Communications in Computer and Information Science 2673

Series Editors

Rationale

The CCIS series is devoted to the publication of proceedings of computer science conferences. Its aim is to efficiently disseminate original research results in informatics in printed and electronic form. While the focus is on publication of peer-reviewed full papers presenting mature work, inclusion of reviewed short papers reporting on work in progress is welcome, too. Besides globally relevant meetings with internationally representative program committees guaranteeing a strict peer-reviewing and paper selection process, conferences run by societies or of high regional or national relevance are also considered for publication.

Topics

The topical scope of CCIS spans the entire spectrum of informatics ranging from foundational topics in the theory of computing to information and communications science and technology and a broad variety of interdisciplinary application fields.

Information for Volume Editors and Authors

Publication in CCIS is free of charge. No royalties are paid, however, we offer registered conference participants temporary free access to the online version of the conference proceedings on SpringerLink (http://link.springer.com) by means of an http referrer from the conference website and/or a number of complimentary printed copies, as specified in the official acceptance email of the event.

CCIS proceedings can be published in time for distribution at conferences or as post-proceedings, and delivered in the form of printed books and/or electronically as USBs and/or e-content licenses for accessing proceedings at SpringerLink. Furthermore, CCIS proceedings are included in the CCIS electronic book series hosted in the SpringerLink digital library at http://link.springer.com/bookseries/7899. Conferences publishing in CCIS are allowed to use our online conference service (Meteor) for managing the whole proceedings lifecycle (from submission and reviewing to preparing for publication) free of charge.

Publication process

The language of publication is exclusively English. Authors publishing in CCIS have to sign the Springer CCIS copyright transfer form, however, they are free to use their material published in CCIS for substantially changed, more elaborate subsequent publications elsewhere. For the preparation of the camera-ready papers/files, authors have to strictly adhere to the Springer CCIS Authors' Instructions and are strongly encouraged to use the CCIS LaTeX style files or templates.

Abstracting/Indexing

CCIS is abstracted/indexed in DBLP, Google Scholar, EI-Compendex, Mathematical Reviews, SCImago, Scopus. CCIS volumes are also submitted for the inclusion in ISI Proceedings.

How to start

To start the evaluation of your proposal for inclusion in the CCIS series, please send an e-mail to ccis@springer.com

Yi Yu · Haiwei Pan · Hongzhi Wang · Chen Yu ·
Qilong Han · Haiyi Liu · Xianhua Song ·
Zeguang Lu
Editors

Data Science

11th International Conference of Pioneering Computer Scientists, Engineers and Educators, ICPCSEE 2025
Hiroshima, Japan, September 19–21, 2025
Proceedings

Springer

Editors
Yi Yu
Hiroshima University
Higashihiroshima, Japan

Haiwei Pan
Harbin Engineering University
Harbin, China

Hongzhi Wang
Harbin Institute of Technology
Harbin, China

Chen Yu
Huazhong University of Science
and Technology
Wuhan, China

Qilong Han
Harbin Engineering University
Harbin, China

Haiyi Liu
Jiangsu Second Normal University
Nanjing, China

Xianhua Song
Harbin University of Science and Technology
Harbin, China

Zeguang Lu
National Academy of Gun Ding Institute
of Data Science
Beijing, China

ISSN 1865-0929 ISSN 1865-0937 (electronic)
Communications in Computer and Information Science
ISBN 978-981-95-2565-2 ISBN 978-981-95-2566-9 (eBook)
https://doi.org/10.1007/978-981-95-2566-9

This Springer imprint is published by the registered company Springer Nature Singapore Pte Ltd.
The registered company address is: 152 Beach Road, #21-01/04 Gateway East, Singapore 189721, Singapore

Preface

As the chairs of the 11th International Conference of Pioneering Computer Scientists, Engineers and Educators 2025 (ICPCSEE 2025, originally ICYCSEE), it is our great pleasure to welcome you to the conference proceedings. ICPCSEE 2025 was held in Hiroshima shi, Japan, September 19–21 2025, hosted by China Computer Federation and ICPCSEE Academic Committee, CCF - Technical Committee on Computer Applications, Hiroshima University, and National Academy of Guo Ding Institute of Data Sciences. The goal of this conference series is to provide a forum for computer scientists, engineers and educators.

This conference attracted 135 paper submissions. After the hard work of the Program Committee, 41 papers were accepted to be presented at the conference, with an acceptance rate of 30.37%. Among them, 1 short paper of fewer than 10 pages was accepted. The conference invited computer scientists, engineers and educators to submit original research, application and education papers on topics of Data Science and its Applications. The accepted papers cover a wide range of areas organized into a Research Track, Industry Track, Education Track and Demo Track, including Education Using Big Data, Intelligent Computing or Data Mining, etc; Education Research, Methods and Materials for Data Science and Engineering; Machine Learning for Data Science; Basic Theory and Techniques for Data Science; Application of Blockchain; Big Data Management and Applications; Industrial Digitization and Applications; Applications of Data Science; Data Security and Privacy; Data-Driven Scientific Research; and Novel Methods for Data Analytics.

We would like to thank all the Program Committee members for their hard work in completing the review tasks. There were at least 3 double-blind reviewers for each manuscript, and each reviewer reviewed no more than 3 manuscripts. Their collective efforts made it possible to attain quality reviews for all the submissions within a few weeks. Their diverse expertise in each research area helped us to create an exciting program for the conference. Their comments and advice helped the authors to improve the quality of their papers and gain deeper insights.

We thank the team at Springer, whose professional assistance was invaluable in the production of the proceedings. A big thanks also to the authors and participants for their tremendous support in making the conference a success.

Besides the technical program, this year ICPCSEE offered different experiences to the participants. We hope you enjoyed the conference.

August 2025

Yi Yu
Haiwei Pan
Hongzhi Wang
Chen Yu
Qilong Han
Haiyi Liu
Xianhua Song
Zeguang Lu

Organization

Honorary Chair

Hai Jin	Huazhong University of Science and Technology, China

General Chairs

Yi Yu	Hiroshima University, Japan
Haiwei Pan	Harbin Engineering University, China

Program Chairs

Chen Yu	Huazhong University of Science and Technology, China
Qilong Han	Harbin Engineering University, China
Xiaoping Li	Guangdong University of Technology, China

Local Chairs

Suhua Tang	University of Electro-Communications, Japan
Junjun Zheng	Hiroshima University, Japan

Publication Chairs

Xianhua Song	Harbin University of Science and Technology, China
Haiyi Liu	Jiangsu Second Normal University, China

Chairman

Hongzhi Wang	Harbin Institute of Technology, China

Vice President

Qilong Han	Harbin Engineering University, China

Secretary General

Zeguang Lu	National Academy of Guo Ding Institute of Data Science, China

Under Secretary General

Xiaoou Ding	Harbin Institute of Technology, China

Secretaries

Dan Lu	Harbin Engineering University, China
Fa Yue	National Academy of Guo Ding Institute of Data Science, China

Executive Members

Cham Tat Huei	UCSI University, Malaysia
Xiaoju Dong	Shanghai Jiao Tong University, China
Lan Huang	Jilin University, China
Ying Jiang	Kunming University of Science and Technology, China
Weipeng Jiang	Northeast Forestry University, China
Min Li	Central South University, China
Junyu Lin	Institute of Information Engineering, CAS, China
Xia Liu	Hainan Province Computer Federation, China
Rui Mao	Shenzhen University, China
Qiguang Miao	Xidian University, China
Haiwei Pan	Harbin Engineering University, China
Pinle Qin	North University of China, China
Xianhua Song	Harbin University of Science and Technology, China
Guanglu Sun	Harbin University of Science and Technology, China

Steering Committee

Qiguang Miao	Xidian University, China
Haiwei Pan	Harbin Engineering University, China
Pinle Qin	North University of China, China
Zheng Shan	PLA Information Engineering University, China
Guanglu Sun	Harbin University of Science and Technology, China
Jie Tang	Tsinghua University, China
Tian Feng	Institute of Software, Chinese Academy of Sciences, China
Tao Wang	Peking University, China
Hongzhi Wang	Harbin Institute of Technology, China
Xiaohui Wei	Jilin University, China
lifang Wen	Beijing Huazhang Graphics & Information Co., Ltd, China
Liang Xiao	Nanjing University of Science and Technology, China
Yu Yao	Northeastern University, China
Xiaoru Yuan	Peking University, China
Yingtao Zhang	Harbin Institute of Technology, China
Yunquan Zhang	Institute of Computing Technology, Chinese Academy of Sciences
Baokang Zhao	National University of Defense Technology, China
Min zhu	Sichuan University, China
Liehuang Zhu	Beijing Institute of Technology, China

Committee Members

Witold Abramowicz	Poznań University of Economics and Business, Poland
Chunyu Ai	University of South Carolina Upstate, USA
Jiyao An	Hunan University, China
Ran Bi	Dalian University of Technology, China
Yi Cai	South China University of Technology, China
Zhipeng Cai	Georgia State University, USA
Zhao Cao	Beijing Institute of Technology, China
Wanxiang Che	Harbin Institute of Technology, China
Chunyi Chen	Changchun University of Science and Technolgoy, China
Hao Chen	Hunan University, China
Wei Chen	Beijing Jiaotong University, China

Xuebin Chen	North China University of Science and Technology, China
Yueguo Chen	Renmin University, China
Siyao Cheng	Harbin Institute of Technology, China
Byron Choi	Hong Kong Baptist University, China
Vincenzo Deufemia	University of Salerno, Italy
Dianxuan Gong	North China University of Science and Technology, China
Jianrui Ding	Harbin Institute of Technology, China
Xiaofeng Ding	Huazhong University of Science and Technology, China
Hongbin Dong	Harbin Engineering University, China
Lei Duan	Sichuan University, China
Xiping Duan	Harbin Normal University, China
Xiaolin Fang	Southeast University, China
Ming Fang	Changchun University of Science and Technology, China
Xiaolin Fang	Southeast University, China
Jianlin Feng	Sun Yat-sen University, China
Jing Gao	Dalian University of Technology, China
Yu Gu	Northeastern University, China
Meng Han	Georgia State University, USA
Qi Han	Harbin Institute of Technology, China
Qinglai He	Arizona State University, USA
Leong Hou	University of Macau, China
Wei Hu	Nanjing University, China
Hao Huang	Wuhan University, China
Lan Huang	Jilin University, China
Bin Jiang	Hunan University, China
Feng Jiang	Harbin Institute of Technology, China
Cheqing Jin	East China Normal University, China
Hanjiang Lai	Sun Yat-sen University, China
Shiyong Lan	Sichuan University, China
Cuiping Li	Renmin University of China, China
Hua Li	Changchun University of Science and Technolgoy, China
Hui Li	Xidian University, China
Jianjun Li	Huazhong University of Science and Technology, China
Min Li	Central South University, China
Mingzhao Li	RMIT University, Australia
Peng Li	Shaanxi Normal University, China

Qiong Li	Harbin Institute of Technology, China
RongHua Li	Shenzhen University, China
Xiaofeng Li	Sichuan University, China
Xiaoping Li	Guangdong University of Technology, China
Zheng Li	Sichuan University, China
Zhixu Li	Soochow University, China
Zhixun Li	Nanchang University, China
Guanfeng Liu	Macquarie University, Australia
Hailong Liu	Northwestern Polytechnical University, China
Yan Liu	Harbin Institute of Technology, China
Yanli Liu	Sichuan University, China
Binbin Lu	Sichuan University, China
Junling Lu	Shaanxi Normal University, China
Zeguang Lu	National Academy of Guo Ding Institute of Data Sciences, China
Jizhou Luo	Harbin Institute of Technology, China
Mohan Li	Jinan University, China
Tiezheng Nie	Northeastern University, China
Haiwei Pan	Harbin Engineering University, China
Fei Peng	Hunan University, China
Jialiang Peng	Norwegian University of Science and Technology, Norway
Jianzhong Qi	University of Melbourne, Australia
Shaojie Qiao	Southwest Jiaotong University, China
Qingliang Li	Changchun University of Science and Technology, China
Zhe Quan	Hunan University, China
Yingxia Shao	Peking University, China
Wei Song	North China University of Technology, China
Minghui Sun	Jilin University, China
Yanan Sun	Oklahoma State University, USA
Guanghua Tan	Hunan University, China
Xifeng Tong	Northeast Petroleum University, China
Yongxin Tong	Beihang University, China
Vicenç Torra	Umeå University, Sweden
Chaokun Wang	Tsinghua University, China
Chunnan Wang	Harbin Institute of Technology, China
Dong Wang	Hunan University, China
Hongzhi Wang	Harbin Institute of Technology, China
Jiapeng Wang	Harbin Huade University, China
Jinbao Wang	Harbin Institute of Technology, China
Peng Wang	Fudan University, China

Xiaoling Wang	East China Normal University, China
Xin Wang	Tianjin University, China
Yingjie Wang	Yantai University, China
Yongheng Wang	Hunan University, China
Huayu Wu	Institute for Infocomm Research, Singapore
Yan Wu	Changchun University of Science and Technology, China
Sheng Xiao	Hunan University, China
Jianqiu Xu	Nanjing University of Aeronautics and Astronautics, China
Jing Xu	Changchun University of Science and Technology, China
Ying Xu	Hunan University, China
Yaohong Xue	Changchun University of Science and Technology, China
Xuwei Li	Sichuan University, China
Mingyuan Yan	University of North Georgia, USA
Gaobo Yang	Hunan University, China
Zhenguo Yang	Guangdong University of Technology, China
Lei Yang	Heilongjiang University, China
Ning Yang	Sichuan University, China
Xiaochun Yang	Northeastern University, China
Yajun Yang	Tianjin University, China
Bin Yao	Shanghai Jiao Tong University, China
Xiufen Ye	Harbin Engineering University, China
Yuxin Ye	Jilin University, China
Dan Yin	Harbin Engineering University, China
Minghao Yin	Northeast Normal University, China
Zhou Yong	China University of Mining and Technology, China
Lei Yu	Georgia Institute of Technology, USA
Ye Yuan	Northeastern University, China
Kun Yue	Yunnan University, China
Yuwei Peng	Wuhan University, China
An Zeng	Guangdong University of Technology, China
Donghuo Zeng	KDDI Research, Japan
Boyu Zhang	Utah State University, USA
Dongxiang Zhang	University of Electronic Science and Technology of China, China
Huijie Zhang	Northeast Normal University, China
Kejia Zhang	Harbin Engineering University, China
Lichen Zhang	Shaanxi Normal University, China

Wenjie Zhang	University of New South Wales, Australia
Xiao Zhang	Renmin University of China, China
Xiaowang Zhang	Tianjin University, China
Yingtao Zhang	Harbin Institute of Technology, China
Yonggang Zhang	Jilin University, China
Yu Zhang	Harbin Institute of Technology, China
Bihai Zhao	Changsha University, China
Jian Zhao	Changchun University, China
Qijun Zhao	Sichuan University, China
Xiaohui Zhao	University of Canberra, Australia
Jiancheng Zhong	Hunan Normal University, China

Contents

Application of Blockchain

Big Data Management and Applications

Industrial Digitization and Applications

Applications of Data Science

Data Security and Privacy

Data-driven Scientific Research

Novel Methods for Data analytics

Education Using Big Data, Intelligent Computing or Data Mining, etc.

Analysis of the Impact of the Information Environment on Students' Self-directed English Learning Behavior and Psychology

Xuemin Du(✉)

Sanya Aviation and Tourism College, Fenghuang Road 218#, Sanya, China
minxuedu@163.com

Abstract. This study constructs a theoretical and analytical framework for examining the interplay between the information environment and autonomous English learning by integrating Zimmerman's self-regulated learning theory with the Technology Acceptance Model (TAM). This integrated framework reveals the underlying connections between learners' self-regulatory processes and their acceptance of technological tools. Using vocational college students as the target population, this study employs both quantitative and qualitative methods to investigate the roles of the information environment, autonomous learning behaviors, learning psychology, and cross-cultural communication competence. The results demonstrate that the information environment significantly and positively influences both autonomous learning behaviors and learning psychology. Furthermore, cross-cultural communication competence is found to mediate this relationship. On the basis of these findings, several strategies are proposed: educational institutions should enhance intelligent resource databases and promote the development of AI-assisted learning systems; educators are encouraged to strengthen their competence in digital instructional design; and learners should be supported in improving their metacognitive strategies and ability to integrate technological tools effectively.

Keywords: Information-based environment · Autonomous English learning behavior · Autonomous English learning psychology · Technology acceptance model (TAM)

1 Introduction

Since 2021, China has issued fifteen policies aimed at advancing the digitalization of education, significantly accelerating the implementation of national strategies for educational informatization [1]. The rapid surge of informatization has brought about unprecedented transformations in student learning. For learners, the information-based environment has greatly expanded access to English learning resources and diversified learning channels. This environment offers an abundance of English learning materials, innovative tools, and highly interactive platforms, creating more accessible and flexible learning conditions. Consequently, English teaching and learning are no longer confined

Y. Yu et al. (Eds.): ICPCSEE 2025, CCIS 2673, pp. 3–24, 2026.
https://doi.org/10.1007/978-981-95-2566-9_1

to the traditional classroom or the teacher's presence. With the fast-paced development of information technology and the rise of multimodal learning contexts, students' modes of perceiving and interpreting knowledge have also increasingly diversified.

However, the information-based environment has also introduced a range of challenges and uncertainties into English education. The overwhelming volume and complexity of online information can obscure learners' goals. Furthermore, the modes of interaction within virtual learning communities may impair language expression, hinder emotional communication, or even contribute to learning anxiety and psychological dependence [2].

2 Significance of the Study

This study conducts a systematic and in-depth analysis of students' use of information-based environments, autonomous English learning behaviors, and learning psychology. It aims to uncover patterns and mechanisms underlying changes in learner behavior and psychological states under informatized conditions. By identifying both positive and negative influences, this study proposes effective strategies for enhancing students' autonomous English learning. These findings are expected to provide scientific support and practical guidance for optimizing English education models and improving learning outcomes. Moreover, this research contributes to the application of smart education in the English language domain by strengthening both behavioral autonomy and positive psychological development among learners.

3 Research Review on the Information-Based Environment and Autonomous English Learning

3.1 The Information-Based Environment

An information-based environment refers to a setting in which information technologies are widely applied. This environment typically relies on computer networking or other digital technologies and encompasses hardware infrastructure, software systems, cultural settings, and organizational structures [3]. Across various sectors, informatization has become a significant driver of progress. In the field of education, iterative upgrades to information-based environments have transformed how students learn, access knowledge, communicate, and solve problems [4].

With respect to its impact on learning outcomes, empirical research has yielded mixed results. Some studies assert that open and flexible digital learning environments enhance educational performance. For example, Wang found through experimental methods that student engagement and academic performance were significantly correlated in a desktop virtual reality environment [5]. Similarly, Cai Su argued that integrating augmented reality (AR) into hybrid instructional scenarios can effectively improve student learning [6]. Zheng Xuxiang et al., through a meta-analysis of 61 empirical studies over the past decade, concluded that digital learning environments significantly impact learning outcomes among vocational college students [7].

Conversely, other research suggests that information-based environments do not consistently lead to improved learning outcomes. For example, Xu Jihong et al. reported no significant differences in learning outcomes when digital media formats were varied in online settings [8]. Wrzesien's comparative study also revealed that virtual classrooms do not outperform traditional classrooms in terms of learning effectiveness [9].

Furthermore, researchers have explored whether digital technologies inherently enhance student learning. Deok-Ho Jang et al., through a meta-analysis, reported that digital textbooks significantly increase learners' motivation [10]. In contrast, Li Gao et al. argued that large-scale digital learning might induce anxiety, stress, loneliness, and burnout, potentially compromising mental and physical well-being [11].

Currently, the development of information-based environments emphasizes the deep integration of educational innovation and technology [4]. Artificial intelligence (AI), data-driven methods, and human-centered design are increasingly central. Such environments possess features—personalization and intelligence—not found in traditional learning contexts. Blended learning, which combines online and offline modalities, has become a key direction for educational reform. The construction of informatized learning environments allows students to benefit from the complementary strengths of both modalities, enhancing learner autonomy and motivation.

3.2 Learners' Autonomous Learning Behavior and Psychology

Self-determination theory (SDT), proposed by Ryan and colleagues in the 1980s, provides a motivational framework for understanding autonomous human behavior [12]. It explores how internal and external factors influence individual decision-making. On this theoretical foundation, researchers have conducted empirical studies verifying the role of self-determination in students' learning motivation, psychology, and behavior. For instance, Vansteenkiste et al. demonstrated that intrinsic goals partially mediate the effect of autonomy-supportive contexts on academic performance [13].

Zimmerman, in his work Self-Regulated Learning and Academic Achievement: An Overview, defines self-regulated learning as learners' ability to actively control their learning processes through metacognitive, motivational, and behavioral strategies [14]. His model comprises three cyclical phases—forethought, performance, and self-reflection—emphasizing psychological mechanisms such as self-efficacy, goal orientation, and attribution. Similarly, Pang Weiguo asserts that autonomous learning entails setting personal goals, formulating learning plans, engaging in self-monitoring and feedback during the learning process, and conducting postlearning evaluations and remedial actions. Without these elements, learning cannot be considered truly autonomous [15].

3.3 Relationship Between the Information-Based Environment and Autonomous English Learning

Scholars have argued that English education is undergoing a profound transformation in today's information-based environment. This environment provides abundant resources and diverse teaching approaches to the field of English language learning. A wide array of materials—such as online courses, English-language films, and audio books—greatly expand learners' access to knowledge. The integration of multimedia tools makes teaching more dynamic and engaging, as visual, audio, and video elements enhance students' interest in and comprehension of the English language. Online learning platforms remove temporal and spatial barriers, allowing learners to set their own schedules and pursue personalized learning paths.

Nonetheless, the information-based environment also presents challenges. The vast and disorganized nature of available content can lead to confusion and difficulty in identifying valuable materials. Overreliance on electronic devices may impair students' critical thinking and concentration.

4 The Impact of the Information-Based Environment on Students' Autonomous English Learning Behavior and Psychology

4.1 Research Objectives

This study aims to conduct an in-depth investigation into the mechanisms through which the information-based environment influences students' autonomous English learning behavior and psychological states. Specifically, it examines the mediating role of intercultural communication competence in this relationship to elucidate how the digital environment affects the autonomous learning process. On the basis of these empirical findings, this study proposes effective strategies for enhancing students' autonomous English learning capabilities, mitigating potential negative psychological effects, and cultivating strategic self-directed learning behaviors. Additionally, the research aims to provide a theoretical foundation and practical guidance for educators engaged in English instruction within informatized contexts, thereby contributing to the optimization of teaching models and the improvement of instructional quality.

4.2 Research Content

The research is structured around the following objectives:

A. To analyse the intrinsic relationship between the information-based environment and students' autonomous English learning behavior and psychology and to assess the current status and existing challenges in the development of autonomous learning competence under informatized conditions.
B. To explore the mediating role of intercultural communication competence in the relationship between the information-based environment and students' autonomous learning behavior and psychology and to examine differences across various student groups.
C. To propose targeted strategies for improving students' autonomous English learning abilities to promote their language competence within the context of the information-based environment.

4.3 Research Hypotheses

The core hypotheses of this study are as follows:

H1: The information-based environment has a direct and positive effect on autonomous English learning behavior. (Path: Information-Based Environment → Learning Behavior).
H2: The information-based environment has a direct and positive effect on students' English learning psychology. (Path: Information-Based Environment → Learning Psychology).
H3: Intercultural communication competence mediates the relationship between the information-based environment and autonomous learning behavior. (Path: Information-Based Environment → Intercultural Competence → Learning Behavior).
H4: Intercultural communication competence mediates the relationship between the information-based environment and learning psychology. (Path: Information-Based Environment → Intercultural Competence → Learning Psychology).

4.4 Research Subjects

The participants of this study were students at College S, a comprehensive vocational institution with approximately 10,000 students across 32 academic programs. Stratified random sampling was employed on the basis of academic year (first to third year) and major (humanities, sciences, engineering, arts, and others). Owing to the low response rate among third-year students engaged in off-campus internships, the final valid sample comprised 322 students, with a greater proportion of underclassmen. This sampling bias may have led to an emphasis on characteristics prevalent among lower-year students; future research should aim to expand the sample to upper-year students to ensure greater generalizability.

The demographic breakdown of the sample is as follows: 103 males (32%) and 219 females (68%); 250 first-year students (77.6%), 61 second-year students (18.9%), and 11 third-year students (3.5%). In terms of disciplinary distribution, 65 students (20.19%) were from humanities, 23 (7.14%) were from sciences, 8 (2.48%) were from engineering, 62 (19.25%) were from arts, and 164 (50.95%) were from other disciplines.

In addition to the questionnaire, qualitative data were collected through one-on-one interviews with five students from different majors. The participants were selected according to the principles of sample diversity and their willingness to be interviewed.

4.5 Research Questions

The study aims to address the following research questions:

What is the current status of students' autonomous English learning behavior and psychological states within the information-based environment?

Are there significant differences in autonomous English learning behavior and psychology across students with different background variables?

What are the primary factors influencing students' autonomous English learning ability, and do these factors have positive or negative effects?

On the basis of the analysis of the current situation and influencing factors, what strategies can be proposed to improve students' autonomous English learning competence in the information-based environment?

4.6 Research Methods

In alignment with the research objectives, this study employs a combination of the literature review method, survey method, and case study method.

Literature Review
During the preparatory phase, a comprehensive literature review was conducted on key concepts, including the information-based environment, autonomous English learning behavior, learning psychology, and intercultural communication competence. The theoretical foundation of this study is based on Zimmerman's (2000) model of self-regulated learning, which categorizes learning behaviors into planning, monitoring, and evaluation phases. This approach is integrated with the technology acceptance model (TAM), which focuses on perceived ease of use and usefulness, and self-determination theory (SDT), which examines how the information-based environment supports learners' autonomy and sense of competence. These frameworks inform the proposed conceptual model linking the information-based environment, learning behavior, learning psychology, and intercultural competence, serving as a basis for the study's design and analysis.

Survey Method
Quantitative data were collected through a self-designed questionnaire developed by adapting and extending existing validated scales on autonomous English learning behavior and learning psychology. The questionnaire covered multiple dimensions, including students' use of digital tools, emotional responses, and perceived learning outcomes in informatized settings. The instrument was pretested and revised to ensure reliability and validity. It was distributed to a wide range of students to obtain a representative dataset.

Case Study Method
A vocational college with a well-established informed English teaching system was selected for case analysis. The study examined initiatives undertaken by the institution, such as the implementation of intelligent learning platforms and the integration of multimedia teaching resources. The observations focused on students' autonomous learning behaviors within this environment and the corresponding psychological responses. Representative students were interviewed to explore their experiences, challenges, and emotional shifts during autonomous English learning. These qualitative insights complemented the survey data and provided a deeper understanding of the relationships among various elements of informatized learning. The triangulation of data from the literature, surveys, and case studies offered a robust empirical foundation for proposing targeted strategies to enhance autonomous English learning competence.

4.7 Questionnaire Design

The questionnaire used in this study consists of three sections: (1) respondents' demographic information, (2) students' engagement with the information-based environment and autonomous English learning, and (3) students' intercultural communication competence. The design of the questionnaire was informed by the learning characteristics and experiences of the respondents, as well as their potential interpretations of the constructs under study. Efforts were made to ensure that each item accurately reflected the relevant dimension.

The initial version of the questionnaire comprised 45 items. A pilot study with 30 students was conducted, and 8 items with corrected item–total correlation (CITC) values below 0.4 were removed (e.g., "I am not interested in foreign festivals," which was excluded because of low correlation with the construct of intercultural attitudes). The final version of the questionnaire included 39 items. All the items were aligned with the study's theoretical framework—for example, Item 9 ("I monitor my learning process using digital platforms") corresponds to the self-monitoring component in Zimmerman's self-regulated learning model, and Item 5 ("The resources on digital platforms are updated frequently") reflects the perceived usefulness dimension from the Technology Acceptance Model (TAM).

The respondents were instructed to mark (√) the box at the intersection of each row (item description) and column (degree of agreement) that best matched their own learning situation.

Section One: Demographic Information
This section gathered basic data such as gender, academic year, and major. These variables were used in subsequent analyses to identify potential effects on students' autonomous English learning behavior and psychological characteristics.

Section Two: Information-Based Environment and Autonomous English Learning
This section assessed students' use of digital platforms and learning resources, as well as their autonomous learning practices. It consists of 15 items and uses a five-point Likert scale ranging from "Strongly Disagree" to "Strongly Agree," which are numerically coded from 1–5. Additionally, response rates were categorized into the following agreement levels for easier analysis: $\leq$10%, ~30%, ~55%, ~80%, and $\geq$90%.

Items 4–7 focused on the use of digital platforms and learning resources, representing the information-based environment dimension. Higher scores indicate greater availability and perceived usefulness/ease of use, which is consistent with TAM (e.g., Item 4: "Abundant resources"; Item 5: "Frequent updates"; Item 6: "Anytime-anywhere learning"; Item 7: "Efficient resource search").

Items 8–13 addressed Autonomous Learning Behavior, covering phases such as planning, execution, monitoring, and evaluation from Zimmerman's self-regulated learning theory (e.g., Item 8: "Setting learning plans"; Item 9: "Monitoring the learning process"; Item 10: "Tool integration"; Item 11: "Collaborative learning"; Item 12: "Use of AI tools"). Higher scores reflect more proactive learning behavior.

Items 14–18 measured learning psychology, reflecting constructs such as self-efficacy and motivation (e.g., Item 13: "Preference for digital instruction"; Item 14: "Confidence in learning"; Item 15: "Positive mindset"; Item 16: "Overcoming difficulties"; Item 17: "Sense of accomplishment"). Higher scores indicate more positive psychological states in English learning.

Section Three: Intercultural Communication Competence
This section was developed on the basis of three core dimensions—knowledge, skills, and attitudes—to assess students' intercultural communication competence.

The knowledge component included an understanding of both home and foreign cultures, encompassing history, society, and values.

The attitude dimension addressed global awareness, cultural sensitivity, and communicative motivation, including aspects such as openness to cultural differences, perspectives on cultural diversity, and the willingness to engage in intercultural communication.

The skills component examined communicative strategies and adaptive capabilities, including language skills, reflection, and flexibility.

Each item in this section used the same five-point Likert scale as the previous section, with values ranging from 1 ("Strongly Disagree") to 5 ("Strongly Agree"). The items were categorized and coded as follows:

Items 19–24: Intercultural knowledge—Higher scores reflect stronger knowledge of intercultural content.
Items 25–35: Intercultural skills—Higher scores reflect a stronger ability to apply intercultural communication strategies.
Items 36–39: Intercultural Attitudes—Higher scores reflect more positive intercultural dispositions.

The mean score across these items represents each student's overall intercultural communication competence. A higher overall score indicates stronger competence.

The detailed scoring rubric for all the items is presented in Table 1.

Table 1. Question option assignments

Options	10%	30%	55%	80%	90%
4–39	≤10% agreement	Approximately 30% agreement	Approximately 55% agreement	Approximately 80% agreement	≥90% agreement

4.8 Survey Results and Analysis

Data for this study were collected via a self-developed questionnaire titled "Survey on the Relationship between the Information-Based Environment and Autonomous English Learning Behavior and Psychology". The survey was administered and collected via the Wenjuanxing platform (Questionnaire Star). A total of 323 students participated in the questionnaire, and after one invalid response was removed, 322 valid responses were retained, yielding an effective response rate of 99.69%.

The data were processed and scored via SPSS 22.0 statistical software. To assess the quality of the measurement instrument:

Exploratory factor analysis (EFA) was conducted to evaluate the construct validity of the scale.

Cronbach's alpha coefficient was used to assess the internal consistency (reliability) of the instrument.

Descriptive statistics were applied to report the means and standard deviations of the three subdimensions of students' intercultural communication competence.

One-way analysis of variance (ANOVA) was used to analyse whether significant differences existed in autonomous English learning behavior and psychological states among students of different academic years and majors (Table 2).

Table 2. Sample Composition of the Actual Survey

Question	Option	Frequency (person)	Percentage (%)
Gender	Male	103	31.99%
	Female	219	68.01%
Grade:	Freshman	250	77.64%
	Sophomore	61	18.94%
	Juniors	11	3.42%
Major	Humanities-related majors	65	20.19%
	Science-related majors	23	7.14%
	Engineering-related majors	8	2.48%
	Arts-related majors	62	19.25%
	Other	164	50.93%

Reliability Analysis

Table 3. Sample Composition of the Actual Survey

Number of Items	Sample Size	Cronbach α Coefficient
39	322	0.983

Cronbach's reliability analysis, also known as Cronbach's alpha, is a statistical method used to measure the internal consistency of a questionnaire or scale. Generally, a Cronbach's alpha coefficient greater than 0.9 indicates excellent internal consistency. As shown in the table above, the reliability coefficient is 0.982, which exceeds the 0.9 threshold, demonstrating that the data collected in this study are highly reliable. Furthermore, all the corrected item–total correlation (CITC) values exceeded 0.4, indicating strong correlations among the items and further confirming the scale's satisfactory reliability (Table 3).

Validity Analysis

Table 4. Sample Composition of the Actual Survey

KMO value		0.964
Bartlett Sphericity Test	Approximate chi-square	15838.202
	df	741
	p value	0.000

The Kaiser–Meyer–Olkin (KMO) value ranges between 0 and 1. A KMO value closer to 1 indicates strong partial correlations among variables, suggesting that the data are well suited for factor analysis. Conversely, a KMO value closer to 0 indicates weak partial correlations, making the data less appropriate for such analysis. If the significance level (P value) of Bartlett's test of sphericity is less than the standard threshold (typically 0.05), it implies that correlations exist among variables, thus justifying the use of factor analysis. According to the table above, the KMO value is 0.964, which exceeds the 0.8 threshold, and the significance level is below 0.005. These results indicate that the questionnaire has excellent construct validity (Table 4).

Descriptive Analysis

Basic descriptive statistics of the information-based environment and students' autonomous English learning behavior and psychology.

The questionnaire items were scored on a five-point scale: a score of 1 was assigned for responses indicating 10% or less agreement; 2 for approximately 30% agreement; 3 for approximately 55% agreement; 4 for approximately 80% agreement; and 5 for 90% or more agreement. A higher score indicates a stronger role of the information-based English learning environment in promoting students' autonomous learning and reflects greater effectiveness in autonomous English learning.

A score of 3 is considered the median value, indicating a moderate impact of the information-based environment on students' autonomous learning. In this study, a score of 4 was used as the benchmark for a favourable level of informatization support. Scores above 4 suggest that digital platforms and learning resources are sufficiently abundant and contribute positively to students' autonomous English learning outcomes.

The descriptive statistical analysis results for students' autonomous learning behavior and psychological state within the information-based environment are presented in Table 5.

Table 5. Descriptive Statistics of the Information-Based Environment and Students' Autonomous English Learning Indicators (n = 322)

Items	Sample Size	Min	Maximum	Mean	Standard Deviation	Median
Information-Based Environment	322	1.000	5.000	4.022	0.793	4.000
Autonomous Learning Behavior	322	1.000	5.000	3.977	0.781	4.000
Autonomous Learning Psychology	322	1.000	5.000	4.069	0.738	4.000

As shown in Table 5, all three indicators exhibit an upper-middle level of performance, with mean values close to 4. This indicates a generally positive trend across the dataset. The results suggest that participants hold favourable perceptions of the information-based environment, demonstrate proactive autonomous learning behaviors, and maintain positive psychological states. Among the three dimensions, autonomous learning psychology scored the highest, with particularly strong performance in the upper quartile (P75 = 5.000), indicating a notably positive psychological attitude toward autonomous learning.

The detailed results of the students' autonomous English learning behavior and psychological responses are presented in Table 6.

Table 6. Descriptive Statistics of Selected Indicators of the Information-Based Environment and Students' Autonomous English Learning (n = 322)

Item	Sample Size	Minimum	Maximum	Mean	Standard Deviation	Median
There is an abundance of online English learning resources that help me learn English more effectively	322	1.000	5.000	3.991	0.922	4.000
The English learning resources provided on digital platforms are updated frequently	322	1.000	5.000	4.009	0.870	4.000
Using digital resources allows me to study English anytime and anywhere	322	1.000	5.000	4.062	0.880	4.000

(*continued*)

Table 6. (*continued*)

Item	Sample Size	Minimum	Maximum	Mean	Standard Deviation	Median
When I search online for resources, I can always find websites or tools that support my English learning	322	1.000	5.000	4.025	0.875	4.000
I am proficient in using digital learning devices (e.g., laptop, tablet) and software for English learning	322	1.000	5.000	3.947	0.934	4.000
I use digital platforms to self-monitor and evaluate the quality of my English learning process	322	1.000	5.000	3.922	0.926	4.000
I consciously use additional functions of digital tools to support my English learning	322	1.000	5.000	4.025	0.853	4.000
I try to integrate different digital learning tools into my English learning	322	1.000	5.000	3.984	0.888	4.000
In online learning, I use WeChat, QQ, or other social media to communicate with peers for English learning	322	1.000	5.000	4.028	0.880	4.000
I frequently use large language models (e.g., Wenxin Yiyan, Lanxin XiaoV, ChatGPT) to support my English learning	322	1.000	5.000	3.957	0.913	4.000

(*continued*)

Table 6. (*continued*)

Item	Sample Size	Minimum	Maximum	Mean	Standard Deviation	Median
I prefer that teachers use digital resources to deliver English instruction	322	1.000	5.000	4.025	0.842	4.000
The information-based environment increases my confidence in learning English	322	1.000	5.000	4.075	0.794	4.000
I have a positive mindset toward the abundance of English learning resources in the information-based environment	322	1.000	5.000	4.081	0.805	4.000
English learning in an informatized environment helps me overcome difficulties and barriers	322	1.000	5.000	4.050	0.826	4.000
When I make progress in English learning in an information-based environment, I feel proud and satisfied	322	1.000	5.000	4.115	0.799	4.000

On the basis of the results from both the questionnaire data and the student interviews, it is evident that students have access to a relatively wide range of digital platforms and English learning resources. These resources are frequently updated, offer a variety of functionalities, and reflect a well-developed information-based learning environment that meets students' English learning needs. While items such as "Using digital resources enables me to learn English anytime and anywhere" ($M = 4.062$) received high scores, items such as "Integration of digital tools" ($M = 3.984$), "Self-monitoring and evaluation" ($M = 3.922$), and "Support from large AI models in English learning" ($M = 3.984$) scored below 4. These results suggest that students exhibit relatively weak metacognitive strategies. This interpretation is further supported by face–to–face interview data. For example, Student A mentioned, "I've used several apps, but they're not very efficient," whereas Student CSY stated that although they use multiple smart platforms, "my grades haven't improved." These comments highlight a clear contradiction: while the informatized environment offers a wealth of tools, students' English learning strategies remain underdeveloped. There is a pressing need to enhance students' abilities in resource filtering and process-oriented learning management.

Case Interview: "On the Information-based Environment and Autonomous English Learning"
Date: March 20, 2025 Location: Classroom, College S
Interviewee: CSY
Background: First-year students, Arts major
Q: What digital tools or resources do you commonly use for autonomous English learning? How do you usually access these resources? How much time do you spend on them daily?
A: I use the Youdao translator and Doubao. I generally rely on learning resources recommended by the school. I spend approximately 30 min a day using informatized English learning resources.
Q: Do you set your own learning plans for English? Does the information-based environment help you plan or adjust your learning more effectively?
A: I mostly use informatized tools to practice English listening. The digital environment helps me make reasonable plans for autonomous English learning.
Q: When you encounter difficulties in English learning, do you turn first to informatized methods for solutions? Compared with asking teachers for help, which do you prefer?
A: I use informatized tools to solve problems, such as grammar issues. However, I prefer to ask teachers because they can apply their knowledge more effectively to help me. Informatized tools do not feel very personalized, which sometimes frustrates me.
Q: During autonomous learning, do you feel that the information-based environment gives you more freedom and flexibility? Have you ever experienced low learning efficiency due to a lack of supervision?
A: Learning in an informatized environment is indeed more flexible and autonomous, but it is not as targeted as learning with a teacher. The environment gives me more freedom, but the efficiency is not high.

Analysis of the Relationship between Gender and Students' Autonomous English Learning Behavior, Learning Psychology, and Intercultural Communication Competence.

An analysis of variance (ANOVA) was conducted using gender as the independent variable and students' autonomous English learning ability and intercultural communication competence as the dependent variables. The aim was to examine whether significant differences exist in these abilities between male and female students. The detailed results of the analysis are presented in Table 7.

Table 7. ANOVA of Students' Autonomous English Learning Behavior, Learning Psychology, and Intercultural Communication Competence by Gender, Academic Year, and Major

	Gender (Mean ± SD)		F	p
	Male(n = 103)	Female(n = 219)		
Autonomous Learning Behavior	3.96 ± 0.88	3.99 ± 0.73	0.092	0.762
Autonomous Learning Psychology	4.02 ± 0.88	4.09 ± 0.67	0.578	0.448
Intercultural Communication Competence	4.01 ± 0.84	3.97 ± 0.65	0.217	0.642
$p < 0.05$ ** $p < 0.01$				

As shown in the table above, there are no significant gender differences in students' autonomous English learning behavior, learning psychology, or intercultural communication competence; the responses of the male and female participants were largely consistent across all three dimensions. This finding contrasts with previous studies that have suggested that female students exhibit stronger autonomous learning abilities. A possible explanation is that vocational college students tend to use digital tools in similar ways, reducing gender-based differences in learning behavior. Additionally, the relatively high proportion of female respondents in the sample may have diluted any potential gender effect, contributing to the lack of statistically significant differences (Table 8).

Table 8. ANOVA of Students' Autonomous English Learning Behavior, Learning Psychology, and Intercultural Communication Competence by Academic Year

Variable					
	Academic Year (Mean ± SD)			F	p
	First Year (n = 250)	Second Year (n = 61)	Third Year (n = 11)		
Autonomous Learning Behavior	3.97 ± 0.78	4.09 ± 0.76	3.56 ± 0.74	2.204	0.112
Autonomous Learning Psychology	4.06 ± 0.72	4.16 ± 0.82	3.71 ± 0.76	1.760	0.174
Intercultural Communication Competence	3.97 ± 0.71	4.05 ± 0.78	3.81 ± 0.45	0.634	0.531
* $p < 0.05$ ** $p < 0.01$					

As shown in the table above, there are no statistically significant differences ($p > 0.05$) in students' autonomous English learning behavior, learning psychology, or intercultural communication competence across different academic years. This finding indicates that students at various year levels exhibit consistent patterns in their autonomous learning behavior, learning psychology, and intercultural communication competence (Table 9).

Table 9. ANOVA of Students' Autonomous English Learning Behavior, Learning Psychology, and Intercultural Communication Competence by Major

Results of Analysis of Variance							
	Major (Mean ± SD):					F	p
	Humanities (n = 65)	Sciences (n = 23)	Engineering (n = 8)	Arts (n = 62)	Other (n = 164)		
Autonomous Learning Behavior	4.04 ± 0.70	4.39 ± 0.78	3.98 ± 0.88	3.88 ± 0.77	3.93 ± 0.80	2.147	0.075
Autonomous Learning Psychology	4.17 ± 0.68	4.40 ± 0.74	4.00 ± 1.00	4.04 ± 0.63	4.00 ± 0.78	1.871	0.115
Intercultural Communication Competence	4.12 ± 0.60	4.27 ± 0.75	3.92 ± 1.06	3.97 ± 0.63	3.90 ± 0.75	2.138	0.076

* $p < 0.05$ ** $p < 0.01$

As shown in the table above, there are no significant differences across majors in terms of students' autonomous English learning behavior, learning psychology, or intercultural communication competence. The results indicate a high level of consistency among students from different academic disciplines. One possible explanation is that the classification of majors was not sufficiently detailed, and participants who selected "Other majors" accounted for 50.93% of the total sample, potentially diluting any observable differences.

Hierarchical Regression Analysis

Table 10. Hierarchical Regression Analysis of the Information-Based Environment, Students' Autonomous English Learning Behavior, and Intercultural Communication Competence

Results of Hierarchical Regression Analysis (n = 322)															
	Model 1					Model 2					Model 3				
	B	Standard Error	t	p	β	B	Standard Error	t	p	β	B	Standard Error	t	p	β
Constant	4.183**	0.237	17.647	0.000	-	0.917**	0.183	4.997	0.000	-	0.072	0.171	0.422	0.673	-
Gender	0.032	0.094	0.342	0.732	0.019	0.021	0.053	0.391	0.696	0.012	0.039	0.045	0.868	0.386	0.023
Academic Year	−0.053	0.087	−0.603	0.547	−0.034	−0.110*	0.049	−2.230	0.026	−0.072	−0.075	0.042	−1.797	0.073	−0.049
Major	−0.052	0.028	−1.885	0.060	−0.107	−0.026	0.016	−1.663	0.097	−0.053	−0.005	0.013	−0.356	0.722	−0.010
Information-Based Environment						0.811**	0.031	26.055	0.000	0.824	0.516**	0.037	14.033	0.000	0.524
Intercultural Communication Competence											0.471**	0.041	11.442	0.000	0.430
R 2	0.012					0.685					0.778				

(continued)

Table 10. *(continued)*

Results of Hierarchical Regression Analysis (n = 322)															
	Model 1					Model 2					Model 3				
	B	Standard Error	t	p	β	B	Standard Error	t	p	β	B	Standard Error	t	p	β
Adjusted R^2	0.002					0.681					0.774				
F value	F (3,318) = 1.246,p = 0.293					F (4,317) = 172.637,p = 0.000					F (5,316) = 220.900,p = 0.000				
ΔR 2	0.012					0.674					0.092				
ΔF value	F (3,318) = 1.246,p = 0.293					F (1,317) = 678.838,p = 0.000					F (1,316) = 130.925,p = 0.000				
Note: Dependent Variable = Autonomous Learning Behavior															
* p < 0.05 ** p < 0.01															

As shown in Table 10, a linear regression analysis was conducted using gender, academic year, and major as independent variables and autonomous learning behavior as the dependent variable. The R^2 value of Model 1 is 0.012, indicating that these three demographic variables account for only 1.2% of the variance in autonomous learning behavior. The F test for the model yielded F = 1.246 with $p > 0.05$, suggesting that the model is not statistically significant. In other words, gender, academic year, and major do not significantly affect autonomous learning behavior. Therefore, further analysis of the impact of individual predictors is not warranted, and the analysis for Model 1 concludes here.

In Model 2, the information-based environment variable was added to the existing model. The F value became statistically significant ($p < 0.05$), indicating that the addition of this variable substantially improved the model's explanatory power. The R^2 value increased from 0.012 to 0.685, implying that the information-based environment explained 67.4% of the variance in autonomous learning behavior. Specifically, the regression coefficient for the information-based environment is 0.811, and this effect is statistically significant ($t = 26.055$, $p = 0.000 < 0.01$), suggesting a strong and positive influence of the information-based environment on autonomous learning behavior.

In Model 3, intercultural communication competence was added on top of Model 2. The change in the F value was again statistically significant ($p < 0.05$), indicating that this additional variable further enhanced the model. The R^2 value rose from 0.685 to 0.778, suggesting that intercultural communication competence contributes an additional 9.2% to the explanatory power of the model. More specifically, the regression coefficient for intercultural communication competence is 0.471, and the result is statistically significant ($t = 11.442$, $p = 0.000 < 0.01$), indicating a strong positive effect on students' autonomous learning behavior.

Table 11. Hierarchical Regression Analysis of the Information-Based Environment, Students' Autonomous Learning Psychology, and Intercultural Communication Competence

	Model1	Model2	Model3
Constant	4.232** (18.944)	1.365** (6.968)	0.254 (1.566)
Gender	0.073 (0.818)	0.063 (1.101)	0.086* (2.036)
Academic Year	−0.052 (−0.634)	−0.103 (−1.946)	−0.056 (−1.426)

(continued)

Table 11. *(continued)*

	Model1	Model2	Model3
Major	−0.059* (−2.264)	−0.036* (−2.159)	−0.008 (−0.640)
Information-Based Environment		0.712** (21.411)	0.323** (9.297)
Intercultural Communication Competence			0.620** (15.916)
Sample Size	322	322	322
R 2	0.018	0.599	0.777
Adjusted R^2	0.009	0.593	0.774
F value	F (3,318) = 1.928,p = 0.125	F (4,317) = 118.135,p = 0.000	F (5,316) = 220.400,p = 0.000
ΔR^2	0.018	0.581	0.179
ΔF value	F (3,318) = 1.928,p = 0.125	F (1,317) = 458.437,p = 0.000	F (1,316) = 253.326,p = 0.000

Note: Dependent Variable = Autonomous Learning Psychology

p < 0.05, p < 0.01,t values are in parentheses

As shown in Table 11, the R^2 value of Model 1 is 0.018, indicating that gender, academic year, and major account for only 1.8% of the variance in students' autonomous learning psychology. The F test for this model yields F = 1.928, with p > 0.05, suggesting that the model is not statistically significant. In other words, gender, academic year, and major do not have a significant effect on students' autonomous learning psychology; thus, no further analysis of individual predictors in this model is warranted.

In Model 2, the inclusion of the information-based environment leads to a statistically significant change in the F value ($p < 0.05$), indicating that this variable significantly improves the model's explanatory power. The R^2 value increases from 0.018 to 0.599, meaning that the information-based environment explains 58.1% of the variance in autonomous learning psychology. Specifically, the regression coefficient for the information-based environment is 0.712, with a significant t value of 21.411 ($p = 0.000 < 0.01$), suggesting a strong positive influence on students' learning psychology.

In Model 3, intercultural communication competence is added. The F value again shows statistical significance ($p < 0.05$), indicating further model improvement. The R^2 value increases from 0.599 to 0.777, demonstrating that intercultural communication competence accounts for an additional 17.9% of the variance in autonomous learning psychology. The regression coefficient for intercultural communication competence is 0.620, with a t value of 15.916 ($p = 0.000 < 0.01$), confirming its significant positive impact on students' learning psychology.

As shown in Table 10, after controlling for gender, academic year, and major, the direct effect of the information-based environment on autonomous learning behavior remains significant ($\beta = 0.824$, $p < 0.01$). However, after the introduction of intercultural communication competence, the effect of the information-based environment is reduced ($\beta = 0.524$, $p < 0.01$), whereas intercultural communication competence itself has a significant effect ($\beta = 0.430$, $p < 0.01$), indicating that it plays a partial mediating role between the information-based environment and autonomous learning behavior. Similarly, Table 11 shows that intercultural communication competence also partially mediates the relationship between the information-based environment and autonomous learning psychology.

The results of the hierarchical regression analysis indicate that intercultural communication competence accounts for 9.2% of the effect of the information-based environment on autonomous learning behavior and 17.9% of the effect on learning psychology. This suggests that intercultural communication competence indirectly facilitates autonomous learning through a chain of knowledge acquisition – skill application – psychological reinforcement, aligning with Byram's (2001) intercultural competence model of knowledge – attitudes – skills.

5 Strategies for Enhancing Students' Autonomous English Learning Ability in the Information-Based Environment

The continuous advancement of information technologies has led to the constant evolution of informatized learning environments. While college students are increasingly engaging with digital platforms and resources for English learning, the abundance and complexity of available materials often leave some students feeling overwhelmed. Many students are not proficient in using digital learning tools for autonomous English study, and most fail to establish clear learning goals when these tools are utilized, leading to reduced focus and learning inefficiencies. Moreover, students' strategic planning and self-monitoring abilities in autonomous learning need further improvement.

Emerging technologies such as artificial intelligence (AI) offer promising opportunities for enhancing English learning by enabling personalized learning material recommendations through big data, facilitating self-evaluation, and assisting in goal setting. However, emotional responsiveness in human–machine interactions remains underdeveloped, and technical support for students' decision-making and self-assessment strategies within the information environment is still lacking.

On the basis of the findings of this study, the following strategies are proposed across three dimensions: institutional informatization and English education reform, the enhancement of teachers' digital literacy, and the promotion of students' autonomous learning awareness.

5.1 Institutional Strategies: Optimizing Smart Resource Repositories and Deploying AI-Supported Systems

First, it is essential to upgrade hardware facilities and diversify software resources. Despite the influx of English learning products and platforms into the educational field, regulatory standards remain underdeveloped, and quality control is insufficient. Schools should build unified digital platforms to meet the growing demands of smart campus development. Integrating educational systems into hyperconvergent platforms and constructing high-quality English resource libraries can support data collection on teacher–student interactions and student learning progress.

Second, institutions should address tool usability and resource compatibility. By aggregating high-quality English learning resources at various levels, schools can create institution-specific digital libraries that support students' diverse and personalized learning objectives. Selecting appropriate digital resources and tools and integrating data on group and individual learning profiles can enhance the effectiveness and efficiency of autonomous English learning. Schools should actively introduce AI-powered English learning support systems (e.g., virtual reality English scenarios and intelligent speech-practice tools) to provide personalized learning paths and real-time feedback.

To address students' low proficiency in using smart tools and limited engagement with AI models, institutions should establish "digital learning workshops" that offer specialized training, such as AI tool mastery and cross-platform collaborative learning. These sessions may include tutorials on platform functionalities and integration with intelligent recommendation systems. In response to significant disparities in intercultural competence across majors, subject-specific cultural resource modules (e.g., embedded cultural case databases) should be developed to enhance students' cultural sensitivity and communication skills.

5.2 Teacher Strategies: Strengthening Digital Instructional Design Competence

The integration of informatization into education imposes new demands on teachers. As instructional designers and facilitators of student learning, English teachers play crucial roles in shaping learners' information literacy and autonomous learning capabilities. To guide students effectively in using online resources for self-directed learning, teachers themselves must possess solid digital competencies, including the ability to develop and organize educational content through information technologies.

Therefore, English teachers should update their pedagogical perspectives, increase their sensitivity to digital learning environments, and strengthen their digital instructional design skills. They should stay informed about technological advancements and incorporate the development of students' digital literacy into English instruction.

To address students' weak self-monitoring, teachers can design integrated tool-based learning tasks (e.g., using a dictionary app, grammar checker, and online collaboration platform for writing). Learning platforms can be used to analyse task duration and tool-switching frequency, helping identify learning bottlenecks and enabling personalized guidance. To mitigate information overload anxiety, microskills such as "resource filtering and goal setting" can be embedded into lessons. Through case-based instruction, teachers can demonstrate how to use AI tools (e.g., Doubao, DeepSeek) to generate personalized learning plans and reduce the aimlessness often observed in autonomous learning.

5.3 Student Strategies: Enhancing Metacognitive Strategies and Tool Integration Skills

On the basis of the theoretical framework of "information-based environment–learning behavior–learning psychology–intercultural competence", students can adopt the following strategies to optimize learning processes and improve overall competence.

First, students should restructure their learning ecosystems and behavior patterns by leveraging informatized environments. They should use technological tools to integrate and organize fragmented learning resources (e.g., categorizing materials by theme), adopt problem-based learning (PBL) strategies, and engage in group collaboration to enhance communication skills. By maintaining learning behavior journals, students can reflect on cognitive biases in intercultural interactions. Simultaneously, psychological self-regulation should be strengthened—by breaking down learning goals to build self-efficacy, managing anxiety through metacognitive strategies, and utilizing emotional support functions embedded in platforms to alleviate negative emotions.

Second, students should deliberately cultivate intercultural competence and adapt their strategies flexibly. This can involve learning about cultural differences (e.g., comparing Chinese and Western etiquette, watching intercultural documentaries), participating in virtual cultural events, and preparing culturally relevant questions for conversations with international peers. After each interaction, students summarize and reflect on the experience. Through continuous adaptation, students can create a positive feedback loop among learning tools, learning behavior, emotional states, and intercultural competence—ultimately fostering proactive, adaptive learning in the information-based environment.

6 Limitations and Future Directions

This study has several limitations. First, the sample was drawn from a single vocational college, with a large proportion of lower-year students (58%), which may affect the generalizability of the findings. Future research could include samples from undergraduate institutions for comparison. Second, the data were collected through questionnaires and interviews, which may be subject to social desirability bias. Future studies should incorporate learning platform log data (e.g., login duration, frequency of resource usage) or eye-tracking experiments for behavioral validation.

Moreover, the mediating mechanism of intercultural communication competence should be further tested via structural equation modelling (SEM) to more accurately reveal the pathways through which the informatized environment influences autonomous learning. These limitations will serve as directions for future research.

7 Conclusion

Information technologies have transformed the landscape of English education by reshaping instructional settings, teaching practices, and learning models. While these technologies can empower education, they also introduce negative side effects. Students may feel lost in a sea of disorganized learning materials or lack clear goals. Furthermore, current human–computer interactions lack emotional depth, which may lead to learning fatigue, confusion, or other negative psychological outcomes.

As AI technologies continue to develop, the informatized learning environment becomes more complex. Although AI is indispensable for broadening students' global outlook and improving learning efficiency, relying solely on information technologies for autonomous English learning presents many challenges—for example, machine learning lacks humanistic understanding. While information technologies can reshape pedagogy, they must not replace the value of interpersonal communication. The complexities and uncertainties of real-life interactions cannot be fully replicated by human–computer interfaces or virtual worlds.

Therefore, during this critical phase of educational informatization, society, institutions, and educators must work collaboratively to build a student-centered, humanistic, and ethically grounded informatized educational environment. By guiding the positive integration of new technologies into English education and exploring context-sensitive practical approaches, we can ensure meaningful support for the enhancement of students' autonomous English learning outcomes.

References

1. Ministry of Education of the People's Republic of China. Ten-Year Development Plan for Educational Informatization (2011–2020) (2012). http://old.moe.gov.cn/publicfiles/business/htmlfiles/moe/s3342/201203/133322.html. Accessed 12 Sept 2019
2. Li, X., Zhang, T., Liu, M.: A study on the influencing factors of learning anxiety in virtual learning communities. Distance Educ. J. **34**(6), 55–62 (2016)
3. He, K.: New developments in the theoretical research of educational informatization in China. China Educ. Technol. (1), 1–19 (2011)
4. Tang, Y., Ru, L., Fan, J., et al.: Personalized learning path planning based on learner profile modelling. E-Educ. Res. **40**(10), 53–60 (2019)
5. Wang, C., Xu, P., Hu, Y.: The effects of desktop virtual reality learning environments on student engagement and academic performance: based on multimodal data. Open Educ. Res. **27**(3), 112–120 (2021)
6. Cai, S., Zhang, H., Xue, X., Wang, T., Wang, P., Zhang, Z.: Application cases of augmented reality (AR) in teaching: a review. China Educ. Technol. (3), 1–9 (2017)
7. Zheng, X., Zhang, Y.: The impact of digital learning environments on vocational students' learning outcomes: a meta-analysis based on 61 empirical studies. Contemp. Vocat. Educ. (2), 74–83 (2024). https://doi.org/10.16851/j.cnki.51-1728/g4.20240322.009
8. Xu, J., Tian, Z.: A comparative study of media effects in online learning environments. Open Educ. Res. (6), 74–76 (2006)
9. Wrzesien, M., Raya, M.A.: Learning in serious virtual worlds: evaluation of learning effectiveness and appeal to students in the E-Junior project. Comput. Educ. **55**(1), 178–187 (2010)
10. Jang, D.H., Yi, P., Shin, I.S.: Examining the effectiveness of digital textbook use on students' learning outcomes in South Korea: a meta-analysis. Asia-Pac. Educ. Res. **25**, 57–68 (2016)
11. Li, G., Song, J.: Practices and insights of digital transformation in education from an international perspective. People's Educ. (19), 71–74 (2022)
12. Deci, E.L., Ryan, R.M. (eds.): Handbook of Self-Determination Research. The University of Rochester Press, New York (2002)
13. Vansteenkiste, M., Simons, J., Lens, W., Sheldon, K.M., Deci, E.L.: Motivating learning, performance, and persistence: the synergistic effects of intrinsic goal contents and autonomy-supportive contexts. J. Pers. Soc. Psychol. **87**(2), 246–260 (2004). https://doi.org/10.1037/0022-3514.87.2.246
14. Zimmerman, B.J.: Attaining self-regulation: a social cognitive perspective. In: Boekaerts, M., Pintrich, P., Zeidner, M. (eds.) Handbook of Self-Regulation, pp. 13–39. Academic Press, San Diego (2000)
15. Pang, W.: On students' autonomous learning. J. East China Normal Univ. (Educ. Sci. Ed.) (2), 78–83 (2001)

Research on the Integration of Ideological and Political Education into Cybersecurity Postgraduate Courses: A Big Data-Driven Approach

Long Zhang(✉), Bing Xia, Ying Xing, Qinghao Zhu, and Mingyang Ji

Zhongyuan University of Technology, Zhengzhou 451191, China
zhanglong@zut.edu.cn

Abstract. This paper investigates the integration of ideological and political education within cybersecurity postgraduate curricula. Through systematic analysis of pedagogical challenges and the integrated application of big data analytics, AI-enhanced pedagogy, and data mining methodologies, we propose an innovative framework to synergize technical training with ethical cultivation. This approach aims to develop high-quality cybersecurity professionals equipped with moral integrity, technical expertise, and strategic compliance awareness, thereby advancing the cybersecurity industry's sustainable development.

Keywords: Cybersecurity Talent Development · Ideological–Political Curriculum Integration · AI-Driven Pedagogical Framework · Ethical–Technical Competency Modelling · Cybersecurity Education Analytics

1 Introduction

With the rapid development of information technology, the field of cybersecurity has become crucial for national security and social stability. Postgraduate education in cybersecurity shoulders the responsibility of cultivating high-level professionals. However, in current education, there is a lack of effective integration of ideological and political education into cybersecurity postgraduate courses [1]. For example, in the teaching of network attack and defense technology, the separation between technical knowledge and ethical norms is prevalent [2]. To address this issue, exploring new educational models and leveraging advanced technologies such as big data, intelligent computing, and data mining is highly important [3].

2 Current Situation of Ideological and Political Education in Cybersecurity Postgraduate Courses

2.1 Weak Integration with Professional Courses

In the current education of cybersecurity postgraduate programs, ideological and political education and professional courses often operate independently. Professional curricula predominantly emphasize technical competencies, whereas ideological-political

Y. Yu et al. (Eds.): ICPCSEE 2025, CCIS 2673, pp. 25–36, 2026.
https://doi.org/10.1007/978-981-95-2566-9_2

education remains confined to traditional pedagogical frameworks, lacking organic integration. For example, in the teaching of network attack and defense technology, teachers rarely explore ideological and political elements, resulting in a fragmented dichotomy that hinders students from integrating ideological-political knowledge with professional practice.

2.2 Insufficiently Targeted Content

“Existing ideological-political content inadequately addresses domain-specific characteristics of the cybersecurity field [4]. Cybersecurity has its own industry norms, professional ethics, and security ethics issues. However, in ideological and political education, these aspects are not well represented. For example, in the teaching of network security technologies, there is a lack of in-depth guidance on how to use technology correctly and avoid abuse, which may lead to students’ lack of correct value judgment in complex cybersecurity situations.

2.3 Lack of Innovative Teaching Methods

The teaching methods of ideological and political education are relatively traditional and rely mainly on lectures. This single teaching method has difficulty stimulating students’ learning interest and initiative, especially for postgraduate students in cybersecurity, who have strong self-learning and practical needs. When ideological and political education is integrated into professional courses, modern educational technology and diverse teaching methods are not fully utilized and fail to meet the diverse learning needs of students.

2.4 Imperfect Evaluation System

The current curriculum assessment system focuses mainly on the assessment of professional knowledge and skills and lacks scientific and effective methods for evaluating the effectiveness of ideological and political education. The proportion of ideological and political education in curriculum assessment is low, and the assessment methods are single, mostly relying on simple theoretical assessments. This approach cannot comprehensively and accurately reflect students’ ideological and political literacy and practical abilities, reducing their attention to ideological and political education. The current situation of ideological and political education in cybersecurity courses is shown in Table 1.

Table 1. Current situation analysis

Problem Dimension	Specific Manifestations	Impact
Weak Integration with Professional Courses	Ideological and political education and professional courses are separate; for example, cybersecurity teaching rarely incorporates ideological elements	Students find it difficult to integrate ideological and political knowledge with professional practice
Insufficiently Targeted Content	The ideological and political content does not consider the characteristics of the cybersecurity field, lacking guidance on technology use	Students may make inaccurate value judgments
Lack of Innovative Teaching Methods	Rely on lectures and use few modern teaching techniques	It is hard to stimulate students' learning interest and initiative
Imperfect Evaluation System	Assessments prioritize professional knowledge, downplay ideological/political elements, and use limited methods	It is difficult to comprehensively reflect students' ideological and political qualities and abilities

3 Significance of Integrating Ideological and Political Education into Cybersecurity Postgraduate Courses

3.1 Improving Postgraduate Comprehensive Quality

Cybersecurity postgraduate students need to not only master solid professional knowledge but also possess noble moral sentiment and social responsibility. By integrating ideological and political education into courses, socialist core values, network ethics, and other content are incorporated into professional teaching. This can guide students to establish correct worldviews, outlooks on life, and values, enhance their moral qualities and professional ethics, and cultivate well-rounded cybersecurity talent with both ability and integrity.

3.2 Promoting Discipline Development

Cybersecurity is an interdisciplinary field involving computer science, law, psychology, sociology, and other disciplines. The integration of ideological and political education can strengthen communication and integration among different disciplines, promote interdisciplinary innovation, and drive the development of the cybersecurity discipline. For example, in the study of network security policies, the integration of legal and ethical knowledge from different disciplines can deepen students' understanding of the subject.

3.3 Maintaining National Network Security

As an important part of national security, cybersecurity is directly related to national sovereignty, security, and development interests. Through ideological and political education, students' awareness of national security and the law can be enhanced. They can develop a profound understanding of the role of cybersecurity in safeguarding national stability and fostering economic development and thus consciously safeguarding national network security.

3.4 Creating a Good Network Environment

The network space is the common spiritual home of hundreds of millions of people. By integrating ideological and political education into courses, students can be guided to establish correct moral network concepts; resist bad behaviors such as network violence, rumour spread, and privacy infringement; and create a clean and upright network environment [5]. The significance of the integration is shown in Table 2.

Table 2. Significance of Integration

Significance Dimension	Specific Elaboration
Improving Postgraduate Comprehensive Quality	Incorporate core values to cultivate well-rounded cybersecurity talents
Promoting Discipline Development	Strengthen interdisciplinary communication and promote innovation
Maintaining National Network Security	Enhance students' awareness of national security and law to safeguard national cyberspace
Creating a Good Network Environment	Guide students to establish correct network ethics and resist bad online behaviors

4 Strategies for Integrating Ideological and Political Education into Cybersecurity Postgraduate Courses with the Aid of Advanced Technologies

4.1 Reconstructing the Curriculum System with the Help of Big Data

Using big data technology, we analyse the current situation of ideological and political education in cybersecurity postgraduate courses. Specialized cybersecurity ethics curricula, including "Cybersecurity Ethics & Professional Conduct" and "Cybersecurity and National Security Strategy", should be established. These courses systematically expound on moral norms, professional regulations, national strategic requirements, and international cooperation and competition in the cybersecurity field. For example, by mining big data on international cybersecurity incidents, we can select typical cases such as the Snowden incident and China's successful cybersecurity protection cases for in-depth analysis in courses, enabling students to consider the complexity and importance of cybersecurity issues from an ideological and political perspective [6].

Moreover, we reexamine the core courses of cybersecurity. Through data mining techniques, ideological and political education points are extracted from each course. In the "Network Attack and Defense Technology" course, in addition to teaching technical principles, data-driven analysis is used to add content about the legality and moral boundaries of network attacks and the importance of defense technology for social stability. The teaching syllabus and content of professional courses should be adjusted according to these ideological and political education points, ensuring the organic integration of ideological and political elements and professional knowledge.

4.2 Innovating Teaching Methods with Intelligent Computing

Intelligent computing can be used to innovate teaching methods. The latest cases in the field of cybersecurity, including positive and negative cases, are collected through data collection and analysis. Positive cases (e.g., successful cyber defense operations) and negative cases (e.g., data breaches caused by security flaws) are utilized for case-based teaching. For example, intelligent computing algorithms can be used to analyse the ethical and legal issues in these cases and present them to students. Students should discuss and analyse these cases in class, encourage them to express different views, and use intelligent tutoring systems to guide and summarize the discussions, helping students deepen their understanding of ideological and political content and improving their ability to solve practical problems with ideological and political knowledge. Figure 1 shows the intelligent computing-enhanced pedagogical workflow, comprising case mining, algorithmic analysis, and interactive tutoring modules.

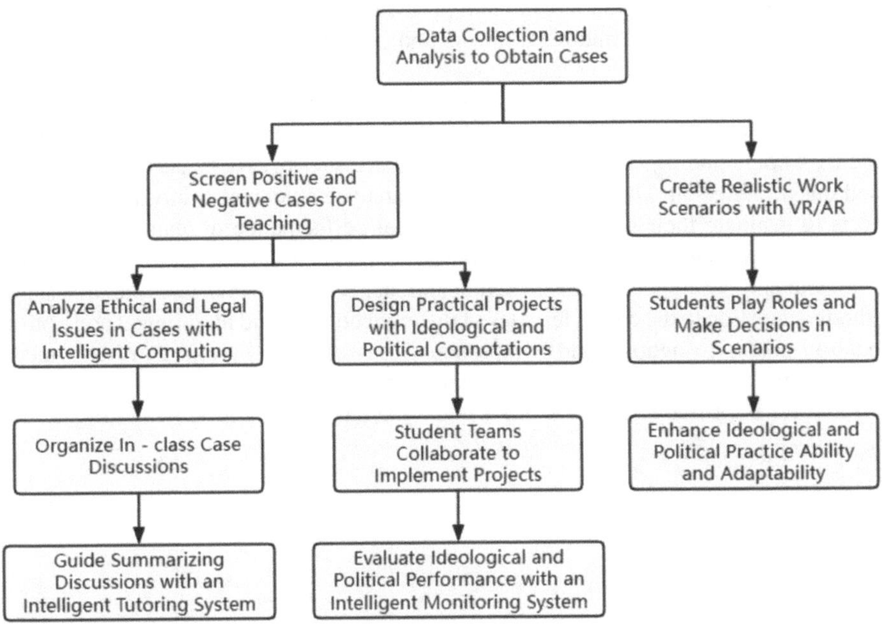

Fig. 1. Flowchart of the application of intelligent computing to innovate teaching methods

Practical projects with ideological and political connotations of cybersecurity should be designed. For example, intelligent computing can be used to simulate a network security protection system design project. In this project, students are required to follow moral and legal norms and consider factors such as user privacy protection and social impact. During the project implementation process, students can improve their ideological and political levels in practice through team collaboration. Teachers can use intelligent monitoring systems to evaluate and provide feedback on students' ideological and political performance in the project.

Realistic work scenarios in the field of cybersecurity, such as network security emergency response centers and internal decision-making scenarios of network security enterprises, are created with the help of virtual reality (VR) and augmented reality (AR) technologies, which are based on intelligent computing. Students play different roles in these scenarios, face various cybersecurity problems, and make decisions using ideological and political knowledge, improving their ideological and political practical ability and adaptability in complex network environments [7].

4.3 Perfecting the Evaluation System with Data Mining

Data mining technology can be applied to improve the evaluation system. The effectiveness of ideological and political education should be incorporated into the curriculum assessment scope [8]. Mine data on students' performance in classroom discussions, such as their degree of participation, rationality, and innovation of viewpoints, as well as their performance in ideological and political assignments, such as the quality of case analysis and understanding depth of network security professional ethics. Additionally, data on students' ideological and political behaviors in practical projects, such as compliance with moral and legal norms and consideration of social impact, should be analysed [9].

In addition to traditional written examinations, data-driven methods are used to adopt multiple assessment methods. For example, interviews were conducted to understand students' understanding of ideological and political issues in cybersecurity through data-based question design. Data mining is used to analyse students' performance in group reports to evaluate their ideological and political performance in teamwork. Through a data-based evaluation of practical operations, we check whether students implement ideological and political concepts in actual actions. By reasonably selecting assessment methods according to the characteristics of different courses and ideological and political education goals, a scientific and comprehensive assessment system can be formed to accurately reflect the effectiveness of students' ideological and political education. The flowchart for improving the evaluation system via data mining is shown in Fig. 2.

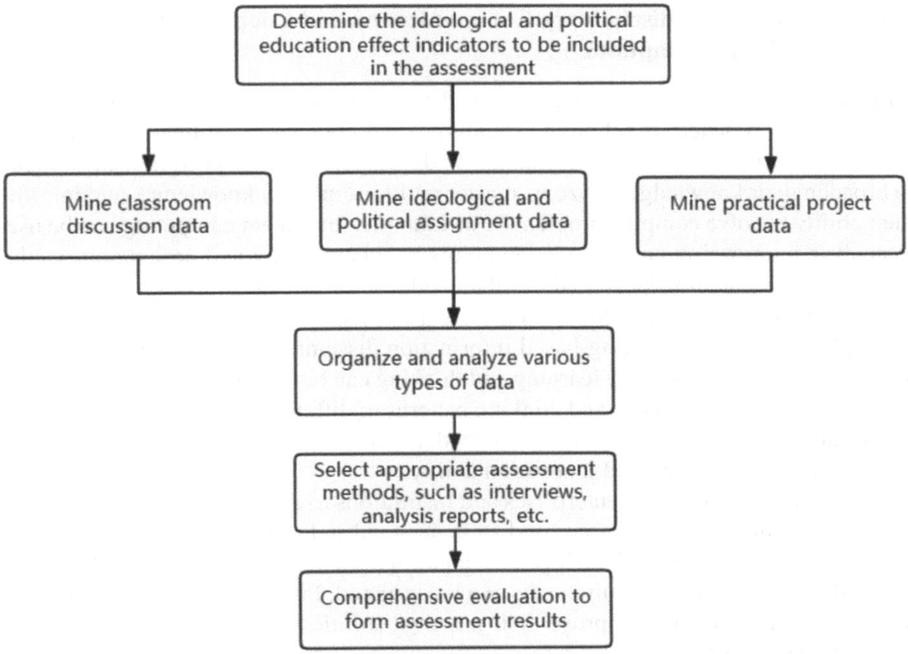

Fig. 2. Flow chart for improving the evaluation system via data mining

4.4 Strengthening the Teaching Staff with the Support of Big Data

Big data is used to analyse the ideological and political education needs of teachers in the field of cybersecurity. Teachers should be encouraged to participate in ideological and political education training courses, including ideological and political theories, educational methods, and curriculum design. The training content can cover Marxist basic principles, the theoretical system of socialism with Chinese characteristics, and Xi Jinping Thought on Socialism with Chinese characteristics for a new era. Through big data-based resource integration, ideological and political education experts, policymakers, and industry experts in the field of cybersecurity can be invited to give lectures and share experiences, broadening teachers' ideological and political horizons and improving their ideological and political education capabilities.

A cooperation mechanism between cybersecurity professional teachers and ideological and political teachers should be established with the help of big-data-enabled communication platforms. Encourage them to jointly conduct teaching research and practice in ideological and political education. Through regular online and offline teaching seminars, big-data-driven data analysis can be used to explore ideological and political education elements jointly, design teaching plans, and analyse teaching effects. Promote cross-teaching and mutual evaluation activities among teachers, enabling professional teachers to learn the teaching methods of ideological and political teachers and ideological and political teachers to understand the content of cybersecurity professional courses, thus improving the overall teaching quality.

4.5 Promoting Interdisciplinary Communication and Cooperation with the Aid of Intelligent Computing

AI-driven convergence should be leveraged to synthesize knowledge from computer science, jurisprudence, psychology, sociology, and cognate disciplines in curriculum settings. Interdisciplinary courses should be developed to guide postgraduate students to broaden their knowledge horizons, master multidisciplinary knowledge, and improve their ability to solve complex problems. For example, intelligent computing can be used to analyse interdisciplinary knowledge points in cybersecurity and design courses that combine technical knowledge with legal and ethical knowledge.

Experts from different fields are regularly invited to give lectures and hold seminars. Through intelligent computing-based information dissemination platforms, postgraduate students' interdisciplinary learning and thinking can be promoted. For example, they analyse the research methods and thinking patterns of different disciplines through intelligent data analysis and introduce them to students during seminars, cultivating students' interdisciplinary thinking and innovation ability.

Cooperate with relevant enterprises and institutions to establish a collaborative education platform. Big data can be used to understand industry needs and development trends, and with the help of intelligent computing, practical projects and training programs that meet industry requirements can be designed. Provide practical opportunities for postgraduate students, improve their practical abilities and professional ethics, and lay a foundation for their future career development.

5 Feasibility Analysis

5.1 Policy Support

The state attaches great importance to the cultivation of cybersecurity talent and the construction of ideological and political education in courses. The Ministry of Education has issued a series of guiding opinions on integrating ideological and political education into various courses. The implementation of the national cybersecurity strategy requires many high-quality cybersecurity professionals with both moral and professional qualities. This project is in line with national policy guidance and can obtain strong policy support [10].

5.2 Teaching Staff

The applicant unit has high-level teaching staff with rich teaching and scientific research experience. Through ideological and political training and interdisciplinary cooperation, professional teachers can quickly master ideological and political education methods and integrate ideological and political elements into professional courses. The team of ideological and political teachers can provide theoretical guidance and teaching support for the integration of ideological and political education into cybersecurity postgraduate courses. Cooperation between the two types of teachers can ensure the scientificity and effectiveness of ideological and political education content and improve the quality of curriculum ideological and political constructions [11].

5.3 Teaching Resources

The cybersecurity major has a certain number of textbooks and rich online teaching resources. By using big data to mine and integrate these resources, ideological and political elements can be added. The college has forged robust partnerships with leading cybersecurity firms, which provide substantial support for practical teaching and students' internships. The resources of enterprises in professional ethics education and practical skills training can be organically combined with ideological and political education, enriching the practical teaching resources of curriculum ideological and political construction.

5.4 Student Demand

Postgraduate students in the field of cybersecurity generally recognize the importance of ideological and political education for their own development and have a high acceptance of integrating ideological and political content into professional courses. Through reasonable teaching design and vivid teaching cases, which can be optimized via data-driven methods, students' learning interest can be stimulated. Postgraduate students have strong self-learning and problem-solving abilities. With the guidance of teachers, they can deeply understand ideological and political education content and combine it with professional knowledge. At the same time, diversified teaching methods and practical activities can meet the learning needs of different students and promote their effective acceptance of curriculum ideological and political education.

6 Expected Results and Benefits

6.1 Expected Results

The ideological and political elements related to "a clean and upright atmosphere" were embedded in the course syllabus. For example, in the "Cybersecurity Laws and Regulations" course, the legal responsibility of maintaining a clean and upright network space is emphasized. In the "Network Attack and Defense Technology" course, students are guided to establish correct values to ensure that technology is used for just purposes. These revised syllabuses will become normative documents for guiding ideological and political teaching.

To develop teaching courseware that integrates ideological and political content for professional cybersecurity courses. In the courseware, ideological and political elements are presented in various forms, such as pictures, videos, and texts, while professional knowledge is taught. For example, when explaining network encryption technology, videos of national policies are inserted to encrypt and protect critical information infrastructure to enhance students' understanding of national cybersecurity strategies and their sense of responsibility.

Publish 1–2 research papers on the construction of ideological and political education in courses in high-level educational and teaching journals. These papers will conduct in-depth research on the theoretical basis, practical models, and effectiveness evaluation of curriculum ideological and political construction, share the innovative ideas and practical experiences of this project in the construction of ideological and political education in courses oriented by "a clean and upright atmosphere", and provide references for the curriculum ideological and political construction of related majors in other universities.

6.2 Implementation Scope

This project will cover the core courses of the cybersecurity major, including but not limited to network security basics, network attack and defense, cryptography, and cybersecurity laws and regulations.

All teachers teaching cybersecurity professional courses will be involved in this project. Through teacher training, seminars, and other activities, each teacher can understand and master the goals and methods of curriculum ideological and political constructions and can effectively implement ideological and political teaching in their own courses [12].

The direct beneficiaries of this project are all postgraduate students majoring in cybersecurity. Through curriculum ideological and political constructions, students' professional qualities and moral responsibilities in the field of cybersecurity can be cultivated, enabling them to uphold the concept of "a clean and upright atmosphere" in their future careers and contributing to the cause of national cybersecurity [13].

6.3 Benefits

For students through curriculum ideological and political constructions, they deeply understand the importance of "a clean and upright atmosphere" in the network space and enhance their recognition of national cybersecurity strategies and network space moral norms. Students' patriotism, social responsibility, and professional ethics should be cultivated, enabling them to abide by laws, regulations, and moral standards in network space behavior and resist bad information and behaviors. The integration of ideological and political elements and professional courses helps students understand the mission and value of the cybersecurity major from a more macroscopic perspective. It also cultivates students' innovation ability and critical thinking, guiding them to comprehensively consider technical, ethical, legal, and other factors when facing complex cybersecurity problems and making correct decisions. This improves students' comprehensive qualities and enhances their competitiveness in the job market, providing high-quality professionals for the national cybersecurity field.

Teachers. Participating in curriculum ideological and political construction projects requires teachers to explore ideological and political elements in detail and explore methods of integrating them with professional knowledge. This will prompt teachers to update their teaching concepts, improve teaching methods, and enhance teaching design capabilities. Through participating in training and seminars, teachers can learn from each other and further improve their ideological and political teaching levels. This project provides rich materials and practical platforms for teachers to carry out teaching research. Teachers can conduct in-depth research on problems and experiences in curriculum ideological and political construction, publish teaching research papers, and enhance their academic status in the fields of education and teaching while also contributing to the teaching reform and development of the school.

For the School and the Major. By implementing high-quality curriculum ideological and political construction projects, the school can form characteristics in the cultivation of postgraduate students majoring in cybersecurity and improve the school's reputation in the education field and society. It demonstrates a school's ability and determination to cultivate high-quality professionals with both moral and professional qualities, attracting more high-quality students and social resources [14]. Curriculum ideological and political constructions help enhance the connotational construction of the cybersecurity major. This makes professional training objectives more in line with national strategic needs and social development requirements, enhances the competitiveness and influence of the major, and promotes the optimization and upgrading of the major in terms of the curriculum system, teaching staff, and teaching resources, driving the major to develop to a higher level.

For Society. Cultivating cybersecurity professionals with the concept of "a clean and upright atmosphere" will help purify the network environment and ensure national network security. These professionals become the backbone of the cybersecurity industry after they enter society; play an active role in various links, such as cybersecurity technology research and development, management, and services; and contribute to the construction of a healthy, safe, and orderly network space. The implementation and result promotion of this project will, to a certain extent, guide the value orientation of talent cultivation in the field of cybersecurity in society, prompt the entire industry to pay more attention to the ideological and political education of talent, create a good industry atmosphere, and promote the healthy and sustainable development of the cause of cybersecurity [15].

7 Conclusions

Integrating ideological and political education into cybersecurity postgraduate courses is an important measure to meet the needs of the development of the cybersecurity industry and cultivate high-quality professionals. With the help of advanced technologies such as big data, intelligent computing, and data mining, we can effectively address the existing problems in the integration process, improve the effectiveness of ideological and political education, and achieve the coordinated development of ideological and political education and professional education. This project has important research significance and high feasibility. Continuous exploration and practice are expected to provide valuable experience and reference for the curriculum's ideological and political construction of the cybersecurity major and other related majors and to contribute to the long-term development of the nation.

References

1. Qin, T., Liu, Y., Zhang, H., et al.: Wireless network-based evaluation of ideological education management in colleges using AI and big data. IEEE Trans. Learn. Technol. **15**(4), 502–512 (2022)
2. Wu, C., Ma, L.: Machine learning-driven online learning model for political education: a case study of cybersecurity ethics. J. Educ. Data Mining **14**(3), 89–103 (2022)

3. Chen, H., Watanabe, T.: Blockchain-based curriculum design for cybersecurity ethics education. In: Proceedings of the 2024 IEEE Global Engineering Education Conference (EDUCON), pp. 1–6 (2024)
4. National Institute of Standards and Technology (NIST). NICE cybersecurity workforce framework 2.0: Integrating ethics into technical competencies (NIST Special Publication 800-181 Rev. 2). U.S. Department of Commerce (2023)
5. European Union Agency for Cybersecurity (ENISA). GDPR compliance in cybersecurity education: Guidelines for higher institutions (ENISA Technical Report No. 2023-07) (2023). https://www.enisa.europa.eu/publications/gdpr-compliance-in-cybersecurity-education. Accessed 18 July 2024
6. ISO/IEC 27032:2023 Information technology — Security techniques — Guidelines for cybersecurity education in critical infrastructure. International Organization for Standardization (2023)
7. Smith, J., Zhou, W. Virtual reality simulations for teaching cybersecurity ethics: a multirole decision-making approach. ACM Trans. Comput. Educ. **24**(2), Article 8 (2024)
8. Brown, T., Garcia, M.: Gamified learning in cybersecurity: embedding ethical scenarios in adversarial machine learning labs. J. Cybersecur. Res. **12**(1), 45–60 (2023)
9. Yin, J., Yuan, X.: Precision teaching in blended learning environments: a case study of cybersecurity course design. Comput. Educ. **185**, Article 104531 (2022)
10. Zheng, Q., Wang, L., Li, X.: Curriculum ideology-politics integration in "information security" courses: a case study of SM4 cryptographic algorithm. IEEE Trans. Educ. **66**(2), 234–245 (2023)
11. Rong, Y., Gang, Z.: Evaluating political education strategies via AI-driven data mining: a longitudinal study of cybersecurity students. Educ. Technol. Soc. **25**(4), 112–124 (2022)
12. Johnson, R., Chen, L.: Federated learning for cross-institutional analysis of cybersecurity education effectiveness. In: Proceedings of the AAAI Conference on Artificial Intelligence, vol. 38, pp. 134–141 (2024)
13. ETSI Cybersecurity Committee. 5G security education framework: Aligning technical training with ethical governance. (ETSI White Paper No. 48) (2024). https://www.etsi.org/deliver/etsi_wp/48_09999.pdf. Accessed 18 July 2024
14. Liu, X., Zhang, Q.: Longitudinal tracking of ethical decision-making in cybersecurity practicums: a natural language processing approach. ACM SIGCSE Bull. **53**(4), 112–125 (2025)
15. United States Cyber Command (USCYBERCOM). Annual assessment of cybersecurity education outcomes: Metrics for ethical competency (Technical Report CYBERCOM-2024-001) (2024)

A Bibliometric Analysis of the Current Status and Trends in Higher Vocational Education Research

Shanshan Li[1], Kebu Wang[1], Xia Liu[2](✉), Yuge Gao[1], and Hongfeng Zhang[2]

[1] Sanya Aviation & Tourism College, Sanya 572000, China
[2] Faculty of Humanities and Social Sciences, Macao Polytechnic University, Macao, China
p2315160@mpu.edu.mo

Abstract. This study employs bibliometric methods to conduct an empirical analysis of 53 vocational education projects approved under the National Education Science Planning (NESP) Program between 2020 and 2024 in higher vocational institutions (HVI) across China. The aim is to reveal the current state, developmental trajectory, key characteristics, and future directions of research in higher vocational education. The findings indicate that the number and categories of approved projects are significantly influenced by national policy orientations, exhibiting notable fluctuations and overall imbalances in distribution. The geographical distribution of project-leading institutions is uneven, and the distinctive characteristics of higher vocational colleges have yet to be fully demonstrated. While the research themes increasingly align with national vocational education policies, the construction of research teams remains weak. Based on the structural analysis of the approved projects, it is recommended that HVI focus on type-based positioning, coordinate research capacity with regional development needs, optimize the allocation of research resources, concentrate on key tasks such as quality improvement and excellence cultivation, and enhance the research motivation of young faculty members. These efforts are essential for advancing vocational education research toward a new stage of high-quality development.

Keywords: Vocational education research · National Education Science Planning (NESP) · quality improvement and excellence cultivation · high-quality development

The National Education Science Planning (NESP) Program constitutes a vital component of educational research in China. As a national-level research initiative, it encompasses 15 disciplinary areas, including fundamental theories of education, educational psychology, higher education, and vocational and technical education (VTE). Among these, research projects in VTE hold a particularly significant position within the overall framework, as they effectively reflect the research priorities and developmental level of the vocational education sector in China [1].

Since the implementation of the 13th Five-Year Plan, the Chinese government has consistently prioritized vocational education as a strategic instrument for promoting

Y. Yu et al. (Eds.): ICPCSEE 2025, CCIS 2673, pp. 37–57, 2026.
https://doi.org/10.1007/978-981-95-2566-9_3

socio-economic development. It has provided unprecedented policy support and institutional guarantees, positioning vocational education as a key area of national education reform. A series of policy documents have been issued to promote institutional innovation and qualitative enhancement in vocational education. Against this policy backdrop, the reform and development of vocational education—particularly within higher vocational colleges—have achieved remarkable progress, making it a highlight of China's broader educational reform efforts. As a core component of the vocational education system, higher vocational institutions (HVI) play a critical role in advancing the development of VTE nationwide.

This study conducts a statistical and analytical examination of VTE projects undertaken by HVI that were approved under the NESP Program between 2020 and 2024. By integrating quantitative data with scholarly interpretations of relevant policy documents, the study aims to systematically explore the current research landscape, developmental characteristics, and evolutionary trajectory of vocational education research in higher vocational colleges during this period. Based on this analysis, the paper further offers both theoretical and practical recommendations regarding the future directions of VTE research in China.

1 Research Design

1.1 Data Sources and Research Subjects

The data for this study were obtained from the official website of the Office of the NESP Leading Group, specifically focusing on VTE projects approved for HVI between 2020 and 2024 [2]. Among these, only the projects approved for higher vocational colleges were selected as the research subjects, in order to ensure the authority and relevance of the data.

1.2 Methods, Tools, and Analytical Content

This study employed homogeneous sampling, a type of purposeful sampling, which is generally used when the research aims to focus on a specific and uniform subgroup. In addition, bibliometric analysis was conducted, supplemented by various analytical tools including Cite Space, Usenet, VOS viewer, and SPSS. These tools were used to analyze the number and categories of approved projects, the responsible institutions and their geographic distribution, institutional types and characteristics, as well as to identify and categorize research hotspots associated with the sampled projects. Descriptive statistical methods were applied to interpret the findings.

2 Analysis of the Current Status of Vocational Education Projects in Higher Vocational Institutions

2.1 Analysis of the Number and Categories of Approved Projects

2.1.1 Quantitative Analysis

As shown in Table 1, a total of 2,473 projects were approved under the NESP Program between 2020 and 2024. Among them, 187 projects were categorized under VTE, accounting for 7.56% of the total, with an average of 37.4 projects approved annually. Of these, 53 projects were undertaken by HVI, representing 2.14% of the total, with an annual average of 10.6 projects.

Table 1. Overview of vocational and technical education (VTE) Projects Approved under the National Education Science Planning (NESP) Program (2020–2024)

Year	Total Approved Projects Nationwide	VTE Projects	Proportion of VTE Projects (%)	VTE Projects Approved for HVI	Proportion of Higher Vocational VTE Projects (%)
2020	437	30	6.85%	12	40.00%
2021	453	28	6.18%	10	35.71%
2022	584	58	9.93%	17	29.31%
2023	525	42	8.00%	11	26.19%
2024	474	29	6.12%	3	10.34%
Total	2473	187	7.56%	53	2.14%

Based on the data in Table 1, a trend graph illustrating the number of approved VTE projects and those specifically undertaken by HVI was generated (see Fig. 1). Chronologically, 2022 appears to mark a turning point in the number of project approvals. According to the annual approval data, the trend can be roughly divided into two phases [3].

The first phase (2020–2022) was characterized by a slight upward fluctuation in project numbers. During this period, the average annual number of approved VTE projects was 38.7, with HVI averaging 13 approved projects per year.

The second phase (2022–2024) showed a sharp decline in project numbers. The total number of VTE projects decreased by 44.44%, while those approved for HVI dropped by 45.45%. Overall, this downward trend is largely consistent with the general trend observed in nationwide project approvals during the same period.

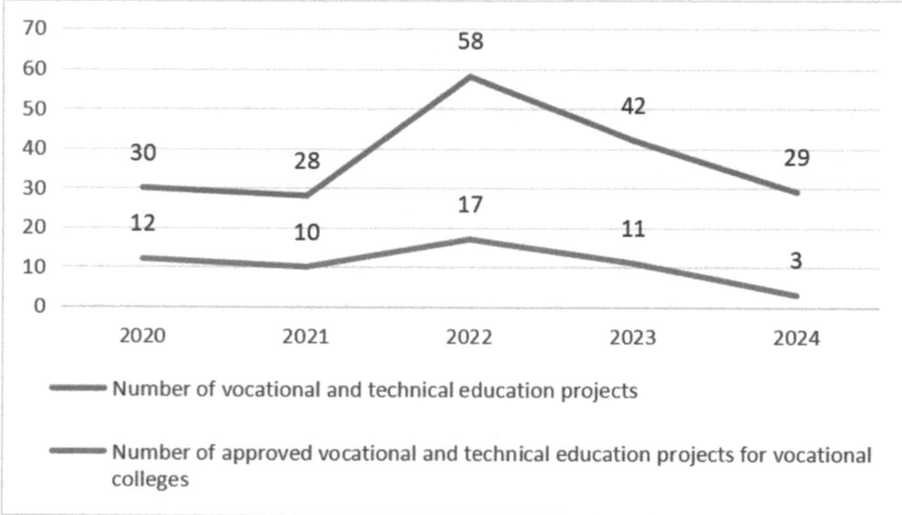

Fig. 1. Annual Trends in the Number of Approved Projects

2.1.2 Category Analysis

Between 2020 and 2024, VTE projects encompassed two primary categories and seven subcategories. The first primary category consists of education research projects funded by the National Social Science Fund and approved by the Office of the NESP Leading Group, including five subcategories: National Major, National Key, National General, National Youth, and Western Region projects [4]. The second category comprises projects approved by the Ministry of Education (MOE) through the same office, with two subcategories: MOE Key Projects and MOE Youth Projects [3]· Notably, Western Region projects were only approved for HVI in 2022. A detailed breakdown of the VTE projects approved for HVI is presented in Table 2.

Table 2. Detailed Breakdown of vocational and technical education (VTE) Projects

Year	National Major Project	National Key Project	National General Project	National Youth Project	Western Region Project	MOE Key Project	MOE Youth Project	Number of Projects Approved for HVI
2020	1	0	4	0	0	5	2	12
2021	0	0	3	0	0	5	2	10
2022	0	0	2	0	1	10	4	17
2023	0	0	6	0	0	3	2	11
2024	0	0	2	0	0	0	1	3
Total	1	0	17	0	1	23	11	53

(continued)

Table 2. (*continued*)

Year	National Major Project	National Key Project	National General Project	National Youth Project	Western Region Project	MOE Key Project	MOE Youth Project	Number of Projects Approved for HVI
Proportion (%)	1.89%	0.00%	32.08%	0.00%	1.89%	43.40%	20.75%	/

Between 2020 and 2024, HVI were granted one National Major project, accounting for 1.89% of the approved VTE projects; 17 National General projects, accounting for 32.08%; one Western Region project, accounting for 1.89%; 23 MOE Key projects, accounting for 43.40%; and 11 MOE Youth projects, accounting for 20.75%. It is evident that during this period, the majority of projects approved for HVI were MOE Key projects, with National General and MOE Youth projects also representing a significant proportion. Notably, only one National Major project was approved in 2020, highlighting three implications: first, competition for National Major projects is intensely fierce; second, the ability to secure such projects depends heavily on the institutions' research capacity; third, at the current stage, HVI generally exhibit relatively weak research strength and limited foundational capacity, rendering them largely incapable of undertaking National Major and National Key projects. Furthermore, from a proportional perspective, MOE Key and Youth projects constitute nearly two-thirds of the total projects approved for HVI, indicating that projects funded by the MOE are relatively more accessible for these institutions. A line chart illustrating the number of various types of VTE projects approved for HVI was plotted, showing MOE Key projects and National General projects as the two peaks. The overall trend of the line chart displays an M-shaped pattern, as shown in Fig. 2.

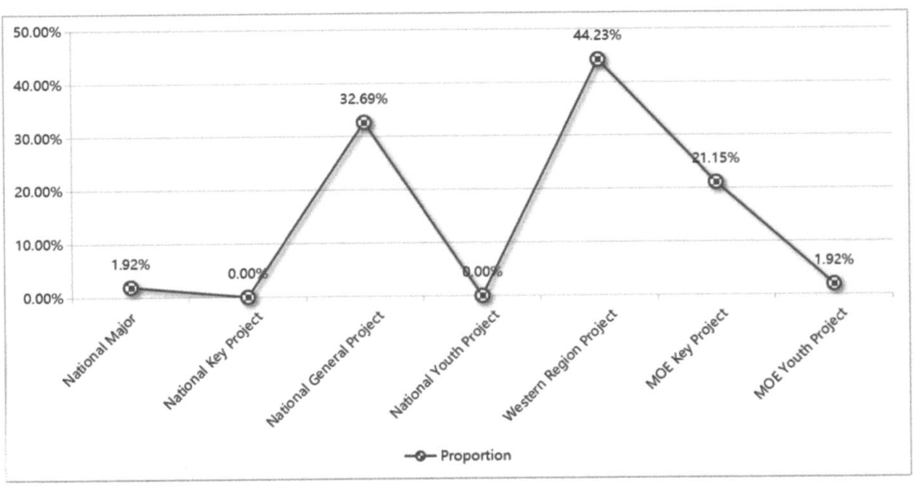

Fig. 2. Proportional Distribution of Subcategories within Vocational Education Projects

2.2 Analysis of Project-Responsible Institutions and Their Regional Distribution

2.2.1 Analysis of Responsible Institutions

From the perspective of project-responsible institutions, a total of 53 HVI were approved for VTE projects during 2020–2024. Based on the number of approved projects, five institutions—Wenzhou Vocational & Technical College, Shandong Business Vocational College, Shandong Animal Husbandry and Veterinary Vocational College, Shandong Polytechnic Vocational College, and Xiangyang Vocational & Technical College—each received approval for two projects, while the remaining 43 institutions each received approval for one project. Projects approved for institutions with two grants are summarized accordingly. As shown in Table 3, MOE Key projects continue to represent the largest share, with 5 projects accounting for 60.0%, followed by National General projects with 3 projects accounting for 30.0%. This further confirms that MOE Key and National General projects remain the primary channels through which HVI secure project approvals.

Table 3. Institutions Awarded Two or More Vocational Education Projects

Institution Name	National General Project	MOE Key Project	MOE Youth Project	Western Project	Total
Wenzhou Vocational & Technical College	1	1	0	0	2
Shandong Vocational College of Commerce	1	1	0	0	2
Shandong Vocational College of Animal Science and Veterinary Medicine	0	1	1	0	2
Shandong Polytechnic Vocational College	1	1	0	0	2
Xiangyang Vocational and Technical College	0	2	0	0	2
Subtotal	3	6	1	0	10

According to Table 3, three HVI in Shandong Province were collectively approved for six projects, one institution in Wenzhou was approved for two projects, and one institution in Hubei Province was approved for two projects. It is evident that these three provinces, known for their relatively advanced vocational education systems, demonstrated notable achievements in vocational education during the 2020 to 2024 period. Moreover, between 2018 and 2020, these provinces were respectively listed by the MOE as regions with significant vocational education reform outcomes and were recommended for incentive support.

2.2.2 Analysis of the Regional Distribution of Responsible Institutions

Based on the geographic locations of the project-responsible institutions, during the period 2020 to 2024, HVI in 17 out of 34 provinces (autonomous regions and municipalities directly under the central government) nationwide—excluding Hong Kong, Macao, and Taiwan—were approved for VTE projects. The specific numbers and proportions are illustrated in Fig. 3. However, HVI in 17 provinces (autonomous regions and municipalities), including Jilin, Heilongjiang, Anhui, Gansu, Inner Mongolia, Hainan, Yunnan, Guizhou, Jiangxi, Henan, Sichuan, Ningxia, Xinjiang, and Qinghai, have never been approved for such projects.

Further analysis reveals significant disparities among provinces with an evident hierarchical pattern. HVI in Zhejiang, Shandong, Jiangsu, and other provinces received above-average approvals, exceeding 9.67 projects, accounting for 11.32% to 24.53% of the total. Tianjin, Hunan, Hubei, and Chongqing each had more than three approved projects; these provinces are generally economically developed with relatively high education levels, making them important regions for vocational education research [5]. Provinces such as Heilongjiang, Tianjin, Fujian, Beijing, Henan, Jiangxi, Liaoning, Tibet, and Chongqing had only one approved project each. The remaining provinces, particularly those with no approved projects, tend to have slower economic development and lower overall levels of vocational education development, confirming the close relationship between vocational education and regional economic and social development.

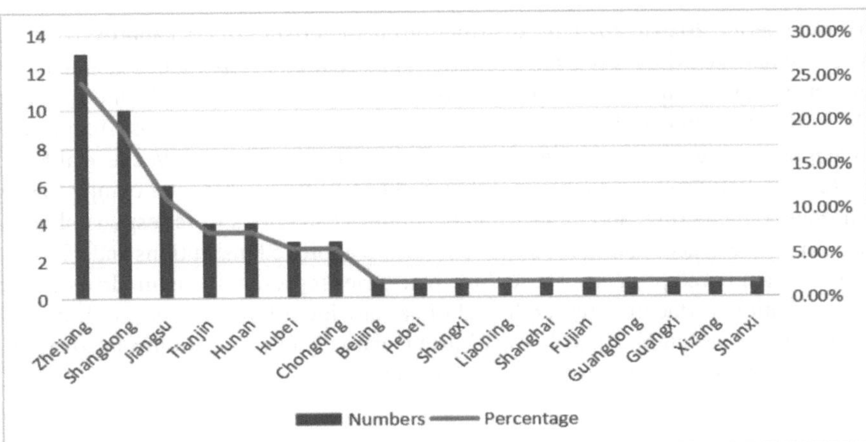

Fig. 3. Provincial Distribution of Responsible Institutions for Approved Projects

On the other hand, according to the traditional regional classification standard by the National Bureau of Statistics [6], which divides China into the Eastern (11 provinces), Central (8 provinces), and Western (12 provinces) regions, the distribution of the 53 approved VTE projects for HVI under the NESP Program is as follows: 39 projects (73.58%) in the Eastern region, 8 projects (15.09%) in the Central region, and 6 projects (11.31%) in the Western region, as shown in Table 4. Table 4 indicates that HVI in the Eastern region hold an absolute advantage in the number of approved VTE projects. However, the strength of HVI in the Central region, particularly in Hunan and Hubei provinces, should not be underestimated, as they have even surpassed regions such as Shanghai, Tianjin, Fujian, and Beijing. In the Western region, only HVI in Tibet, Chongqing, Guangxi, and Shanxi have received project approvals.

Table 4. Regional Distribution of Responsible Institutions for Funded Projects

Year	Eastern Region	Central Region	Western Region	Total
2020	8	2	2	12
2021	9	0	1	10
2022	13	2	2	17
2023	7	3	1	11
2024	2	1	0	3
Total	39	8	6	53
Proportion	73.58%	15.09%	11.32%	/

2.3 Analysis of the Categories and Nature of Responsible Institutions for Approved Projects

2.3.1 Analysis of Institutional Categories

During the period 2020 to 2024, the distribution characteristics of HVI approved for VTE projects under the NESP Program primarily consisted of comprehensive, science and engineering, and finance and economics institutions. There were significant differences in the number of approved projects among different types of institutions. According to the category of the responsible institutions (based on data from the Vocational Education Reform Network), as shown in Table 5, 44 institutions from the comprehensive, science and engineering, and finance and economics categories were responsible for 48 approved projects, accounting for 91.67% of the 48 approved institutions and 90.57% of the total 53 approved projects. These three categories represent the main driving forces behind vocational education research in HVI nationwide [6]. In contrast, HVI specializing in medicine, agriculture, and forestry hosted relatively few education-related projects, indicating that their primary focus lies outside the discipline of VTE.

Table 5. Funded Project Distribution by Type of Institution

Type of Institution	Number of Institutions	Proportion (%)	Number of Funded Projects	Proportion (%)
Comprehensive	18	41.67%	21	44.83%
Science and Engineering	17	36.11%	17	33.33%
Finance and Economics	9	11.11%	10	12.64%
Medical and Pharmaceutical	3	5.56%	3	4.60%
Agricultural	2	4.17%	2	3.45%

2.3.2 Analysis of Institutional Nature

Based on the list of "Double High Plan" institutions released by the MOE in 2019, the responsible institutions for the projects were categorized into Double High institutions and regular institutions. Among the 53 approved HVI, 20 were designated as Double High institutions, while 33 were regular institutions. The 19 Double High institutions were awarded 9 projects, including 1 National Major Project, 7 National General Projects, 9 MOE Key Projects, and 3 MOE Young Scholars Projects. The 34 regular institutions received 34 projects, comprising 10 National General Projects, 1 National Western Project, 14 MOE Key Projects, and 8 MOE General Projects.

Further analysis reveals that among the Double High institutions, four schools were awarded two or more projects, whereas among the regular institutions, only one school received two or more projects. The National Major Project titled "Research on the Characteristics of Vocational Education Types and the Construction of the 'Dual-Track' and 'Dual-Path' Systems with General Education," led by scholar Sun Shanxue and approved in 2020, is the only VTE project granted to higher vocational scholars during the 2020 to 2024 period.

This reflects several points: First, it demonstrates the strong research foundation and capability of the team led by Sun Shanxue; Second, it highlights the breakthrough achieved by Beijing Electronic Science and Technology Vocational College, a Double High institution where Sun Shanxue is based, in securing a National Major Project; Third, it underscores the precise understanding by the project team and their institution of national vocational education policies, especially the "Implementation Plan for National Vocational Education Reform." In the grand journey of vocational education reform and development in contemporary China, there is an urgent need for a group of outstanding higher vocational scholars to make significant theoretical and practical contributions.

3 Historical Evolution of Approved Projects on Vocational and Technical Education in Higher Vocational Colleges

Using Cite Space and VOS viewer software; a keyword analysis was conducted on 53 research projects on vocational and technical education (primarily consisting of 359 publicly published journal articles available from CNKI) to outline the research hotspots in vocational and technical education in higher vocational colleges.

3.1 Frequency Dimension

3.1.1 Keyword Frequency

The frequency of keywords can reflect the research focus of vocational education projects. The higher the frequency is, the greater the attention researchers pay to the research content represented by the keyword, and vice versa. As shown in Table 6, during the 2020 to 2024 period, keywords such as vocational education, higher vocational colleges, vocational education, vocational colleges, and craftsmanship had higher frequencies, indicating that research on vocational education in practice and application and in response to the theme of the times was the main content of project research. This specifically involves many hot topics in vocational education research, such as school-enterprise cooperation, the integration of industry and education, mixed ownership reform, vocational education governance, and modern apprenticeship, with research hotspots closely following the theme of the time.

Table 6. Top 18 Keyword Frequencies of National Education Science Planning Vocational Education Project Research Results

Number	Frequency	Centrality	Keywords	Number	Frequency	Centrality	Keywords
1	24	0.31	Vocational Education	10	4	0.35	Skilled personnel
2	18	0.21	Higher Vocational Colleges	11	4	0.08	Corporate governance structure
3	13	0.37	Higher Vocational Education	12	4	0.08	Pathway
4	12	0.49	Vocational Colleges	13	3	0.09	Higher Vocational Education
5	12	0.04	Craftsmanship spirit	14	3	0.03	New Era
6	9	0.59	Mixed ownership	15	3	0.04	The Belt and Road Initiative
7	6	0.38	Integration of Industry and Education	16	3	0	Mixed ownership vocational colleges

(continued)

Table 6. (*continued*)

Number	Frequency	Centrality	Keywords	Number	Frequency	Centrality	Keywords
8	4	0.14	School-enterprise cooperation	17	3	0.04	Modern Apprenticeship
9	4	0	Academic drift	18	3	0.02	Targeted poverty alleviation

3.1.2 Keyword Clustering and Cooccurrence Map

As shown in Fig. 4, the log-likelihood rate (LLR) algorithm was used to analyse the keywords, generating a knowledge spectrum map of keyword co-occurrence and extracting eight main clusters. From the image, the larger the coverage area of the cluster is, the more prominent its feature of broad research and an interwoven network, which facilitates the further tracing of hotspots. Each node in the keyword graph represents a label, with the node's size representing the frequency with which the keyword has been used. The larger the node size is, the more frequently the keyword has been used. In Fig. 4, the module quality (Modularity Q) is 0.8253, which is greater than the critical value of 0.3, indicating that the community structure partitioned by the cooccurring words is significant. The silhouette value (mean silhouette) is 0.9535, which is greater than the critical value of 0.7, indicating that the clustering is convincing. As listed in sequence, the clusters are #0 vocational education, #1 higher vocational education, #2 higher vocational education, #3 vocational colleges, #4 education modes, #6 craftsmanship, #7 countermeasure suggestions, and #8 mixed-ownership vocational colleges.

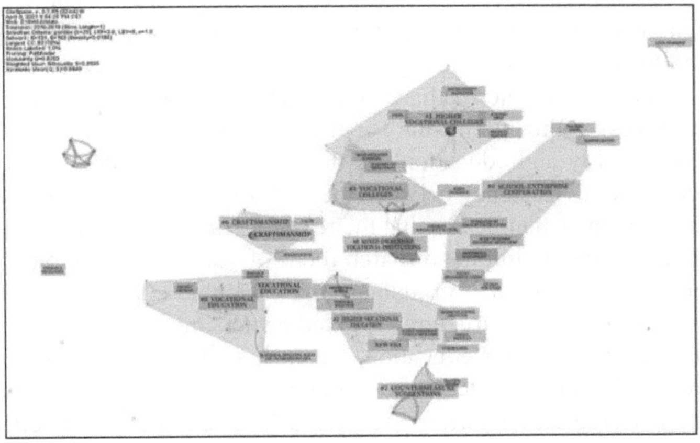

Fig. 4. Keyword clustering map

As shown in Fig. 5, the two traditional hotspots of "vocational education" and "higher vocational colleges" are still at the centre of research, whereas "modern apprenticeship," "craftsmanship," "integration of industry and education," "building moral character" and "professional cluster construction" are the central nodes of the research network during the 2020 to 2024 period, which is directly related to the release of policy documents such as the Action Plan for the Innovative Development of Higher Vocational Education (2015--2018), the National Implementation Plan for Vocational Education Reform and the National Pilot Implementation Plan for the Integration of Industry and Education, making the above five keyword hotspots during the during the period from 2020 to 2024.

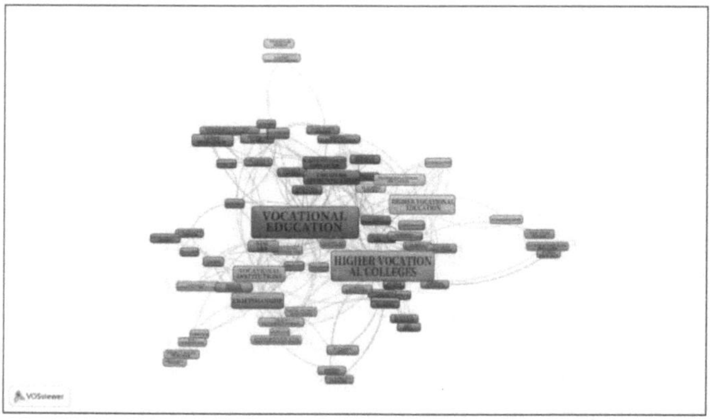

Fig. 5. Co-occurrence network of keywords

3.1.3 Network Relationship of Keywords

Keywords that appeared more than twice were selected to generate the co-occurrence matrix. As shown in Fig. 6, the matrix was imported into Usenet for network relationship analysis, and a network relationship diagram was generated. During the 2020 to 2024 period, higher vocational colleges formed three key research areas in the field of vocational and technical education. The first is the traditional research area, with "vocational education" as the intermediary centre. The second is the research area, with "vocational colleges" as the intermediary centre. The third area is the research area, with "integration of industry and education" as the intermediary centre. These three areas have a high degree of consistency between the research hotspots and policy hotspots of the researchers, which also indicates that the researchers have maintained a high degree of stable attention to the traditional research focus of vocational education. In addition, "school-enterprise cooperation," "craftsmanship," "Double High Plan," "Luban Workshop," "corporate governance structure," "mixed ownership," "modern apprenticeship," "professional clusters," "labour education," "1 + X certificate," "Double High Plan," "Precise poverty alleviation,", etc., have gradually entered the research field's vision. Generally, the research topics during the period from 2020 to 2024 reflect a continued focus on traditional topics and a close observation of new policies in vocational education.

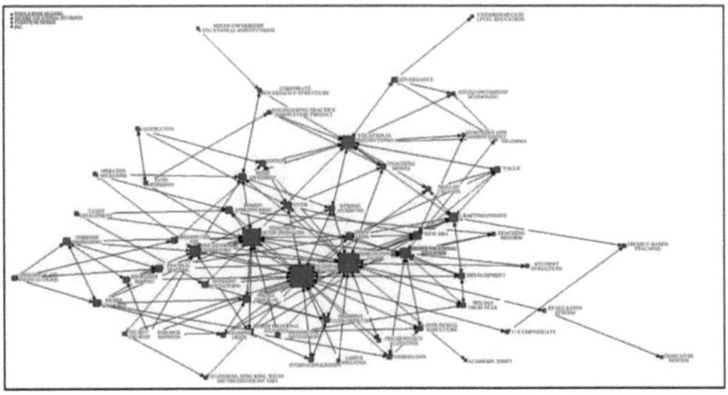

Fig. 6. Network relationship diagram of keywords

3.2 Time Span

3.2.1 Multidimensional Scaling Analysis

Researchers have conducted a series of studies on a theme related to vocational education, forming a distinct discipline and timeline of research. To create a knowledge spectrum map of research results on vocational and technical education topics in higher vocational colleges during the period from 2020 to 2024. a co-occurrence matrix was used to generate a dissimilarity matrix, and a multidimensional scaling analysis was conducted via SPSS 21.0. Figure 7 shows the map divided into four quadrants in counterclockwise order. Keywords such as "mixed-ownership school," "mixed-ownership vocational colleges," "talent training," "industrial structure," and "Guangdong-Hong Kong-Macao Greater Bay Area" are in the first quadrant, indicating that they are the centre of the network of vocational and technical education research in Chinese higher vocational colleges and have a tight internal student structure. Keywords such as "undergraduate level vocational education," "mixed-ownership," "corporate governance structure," "evaluation system," and "engineering practical innovation project" are in the second quadrant, indicating that they have a loosely structured internal network with research potential that needs further exploration. Keywords such as "Luban Workshop," "craftsmanship," "precise poverty alleviation," and "labour education" are in the third quadrant, indicating that they have a tight internal student structure and have already been studied by relevant organizations and individuals. Keywords such as "professional cluster construction," "integration of industry and education," "Double High Plan," "intelligent manufacturing," "internationalization," and "combined engineering" are in the fourth quadrant, indicating that they are on the edge of the research network and have not yet received much attention from researchers (Fig. 7).

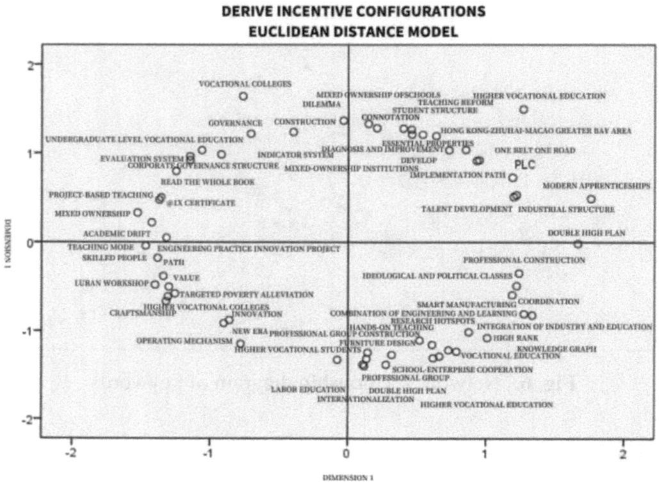

Fig. 7. Keyword knowledge spectrum map

3.2.2 Timeline Spectrum

Figure 8 shows the chronological development order of each cluster and the year the research hotspots were founded, as well as which other research hotspots were involved in the development of each cluster. The research data began in 2020 and have been used to develop macro studies of policies regarding vocational education, including industrial transformation and upgrading, industry-education alliances, education ecology, the Guangdong–Hong Kong–Macao Greater Bay Area, education, education supply-side reform, and industrial restructuring; on the other hand, it has developed mixed ownership, industrial colleges, precise poverty alleviation, the community, the governance system, and the corporate governance structure, reflecting concerns for development reform and governance in the field of vocational education. In addition, the focus on school-enterprise cooperation, professional development, talent training and professional cluster construction reflects the focus on the reform of education and teaching in higher vocational colleges and universities; the focus on professional skills, striving for perfection, skill competitions, and skilled talent reflects the concern of the academic community for the need to cultivate more highly skilled talent and great artisans for Chinese manufacturing since entering the new era and is a reflection of practitioners' professional value orientation and professional ethics. It reflects the professional values, professional ethics and professional quality of practitioners. This study focuses on the hot issues of modern vocational education in light of the problems and challenges that Chinese economic and social development faced during the new era. A change in the governance system of vocational education in a problem-oriented manner is recommended, as well as a realistic option of building modern vocational education with socialist Chinese characteristics.

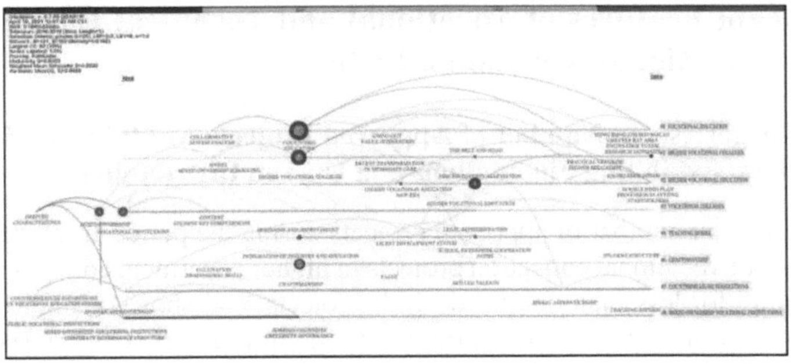

Fig. 8. Keyword timeline spectrum

3.3 Evolution of Popular Terms

As shown in Fig. 9, a scatter diagram of the evolution of the 15 high-frequency buzzwords for higher education institutions each year during the period from 2020 to 2024 was drawn. In the diagram, the cross represents the keyword's position, and the circle represents the frequency of the keyword. The lower the cross is, the higher the ranking; the larger the circle is, the higher the frequency. It is obvious that during the 2020 to 2024 period, some high-frequency keywords received more attention; for example, "vocational colleges" and "craftsmanship" have received attention from academics for five consecutive years, and "vocational education" and "higher vocational colleges" have received more attention for five consecutive years, " again proving that the above four keywords have always been the core of scholars' research on higher vocational institutions. Other keywords, such as "modern apprenticeship," "corporate governance," and "mixed ownership," have continued to attract academic attention since 2020, whereas "integration of industry and education," "school-enterprise cooperation," "precise poverty alleviation," and "education and training" have continued to attract academic attention. The emergence of "precise poverty alleviation," "student structure," "double high plan," and "Lupin workshop" also confirms the trend of following the changes in national vocational education policies.

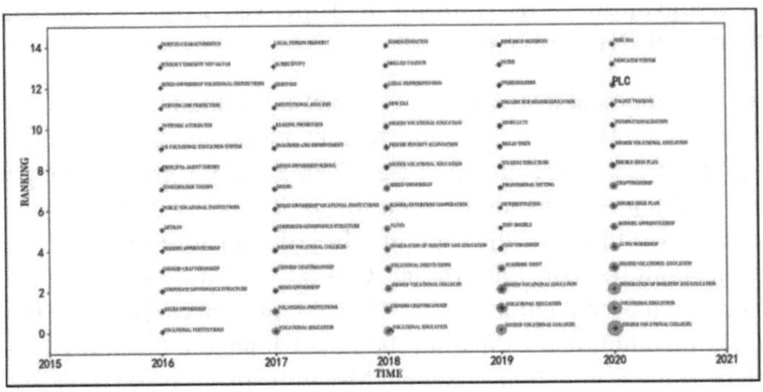

Fig. 9. Scatterplot of keyword evolution over time

4 Significant Features of Vocational and Technical Education Projects in Higher Vocational Institutions

4.1 The Categories and Number of Approved Projects Are Influenced by Policy Directives, Exhibiting Fluctuations and Uneven Distribution

In terms of the total number of approved projects during the period from 2020 to 2024, the proportion of vocational education projects in higher vocational colleges is relatively low, accounting for less than one-third of all projects, and shows a slight downwards fluctuation over time. This can be attributed to several factors. On the one hand, the overall research capacity of higher vocational colleges is comparatively weak. In contrast to undergraduate institutions, vocational colleges place less emphasis on research, and their research foundations are also underdeveloped. As a result, some researchers prioritize teaching over research. On the other hand, research teams at these institutions have not yet formed stable academic communities and lack experienced experts and scholars. Team collaboration is weak, and there is insufficient systematic training in research methodologies, leading to a misalignment between project application directions and policy requirements. Consequently, there is a pressing need to strengthen problem awareness and the application of methodologies.

With respect to project categories, the success rate for key Ministry of Education (MOE) projects are relatively high, accounting for more than half of the total, whereas general national projects make up approximately one quarter of the total. This phenomenon can be attributed to the strict age restrictions for youth projects, as well as the relatively underdeveloped research foundation of younger faculty members. The overall quality of research teams requires improvement, reflecting the insufficient investment in fostering young research talent in higher vocational colleges. Additionally, the distribution and quantity of approved projects exhibit cyclical fluctuations: the number of general national projects first decreased, then increased, and then decreased again; the MOE key projects show a "reverse V" shaped fluctuation, whereas the MOE general projects display a "W" shaped trend. These patterns indicate that the annual variations in different project categories do not follow a consistent trend.

4.2 Uneven Geographical Distribution of Approved Projects and Lack of Institutional Distinctiveness

In terms of the geographical distribution of project-initiating institutions, there is a clear trend of "Eastern regions leading, Central regions moderate, and Western regions lagging." Most projects are concentrated in East China and Central China, particularly in provinces such as Zhejiang, Shandong, Jiangsu, and Tianjin, with these regions accounting for more than two-thirds of the total number of approved projects. The combined number of projects in the two Eastern and Central provinces (Tianjin, and Hunan) is nearly equivalent to the total number in the Western provinces. Notably, provinces such as Shanghai, Jilin, Anhui, and Hainan in the East China and Central China, as well as Gansu, Qinghai, Yunnan, Guizhou, and Inner Mongolia in the West China, did not receive any approved projects during the period from 2020 to 2024. This disparity reflects the differences in research resources, institutional development, and financial investment across regions.

With respect to the type and nature of institutions, comprehensive, engineering, and business-related vocational colleges dominate the approval of projects, with these three types of institutions collectively accounting for nearly 90% of the total. Specifically, among the comprehensive colleges, institutions such as Wenzhou Vocational and Technical College, along with 2 other institutions, each received 2 projects, which represents nearly half of the total projects awarded to comprehensive colleges. Among the engineering colleges, only 2 institutions, including Shandong University of Technology Vocational College, received 2 projects. Among business colleges, only 2 institutions, Shandong Commercial Vocational and Technical College, received 2 or more projects. Notably, some vocational colleges did not sufficiently focus on their specialized or distinctive fields when they applied for projects, resulting in research topics that failed to fully highlight the unique features of the institutions. This lack of focus undermines the role of research in demonstrating institutional strengths and limits overall development in this regard.

4.3 Research Directions Gradually Align with National Vocational Education Policies, but the Strengthening of Research Teams is Urgent

In terms of research hotspots, although emerging topics are receiving increasing attention, traditional issues continue to dominate the core of vocational education research. This finding indicates that vocational education research is still constrained by conventional problems in certain areas. However, as national vocational education policies guide the development of research, new research topics and focal points are increasingly aligned with the direction set by these policies, reflecting the academic community's quick response to policy shifts. For example, there has been a gradual increase in research that is both problem-oriented and practice-oriented, with a stronger focus on addressing practical issues closely related to policy, thereby aligning research more closely with national policy objectives [7].

However, in terms of collaboration within research teams, while vocational colleges have begun to establish cooperative models, with core scholars leading research in specific disciplines and fields, interdisciplinary and interinstitutional collaboration remains limited. Specifically, school-enterprise cooperation and government-school collaboration are relatively rare, and collaborative mechanisms are still underdeveloped [8]. The underlying reasons for these challenges are multifaceted. First, vocational college faculty members face significant teaching workloads, which limit their research capacity and scientific literacy. This leads to a lower rate of transformation of research outcomes, particularly those that could benefit teaching. Second, the overall low number of research projects and the insufficient quality of research results hinder the effective application of research outcomes in the vocational education sector [8]. Therefore, vocational colleges are recommended to enhance the construction of research platforms, promote the establishment of school-enterprise collaboration mechanisms, and improve the research capabilities of faculty members. Additionally, it is important to align research efforts with local and regional development needs, build high-quality research teams based on a stable academic community, and strengthen cross-disciplinary and cross-field cooperation and innovation. This will contribute to the overall enhancement of research teams' effectiveness and the development of vocational education research [9].

5 Future Directions of Vocational and Technical Education Research Project Proposals in Higher Vocational Colleges.

5.1 Coordinating Research Resources in Vocational Colleges and Establishing Innovation Platforms for Young Researchers

From the perspective of the number of research projects and their changing trends, the National Education Science Planning Leadership Group has shown significant support for vocational colleges' research projects. However, the overall trend reflects a fluctuating pattern of decline, rise, and decline. This phenomenon indicates that vocational colleges' research levels in vocational education are still insufficient and require further strengthening. Therefore, the following recommendations are proposed:

On the one hand, moderately increasing attention and resource allocation for research project approval in vocational colleges is suggested, clearly highlighting the potential and development space for research in vocational education within this specific educational type. On the other hand, in addition to prioritizing the approval of national-level projects, it is necessary to increase funding support for National Youth Projects and the Ministry of Education Youth Projects. Furthermore, vocational colleges are recommended to place more emphasis on building their research teams. This can be achieved by increasing investments in research funding, aligning with national policies such as the "Quality Improvement and Excellence Action Plan," and actively attracting high-level research talent. Particular attention should be given to cultivating a cohort of young and middle-career research leaders. This approach will not only enhance the research capabilities and career development of young teachers but also motivate their research enthusiasm, ultimately driving the overall advancement of research teams in vocational colleges.

5.2 Adjusting the Allocation of Research Resources to Narrow Regional Disparities and Strengthening the Demonstration Role of "Double-High" Institutions

Given the significant regional imbalance in the distribution of research projects, particularly the concentration in the eastern region, continuing to support the advantageous regions in the east while also actively promoting a research bias toward the central and western regions to bridge the regional gap is recommended. Additionally, "Double High" institutions, as top-tier vocational colleges, should play a more prominent role in leading and demonstrating vocational education research. Specifically, the following actions are suggested:

First, it is essential to rationally allocate research resources and balance the distribution of projects across provinces, with particular emphasis on channelling resources toward the western regions. This would help unleash the research potential in the western areas and reduce regional disparities. Simultaneously, efforts should be made to promote mechanisms for cross-regional joint project applications, breakdown regional barriers, strengthen collaboration and exchanges between institutions, and improve overall research standards.

Second, the demonstration function of "Double High" institutions should be reinforced. Although, with the exception of Beijing Electronic Vocational College, most "Double High" institutions have not received approval for national key projects, only 10.15% of "Double High" institutions successfully received project approval during the period from 2020 to 2024, indicating that their research advantages have not been fully realized. Therefore, it is recommended that "Double High" institutions leverage their own strengths, integrate the characteristics of their institutions with regional needs, and build academic communities cantered around their key disciplines. They should actively apply for research projects and lead other colleges in improving research standards, thus promoting the overall development of research in vocational colleges.

5.3 Focusing on the Research Hotspots and Difficulties of Higher Vocational Colleges to Accurately Grasp Policy Trends and Effectiveness

Research on vocational education in China's higher education institutions has overcome these hurdles and is entering a new stage of quality improvement and excellence. The period from 2020 to 2024 is a five-year period for vigorously developing higher vocational education in China. During the 14th Five-Year Plan period, there was an urgent need for higher vocational education in China to promote the goal of "Chinese characteristics and the world level" through the Chinese model and the Chinese road [10], with a focus on nurturing morality and fostering talent, enrolment and examination systems, the reform of three education systems, the integration of industry and education, and the ability of governance, etc. [11], to effectively grasp the direction of scientific research development and reform of higher vocational education in China, pay attention to the hot and difficult points in the current research, clarify the next direction, highlight the timeliness of research, and promote higher vocational education into the stage of high-quality development of improving quality and cultivating excellence.

First, in the context of nine ministries and commissions' joint promotion of the Action Plan for the Equipping of Vocational Education System (2020--2023), a number of new pathways, new experiences, and new achievements in the construction of "vocational education highlands" will be presented, providing experiences and models that can be replicated and promoted for China's modern vocational education system and providing theoretical support for better grasping the laws of vocational education development with Chinese characteristics. It also provides theoretical support for better grasping the law of vocational education development with Chinese characteristics [12].

Second, integrating industry and education helps deepen collaborative innovation and provides reference and guidance for forming vocational education groups (alliances), industry-education docking spectrum maps and new apprenticeship systems [13].

Third, it aims to achieve rapid development in system construction, enhance the adaptability of higher vocational education, make new breakthroughs in modernizing the governance system and governance capacity of higher vocational education [14], and promote the modernization of the governance system and governance capacity with system construction at the core [15].

Fourth, we should take major reform projects such as the "Double High School Plan" as a starting point to deepen education and teaching reforms; provide necessary guarantees in building innovative teams of dual-teacher teachers, talent training models and the "1 + X" certificate system [16]; and explore the possibility of improving the quality and excellence of higher vocational education and adding value to it. The aim is to explore how to improve the quality of higher vocational education, increase its value-added capacity and develop it sustainably [17].

Through the above measures, higher vocational institutions should fully leverage policy guidance opportunities, closely align with national strategic goals, focus on key hotspots and challenges in research, and gradually drive vocational education research into a new stage of high-quality development.

Funding. Project supported by Sanya Aviation & Tourism College, Project number: SATC2023JG-09." The updated funding information should read: "Funding: Project supported by the Education Department of Hainan Province, Project number: Hnjg2025ZC-177. Project supported by Sanya Aviation & Tourism College, Project number: SATC2023JG-09.

References

1. Liu, X., Chen, L., Liu, K., et al.: The status, evolution, characteristics, and dimensions of educational research in higher vocational institutions — an empirical analysis based on national education science "13th five-year plan" vocational education projects. Educ. Sci. Forum **36**, 36–45 (2021)
2. Wu, H.: The status and hotspots of vocational and technical education research in china in the last decade — based on the analysis of national education science planning project data from 2010 to 2019. Educ. Sci. Forum **30**, 15–21 (2020)
3. Liu, J., Zhong, J.: Quantitative research and advancement paths of vocational education projects in china — based on the analysis of national education science planning projects from 2018 to 2022. Educ. Vocat. Educ. **16**, 11–17 (2023). https://doi.org/10.13615/j.cnki.1004-3985.2023.16.015
4. Lin, N.: Empirical analysis of vocational education research in china under the type-based education background — based on the statistical data of national education science planning projects from 2015 to 2019. Vocat. Tech. Educ. **41**(32), 76–80 (2020)
5. Wang, X., Zhou, X., Li, L., et al.: Statistical analysis of higher vocational education research papers in Chinese universities in 2019 — based on the statistical data of 18 Chinese core educational journals. Chin. High. Educ. Res. **328**(12), 96–102 (2020)
6. Wang, Q.: The basic status and development vision of basic education research projects in china — based on the quantitative analysis of national education science planning projects from 2001 to 2020. Educ. Theory Pract. **42**(23), 3–8 (2022)
7. Zhou, X., Wang, X., Liu, Z.: Progress and characteristics of vocational education research in china — statistical analysis of vocational education research papers in 2022. Chin. High. Educ. Res. **12**, 101–108 (2023). https://doi.org/10.16298/j.cnki.1004-3667.2023.12.14
8. Zhou, X., Wang, X., Liu, Z.: Research progress and hotspots in vocational education in China — statistical analysis of vocational education research papers in 2023. Chin. High. Educ. Res. **07**, 95–108 (2024). https://doi.org/10.16298/j.cnki.1004-3667.2024.07.14

9. Wang, Y.: Comprehensive enhancement of vocational education research to contribute to high-quality development — based on the "2019 China vocational education research development report". China Vocat. Tech. Educ. 5–15 (2020)
10. Fo, Z., Li, J., Yang, H.: Improving teacher research capacity in higher vocational institutions from the perspective of collaborative management. Educ. Vocat. Educ. **935**(7), 72–77 (2019)
11. Liu, X., Chen, L., Liu, K., et al.: Research on the evaluation of research capacity in higher vocational institutions — a case study of the research achievements of Hainan's higher vocational institutions in 2019. Vocat. Tech. Educ. **41**(33), 28–34 (2020)
12. Zhou, J., Chen, Z.: A review of China's higher vocational education during the "13th five-year plan" and outlook for the "14th five-year plan." China Vocat. Tech. Educ. **770**(10), 18–22 (2021)
13. Yu, Y., Zhu, H.: Building national vocational education innovation development high grounds under the quality improvement and excellence action plan. Educ. Vocat. Educ. **984**(8), 5–11 (2021)
14. Le, C., Wang, Q.: Overview of vocational and technical education theoretical research in China in 2020 — based on statistical analysis of the "vocational and technical education" journal in Renmin University copying and reprints. Vocat. Educ. Forum **37**(2), 6–16 (2021)
15. Chen, Q.: The starting point, difficulties, and breakthroughs of enhancing the adaptability of vocational education under the quality improvement and excellence background. Educ. Vocat. Educ. **987**(11), 5–12 (2021)
16. Zhao, X., Gao, H.: The fundamental guarantee for improving quality and excellence in vocational education: advancing the modernization of governance systems and governance capabilities with institutional development at the core. Vocat. Tech. Educ. **41**(36), 12–17 (2020)
17. Yi, Y., Feng, T.: Research on the construction of high-level vocational teacher teaching innovation teams under the quality improvement and excellence action plan. Educ. Vocat. Educ. **985**(9), 76–82 (2021)

An Empirical Study on the Influential Factors of Improving the Quality of Teaching in the Civics Classroom in Higher Vocational Colleges—Take Sanya Aviation and Tourism College as an Example

Zhi-peng Ou[1(✉)], Zhang Han[1], and Ngan Iek Hang[2]

[1] Sanya Aviation and Tourism College, Sanya 572000, Hainan, China
hainanozp@126.com

[2] Faculty of Humanities and Social Sciences, Macao Polytechnic University, Macao 999078, China
p2209551@mpu.edu.mo

Abstract. The Civics course is an important part of the current teaching in colleges and universities, and it is very necessary to analyze the factors affecting the teaching effect of the Civics course. Based on 262 questionnaires, a model of factors influencing the teaching effect of Civics courses was established using structural equation modeling with the course Introduction to Socialist Thought with Chinese Characteristics in the Xi Jinping Era as an example. By analyzing the results of the final model, it is concluded that the teacher aspect, the student aspect, the course content and the teaching method are the most important factors influencing the teaching effect.

Keywords: Civics courses · Influencing factors · Structural equation modeling

1 Introduction

The Civics courses mainly refer to a series of courses with the core objective of ideological and political education for students, to establish a correct political direction: to guide students to firmly believe in Marxism, to have faith in socialism with Chinese characteristics, and to enhance their confidence in the leadership of the Communist Party of China (CPC), so as to make students maintain a high degree of consistency with the Central Committee of the CPC ideologically, politically, and in terms of their actions. Cultivating good moral character: to cultivate students' core socialist values, equipping them with moral qualities such as patriotism, dedication, honesty and friendliness, improving their ability to practise morality, and promoting their all-round development and healthy growth. Enhancing awareness of the rule of law: to make students learn about the law, enhance their concept of the rule of law, consciously abide by laws and regulations, learn to use legal weapons to safeguard their lawful rights and interests, and

Y. Yu et al. (Eds.): ICPCSEE 2025, CCIS 2673, pp. 58–72, 2026.
https://doi.org/10.1007/978-981-95-2566-9_4

be trained to become modern citizens with the spirit of the rule of law. Enhancing the sense of social responsibility: to make students realize their responsibility and mission as a member of the society, care about national events and hot issues in the society, actively participate in social practice, and contribute their strength to the development of the society. At present, students at the vocational and tertiary level mainly study courses such as Mao Zedong Thought and the Theoretical System of Socialism with Chinese Characteristics as well as Introduction to Xi Jinping's Thought on Socialism with Chinese Characteristics for a New Era. How to make the Civics class lively and interesting, easy for students to accept, which requires education work to pay attention to the teaching effect of the Civics course, and research to promote the construction of the Civics course in colleges and universities. There are many factors that can influence the teaching effect of a certain teaching mode, and many of them cannot be measured directly, thus increasing the difficulty of teaching effect evaluation. Therefore, it is necessary to analyze the factors influencing the teaching effect of Civics courses, analyze the difficulties of Civics courses in depth, enhance the quality of Civics courses, and provide reference for future Civics courses.

For analyzing the influencing factors of teaching effectiveness, several scholars have used different research methods to study different teaching models. For example, Song Pisheng [1] used regression analysis and reliability analysis to identify the more significant influencing factors in project teaching from five groups such as project teachers' resources; Jiang Yurong et al. [2] used hierarchical analysis to construct a model of the influencing factors of online teaching effectiveness and formed a multi-hierarchical structure of online teaching effectiveness from four aspects, including teachers' factors, on the basis of the research literature; Zeng Liwen [3], in the context of epidemic control, analyzed the influence of online teaching implementation by constructing a multi-linear regression model from teaching and learning perspectives and analyzed the influence of online teaching implementation by constructing a multivariate linear regression model from teaching and learning perspectives; Zhao Yunhe et al. [4] used factor analysis to analyze the main factors influencing the effectiveness of computer-assisted instruction from the students' point of view, and provided references for teaching reform.

Apart from studying the effects of different modes of teaching, scholars at home and abroad have also analyzed and researched the factors influencing the effectiveness of teaching from different disciplines and other directions. For example, Wu Sha et al. [5] analysed the influencing factors of the teaching effect of medical immunology course from the perspective of students' own factors by adopting a questionnaire survey; Wang L [6] constructed a language artificial intelligence teaching model by adopting improved machine learning algorithms, and optimised the root cause analysis algorithms based on MCTS, so as to improve the effect of linguistics teaching; Liu LL et al. [7] combined with the current status of bilingual teaching of logistics in Yunnan universities, analysed the factors influencing the efficiency of bilingual teaching of logistics, and established an explanatory structure model (ISM) to analyse the correlation between the factors.

In summary, scholars have analyzed the influencing factors of teaching effectiveness in a more systematic way, but few scholars have studied the main factors influencing the teaching effectiveness of Civics courses in the context of the Civics course, taking Xi Jinping's Introduction to Socialist Thought with Chinese Characteristics in the New Era course as an example. Through reviewing a large amount of literature, four potential influencing factors such as teachers' factors, students' factors, course content and teaching methods, school and social environment were identified to establish a model of influencing factors on the teaching effectiveness of Civics courses, and the questionnaires and models were further examined, analysed and corrected with the help of SPSS and AMOS software, so as to improve the accuracy and reliability of the research results, and ultimately, to propose the corresponding improvement measures based on the results of the research.

2 Proposed Factors Influencing the Effectiveness of Teaching Quality Improvement in Civics Courses

The content of this essay is to study the main factors influencing the teaching effectiveness of the Civics course in the context of the Civics course, taking Xi Jinping's Introduction to Socialist Thought with Chinese Characteristics in the New Era course as an example. Through reviewing a large amount of literature, four potential influencing factors are identified: teacher factors, student factors, course content and teaching methods, and school and social environment.

Course content and teaching methods. Attractiveness and practicality of the course content: whether the course content is close to the actual life of the students and whether it is contemporary and practical is an important factor influencing the interest and effect of students' learning. Combining the Civics courses with real life and hot issues in the society can improve the attractiveness of the courses.

Innovative teaching methods: The traditional teaching method of 'filling the class with knowledge' is likely to lead to a decrease in students' motivation to learn. Innovative methods such as online and offline mixed teaching modes and project-based teaching can provide students with more diversified learning opportunities and promote their active participation. Accordingly, the hypothesis is formulated as A1: There is a significant positive effect of course content and teaching methods on teaching effectiveness.

The teacher aspect. Teaching methods and skills: Whether a teacher's teaching methods are flexible and varied, and whether they can effectively stimulate students' interest and participation in learning are important factors influencing the effectiveness of teaching. For example, the use of case studies, discussions and other highly interactive teaching methods can better guide students' thinking and participation.

Qualities of Teachers: Teachers' theoretical skills, language expression ability and teaching attitude will directly affect students' learning effect. Teachers need to have solid professional knowledge and a good teaching attitude in order to better impart the content and requirements of the Civics course. Accordingly, the hypothesis is formulated as A2: There is a significant positive effect of teacher aspect on course content and teaching methods.

The student aspect. Learning Motivation and Attitude: students' learning motivation and attitude is one of the key factors influencing their learning effectiveness. Students who are interested in the Civics courses tend to be more likely to be engaged in learning and thus achieve better learning results.

Knowledge background and ability: Students' knowledge structure and learning ability will also affect their understanding and mastery of the Civics courses. Students with different majors and backgrounds may have different performance and needs when studying Civics courses. Accordingly, the hypothesis is formulated as A3: There is a significant positive influence of student aspect on course content and teaching methods.

School and social environment. School support and resource allocation: the degree of importance the school attaches to the Civics courses, the allocation of teaching resources and the creation of the teaching environment will all affect the teaching effect. Good teaching facilities and rich teaching resources can provide strong support for the teaching of the Civics courses.

Influence of family and society: The educational environment of the family and the comprehensive quality of the parents, and the importance society attaches to Civics education will also have an impact on the learning effect of students' Civics courses. Family support and good social atmosphere can provide good external conditions for students' Civics learning. Accordingly, the hypothesis is formulated as A4: There is a significant positive influence of school and social environment on the students' aspects.

From a theoretical point of view, in the teaching of Civics courses, the school and society are the main driving forces, teachers are the main implementers, and students are the main participants. Through reasonable and perfect course contents and teaching methods, we can realize the unity of knowledge transmission, ability cultivation and value guidance, so as to achieve the goal of cultivating morality and enhancing students' comprehensive ability. Generally speaking, the main factors affecting the effectiveness of course teaching are: (1) the school and social environment and teachers have an interactive relationship between the course content and teaching methods and so on to influence the effect of student acceptance, between students and teachers will change the teacher's teaching methods and teaching attitudes; (2) Students' motivation and attitudes will change their learning styles and enhance their learning ability, thus influencing the effectiveness of teaching Civics courses; (3) Teachers' ability to teach and students' attitudes towards learning will influence the implementation of the curriculum and hence the effectiveness of teaching and learning. Taking into account the characteristics of Introduction to Socialist Thought with Chinese Characteristics in the New Era of Xi Jinping, this essay constructs a conceptual model consisting of four factors, namely, student aspects, teacher aspects, school and social environments, and course contents and teaching methods, as shown in Fig. 1.

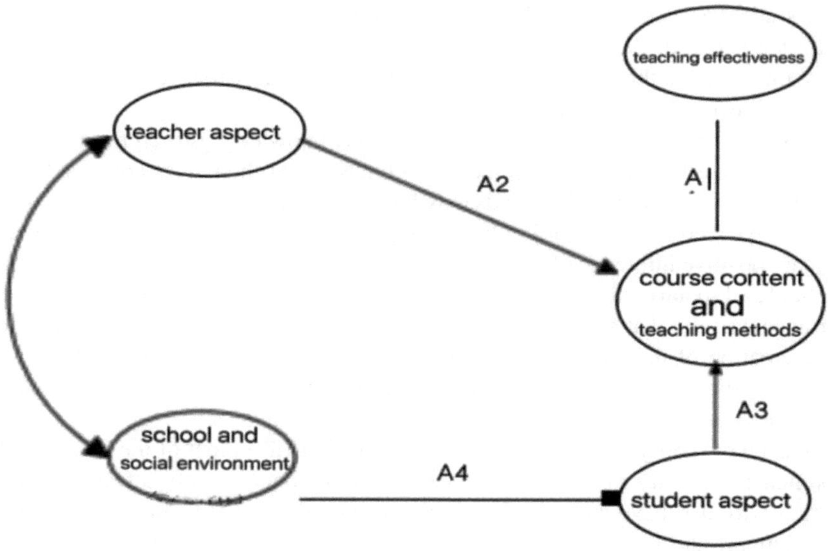

Fig. 1. Conceptual Model of Influencing the Teaching Effectiveness of Civics Courses

3 Research Methodology and Data Collection

3.1 Methodology for Structural Equation Modelling

Structural Equation Model (SEM) is a statistical analysis tool for analyzing the relationship between variables. The relationship between variables and potential variables cannot be directly observed and needs to be obtained by using structural models, while the relationship between potential variables and measured variables can be directly observed by using measurement models, and structural models and observation models are the two components of structural equation modelling. Structural equation modelling is a method of multi-statistical analysis that not only tests some unpredictable abstract concepts, but also parameterizes models of causal relationships between independent and dependent variables. Structural equation modelling is one of the most prominent research methods across multiple disciplines [8], which can be applied to the field of pedagogy in addition to being widely used in the fields of management science and social science. This essay applies structural equation modelling to the study of teaching effectiveness in Civics courses to explore the important factors influencing teaching effectiveness.

3.2 Questionnaire Design and Survey

Data were collected in the form of questionnaires, and the target respondents were students of Sanya Aviation and Tourism Vocational College. A total of 282 questionnaires were returned, of which 262 were valid, with a validity rate of 92.91%. In order to have a more comprehensive understanding of the influence of teachers, students, course content and teaching methods, and the school and social environment on the effectiveness of teaching and learning, a five-point scale was used for the question items (1 for ‘completely disagree’, 2 for ‘disagree’, 3 for ‘neutral’, 4 for ‘neutral’, 5 for ‘disagree’, and 6 for ‘neutral’). The question items were measured using a five-point scale (1 for ‘completely disagree’, 2 for ‘disagree’, 3 for ‘neutral’, 4 for ‘agree’, and 5 for ‘completely agree’). The measurements of teachers were referred to the study of Yang Yuefeng et al. [9] and revised; the measurements of students were referred to the study of Xian Fengying et al. [10] and revised; the measurements of course content and teaching methods were referred to the study of Zou Qi et al. [11] and revised; the measurements of school and social environment were referred to the study of Ma Yunpeng et al. [12] and designed independently; the measurements of teaching effectiveness were developed independently in terms of the perceived validity of the teaching effectiveness. The questionnaire covers the five factors mentioned above, namely, teacher aspects, student aspects, course content and teaching methods, school and social environment, and teaching effectiveness, and students are asked to answer truthfully according to their personal circumstances. The potential and observed variables of the model and their meanings are shown in Table 1.

4 Empirical Analysis

4.1 Reliability and Validity Tests of the Questionnaire

In order to show the reliability and consistency of the test results, the questionnaire data need to be tested for reliability and validity before the empirical analysis. Cronbach’s alpha was used to judge the reliability of the questionnaire, and the results showed that the Cronbach’s alpha for the total scale was 0.886, and the Cronbach’s alpha for each subscale was greater than 0.7, as shown in Table 1.

Table 1. Variable Measures and Reliability Test Results

latent variable	observed variable	Meaning of observed variables	Cronbach’s α
Teacher aspect	TA1	Particularly satisfied with the way the teacher taught	0.781
	TA2	The teacher’s teaching process is clear and can well integrate the knowledge of Civics with professional knowledge	
	TA3	Teachers have a high theoretical level and a good understanding of curriculum Civics	
	TA4	Teachers treat the teaching of Civics in the programme seriously and responsibly	

(*continued*)

Table 1. (*continued*)

latent variable	observed variable	Meaning of observed variables	Cronbach's α	
Student aspect	SA1	I look forward to my Xi Jinping Introduction to Socialist Thought with Chinese Characteristics in the New Era class every week	0.792	
	SA2	Previously had a good knowledge of the Civics course and understood the purpose of the Civics course		
	SA3	Motivated learner, able to prepare and revise well		
	SA4	Learning objectives are clear, and they know what they can achieve after learning Civics through the course.		
	SA5	Ability to reflect and summarize		
School and social environment	TS1	The school attaches great importance to the Civics Course, and often the leaders of the party committee of the school come to the class	0.736	0.886
	TS2	The school's party leaders are very aware of the specific learning situation of the students		
	TS3	Teachers will provide targeted teaching in accordance with the requirements of the school's party committee		
	TS4	Parents will discuss national and international hot topics with students		
Course content and teaching methods	CD1	Teaching content closer to students' reality	0.765	

(*continued*)

Table 1. (*continued*)

latent variable	observed variable	Meaning of observed variables	Cronbach's α	
	CD2	The evaluation system of the Civics courses is reasonable and can promote teaching well		
	CD3	The way the course is assessed makes sense		
	CD4	The teaching system can effectively carry the teaching objectives and tasks of the Civics courses		
	CD5	The curriculum has a clear objective of cultivating moral integrity and nurturing all-round talents		
teaching effectiveness	TE1	Through the course, I can master the knowledge of Civics well	0.708	
	TE2	Through the course, my general competence has been greatly enhanced		
	TE3	Through the course, I am very satisfied with the effectiveness of the teaching of the Civics courses		
	TE4	Through the course Civics, my code of conduct and moral values have changed considerably		

It is concluded that the questionnaire designed is reasonable and the data collected has a high reliability. Validity reflects the accuracy and usefulness of the test. The structural validity of the questionnaire is expressed by Bartlett's ball test and KMO, and the research results show that the KMO value of the total scale is 0.830, and the significance probability of Bartlett's ball test is 0.000. The structural validity of the questionnaire designed in this essay is in line with the requirements, and the content of the questionnaire can effectively reflect the real situation. After completing the reliability test, the scale was subjected to principal component analysis and the questionnaire was re-dimensioned. The initial eigenvalue greater than 1 is five items, which can reduce the dimension of the measurement index to five components, and the total equation interpretation rate is 66.775% (>60%), that is, the five common factors extracted can effectively reflect the 22 variables of teaching effectiveness. The five components obtained from the analysis of the rotated component matrix were regrouped and named. The rotated component matrix is shown in Table 2. The analysis of the rotated component matrix reveals that the attribution of each question item after regrouping is almost the same as the original hypothesis, therefore, the grouping of the original hypothesis is still used for the analysis.

Table 2. Rotated Component Matrix

Measurement indicators	element				
	1	2	3	4	5
TA1	0.315	**0.636**	0.113	0.227	−0.301
TA2	−0.157	**0.796**	0.101	0.064	0.255
TA3	0.128	**0.777**	0.146	0.224	0.137
TA4	0.173	**0.716**	0.175	−0.055	0.213
SA1	**0.247**	−0.136	0.099	0.166	0.114
SA2	**0.742**	0.266	0.069	0.077	−0.028
SA3	**0.797**	−0.034	0.162	0.059	0.233
SA4	**0.866**	−0.029	0.102	0.162	0.133
SA5	**0.666**	0.222	0.172	0.187	0.263
TS1	0.174	0.090	**0.695**	−0.214	0.197
TS2	0.334	0.378	**0.650**	0.150	−0.126
TS3	0.366	0.239	**0.676**	0.117	0.004
TS4	−0.196	−0.002	**0.724**	0.230	0.215
CD1	0.311	0.301	0.423	0.295	**0.302**
CD2	0.210	0.195	0.337	0.499	**0.251**
CD3	0.093	0.173	0.295	0.068	**0.650**
CD4	0.177	0.354	0.068	0.179	**0.669**
CD5	0.397	0.173	0.150	0.296	**0.327**
TE1	0.208	−0.037	0.066	**0.359**	0.683
TE2	0.035	0.081	0.257	**0.717**	0.135
TE3	0.164	0.074	−0.068	**0.752**	0.058
TE4	0.121	0.191	−0.014	**0.606**	0.368

4.2 Model Fitting and Correction

The initial model Sl was fitted using AMOS24.0 and the fitting metrics are shown in Table 3. From Table 3, it can be seen that the initial model does not meet the requirements of RMSEA, AGFI and NFI. Therefore, it is necessary to modify the initial model to determine the optimal model. Through the comparison and analysis of the modified models, the fitting indexes of the modified final model S2 basically meet the requirements.

Table 3. Structural equation model fit indices

	χ^2/v	RMSEA	GFI	AGFI	NFI	CFI	TLI	IFI
Acceptable range	<3	<0.08	>0.8	>0.8	>0.8	>0.8	>0.8	>0.8
initial model S1	2.121	0.083	0.806	0.759	0.717	0.824	0.801	0.828
final model S2	1.682	0.065	0.859	0.814	0.797	0.904	0.885	0.907

The convergent and discriminant validity analyses of the final model S2 are shown in Table 4. As can be seen from Table 4, the values of combined validity (CR) at each variable level are all greater than the recommended value of 0.7, and the average variance extracted (AVE) can reach the minimum required value, which indicates that there is a better consistency within the model constructs, and there is a high degree of convergent validity between the measurement models. Convergent validity analyzes the consistency within the dimensions of the variables, while discriminant validity analyzes the dissimilarity between the dimensions of the variables. In Table 4, the values of the mean variance extracts of each variable (bolded diagonal values) are larger than the squared Pearson's correlation coefficients of the corresponding rows and columns, which indicates that the measurement models have high discriminant validity. All the fitting indexes of the final model meet the requirements, and also have good convergent and discriminant validity, which indicates that the measurement model has high reliability and validity after fitting and correction.

Table 4. Convergent and discriminant validity of the model

variant	convergent validity				discriminant validity		
	CR teaching effectiveness	AVE	teacher-student interaction	Teacher aspect	Student aspect	Course implementation	
teacher-student interaction	0.788	0.488	**0.488**				
Teacher aspect	0.829	0.549	0.066	**0.549**			
Student aspect	0.735	0.438	0.072	0.058	**0.438**		
Curriculum implementation	0.708	0.378	0.080	0.150	0.143	**0.378**	
teaching effectiveness	0.772	0.409	0.055	0.102	0.098	0.216	**0.409**
recommendation value	>0.7	>0.36		Diagonal values should be greater than the square of the correlation coefficients of the corresponding rows and columns			

4.3 Path Factor Analysis

In this essay, structural equation modelling analysis was carried out using AMOS 24.0 software to test the research model hypotheses. From the results of the analysis, it can be seen that the results of the four hypotheses have been verified, and the specific results are shown in Table 5. Afterwards, the standardized factor loadings of the output model were analyzed, and the model was further adjusted to finally obtain the model of influencing the effectiveness of teaching Civics courses as shown in Fig. 2.

Further comparison of the standardized path coefficients reveals that curriculum implementation has a direct influence on teaching effectiveness, with a coefficient of 0.73; teacher aspects, student aspects and teacher-student interactions indirectly influence teaching effectiveness by influencing curriculum implementation, with coefficients of 0.59, 0.31 and 0.47 respectively.

Table 5. Results of testing research hypotheses

causal relationship	Estimate	C.R.	*P*	Verification results
Student aspect ← teacher-student interaction	0.873	3.766	***	support
Curriculum implementation ← Teacher aspect	0.843	5.751	***	support
Curriculum implementation ← Student aspect	0.332	3.657	***	support
teaching effectiveness ← Curriculum implementation	0.682	5.972	***	support

Note: *** indicates P less than 0.001

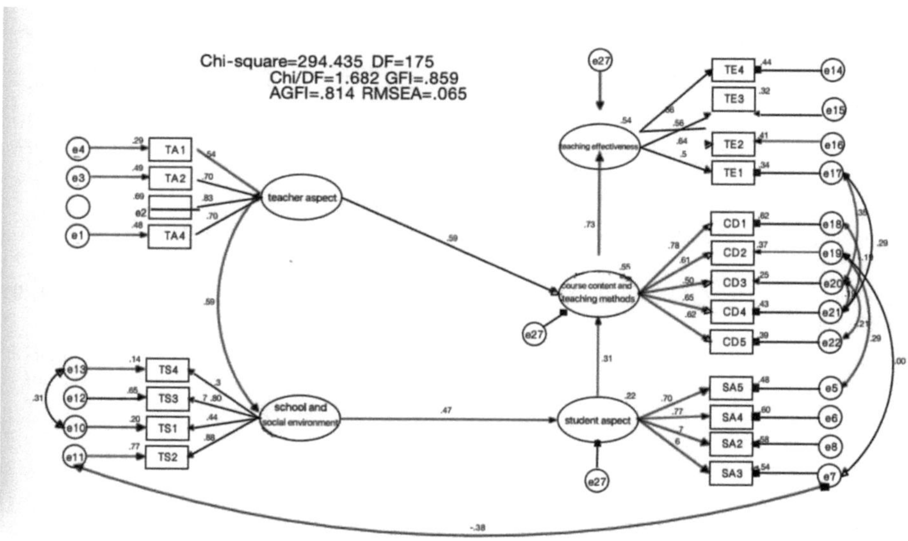

Fig. 2. Influence model of teaching effectiveness of Civics courses

4.4 Analysis of Conclusions

The analysis of the model of influence on the teaching effect of Civics courses can be concluded that the strongest influence on the teaching effect is exerted by the course content and teaching methods, which indicates that the course implementation process plays an extremely important role in the teaching effect of Civics courses, and the more adequate the preparation before the implementation of the course, and the better and more reasonable the system and assessment in the process of the implementation, the better the teaching effect of Civics courses is. Both students and teachers indirectly influence the teaching effect through the implementation of the programme. Compared with students, teachers have a higher influence on the implementation of the programme, and their personal quality, personal ability and teaching attitude will have a significant impact on the implementation of the programme, which in turn influences the teaching effect of Civics courses; in addition to this, the school and the social environment indirectly influence the implementation of the programme by influencing the students' enthusiasm for learning and their attitude towards learning, which is also an important factor that influences the teaching effect.

The teaching of Civics courses requires professional teachers to have higher teaching ability, and the teachers' educational concepts, values and guidance to students are particularly important, and the differences in teachers' personal concepts and abilities will significantly influence the implementation of the courses, thus influencing the teaching effect of Civics courses. On the whole, the existing Civics courses are still in the stage of needing further improvement, and it is not easy to achieve good teaching results. This requires teachers to communicate more with students, understand the actual needs of students, and promptly adjust the teaching methods and curriculum design to achieve the best teaching results.

The studies in the existing literature usually take both the teacher aspect and the student aspect as important factors influencing the teaching effect. Based on the characteristics of teaching Civics courses and the specifics of teaching practice, the teacher aspect is taken as the main factor influencing the teaching effect of Civics courses, and the student aspect is taken as a secondary factor. In addition, this essay also investigates the influence of course content and teaching methods, school and social environment on the teaching effect of Civics courses, and through the study, it is found that course content and teaching methods, school and social environment are all important factors influencing the teaching effect of Civics courses.

5 Conclusion

This essay uses structural equation modelling to analyze the factors influencing the Civics courses, and the results show that the teacher aspect, the student aspect, the course content and the teaching method are the most important factors influencing the teaching effect. Based on the above conclusions, we put forward suggestions and improvement measures for Civics courses teaching from the following three aspects.

5.1 Strengthening Teachers' Teams and Building the Main Positions of Civics Courses

Teachers play a major role in the teaching process of the Civics courses, and the effect of the Civics courses is closely related to the teaching ability of the teachers. Teachers' knowledge level, teaching commitment, teaching attitude, attention to students, etc. will subconsciously influence students' learning, and the good charisma and teaching attitude shown by teachers in teaching will leave a positive impression in students' minds and promote students' all-round development. It is necessary to continuously strengthen the preaching of the Civics Course, gradually improve the knowledge and understanding of the Civics courses by teachers, help teachers better understand the connotation and logic of the Civics course, master the content system of ideological and political education, and continuously improve their ideological and political literacy, so as to make teachers have good ideological and political education ability. In the wave of education reform and development in the new era, we will continue to strengthen the construction of the teaching team, improve teachers' ideological and political literacy and professional teaching ability in all aspects, and create a high-quality teaching team with excellent politics and excellent skills through a multi-pronged approach from the selection of talents, the improvement of the training system to the innovation of incentive mechanism, so as to build a solid ideological and political curriculum and let the ideological and political education take root in campuses and flourish, and strengthen the foundation of the new man of the times who develops comprehensively in morality, intelligence, physical fitness, aesthetics, and labour.

5.2 Comprehensive Understanding of Students' Needs and Enhancement of Students' Civics Learning Effect

Perceiving students' needs in depth, making every effort to enhance the effectiveness of Civics learning, and letting Civics education become the light that illuminates students' growth. Perceiving students' needs in all directions and exerting efforts from multiple angles, so that Civics learning can take root in students' hearts and blossom as the flower of thoughts. We will accurately grasp the needs of students, strengthen the effect of Civics learning with personalised strategies, and help students to build the foundation of faith and take up the mission of the times. In the journey of Civics education, adhering to the educational concept of teaching students according to their aptitude, we comprehensively and deeply understand the diversified needs of students at different stages of growth and in different life backgrounds through diversified communication channels, such as classroom interactions, post-course exchanges, questionnaire surveys, and psychological counselling, etc. We have been able to accurately grasp students' ideological dynamics and interests, as well as their needs and interests. We can accurately grasp the students' thinking dynamics, interests and confusions, and based on this, we can skillfully integrate theoretical knowledge and life examples, innovate teaching methods, optimise teaching content, and carefully create a Civics classroom that meets the needs of the students, so as to effectively enhance the students' Civics learning effect, stimulate the students' enthusiasm for actively exploring the Civics knowledge, and really make Civics education into the minds and hearts of the students to lead them to set up a correct outlook on the world, life, and values.

5.3 Focus on Curriculum Implementation and Efficiently Promote the Construction of the Civics and Political Science Programme

Before the implementation of the courses, it is necessary to clarify the objectives of the course construction, establish a perfect evaluation system for the Civics courses, analyse the specific conditions of the courses and the students, and set up a reasonable assessment method. In the process of course implementation, the content of the Civics courses should be combined with the specific teaching contents in the classes, the contents of the Civics courses should be combined with the majors of the students taught, and attention should be paid to the details of the teaching, so as to strengthen the implementation of the concept of Civics courses, and to achieve the all-round development of the students. In the critical period of education reform, educators need to focus on every detail of the implementation of the curriculum. From the precise setting of teaching objectives, to the reasonable selection of teaching methods, to the scientific construction of teaching evaluation, the whole process of control, do not miss any factor that influences the quality of teaching. In the construction of the Civics courses, we ensure that each Civics class can give full play to its nurturing value with a rigorous and pragmatic attitude and a professional and responsible spirit. By creating diversified teaching modes, such as group discussion, practical research, case study, etc., we stimulate students' interest in learning and improve the attractiveness of the courses. At the same time, we actively integrate all kinds of educational resources, unite the power of school, family and society to form a strong synergy of collaborative education, efficiently promote the construction of the Civics course in a lasting way, lay a solid ideological and moral foundation for the lifelong development of students, and let the light of Civics education illuminate the students in every step of their journey.

Acknowledgement. Project supported by Sanya Aviation and Tourism College. The ideological and political special project number: SATC2023SZ-04.

References

1. Song, P.: Analysis of influencing factors of project teaching effect in colleges and universities: an empirical test based on economics and management disciplines. Mod. Educ. Sci. (9), 71–76, 91 (2018)
2. Jiang, Y., Meng, F.: Research on factors influencing the effectiveness of adult online teaching based on AHP. China Adult Educ. (24), 28–31 (2020)
3. Zeng, L.: Analysis of the effectiveness and influencing factors of online teaching in the context of epidemic prevention and control: a survey based on universities in Guangdong Province. High. Educ. Explor. (7), 85–91 (2020)
4. Zhao, Y., Wang, L., Ma, L.: Analysis of factors influencing the effectiveness of computer-assisted linear algebra teaching. J. Math. Educ. **22**(5), 73–77 (2013)
5. Wu, S., Zhou, J., Liu, Y., et al.: Analysis of teaching effect of medical immunology online course based on students' own factors. Chin. J. Immunol. **36**(18), 2293–2299 (2020)
6. Wang, L.: Role of machine learning and artificial intelligence algorithms for teaching reform of linguistics. J. Intell. Fuzzy J. Intell. Fuzzy Syst. **40**(2), 3251–3262 (2021)

7. Liu, L.L., Xia, Y.C.: Study on factor analysis of bi-lingual teaching for logistics in Yunnan's universities based on interpretive structural model. In: Proceedings of the 2015 4th International Conference on Social Sciences and Society (ICSSS 2015), Pt 2. Paris (2015)
8. Shen, Y.: Analysis of teaching effectiveness in colleges and universities based on structural equation modelling. J. High. Educ. (10), 72–74 (2015)
9. Yang, Y., Li, Z., Xu, X.: Evaluation of classroom case teaching effect based on structural equation modelling: an example of economic law course. Res. High. Finan. Educ. **16**(3), 61–65 (2013)
10. Sin, F., Wang, X.: Evaluation of WPBL learning outcomes based on structural equation modelling. China Health Stat. **32**(1), 88–90, 94 (2015)
11. Zou, Q., Fan, L., Zhang, E., et al.: Analysis of satisfaction and influencing factors of practical teaching of 'curriculum Civics' in colleges and universities: an empirical analysis based on structural equation modelling. J. Beijing Univ. Chem. Technol. (Soc. Sci. Ed.) (2), 90–97 (2021)
12. Ma, Y., Jin, X., Zhang, Z.: Review and prospect of 20 years of research on curriculum implementation in China. Educ. Res. Exp. (5), 38–44 (2019)

A Study on English-Reading Anxiety of Science and Engineering Students in Hainan Vocational Colleges

Han Wu(✉)

Sanya Aviation and Tourism College, Sanya 572000, Hainan, China
516095898@qq.com

Abstract. With the acceleration of globalization and the rapid economic growth in Hainan province, English-language learning is becoming increasingly important for students from Hainan vocational college. As an important part of English language skills, English reading ability is affected by many factors, among which anxiety is an crucial element which will hinder students' reading level. Therefore, this study focuses on the group of science and engineering students in Hainan vocational colleges. A total of 244 science and engineering students from a higher vocational college in Sanya were chosen as the study objects. The level English reading anxiety and its correlation with English scores were explored in detail through a questionnaire survey. The results show that there is a certain degree of English reading anxiety among higher vocational science and engineering students in Hainan, which is manifested mainly in the strangeness of reading materials, the lack of vocabulary and the lack of confidence in reading comprehension. Further correlation analysis revealed that English reading anxiety is significantly negatively correlated with students' English scores; that is, the reading anxiety level has the negative correlation with students' English scores. These findings indicate that reading anxiety hinders the improvement of students' English reading ability and English learning performance to some extent. This study also analyses various factors that lead to reading anxiety, such as teaching methods, learning strategies, individual psychology, etc., and proposes targeted mitigation strategies accordingly, aiming to provide a reference to reform science and engineering students' English teaching in Hainan vocational colleges. Through above methods, teachers can help students relieve reading anxiety so as to promote the progress of English reading ability and the overall English level to better adapt to the requirements of higher vocational education for English application ability.

Keywords: Science and Engineering Students · English Reading Anxiety · Strategies

1 Introduction

With the accelerating process of globalization in contemporary society, English, as the main language of international communication, has played a vital role in various fields. For students in higher vocational colleges, having good English reading ability is not

Y. Yu et al. (Eds.): ICPCSEE 2025, CCIS 2673, pp. 73–84, 2026.
https://doi.org/10.1007/978-981-95-2566-9_5

the need for students' career learning but an important foundation for future career development and international communication. However, in the actual English learning process, many students face different degrees of English reading anxiety, which seriously affects their English learning as well as learning enthusiasm.

As an important free trade port in China, Hainan has increasingly frequent international exchanges in economy, culture, education and other fields, and the demand for English talent is increasingly urgent. As an important base for training high-quality technical talent, higher vocational colleges take the heavy duty of transporting a large number of professional talents for the Hainan's future development. China's higher vocational colleges who play an very important role in our country's future development, their science and engineering students often have high requirements for English reading ability in their professional learning. Both reading professional literature, understanding international frontier technology, and future career development require a solid foundation of English reading. However, owing to the comprehensive effects of many factors, anxiety exists in the English reading learning process of science and engineering students at Hainan vocational colleges and needs to be further studied and solved.

English reading anxiety is a complex emotional reaction that involves many factors, such as students' mental state, learning strategies, and teaching environments. From a psychological point of view, anxiety affects the cognitive processing of students and reduces their learning efficiency and effectiveness. In English reading learning, anxiety may lead to difficulty in understanding reading materials, slow down reading speed, and decline in memory ability, thus forming a vicious cycle and further aggravating students' anxiety. In addition, English reading anxiety may also be closely related to students' self-confidence, learning motivation, learning attitudes, etc., and affect students' overall interest in and involvement in English learning.

The educational environment and student groups of Hainan higher vocational colleges have particularities. On the one hand, Hainan has relatively limited educational resources, especially English teaching, which may face problems such as insufficient teachers and imperfect teaching facilities, all of which may have an impact on student's learning. On the other hand, most of the vocational college students from Hainan are from local areas, their basic knowledge for English is relatively weak, and their learning habits and learning methods also have certain limitations. Science and engineering students tend to spend more time on the cultivation of practical operation and technical skills in professional learning and pay less attention to and invest less time in English learning, which also aggravates their anxiety in English reading learning to a certain extent.

Internationally, Saito (1999) developed the Foreign Language Reading Anxiety Scale (FLRAS) to study the reading anxiety of foreign language learners. Sellers (2000) investigated the reading anxiety of college foreign language learners and reported that learners with high reading anxiety often had more difficulty completing reading tasks and had lower reading scores than those without anxiety. At present, research on English reading anxiety has made some achievements in China. Ma Dongmei (2010) reviewed recent domestic and foreign research on the three important nonlinguistic factors that affect foreign language reading comprehension-reading motivation, reading anxiety, and reading strategies-and looked forward to the development direction of future research. An

Weiyu (2012) investigated some vocational college students' English reading anxiety from Tianjin by means of the English reading anxiety scale. Based on the survey results, he proposed effective countermeasures to relieve the reading anxiety to vocational college students and improve their English reading level. Tang Weiling (2014) conducted a study on reading anxiety and its correlation with English scores of English majors in different language-level groups. Lin Ying (2022) conducted a questionnaire survey to explore the correlations among reading anxiety, strategies and academic performance. Zhao Songli and Wei Qiwei (2023) conducted a questionnaire survey among 561 high school students from 3 senior middle schools in Qiannan Buyi and Miao Autonomous Prefecture and conducted semistructured interviews with 19 students according to the interview outline. This paper probes the causes of anxiety in English reading for different genders and nationalities and proposes relevant coping strategies according to the investigation results. Zhu Tian and Kang Zhennan (2023), through a literature review, analysed the problem of students' anxiety in the teaching of high school English reading test questions and proposed corresponding teaching strategies to alleviate students' anxiety. In particular, they explored the specific application of meta-cognitive strategies in alleviating test question anxiety, with the aim of improving the English reading comprehension ability of high school students. Zhu Tian and Xu Ludan (2023) explored the specific application of the CLIL teaching concept in college English reading teaching to alleviate reading anxiety, with the aim of enhancing college students' enthusiasm for English reading. Zhao Songli and Wei Qiwei (2023) conducted a questionnaire survey and interviews among students from three senior high schools in Qiannan Buyi and Miao Autonomous Prefecture to understand the differences in English reading anxiety among high school students of different genders and ethnic groups and the specific manifestations of English reading anxiety among high school students and provided corresponding strategies on the basis of the research results. On the basis of Saito's Foreign Language Reading Anxiety Scale and Hu Qiao's Learning Burnout Scale, Wang Danyang and Zhou Qilin (2024) investigated English reading anxiety and learning burnout, the relationship between English reading anxiety and learning burnout, and the relationship between English reading anxiety and learning burnout in high school students. At present, research on English reading anxiety is relatively mature in China, but relatively few studies have been conducted on English reading anxiety among science and engineering students in Hainan vocational colleges. The purpose of this study is to investigate and analyse the current situation of English reading anxiety among science and engineering students in Hainan higher vocational colleges, explore its causes, and propose corresponding solutions to provide a theoretical basis and practical guidance for English teaching reform in Hainan higher vocational colleges, help science and engineering students overcome English reading anxiety and improve their English reading ability. To cultivate more high-quality technical talent with good English literacy for the construction of the Hainan Free Trade Port.

In the research process, 244 science and engineering students from Sanya Air Tourism Vocational College were surveyed via a questionnaire to understand their English reading anxiety level and its correlation with their English scores. According to the survey results and the actual situation of Hainan vocational colleges, this paper discusses how to effectively relieve students' English reading anxiety and improve their

English learning effect and learning experience by optimizing teaching methods, improving the teaching environment and strengthening psychological counselling. This study not only has important theoretical significance but also has strong practical guiding value and plays a positive role in promoting the development of English teaching and students' overall development in Hainan higher vocational colleges.

2 A Study on English-Reading Anxiety of Science and Engineering Students in Hainan Vocational Colleges

2.1 Research Subjects

A total of 244 freshmen majoring in aviation mechanical and electrical equipment maintenance, aviation maintenance and new energy vehicle manufacturing from Sanya Aviation and Tourism College were studied.

2.2 Research Questions

Current English reading anxiety of first-year science and engineering students in higher vocational colleges;

Differences of science and engineering students' English reading anxiety according to different majors and different places of origin inside and outside Hainan Province;

Correlation between students' English reading anxiety and final English exam scores.

2.3 Research Methods and Tools

A questionnaire survey was adopted in this study. The questionnaire questions were derived from the Foreign Language Reading Anxiety Scale (FLRAS) figure out by Saito in 1999. It consists of 20 items. A 5-point Likert scale was used to score the listed questions from "strongly disagree" to "strongly agree", and the score range for each question was 1–5 points. This scale mainly involves emotional attitudes (confidence, confusion, frustration, timidity, etc.) and the perceptual measurement of reading comprehension levels. The questions also include topics related to reading, grammar, pronunciation, translation, culture, ideology, and history of the target language.

2.4 Data Analysis of the Questionnaire

In this research, Students were given the questionnaire through Questionnaire Star. After date cleaning, There were 244 valid questionnaires. All the data in the questionnaire was calculated and analysed via SPSS 22.0. The final results of the data analysis were got.

Reliability and Validity Test. Although this questionnaire is mature and its reliability and validity have been recognized by many experts and scholars, reliability analysis of the questionnaire is still carried out to ensure its reliability. Through SPSS22,0 analysis, you will find the data in Table 1 that the questionnaire's Cronbach coefficient is 0.924 which means that the reliability of the scale is good. As shown in Table 2, the KMO value of the scale is 0.926, which is greater than 0.9. As well, the Bartlett sphericity test's sig value's is 0.000, which is less than 0.000. All these data show that the scale has good validity.

Table 1. Statistics of Reliability

Cronbach'Alpha	N of Items
.924	20

Table 2. KMO and Bartlett test

Kaiser-Meyer-Olkin Measure of Sampling Adequacy.		.926
Bartlett's Test of Sphericity	About chi-square	3029.487
	df	190
	Sig.	.000

The Current Situation of English Reading Anxiety to Science and Engineering Students in Hainan Higher Vocational Colleges

Overall Anxiety. The overall anxiety to science and engineering students from a higher vocational college in Sanya has been analyzed according to the data collected via SPSS 22.0. As the data shown in Table 3, and the mean value was 68.7910. The minimum average of this scale is 20 points, and the maximum average is 100. The closer the average is to 100, which means the English reading anxiety is higher for students. The average value of 68.7910 is close to 70, indicating that the students' current English reading anxiety is above the medium level or higher. Reading ability is very important to science and engineering students. When they read technical English materials, the lower reading ability will cause students' difficulty in understanding the meaning of there materials. When they can't understand the meaning of these materials, they may not be competent for their future job. A lot of graduate can not find a satisfied job or get a smooth promotion due to their poop English ability. It will seriously hinder their future career development. Reading anxiety as one of the most important factor will be crucial that will hinder students' reading improvement.

Table 3. Descriptive statistics

	N	Minimum	Maximum	Mean	Standard deviation
Total value	244	20.00	100.00	68.7910	14.19418
Valid N (listwise)	244				

English Reading Anxiety in Different Dimensions. Questions 1, 2, 3, 6, 7, 8, 9, 11, and 15 in the questionnaire investigated the influence of students' vocabulary and grammar mastery on students' English reading anxiety level. In terms of the survey results in Table 4, the average value is 3.4950, which is close to 3.5, indicating that students' vocabulary and mastery of grammar gives a moderate effect to the level of students

English reading anxiety. The more vocabulary students have, the more proficient they are in grammar, and the lower their reading anxiety level will be.

Table 4. Descriptive statistics

	N	Minimum	Maximum	Mean	Standard deviation
The impact of vocabulary and grammar	244	1.00	5.00	3.4950	.85624
Valid N (listwise)	244				

Questions 4, 12, 13, 14, 16, 17, and 18 of the questionnaire investigated the influence of students' emotional state on their level of English reading anxiety. The more positive students' attitudes towards reading are, the lower their anxiety level in English reading. According to the survey results in Table 5, the average value is 3.3630, which is close to 3.4. This shows that the students are not confident in reading.

Table 5. Descriptive statistics

	N	Minimum	Maximum	Mean	Standard deviation
The impact of emotional state	244	1.00	5.00	3.3630	.69905
Valid N (listwise)	244				

Questions 5, 10, 19, and 20 of the questionnaire investigated the influence of students' mastery of cross-cultural background knowledge on English reading anxiety. According to the survey results in Table 6, the average value is 3.4488, indicating that students' cross-cultural background knowledge reserve has a moderate effect on their English reading anxiety level. The richer the students' cultural background knowledge is, the lower their reading anxiety level will be.

Table 6. Descriptive statistics

	N	Minimum	Maximum	Mean	Standard deviation
The impact of cultural background	244	1.00	5.00	3.4488	.83853
Valid N (listwise)	244				

Differences of English Reading Anxiety for Science and Engineering Students who are from Different Regions and Majors. Independent sample T test is conducted to check whether there is significant difference for science and engineering students who are form different regions. Meanwhile, one-way ANOVA is given to check the significant differences for students who are from different majors.

Independent Sample T Test. According to Table 7, the result of the independent sample t test is shown according to the different places of students' origin. From the data, you will find whether it exists differences in English reading anxiety for students who are from different places of origin. Based on the research, 244 students were divided into two groups. In group one, students are all from Hainan Province. In group 2, students are all from outside Hainan Province. The significance values in terms of total anxiety, the lexical grammar dimension, the emotional factor dimension, and the cultural factor dimension were 0.009, 0.010, 0.041, and 0.042, respectively, and their values were all lower than 0.05. These findings indicate that the difference of English reading anxiety for science and engineering students who are from different places is significant. According to the group statistics in Tables 7 and 8, the overall English reading anxiety level of mainland students is 67.0433, whereas that of Hainan students is 72.0000, indicating that students who are from Hainan are more anxious than that of students from other provinces when they are reading English materials, which is due to the backwater of basic education in Hainan and the poor basic knowledge of students in junior and senior high schools. From the perspective of different dimensions, Hainan students are more anxious about vocabulary and grammar than non-provincial students are, are less confident in reading, and are more anxious about cultural background knowledge during reading.

Table 7. Independent-samples T test

	Levene's test forequality of variances		T test for whether the mean is equal						
	F	Sig.	T	df	Sig. (2-tailed)	Mean difference	Standard error	95% Confidence Interval	
								Lower Bound	Upper Bound
Overall anxiety	.243	.622	−2.637	242	.009	−4.9557	1.87919	−8.6573	−1.25404
			−2.678	182.556	.008	−4.9557	1.85069	−8.6071	−1.30421
Vocabulary and grammar anxiety	.211	.646	−2.602	242	.010	−.29505	.11340	−.51843	−.07167
			−2.637	181.642	.009	−.29505	.11188	−.51579	−.07430
Emotional anxiety	.058	.809	−2.053	242	.041	−.19105	.09306	−.37437	−.00774
			−2.124	192.413	.035	−.19105	.08996	−.36849	−.01361
Cultural background anxiety	.564	.453	−2.158	242	.032	−.24073	.11153	−.46042	−.02104
			−2.175	178.572	.031	−.24073	.11068	−.45914	−.02231

Table 8. Group statistics

	City	N	Mean	Standard deviation	Standard error mean
Overall anxiety	Outside Hainan	158	67.0443	14.27141	1.13537
	Hainan	86	72.0000	13.55338	1.46150
Vocabulary and grammar anxiety	Outside Hainan	158	3.3910	.85954	.06838
	Hainan	86	3.6860	.82114	.08855
Emotional anxiety	Outside Hainan	158	3.2957	.72092	.05735
	Hainan	86	3.4867	.64276	.06931
Cultural background anxiety	Outside Hainan	158	3.3639	.83970	.06680
	Hainan	86	3.6047	.81841	.08825

One-Way ANOVA for Students with Different Majors. An ANOVA was conducted on students from three different majors. Table 9 shows that the significance values of the total reading anxiety level, reading anxiety level of vocabulary and grammar, reading anxiety level of emotional factors and reading anxiety level of cultural background knowledge of students from three different majors were 0.000, 0.000, 0.001 and 0.012, respectively, which were all less than 0.05. The overall reading anxiety levels of the three majors, the reading anxiety levels of vocabulary and grammar, the reading anxiety levels of emotional factors and the reading anxiety levels of cultural background knowledge are significantly different. The overall reading anxiety level of electronic majors is higher than that of other mechanical and electrical majors and new energy majors. In different dimensions, the reading anxiety level of electronic majors is higher than that of the other two majors in terms of vocabulary, grammar, emotional factors and cultural background knowledge. The main reason for these results is that students from electronic majors have poor English foundations. They have less vocabulary than the other two majors do. When they read English materials, they feel more anxious.

Correlation Analysis of Students' English Reading Anxiety Levels and Final Exam Scores. As shown in Table 10, between the final scores and the overall anxiety for students in English reading, the Pearson correlation coefficient is −0.013, $P = 0.00 <$ 0,05. In conclusion, students' final exam scores and their English reading anxiety' the correlation is negative. Similarly, the Pearson correlation coefficients for students' final grades and students' vocabulary and grammar reading anxiety and cultural background knowledge reading anxiety are −0.062 and −0.61, respectively ($P = 0.00 <$ 0,05). In conclusion, students' final exam scores are negatively correlated with their reading anxiety in terms of vocabulary and grammar and their cultural background knowledge anxiety. In contrast, between students' final grades and reading affective anxiety, the Pearson correlation coefficient is 0.102 ($P = 0.00 < 0.05$), indicating that students' final grades are positively correlated with reading affective anxiety, teacher-student interactions and negative attitudes toward English classes.

Table 9. ANOVA

	Quadratic sum	df	Mean square	F	Sig.
Overall anxiety	3418.508	2	1709.254	9.045	.000
	45539.832	241	188.962		
	48958.340	243			
Vocabulary and grammar anxiety	12.395	2	6.197	9.010	.000
	165.760	241	.688		
	178.154	243			
Emotional anxiety	6.318	2	3.159	6.771	.001
	112.429	241	.467		
	118.747	243			
Cultural background anxiety	6.184	2	3.092	4.525	.012
	164.676	241	.683		
	170.860	243			

Table 10. Correlation analysis

		Final sccore	Overall Anxiety	vocabulary and grammar anxiety	emotional anxiety	cultural background anxiety
Final score	Pearson Correlation	1	−.013**	−.062**	.102**	−.061**
	Sig.(2-tailed)		.000	.000	.000	.000
	N	244	244	244	244	244

**. Correlation was significant at 0.01 level (two-tailed)

3 Strategies for Alleviating Science and Engineering Students' English-Reading Anxiety in Hainan Vocational Colleges

Based on the above data analysis, the research proposes relevant strategies to relieve Hainan senior vocational science and engineering students' reading anxiety.

3.1 From the Aspects of Teaching Methods

Teachers should optimize reading teaching content and select English reading materials related to science and engineering, such as science and technology news, professional popular science articles, etc., that are closely related to students' lives and interests so that students can understand the value of English in reading and increase their interest in reading. Meanwhile, teachers should also control the difficulty of the reading materials according to the students' English level, so as to avoid materials that are too difficult or too easy, which will lead to anxiety or loss of interest. By cultivating reading skills and teaching reading strategies, such as quick scanning, skimming, and intensive reading, students can choose appropriate reading methods according to different reading purposes. For example, when reading science and technology news, they can first read the title and the topic sentence of each paragraph to get the main idea of content. In addition, they can also perform intensive reading of the parts they are interested in. Teachers trains students how to to guess words meaning and encouraging them to predict the new words according to the context, word formation, etc., rather than over reliance on dictionaries, so as to improve students' reading speed and comprehension. Innovative teaching methods such as group cooperative learning, organizing students to have group discussions and cooperative reading should be used in daily English class so that students can inspire each other and enhance confidence in communication. For example, students can be divided into groups to read a scientific article together, with each group member responsible for a different part and then summarizing the discussion. Multimedia assisted teaching is also important. The use of multimedia resources, such as English movies and documentaries, so that students can improve their listening and reading ability.

3.2 From the Aspects of Psychological Counseling

A positive attitude should be established. Self-confidence is one of the positive attitudes for English reading. It can be enhanced through encouraging students, giving positive feedback and other ways to help students build reading confidence. For example, when students make progress in reading, teachers should promptly praise their efforts and results. Expectations for reading should be adjusted. Students should be guided to set learning goals reasonably, therefore, anxiety caused by high expectations can be avoided. To relieve anxiety and catharsis, students are encouraged to vent and relieve anxiety through sports, writing diaries, talking to friends and other ways. Relaxation training should be conducted to students. These relaxation techniques include deep breathing and meditation to help them adjust their mindset before reading.

3.3 From the Aspects of Study Habits

For private vocational college students, they haven't developed good study habits for a long time. In order to relieve students' English reading anxiety, independent learning ability should be cultivated. They can make their own learning plans according their own English reading ability, including daily reading time and reading amount, so that they can develop good learning habits. For students, independent reading should be encouraged. They can use their spare time to read independently. The more they practice, the better study habits can be developed. Once the good reading habit has been developed, the scope of knowledge will be broadened and their reading ability can be improved. Vocabulary is crucial for students reading, and a good habit for words learning will help students accumulate more vocabulary. When students accumulate enough words, their reading anxiety can be well relieved. Vocabulary can be mastered through multiple channels, such as reading, dictating, memory software and other methods. Regular review and consolidation are also important for students to memorize vocabulary regularly.

3.4 From the Aspects of Extracurricular Activities

In order to relieve students' English reading anxiety, English corners or reading clubs can be conducted. Students can talk about their favorite books or articles to others, exchange English reading experiences and share their learning experiences in a relaxed atmosphere so that their interests for reading can be developed. In addition, some experts can be invited to deliver lectures to students. They can introduce reading skills, cultural background and other knowledge to improve their reading skills and enrich their cultural background. At last, students are encouraged to participate in English reading competitions or writing competitions to stimulate their learning motivation. In summary, after the competition, students can summarize their experience and lessons to further improve their reading ability.

4 Conclusion

Through the investigation and research of the English reading anxiety of science and engineering students in Hainan vocational colleges, we gain a deep understanding of the current situation and causes of anxiety faced by students when they read all kinds of English materials. When conducting the research, we realize that the alleviation of English reading anxiety does not occur overnight but requires the joint efforts of teachers, students and schools. Future research could further pursue the long-term influence of different teaching methods and how to better combine mental health education with English reading teaching. Through continuous research and practice, we hope to create a more relaxed and efficient learning environment for Hainan vocational science and engineering students, help them make greater progress in English reading learning, and lay a solid foundation for their future development.

References

Wang, D., Zhou, Q.: A study on the correlation between English reading anxiety and learning burnout in high school students. J. Overseas English **01**, 185–187 (2024)

Zhao, S., Wei, Q.: A survey of the causes and coping strategies of English reading anxiety in senior high school students based on different genders and nationalities. J. Kaili Coll. **41**(02), 91–100 (2023)

Tang, W.: A preliminary study on English reading anxiety of English majors. J. Xiangnan Univ. **35**(03), 63–66+70 (2014)

Jiang, D., Li, H.: Cooperative learning and relieving oral English anxiety. J. Univ. Shanghai Sci. Technol. (Soc. Sci. Ed.) **35**(1), 36–40 (2013)

An, W.: Correlation analysis of English reading anxiety and reading strategy reading achievement of vocational college students. J. Chin. Adult Educ. (20), 100–103 (2012)

Ma, D.: A review of research on nonlinguistic factors affecting foreign language reading comprehension. J. Educ. Dev. (04), 69–72 (2010)

Saito, Y.: Foreign language reading anxiety. Mod. Lang. J. (83), 202–206 (1999)

Sellers, V.: Anxiety and reading comprehension in Spanish as a foreign language. Foreign Lang. Ann. (5), 512–518 (2000)

Zhu, T., Kang, Z.: An exploration of the impact of meta-cognitive strategies on high school students' English reading anxiety. J. English Square **22**, 125–128 (2023)

Zhu, T., Xu, L.: Research on the influence of CLIL teaching concept on college students' english reading anxiety. J. Overseas English **12**, 86–88 (2023)
Lin, Y.: A study on the correlation among college students' English reading anxiety, strategies and grades. J. Taiyuan Urban Vocat. Coll. **02**, 155–157 (2022)
Wu, H.: A study of vocational college students' anxiety in english vocabulary learning. In: Wang, Y., Zhu, G., Han, Q., Zhang, L., Song, X., Lu, Z. (eds.) Data Science. ICPCSEE 2022. Communications in Computer and Information Science, vol. 1629, pp. 412–426. Springer, Singapore (2022)
Li, X.: SPSS 22.0 statistical analysis from entry to the master. Electronic Industry Press, Beijing (2015)
Zhu, Y.: An empirical study on the impact of collaborative learning on english reading anxiety and reading performance. J. Mudanjiang Coll. Educ. (03), 67–69 (2021)
Mo, L.: Educational psychology. Guangdong Higher Education Press, Guangzhou (2002)
Xiang, Y., Shi, M.: Research on english reading anxiety of vocational college students. J. Lanzhou Vocat. Tech. Coll. **38**(14), 48–50 (2022)
Dong, L., Liu, M.: The development trajectory of foreign language reading anxiety and the application of reading strategies: based on the latent variable growth model. Foreign Lang. Teach. (04), 134–144+150 (2022)
Quan, Z.: Research on the influence of multimodal reading teaching mode on students' english reading anxiety. J. Inner Mongolia Univ. Finance Econ. **23**(03), 50–53 (2022)
Fan, H.: Investigation and research on english reading strategy training for junior high school students. J. English Teach. **21**(02), 27–31 (2021)
Jia, W.: Analysis of the application practice of cooperative learning theory in college english reading teaching. J. Overseas English (24), 136–138+141 (2024)

A Study on the English Listening Anxiety of Students in Private Vocational Colleges

Han Wu(✉)

Sanya Aviation and Tourism College, Sanya 572000, Hainan, China
516095898@qq.com

Abstract. With the world has become the global village, communication between countries is becoming increasingly common. As an important tool for international communication, English has become increasingly important. However, for students in private vocational colleges, English listening anxiety is a common phenomenon, which not only affects the learning ability of students but also restricts the improvement of their comprehensive English ability. In this study, 242 first-year students of a private vocational college in Hainan, were used as the study objects, and the level, causes and influencing factors of their English listening anxiety were deeply analyzed via a questionnaire survey. Strategies to alleviate students' English listening anxiety were discussed from the three aspects of teachers, students and schools. The conclusions of this study have certain theoretical and practical significance for improving the the teaching state for English listening to college students from private higher colleges. Meanwhile, It can also enhancing the learning effect of students to English. The comprehensive use of various strategies can effectively relieve students' English listening anxiety, improve their English listening level and learning enthusiasm, and make great contributions to the future career development for them. However, owing to limitations in terms of research time and conditions, this study has several limitations. Future studies can further expand the research scope, increase the diversity and quantity of research samples, and adopt a combination of multiple research methods to conduct a more in-depth discussion on this issue.

Keywords: Private Higher Vocational Colleges · English Listening Anxiety · Cause of Formation · Suggestions

1 Introduction

Given the surge in globalization, the importance of English as the core language of international communication is self-evident. For private vocational college students, English learning is not only a key skill for improving their comprehensive quality and broadening employment channels but also a necessary quality to adapt to future social development and integrate into the multicultural environment. Emotional factors play an important role in the process of English learning, and language anxiety is one of the important emotional factors that restricts the learning effect in second language acquisition. In English study, there are four basic skills, listening, speaking, reading and

Y. Yu et al. (Eds.): ICPCSEE 2025, CCIS 2673, pp. 85–97, 2026.
https://doi.org/10.1007/978-981-95-2566-9_6

writing. For these four skills, listening may have the most problems, that is, the highest anxiety. However, in many aspects of English learning, listening teaching has always been a major problem for students, and the English listening anxiety caused by it has become a bottleneck restricting the improvement of students' English ability, which is worthy of in-depth exploration and analysis.

The concept of English listening anxiety has long been a concern in the field of language learning. It is not a simple nervous emotion but a complex emotional experience, covering anxiety, fear, tension, frustration and other negative emotions. These emotions interweave with each other, similar to an invisible net, which tightly restricts the performance and advance for private higher vocational college students in the process for English listening learning. Because of their special learning background and psychological state, private vocational students are more likely to fall into the vortex of anxiety. Compared with ordinary college students, private vocational college students have relatively weak English foundations at the time of admission, and there are great differences in their learning habits and learning methods. In the context of English listening, they often feel powerless because of the lack of vocabulary, grammar knowledge, lack of a language environment and many other factors. This enormous gap between the lack of ability and the learning goal can easily cause students' anxiety. They worry that they cannot understand the content of listening materials, are afraid of answering questions in class and being laughed at by classmates, and even fear English listening, thus falling into a vicious circle.

From a psychological perspective, private vocational college students are in the critical transition stage from adolescence to early adulthood; during this period, their self-awareness is gradually awakening, and their self-evaluations and expectations are also increasing. In English listening learning, once they encounter setbacks, they can easily attribute this failure to their own lack of ability and then self-doubt and denial. This psychological state not only affects their enthusiasm for English listening but also may affect the entire English learning process and even their self-confidence and mental health negatively. In addition, the teaching resources for some private vocational colleges are limited, as well as the teaching facilities may not be comparable to those of public colleges. In English listening teaching, the lack of sufficient language practice opportunities and personalized guidance also makes it difficult for students to obtain effective solutions when facing listening problems, which further aggravates their anxiety.

Given the importance of English listening anxiety for private higher college students, the method to overcome the anxiety should be figure out. With the deepening of educational psychology and linguistics research, more and more scholars have started to focus on this problem. In foreign countries, foreign language listening anxiety was first proposed by Horwitz et al. in 1986. Kim (2000) developed the English listening anxiety scale on the basis of the English Classroom Anxiety Scale to test the listening anxiety level of English learners. Later, an increasing number of foreign scholars conducted in-depth research on the causes of English listening anxiety and its solution strategies. John Field (2019) noted the specific causes of English listening anxiety. He also put forward that English listening anxiety is one of the important factors hindering the improvement of English proficiency in English learners. In China, Liu Hong (2008) conducted a systematic empirical analysis on the listening anxiety of 189 non-English majors at universities for nationalities and investigated the listening anxiety of English learners, as well as the relationships among listening anxiety, language learning anxiety as well as listening level. Chen Yanjun and Liu Dejun (2010) conducted an in-depth study on the causes of college students' English listening anxiety. Meanwhile, some corresponding suggestions have been given to alleviate the listening anxiety for students. Zhang Fengmei and Wang Lan (2010) used quantitative analysis to study English anxiety and listening anxiety in undergraduate students and their correlation with CET-4 scores. Liu Hang and Zhang Xiaojun (2011) investigated the factors of English listening anxiety among 152 first-year undergraduate students of four non-English majors at a science and technology university by means of a questionnaire survey and group interviews and proposed corresponding strategies according to the survey results. Zhao Linlin, Zhang Yao, and Fang Wei (2018) used nursing undergraduates as research objects to understand the causes of their medical English listening anxiety and evaluate the effect of the implementation of learning strategies on their English listening anxiety. Chai Xin (2022) used Grade 7 and Grade 8 students at Z Middle School as research subjects and conducted a survey about the current situation, causes of English listening anxiety among rural junior high school students. In the end some useful suggestions had been given to deal with the situation. Ji Xinyan (2022) focused on how to use note-taking strategies to alleviate anxiety in English listening. Lin Min (2022) explored and analyzed the effects of English listening anxiety to the students in primary school via questionnaires. On the basis of the survey results, corresponding methods and strategies to alleviate the listening anxiety of primary school students were proposed. Wang Qiaoling (2023) conducted a survey on the hearing anxiety of 113 college students through the Hearing Anxiety Scale (FLLAS). Pan Liu and He Mingyu (2024) conducted an exploratory factor analysis and difference analysis on the second language listening anxiety of students majoring in business English in the form of a questionnaire survey. Furthermore, corresponding policies are proposed from both the aspects of teachers' listening teaching and students' listening learning to reduce students' second language listening anxiety in business English learning. In recent years, research on English listening anxiety has continued to expand. However, compared with most studies that tend to focus on students in some other schools, relatively fewer studies have focused on English listening for students who are from private vocational colleges.

Focusing on a specific group of private vocational college students, this study takes 242 first-year students from a private vocational college in Sanya as the research objects and adopts the method of questionnaire survey to explore the status quo, causes and coping strategies of their English listening anxiety in depth, aiming to provide empirical evidence for the improvement of English listening teaching, help students eliminate anxiety and improve their English listening level. This promotes the development of their overall English ability. Through the systematic study of this problem, we can not only provide useful references for the reform of English teaching in private vocational colleges but also give more useful experience for the future research related to such topic.

2 English Listening Anxiety Among Students in Private Vocational Colleges

2.1 Study Subjects

This study takes 242 non-English majors from a private higher vocational college in Sanya as the the objects of study. All the students are male students and freshmen, who majored in aviation mechanical and electrical equipment maintenance, avionics equipment maintenance and new energy vehicle manufacturing.

2.2 Study Questions

The current situation of English listening anxiety for students who are from private vocational colleges and its causes;

Differences of English listening anxiety for students who are form different majors and different places of origin inside and outside Hainan Province;

The correlation between students' English listening anxiety and their Level A English test scores.

2.3 Study Methods and Tools

A questionnaire survey was adopted in this study. The questionnaire for the study is the Foreign Language Listening Anxiety Scale, developed by Korean scholar Kim in 2000, which is specifically used to measure the anxiety of foreign language learners in the process of listening to English materials. The questionnaire consists of 33 questions and a 5-point Likert scale is used, which ranged from 'strongly disagree' to 'strongly agree'. The score varies from a minimum of 1 to a maximum of 5. People who are tested with higher scores, which means higher anxiety levels. In this questionnaire, 33 questions can be divided into four dimensions, namely, the emotional dimension, the listening ability dimension, the listening material dimension and the external environment dimension. Through a questionnaire survey, we can further clarify the current situation English listening anxiety for students who are form private vocational colleges and further explore its causes.

2.4 Study Results Analysis

In this study, this questionnaire was mainly distributed online, and the students answered it online. A total of 242 valid questionnaires were collected, and the data obtained were analyzed via SPSS 22.0.

Testing of Reliability and Validity. First, a reliability analysis of the questionnaire was carried out to further clarify whether it has good reliability in testing the situation of English listening anxiety for students who are from a private vocational colleges. The figure of Cronbach coefficient can indicate whether the questionnaire has a good reliability. If it is greater than 0.8, it indicates that the overall reliability of the questionnaire is very good. As is shown in Table 1, the Cronbach coefficient of the scale is

0.963. It indicate that the questionnaire has high reliability. Second, the KMO value of the questionnaire is 0.951. Meanwhile the P value of the Bartlett sphericity test is 0.000. Whereas the commonly used KMO metric is above 0.9, and the P value of the Bartlett sphericity test is less than 0.01, which means that the validity of the scale is good and is highly proper for factor analysis (Table 2).

Table 1. The statistics of Reliability

Cronbach'Alpha	N of Items
.963	33

Table 2. KMO and Bartlett test

Kaiser-Meyer-Olkin Measure of Sampling Adequacy.		.951
Bartlett's Test of Sphericity	About chi-square	5722.928
	df	528
	Sig.	.000

The Current Situation of English Listening Anxiety for Students in Private Higher Vocational Colleges

Overall Anxiety. First, Descriptive statistical analysis has been carried out to the 33 questions of the scale, and the average value obtained, as shown in Table 3, was 3.384, indicating that the subjects generally had a moderate degree of listening anxiety. Generally, FLLAS scale scores range from 1 to 5 points, and the average of 3.384 is at a medium level, indicating that students have some anxiety when they listen to English materials, but it has not reached a serious degree.

Table 3. Descriptive statistics

	N	Minimum	Maximum	Mean	Standard deviation
Total value	242	1.00	5.00	3.3843	.71165
Valid N (listwise)	242				

English Listening Anxiety in Different Dimensions. Questions 6, 10, 14, 17, 20, 22, and 32 of the questionnaire discussed the influence of emotional factors on the English listening anxiety level of students in private vocational colleges from the emotional dimension. In the process of listening to English, if the students are very relaxed and very confident in themselves, their anxiety level is low; in contrast, if they are very nervous and not confident in themselves, the will feel more anxiety. As is shown in Table 4, the mean value for emotional dimension in the scale is 3.2255, which means that the tested students believe that their nervousness and lack of confidence in the process of listening to English will lead to a medium level of anxiety when listening to English materials.

Table 4. Descriptive statistics

	N	Minimum	Maximum	Mean	Standard deviation
Emotional Factor	242	1.00	5.00	3.2255	.65736
Valid N (listwise)	242				

Questions 7, 8, 9, 11, 12, 13, 15, 19, 21, 23, 28, and 33 of the questionnaire address the dimension of listening ability to explore the influence of the students' listening ability on English listening anxiety. According to Table 5, the average value of questions in this dimension of the questionnaire is 3.3833. The students tested had average listening ability, which caused moderate English listening anxiety. The tested students come from private vocational colleges, and their English foundation is weak. Some students even have no English foundation, so their English listening ability is poor, which easily causes English listening anxiety.

Table 5. Descriptive statistics

	N	Minimum	Maximum	Mean	Standard deviation
Listening Ability	242	1.00	5.00	3.3833	.80168
Valid N (listwise)	242				

Questions 1, 2, 3, 4, 5, 16, 18, 24, 29, 30, and 31 of the questionnaire discussed the influence of listening materials on students' English listening anxiety from the perspective of listening materials. According to Table 6, the average number of questions in this dimension of the questionnaire is 3.5203. The type and difficulty of listening materials cause moderate English listening anxiety in students, and the impact on listening anxiety is greater than that on students' listening ability. Given the weak English listening ability for students who are from private higher vocational colleges, listening to obscure English materials increase students' listening anxiety, but choosing easy and interesting materials can alleviate students' anxiety.

Questions 25, 26 and 27 of the questionnaire address the impact of environmental factors on the English listening anxiety of students. According to the data in Table 7, the mean value of the number of questions in this dimension of the questionnaire is 3.2603, indicating that different scenarios in which students listen to English can also cause

Table 6. Descriptive statistics

	N	Minimum	Maximum	Mean	Standard deviation
Listening Materials	242	1.00	5.00	3.5203	.76640
Valid N (listwise)	242				

moderate anxiety in English listening. Students will feel nervous when communicating with others face to face or on the phone, which will cause anxiety about listening. Over the years, students have mostly learned English for the purpose of taking the test and have less oral expression practice. Especially in real-life scenarios, students are more ashamed to express themselves, and they will be nervous when they hear others speak English, thus aggravating English listening anxiety.

Table 7. Descriptive statistics

	N	Minimum	Maximum	Mean	Standard deviation
Environmental factors	242	1.00	5.00	3.2603	.85170
Valid N (listwise)	242				

The Correlation between Each Dimension of the English Listening Anxiety Scale and Total Listening Anxiety. According to Table 8, A significant positive correlation has been shown among the factors causing English listening anxiety and overall anxiety. The correlation coefficients between total anxiety and self-evaluation, listening ability, listening materials and environmental factors were 0.893, 0.968, 0.949, and 0.808, respectively, and all P = 0.000, which is less than 0.01, which indicates that is a significant positive correlation among each other. That is, the lower the student's self-evaluation is, the weaker the listening ability, the more difficult the listening materials, and the more unfamiliar the environment, the greater the level of English listening anxiety.

Table 8. Correlation analysis

		Emotional Factor	Listening Ability	Listening Materials	Environmental factors
Total Value	Pearson Correlation	.893**	.968**	.949**	.808**
	Sig.(2-tailed)	.000	.000	.000	.000
	N	242	242	242	242

**. Correlation was significant at 0.01 level (two-tailed)

Differences of Private Higher Vocational College Students' English Listening Anxiety according to Different Regions and Majors. Independent T test was conducted on the data of the questionnaire to explore whether there was difference of English listening anxiety for private higher vocational college students according to different regions via SPSS22.0. One-way ANOVA was also carried out to determine whether there was differences in English listening anxiety among students from majors.

Independent Sample t Test for Students from Different Regions. Because the test students are currently studying in Sanya and there are differences in education levels between Hainan and mainland China, they were divided into two groups: mainland China and Hainan local students. The differences in English listening anxiety levels among the subjects were analyzed. According to Table 9, The Sig. values for listening anxiety caused by factors such as total anxiety, self-evaluation, listening ability, listening materials, and their environment are 0.015, 0.026, 0.027, 0.033, and 0.002, respectively, all of which are less than 0.05. This finding indicates that the significant differences has been shown in total English listening anxiety levels between Hainan local students and mainland China students. Meanwhile, the differences also exists in listening anxiety caused by various factors. According to Table 9, the overall state of listening anxiety for students who are from Hainan province is greater than that among students in mainland China, and the English listening anxiety caused by various factors is more severe among students in Hainan than among students in mainland China. Hainan needs to further improve the level of basic education, especially to enhance students' English proficiency (Table 10).

Table 9. Independent-samples T test

	Levene's test forequality of variances		T test for whether the mean is equal						
	F	Sig.	T	df	Sig. (2-tailed)	Mean difference	Standard error	95% Confidence Interval	
								Lower Bound	Upper Bound
Overall Anxiety	1.640	.202	−2.499	240	.013	−.23633	.09456	−.42260	−.05007
			−2.468	168.983	.015	−.23633	.09575	−.42536	−.04731
Emotional factor	2.273	.133	−2.274	240	.024	−.19906	.08753	−−.37149	−.02663
			−2.243	168.486	.026	−.19906	.08873	−.37422	−.02389
Listening ability	.277	.599	−2.245	240	.026	−.23971	.10678	−.45006	−.02937
			−2.237	173.519	.027	−.23971	.10716	−.45122	−.02821
Listening materials	1.612	.205	−2.180	240	.030	−.22271	.10214	−.42391	−.02150
			−2.152	168.784	.033	−.22271	.10347	−.42698	−.01844
Environmental factor	.906	.342	−3.205	240	.002	−.35977	.11225	−.58089	−.13865
			−3.206	175.528	.002	−.35977	.11221	−.58122	−.13832

One-way ANOVA for Private Higher Vocational College Students with Different Majors. In this questionnaire survey, the participants were from four different majors. Therefore, to understand whether there are differences in English listening anxiety levels for private higher vocational college students from different majors, **ANOVA** was conducted on the questionnaire data. According to Table 11, the variances of the total English listening anxiety level, self-evaluation listening anxiety dimension, listening ability anxiety dimension, listening material anxiety dimension, and environmental factor anxiety

Table 10. Group statistics

	City	N	Mean	Standard deviation	Standard error mean
Overall Anxiety	Outside Hainan	156	3.3003	.69301	.05549
	Hainan	86	3.5366	.72368	.07804
Emotional Factor	Outside Hainan	156	3.1548	.64072	.05130
	Hainan	86	3.3538	.67139	.07240
Listening Ability	Outside Hainan	156	3.2981	.79159	.06338
	Hainan	86	3.5378	.80131	.08641
Listening Materials	Outside Hainan	156	3.4411	.74822	.05991
	Hainan	86	3.6638	.78241	.08437
Environmental Factor	Outside Hainan	156	3.1325	.83617	.06695
	Hainan	86	3.4922	.83506	.09005

dimension of the students from the four different majors were P values of 0.021, 0.055, 0.041, 0.014, and 0.022, respectively, all of which were <0.05. These findings indicate that there are significant differences in listening anxiety levels, self-evaluations of listening anxiety levels, listening ability anxiety levels, listening material anxiety levels, and the influence of environmental factors on listening anxiety among students from different majors. Students from different majors have different English foundations and requirements for English listening. For majors that focus on cultivating students' reading and writing abilities, students' anxiety about English listening is reduced. Among these students, some with excellent academic performance, because they care about their English grades and have requirements for English learning achievements, have a higher level of anxiety about listening than those with weak foundations and no requirements for English learning. Different majors, such as mechanical and electrical engineering and electronics, have higher requirements for students' English listening skills, so students' anxiety about their English listening proficiency is relatively high. However, for the new energy vehicle manufacturing major, since the requirements for students' listening proficiency are lower than those of the other two majors are, the requirements for students' English listening proficiency are slightly lower.

Correlation Analysis between Students' Level A Test Scores and the Level of English Listening Anxiety. The College Level A English test is an important English exam that vocational college students can take on campus, and it is also a crucial English exam for many majors in their future job hunting. As a result, study the correlation between students' Level A English test scores and listening anxiety levels will be essential. A correlation analysis was conducted between the subjects' Level A English test scores and the students' overall anxiety level in English listening. According to Table 12, The Pearson correlation coefficient between the Level A English test scores and the overall anxiety level in English listening is −0.136, and the significant P value is $0.000 < 0.05$, indicating a negative correlation between the students' Level A English test scores and

Table 11. ANOVA

	Quadratic sum	df	Mean square	F	Sig.
Overall Anxiety	4.886	3	1.629	3.309	.021
	117.167	238	.492		
	122.054	241			
Emotional factor	3.263	3	1.088	2.566	.055
	100.880	238	.424		
	104.143	241			
Listening ability	5.255	3	1.752	2.786	.041
	149.634	238	.629		
	154.890	241			
Listening materials	6.189	3	2.063	3.627	.014
	135.368	238	.569		
	141.557	241			
Environmental factor	6.897	3	2.299	3.258	.022
	167.924	238	.706		
	174.821	241			

their English listening anxiety level. The higher the level of English listening anxiety among students is, the lower their Level A English test scores. Moreover, for four different dimensions, there is a negative correlation between students' Level A English test scores and their performance.

Table 12. Correlation analysis

		Overall Anxiety	Emotional Factor	Listening Ability	Listening Materials	Environmental Factor
Level A test score	Pearson Correlation	−.136	−.114	−.146	−.102	−.122
	Sig.(2-tailed)	.000	.000	.000	.000	.000
	N	67	67	67	67	67

**. Correlation was significant at 0.01 level (two-tailed)

3 Methods and Suggestions for Relieving Private Higher Vocational College Students' English Listening Anxiety

Based on the analysis for the above data, students show moderate degree of anxiety in their English listening at present. The factors of students' self-evaluations, listening ability, listening materials and listening environment can all cause students' listening anxiety. On this basis, the following suggestions are proposed to alleviate students' English listening anxiety.

3.1 From the Aspects of Teachers

By improving the teaching method, choosing the appropriate English materials and teaching mode in terms of the actual situation for private higher vocational college students, teachers can start with simple and easy-to-understand content, gradually increase the difficulty, and help students build confidence. Rich teaching methods, combined with multimedia resources, such as English movie clips, interview programs, and talk shows, enable students to feel the practical application of English in a relaxed atmosphere. Moreover, teachers can use these materials to analyze and help students understand vocabulary, grammar and cultural background. It adopts a variety of listening training methods, combining intensive listening and extensive listening. In intensive listening, students can understand word-by-word and sentence-by-sentence, whereas in extensive listening, students can focus on the general idea of the text. It is also possible to combine recording with discussion, performance and group competition to increase students' participation and interest. To create a relaxed and pleasant classroom atmosphere, teachers should maintain a positive mood and friendly attitude and use humorous language and nonverbal behaviors (such as smiling and eye contact) to relieve students' tension. The use of encouraging language in class provides recognition and praise for students' performance. To evaluate students' performance correctly, teachers should let students know that listening to anxiety is a normal phenomenon and avoid interrupting their thinking and increasing anxiety due to excessive error correction. In the evaluation, encouragement should be given priority, and students' progress and advantages should receive more attention. For students' mistakes, indirect correction can be adopted, such as first affirming students' efforts and then pointing out that local teachers need to know students' learning situations and problems in time and provide targeted feedback and suggestions. At the same time, teachers can use the time after class to communicate with students individually to help them solve difficulties in learning.

3.2 From the Aspects of Students

Students should understand that listening to anxiety is universal and should not deny their ability because of anxiety. Individuals believe that through hard work and practice, they can gradually overcome anxiety and improve their listening level. Make reasonable learning goals according to your actual situation. Goals should be specific, measurable and achievable, such as listening to a piece of English news every day, completing a listening exercise once a week, etc. By achieving these small goals, one can gradually accumulate a sense of accomplishment and enhance one's self-confidence. Students who are from private higher vocational colleges, especially those from Hainan, have poor basic knowledge of English. Therefore, the keys to improving listening comprehension are to improve the English foundation, expand vocabulary and master basic grammar knowledge. Students can consolidate their English foundation by reciting words, reading English articles, and doing grammar exercises. Students in private schools have not developed good learning habits over the years, so it is necessary to cultivate students' ability to make their own learning plans and arrange their learning time reasonably. Moreover, students are encouraged to learn to summarize and reflect, find their own problems in time and make improvements. Students should also master certain learning strategies. Before listening to practice, they can relieve their tension by taking deep breaths and listening to relaxing music. Keep a good attitude and believe that you can complete the listening task. Master certain listening skills, such as reading the question first and then listening to the tape, grasping the key information, and learning to predict the content. These skills can help students better cope with listening materials and improve the listening effect. Students should also actively participate in extracurricular activities, such as English corners, English speech contests, English song contests, etc. These activities can provide more opportunities for practising students' listening, reading,speaking and writing, assist students improve their listening and speaking ability, and help them build confidence for English learning.

3.3 From the Aspects of Colleges

Students in vocational colleges spend most of their time studying and living in schools. Schools can enrich teaching resources, provide more online learning resources, and provide students with diverse learning materials. Moreover, schools can also plan special English corners and learning exchange rooms to provide students with places for communication and learning. Hold more oral activities to create more opportunities for students to show themselves so that they can gradually overcome psychological anxiety and bold expression so that their listening and speaking ability can be improved. Given the uneven English level of students in private vocational colleges, small classes can be taught at different levels to teach students according to their aptitude.

4 Conclusion

The listening anxiety level for students who are from private higher vocational colleges restricts their English listening skills. To relieve the English listening anxiety for students in private higher vocational college, long-term attention and effort are needed, and joint efforts and cooperation among teachers, students and schools are needed to achieve good results. It is hoped that through discussion and analysis in this study, more educators can pay attention to this problem and make joint efforts to advance English learning effect so as to promote the cultivation of comprehensive quality of students from private vocational colleges.

References

Liu, H.: An empirical study on listening anxiety in college English classes in ethnic colleges – a case study of Guizhou University for Nationalities. J. Guizhou Normal Univ. (Soc. Sci. Ed.) (04), 115–119 (2008)

Chen, Y., Liu, D.: An empirical study on the causes and coping strategies of English listening anxiety. J. Foreign Lang. **26**(01), 141–144 (2010)

Zhan, F., Wang, L.: A study on the relationship between foreign language anxiety, listening anxiety and CET-4 achievement. J. Vocat. Tech. Educ. Forum (24), 59–61 (2010)

Liu, H., Zhang, X.: An empirical study on English listening anxiety factors of non-English majors in college. J. Educ. Vocat. (35), 182–184 (2011)

Zhao, L., Zhang, Y., Fang, W.: An evaluation of the causes and learning strategies of medical English listening anxiety in nursing undergraduates. PLA Nurs. J. **35**(14), 46–49 (2018)

Kim, J.: Foreign Language Listening Anxiety: Study of Korea Students Learning English. Doctoral Dissertation. The University of Texas at Austin (2000)

John, F.: Second Language Listening: Current Ideas, Current Issues. Cambridge University Press, Cambridge (2019)

Pan, L., He, M.: Investigation and countermeasure research on second language listening anxiety of business english students. J. Overseas English (14), 113–115 (2024)

Wang, Q.: Research on the current situation of english listening anxiety among college students. J. Overseas English (13), 126–128 (2023)

Chai, X.: Research on the causes and countermeasures of english listening anxiety among rural junior high school students. J. Overseas English (16), 171–173 (2022)

Ji, X.: A case study on note-taking strategies for treating listening anxiety among english learners. J. Heihe Univ. **13**(07), 10–12 (2022)

Lin, M.: The influence of anxiety in english listening among primary school students on their academic performance. Jo. English Square (08), 124–129 (2022)

Wu, H.: A study on english classroom learning anxiety of private vocational college students. In: Yu, Z., et al. (eds.) Data Science. ICPCSEE 2023. Communications in Computer and Information Science, vol. 1880, pp. 393–407. Springer, Singapore (2023)

Pei, T., Lu, Z.: A meta-analysis study on the influence effect of metacognitive teaching mode on second language listening ability. Foreign Lang. Bimonthly **46**(04), 128–136 (2023)

Su, L.: The efficacy of multimodal input on ESL learners' listening and speaking abilities. Sino-US English Teach. **19**(08), 263–269 (2022)

Li, X.: Research on the characteristics of english listening anxiety among non-english major college students. J. Harbin Univ. **46**(05), 87–91 (2025)

Qin, X.: Quantitative Data Analysis in Foreign Language Teaching Research. HuaZhong University of Technology Press, Wuhan (2003)

Wang, Z., Jiang, Y.: Research on the current situation and relationship of english learning anxiety, learning burnout and self-efficacy among college students. J. Zhejiang Ocean Univ. (Humanit. Sci.) **38**(02), 52–59 (2021)

Ronald, D.: SPSS Demystified. China Renmin University Press, Beijing (2019)

Zhao, Q.: Research on the teaching mode of listening and speaking in english classrooms driven by artificial intelligence. English Teach. **25**(06), 18–20 (2025)

Education Research, Methods and Materials for Data Science and Engineering

Construction Method and Practice for the Curriculum Ideological and Political System of Engineering Courses at Military University Under the Education Philosophy of "Politics + Military + Technology"

Qingzheng Xu(✉) and Na Wang

College of Information and Communication, Information Support Force Engineering University, Wuhan 430019, China
xuqingzheng@hotmail.com

Abstract. The teaching reform and practice of curriculum ideological and political education in engineering courses at military universities is a critical measure for implementing the military education policy proposed by President Xi Jinping in the new era. The education philosophy of "politics + military + technology" of the National University of Defense Technology is the key to nurturing high-quality and specialized military talent. Taking the *Fundamentals of Artificial Intelligence* as an example, the inherent advantages of engineering courses at military universities are discussed in this paper. A comprehensive approach to construct the curriculum ideological and political system is then proposed, encompassing excavating ideological and political elements, reorganizing the knowledge system, and exploring teaching methods. The goal of this approach is to address fundamental issues such as integrating what, where and how to integrate. Furthermore, preliminary teaching practices have achieved satisfactory results.

Keywords: Curriculum Ideological and Political Education · Military University · Education Philosophy · Ideological and Political Elements · Teaching Method

1 Introduction

The curriculum of ideological and political education is an educational concept with distinct Chinese characteristics and the spirit of the times [1]. It is generally believed that this term was first proposed by some universities in Shanghai in 2014. Its original intention was to explore solutions to the two-skin phenomenon between ideological and political courses and general education courses, as well as professional courses in universities [2]. Since the 18th National Congress of the Communist Party of China, General Secretary Xi has presided over several important meetings, such as the National Conference on Ideological and Political Work in Colleges and Universities, the National Education Conference, and the Symposium on Teachers of Ideological and Political

Y. Yu et al. (Eds.): ICPCSEE 2025, CCIS 2673, pp. 101–114, 2026.
https://doi.org/10.1007/978-981-95-2566-9_7

Theory Courses in Schools, and then made a series of important instructions emphasizing the need to strengthen ideological and political education in colleges and universities. These significant speeches pointed out the direction and provided fundamental guidelines for advancing the development of curriculum ideological and political education in colleges and universities. In May 2020, the "Guidelines for the Curriculum Ideological and Political Construction in Higher Education Institutions" issued by the Ministry of Education of the People's Republic of China clearly stated that the overall goal of curriculum ideological and political construction is to address the fundamental issues of who to cultivate, how to cultivate, and for whom to cultivate. Centering on the core of comprehensively improving talent cultivation capabilities aims to fully promote and form a widespread consensus on the concept of curriculum ideology and politics in all colleges and universities, as well as all disciplines and majors. Furthermore, teachers' awareness and abilities are enhanced to engage in curriculum ideology and politics construction, a sound institutional mechanism is then established for collaborative promotion, and the effectiveness of moral education in colleges and universities is ultimately improved [3].

In November 2019, President Xi clearly proposed a military education policy for the new era at the opening ceremony of centralized training for military academy presidents. It aims to adhere to the Party's absolute leadership over the military, serve the goal of building a strong military, establish morality and cultivate people, cultivate talent for the war, and cultivate high-quality, professional, and new-type military talents with both ability and moral integrity. This policy scientifically answers the critical issues of the times for military academies, namely, "for whom to cultivate, how to cultivate, and who to cultivate". It can provide fundamental guidelines and point out the direction of action for military academy education in the new era [4]. Therefore, how to construct a curriculum ideological and political system and form a teaching mode that organically integrates ideological and political elements with professional courses under military education policy for the new era has become an important and urgent problem to solve.

The digital transformation of curriculum ideological and political education was analysed, and a digital assessment model was subsequently established on the basis of the correlation matrix [5]. Tian studied the coupling of the five links of ADDIE and the five essential elements of the ideological and political classroom teaching model [6]. On the basis of the ADDIE model, the curriculum system of "analysis-design-development-implementation-evaluation" was established. After investigating 1,140 college students at 10 universities, the teaching quality evaluation model and index system of ideological and political education were also established on the basis of the IPO method [7]. Recently, curriculum ideological and political education reforms have been explored and implemented in some majors, such as foreign language [8], physics [9], computer science [10, 11], etc.

The education philosophy of the National University of Defense Technology is "politics + military + technology", which refers to cultivating students' high ideological and political consciousness, solid military training qualities, and excellent scientific and technological innovation abilities. This talent cultivation concept fully considers the development and construction needs of the country, society, and military, vividly reflects the comprehensiveness and modernity of higher education, and is highly important for cultivating high-quality, professional, and new-type military talent in the new era.

2 Course Overview

The *Foundations of Artificial Intelligence* course is a generally foundational elective course for engineering majors such as the communication engineering major at the National University of Defense Technology, China. It has a weight of 3 credits and spans 48 h, including 38 h of lectures and 10 h of practice. The course forms an organic whole with subject fundamentals electives and professional electives and lays the foundation for practical teaching subjects such as internship practice, graduation design, and discipline competitions. Students can cultivate their comprehensive qualities, broaden their career development paths, and effectively support the cultivation of high-quality new military talent. The teaching goal of this course is to enable students to understand AI's (artificial intelligence) research scope, application fields, and development directions; master the basic concepts, principles, and methods of AI; and develop problem-solving abilities. As a result, it can lay the foundation for theoretical learning and comprehensive practice in AI-related courses for initial professional education.

2.1 Course Content

To construct a curriculum content system that conforms to the discipline development law and presents a relatively complete artificial intelligence knowledge system structure for students, it is organized according to a "three-horizontal, three-vertical" framework, as shown in Fig. 1 [12]. "Three-horizontal" means that, following the migration sequence of hotspots in the history of artificial intelligence, the main content of this course is divided into three parts: reasoning, knowledge, and learning. The term "three-vertical" refers to the gradual deepening and upgrading of the basic concepts, algorithm models and application designs within each chapter. This structure of chapter organization can reflect the universal law of human understanding of new things.

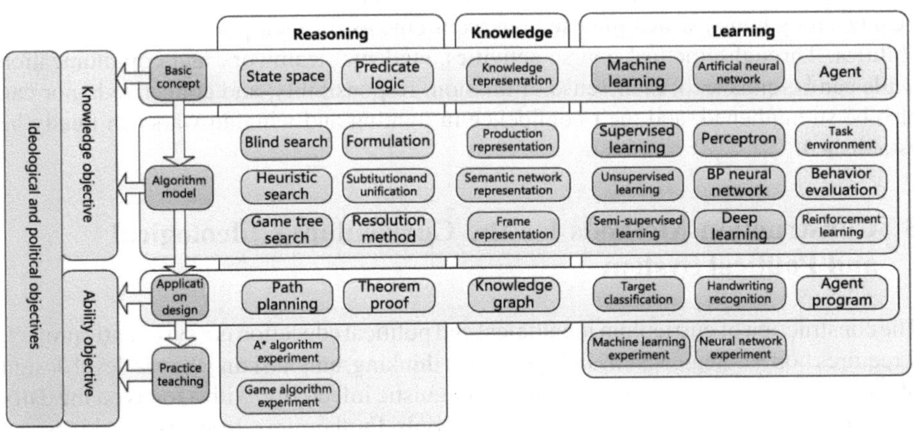

Fig. 1. Knowledge architecture of the *Foundations of Artificial Intelligence* course.

2.2 Foundation of Ideological and Political Education

As a typical representative of engineering courses at military universities, the *Fundamentals of Artificial Intelligence* course possesses numerous advantages for carrying out ideological and political education reforms and practices.

First, it has a broad audience among students and has a significant impact on their subsequent development. As a generally foundational course for engineering majors, this course is taught to junior or senior students, serving as a crucial "bridge" between academic learning and position competency. In conjunction with ideological and political courses, second classrooms, and graduation education, this course can create a synergistic educational effect, bridging the "last mile" of students' professional education and supporting their initial appointment and long-term development in the military.

Second, the course covers a wide range of topics, which contain rich ideological and political materials. This course is comprehensive and foundational. Its content encompasses state space search, logical reasoning, knowledge representation, machine learning, and neural networks. The extensive teaching content is imbued with abundant ideological and political materials, ranging from traditional culture and patriotic feelings to obey the Party's directives and spirit of militancy, as well as from the basic laws of dialectics to scientific and innovative thinking.

Third, AI technology has undergone rapid updates and deep integration with disciplines and majors. Currently, information technology is rapidly transitioning from "Internet+" to "AI+". AI technologies such as deep learning and generative AI profoundly impact various aspects of the economy, society, national defense, and military. This course can help students enhance their confidence and identification with their chosen disciplines and majors, gradually plan clear personal professional goals, and cultivate good habits of autonomous learning, creative learning, and lifelong learning.

Finally, this course fully integrates theory with practice, which can improve students' practical and innovative abilities. A typical characteristic of engineering courses is the deep integration of theory with practice, often including varying amounts of practical teaching hours, which impose high requirements for their practical and innovative abilities. Through practical course activities, students' teamwork and communication skills can be enhanced, their sense of mission, responsibility, and collective honor can also be strengthened, and their confidence in fighting and winning wars can finally be established.

3 Construction Methods for the Curriculum's Ideological and Political System

The construction of curriculum ideological and political education is a systematic project. It requires both strategic vision and systematic thinking, emphasizing the top-level design of courses, and involves keen intuition and linguistic infectivity, with a focus on the daily accumulation of ideological and political materials. Furthermore, teachers should respect the basic laws of education and teaching, adhere to the original aspirations and bottom lines, adapt to the needs of the times, and constantly innovate teaching methods for curriculum ideological and political education. To construct a good curriculum ideological and political system, three fundamental problems must be scientifically solved: what to integrate, where to integrate, and how to integrate.

3.1 Multipronged Approach to Deeply Mining Curriculum Ideological and Political Elements

Its goal is to solve the "what to integrate" problem. Mining diverse curriculum ideological and political elements is the key to constructing and practicing curriculum ideological and political education. Chen suggested that the goals of the ideological and political education of specialized courses generally include two categories: general education goals and professional value goals [13]. On this basis, Liu added engineering education goals, such as innovative thinking, team collaboration, engineering ethics, and ecological protection [14]. Given the uniqueness of military universities, which is based on the philosophy of "politics + military + technology" of the National University of Defense Technology, the goals of the ideological and political education of military universities can be divided into three categories: politics goals, military goals, and technology goals. Each category may contain some specific ideological and political elements, effectively addressing the "what to integrate" problem.

The goal of politics is to cultivate students' high ideological and political awareness and attempt to answer "what a comprehensive individual looks like". Specifically, it can include traditional culture, patriotism, core socialist values, the "Four Confidences" (confidence in the path, theory, system, and culture of socialism with Chinese characteristics), the "Three Views" (worldview, outlook on life, values), and mental health. The military goal aims to cultivate students' solid military training qualities and attempt to answer "what a revolutionary soldier looks like". It can include obeying the Party's directives, the spirit of militancy, the strong sense of professional mission, awareness of hardships, excellent traditions, job ethics, awareness of security and confidentiality, and the core values of revolutionary soldiers. The technology goal aims to cultivate students' outstanding technological innovation abilities and attempt to answer "what a new-era student looks like". Specifically, it can include dialectical materialism, scientific thinking (systematic thinking, critical thinking, reverse thinking, etc.), craftsmanship spirit, exploratory and innovative spirit, teamwork spirit, engineering ethics, and entrepreneurial spirit.

3.2 Reviewing the Past Leads to New Insights and Comprehensively Organizing the Course Knowledge System

Its goal is to solve the "where to integrate" problem. Clarifying a complete knowledge system and constructing specific knowledge units are the foundations for carrying out the construction and practice of curriculum ideological and political education. On the basis of the "Teaching Plan of the *Fundamentals of Artificial Intelligence*" and practical teaching experiences, the teaching team reorganized the knowledge points for each chapter, effectively solving the "where to integrate" problem. As shown in Fig. 2, taking Chapter 2, "State space search," as an example, this chapter includes three subsections: overview, blind search, and heuristic search, encompassing a total of 31 knowledge points.

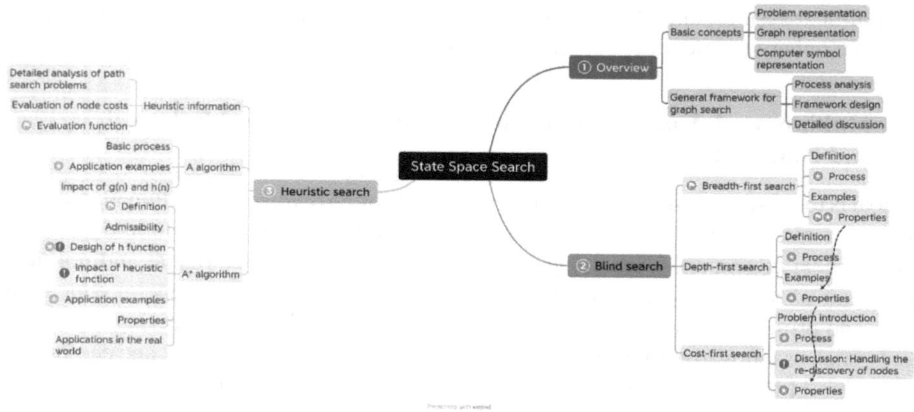

Fig. 2. Knowledge Points of Chapter 2 "State space search".

3.3 Actively Exploring Teaching Methods for Curriculum Ideological and Political Education

Its goal is to solve the "how to integrate" problem. Exploring novel teaching methods for curriculum ideological and political education involves building a "bridge" between ideological and political elements and the knowledge system to achieve their organic integration. Essentially, mining the curriculum's ideological and political elements and organizing the knowledge system both serve the purpose of achieving classroom teaching effects and cultivating students' qualities and abilities. Therefore, it is particularly important to organically integrate ideological and political elements into teaching content. This integration is not about indiscriminate infusion, forced implantation, mechanical copying, or mimicking but rather about being embedded in knowledge, culture, and activities, such as saltwater, moisturizing silently [15]. Therefore, teachers must carefully design each segment before class, grasp the inherent logicality of the whole course, follow the general laws of students' ideological and conceptual development, and consider the uniqueness of new-era military university students. To establish morality, cultivate people, and cultivate talent for the war, the "Politics + Military + Technology" education philosophy is organically integrated into teaching. Through many teaching practices, the teaching team has summarized the following methods for integrating ideological and political elements, achieving good teaching results.

First, there is the case-based teaching method. By utilizing practical cases, especially in the military field, theoretical knowledge can be combined with engineering applications, and the rich ideological and political elements contained in actual cases can then be mined, cultivating students' innovative thinking and value judgment abilities. Within the chapter "Machine learning", on the basis of the actual case of fighting against COVID-19, we can conduct theoretical knowledge instruction and ideological and political education from multiple perspectives. For example, decision tree algorithms can assess the four levels of COVID-19, medical image analysis technology can perform automatic screening for bronchial diseases, and speech recognition technology can carry out screening of key populations, publicity and education, and contactless data collection. These real cases can increase students' awareness of the severity of COVID-19 and help them recognize that the most powerful weapon for overcoming disease is scientific development and technological innovation. These cases are also helpful for cultivating their scientific spirit of respecting science, seeking truth, being pragmatic, and being pioneering and innovative. As a result, we can guide them to learn skills well, serve the country, and inspire them to aspire to strengthen the military and take on responsibilities.

The second method is the heuristic teaching method. Under the guidance of teachers, all students are encouraged to explore unknown areas independently and autonomously construct their knowledge system. During this process, ideological and political education is subtly incorporated into the curriculum. For example, when explaining the basic concepts of state space search, teachers can introduce examples such as Alice's Adventures in Wonderland maze and the nine-level maze of the Monster Tower. By first studying the representation method of specific problems such as maze problems and then guiding them to summarize the representation method of general search problems, students can deeply understand the scientific research approach that moves from specific observations to broader generalizations. Therefore, this approach can increase confidence and determination in scientific research.

The third category includes problem identification and analysis teaching methods. The teachers can set an open question or a set of sequentially deepening questions around the course content. The goal of this approach is to stimulate students' active thinking and exploration and promote the exchange of opinions and in-depth discussions. For example, in the chapter "Overview of artificial intelligence", the classic open question is "When will artificial intelligence surpass human intelligence?" or "Is the emergence of superintelligent AI a good thing?". The goal of these questions is to guide students in the use of natural dialectics to understand and analyse AI development trends and humanity's attitudes and then enable them to comprehend and establish a correct scientific worldview. Another example is that, in the chapter "Advanced topics of artificial intelligence", the teaching team uses the US Air Force's ALPHA air combat system as a case study. Some strongly related questions are set and discussed, such as how AI technology can empower military warfare, how China's military can develop intelligent air combat systems with independent intellectual property rights, and what decision-making authority or autonomy intelligent unmanned aerial vehicles should have. These problems can encourage students to pay attention to the logical relationship between technology progress and military reforms, cultivate their scientific literacy for independent innovation, and consider the ethics of artificial intelligence in warfare. As a result, it will ultimately strengthen their sense of responsibility as revolutionary soldiers.

Fourth, the project-driven teaching method is used. Engineering courses often include practical teaching sessions. It involves both developing small programs independently and conducting secondary development on domestic company research platforms. Through a series of practical projects, students can experience the satisfaction of solving practical problems via theoretical knowledge, which can both stimulate their enthusiasm for scientific research and simultaneously achieve professional skill and ideological and political education. Furthermore, all students need to independently complete algorithm design and code writing, prohibit copying of others' code, and submit high-quality experimental reports. These requirements can cultivate honesty and the craftsmanship spirit of striving for excellence.

The fifth method is the homework assignment-based teaching method. In addition to classroom teaching activities, assigning homework or thinking questions after class is also an important approach for effectively integrating ideological and political elements into professional courses. For example, for Chapter 6, "Knowledge representation and reasoning", the questions are designed such as "Use the semantic networks and frames to represent the following knowledge: The National University of Defense Technology is a national 'Double First-Class' university. The school motto is 'Cultivating Virtue and Expanding Knowledge, Strengthening the Military and Revitalizing the Country'. It was formerly known as the Military Engineering Institute of Harbin established in 1953", enhancing students' sense of collective honor and mission to dedicate themselves to national defense. Another possible question is "Using semantic networks and frames to represent the following knowledge: the average elevation of the Jiajin Mountains is 3,437 m, with the mountain pass at an elevation of 1,300 m. In addition, Daxue Mountain at an elevation of 7,556 m, covered with snow all year round", allowing students to appreciate the bravery and firm beliefs of their revolutionary predecessors and to feel and strive to inherit the great spirit of the Long March.

4 Construction Practice for the Curriculum Ideological and Political System

4.1 A Teaching Practice Case

On the basis of the aforementioned construction methods for the curriculum ideological and political system, an initial ideological and political education system for the professional *Fundamentals of Artificial Intelligence* course is designed, and part of the content is illustrated in Table 1.

Table 1. Ideological and political system for the *Fundamentals of artificial intelligence* course (excerpt).

Chapter	Knowledge point	Ideological and political element	Ideological and political objective	Teaching method	Teaching strategy and expected outcomes
2.1 Overview of state space search	Basic concepts	Scientific thinking: Inductive reasoning	Technology objective	Heuristic method	First, investigate specific problems such as the maze problem, and then generalize and abstract these studies to formulate a general approach to representing search problems.

(*continued*)

Table 1. *(continued)*

Chapter	Knowledge point	Ideological and political element	Ideological and political objective	Teaching method	Teaching strategy and expected outcomes
2.2 Blind search	Breadth-first search	Innovative spirit	Technology objective	Heuristic method	To better understand the nuances between general search and breadth-first search, we will delve into their core principles. This comparison can not only highlight the technical differences but also inspire learners to rethink innovation, bolster their confidence in innovation, and encourage them to pursue innovative outcomes.

(continued)

Table 1. (*continued*)

Chapter	Knowledge point	Ideological and political element	Ideological and political objective	Teaching method	Teaching strategy and expected outcomes
		Mental health	Politics objective	Problem identification and analysis method	First, propose an open question: How can we correctly understand the evaluation? Then, encourage deep thinking and active discussion among learners from various perspectives, including the significance of evaluation, the selection of evaluation metrics, and the application of results. Therefore, cognitive misconceptions will be eliminated and psychological barriers will be overcome. The teachers should provide timely summaries to reinforce key insights.

(*continued*)

Table 1. (*continued*)

Chapter	Knowledge point	Ideological and political element	Ideological and political objective	Teaching method	Teaching strategy and expected outcomes
	Blind search	Professional mission	Military objective	Homework assignment-based method	Using the scenario of delivering supplies to an island as a background, analyse the limitations of blind search algorithm. Through this analysis, guide learners to understand the importance of developing more efficient and intelligent algorithms for military logistics operations. Emphasize the need for students to cultivate a strong commitment to strengthen the military and protect the country, and a profound sense of mission and responsibility.

(*continued*)

Table 1. (*continued*)

Chapter	Knowledge point	Ideological and political element	Ideological and political objective	Teaching method	Teaching strategy and expected outcomes
2.3 Heuristic search	Heuristic information	Outlook on life	Politics objective	Heuristic method	Examine the features of heuristic search algorithms, which incorporate problem-specific knowledge to guide the search process more efficiently. Through this analysis, guide students to develop a correct outlook on life, emphasizing both personal effort and adherence to the right direction of progress.
	A* algorithm	Innovative spirit	Technology objective	Heuristic method	Revisit the historical story of Nils Nilsson, the founder of the A* algorithm, to inspire students with the spirit of 'A young calf unafraid of tigers'. In addition, encourage them to take scientific pioneers as role models, boldly innovate, and pursue progress with determination.

4.2 Teaching Outcomes

Teaching practices related to this topic have been attempted in the classroom, and satisfactory results have been achieved. On the one hand, students have become more active in asking questions and engaging in interactive discussions during class, and they are more proactive in learning new knowledge and skills after class. On the other hand, survey results and individual interviews indicate that they generally approve of the curriculum ideological and political education reforms and are attempting to digest and accept the "Politics + Military + Technology" education philosophy that the teachers aim to convey.

The teaching outcomes were highly encouraging: the course achieved a 100% pass rate for the first time, with the excellence rate increasing by 5% compared with previous academic years. Furthermore, the course received an "outstanding" rating in the university's classroom teaching quality evaluation.

5 Conclusion

In carrying out curriculum ideological and political education reform and practice in engineering courses at military universities, it is necessary to fully explore the ideological value and spiritual power embedded in course knowledge while considering the integrated development needs of military intelligence and professional talent cultivation. Our ultimate goal is to subtly implement the education philosophy "Politics + Military + Technology". In this manuscript, a comprehensive approach was proposed to construct the curriculum ideological and political system, and its feasibility was validated via the *Fundamentals of Artificial Intelligence* course as a case study. In fact, the implementation strategies of other courses should be developed and compared to enhance their generalizability.

Currently, the curriculum ideological and political education reform in the *Fundamentals of Artificial Intelligence* course is still in its exploratory stage. Various practical and theoretical challenges deserve in-depth research, including the diversity of sources of ideological and political elements, the updating and iteration of the knowledge system, the matching degree between ideological and political teaching methods and the "profiles" of students, and the assessment of the effectiveness of curriculum ideological and political education.

Acknowledgements. This work was funded by the Education Science Planning Project of Hubei Province (Grant no. 2024GB424) and the Undergraduate and Professional Education Teaching Research Project of the National University of Defense Technology (Grant no. U202320).

Disclosure of Interests. The authors have no competing interests to declare that are relevant to the content of this article.

References

1. Liu, T.: The challenges of ideological and political education in higher education institutions in the era of big data and its path analysis. Appl. Math. Nonlinear Sci. **9**(1), Article number: 20230531 (2024)

2. Wang, R.L., Wang, J.: The concept connotation, driving force, and implementation strategy of curriculum ideological and political in primary and secondary school. Moral Educ. China **19**(8), 15–20 (2024). (in Chinese)
3. The Ministry of Education of the People's Republic of China. Guidelines for the curriculum ideological and political construction in higher education institutions (2024). http://www.moe.gov.cn/srcsite/A08/s7056/202006/t20200603_462437.html?eqid=b04748c500024fb1000000036447884a. (in Chinese)
4. Xi, J.P.: Implement the military education policy for a new era, deepen the reform and innovation of military academy, and cultivate high-quality and specialized new-type military talents of have both ability and moral integrity. People's Daily (2019). (in Chinese)
5. Yang, X.B.: Digital assessment of curriculum ideological and political education in higher vocational institutions' courses based on correlation matrix. Appl. Math. Nonlinear Sci. **9**(1), Article number: 20242823 (2024)
6. Tian, B., Xie, Z.G.: Thoughts and design of ideological and political construction based on the ADDIE model. Appl. Math. Nonlinear Sci. **8**(2) (2023)
7. Song, L.L.: Curriculum construction practice of ideological and political education based on the IPO model. Appl. Math. Nonlinear Sci. **8**(2), 2817–2826 (2023)
8. Kong, L.P.: Optimization analysis of the path of ideological and political construction in higher vocational English courses based on intelligent data analysis. Appl. Math. Nonlinear Sci. **9**(1) (2024)
9. Liu, Y., Hu, H.X.: Practice of ideological and political construction of college physical education curriculum from the perspective of moral education based on 'Integration of inside and outside class'. In: International Conference on Measuring Technology and Mechatronics Automation, 28–29 February 2020, Phuket, Thailand, pp. 1100–1103 (2020)
10. Gou, P.Z., Zhang, W.X., Han, Y.Y.: Research on teaching of ideological and political education in computer foundation based on computational thinking. In: International Conference on Computer Science & Education, 17–21 August 2021, Lancaster, United Kingdom, pp. 427–430 (2021)
11. Sun, B., Dong, Y.L.: Construction and implementation of ideological and political curriculum system for computer science and technology major. In: International Conference on Computer Science & Education, 17–21 August 2021, Lancaster, United Kingdom, pp. 452–455 (2021)
12. Liu, W., Li, M., Chen, S.F., Xie, H.B.: Fundamentals of Artificial Intelligence. Science Press, Beijing, China (2024). (in Chinese)
13. Chen, L.: Discussion on the reform of urban planning principles courses based on curriculum ideological and political. J. Yancheng Teach. Univ. (Humanit. Soc. Sci. Ed.) **39**(6), 113–116 (2019). (in Chinese)
14. Liu, H.L., Li, J., Li, Y.J., Li, H.Y.: Ideological and political teaching system construction of engineering major based on engineering education accreditation. J. High. Educ. Res. **43**(3), 86–91 (2020). (in Chinese)
15. Sha, N., Guo, M.X., Xie, W., Xu, K., Lu, L.: Research on construction of ideological and political element library in thee 'anti-jamming technology and application in communications' course. J. High. Educ. Res. **44**(4), 89–92 (2021). (in Chinese)

A Brief Review of Generative Artificial Intelligence in Higher Education

Qingzheng Xu(✉) and Na Wang

College of Information and Communication, Information Support Force Engineering University, Wuhan 430019, China

xuqingzheng@hotmail.com

Abstract. The advent of generative artificial intelligence (GAI) has resulted in significant transformations across diverse sectors, with higher education being a notable beneficiary. The integration of GAI into higher education marks a paradigm shift in the methodologies employed by educational institutions for teaching and learning. This review commences by furnishing a definition of the GAI and introducing its fundamental and core technologies. It then proceeds to comprehensively encapsulate the prevalent applications of GAI within higher education, zeroing seven pivotal aspects: personalized learning experiences, automated content generation, virtual teaching assistants, enhancing academic communication, assessment and feedback mechanisms, data analytics, and faculty support and development. Additionally, this review delves into the ethical dilemmas posed by the GAI era, highlighting academic integrity, data privacy and security, algorithmic bias and fairness, and transparency and accountability. Ultimately, this review concludes by suggesting several invaluable research avenues for other researchers to consider and draw inspiration from.

Keywords: Generative artificial intelligence · Higher education · Teaching and learning · Content generation · Curriculum reform

1 Introduction

The rise of generative artificial intelligence (GAI) is driving transformative changes in higher education, reshaping teaching and learning. As institutions seek innovative ways to enhance educational outcomes, the GAI has emerged as a powerful tool for personalizing learning, streamlining content creation, and providing real-time support through virtual teaching assistants.

Studies highlight the GAI's potential to tailor learning experiences to diverse student needs, fostering individualized learning pathways. This adaptability is crucial in modern education, where varied learning styles require innovative instructional strategies. GAI also enhances academic efficiency by automating content generation, enabling educators to develop personalized materials while reducing administrative burdens. This allows instructors to focus on interactive teaching and student engagement, boosting both productivity and learning outcomes.

Y. Yu et al. (Eds.): ICPCSEE 2025, CCIS 2673, pp. 115–129, 2026.
https://doi.org/10.1007/978-981-95-2566-9_8

Moreover, the GAI revolutionizes assessments by enabling adaptive evaluation methods. Automated grading and real-time feedback enhance efficiency while personalizing student evaluations. While intelligent algorithm design affects effectiveness, such systems hold promise for improving engagement and academic success.

Despite its advantages, GAI integration presents challenges, including data privacy concerns, intellectual property issues, and academic integrity risks. Ensuring responsible use requires clear policies and proactive measures to prevent misuse. A collaborative approach among educators, policymakers, and developers is essential to establish best practices.

2 Background

2.1 Definition of the GAI

GAI represents a transformative branch of GAI focused on synthesizing novel content through machine learning, spanning text, images, music, and multimodal outputs. Unlike traditional GAI systems, which prioritize rule-based analysis or decision-making, GAI emphasizes creativity by identifying latent patterns in training data to generate original artifacts that emulate human-like innovation.

The core of the GAI is its data-driven learning mechanism. Models trained on extensive textual corpora, for example, internalize linguistic patterns and semantic relationships, enabling contextually coherent text generation [1]. This capability extends to complex creative tasks, such as composing emotionally resonant music or producing stylistically consistent visual art, demonstrating hierarchical creativity from basic replication to advanced conceptualization.

In education, the adaptability of the GAI enables personalized learning experiences through dynamic content generation tailored to individual needs [2]. Interactive interfaces, including conversational agents, further enhance engagement by simulating human-like dialogue, with efficacy contingent on dataset quality, prompt engineering, and user-centered design principles.

The GAI thus redefines the role of the GAI by integrating generative and analytical capabilities. While its potential to revolutionize fields such as education is undeniable, rigorous evaluation of technical boundaries, ethical implications, and domain-specific constraints remains critical for responsible deployment.

2.2 Key Technologies of GAI

The educational integration of GAI hinges on three interconnected technological pillars: large language models (LLMs), natural language processing (NLP), and computational infrastructure.

Large Language Models: Transformer-based architectures such as GPT and BERT have redefined machine–language interactions through self-attention mechanisms, enabling real-time text processing for applications such as automated essay scoring and adaptive feedback generation [3]. Trained on multimodal datasets, these models capture linguistic nuances—from syntax to pragmatics—producing outputs indistinguishable from human-authored content in controlled contexts.

Natural Language Processing: Enhanced semantic parsing and discourse modelling now support granular educational analytics. Sentiment analysis identifies learner disengagement, whereas entity recognition extracts conceptual gaps from student submissions [4]. Such advancements underpin intelligent tutoring systems capable of context-aware dialogue, bridging the historic limitations of the conversational GAI in pedagogical settings.

Computational Infrastructure: Cloud-edge computing convergence reduces latency and costs, democratizing GAI access. Institutions leverage distributed frameworks to deploy resource-intensive models without prohibitive hardware investments, enabling small-scale educators to implement personalized learning tools [5].

3 Applications of the GAI in Higher Education

3.1 Personalized Learning Experiences

Data-Driven Adaptation. The GAI analyses behavioral data (e.g., engagement patterns, performance metrics) to map individual learning trajectories. Machine learning models identify critical success factors, such as optimal content sequencing and pacing, enabling real-time adjustment of material difficulty. For example, students encountering conceptual hurdles receive scaffold resources, whereas high achievers access advanced challenges, maintaining cognitive flow [6].

Dynamic Content Generation. Beyond adaptive pacing, GAI synthesizes multimodal content aligned with learners' interests and goals. By correlating curricular objectives with personal preferences (e.g., linking calculus problems to engineering case studies), it enhances relevance and intrinsic motivation [7].

Proactive Intervention. Early warning systems leverage natural language processing to detect disengagement signals from forum posts, submission delays, or interaction dips. Predictive models trigger targeted interventions from automated nudges to instructor alerts, reducing dropout risk by preemptively addressing academic or motivational gaps [8].

Real-Time Feedback. The GAI replaces periodic assessments with continuous feedback loops. Writing assistants, for example, provides stylistic and logical critiques during the drafting phase, cultivating iterative improvement. This mirrors expert tutoring dynamics, strengthening self-regulated learning skills [9].

Gamified Engagement. By integrating game mechanics (e.g., progress badges, leaderboards), the GAI can personalize motivational incentives. Competitive learners receive challenge-based quests, whereas collaboration-oriented students engage in GAI-mediated group tasks, diversifying engagement strategies without compromising pedagogical rigor [10].

3.2 Automated Content Generation

A key benefit of the GAI is its ability to produce customized educational materials tailored to diverse student needs. For example, it can analyse curriculum requirements to generate lecture notes aligned with specific learning objectives. Additionally, the GAI can create formative assessments, such as quizzes and exams, adapting difficulty levels to suit different student groups. By leveraging data-driven insights, these tools design personalized assessments that balance challenge and engagement, ensuring that students are neither overwhelmed nor understimulated. For example, the GAI can develop quizzes from existing resources, offering relevant and engaging content for learners [11].

Beyond basic content generation, GAI enriches learning materials by incorporating semantic roles and contextualized examples, fostering deeper comprehension and real-world application [12]. This approach enhances student engagement and knowledge retention.

As the GAI evolves, its adaptive capabilities improve, enabling the creation of materials that reflect the dynamic nature of academic discourse [13]. By analysing student performance and feedback, the GAI refines content to address learning gaps, enhancing educational effectiveness. Learning analytics tools further empower educators by providing insights into material effectiveness and guiding resource refinement.

3.3 Virtual Teaching Assistants

Virtual teaching assistants (VTAs) powered by GAI are revolutionizing higher education by streamlining course management, enhancing communication, and personalizing learning experiences. These GAI-driven tools are transforming traditional educational support, enabling more efficient and responsive interactions between students and instructors.

One of the key functions of VTAs is managing course materials. By organizing and distributing essential resources such as syllabi, lecture notes, and multimedia content, VTAs ensure seamless access for students. Equipped with NLP, they can instantly respond to student inquiries by retrieving relevant documents or links, reducing the administrative workload for educators [14]. Beyond material management, VTAs facilitate communication by addressing common student queries on assignments, scheduling, and course policies. When they are available beyond office hours, they provide timely assistance, improving student engagement. Additionally, by analysing communication patterns, VTAs refine their responses to increase clarity and relevance, making them particularly valuable in diverse classrooms where students have varied proficiency levels and learning needs [15].

VTAs also leverage data analytics to personalize learning. By analysing students' past performance, they offer tailored recommendations, ensuring access to resources aligned with individual learning paths. This personalization improves academic outcomes and helps instructors identify areas where students may struggle. Moreover, in large classes where individualized attention is challenging, VTAs systematically manage student queries, alleviating instructor workload while maintaining effective support [16].

Another critical function of VTAs is data-driven insight generation. By monitoring student interactions, they provide valuable analytics that inform curriculum improvements and teaching strategies. Identifying common misconceptions allows educators to proactively address knowledge gaps, leading to more effective instruction [17].

As student populations grow increasingly diverse, VTAs also contribute to inclusivity. They can offer translations and context-specific explanations, improving accessibility for international students and those with learning challenges. This fosters a more equitable educational environment, ensuring broader access to quality learning opportunities [18].

3.4 Enhancing Academic Communication

GAI is transforming academic communication in higher education by improving language translation and writing assistance. These tools benefit both native and nonnative English speakers, fostering a more inclusive learning environment. In language translation, the GAI provides real-time support for students struggling with academic vocabulary and complex structures. Unlike traditional translation methods, GAI leverages NLP to deliver accurate, context-aware translations. Machine translation platforms help students craft clearer texts by offering instant feedback on linguistic choices, enabling both the refinement of writing and deeper language comprehension [19]. For writing assistance, GAI-powered tools such as ChatGPT offer grammar corrections, style enhancements, and sentence suggestions—key for mastering academic writing. Research indicates that GAI-driven feedback significantly improves writing proficiency by helping students learn from their mistakes in real time, promoting academic growth in a supportive environment [20].

Beyond individual writing support, GAI enhances collaborative learning through peer assessment platforms, streamlining the review process and fostering constructive student interactions. Studies have shown that peer assessment benefits students across different skill levels, providing valuable insights that improve writing quality [21]. The GAI also addresses challenges faced by multilingual students, who may struggle in traditional academic settings. By analysing writing samples, the GAI delivers targeted feedback on common language issues specific to different linguistic backgrounds. Additionally, it bridges the gap between receptive and expressive language skills, assisting students who understand academic content but struggle to articulate their thoughts effectively.

3.5 Assessment and Feedback Mechanisms

The integration of the GAI in higher education has revolutionized assessment and feedback through automated grading and real-time feedback. These innovations address the growing need for efficiency and consistency in evaluating student performance. Automated grading alleviates educators' workload while ensuring fair and objective assessments. Traditional grading is time-consuming, particularly for large classes, whereas GAI-driven systems use NLP and machine learning to evaluate submissions swiftly and accurately. Studies have shown that GAI-based grading can assess essays on the basis of semantic meaning rather than simple keyword matching, providing more meaningful feedback and allowing instructors to focus on deeper student engagement [22]. Real-time feedback mechanisms further enhance learning by offering instant assessments that improve retention and outcomes. GAI-powered educational platforms adapt to student progress, providing tailored insights and highlighting areas for improvement. Research indicates that instant feedback boosts student engagement and performance, fostering a more interactive learning experience [23].

The GAI also enriches formative assessments by generating personalized feedback on specific errors or misconceptions. Studies have demonstrated that students appreciate receiving instructive hints and performance analytics, particularly in technical subjects such as programming. This combination of automated grading with detailed feedback cultivates a supportive and dynamic learning environment [24]. Moreover, GAI-enhanced feedback helps educators identify common challenges by analysing student submission patterns, enabling curriculum adjustments that improve teaching strategies. Research on GAI-driven tools reveals their ability to provide immediate, detailed feedback, allowing students to learn from mistakes in real time and strengthening their subject mastery [25]. Additionally, GAI-driven platforms seamlessly integrate into learning management systems, streamlining grading and feedback without disrupting established workflows. This accessibility encourages instructors to adopt GAI technology, enhancing both teaching and learning experiences [26].

3.6 Data Analytics

GAI-powered data analytics serve as a vital tool in modern education by extracting actionable insights from vast educational datasets. By analysing data from learning management systems, assessments, and student interactions, institutions can gain a comprehensive understanding of learning behaviors and patterns.

GAI enables predictive analytics to detect early warning signs of academic struggles, such as low attendance, late submissions, and declining assessment scores. This allows educators to intervene proactively, providing personalized support and fostering adaptive learning environments. Educational data mining further refines this process by tracking student engagement with learning materials, helping educators optimize content delivery and instructional methods [27].

In addition to quantitative data, the GAI also analyses qualitative inputs, such as student feedback from open-ended surveys. These insights reveal challenges in various learning environments, including online and hybrid settings, guiding curriculum improvements to enhance engagement and satisfaction [28]. The significance of GAI-driven analytics became especially clear during the shift to digital learning, accelerated by the COVID-19 pandemic. By enabling real-time monitoring of student performance, the GAI supports the continuous refinement of online learning strategies [29].

On an institutional level, large-scale data analysis helps identify systemic issues, informing strategic decisions on program development and resource allocation. Institutions leveraging GAI effectively create a feedback loop that continuously improves teaching methodologies and learning materials on the basis of empirical evidence [30].

3.7 Faculty Support and Development

The GAI enables personalized lesson planning by analysing student demographics, performance data, and learning preferences. Educators can input specific parameters, prompting the GAI to generate tailored lesson plans incorporating diverse pedagogical strategies. This personalization enhances teaching effectiveness and better addresses varied learning needs within a single classroom [31]. Beyond lesson planning, GAI aids in the creation of instructional materials. It automates the development of assessments, multimedia content, and supplementary resources aligned with course objectives, reducing preparation time. Additionally, GAI-based data analytics capabilities provide insights into student performance, helping educators refine their teaching strategies in real time.

The GAI further supports continuous professional growth through personalized training recommendations. By tracking faculty engagement and development areas, these systems suggest relevant workshops and instructional innovations, ensuring that educators stay updated with evolving teaching methodologies. A key advantage of the GAI is reducing administrative burdens. By automating grading, scheduling, and student communications, faculty gain more time for direct interaction and instructional improvement, ultimately increasing student satisfaction and retention [32]. The GAI also improves assessment design by enabling adaptive evaluations that adjust in real time on the basis of student progress. This promotes personalized learning experiences and helps educators refine their instructional approaches [33].

4 Ethics Challenges in the Age of the GAI

4.1 Academic Integrity

A key issue of academic integrity is the ease with which students can submit GAI-generated work on their own. This not only violates ethical standards but also devalues the effort and critical thinking essential to academic growth. Effective plagiarism detection systems are increasingly vital to combat these new forms of misconduct [34]. Additionally, GAI risks diminishing students' engagement with learning, encouraging passive consumption over active inquiry and problem solving. Overreliance on GAI tools could erode creativity and intellectual independence, which are core pillars of higher education [35].

In addition to student practices, faculty and institutions face new challenges in maintaining academic integrity. The obscured authorship of written work complicates evaluation, fostering skepticism and straining educator–student trust. Thus, institutions must prioritize academic integrity education, equipping both students and educators with strategies to navigate GAI's role in learning [36].

Proactive institutional measures are essential, including clear policies on GAI use, updated plagiarism guidelines, and curricula that emphasize ethical writing and originality. GAI tools should be integrated as learning aids rather than replacements for genuine effort [37]. Furthermore, the definition of plagiarism may need reevaluation, as GAI blurs the line between human- and machine-generated content. Rather than solely policing misconduct, educators should instill a deeper appreciation for intellectual honesty [38].

4.2 Data Privacy and Security

As GAI technologies gain traction in higher education, concerns over data privacy and security become increasingly critical. GAI-driven personalization aggregates user data to enhance learning, but this approach raises significant privacy concerns. Detailed tracking of user interactions may inadvertently reveal personally identifiable information. Inadequate security measures can lead to unauthorized access, with severe consequences for affected individuals. The increasing number of data breaches in education underscores the need for stringent security protocols [39].

Additionally, evolving regulations impose strict data protection requirements. Institutions must ensure compliance to uphold users' rights over their data, as violations can result in heavy penalties, reputational damage, and financial losses. Research highlights the increasing risks of data breaches, emphasizing the urgent need for robust governance policies.

Data breaches are not just technical failures; they reflect an institution's ethical responsibility to protect personal data. High-profile incidents reveal vulnerabilities in cybersecurity practices, exposing institutions to financial and reputational harm. Many struggle with implementing adequate protection, making them susceptible to attacks [40].

Educational data, encompassing academic records and personal details, are particularly sensitive. Mishandling such information can increase institutional liability. Proactive security assessments and ongoing adaptation to emerging threats are essential for maintaining data integrity. The intersection of emerging technologies and existing privacy regulations presents additional challenges. Institutions must adapt their privacy strategies to align with advancements in GAI while maintaining compliance with legal frameworks. Effective risk mitigation involves implementing strong encryption, data loss prevention strategies, and staff training on cybersecurity best practices.

4.3 Algorithm Bias and Fairness

Another key issue of the GAI is its potential for discrimination on the basis of race, gender, socioeconomic status, or ability. If trained on historically biased datasets, these systems may reinforce disparities, disadvantaging underrepresented students in areas such as educational pathways, resource access, and feedback mechanisms. Similar concerns have been raised regarding biases in other educational algorithms, whereas nonrepresentative data lead to discriminatory outcomes [41]. Bias in GAI-driven assessments further complicates fairness. Algorithms may favour certain traits, penalizing students who do not align with implicit standards. For example, standardized assessments influenced by institutional history may fail to reflect students' true abilities, undermining academic integrity and equitable evaluation [42].

To mitigate these risks, proactive strategies are essential. Diversifying training datasets can enhance fairness by encompassing varied demographics and experiences. Transparency in algorithm design, data collection, and decision-making fosters trust and accountability. Ethical frameworks should guide GAI development, ensuring that biases are addressed from the outset rather than reactively [43]. Continuous monitoring and audits are also crucial. Regular assessments help identify unintended biases, promoting fairer educational practices. Ethical design principles should be embedded in GAI development, with developers trained to recognize and mitigate bias. A participatory approach involving students, educators, and ethicists can align GAI tools with educational values and needs.

4.4 Transparency and Accountability

The transparency of GAI ensures that stakeholders understand how these technologies function, fostering trust and institutional responsibility. Transparency encompasses clarity in data usage, algorithms, and decision-making criteria. It enables students, educators, and policymakers to assess the GAI's fairness and contest its decisions, reinforcing accountability. A lack of transparency can exacerbate biases, as users cannot determine whether ethical considerations are applied [44]. Providing access to training data can help identify and mitigate such biases, ensuring equitable educational outcomes.

Accountability requires institutions to justify their use of the GAI, mitigate potential harm, and align GAI applications with educational and ethical standards. Establishing clear guidelines on GAI deployment, data collection, and intended outcomes is essential. Robust governance frameworks must oversee GAI implementation, ensuring compliance with ethical norms and legal standards. Algorithm decision-making further complicates accountability, particularly in student evaluations and resource management. Without insight into how factors such as attendance or participation are weighted, students and faculty may struggle to accept GAI-generated outcomes. Transparency in these processes is key to maintaining fairness and institutional credibility [45].

The fostering of transparency and accountability requires inclusive engagement with educators, administrators, students, and parents. Collaborative discussions can help shape GAI practices that reflect diverse needs and values, preventing the GAI from being developed in isolation from human judgment. As technology evolves, accountability frameworks must remain adaptable. Higher education institutions should continuously refine their oversight mechanisms, ensuring that ethical integrity keeps pace with technological advancements. By prioritizing transparency and accountability, institutions can harness the GAI's potential while safeguarding educational equity and trust [46].

5 Discussion and Future Research Opportunities

5.1 Impact on Curriculum Development

A key benefit of the GAI in curriculum development is its ability to tailor learning experiences on the basis of individual abilities and interests. Unlike traditional one-size-fits-all models, GAI leverages algorithms and data analytics to create customized educational pathways that adjust in real time. Studies have shown that GAI-driven adaptive learning paths align with students' goals by continuously analysing user data and learning preferences [47].

GAI further enhances instructional design by enabling educators to refine curricular content via data-driven insights. Automated content generation fills gaps in underrepresented topics and elaborates on complex concepts on the basis of student assessments, ensuring relevance and comprehensiveness [48]. Additionally, learning analytics powered by the GAI also inform curriculum adaptation, helping institutions identify trends, strengths, and weaknesses. This evidence-based approach supports informed decision-making regarding instructional strategies, resource allocation, and curriculum focus [49].

As the GAI automates administrative and assessment tasks, educators can shift towards strategic roles emphasizing mentorship, facilitation, and personalized support. This transformation encourages innovative teaching practices and deeper student engagement [50].

5.2 Evolving Roles of Educators

As GAI tools become essential, educators now guide students in navigating GAI-enhanced learning environments rather than merely delivering content. Therefore, educators' roles are transferred from traditional knowledge providers to facilitators of student-centered learning. This shift requires educators to focus on cultivating critical thinking, creativity, and problem-solving skills. GAI enables personalized learning, allowing students access to tailored resources and adaptive environments. Educators play a crucial role in ensuring that students engage critically with GAI-generated content rather than passively consuming it, fostering deeper understanding and active participation.

The evolving landscape also redefines educators as mentors and learning coaches, promoting student autonomy and self-directed learning. With the GAI handling administrative and content-generation tasks such as grading and instructional material creation, educators can allocate more time to mentorship, meaningful interactions, and personalized feedback, enhancing both academic and professional development. Additionally, the impact of the GAI necessitates interdisciplinary curriculum design, encouraging collaboration across fields to keep learning materials relevant and aligned with technological advancements. This integration ensures that students are better prepared for future workforce demands.

To effectively integrate GAI, educators must continuously develop their technological proficiency, committing to lifelong learning and modelling responsible GAI use. Professional development is essential for mastering emerging tools and fostering ethical technology adoption among students. Another critical role for educators is addressing academic integrity challenges posed by GAI-generated content. Traditional assessment methods must evolve to emphasize process, critical thinking, and creativity over rote memorization. Innovative assessment strategies can uphold originality while leveraging the GAI's potential. Furthermore, educators, as researchers, play a vital role in analysing its effectiveness in teaching, refining methodologies, and sharing best practices. This research-driven approach ensures continuous improvement and contributes to the broader educational discourse on the role of the GAI in academia.

5.3 Institutional Policy Planning

As the GAI evolves, institutions must establish governance frameworks and faculty development programs to ensure its effective and ethical implementation. Institutional policies must first address the ethical challenges of the GAI, including academic integrity, plagiarism, data security, and system transparency. Clear guidelines should define acceptable use, ensuring that faculty and students understand GAI's implications for learning and assessment. Policy development should involve collaboration between IT departments, faculty, and administrations to foster a shared understanding of the role and risk of the GAI [51]. A strong governance framework is essential for strategic planning. Institutions must adopt adaptable policies that align GAI initiatives with broader educational goals while accommodating departmental needs. This requires stakeholder engagement and systematic resource allocation to optimize academic performance and operational efficiency [52].

Faculty development is equally critical. Effective GAI integration demands training beyond technical proficiency, incorporating pedagogical strategies that enhance student engagement. Professional development programs should equip faculty with the ability to leverage the GAI for personalized learning and curriculum design.

Finally, given the rapid evolution of the GAI, policies and strategies should be regularly reviewed and refined on the basis of emerging trends and faculty–student feedback. An iterative approach fosters continuous improvement and ensures responsiveness to real-world applications. Institutions should also support research on the GAI's impact on learning and pedagogy. Encouraging evidence-based studies will inform best practices and policy refinements, strengthening the role of the GAI in academia [53].

5.4 Equity and Accessibility

A key concern of the GAI is addressing disparities in socioeconomic status, disability, and geographical location. Marginalized groups already face barriers to educational resources, and without careful management, GAI may widen these gaps. For example, while AI can streamline processes and personalize learning, students lacking access to technology or digital literacy may be left behind.

Ensuring accessibility requires GAI-driven tools that accommodate diverse learning needs, particularly for students with disabilities. Adaptive learning systems and assistive technologies must be thoughtfully designed to foster inclusivity. Additionally, digital equity extends beyond hardware access—it necessitates comprehensive training and support to equip all students with the skills to navigate GAI-enhanced learning environments effectively [54].

Ethical concerns, particularly algorithmic bias, must also be addressed. GAI systems trained on historical data risk perpetuating inequalities if not regularly audited. Institutions should implement safeguards to prevent bias and discrimination, ensuring that GAI tools support rather than hinder marginalized students [55].

Transparency is another critical factor. Students must understand how GAI tools influence grading and content recommendations, fostering trust and engagement. Institutions should also establish policies that prioritize equity, set clear inclusion objectives, and regularly assess progress to ensure that the GAI enhances educational opportunities rather than reinforcing existing inequities [56].

6 Conclusion

The integration of GAI into higher education has catalyzed a paradigm shift in pedagogy, institutional workflows, and ethical frameworks. Our synthesis reveals three critical dimensions:

(1) Personalized learning and pedagogical innovation. GAI enables hyperpersonalized education through adaptive content delivery and coideation tools (e.g., DeepSeek, ChatGPT), enhancing knowledge acquisition through alignment with individual cognitive profiles. However, its efficacy depends on faculty upskilling to harness these technologies strategically—transforming educators into curators of GAI-augmented learning ecosystems.
(2) Efficiency–ethics tension. While GAI streamlines automated pedagogical content generation and assessment, it risks propagating dataset biases and compromising academic integrity. To preserve authenticity, institutions must develop hybrid evaluation frameworks that balance GAI-assisted grading with human oversight.
(3) Ethical imperatives and research frontiers. Emerging challenges, such as data privacy violations, algorithmic opacity, and disciplinary equity gaps, demand proactive governance. The current literature disproportionately focuses on STEM/business applications, neglecting domain-specific impacts in humanities and regional contexts. Future research should prioritize cross-disciplinary efficacy studies, culturally responsive implementation models, and educator-GAI codesign frameworks.

The path forward requires reimagining faculty roles through continuous professional development, emphasizing the GAI's creative application and critical limitations. Only through interdisciplinary collaboration and ethical vigilance can higher education harness GAI's transformative potential without compromising its humanistic mission.

Acknowledgements. This work was funded by the Education Science Planning Project of Hubei Province (Grant no. 2024GB424) and the Undergraduate and Professional Education Teaching Research Project of the National University of Defense Technology (Grant no. U202320).

Disclosure of Interests. The authors have no competing interests to declare that are relevant to the content of this article.

References

1. Katsamakas, E., Sánchez-Cartas, J. M.: Generative artificial intelligence, content creation, and platforms. J. Ind. Competition Trade **24**, Article: 19 (2024)
2. Guo, J., Ma, Y., Li, T., Noetel, M., Liao, K., Greiff, S.: Harnessing artificial intelligence in generative content for enhancing motivation in learning. Learn. Ind. Differ. **116**, Article: 102547 (2024)
3. Ren, M.: Advancements and applications of large language models in natural language processing: a comprehensive review. Appl. Comput. Eng. **97**, 55–63 (2024)
4. Rongali, S.K.: Natural language processing (NLP) in artificial intelligence. World J. Adv. Res. Rev. **25**(1), 1931–1935 (2025)
5. Chandrika, G.: Natural language processing (NLP). Int. J. Res. Appl. Sci. Eng. Technol. **12**(6), 1092–1095 (2024)

6. Zaharuddin, Yu, C., Yao, G.: Enhancing student engagement with AI-driven personalized learning systems. Int. Trans. Educ. Technol. **3**(1), 1–8 (2024)
7. Zharmukhanbetov, S., Singh, C. P.: Enhancing flipped classroom engagement and personalized learning through AI-powered adaptive content delivery. In: International Conference on Technological Advancements in Computational Sciences, Tashkent, Uzbekistan, 1–3 November 2023 (2023)
8. Muin, S.M.M., et al.: Predicting academic student performance based on e-learning platform engagement using learning management system data. Int. J. Recent Innov. Trends Comput. Commun. **11**(9), 1859–1866 (2023)
9. Staikopoulos, A., OKeeffe, I., Yousuf, B., Conlan, O., Walsh, E., Wade, V.: Enhancing student engagement through personalized motivations. In: International Conference on Advanced Learning Technologies, Hualien, Taiwan, 6–9 July 2015 (2015)
10. Krishnapriya, S., Malathy, R., Banu, A.S., Gomathi, R.D., Vishnupriyan, T., Kavitha, S.: Collaborative learning with IoT: enhancing student engagement. In: International Conference on Advanced Computing and Communication Systems, Coimbatore, India, 14–15 March 2024 (2024)
11. Tamura, Y., Takase, Y., Hayashi, Y., Nakano, Y.: Generating quizzes for history learning based on Wikipedia articles. In: International Conference on Learning and Collaboration Technologies, Los Angeles, CA, USA, 2–7 August 2015 (2015)
12. Trandabat, D.: Enriching learning materials with semantic roles. Procedia Comput. Sci. **112**, 288–295 (2017)
13. Huang, W., Brown, G.T.L., Stephens, J.M.: How technology assists the feedback process in a learning environment: a review. In: International Conference on Advanced Learning Technologies, Orem, UT, USA, 10–13 July 2023 (2023)
14. Kamal, Z.H., Cooper-Stachowsky, M.: First impressions on supporting students with a course-aligned, AI-powered virtual teaching assistant (TA), Oliver. In: Canadian Engineering Education Association Conference, Edmonton, Alberta, Canada, 15–19 June 2024 (2024)
15. Liu, J., Zhang, L., Wei, B., Zheng, Q.: Virtual teaching assistants: technologies, applications and challenges. Humanity Driven AI, pp. 255–277. Springer, Cham (2021)
16. Sandhu, M., Patel, J.J., Jena, A., Kiran, P.R., Jadhav, R.S., Abraham, G.: AI-powered virtual health assistants: bridging gaps in patient care. J. Electr. Syst. **20**(10s) (2024)
17. Reid, J.W., Weigel, E.G.: Examining perspectives of teaching among biology teaching assistants. J. Coll. Sci. Teach. **51**(3), 67–77 (2022)
18. Quicke, J.: Teaching assistants: students or servants? Forum **45**(2), 71–74 (2003)
19. Cameron, C., Deming, S.P., Notzon, B., Cantor, S., Broglio, K.R., Pagel, W.: Scientific writing training for academic physicians of diverse language backgrounds. Acad. Med. J. Assoc. Am. Med. Coll. **84**(4), 505–510 (2009)
20. Feng, H.: Real-time language assistance in teaching academic english writing to graduate students based on language intelligence technology (ChatGPT). In: Eurasian Conference on Educational Innovation, Bangkok, Thailand, 26–28 January 2024 (2024)
21. Li, L.: How do students of diverse achievement levels benefit from peer assessment? Int. J. Sch. Teach. Learn. **5**(2), Article: 14 (2011)
22. Ye, X., Manoharan, S.: Providing automated grading and personalized feedback. In: International Conference on Artificial Intelligence, Information Processing and Cloud Computing, Sanya China, 19–21 December 2019, pp. 1–5 (2019)
23. Mane, P.A., Jagtap, S.: AI-enhanced educational platform for personalized learning paths, automated grading, and real-time feedback. Int. J. Sci. Res. Eng. Manag. **8**(12), 1–9 (2024)
24. Lee, H.H.: Effectiveness of real-time feedback and instructive hints in graduate CS courses via automated grading system. In: ACM Technical Symposium on Computer Science Education, Virtual Event USA, 13–20 March 2021, pp. 101–107 (2021)

25. Gambo, I., Abegunde, F.-J., Gambo, O., Ogundokun, R.O, Babatunde, A.N., Lee, C.-C.: GRAD-AI: an automated grading tool for code assessment and feedback in programming course. Educ. Inf. Technol. (OnlineFirst)
26. Aggarwal, A., Vasquez, A.: Unison live: automated feedback, grading, and analytics LTI application. In: Koli Calling International Conference on Computing Education Research, Koli, Finland, 17–20 November 2022, Article: 34 (2022)
27. Asif, R., Merceron, A., Ali, S.A., Haider, N.G.: Analyzing undergraduate students' performance using educational data mining. Comput. Educ. **113**, 177–194 (2017)
28. Blobstein, A., et al.: #let's-discuss: analyzing students' use of emoji when interacting with course readings. Int. J. Artif. Intell. Educ. (OnlineFirst)
29. Lenhart, C., Bouwma-Gearhart, J.: STEM faculty instructional data-use practices: informing teaching practice and students' reflection on students' learning. Educ. Sci. **11**(6), Article: 291 (2021)
30. Cheng, L.-C., Chu, H.-C., Shiue, B.-M.: An innovative approach for assisting teachers in improving instructional strategies via analyzing historical assessment data of students. Int. J. Distance Educ. Technol. **13**(4), 40–61 (2015)
31. Morrow, L.W., Morrow, S.A.: Instructional focus: What are we teaching students? Preventing Sch. Fail. Altern. Educ. Child. Youth **35**(1), 33–36 (1990)
32. Lee, E.-J., Son, Y.-A.: A case study and consultation on instructional design process for integrated science lesson by secondary science teachers - focus on the integrated science five-step instructional design for integrated science lesson. J. Korean Assoc. Res. Sci. Educ. **33**(1), 208–227 (2013). (in Korean)
33. Tedick, D.J., Lyster, R.: Module and lesson instructional design. Scaffolding Language Development in Immersion and Dual Language Classrooms, 1st edn. Taylor & Francis Group (2019)
34. Ismail, I., Jabri, U.: Academic integrity: preventing students' plagiarism with TURNITIN. Edumaspul: Jurnal Pendidikan **7**(1), 28–38 (2023)
35. Jamieson, S., Howard, R.M.: Rethinking the relationship between plagiarism and academic integrity. Int. J. Technol. High. Educ. **16**(2), 69–85 (2019)
36. Joyce, D.: Academic integrity and plagiarism: Australasian perspectives. Comput. Sci. Educ. **17**(3), 187–200 (2007)
37. Salunkhe, R., Kasar, N., Bainade, S., Pawar, V.: Enhancing academic integrity: plagiarism detection with machine learning. Int. J. Sci. Res. Eng. Manag. **8**(4), 1–6 (2024)
38. Kumar, R., Pratap, B., Pateria, R.K.: An effect on academic integrity due to plagiarism: a review. Libr. Prog. (Int.) **42**(1), 206–219 (2022)
39. Gupta, I., Gusain, A., Mahajan, K., Rais, F., Chaudhary, R.: Data security and data breaches. Int. J. Eng. Appl. Sci. Technol. **8**(2), 174–177 (2023)
40. Gomathy, C.K., Geetha, V., Aravind, S., Surya, V.V.: Ethical implications of cybersecurity breaches and data privacy violations. Int. J. Sci. Res. Eng. Manag. **8**(9), 1–6 (2024)
41. Грицак, Н, Ісаєва, О: Generation z students: educational issues. Probl. Mod. Teach. Train. **1**(21), 64–72 (2020). (in Russian)
42. Fevre, R., Grainger, H., Brewer, R.: Discrimination and unfair treatment in the workplace. Br. J. Ind. Relat. **49**(s2), s207–s235 (2011)
43. Straton, R.G.: Ethical issues in evaluating educational programs. Stud. Educ. Eval. **3**(1), 57–66 (1977)
44. Kaminski, M.: Understanding transparency in algorithmic accountability. The Cambridge Handbook of the Law of Algorithms, pp. 121–138. Cambridge University Press (2020)
45. Otu, J.E., Ushle, M.A., Attah, F.M., Owan, E.J.: Applying accountability and transparency in the management of tertiary institutions in Nigeria. Sophia Afr. J. Philos. **10**(2) (2008)
46. Vedaschi, A.: Transparency and accountability versus secrecy in intelligence operations: an Italian case study. Accountability and the Law, 1st edn. Taylor & Francis Group (2021)

47. Joudieh, N., Eteokleous, N., Champagnat, R., Rabah, M., Nowakowski, S.: Employing a process mining approach to recommend personalized adaptive learning paths in blended-learning environments. In: International Conference in Open & Distance Learning, Athens, Greece, November 2023, pp. 15–37 (2023)
48. Thimmanna, A.V.N.S., Naik, M.S., Radhakrishnan, S., Sharma, A.: Personalized learning paths: adapting education with AI-driven curriculum. Eur. Econ. Lett. **14**(1), 31–40 (2024)
49. Şuşnea, E.: How big data analytics will reshape E-learning? In: International Conference eLearning and Software for Education (2018)
50. Bishop, P., et al.: Teacher roles in personalized learning environments. Elem. Sch. J. **121**(2) (2020)
51. Stein, C.D., Eisenberg, E.S., O'Donnell, J.A., Spallek, H.: What dental educators need to understand about emerging technologies to incorporate them effectively into the educational process. J. Dent. Educ. **78**(4), 520–529 (2014)
52. Leardini, C., Rossi, G., Moggi, S.: Governance and strategic planning. Board Governance in Bank Foundations: The Italian Experience, pp. 71–79. Springer, Heidelberg (2014)
53. Voorhees, R.A.: Institutional research's role in strategic planning. N. Dir. High. Educ. **141**, 77–85 (2008)
54. Mitre, X., Zeneli, M.: Using AI to improve accessibility and inclusivity in higher education for students with disabilities. In: International Conference on Information Technology Based Higher Education and Training, Paris, France, 6–8 November 2024, pp. 1–8 (2024)
55. Azevedo, L., Mallinson, D.J., Wang, J., Robles, P., Best, E.: AI policies, equity, and morality and the implications for faculty in higher education. Public Integrity (OnlineFirst)
56. Jones, K.M.L., Afnan, T.: "For the benefit of all students": student trust in higher education learning analytics practices. Proc. Assoc. Inf. Sci. Technol. **56**(1), 682–683 (2019)

Reform and Practice of a Large Language Model Method Based on Human Feedback in Algorithm Analysis and Design of a Course

Kejia Zhang[1], Haiwei Pan[1], Zhiqiang Ma[1], Shaoqiang Zhu[1], Yingxin Qin[1], and Lan Zhang[2](✉)

[1] College of Computer Science and Technology, Harbin Engineering University, No. 145, Nantong Street, Nangang District, Harbin, Heilongjiang, China
panhaiwei@hrbeu.edu.cn

[2] Computer and Control Engineering, Northeast Forestry University, No.26 Hexing Road, Xiangfang District, Harbin, China
panhaiwei@hrbeu.edu.cn

Abstract. With the continuous development of artificial intelligence technology, large-scale language models have demonstrated significant potential across various fields. In education, an increasing number of methods leverage large-scale language models to enhance educational quality, introducing new ideas and opportunities for reform. However, training a large language model with substantial professional knowledge to meet teaching needs incurs high labor costs. The fine-tuning approach based on human feedback alignment can significantly lower these model labor costs. Consequently, this article thoroughly investigates the application of this large prediction model method, which is rooted in human feedback alignment, within the educational reform of algorithm analysis and design courses and examines its impact on teaching effectiveness and students' learning experiences.

Keywords: Computer Science · Large Language Model · Human Feedback · Teaching Reform

1 Introduction

With the rapid development of large language models, more fields have begun to apply large language models [1] to solve practical problems [2–4]. Since the development of the AI era, the "large language model + education" teaching model has emerged [4–8]. The application of large language models has also become an important tool for improving teachers' teaching efficiency and quality.

The algorithm analysis and design course is a core course for computer science majors [9]. The course requires students to have certain algorithm analysis capabilities

This work is supported by Harbin Engineering University's 2021 Education Reform Project "How to Make Computer Theory Teaching Serve Employment" (Grant No. JG2021B0609).

Y. Yu et al. (Eds.): ICPCSEE 2025, CCIS 2673, pp. 130–139, 2026.
https://doi.org/10.1007/978-981-95-2566-9_9

and be proficient in using algorithm design techniques and tools, which is more targeted for the cultivation of students' professional skills [10]. By learning basic knowledge in the algorithm analysis and design course, students can lay a solid professional foundation and lay the foundation for subsequent professional courses. At present, most algorithm analysis and design courses are based on knowledge points in traditional textbooks. However, with the development of computer-related technologies, the design methods and analysis skills of algorithm analysis and design courses cannot keep pace with the times. The basic content of the course is abstract, and the algorithm logical thinking requirements are high. It is difficult for students to find a learning path that uses their characteristics and lacks independent thinking. In the process of teaching a course, teachers have a single teaching method, and it is difficult to design personalized, customized teaching plans for each student to meet the students' professional needs.

In the process of teaching practice, coupled with reinforcement learning from human feedback (RLHF) [11, 12], this paper uses the "big language model+" model to propose an improvement plan for the teaching design of the algorithm analysis and design course by using a more personalized educational method, namely, the "large Language Teaching Model based on Student Feedback" (LTMSF). This emphasizes improving students' hands-up ability, strengthening their analytical ability for different practical problems, and better mastering the complex theoretical knowledge points in the course, thereby improving the teaching quality and teachers' work efficiency. This paper discusses the effect of teaching reform on the improvement of algorithm analysis and design courses and explores the difficulties and preliminary solutions in practical applications of the teaching model of the large language model RLHF.

2 Instructional Design of Large Language Models on the Basis of Human Feedback

The human feedback model method refers to incorporating human feedback information into the model training process to achieve better model performance and output results that are more in line with human needs. In the field of education, introducing human teaching feedback information into the training of large-scale language models can improve the performance of the model in different teaching scenarios [4].

2.1 A Subsection Sample

The training method for a qualified student in the teaching process should be combined with actual enterprise product knowledge to improve the students' ability to analyse practical problems. The specific process is shown in Fig. 1. First, students learn courses and pass exams. Students have certain general course knowledge and the ability to analyse and write algorithms. However, since the knowledge points in textbooks are relatively fixed and highly abstract, students cannot solve practical problems. At this time, the school arranges internships and training in enterprises to supplement the professional application knowledge of products. Since different enterprises and products are quite different, students will gradually have different knowledge tendencies during internships and training. Students can initially solve some practical problems, but due to a lack of practical experience, some errors can occur. Usually, the instructor builds a feedback system to provide students with modification suggestions. Through the process of practice and feedback, students continuously improve their professional skills and gradually grow into qualified programmers.

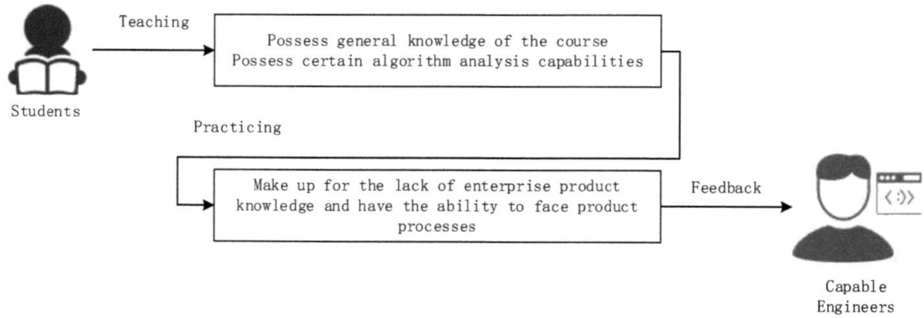

Fig. 1. Students' professional skills learning route

Similar large-scale language-based models also need to be fine-tuned with professional knowledge. Unlike the fine-tuning of downstream tasks of previous large language models, the course evaluation fine-tuning model built with RLHF is more in line with the application direction of the "Algorithm Analysis and Design" course. The specific process is shown in Fig. 2. The model is trained with a large amount of data in the cloud. Owing to the large amount of data and the large demand for computing resources, this step of pretraining does not require teacher participation, and pretrained models are usually used. Mature large model parameters can usually answer some questions, but the model lacks professional knowledge and lacks knowledge of the details in the algorithm analysis and design course; thus, teachers are required to fine-tune the results of downstream tasks. Unlike the manual fine-tuning method of the previous large-scale language model, the fine-tuning method based on RLHF fine-tunes the model by building an expert model and a reward model. The domain dataset that the input model needs to learn (i.e., the knowledge of the algorithm analysis and design course) enables the model to have general knowledge and certain course expertise. However, the model at this time will have some errors that do not conform to human analysis methods, and the output still has shortcomings. At this time, we follow the students' learning path and build a practice and feedback mechanism. Teachers do not need to adjust many model outputs. Instead, they use the built reward model to score and fine-tune the model output, which greatly improves work efficiency. RLHF-based model fine-tuning uses a specific dataset to train the reward model to introduce human preference information and reduce losses. The model uses different human values to reduce training losses, so the resulting model is more in line with students' way of thinking, and students can more easily understand the course content.

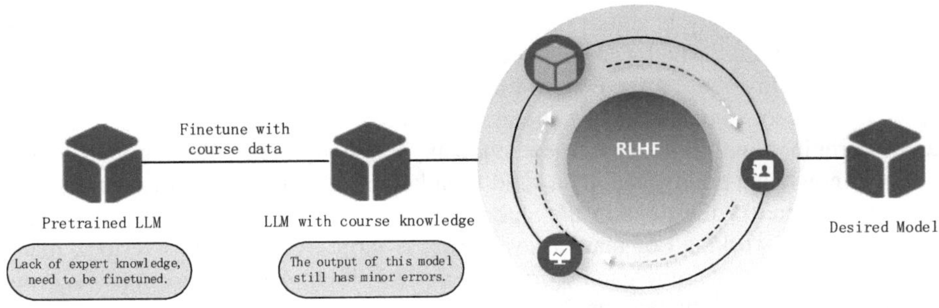

Fig. 2. Fine-tuning method of the large-course model that is based on RLHF. First, we use an existing large language model (LLM) as the backbone; then, we finetune this LLM with domain data, for example, the course data of Algorithm Analysis and Design. Next, we use the RLFH algorithm coupled with a reward function from inverse reinforcement learning to obtain the LTMSF model.

2.2 Using Large Language Models to Improve Students' Learning Efficiency

The large language model constructed by teachers for algorithm analysis and design can improve students' learning efficiency. In the process of students' learning, the knowledge map of the algorithm analysis and design course can be obtained through the large language model. Compared with the traditional knowledge construction system based on data search, the knowledge map obtained by the large language model can be more inclined to the personalized questions of different students. Different students have different needs and purposes for knowledge. When students use a large language model, they can have different preferences for questions. The model can generate different knowledge maps and related content according to students' needs, thereby improving students' interest and efficiency in learning. In the past teaching process, teachers needed to have different teaching methods for different students to improve students' subjective initiative in learning, but this usually required an enormous workload, and teachers do not know enough about students, so it is difficult to analyse and teach each student individually. Using RLHF to build a large language model can replace part of the teacher's work. The model generates a knowledge framework for different students and improves students' learning efficiency.

Practice and feedback are usually difficulties in students' learning in algorithm analysis and design courses. When facing different practical application problems, although the ideas and basic concepts are similar, the specific practical details are not fixed, which makes it difficult for students to obtain specific analysis and guidance from teachers. Moreover, different practical problems have various solutions, and students cannot directly apply reference answers to judge their algorithms. Teachers promote the efficiency of students' self-study by building a large language model of the course. Students can use the mature course large language model to judge the results of their homework and algorithms and further adjust their algorithms on the basis of the results of model feedback. Students can discover more algorithm structures through a large language model to broaden their design ideas.

The algorithm analysis and design course has strong professionalism and practical application scenarios, which can enhance students' subjective initiative in learning and improve the quality of teaching. By building a large language model, a more personalized self-study plan for students can be generated. In addition to teaching, the course content can be combined with practice to meet students' self-study needs, deepen their ability to apply knowledge points, and lay a solid foundation for students' future career paths.

The open-ended questions in the algorithm analysis and design course have many details, and the workload of manual evaluation is large. It is easy to overlook many details, so the error rate of manual scoring is relatively high. An assessment and scoring system using a large language model can greatly reduce the workload of teachers and conform to the evaluation habits of teachers. For the evaluation of open-ended questions in the algorithm analysis and design course, a relatively fair scoring standard is specified to improve the work efficiency of teachers. Moreover, the model also provides specific evaluations in addition to scores. Compared with the traditional scoring system, the scoring method based on the large language model can better highlight the evaluation details. Teachers can better grasp the students' learning situation, provide in-depth explanations of the key points and difficulties, and improve their teaching efficiency. At the same time, teachers can use a large language model to build a scoring scheme that is more in line with professional skills, overcomes the limitations of teaching, and expands teachers' teaching ideas.

2.3 The Use of Large Language Models to Reform Teaching Methods

With the development of the computer industry, students need not only to gain professional knowledge in class but also to improve their professional skills in the application field to gain an advantage in fierce professional competition in the future. Therefore, when teachers teach algorithm design and analysis courses, they need to highlight the improvement of students' ability to solve practical problems. However, owing to the limitations of the book compilation process, many practical application examples are difficult to reflect in textbooks. Moreover, with the advancement of science and technology, many new algorithm analyses and design schemes have emerged, and teachers' courses need to keep pace with the times. These needs require considerable manpower and considerable professional knowledge. Since teachers themselves have little experience in corporate work, it is difficult for them to face the actual needs of enterprises during the teaching process. Therefore, building a large model on the basis of professional needs can compensate for the lack of practical links in the teaching process. For students' engineering needs, the large language model based on RLHF fine-tuning can be used as an additional way for students to acquire knowledge. Students can build their learning path through a large language model. Teachers can gradually build a large language model with professional knowledge through sufficient professional knowledge to meet students' professional needs. The course teaching method can be reformed in the following three ways:

Course content optimization In traditional algorithm analysis and design courses, students usually face challenges in terms of content depth and breadth. To address these issues, the course content is optimized via a large language model that is based on human feedback alignment. The model analyses common problems in existing textbooks and teaching materials, such as too many difficult points and unclear explanations, and recommends improvement measures on the basis of student feedback. The teacher's teaching process can add modular design to decompose the course content into several modules, each focusing on a core topic, such as time complexity analysis, sorting algorithms, and graph algorithms. A large language model helps teachers organize these topics more clearly and provides detailed explanations and examples for each module. Moreover, adaptive learning paths are constructed for different types of students. On the basis of the student's learning progress and mastery, the model generates personalized learning path recommendations. It adjusts the difficulty and depth of the learning content on the basis of student feedback to ensure that each student can master the course content at a pace that suits it.

Assignment and assessment reform Traditional homework and assessment methods often focus on standardized questions and tests, which make it difficult to assess students' actual abilities accurately. To solve this problem, a model-based intelligent homework and assessment system was introduced. Homework can be dynamically generated in teaching tasks, and homework can combine knowledge points with practical applications to the greatest extent possible. **The dynamic homework generation system** is a large language model that generates personalized homework questions on the basis of the course content and the current mastery of students. These questions not only involve basic knowledge but also include practical applications and complex problems to promote the improvement of students' comprehensive abilities and ensure that students can apply what they have learned. Moreover, an **instant feedback mechanism** can be built on the basis of the large language model; that is, by analysing students' homework submissions through the model, the system can provide timely and detailed feedback. This feedback includes suggestions for problem-solving steps, tips for common errors, and improvement strategies to help students quickly understand and correct errors.

Teaching interaction and support In traditional teaching, the interaction between teachers and students is often limited by class time and the number of students. The introduction of large language models provides a new way of teaching interaction and enhances students' learning support. Through a large language model, teachers can build **intelligent tutoring assistants**. The model is used as a virtual tutoring assistant to provide students with 24 h/7 d of online help. When students encounter problems, they can obtain instant answers and suggestions through the model so that they can also obtain effective learning support outside of class, reduce the delay of students' knowledge, quickly respond to students' problems, and reduce the number of students' repeated mistakes caused by untimely corrections. At the same time, based on large language model teaching, classroom interaction is enhanced. In the classroom, teachers use real-time data generated by the model and student feedback to adjust teaching strategies. For example, the model can analyse students' mastery of a certain topic in real time, and teachers can adjust the depth and focus of the explanation accordingly to improve the pertinence and efficiency of classroom teaching.

Enrichment of learning resources To enrich students' learning resources, auxiliary materials and reference materials generated by the large language model are incorporated into the course system. To generate additional learning materials, the model can automatically generate supplementary reading materials, exercises, and actual cases, providing more learning resources to help students better understand and master the course content. Moreover, an intelligent recommendation system can be built on the basis of a large language model for different students. On the basis of students' learning records and feedback, the model can recommend relevant academic papers, video lectures, and online courses to help students expand their knowledge and deepen their understanding of complex concepts.

Through the above reforms, courses can better meet the learning needs of students and improve the quality and efficiency of teaching. Model-driven content optimization and evaluation methods not only improve the relevance of the courses but also enhance the learning experience of students.

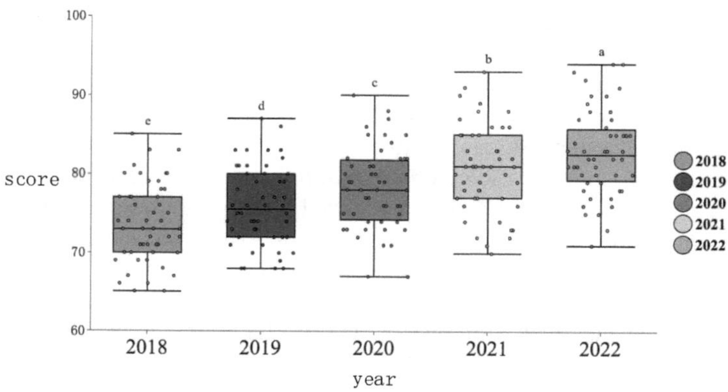

Fig. 3. Comparison of students' midterm exam scores before and after the teaching reform

3 Effects of Teaching Reform

3.1 Student Performance Analysis

To evaluate the impact of the curriculum reform on student learning, this paper compared student grades and performance before and after the reform. By analysing students' test scores, homework scores, and project completion in detail, this paper can more accurately measure the effectiveness of the reform.

A comparison of the results is shown in Fig. 3, which shows the comparison of students' test scores before and after the reform. The students' average midterm and final exam scores increased by 15% and 12%, respectively. The improvement in grades shows that the optimization of course content and personalized homework design based on the RLHF large language model effectively enhanced students' understanding and application capabilities.

Improving the operation quality is shown in Fig. 4. After the reform, the standard deviation of homework scores significantly decreased, and students' regular homework scores improved steadily, which shows that the introduction of a personalized and customized homework mechanism can improve students' participation and homework completion accuracy.

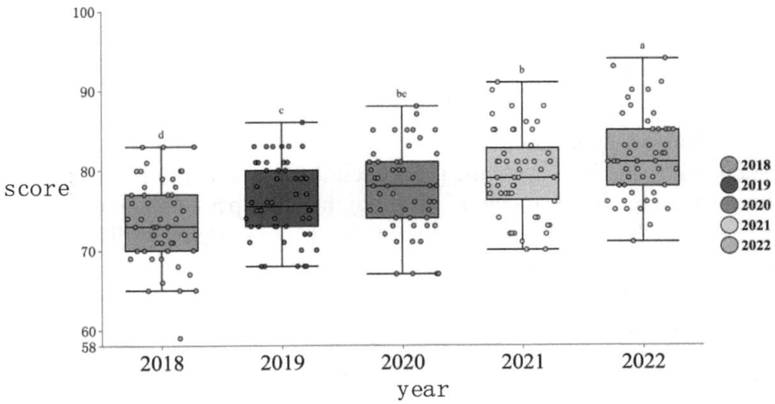

Fig. 4. Comparison of students' homework scores before and after the teaching reform

3.2 Learning Experience Survey

To gain a deeper understanding of students' acceptance of and satisfaction with the new teaching methods, this paper conducted a comprehensive questionnaire survey and interviews. These surveys enabled teachers to obtain quantitative and qualitative feedback on curriculum reform.

Questionnaire results: The survey revealed that 85% of the students were satisfied with their personalized learning paths and intelligent tutoring assistants. They believe that these measures have improved learning efficiency and interest. In terms of homework feedback, 72% of the students said that instant feedback helped them correct mistakes faster and improve their grades.

Interview feedback: In the interviews, the students generally believed that the modularity and real-time interaction of the course content enhanced their understanding and retention. Many students mentioned that the virtual tutoring assistants became indispensable support tools in their learning process.

3.3 Model Effect Evaluation

Evaluate the practical effectiveness of large language models in course improvement, including their impact on course design, teaching interactions, and student support.

Accuracy of content generation: The content and assignments generated by the model showed high accuracy and relevance in the teaching process. Teacher feedback revealed that the suggestions and materials provided by the model were practically helpful for classroom teaching and could be adjusted in time to meet students' needs.

Analysis of interactive effects: By analysing the interaction records between the model and students, the model can effectively identify and respond to students' common questions, which improves the interactivity and pertinence of learning. However, the model still has limitations when dealing with complex problems and needs further optimization.

3.4 Suggestions for Improvement

On the basis of the results of the effect evaluation, this paper proposes some improvement suggestions to further optimize the effect of curriculum reform:

Model optimization enhances the model's ability to handle advanced algorithmic problems, ensuring that it can provide more accurate and in-depth answers.
Strengthening the feedback mechanism strengthens the personalization of immediate feedback to ensure that it provides appropriate advice to students of different levels.
Adjustment of teaching methods is based on student feedback, teaching methods and materials are adjusted, and more interactive sessions and practical cases are added to further enhance students' learning experience.

4 Conclusion

In the past five years of curriculum reform practices, this article has systematically improved the "Algorithm Analysis and Design" course by introducing a large language model method based on human feedback alignment. This reform has not only brought about significant changes in teaching methods and assessment strategies but also significantly improved students' overall assessment results. It is clear from the data that students' scores on the midterm and final exams improved significantly. In addition, through further analysis of the data, this study revealed that students' learning interest and participation were also significantly enhanced. This change can be attributed to the application of large language models based on human feedback alignment in teaching, which not only provides personalized learning suggestions but also promotes a more interactive and dynamic classroom environment. Student feedback indicated that they had a deeper understanding of the course content and were satisfied with the improvement in their problem-solving skills.

Overall, the application of the large language model method based on human feedback alignment in the "Algorithm Analysis and Design" course has demonstrated its potential in the field of education. The curriculum reform not only improved students' academic performance but also enhanced their learning experience and sense of participation. In the future, this method will continue to be optimized to further improve teaching effectiveness and student learning outcomes.

References

1. Kaplan, J., McCandlish, S., Henighan, T., et al. Scaling laws for neural language models. arXiv preprint arXiv:2001.08361 (2020)
2. Yao, S., Yu, D., Zhao, J., et al.: Tree of thoughts: deliberate problem solving with large language models. Adv. Neural. Inf. Process. Syst. **36**, 11809–11822 (2023)
3. Thirunavukarasu, A.J., Ting, D.S.J., Elangovan, K., et al.: Large language models in medicine. Nat. Med. **29**(8), 1930–1940 (2023)
4. Kasneci, E., Seßler, K., Küchemann, S., et al.: ChatGPT for good? On opportunities and challenges of large language models for education. Learn. Individ. Differ. **103**, 102274 (2023)
5. Jeon, J., Lee, S.: Large language models in education: a focus on the complementary relationship between human teachers and ChatGPT. Educ. Inf. Technol. **28**(12), 15873–15892 (2023)

6. Jowsey, T., Stokes-Parish, J., Singleton, R., et al.: Medical education empowered by generative artificial intelligence large language models. Trends Mol. Med. **29**(12), 971–973 (2023)
7. Killian, C.M., Marttinen, R., Howley, D., et al.: "Knock, Knock… Who's there?" ChatGPT and artificial intelligence-powered large language models: reflections on potential impacts within health and physical education teacher education. J. Teach. Phys. Educ. **42**(3), 385–389 (2023)
8. Hoi, V.N., Mu, G.M.: Perceived teacher support and students' acceptance of mobile-assisted language learning: evidence from Vietnamese higher education context. Br. J. Edu. Technol. **52**(2), 879–898 (2021)
9. Zhang, B., Zheng, Y., Liu, Q., et al.: On the practice of 'algorithm analysis and design' course based on blackboard platform. In: International Conference on Education, Management and Computing Technology (ICEMCT-15), pp. 1438–1442. Atlantis Press (2015)
10. Sen, S.: Algorithm Analysis and Design (2005)
11. Liu, F.: Learning to summarize from human feedback. In: Proceedings of the 58th Annual Meeting of the Association for Computational Linguistics, pp. 583–592 (2020)
12. Bai, Y., Jones, A., Ndousse, K., et al.: Training a helpful and harmless assistant with reinforcement learning from human feedback. arXiv preprint arXiv:2204.05862 (2022)

Teaching Reform and Practice of Algorithm Design and Analysis Course

Kejia Zhang[1], Haiwei Pan[1], Zhiqiang Ma[1], Shaoqiang Zhu[1], Yingxin Qin[1], and Lan Zhang[2](✉)

[1] College of Computer Science and Technology, Harbin Engineering University, No. 145, Nantong Street, Nangang District, Harbin, Heilongjiang, China

[2] Computer and Control Engineering, Northeast Forestry University, No.26 Hexing Road, Xiangfang District, Harbin, China

zhanglan@nefu.edu.cn

Abstract. This paper reforms the shortcomings and difficulties in the teaching process of the "Algorithm Design and Analysis" course. The knowledge graph optimizes the teaching content, abandons the traditional teaching method, captures the direction of talent demand, adjusts the class time allocation, reorganizes the assessment method, focuses on practical hands-on ability, and designs a multistage teaching quality evaluation system to promote the overall improvement of teaching quality. The practice of course reform has proven that the "Algorithm Design and Analysis" course has achieved good teaching results after a series of teaching reform measures.

Keywords: Algorithm Design · Analysis · Teaching Reform · Talent Cultivation

1 Introduction

As an emerging interdisciplinary subject, computer science systematically studies the foundation of information and computational theory, laying the foundation for the basic theories of discrete mathematics, algorithm analysis and design, data structure, and other disciplines [1, 2]. It insists on the coexistence of theory and practice and continuously cultivates talent ability to solve complex problems in scientific and engineering computing. However, in the current educational context, subject education on the basis of basic theory has always been the focus and difficulty of computer teaching. The teaching mode is single, and the theoretical content is rich and difficult to understand, which makes students feel deeply powerless and fearful when facing such courses. Therefore, to comply with the national talent training strategy and break traditional teaching barriers, course teaching must be reformed.

As a core course in computing science, Algorithm Design and Analysis integrates and expands the contents of advanced mathematics, discrete mathematics, data structure, and programming foundation. It is also a key tool that has led the construction

This work is supported by Harbin Engineering University's 2021 Education Reform Project "How to Make Computer Theory Teaching Serve Employment" (Grant No. JG2021B0609).

Y. Yu et al. (Eds.): ICPCSEE 2025, CCIS 2673, pp. 140–149, 2026.
https://doi.org/10.1007/978-981-95-2566-9_10

of emerging fields such as big data and artificial intelligence [1, 3–5]. Starting from the basic methods and principles of algorithm complexity analysis, this course systematically studies algorithm theory, deeply understands practical problems in computer science and applications, designs algorithms to describe and solve corresponding problems, and continuously cultivates students' algorithm design and complexity analysis, as well as comprehensive quality and ability to solve problems. For a long time, the teaching of Algorithm Design and Analysis has followed the combination of theory and experiments. Theoretical teaching is based mainly on "inculcation" education. The teaching time is tight, and the teaching content is complicated, with dense knowledge points involving various disciplines. Students have shallow basic knowledge and find it difficult to understand, which easily leads to a disconnect in their thinking. In addition, theoretical learning is boring, and the initiative of after-class learning is poor, which easily leads to the problem of "not seeking to understand thoroughly". Experimental methods mostly adopt the "on-the-computer" mode to reproduce classic algorithms. The algorithm application scenarios are unrealistic and cannot coordinate the "supply side" of colleges and universities and the "demand side" of employers. It is very easy to form information islands between student groups and teacher groups and work behind closed doors, resulting in poor hands-on practical skills and a lack of practical ability to draw inferences from one example. School assessment methods are mostly a combination of examinations and practice, and the proportion of practice is relatively small, resulting in students focusing on theoretical knowledge, rote memorization, and learning for examinations. They lack enthusiasm and initiative in learning, which easily leads to the problem of learning without knowing how to use it.

Talent cultivation must be a unified process of educating people and cultivating talent, and educating people is the foundation. Educating people is not a simple "assembly line" operation. It is necessary to understand the characteristics of college students in the new era promptly, understand what students think, want, and seek, and cannot "use one teaching plan to go around the world". We must abandon the "one size fits all" education model. Therefore, this paper takes the teaching of "Algorithm Design and Analysis" carried out by the School of Computer Science and Technology of our university as an example, reforms the traditional teaching content and teaching methods of this course, balances the relationship between teaching and learning, theory and practice, designs a talent training program that is closer to reality, improves students' learning enthusiasm and learning initiative, stimulates students' awareness of independent innovation, breaks cognitive defects and information barriers, improves students' programming and practical abilities, meets their own development and social needs, and truly applies what they have learned.

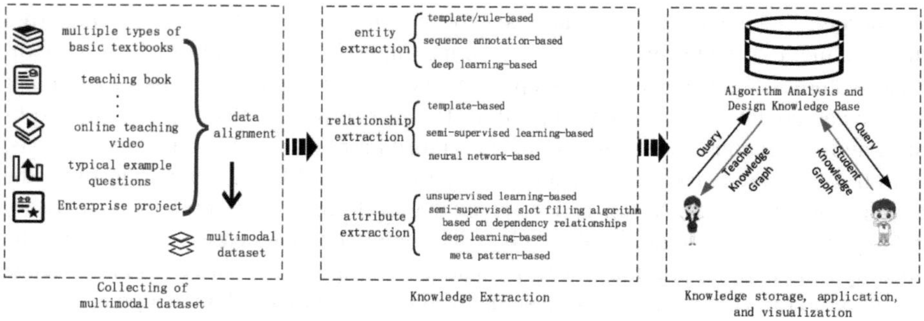

Fig. 1. Algorithm design and analysis knowledge graph construction process diagram

2 Teaching Reform Methods

2.1 Reform of Training Objectives

In addition to the "education-based, student-centered, market demand-oriented, and application-oriented talent training model," the single teaching and assessment model should be changed, the reconstruction of the teaching plan should be diversified, the teaching content should be enriched, and the relationships between teaching and learning, theory and practice should be fundamentally balanced. Continuously deepening the "Algorithm Design and Analysis" course reform, improving students' algorithm design and complexity analysis and corresponding problem-solving comprehensive quality and ability, expanding students' thinking, cultivating students' learning initiative, enthusiasm and innovation ability, eliminating students' inner panic and powerlessness in the face of this course, and truly applying what they have learned.

2.2 Reform of Teaching Content

The traditional teaching content structure of "Algorithm Design and Analysis" is single, and the chapter content is consistent with a certain reference textbook. It can be understood as copying the content of the textbook without deletion of the teaching tool, such as the PPT; however, the algorithms in "Algorithm Design and Analysis" involve previous courses such as discrete mathematics, data structure, and advanced mathematics, with complicated definitions, wide-ranging knowledge points, scattered knowledge, lack of focus, and redundant content; the traditional teaching content lacks algorithm optimization solutions and modern practical application cases, which easily leads to the phenomenon of learning without knowing how to use. Therefore, in response to the above problems, the teaching content has been reformed as follows:

On the basis of the domain knowledge graph algorithm, the knowledge base is constructed, and the relationship between each knowledge point is deeply sorted to achieve the organic connection and optimized integration of knowledge. The traditional "Algorithm Design and Analysis" textbook builds teaching content around problems or algorithms. Sometimes, the content involved is relatively broad, and the span is large, which makes it impossible for students to understand all the knowledge and evolution process of a certain algorithm from shallow to deep. Therefore, on the basis of the knowledge graph, the "Algorithm Design and Analysis" knowledge base is constructed, and the relationships among the syllabus, teaching chapters and teaching content are reasonably organized to help teachers better understand the context and logic of the teaching content and provide students with more systematic and comprehensive learning resources and guidance. The construction process of the knowledge graph of "Algorithm Design and Analysis" is shown in Fig. 1. First, we fully investigated the market and visited familiar companies. We focus on student needs and implement training methods for applied talent. We collect different types of data suitable for teaching, including basic textbooks [3, 6–8], teaching handouts, online teaching videos, typical examples, corporate projects, etc. We filter and streamline the data to retain the integrity of the data points and perform multimodal alignment of text domain images to construct a multivariate dataset. Second, we extract knowledge to construct structured hierarchical relationships and entity relationship sets. Relying on the Neo4j graph database, we build and store knowledge graphs to form a corresponding knowledge base.

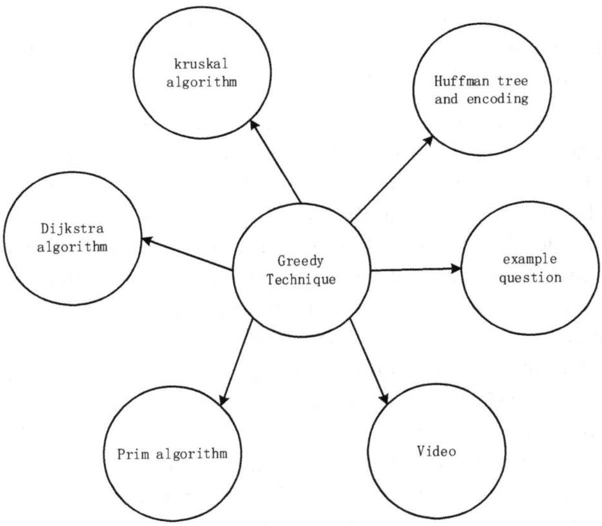

Fig. 2. Greedy technology knowledge graph visualization

The algorithm analysis and design knowledge base is aimed at teachers and student groups. Through the online platform, we perform keyword queries to obtain keyword knowledge graphs. The interface is displayed in the form of nodes. Clicking on a node will display the knowledge points associated with the node information. As shown in Fig. 2, the keywords are the Kruskal algorithm [9], the Dijkstra algorithm [10], the Prim algorithm [11], the Huffman tree and encoding [12], teaching videos, typical examples, and other knowledge points involved in "greedy technology". With the help of knowledge graph-based algorithms[13], the teaching content of the "Algorithm Design and Analysis" course will be more organized and coherent, providing teachers and students with more efficient and systematic learning support and guidance and promoting the continuous improvement of teaching quality and the innovative development of education and teaching. In the process of building an algorithm analysis and design knowledge base, we will grasp the direction of cultivating practical talent, carefully selecting algorithm application cases, deleting unrealistic engineering cases, and making the application scenarios clear to encourage students to achieve a thorough understanding and trigger continuous deduction of thinking.

2.3 Reform of Teaching Methods

In traditional Algorithm Design and Analysis teaching, whether theoretical teaching or practical teaching, the main teaching method is "inculcation". That is, the teacher uses multimedia to explain the theory or practical process, and the students listen. It is just inculcation, lacking interaction between teachers and students and between students and students. The learning process is boring and uninteresting, and the students' status is ignored. Second, in traditional "inculcation" education, the teacher leaves after class, there are no exercises after class, no question-and-answer sessions, etc., and students' after-class learning interaction is poor, which easily leads to the problem of "not seeking to understand thoroughly". Therefore, in response to the above problems, the teaching method is reformed in the following aspects:

- With respect to the traditional teaching model, the algorithm design and analysis knowledge graph are combined, and a new online and offline teaching model is adopted. In the traditional teaching model, the focus is usually on offline teacher teaching, and students listen mainly passively. Even if students are encouraged to preview independently, since the "Computer Design and Analysis" course is relatively unfamiliar to most freshmen, they often have no way to start effective preview; at the same time, the class time is tight, and the teacher cannot explain all the knowledge comprehensively and systematically, which may cause students to fail to keep up with the teacher's thinking in class. Through the algorithm design and analysis knowledge graph, teachers can ask questions or preview requirements offline in advance, and students can use this knowledge graph to search for questions or algorithms online, which helps them build an orderly knowledge structure, mark the knowledge of interest and knowledge points that they do not understand, conduct research and discussion with teachers offline, reduce learning complexity and improve students' initiative in independent learning; students can also review knowledge through the online algorithm design and analysis knowledge graph. Second, teachers can use algorithms to design and analyse knowledge graphs on the basis of the overall qualities of students; tailor lesson plans and explanation content for them; carefully select suitable teaching examples; introduce a teaching and research discussion model; establish group discussions; offline students review and master the recommended examples; address unresolved problems in the classroom; analyse them with teachers and classmates; learn from each other's strengths and weaknesses; and continuously improve problem thinking and implementation plans to truly apply what they have learned.
- The direction of talent demand should be grasped, schools and enterprises should be coordinated, the allocation of class hours should be adjusted, and the time for offline

explanations and engineering practice should be balanced. The practical course of "Algorithm Analysis and Design" should take promoting students' understanding and mastery of algorithm design as the main direction and train students' hands-up ability as the goal. By adopting the linkage between schools and enterprises, "Algorithm Design Analysis" is divided into three parts: classroom teaching, basic algorithm practice, and project practice. Classroom teaching cultivates students' basic knowledge literacy, basic algorithm practice implements application scenarios, and cultivates students' basic ability to independently realize algorithms; project practice is close to actual work needs, guided by enterprise teachers, and plans project progress to ensure that students' learning is in line with industry standards and best practices.

- The assessment method is reorganized, and the proportion of practice is adjusted. In the traditional assessment method, practical activities are only regarded as classroom performance and are included in the regular score, which leads to students focusing on theoretical knowledge and having poor hands-up ability. Therefore, in this reform, we divide the final score into three parts: regular score, final exam, and experiment. Specifically, the regular score accounts for 10% of the total final score, which is based on the completion of homework; the final exam adopts a closed-book exam method, accounting for 40% of the total final score, and examines students' understanding of algorithm theory; the experimental part accounts for 50% of the final assessment, of which basic algorithm practice accounts for 15%; the project accounts for 35%, and is scored by enterprise teachers to implement students' hands-on practice and adapt to society through practical actions.
- A multidimensional and multistage teaching quality evaluation system for algorithm design and analysis is established. Traditional course evaluation is often an overall evaluation after the course is completed. It is unable to implement points and surfaces, effectively collects students' real feelings and specific questions about each stage of online and offline teaching, and the evaluators are often limited to teachers or students of the department, which lacks comprehensiveness.

In this reform, a multidimensional and multistage teaching quality evaluation system is established, which specifically includes the following aspects:

- **Staged evaluation** divides the teaching process of the algorithm design and analysis course into different stages and conducts real evaluation and feedback for each stage. **Preclass preparation** is used to evaluate whether the questions or instructions raised by the teacher before class are attractive and whether they can mobilize students' initiative in independent learning; **classroom teaching** is used to evaluate the quality of the teacher's classroom teaching, whether the knowledge is clear, whether the logic is regular, whether the students can understand, and why the students cannot understand; homework is used to evaluate the difficulty of the design assignment, whether it is highly relevant to the class, and whether it can help students understand the content of the class; **offline discussion** is used to evaluate the form of group discussion and whether it can solve students' difficulties in classroom knowledge.
- **Diverse evaluators** In addition to faculty and students within the department, consideration should be given to introducing external evaluators, such as industry experts, interdisciplinary teachers, or off-campus internship units, to obtain broader and more objective evaluation opinions.

- **The regular feedback mechanism** establishes a regular feedback mechanism, such as collecting students' opinions and suggestions every week or month and adjusting the teaching plan on time.
- **Diversified evaluation tools** combine multiple evaluation tools, such as questionnaires, group discussions, learning logs, and field observations, to obtain more comprehensive and in-depth information.
- **For data analysis and improvement**, the collected data are integrated, quantitative and qualitative analyses are conducted to identify problems and room for improvement, and teaching strategies and methods are adjusted in a timely manner.

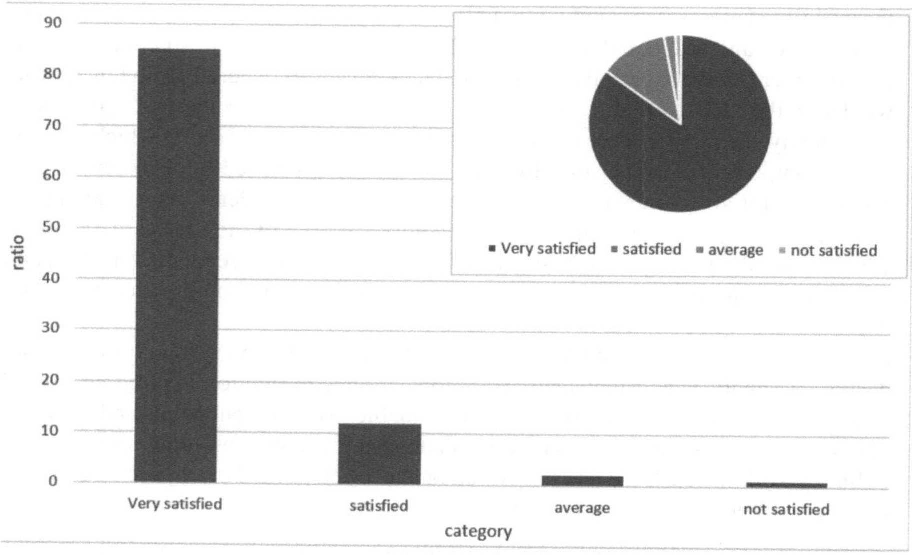

Fig. 3. Survey results of algorithm design and analysis knowledge graph

2.4 Continuous Advancement of Teaching Reform

Teaching reform is a process of continuous improvement and progressive progress. In the contemporary education model and system, teaching reform insists on being "student-oriented", constantly analysing and investigating students' opinions, combining with the background of engineering applications, and constantly cycling through the "evaluation", "feedback" and "improvement" of teaching activities, keeping pace with the times, balancing the relationship between the "supply side" of universities and the "demand side" of enterprises, improving teaching quality, and cultivating comprehensive talent. To this end, in the "Algorithm Analysis and Design" course, we analyse the shortcomings of traditional teaching content and teaching methods; optimize teaching content; listen to students' opinions; improve teaching methods; balance the relationship between "teaching" and "learning" in class; deeply tap students' ability to explore, analyse, and solve problems; improve students' initiative and enthusiasm for independent learning; cultivate students' research interests; reform the assessment method so that students can truly apply what they have learned; improve the teaching quality evaluation system; and gradually promote reform.

Table 1. The average pass rate comparison before and after the reform.

state	class			
	1	2	3	Average Pass Rate
Before the Reform	85	86	89	86.7%
After the Reform	98	95	98	97%

3 Effects of Teaching Reform

For the theoretical course "Algorithm Design and Analysis", the series of teaching reform measures we carried out achieved good teaching results, which are reflected mainly in the following aspects:

- To investigate students' satisfaction with the algorithm design and analysis knowledge graph, 100 students were randomly selected and sent a questionnaire to investigate the algorithm design and analysis knowledge graph. The survey results are shown in Fig. 3. Eighty-five percent of the students are very satisfied with the algorithm design and analysis knowledge graph, and 12% of the students are satisfied. In addition, through the application of the knowledge graph algorithm, the various knowledge points in the textbook are organically connected to form a complete and coherent knowledge network, which can help them better understand and apply the knowledge they have learned; 2% of the students are average, and 1% are dissatisfied. We list the opinions of the last 3% and gradually improve the content and structure of the algorithm design and analysis knowledge graph.
- Improved student attendance and participation. Before the reform, to improve the attendance and participation of students in the algorithm design and analysis class, students were often required to take compulsory assessments such as checking the student list and asking classmates in class questions, which resulted in poor student initiative. After the reform, the student attendance rate increased from 86.7% to 97%, and the attendance assessment method was abolished. During class, students were energetic and focused. They actively expressed their views and solutions to problems and took the initiative to show their self-study results. They actively communicated with teachers about the difficulties they encountered, and their homework completion rate was high.
- Improved students' comprehensive professional ability. Before the reform, students studied the key chapters in the exam, neglected hands-on practical skills, studied for the sake of exams, did not understand the subject thoroughly, and had poor professional ability. After the reform, we compared the comprehensive scores of the three classes of the previous year that did not participate in the reform with the three classes

of the current year. To ensure that the number of students was the same, we selected the top 50 students in each class for comparison. The comprehensive scores (regular scores + final exams + experiments) are shown in Table 1. The pass rate of each class after the reform was higher than that before the reform. Moreover, after the reform, students' enthusiasm for learning significantly improved. Students have broken free from the constraints of exams and no longer study for the sake of "passing". They consider both theory and practice and improve their independent exploration ability, hands-on practical ability, teamwork ability, and communication and expression ability.

- Expanding employment channels for students. After the curriculum reform, the proportion of hands-on practice increased, and the practice content was mostly practical problems currently encountered by enterprises. Students were allowed to understand the employment background and actual application scenarios of algorithms in advance, which prompted them to make employment plans in advance. Moreover, to grasp the direction of talent demand in the social market, schools continuously visit and investigate relevant enterprises, establish good relationships with enterprises, and direct the input of professional talent.
- Teachers' knowledge and teaching methods are constantly updated and improved. While teachers have rich theoretical knowledge, they constantly communicate with enterprises to grasp the latest industry developments and technological trends, constantly absorb the latest knowledge, and gradually update their own theoretical and technical frameworks. Second, through a multidimensional and multistage teaching quality evaluation system, classification is collected and analysed, more practical subject training plans are formulated, and their own teaching and guidance capabilities are continuously improved.

4 Conclusion

As one of the core courses of computer science, Algorithm Design and Analysis uses algorithms to solve practical problems, which is one of the important skills that contemporary computer science students need to master. However, traditional teaching content and teaching methods have certain drawbacks in cultivating comprehensive talent. This paper reforms the traditional teaching content and teaching methods, takes "students as the center" and the needs of "employment units" as the basis, continuously improves the teaching content, reforms the teaching methods, stimulates students' interest in learning, cultivates students' hands-on practical ability, continuously improves students' professional comprehensive ability, and helps students expand employment channels. This study provides a reference for the reform of other courses in the future.

References

1. Wang, R., Zeng, F., Yao, L., et al.: Game-theoretic algorithm designs and analysis for interactions among contributors in mobile crowdsourcing with word of mouth. IEEE Internet Things J. **7**(9), 8271–8286 (2020)
2. Rude, U., Willcox, K., McInnes, L.C., et al.: Research and education in computational science and engineering. SIAM Rev. **60**(3), 707–754 (2018)

3. Goodrich, M.T., Tamassia, R.: Algorithm Design: Foundations, Analysis, and Internet Examples. John Wiley & Sons, Hoboken (2001)
4. Xu, B., Jia, Q., Hsu, L.T.: Vector tracking loop-based GNSS NLOS detection and correction: algorithm design and performance analysis. IEEE Trans. Instrum. Meas. **69**(7), 4604–4619 (2019)
5. Zhang, L., Chakareski, J.: UAV-assisted edge computing and streaming for wireless virtual reality: Analysis, algorithm design, and performance guarantees. IEEE Trans. Veh. Technol. **71**(3), 3267–3275 (2022)
6. Panneerselvam, R.: Design and Analysis of Algorithms. PHI Learning Pvt. Ltd. (2015)
7. Alsuwaiyel, M.H.: Algorithms: Design Techniques and Analysis. World Scientific (2016)
8. Li, N., Hou, J.C., Sha, L.: Design and analysis of an MST-based topology control algorithm. IEEE Trans. Wirel. Commun. **4**(3), 1195–1206 (2005)
9. Kruskal, J.B.: On the shortest spanning subtree of a graph and the travelling salesman problem. Proc. Am. Math. Soc. **7**(1), 48–50 (1956)
10. Dijkstra, E.W.: A note on two problems in connexion with graphs. In: Edsger Wybe Dijkstra: His Life, Work, and Legacy, pp. 287–290 (2022)
11. Prim, R.C.: Shortest connection networks and some generalizations. Bell Syst. Techn. J. **36**(6), 1389–1401 (1957)
12. Huffman, D.A.: A method for the construction of minimum-redundancy codes. Proc. IRE **40**(9), 1098–1101 (1952)
13. Wu, C., Liu, S., Zeng, Z., et al.: Knowledge graph-based multicontext-aware recommendation algorithm. Inf. Sci. **595**, 179–194 (2022)

Using Large Language Models to Promote Vocational Skills Improvement: Reform and Practice of the "Algorithm Design and Analysis" Course in Higher Education

Kejia Zhang[1], Haiwei Pan[1], Zhiqiang Ma[1], Shaoqiang Zhu[1], Xiaoliang Qin[1], and Lan Zhang[2(✉)]

[1] College of Computer Science and Technology, Harbin Engineering University, No. 145, Nantong Street, Nangang District, Harbin, Heilongjiang, China
[2] Computer and Control Engineering, Northeast Forestry University, No.26 Hexing Road, Xiangfang District, Harbin, China
zhanglan@nefu.edu.cn

Abstract. "Algorithm Design and Analysis" is not only one of the important courses in the undergraduate teaching of computer science and technology but also a key part of computer professional skills. In recent years, with the rise and widespread application of big language models, many teaching reform plans have been produced to promote the quality and efficiency of teaching. This paper studies how to refer to software development professional skills standards, investigates the knowledge points of "Algorithm Design and Analysis" courses in other educational institutions, uses cutting-edge core technology big language models to drive the improvement of teaching evaluation methods, improves teaching efficiency, and carries out reforms and practices in teaching content for undergraduate students in computer science.

Keywords: Algorithm Design and Analysis · Large Language Model · Professional Skill Standards · Teaching Reform

1 Introduction

At present, computer information technology is widely used in most fields [1–3]. Developing an efficient program requires not only programmers but also programmers with proper data organization skills and outstanding algorithm analysis capabilities. Some top computer scientists believe in their computer science monographs that algorithms are a form of creative thinking that must be design oriented. As an important course in computer science, the "Algorithm Design and Analysis" course is more targeted at cultivating students' algorithm writing capabilities. An excellent "Algorithm Design and

This work is supported by Harbin Engineering University's 2021 Education Reform Project "How to Make Computer Theory Teaching Serve Employment" (Grant No. JG2021B0609).

Y. Yu et al. (Eds.): ICPCSEE 2025, CCIS 2673, pp. 150–159, 2026.
https://doi.org/10.1007/978-981-95-2566-9_11

Analysis" course should have strong practical applicability and course professionalism, cultivate students' ability to correctly analyse the complexity of algorithms, enable students to analyse the correct algorithms in practical problems, and create a good theoretical foundation for students' algorithm analysis. This is an essential factor for any engineer involved in computer systems, system software, and software development.

In the teaching of algorithm design and analysis, instructors often prioritize the explanation and formal evaluation of algorithmic techniques. The standard instructional approach can become overly mechanical and detached from practical contexts. Case studies are frequently unrepresentative, leading students to perceive a lack of real-world applicability and leaving them ill equipped to confront meaningful problem-solving tasks. This disconnect gives rise to a dual challenge: instructors find it difficult to teach effectively, while students struggle to learn meaningfully. To address this issue, the instructional process should purposefully integrate abstraction, analytical reasoning, modelling, and algorithm construction with the cultivation of computational thinking. By structuring lessons around authentic, problem-driven scenarios, educators can better engage students' curiosity, promote active participation, and enhance their problem-solving capabilities. This approach not only bridges the gap between theory and practice but also fosters intrinsic motivation, enabling students to internalize and apply algorithmic concepts more effectively.

The emergence of large multimodal language models [4] provides new teaching ideas and practical approaches for algorithm design and analysis. The application of large language models [3, 5–8] can effectively reduce the processing cost and time cost of students' exercise. With the use of large language models [9–11], students can learn simple algorithms independently through questions and answers, which improves their learning efficiency. Moreover, large language models have powerful information extraction capabilities [12, 13]. The analysis results of large language models can provide teachers with comprehensive student characteristics and difficult knowledge points in learning and provide more efficient teaching ideas for teachers in the Algorithm Design and Analysis course. Expanding the application scope of multimodal large language models in teaching tasks can promote the digitization and characterization of the teaching of the Algorithm Design and Analysis course and promote improvements in the intelligence of the course. With the continuous updating of multimodal large language model technology, this paper innovates the application scenarios of large language models, connects with students' future professional skills, and aims to cultivate students with strong professional skills to promote the improvement of the quality and efficiency of the teaching of the Algorithm Design and Analysis course.

Algorithm design and analysis constitute a core course in computer science. How to adapt students to new industries and adapt them to international competitors is an important issue in the reform of university education. This paper discusses the content and methods of computer science course construction, promotes the reform of teaching methods through large language models, improves the quality of classroom content, promotes an increase in teachers' work efficiency, improves course evaluation and scoring methods, and cultivates students' stronger comprehensive algorithm ability and application innovation ability.

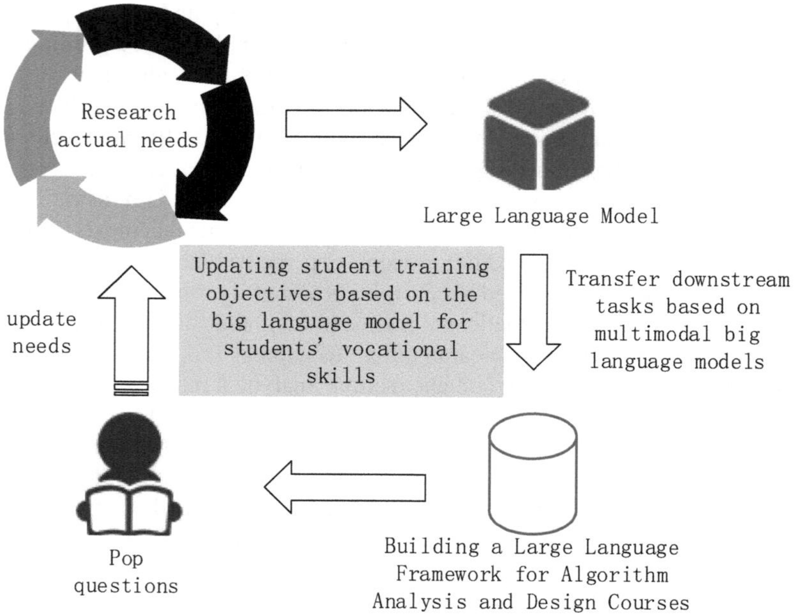

Fig. 1. Reform Plan for Student Training Objectives

2 Teaching Reform Measures Based on the Large Language Model

2.1 The Big Language Model Can Be Used to Reform Student Training Goals

The big language model is used to analyse the needs and current status of the "Algorithm Design and Analysis" course in professional applications. The current actual enterprise demand tasks are counted, and a knowledge base is built on the basis of the big language model analysis. The analysis capabilities of the multimodal big language model are used, and the existing course knowledge points are combined to build an application skill library suitable for students. The specific student training goal reform plan process is shown in Fig. 1.

To prevent the reform teaching objectives from deviating from the teaching content, teachers summarize the knowledge points in the textbooks, take the textbook knowledge points as the starting point, find the demand tasks that will be encountered in practical applications, and build a preliminary student knowledge demand framework by combining the difficulties of students' understanding in previous learning and the demand tasks proposed by enterprises. By using the computer-oriented large language model, the knowledge framework is input into the large language model, and the downstream task branch is constructed. By using the powerful feature extraction ability of the large language model, the parameters of the downstream branch model of the large language model are fine-tuned to obtain a knowledge large language framework for "Algorithm Design and Analysis" so that the framework has a certain knowledge extraction ability for the "Algorithm Design and Analysis" course. After the preliminary model is formed, due to certain limitations of the preliminary investigation, the framework needs to be further updated. Students, enterprise personnel and teachers can ask questions about the framework to obtain new requirements, and teachers update the framework requirements. The updated requirements are input into the large language model again, and the parameters in the framework are gradually iterated and updated. Since more enterprise requirements and student professional skills are included in this process, the updated large language model framework can generate a target training plan for student professional skills.

At present, the application directions and practical needs involved in computer courses are increasing, which places greater requirements on teaching methods and training programs. By using interactive learning between different modal data and a strong model foundation, large language models can improve the efficiency and accuracy of data analysis, so they have great potential for computer applications.

The goal of this course is to cultivate mastery through practical application. Students will learn to describe algorithms by combining programming languages with structured natural-language steps and will gain a clear understanding of algorithmic complexity, accuracy, and correctness. They will be introduced to mathematical methods for deriving and proving time and space complexities and will explore computational models, including the P class, NP class, and NP-complete problems. Using foundational concepts in algorithm design, students address real-world problems, perform linear and in-depth analyses of complex engineering challenges, abstract structures, and design both models and experiments. Emphasis is placed on strengthening computational skills and logical thinking. To support diverse learning levels, the curriculum is organized around problem solving: each module presents a representative problem, interprets the underlying theory, analyses algorithmic strategies, and provides hands-on exercises in algorithm implementation and application.

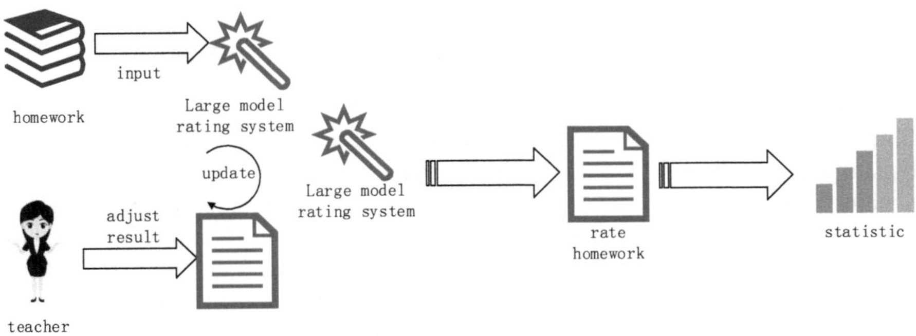

Fig. 2. Curriculum evaluation system based on a large language model

2.2 Using the Big Language Model to Reform the Assessment Method

Since the course is aimed at vocational skills, it is necessary to reform the assessment method of students' courses. A single, simple final exam cannot accurately assess the level of students' vocational skills. Owing to the diversity of vocational skills, teachers have a heavy workload following the traditional marking and grading system, and the assessment of vocational skills usually has no unique conclusions. Therefore, it is necessary to innovate and reform new assessment methods to reduce teachers' workload and improve work efficiency.

This paper constructs a course evaluation system based on a large language model, as shown in Fig. 2. First, the student assignments are input into the large language model scoring system to obtain the corresponding scoring results. The teacher fine-tunes the model output results on the basis of experience to ensure that the model output results meet the evaluation criteria of the "Algorithm Design and Analysis" course, inputs the adjusted results into the model, and updates the model parameters. Owing to the powerful learning ability of the large language model, the number of input examples is usually approximately 20, and the workload is small. After adjustment, the teacher can obtain a large language model scoring system for "Algorithm Design and Analysis". The teacher can input student assignments or classroom tests into the large language model scoring system to obtain student assignment evaluations. The quantitative statistical information of students is obtained through the evaluation generated by the large language model.

The open-ended questions in the "Algorithm Design and Analysis" course have many details, which require many manual evaluations. It is easy to overlook many details, and the manual scoring error rate is high. An assessment and scoring system using a large language model can greatly reduce the workload of teachers and is in line with teachers' evaluation habits. A relatively fair scoring standard is specified for the evaluation of open-ended questions in the "Algorithm Design and Analysis" course.

Teachers have deepened the reform of curriculum evaluation methods by strengthening the management of the teaching process, scientifically evaluating students' learning effects, and guiding students to transition from learning knowledge to ability and quality. They fully understand the learning situation of students, provide timely feedback to students, pay attention to normal learning and accumulation, and evaluate students in an objective way.

2.3 The Use of Large Language Models to Reform Teaching Methods

In the algorithm design and analysis course, the traditional teaching methods include offline classroom teachers and students face-to-face, and teachers teach according to the plan. Teaching courses are taught through plans, teaching calendars, etc. Students are mainly responsible for listening, sometimes answering teachers' questions in class, answering teachers' questions after class and assigning homework to consolidate learning, and real-time operation is relatively simple. The teacher is the protagonist and is responsible for organizing and managing the entire project, and the learning students are passive learners. In this process, students' initiative to ask questions is reduced, teachers lack the ability to observe students' learning progress, and they lack an understanding of the extent to which students have mastered knowledge. Owing to problems such as students' personalities and teaching methods, students are more inclined to use the internet, self-study, and other methods to learn courses.

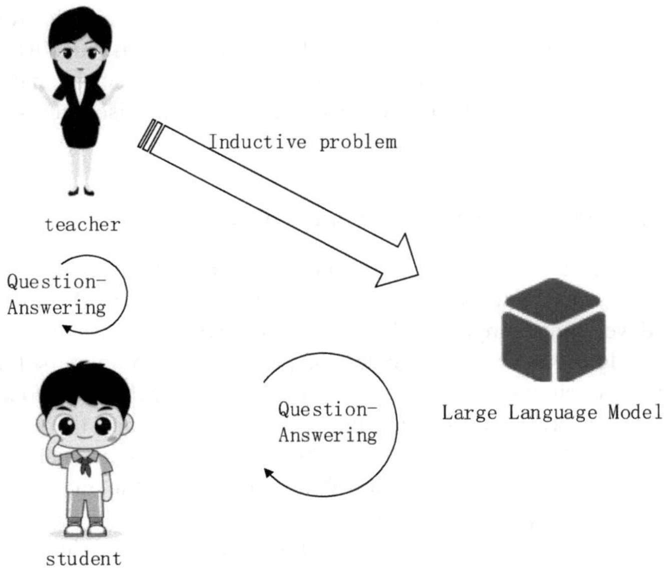

Fig. 3. Student Learning Observation System Based on a Large Language Model

This paper builds a learning observation system for the "Algorithm Design and Analysis" course on the basis of the big language model, as shown in Fig. 3. During the teaching process, students can use the established big language model of the "Algorithm Design and Analysis" course to ask their questions, and the big language model answers the students' questions. In the teaching questioning session, the teacher condenses the questions, guides the students in their careers, and evaluates the knowledge and ability of the students in the learning process. At the same time, the teacher summarizes the knowledge points in the professional skills. The teacher summarizes the questions generated during the teaching time and inputs them into the big language model to improve the professionalism of the big language model. At the same time, students can also use the big language model to ask questions efficiently and provide feedback on their homework for preliminary evaluation. During the whole process, the big language model can participate in discussions between students, monitor students' learning status, guide students in professional knowledge, and perceive students' knowledge mastery and understanding from multiple dimensions, such as student homework, questions, and query methods.

In the online and offline hybrid teaching model, students can spend 20% to 50% of their time online. The premise of hybrid teaching is to use rich course teaching resources, and teachers should prepare relevant teaching materials, including but not limited to teaching videos, relevant courseware, key exercises, etc.

In the initial classroom teaching activities, the teachers spent more time on these activities, and the students listened most of the time. It is difficult for students to concentrate for 45 min, and the content of a lecture is more difficult to understand. If they do not follow the teacher's progress, they will easily feel tired. In online and offline hybrid education, teachers must set aside some time to learn to think, discuss, and answer questions.

Therefore, in teaching design, teachers should choose some open topics for students to discuss, and students can find answers through various methods, such as backtracking and dynamic programming; this approach can also enable students to analyse the problems solved by previous students. The teacher organized group discussions and reviewed and revised the discussion results of each group.

2.4 Continuous Advancement of Teaching Reform

Teaching reform is a long-term and systematic project that cannot be completed overnight. Even successful teaching models need to keep pace with the times, especially in the rapidly developing computer industry. Algorithm design and analysis courses usually teach basic algorithm design strategies and provide some problem-solving methods for each strategy. Each algorithm and solution is different from other course chapters. The idea of solving problems is independent and scattered, and there is an intrinsic connection between them. Most existing textbooks use deterministic algorithms. They master the characteristics of many common algorithms, such as the state planning method of dynamic knapsack problems. Solving such problems is an NP. These deterministic algorithms are no longer used, but modern optimizer calculation methods such as neural networks, genetic algorithms, ant colony algorithms, etc. Parallel algorithms, random algorithms and approximate calculation methods, encryption algorithms and intelligent optimization algorithms developed for new engineers are less common. If the internal teaching level is not optimized and a targeted knowledge system is lacking, teachers cannot link the teaching content with the real problems of today's society.

Table 1. Scoring rules for "Algorithm Design and Analysis"

Assessment Method	Process Assessment				Final Assessment	Value-added evaluation
	Regualr attendance	Operation	Comprehensive design	Online Learning	Final Exam	Competition winners
Score setting	10	20	10	20	40	10
Evaluation subject	teacher	Teacher, Large Language Model	Teacher, Large Language Model	Large Language Model	teacher	teacher
Evaluation method	Online and offline	Online and offline	Online and offline	Offline	Offline	Offline

As teaching progresses, teachers can adjust the parameters and results of the large language model to obtain evaluation standards, knowledge frameworks, and teaching observation systems that meet current students. As the amount of data increases, the knowledge contained in the teaching system, which is based on a large language model, becomes more comprehensive, and the scoring system becomes more objective. Teachers can expand more functions in the system to achieve diversified teaching methods and evaluations for students, in line with students' career needs, meet the employment needs of enterprises, and enhance students' professional skill competitiveness. See Table 1 for details.

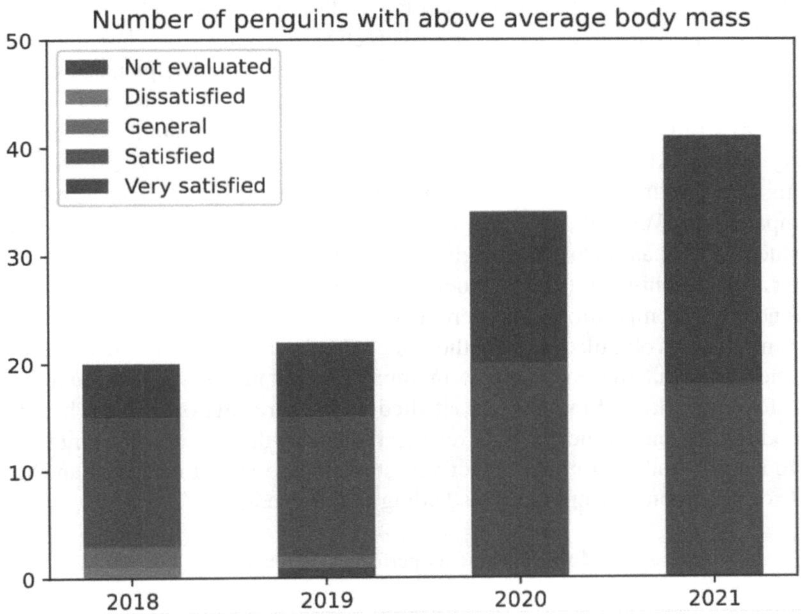

Fig. 4. Student Satisfaction Survey.

In the new engineering context, students are required to fully and thoroughly master the major to meet the needs of the talent market when they graduate and to continuously innovate activities. This is the basic way and method used to ensure that students have a strong ability to deal with complex problems. From the school's innovative experiments to algorithm courses, we encourage outstanding students to apply for creative projects, internet+, participate in software competitions, mathematics competitions, and discover student cooperation projects with scientific research potential for scientific education. Through research and learning, writing scientific articles, and graduation theses, they are greatly improved.

3 Effects of Teaching Reform

The series of teaching reform measures we have implemented for the core courses of our school's School of Software have achieved good teaching results, which are reflected mainly in the following aspects.

Students' satisfaction with teaching has increased significantly, as shown in Fig. 4. Since the beginning of 2020, the proportion of students who are very satisfied or satisfied with teaching has increased significantly, indicating that teaching reform based on a large language model can enhance students' recognition of teaching and increase their learning participation.

The algorithm design and analysis course has achieved good results through multiple mixed teaching practices. Students' enthusiasm for learning is increasing, but fewer people are looking at their phones and sleeping in class. Students' ability to ask questions and use tables has improved. The ability to practice engaging with class and asking or answering fewer questions has increased. The number of participants has increased from 10 to approximately 40, and the ability to think actively has also greatly improved. Students have improved their problem-solving skills and the quality of their course summary reports. After improvements, they can improve their ability to think and solve problems from multiple perspectives and propose different solutions to the same problem and compare them. When the difficulty of the subject remains the same, the average score of the students increases from the original 70 points. Up to 75 points (see Table 2).

After the teaching reform, students actively participated in innovation and entrepreneurship competitions and were able to obtain certain honors. Therefore, after 2020, some students obtained credit in the value-added evaluation to improve their overall scores, indicating that the teaching reform improved their hands-on ability and increased their professional skills. Students are satisfied with the results of mixed education and have good evaluations of the teaching results, indicating that the use of a large language model to reform student training objectives, student assessment methods, and teaching methods can promote students' understanding of knowledge.

Table 2. Student performance records

Assessment Method	Process Assessment				Final Assessment	Value-added evaluation	Total score
	Regualr attendance	Operation	Comprehensive design	Online Learning	Final Exam	Competition awards, etc.	
Score setting	10	20	10	20	40	10	110
2018	8.5	12.3	6.4	15.7	27.8	0	70.7
2019	8.1	12.8	7.3	14.8	28.1	0	71.7
2020	8.8	13.7	7.8	15.1	29.8	2.3	75.2
2021	9.1	13.1	8.1	15.9	30.1	1.9	76.3

4 Conclusion

This paper explores the problems of inefficiency and lack of interactivity of traditional teaching methods, proposes four reform measures for the "Algorithm Design and Analysis" course via the large language model method, and analyses them one by one. Second, the application of new online teaching methods to algorithm design and analysis courses can improve the interaction between students and teachers, students' learning efficiency, and the initiative of independent learning. By using new teaching methods, students' professional level and enthusiasm for algorithm design and analysis courses have significantly improved, and students' performance has significantly improved. Teachers will also have more work to prepare lessons, greatly improving work efficiency. The implementation of a series of teaching reforms has enabled computer major teaching in our school to gradually enter an accelerated development stage.

References

1. Yao, S., Yu, D., Zhao, J., et al.: Tree of thoughts: deliberate problem solving with large language models. Adv. Neural. Inf. Process. Syst. **36**, 11809–11822 (2023)
2. Thirunavukarasu, A.J., Ting, D.S.J., Elangovan, K., et al.: Large language models in medicine. Nat. Med. **29**(8), 1930–1940 (2023)
3. Kasneci, E., Seßler, K., Küchemann, S., et al.: ChatGPT for good? On opportunities and challenges of large language models for education. Learn. Individ. Differ. **103**, 102274 (2023)
4. Kaplan, J., McCandlish, S., Henighan, T., et al.: Scaling laws for neural language models. arXiv preprint arXiv:2001.08361 (2020)
5. Jeon, J., Lee, S.: Large language models in education: a focus on the complementary relationship between human teachers and ChatGPT. Educ. Inf. Technol. **28**(12), 15873–15892 (2023)
6. Jowsey, T., Stokes-Parish, J., Singleton, R., et al.: Medical education empowered by generative artificial intelligence large language models. Trends Mol. Med. **29**(12), 971–973 (2023)
7. Killian, C.M., Marttinen, R., Howley, D., et al.: "Knock, Knock… Who's there?" ChatGPT and artificial intelligence-powered large language models: reflections on potential impacts within health and physical education teacher education. J. Teach. Phys. Educ. **42**(3), 385–389 (2023)
8. Hoi, V.N., Mu, G.M.: Perceived teacher support and students' acceptance of mobile-assisted language learning: evidence from Vietnamese higher education context. Br. J. Edu. Technol. **52**(2), 879–898 (2021)
9. Karpukhin, V., Oguz, B., Min, S., et al.: Dense passage retrieval for open-domain question answering. In: EMNLP, no. 1, pp. 6769–6781 (2020)
10. Nakano, R., Hilton, J., Balaji, S., et al.: Webgpt: browser-assisted question-answering with human feedback. arXiv preprint arXiv:2112.09332 (2021)
11. Law, P.M., Lo, L.Y.H., Endert, A., et al.: Causal perception in question-answering systems. In: Proceedings of the 2021 CHI Conference on Human Factors in Computing Systems, pp. 1–15 (2021)
12. Jiang, Y., Zhang, W., Zhang, X., et al.: Prior knowledge integration via llm encoding and pseudo event regulation for video moment retrieval. In: Proceedings of the 32nd ACM International Conference on Multimedia, pp. 7249–7258 (2024)
13. Shi, M., Liu, F., Wang, S., et al.: Eagle: exploring the design space for multimodal llms with mixture of encoders. arXiv preprint arXiv:2408.15998 (2024)

Implementation Path of Computer Science Competition in the Context of Industry-Education Integration

Yunge Gao[1](✉), Shanshan Li[1], Kebu Wang[1], Xia Liu[2](✉), and Hongfeng Zhang[2]

[1] Sanya Aviation and Tourism College, Sanya 57200, Hainan, China
[2] Faculty of Humanities and Social Sciences, Macao Polytechnic University, Macao 999078, China
p2315160@mpu.edu.mo

Abstract. In the context of industry-education integration, computer science competitions face new opportunities and challenges. Taking College S as an example, this paper analyses the current development status and difficulties faced by discipline competition in this area. On this basis, in combination with the requirements of industry-education integration, four aspects of implementation paths are proposed: improving competition management mechanisms, optimizing competition organization models, innovating competition content design, and enhancing the transformation of competition results. This paper also constructs a computer science discipline competition system in three dimensions: cultivating students' innovative abilities, enhancing teachers' practical teaching levels, and deepening school-enterprise cooperation. The aim is to provide a reference for the conduct of computer science discipline competitions in higher education institutions and to promote the reform and innovation of talent cultivation models.

Keywords: integration of production and education · computer science · subject competition · implementation path · system construction

1 Introduction

The "Implementation Plan for the Empowerment and Enhancement Action of Industry--Education Integration in Vocational Education (2023–2025)" clearly states that deepening industry--education integration will promote effective connections between the education chain, talent chain, industrial chain, and innovation chain. It supports vocational colleges and enterprises in jointly organizing skills competitions, emphasizing that these competitions can drive teaching reforms and innovations in talent cultivation models, making academic competitions an important tool for deepening industry-education integration [1]. The "Opinions on Promoting High-Quality Development of Modern Vocational Education" (2022) also explicitly proposes building an integrated system of "positions, courses, competitions, and certifications." By organically combining "positions," "courses," "competitions," and "certifications," it aims to make academic

Y. Yu et al. (Eds.): ICPCSEE 2025, CCIS 2673, pp. 160–174, 2026.
https://doi.org/10.1007/978-981-95-2566-9_12

competitions a crucial part of talent cultivation [2]. As a vital bridge connecting higher education with industry needs, academic competitions play a significant role in promoting the deep integration and precise alignment of education and industry. Therefore, how to promote industry-education integration through academic competitions and cultivate high-quality applied talent that meets industry demands has become an important topic in current educational reform.

Computer science, as an essential component of engineering education, has seen its practical skills, innovation capabilities, and interdisciplinary integration abilities increasingly challenged due to the rapid development of new-generation information technologies such as big data, cloud computing, and artificial intelligence. However, under the backdrop of industry-education integration, many issues still exist in current computer science competitions. For example, the competition content often diverges from actual industry needs, with low conversion rates for competition outcomes and an underdeveloped mechanism for nurturing talent through these competitions. These problems severely limit the effectiveness and impact of academic competitions in talent cultivation. To address these issues, the "Action Plan for Enhancing Industry-Education Integration in Vocational Education (2023--2025)" proposes optimizing the industry-education integration mechanism and further strengthening the innovation and transformation of academic competitions on this basis [1]. Therefore, universities urgently need to improve competition management mechanisms, optimize competition organizational models, innovate in terms of competition content design, and enhance the conversion of competition results. By doing so, they can explore implementation paths for computer science competitions against the background of industry-education integration, build a discipline-specific competition system with professional characteristics, fully leverage the exemplary and leading role of academic competitions in computer talent cultivation, promote deep integration of industry and education, and enhance the quality of talent cultivation.

This article takes College S as an example. An analysis of the current development status and challenges faced by computer science discipline competitions combines the requirements of industry-education integration to propose implementation paths such as improving competition management mechanisms, optimizing competition organization models, innovating competition content design, and enhancing the transformation of competition outcomes. It constructs a computer science discipline competition system in three dimensions: cultivating students' innovative abilities, improving teachers' practical teaching levels, and deepening school-enterprise cooperation. The aim of this study is to provide a reference for the conduct of computer science discipline competitions in higher education institutions and to promote the reform and innovation of computer science talent cultivation models in the context of industry-education integration.

2 Background and Development Dilemma of Computer Science Competition

2.1 Background of Generation

As the degree of informatization in China continues to rise, new-generation information technologies, such as big data, cloud computing, and artificial intelligence, are developing rapidly. The structure of the information industry is accelerating its adjustment. To cultivate more high-quality, innovative, and versatile IT talent, universities across the board are exploring a new talent cultivation model centred on collaborative education among industry, academia, and research. As an important component of this model, academic competitions have emerged accordingly.

Computer science, as an essential component of engineering education in universities, has always placed great emphasis on the unique role of academic competitions in talent cultivation. Since the 1990s, a series of influential events, such as the National College Students' Electronic Design Competition and the ACM International Collegiate Programming Contest, have been held, significantly advancing the reform and innovation of computer science teaching in higher education institutions. In the 21st century, with the deepening integration of industry and education, increasingly specialized academic competitions, such as the "China College Students' Computer Design Competition," the "College Students' Computer Innovation Works Competition," and the "China University Computer Competition," have emerged. These events have injected new vitality and momentum into the cultivation of computer science talent. Academic competitions not only stimulate students' innovative thinking and practical skills but also provide valuable opportunities for them to connect with real-world industry practices. In this context, competitions have gradually become an integral part of computer science education, playing an irreplaceable role in nurturing high-level innovative talent and promoting the deep integration of education and industry [3].

S College has always regarded academic competitions as one of the key pathways for enhancing the quality of talent cultivation. Since the establishment of the Computer Science Department in 2020, the college has organized students annually to actively participate in national and provincial key discipline competitions. The cumulative number of projects entered exceeds ten, with over 1,200 student participants. For detailed information on participation and awards, see Table 1. These achievements are attributed not only to the college's high interest in academic competitions, treating them as an essential part of innovative talent development and providing ample support in terms of policy and funding but also to the continuous optimization of competition organization models and constant innovation in competition project design. The college emphasized the deep integration of competitions with teaching and research, striving to promote the transformation and application of competition outcomes, thereby achieving a dual promotion of competitions and academic development [4].

Table 1. Competition participation and award-winning situation of computer disciplines at S College since 2020

name of competition	host unit	The level of competition	Number of awards
Hainan Provincial Competition of Chinese College Students Computer Design	Hainan Provincial Department of Education	Class B at provincial level	27
Hainan Provincial Competition of Computer Innovation Works for College Students from Both Sides of the Taiwan Straits and Hong Kong and Macao	Hainan Provincial Department of Education	Class B at provincial level	30
Hainan Provincial Competition of Team Programming Contest for Chinese University Computer Competition	National institutions of higher learning Computer Education Research Association	Class B at provincial level	13
Hainan Provincial Competition of RPA + AI Innovation Challenge for College Students in China Computer Competition	Hainan Provincial Department of Education	Class B at provincial level	4
Hainan Provincial Competition of Chinese University Computer Competition-Artificial Intelligence Creative Competition	Hainan Provincial Department of Education	Class B at provincial level	5
Code Hoof Cup National College Students Programming Competition	Organizing Committee of the Hoof Cup	National Class B	24
Robot Developer Competition	MIIT Talent Exchange Center	National Class B	28
Hainan College Students Low code Programming Competition	Hainan Computer Society	Class C at provincial level	1

(*continued*)

Table 1. (*continued*)

name of competition	host unit	The level of competition	Number of awards
Hainan Vocational College Skills Competition	Human Resources and Social Security Department of Hainan Province	Class A at provincial level	4
National College Computer Competency Challenge	National institutions of higher learning Computer Education Research Association	National Class C	115

2.2 Development Dilemma

Although the discipline competition of computer science in colleges and universities has developed vigorously in recent years, it still faces many challenges and difficulties in the context of industry-education integration.

(1) Some competitive content is disconnected from actual industry needs. Many current computer science competitions remain at a low level, focusing on algorithm design and program optimization, with an emphasis on code skills. This falls short of the practical engineering and innovation capabilities that the industry urgently requires. The connection between competitions and real-world application scenarios is not tight enough, and students often participate merely to win awards. As a result, the role of competitions in enhancing students' practical and innovative abilities is limited [5].

(2) The conversion efficiency of competition outcomes is low. Since competitions primarily aim to assess students' abilities, they often diverge from the teaching content. As a result, the ideas and works generated usually remain at the initial concept or prototype stage or merely satisfy formal requirements, lacking systematicness, completeness, and sustainability. Consequently, realizing their value in practical applications is difficult. Although students invest considerable time and effort into participating in competitions, the insufficient transformation of results fails to ignite their passion for innovation and entrepreneurship.

(3) Competition-driven education mechanisms need to be improved. Currently, many schools do not place enough emphasis on competition, and the management system is inadequate. There is a disconnect between competition and teaching, and the alignment with talent cultivation goals is insufficient. Teachers lack enthusiasm for guiding competitions, students lack motivation to participate, and the exemplary and radiating effects of competitions have not been fully realized.

(4) The depth and breadth of industry-education integration need to be further strengthened. Currently, the interaction between universities and industries in competitions remains superficial. The channels for aligning competition with industry and enterprise needs are not smooth, and there is a lack of a regular mechanism for cooperative exchange. Industrial resources have not been fully utilized, and there is significant

room for improvement in cooperation between universities and enterprises in areas such as talent cultivation and technology transfer [6].

In response to these issues, S College has actively explored and practiced subject competitions against the background of industry-education integration. First, it optimizes competition content design by introducing real projects from enterprises, enabling students to enhance their skills through solving practical problems. Second, it improves process management and mechanisms for converting results into applications, incubating outstanding works and encouraging student innovation and entrepreneurship. Third, it deeply integrates competitions with teaching and research, exploring new models of collaborative training. Fourth, it strengthens cooperation between schools and enterprises, fully utilizing industry resources to provide students with practical platforms. These measures have shown initial success, laying a foundation for further optimizing and refining the implementation path of computer science subject competitions.

3 Design of the Implementation Path of Computer Science Discipline Competition

Since 2020, Institute S has formed teams to participate in competitions. Through years of organizing and participating in these contests, they explored an implementation path for academic competitions characterized by "one improvement, two optimizations, three innovations, and four enhancements," as shown in Fig. 1. "One improvement" refers to improving the competition management mechanism; "two optimizations" refer to optimizing the competition organization model; "three innovations" refer to innovating the design of competition content; and "four enhancements" refer to enhancing the transformation of competition outcomes.

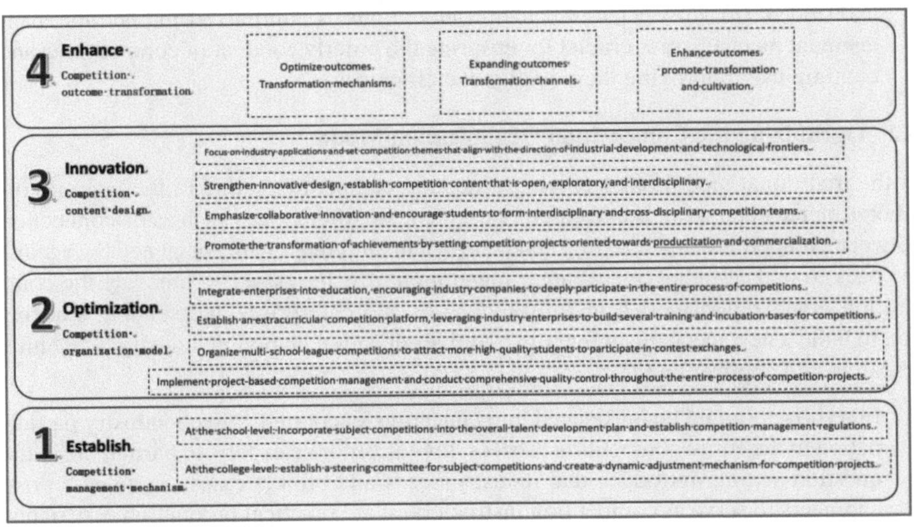

Fig. 1. Implementation path of discipline competition: "one sound, two optimizations, two innovations and four strengthening"

3.1 Improving the Competition Management Mechanism

In the context of industry-education integration, competition in the computer science discipline should give full play to its important role in talent training, so it is necessary to establish a sound competition management mechanism.

(1) Schools should incorporate subject competitions into the overall plan for talent cultivation, formulate competition management methods, clarify the role of competitions in talent development, the organizing entities, the workload calculation for participating teachers, and the connection between student participation and credit. Second, colleges should establish a leadership group for subject competitions, designate a responsible leader, and set up a competition office to handle daily management. The leadership group should develop a competition work plan and coordinate the application of competition projects, faculty allocation, financial support, process management, and summary commendations.
(2) The college should establish a dynamic adjustment mechanism for competition projects, updating the content of competitions as appropriate on the basis of adjustments in academic disciplines, technological trends, and changes in industry and enterprise needs. This will promote the deep integration of competitions with talent cultivation. A quality monitoring and evaluation system for competitions should be established, along with an assessment indicator system for competition projects. Regular assessments should be conducted to track and provide feedback on the effectiveness of competitions and improvements in student capabilities, forming a continuous improvement mechanism for competition projects. An incentive mechanism for competitions should also be established, using competition awards as important criteria for teacher performance evaluations and title promotions and considering participation and award-winning situations as key references for student excellence awards and recommendations for further education, thereby motivating both faculty and students to actively participate in competitions. An optimized competition management mechanism is crucial for ensuring the orderly conduct of competitions and continuously improving their quality and efficiency.

3.2 Optimizing the Competition Organization Mode

In the traditional model of organizing academic competitions, due to the lack of collaboration between schools and enterprises, there are often issues such as disconnection between competition and teaching and insufficient alignment with social needs. Against the background of industry-education integration, it is necessary to innovate the competition organization model, inviting industries and enterprises and sending students out, to build a new mechanism for collaborative education involving industry, academia, research, and application [7].

(1) Enterprises should be introduced into education, encouraging deep industry participation in the entire competition process. Invite corporate experts to participate in the question setting, introduce real industry needs and project cases; invite enterprise engineers to serve as competition instructors, share practical production experience from their companies; and invite senior executives to act as judges, providing optimization suggestions for entries from an application value perspective. Dual-teacher

guidance by both schools and enterprises ensures that students grow rapidly through the development of real projects.

(2) Establish an off-campus competition platform that relies on industry enterprises to build several competition training and incubation bases. Leading enterprises with representative influence in the industry should be selected to establish competitive teams and project incubation centers jointly. Students should be provided with project development venues, software and hardware facilities, financial support, and other supporting conditions to encourage them to transform their ideas into practical productivity. Students' engineering practice capabilities can be enhanced through real enterprise projects, case studies, and project development.

(3) Organize intercollegiate competitions to attract more high-quality students to participate in competition exchanges. Schools should be encouraged to jointly host academic competitions to promote resource sharing, complementary strengths, and the transformation of results, achieving mutual benefits for all parties involved. S College uses important events such as the China University Computer Skills Challenge as platforms to actively organize and host interschool leagues, expanding the influence of academic competitions both on and off campus.

(4) Implement project-based competition management, comprehensively adopting engineering methods and means such as project application, requirement analysis, development implementation, testing, and acceptance to control the quality of the entire process of competition projects. Competition teams should also be organized and implemented according to typical software engineering development models, with project managers and technical leaders appointed, clear role assignments, reasonable resource allocation, enhanced process management, risk control, and performance evaluation. The innovation and optimization of competition organization models need to be based on the requirements of industry-education integration, with schools and enterprises working together to build a continuously improving and healthy competition ecosystem.

3.3 Innovation Competition Content Design

The content design of academic competitions largely determines their appeal and effectiveness in fostering students' innovative abilities. Against the background of industry-education integration, competition content design should closely align with the needs of industries and enterprises, emphasizing the orientation towards complex engineering problems to cultivate students' practical engineering skills and innovation and entrepreneurship capabilities [8].

(1) Focus on industry applications and set competition themes that align with the direction of industrial development and technological frontiers. Closely integrate the development of new-generation information technologies such as big data, artificial intelligence, and blockchain. Competition topics are extracted from industry application scenarios such as smart healthcare, intelligent transportation, and digital twins, guiding students in applying their knowledge and skills to solve real-world problems.

(2) Enhance innovative design by setting up open, exploratory, and interdisciplinary competition content. Focus on new technologies and applications such as the Internet

of Things, cloud computing, big data, and virtual reality to create challenging and innovative projects. Through project-driven competitions, students can be guided to integrate technology innovation and cross-disciplinary technological fusion and design and develop creative system solutions.

(3) Emphasize collaborative innovation and encourage students to form interdisciplinary and interprofessional competition teams. Through teamwork, they can address challenges together, completing the entire process from technological innovation and product design to engineering implementation. Students from different professional backgrounds should be guided to leverage their strengths, work together, and develop their organizational management and teamwork skills.

(4) Promotion of the transformation of results by setting up competition projects oriented towards productization and commercialization. Through these competitions, platforms for showcasing achievements and facilitating their conversion into practical applications can be created. Entrepreneurial mentors should be introduced to provide follow-up incubation support for outstanding competition projects. Encourage students to further develop their competitive ideas by writing business plans, conducting market demand analysis, and building business models. Continuous innovation in the design of competitive content is crucial to maintaining its vitality.

3.4 Strengthening the Transformation of Competition Results

In the context of industry-education integration, university academic competitions should not merely be about "competing for the sake of competing." They must emphasize application-oriented approaches, strengthen alignment with the technological innovation needs of industries and enterprises, and accelerate the transformation and application of competition outcomes. The conversion of competition results is a systematic project that requires coordinated efforts from multiple dimensions, such as mechanisms, platforms, and cultivation.

(1) Improving the mechanism for converting results. Two-level technology transfer centers at the school and college levels should be established, management methods for scientific and technological achievements should be formulated, and matters such as intellectual property rights and the profit distribution of competition outcomes should be clarified. An innovation and entrepreneurship guidance committee should be set up to develop policies supporting the conversion of competition results. A joint research project database with enterprises should be established, outstanding competition projects should be regularly recommended to companies, and the transformation of competition results into real productive forces should be guided.

(2) Expand channels for the transformation of results. An integrated online and offline platform is built to showcase outstanding competition projects on a regular basis. Actively connect with technology transfer institutions, introduce social capital, and utilize angel investment and venture capital to provide financial support for the subsequent incubation of competition outcomes. Encourage students to apply for the Undergraduate Innovation and Entrepreneurship Training Program, further develop their competition projects, and strive for scientific research project funding.

(3) Enhance the cultivation of transformation outcomes. Innovation and entrepreneurship education are integrated throughout the competition process, providing practical

training for participating students in project management and market expansion. A team of mentors specializing in technology, management, and law should be formed to offer guidance on competition projects. Invite corporate mentors to share experiences in business model development and financing roadshows. Establish a special fund for incubating competition outcomes, offering support in terms of venue and funding, to create an accelerator for nurturing these results.

4 Construction of the Computer Science Competition System

In the context of the current digital transformation in education, building a systematic and holistic discipline competition system is crucial for fostering students' innovative capabilities and enhancing their practical skills. To this end, this paper explores three dimensions: the development of a project training ecosystem, the construction of on- and off-campus practice teaching platforms, and the deepening of school–enterprise cooperation. The aim is to promote a comprehensive upgrade and optimization of the discipline competition system, facilitating the organic integration of interdisciplinary learning and practical education [8]. As shown in Fig. 2.

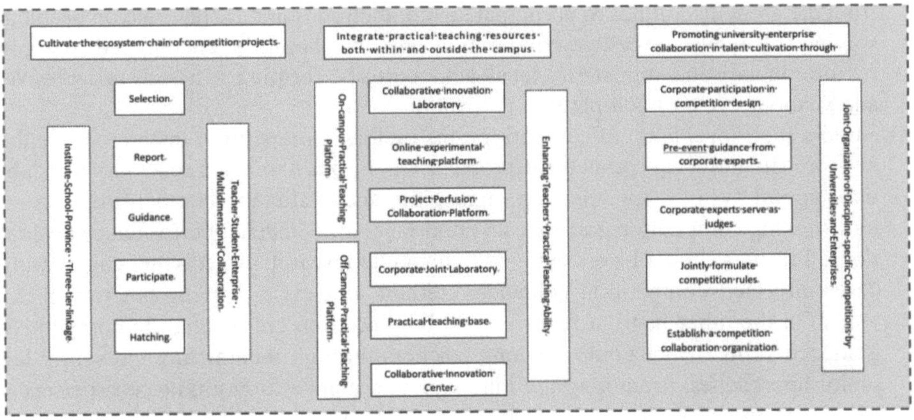

Fig. 2. "3d integration" subject competition system

4.1 To Cultivate Students' Innovation Ability, a Competition Project Cultivation Ecological Chain Should Be Built

The fundamental goal of computer science discipline competitions is to cultivate students' innovative practical abilities. To achieve this goal, it is essential to focus on the need to foster students' innovation capabilities and systematically construct an ecosystem for nurturing competition projects. In the competition project ecosystem centered around cultivating innovation capabilities, collaboration among three levels—university, school, and province—is needed, along with coordinated efforts from multiple stakeholders, including teachers, students, and enterprises. This involves continuous cultivation at various stages, such as topic selection, application, guidance, participation, and incubation, with the aim of realizing the aggregation and mutual promotion of talent, projects, resources, and platforms, as well as innovative iteration [9].

(1) Orienting towards the cultivation of innovative capabilities and optimizing the setting of competition projects. Closely align with the latest technological trends and industry application needs in the field of computing. The competition themes are extracted from directions such as cloud computing, big data, artificial intelligence, blockchain, and virtual reality. Focus on core technical topics such as data analysis, intelligent algorithms, system architecture, and application development to design competition content. Students should be guided to explore the frontier and boldly innovate.
(2) Establish a multilevel and three-dimensional competition project cultivation system. At the college level, teachers should be encouraged to apply for various professional competitions that integrate research projects and teaching activities, organizing selection contests within the college to select the best for participation in university-level competitions. At the university level, a cross-disciplinary competition project incubation platform should be built, providing continuous guidance and resource support for outstanding competition projects, promoting teams of students from different disciplines to participate in competitions and collaborate on problem solving. Moreover, excellent competition projects should be encouraged to apply for provincial- and ministerial-level competitions, forming a tiered, progressive, and virtuous cycle of competition cultivation.
(3) Build a high-level team of competition instructors. Competition instructors are the key force in nurturing competition projects. On the one hand, it is necessary to establish support systems for selecting, managing, and evaluating competition instructors, making competition guidance an important part of teacher performance evaluations. Teachers should be encouraged to integrate research projects and engineering cases into their competition guidance, guiding students in solving practical problems. On the other hand, a focus should be placed on cultivating the competitive guidance skills of outstanding young teachers through mentorship and senior-led junior approaches, promoting the inheritance of competition guidance experience. Moreover, industry resources can be fully utilized by hiring technical experts and innovation mentors from enterprises as student competition instructors, building a dual-teacher competition guidance system.
(4) Emphasize the radiating and leading role of competition projects among lower-grade students. Low-grade students should be encouraged to participate in the development and testing of competition projects, allowing them to absorb cutting-edge technology while assisting their seniors in participating. This helps cultivate their innovative awareness and enables them to enter the competition role earlier. A "seedling support project" can be explored, where project leaders select lower-grade student assistants, fostering a positive mentoring mechanism for experienced leading newcomers. This promotes students' subtle exposure to an innovative atmosphere throughout the entire process of participating in competition projects, experiencing the innovation process, and honing their innovative abilities.

4.2 Build On-Campus and Off-Campus Practical Teaching Platforms with the Aim of Improving Teachers' Practical Teaching Level

Teachers are the leaders in subject competitions and practical teaching. Enhancing teachers' practical teaching skills is crucial for ensuring the quality of competitions and practical teaching. Computer science knowledge updates rapidly and is highly practical, requiring teachers to update their knowledge systems in a timely manner, broaden their engineering practice perspectives, and improve their hands-on skills. Subject competitions serve as a grand stage for honing teachers' practical teaching abilities [10].

(1) Guiding students in competitions allows teachers to stay at the forefront of computer science and understand the latest technological developments; by guiding students in developing comprehensive solutions, they can enhance their engineering practice skills and accumulate project management experience; and by involving students in high-level competitions, they can broaden their academic horizons and establish peer networks. These insights can be transformed into practical teaching wisdom, which feeds back into the classroom to form a synergistic mechanism where competition promotes teaching and learning.

(2) To fully leverage the role of academic competitions in cultivating teachers' practical teaching skills, it is essential to focus on building comprehensive on- and off-campus practical teaching platforms. The construction of on-campus practical teaching platforms should adhere to the principles of resource sharing and platform integration. Taking advantage of academic competitions as an opportunity, experimental teaching resources should be systematically integrated. To overcome barriers related to laboratory ownership, college-level and university-level shared innovation laboratories should be established to provide open and flexible practical environments for both teachers and students. Leveraging technologies such as virtualization and big data, online experimental teaching platforms should be built to enrich high-quality experimental teaching resources, offering personalized and autonomous practical conditions for teachers and students. By tapping into the development needs of competition projects, a project management collaboration platform should be developed to increase teachers' engineering practices and teamwork capabilities.

(3) To make good use of off-campus practical teaching platforms, strengthening the integration of industry and education and collaborating with leading enterprises in the sector to establish joint laboratories, practical teaching bases, and collaborative innovation centers are essential. Teachers and corporate engineers should be guided to engage in topic collaboration, project tackling, and technical training activities, enhancing practical teaching standards through cooperation. Teachers should be encouraged to take up positions in companies for on-the-job training and to participate in major enterprise projects to improve their engineering practice capabilities. Support teachers and students in joining open-source communities and developer conferences organized by industry groups, aligning with the latest industry trends and fostering a habit of open-source contributions. Teachers should be guided to introduce corporate projects and engineering cases into academic competitions, leveraging these competition projects as "training grounds" to enhance teachers' practical teaching skills while guiding students to solve real-world problems.

4.3 Deepening the Cooperation Between Schools and Enterprises as a Means to Promote the Deep Integration of Competition and Talent Training

School–enterprise cooperation is crucial for deepening the integration of industry and education, enhancing the quality of talent cultivation. The computer science discipline has strong practicality and close ties with industries, necessitating a precise alignment between talent development and industrial needs. Competition in this field, characterized by its strong practicality and alignment with industry, stands as the optimal convergence point for school-enterprise collaboration and the integration of industry and education [11].

(1) Engaging in school-enterprise cooperation and jointly organizing academic competitions can introduce real projects and application scenarios from enterprises, allowing students to face industry technical challenges directly and strengthening the cultivation of engineering practice skills. By driving competition through school-enterprise collaboration projects, competition is more closely aligned with industrial needs, significantly enhancing the integration of competition and talent development.
(2) Different stages of competition can involve active participation from enterprises. In the question design phase, enterprise engineers can be invited to participate in setting questions, extracting competition tasks from actual projects and embedding industry needs into the questions. During the preparation phase, technical experts from enterprises can be invited for precomputation guidance, sharing practical industry experience and helping address technical challenges. In the implementation phase, enterprise experts can serve as judges, evaluating entries from perspectives such as technological advancement and application value. In the transformation phase, enterprises can connect with outstanding competition projects to provide subsequent industrial support, allowing the competition outcomes to be tested and improved through practical applications. Through cooperation between schools and enterprises, discipline competitions can be closely integrated with talent cultivation needs and course teaching design, incorporating industry demands throughout the process of topic selection, question development, and entry work, making discipline competitions an important window for assessing the quality of talent cultivation and highlighting its achievements.
(3) To deepen cooperation between schools and enterprises through academic competitions, it is essential to establish flexible and diverse cooperative mechanisms. By relying on industry associations and societies, competition cooperation organizations can be set up by jointly initiating specialized events or categories, with schools and enterprises working together to formulate competition plans and rules. Actively seek support from local governments and industrial parks to build off-campus competition training bases, involving both enterprises and the government in the organization and implementation of competitions. Encourage joint laboratory and practical teaching base collaborations with enterprises, engaging in deep cooperation in areas such as cultivating competition teams and developing competition platforms. Student competition teams should be supported in conducting internships, practical training, and project development at enterprises, reinforcing the integration of competitions with production practices. Improve systems for teachers to take up positions in enterprises and participate in enterprise projects, encouraging teachers

to bring competition projects to enterprises for technology transfer and achieving regularized school–enterprise cooperation.

5 Conclusion

The integration of industry and education is an inevitable path for the reform and development of higher education. Computer science competitions are crucial tools for deepening this integration and enhancing the quality of talent cultivation. This paper takes College S as a case study and analyses the developmental challenges faced by computer science competitions. It systematically outlines the implementation paths of computer science competitions against the background of industry-education integration from aspects such as improving competition management mechanisms, optimizing competition organization models, innovating competition content design, and strengthening the transformation of competition outcomes. The paper also constructs a system of computer science competitions in three dimensions: fostering students' innovative abilities, enhancing teachers' practical teaching skills, and deepening school-enterprise cooperation. These explorations and practices provide reference cases for promoting innovation and reform in computer science competitions in the context of industry-education integration.

Funding:. Project supported by Sanya Aviation & Tourism College, Project number: SATC2023JG-09.

References

1. Notice by the National Development and Reform Commission and Other Departments on Issuing the Implementation Plan for 'the Empowerment and Enhancement Action of Vocational Education through Industry-Education Integration (2023–2025)'. https://www.gov.cn/zhengce/zhengceku/202306/content_6886061.htm. Accessed 5 June 2025
2. The General Office of the CPC Central Committee and the General Office of the State Council Issued the 'Opinions on Promoting the High-Quality Development of Modern Vocational Education'. https://www.gov.cn/zhengce/2021-10/12/content_5642120.htm Accessed 5 June 2025
3. Liu, Z.: Constructing a high-level experimental teaching platform aimed at cultivating innovative talents. China High. Educ. **06**, 20–22 (2009)
4. Zhang, J., Li, J.: Teaching research on computer network technology major based on virtual simulation technology. Theory Pract. Educ. **45**(09), 56–59 (2025)
5. Wang, W.: Exploration of practical teaching in undergraduate discipline competitions for mechanical engineering—a case study of the "Intelligent Plastic Waste Sorting Bin" project. Plast. Ind. **53**(01), 172–173 (2025)
6. Zhan, H., Yan, X., Zhang, Z.: Exploration and practice of the integrated "Course–Competition–Innovation" collaborative education system. Lab. Res. Explorat. **43**(12), 113–116 (2024). https://doi.org/10.19927/j.cnki.syyt.2024.12.023
7. Tang, C., Lei, X., Qing, J.: Issues and countermeasures of provincial vocational college skills competitions from the perspective of regional economic development—an analysis based on competition data from Hunan Province. Vocat. Techn. Educ. **45**(06), 49–56 (2024)
8. Chen, L., Yang, Y., Lin, L., et al.: Implementation path of new engineering experimental competitions under the background of educational digital transformation. Lab. Res. Explorat. **42**(10), 174–180+195 (2023). https://doi.org/10.19927/j.cnki.syyt.2023.10.035

9. Lu, K., Duan, Z., Zhou, C.: Exploration and practice of deep integration of discipline competitions and provincial experimental teaching demonstration centers. Lab. Res. Explorat. **42**(06), 150–154 (2023). https://doi.org/10.19927/j.cnki.syyt.2023.06.030
10. Sun, L., Chen, Z., Zhang, J.: Research on training problems and countermeasures for vocational college skills competitions—based on interviews with participants in the automotive technology competition. Vocat. Techn. Educ. **45**(23), 53–58 (2024)
11. Huang, X.: Analysis of the talent cultivation model for innovation and entrepreneurship based on industry–education integration. J. Nanning Polytech. Univ. 1–7 (2025). https://doi.org/10.19846/j.cnki.nzxb.202503008

Research on the Evaluation of Teaching and Research Ability in Higher Vocational Colleges: Based on the Teaching and Research Achievements of Hainan Higher Vocational Colleges from 2020–2023

Kebu Wang[1], Yunge Gao[1], Xia Liu[2](✉), and Hongfeng Zhang[2]

[1] Sanya Aviation and Tourism College, Sanya 572000, China
[2] Faculty of Humanities and Social Sciences, Macao Polytechnic University, Macao 999078, China
p2315160@mpu.edu.mo

Abstract. This study focuses on 14 higher vocational colleges in Hainan Province. By collecting data from five dimensions—academic papers, research projects, patents, teacher teaching competitions, and teaching achievement awards—between 2020 and 2023, this research analyses the current status of teaching and research at these 14 higher vocational colleges in Hainan. Also evaluated the teaching and research capabilities of these institutions via the entropy value method. The findings reveal that academic papers of high volume and quality are predominantly found in Hainan's higher vocational colleges designated as "Double High" institutions, which possess distinct professional characteristics. With respect to research projects, Hainan's higher vocational colleges conduct scientific research grounded in provincial conditions, with a focus on the construction of the Hainan Free Trade Port and key local industries. In terms of patents, Hainan's higher vocational colleges place greater emphasis on the cultivation of utility model patents. With respect to teaching achievement awards, Hainan has a relatively high selection frequency, with a particular focus on participation from department heads and frontline teachers. In teacher teaching competitions, Hainan's higher vocational colleges utilize their professional characteristics to develop competition projects centered around clusters of majors. On the basis of the results of the data analysis, recommendations are proposed for transforming research paradigms, building high-quality teaching and research teams, establishing a sustainable teaching and research evaluation system, and implementing categorized management for key and general institutions, providing a basis for enhancing teaching and research capabilities at Hainan's higher vocational colleges.

Keywords: Vocational education · Integration of science · education · and technology · Vocational colleges · Teaching and research evaluation

Y. Yu et al. (Eds.): ICPCSEE 2025, CCIS 2673, pp. 175–199, 2026.
https://doi.org/10.1007/978-981-95-2566-9_13

1 Introduction

The 20th National Congress of the Communist Party of China proposed advancing the integration of vocational and general education, the integration of industry and education, and the integration of science and education, clarifying the research orientation of vocational colleges and strengthening the linkages among research, national scientific and technological innovation, and educational and teaching reforms. Against this backdrop, it is necessary to unify teaching and research and clarify the characteristics of teaching and research in vocational education [1, 2]. In 1990, Boyer first introduced the concept of "teaching scholarship," suggesting that teaching and scholarship complement each other. Pan Maoyuan argued that teaching and research activities in application-oriented and vocational colleges should be conducted in tandem.

In the context of performance management, vocational colleges generally face the issue of "disjointed" teaching and research [3]. Regarding the academic level of research, Haapakorpi A. and others have studied methods for evaluating faculty research performance [4–7]; Liu Hong (2019) analysed the research status of vocational colleges from the perspectives of papers, projects, and patents [8–10]. For teacher development evaluation, Xu Qiaoning (2021) analysed the status of teacher development on the basis of the vocational education development index [10], and Wang Bin (2021) developed a teacher development index evaluation model [11].

Most studies evaluate research performance and teaching development as independent topics. This paper argues that the teaching and research capabilities of vocational college faculty should encompass both research capabilities and teaching development levels. Drawing on the evaluation models of Wang Bin (2021) and the research output dimensions proposed by Liu Xia (2020), this study establishes an analytical framework for assessing the teaching and research capabilities of vocational colleges [8, 11]. The analysis is conducted from five aspects: academic papers, research projects, patents, teaching achievement awards, and teaching competitions.

On the basis of data resources from CNKI (China National Knowledge Infrastructure), the Ministry of Education, and other sources, this study focuses on the teaching and research capabilities of faculty in vocational colleges in Hainan. Through a comparative analysis using the above indicators, the entropy method is employed to construct a measurement system for evaluating teaching and research capabilities. The goal is to provide a reference for enhancing the teaching and research abilities of vocational colleges in Hainan and to fill the gaps in current research.

2 Research Design

2.1 Research Subject

As of 2023, there are 14 independent higher vocational colleges in Hainan (hereinafter referred to as "vocational colleges"), as detailed in Table 1. The study focuses on these 14 vocational colleges in Hainan, excluding institutions established for approximately three years or less. The data for this study cover the period from January 1, 2020, to December 31, 2023 and are sourced from various publicly available data dimensions provided by CNKI (China National Knowledge Infrastructure), the Hainan Provincial

Department of Education, the Hainan Provincial Department of Science and Technology, and the Hainan Provincial Social Union, among others.

Table 1. Data sources: National Higher Vocational College Status Data Monitoring Center

Order number	School name	Educational level	Date of establishment	Abbreviations used in the text	Remarks
1	Hainan College of Economics and Business	Higher vocational college	2004	HJM	National double high C level institutions, provincial double high institutions, national backbone higher vocational colleges
2	Hainan Vocational University	Higher vocational college	2000	HZY	Provincial double high school, national demonstration higher vocational college
3	Hainan Vocational College of Political Science and Law	Higher vocational college	2003	HVCPL	Provincial high-level professional group
4	Hainan Collegeof Software Technology	Higher vocational college	2003	HR	Provincial high school and provincial quality higher vocational colleges
5	Hainan College of Foreign Studies	Higher vocational college	2003	WYXY	Provincial high school and provincial quality higher vocational colleges

(*continued*)

Table 1. (*continued*)

Order number	School name	Educational level	Date of establishment	Abbreviations used in the text	Remarks
6	Hainan Provinclal Sports Academy	Higher vocational college	2016	HNTZY	
7	Hainan Technology and Business College	Higher vocational college	2004	HGS	Provincial high-level professional group
8	Hainan Health Management College	Higher vocational college	2016	HJK	
9	Hainan Health Vocational College	Higher vocational college	2021	HWJY	Provincial high-level professional group
10	Hainan Vocational University of Science and Technology	Vocational college	2007	HKJ	Provincial double high school
11	Sanya Aviation and Tourism College	Higher vocational college	2005	HKXY	Provincial double high school
12	SanYa Institute Of Technology	Higher vocational college	2009	LG	Provincial high-level professional group
13	Sanya City College	Higher vocational college	2001	CSXY	Provincial high-level professional groups
14	Hospitality Institute of Sanya	Higher vocational college	2017	ZR	Provincial high-level professional group

2.2 Selection and Construction of Indicators

On the basis of an analysis of related research on the evaluation of university research capabilities and teacher development levels and in conjunction with the current reform trends in higher education evaluation aimed at "breaking the five exclusives" and the practical realities of teaching and research management in vocational colleges, the evaluation of teaching and research capabilities in Hainan's vocational colleges is designed around five dimensions: academic papers, research projects, patents, teaching achievement awards, and teacher teaching competitions. The framework of the evaluation indicators is shown in Table 2.

Table 2. Evaluation index system of teaching and research ability of higher vocational colleges in Hainan

Target layer	Level 1 indicator	Level 2 indicators
Research and teaching ability of higher vocational colleges in Hainan	thesis	Number of core journal papers published
		Number of regular journal papers published
	topic	The number of projects approved by the Education Department of Hainan Province
		Number of projects approved by the National Natural Science Foundation
		Number of projects approved by the Social Science Foundation
	patent	Number of invention patents granted
		Number of utility model patents granted
	Teacher teaching competition	Number of first prize winners
		Number of second prize winners
		Number of third prize winners
	teaching results	Number of first prize winners
		Number of second prize winners

2.3 Research Methods

A combination of quantitative analysis and descriptive analysis is employed to conduct a detailed analysis of the teaching and research capabilities of vocational colleges in Hainan. On the basis of the data analysis, the extreme value treatment method and entropy method are used to construct a measurement system for evaluating the teaching and research capabilities of vocational colleges, followed by a comprehensive assessment of the 14 vocational colleges in Hainan.

2.3.1 Extreme Value Treatment Method

Data standardization includes both nondimensionalization and unification. The purpose of nondimensionalizing the indicators is to eliminate the incommensurability caused by differences in units across various indicators.

The original x_{ij} data used as indicators x'_{ij} in this study are normalized data. x'_{ij} To avoid the occurrence of $= 0$, positive indicators are defined as:

$$x'_{ij} = \frac{x_{ij} - x_{j\min}}{x_{j\max} - x_{j\min}}$$

where i is the evaluation j year number and is the evaluation index. To avoid too small a comprehensive evaluation score, the standardized value is shifted to between 1 and 2, and the standardized value is shown in Table 3.

Table 3. Standardized values of teaching and research ability evaluation indices in Hainan higher vocational colleges

School	Total number of papers	Total number of projects	Total number of patents	Total number of awards in teaching competitions	Total number of awards for teaching achievements
CSXY	1.14	1.00	1.00	1.03	1.17
HGS	1.16	1.21	1.00	1.28	1.50
HZY	1.46	1.59	1.53	1.64	1.67
HKXY	1.35	1.54	1.18	1.67	1.50
HJK	1.04	1.01	1.04	1.03	1.00
HJM	2.00	2.00	1.96	2.00	2.00
LG	1.21	1.24	1.00	1.14	1.50
HR	1.14	1.29	1.42	1.81	1.33
HNTZY	1.00	1.03	1.01	1.00	1.17
WYXY	1.29	1.22	1.00	1.17	1.33
HWJY	1.06	1.12	1.00	1.22	1.17

(continued)

Table 3. *(continued)*

School	Total number of papers	Total number of projects	Total number of patents	Total number of awards in teaching competitions	Total number of awards for teaching achievements
HVCPL	1.35	1.43	1.05	1.67	1.50
ZR	1.25	1.15	1.05	1.39	1.33
HKJ	1.51	1.69	2.00	1.67	1.67

2.3.2 Entropy Method

The entropy method determines the dispersion of indicators through the definition of information entropy. The smaller the value of information entropy is, the greater the dispersion of the indicator, and thus, its impact on comprehensive evaluation (i.e., weight) is greater. If an indicator's value is exactly equal, it has no effect on the comprehensive evaluation. Therefore, information entropy can be used to calculate the weights of each indicator, providing a basis for comprehensive evaluation of multiple indicators.

2.3.2.1 Calculation of the Entropy Value of the Evaluation Index

The E_j information j entropy of the index $K = \frac{1}{\text{In m}}$ is $P_{ij} = \frac{x\prime_{ij}}{\sum_{i=1}^{m} x\prime_{ij}}$ a i constant j; the proportion of the index in the year is the index, then:

$$E_j = -k \sum\nolimits_{i=0}^{m} (P_{ij} \ln P_{ij}) \ldots\ldots$$

2.3.2.2 Weight Calculation of Evaluation Indicators

The θ_j entropy j value (weight) of the index is set as:

$$\theta_j = \frac{1 - E_j}{\sum_{j=1}^{n} (1 - E_j)} \ldots\ldots$$

Among them, j for the first index, the greater the entropy weight is, the greater the impact of the index on the results, and vice versa. Therefore, five first-level indices and 12 s-level indices are calculated, as shown in Table 4.

Table 4. Weights of the teaching and research ability evaluation indices in Hainan higher vocational colleges

Target layer	Level 1 indicator	Variable	Weight of the first level indicator	Level 2 indicators	Variable	Level 2 indicator weight
Research and teaching ability of higher vocational colleges in Hainan	thesis	A_1	0.1536	Number of core journal papers published	B_1	0.5983
				Number of regular journal papers published	B_2	0.4017
	topic	A_2	0.1730	The number of projects approved by the Education Department of Hainan Province	B_3	0.2766
				Number of projects approved by the National Natural Science Foundation	B_4	0.4431
				Number of projects approved by the Social Science Foundation	B_5	0.2804
	patent	A_3	0.4297	Number of invention patents granted	B_6	0.5415
				Number of utility model patents granted	B_7	0.4585

(*continued*)

Table 4. (*continued*)

Target layer	Level 1 indicator	Variable	Weight of the first level indicator	Level 2 indicators	Variable	Level 2 indicator weight
	Teacher teaching competition	A_4	0.1550	Number of first prize winners	B_8	0.4949
				Number of second prize winners	B_9	0.2907
				Number of third prize winners	B_{10}	0.2144
	Teaching achievements won awards	A_5	0.0887	Number of first prize winners	B_{11}	0.8678
				Number of second prize winners	B_{12}	0.1322

2.4 Research Explanation

The construction of an evaluation index system for teaching and research capabilities in higher vocational colleges is a systematic project. The index system and analytical dimensions constructed in this paper are merely an initial exploration. The research results can to some extent truly reflect the actual teaching and research levels of teachers at higher vocational colleges in Hainan. Subsequent studies will be conducted on the basis of the assessment results to investigate the current status, issues, and governance optimization of teaching and research capabilities in Hainan's higher vocational colleges, exploring how to enhance teachers' teaching and research abilities through management.

3 Analysis of Teaching and Research Status in Hainan Vocational Colleges

3.1 Academic Paper Publication

Academic journal papers are a key medium for scientific research and a crucial indicator for measuring research capability. The total number of papers reflects the research activity in Hainan's vocational colleges, whereas the volume of publications in core journals reflects their research strength. This study utilized the "Advanced Search" function of the China National Knowledge Infrastructure (CNKI) to collect academic journal paper data from 14 vocational colleges in Hainan from 2020–2023. The evaluation of research performance is primarily based on two dimensions: the total number of papers published and the number of papers published in authoritative journals.

From 2020 to 2023, the 14 vocational colleges in Hainan published a total of 2,508 papers, with an average of 179 papers per institution. As shown in Fig. 1, the top six colleges published more than the average, with the HJM leading with 595 papers, more than three times the provincial average. Among the eight institutions with fewer than average publications, five had fewer than 100 papers, indicating relatively low research activity.

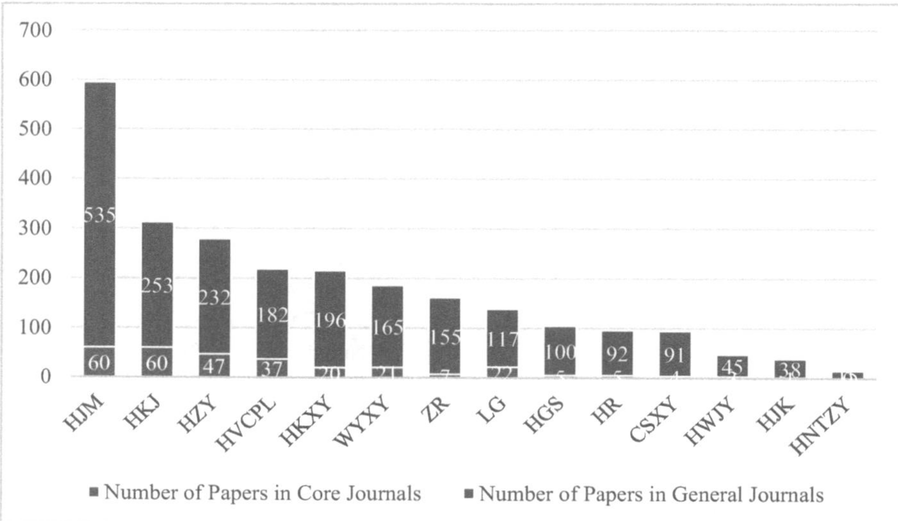

Fig. 1. Number of papers published by 14 higher vocational colleges in Hainan from 2020–2023

3.1.1 Publications in Core Journals

From 2020 to 2023, the 14 vocational colleges in Hainan published a total of 292 papers in core journals, with an average of 21 papers per institution. Seven institutions exceeded the average, five of which were "Double-High" institutions, and HVCPL was a provincial-level high-level professional cluster institution. HKJ and the HJM had the highest publication volumes, with their respective numbers being more than three times the provincial average. ZR increased its publication count to seven papers.

The core journal papers published by HZY are primarily concentrated in Peking University core journals, with HKJ and the HJM accounting for 70% of the provincial publication output. Eight institutions published papers in CSSCI core journals, with the HJM, HKJ, and the HVCPL each publishing more than five papers. Additionally, eight institutions published papers in CSCD core journals, with HKJ leading with 18 papers, which was twice the number published by the HJM.

Research strength is closely related to the overall strength of institutions. HR, HJK, CSXY, and HWJY should strengthen their research efforts to increase their competitiveness and soft power (Fig. 2).

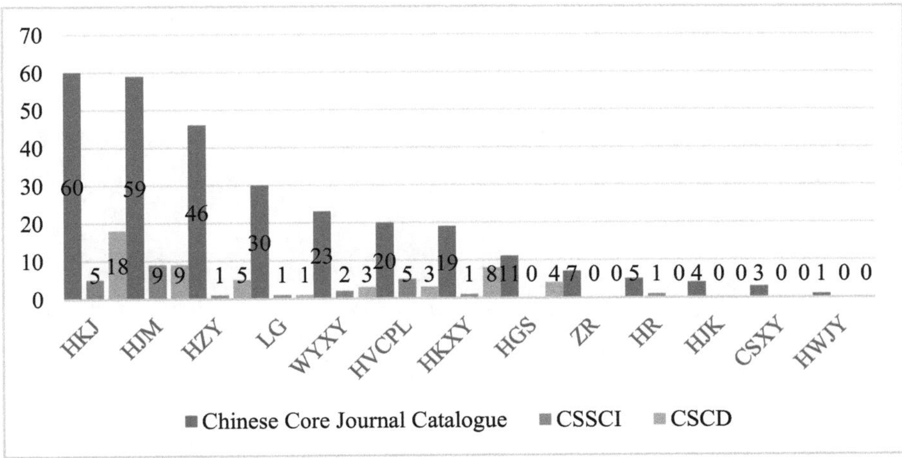

Fig. 2. Total number of articles published in core journals of 14 higher vocational colleges in Hainan from 2020–2023

3.1.2 Publications by Subfields of Specialization

CNKI's Database categorizes specialties into 19 major categories and 99 subcategories on the basis of the Chinese Library Classification System and the Ministry of Education's Directory of Higher Vocational Education (Associate Degree) Specialties (2015), covering various sectors of the national economy. By analysing the publication patterns across different specialties, it is possible to objectively reflect the research activity and competitiveness of institutions in various academic fields.

Since HKJ did not provide relevant data for analysis, it was excluded from the comparison of publications by subfield. Figure 3 displays the top five subfields of specialization in terms of publication volume across 13 vocational colleges in Hainan. Most institutions show the highest publication output in education-related specialties, indicating that vocational colleges place significant emphasis on vocational education reform and the development of educational content. This is particularly evident in the research on vocational education theory, talent cultivation, and the laws of teaching.

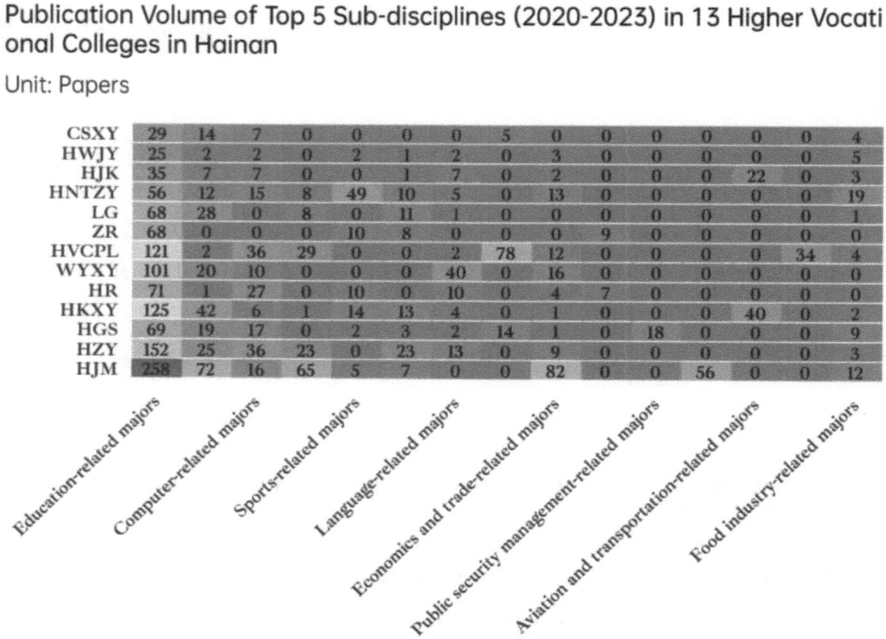

Fig. 3. Number of publications on professional subcategories (top 5) in 13 higher vocational colleges in Hainan from 2020–2023

From the perspective of paper publications, the disciplinary strengths of institutions are closely aligned with the industrial characteristics of the Hainan Free Trade Port. As a "Double-High" institution, the HJM focuses on hot issues in vocational education, with publications in related fields exceeding 50 papers. HZY's publications are concentrated in the fields of engineering, business management, and liberal arts, reflecting its comprehensive nature. HR, WYXY, and HVCPL show strong performance in their respective fields, whereas the HKXY's papers focus on air transport and tourism, and the HNTZY is concentrated in sports-related disciplines. ZR publishes more papers in tourism than in education, indicating its emphasis on developing its hotel and tourism programs.

3.1.3 Analysis of Academic Paper Authors

3.1.3.1 Active Authors in Academic Journals

Academic paper authors are the core of vocational education research, with active authors driving the development of the field. This study analysed the publication patterns of the top 40 first authors from the 14 vocational colleges in Hainan from 2020–2023. The 2,508 papers published by the 14 institutions in Hainan were authored by 500 individuals, with the highest number of papers published by a single author being 13 and the lowest being 2. On average, each author published 3 papers, with 236 authors publishing below the average.

According to Lotka's Law, a small number of scholars have contributed to the majority of the research output. From 2020 to 2023, the first authors who published more than six papers were defined as active authors, totaling 57 individuals, accounting for 9% of the authors. Their combined publications totaled 482 papers, representing 19% of the total number of papers. The average number of papers published by active authors was 8, with 28 from "Double-High" institutions, accounting for 70%. The remaining active authors from other institutions totaled 13, representing 32%. Notably, only one active author was from the HKXY and HR (Table 5).

Table 5. Active authors of papers in 14 higher vocational colleges in Hainan from 2020–202340 people (6 or more papers)

Order number	Author	The school where you are studying	Number of domestic journal papers	Order number	Author	The school where you are studying	Number of domestic journal papers
1 (3 persons)	Lin *	HJM	13	5 (12 persons)	Liu*	HKXY	7
	Li*Jia	HJM	12		Wang*Ping	HKJ	7
	Lin*	HJM	12		Wu*Li	HKJ	7
2 (2 persons)	Jiang*Xiang	WYXY	10		Fu*Jun	HVCPL	7
	Xu*Bin	HZY	10		Chen*	HZY	7
3 (5 persons)	Chen*Yan	HZY	9		Zheng*Ling	HJM	7
	Qin*	HJM	9		Hui*	HJM	7
	Zhu*	HJM	9	6 (14 persons)	Luo*Xiang	HGS	6
	Xia*	WYXY	9		Li*	ZR	6
	Wang*Yan	HJM	9		Guo*Li	ZR	6
4 (6 persons)	Wang*	HKJ	8		Cui*	HKJ	6
	Sun*	ZR	8		Zhang*Fei	ZR	6
	Li*	HJM	8		Ma*Wei	HGS	6
	Li*	HJM	8		Li*	HNTZY	6
	Chen*Li	HGS	8		Li*Mei	HJM	6
	Fu*	HJM	8		Liu*Chun	HGS	6
5 (12 persons)	Song*Jing	HKJ	7		Xu*Te	HJM	6
	Wu*Jin	ZR	7		Li*Min	HZY	6
	Qu*	ZR	7		Hou*Ying	HJM	6
	Chen*Hua	WYXY	7		Li*Lin	HJM	6
	Kang*	HVCPL	7		Feng*Na	HR	6

3.1.3.2 Active Authors in Core Journals

The key to academic quality lies in the level of scholarly work, particularly publications in core journals. In this study, first authors who published more than two papers in core journals from the 14 vocational colleges in Hainan were defined as "active authors in core journals." A total of 19 active authors were identified, with a combined total of 52 papers. The analysis reveals that among the active authors in core journals in Hainan, 53% are male and 47% are female. Among the male authors, 6 have published 3 or more papers. Among the active authors, 4 are professors (21%), 9 are associate professors (47%), and 6 are lecturers (32%). The majority of active authors are between the ages of 40 and 49, with associate professors and higher academic ranks being predominant. These findings suggest that these authors have accumulated significant experience and that their research capabilities have steadily improved. It is recommended that vocational colleges in Hainan provide more growth opportunities for post-80 s teachers to promote the production of high-quality academic outcomes (Table 6).

Table 6. Active authors of core journal papers in 14 higher vocational colleges in Hainan from 2020–2023 (publishing 2 or more papers)

Order number	Active author	Author's unit	Number of articles published	Sex	Professional ranks and titles
1 (1 persons)	Chen*Yuan	LG	5	woman	lecturer
2 (4 persons)	Bai*Jun	HJM	4	man	adjunct professor
	Song*Pei	HKJ	4	man	professor
	Wang*Hai	HVCPL	4	man	adjunct professor
	Kang*	HVCPL	4	man	adjunct professor
3 (3 persons)	Yang*Juan	HJM	3	woman	adjunct professor
	Xia*	WYXY	3	man	adjunct professor
	Li*	ZR	3	man	professor
4 (11 persons)	Li*	HJM	2	woman	adjunct professor
	Guo*Fang	HZY	2	woman	adjunct professor
	He*	HZY	2	woman	lecturer
	Liu*	HKXY	2	man	professor
	Chen*	HKXY	2	man	professor
	Hou*	HKXY	2	woman	adjunct professor
	Lv*Ru	WYXY	2	woman	lecturer
	Chen*	WYXY	2	woman	lecturer
	Jiang*Xiang	WYXY	2	man	lecturer
	Wang*Fa	HVCPL	2	man	lecturer
	Wang*Ping	HKJ	2	woman	adjunct professor

3.2 Analysis of Research Project Approvals

Research projects are important evaluation indicators for scientific research in vocational colleges. This study collects information on vertical research projects in Hainan from

2020–2023, with data sourced from official project approval documents published by the Hainan Provincial Department of Education, Hainan Provincial Social Union, and Hainan Provincial Department of Science and Technology.

3.2.1 Analysis of the Number of Research Project Approvals

From 2020–2023, the number of research projects approved in Hainan shows a clear "tiered effect," as shown in Fig. 4. As shown in Table 4, the 14 vocational colleges in Hainan collectively received approval for 503 research projects, with an average of 40 projects per institution. Institutions with a number of approved projects above the average are considered "active" institutions. From 2020 to 2023, the active institutions with a high number of approved projects include the HJM, HKJ, HZY, the HVCPL, and the HKXY. Notably, most of the active institutions are "double-high" colleges.

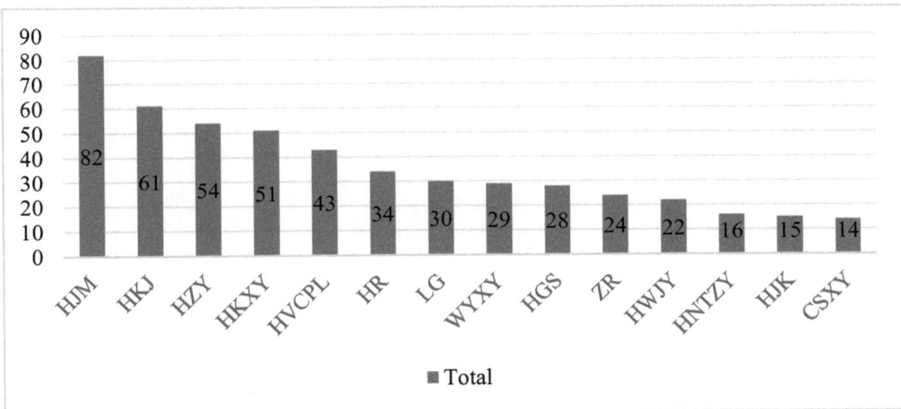

Fig. 4. Total number of projects approved by 14 higher vocational colleges in Hainan from 2020–2023

As shown in Table 7, from 2020–2023, most vocational colleges in Hainan focused primarily on applying for research projects related to educational reform and scientific research approved by the Hainan Provincial Department of Education. However, they exhibited insufficient research capability in the natural sciences, as evidenced by a low number of approved projects in this area.

3.2.2 Analysis of Research Themes by Category

Research projects are categorized into two types: educational reform and scientific research. Educational reform projects have focused mainly on topics related to talent cultivation, the integration of industry and education, curriculum development, the "Double-High" plan, and the 1 + X certification system. These topics form the core content of vocational education theory construction and the development of vocational education reforms. Scientific research projects, on the other hand, are centered primarily around the development of the Hainan Free Trade Port, which extends to areas such as Hainan's four major industries, international exchanges, and the collaborative development of related industries and the economy.

Table 7. Types of projects approved by 14 higher vocational colleges in Hainan from 2020–2023

School	Education Department project	Science and Technology Department project	Social Science Association project	Amount to
HJM	48	14	20	82
HKJ	26	13	22	61
HZY	40	9	5	54
HKXY	46	1	4	51
HVCPL	26	6	11	43
HR	27	4	3	34
LG	20	0	10	30
WYXY	26	0	3	29
HGS	20	2	6	28
ZR	22	0	2	24
HWJY	16	3	3	22
HNTZY	14	0	2	16
HJK	14	1	0	15
CSXY	14	0	0	14

3.3 Patient Achievements

Patents serve as crucial indicators of technological research and innovation in vocational colleges, particularly invention patents. This study collected patent data from 14 vocational colleges in Hainan between 2020 and 2023, covering both invention patents and utility model patents (excluding design patents). During this period, these institutions collectively filed 385 patent applications, of which 266 were granted. Nine institutions received patent authorizations, primarily "Double-High" institutions and science and engineering colleges, indicating a stronger emphasis on technological research and development in these institutions.

The number of patent applications reflects an institution's awareness and enthusiasm for patent activity, whereas the number of authorized patents represents their innovation and practical applicability. HKJ had the highest number of patent applications, totaling 142, more than twice that of other institutions. In terms of patent authorization rates, HKJ recorded 59.15%, the HJM 61.21%, the HR 93.33%, and HZY 89.58%, with science and engineering colleges generally exhibiting higher authorization rates.

With respect to patent types, the HJM focused primarily on invention patents and utility model patents. The HR and HKXY maintained a balanced distribution between patent applications and authorizations, whereas the HZY and the HJM placed greater emphasis on utility model patents (Table 8).

Table 8. Patent application and authorization statistics of 14 higher vocational colleges in Hainan from 2020–2023

School	Patent application			Patent licensing		
	Number of patent applications	Number of invention patent applications	Number of utility model patent applications	Number of patents granted	Number of invention patents granted	Number of utility model patents granted
HJM	116	62	54	71	27	44
HR	45	14	20	42	13	18
HZY	48	11	34	43	9	30
HKJ	142	65	68	84	28	46
HKXY	20	8	11	14	5	8
HVCPL	6	2	4	4	0	4
ZR	4	0	4	4	0	4
HJK	3	2	1	3	2	1
HNTZY	1	0	1	1	0	1
WYXY	0	0	0	0	0	0
HGS	0	0	0	0	0	0
LG	0	0	0	0	0	0
CSXY	0	0	0	0	0	0
HWJY	0	0	0	0	0	0

3.4 Analysis of Vocational College Faculty Teaching Development

Referring to the teacher development index evaluation system constructed by Wang Bin (2021) and other scholars, this study analyses the teaching development of faculty in Hainan vocational colleges using two tertiary indicators as primary dimensions: teaching achievement awards and faculty teaching competitions.

3.4.1 Analysis of Teaching Achievement Awards

The data on teaching achievement awards in Hainan's vocational colleges are sourced from the official announcements of the awarded projects published by the Hainan Provincial Department of Education in 2020 and 2022. The key information of the provincial-level teaching achievement awards in higher education in Hainan includes the project title, contributors, affiliated institutions, academic disciplines, and evaluation results.

3.4.1.1 Analysis of Award Levels

In terms of award levels, Table 9 presents the distribution of teaching achievement awards received by the 14 vocational colleges in Hainan from 2020–2022. A total of 35 awards were granted across these institutions, including 11 first prizes and 24 s prizes, with an average of fewer than three awards per institution. The top six institutions in terms of the number of awards were all "double-high" construction colleges or high-level professional cluster institutions. Notably, the College of Economics and Trade received more than twice the average number of awards.

With respect to the types of achievements recognized, most awards were granted for research in practical exploration, whereas research related to textbooks was limited to a single project, which received a second prize.

Table 9. Teaching achievement awards of 14 higher vocational colleges in Hainan from 2020–2022

Order number	School	First prize	Second prize	Amount to
1	HJM	3	3	6
2	HZY	2	2	4
3	HKJ	2	2	4
4	HGS	1	2	3
5	HVCPL	1	2	3
6	HKXY	2	1	3
7	LG	0	3	3
8	HR	0	2	2
9	WYXY	0	2	2
10	ZR	0	2	2
11	HNTZY	0	1	1
12	CSXY	0	1	1
13	HWJY	0	1	1
14	HJK	0	0	0

3.4.1.2 Analysis of Award-Winning Project Types

Awards can be categorized by institution type and project category. The award-winning units for teaching achievements are divided into three categories: provincial "Double High" construction units, provincial high-level professional cluster construction units, and ordinary higher vocational colleges. According to Table 10, the number of awards won by provincial "Double High" construction units is the highest, accounting for 57.14% of the total, followed by provincial high-level professional cluster construction units, which account for 25.71%. The number of awards won by "Double High" institutions is four times greater than that won by other institutions, with four "Double High" institutions winning first and second prizes and provincial high-level institutions (business and law) winning twice. This finding indicates that "Double High" and high-level professional cluster construction units place greater emphasis on cultivating teaching achievement awards. In terms of award categories, educational outcomes constitute the largest share, exceeding 58%, followed by ideological and political education outcomes and teaching reform outcomes.

Table 10. Distribution of the number of teaching achievement awards won by 14 higher vocational colleges in Hainan from 2020–2022

Unit type	First prize	Second prize	Proportion	Single-digit
Provincial "double high plan" construction unit	9	11	57.14%	6
Provincial high-level professional group construction unit	2	7	25.71%	3
Ordinary higher vocational colleges	0	4	11.43%	4

3.4.1.3 Analysis of Collaboration in Awarded Projects

Collaboration in awarded projects can be categorized into institutional collaboration and contributor collaboration. According to statistics, 21 awarded projects were independently completed by a single institution, accounting for 60% of the total, whereas 14 projects were jointly applied for by two or more institutions, representing 40%. The forms of collaboration include interinstitutional cooperation and multistakeholder collaboration. Interinstitutional cooperation involves partnerships between vocational colleges and various educational institutions, such as undergraduate universities, secondary vocational schools, kindergartens, and primary schools, with four such projects accounting for 11.43% of the total. Multistakeholder collaboration, involving vocational colleges working with enterprises, industries, government agencies, and public institutions, accounted for 13 projects (37.14%), reflecting a new model of industry-education integration.

With respect to the collaboration among project contributors, only one second-prize project was completed independently by a single individual, whereas the largest project, which won a first prize, involved more than 30 contributors. The majority of projects had collaboration teams consisting of 5 to 10 members. The cultivation of teaching achievement awards places greater emphasis on teamwork, with teams comprising enterprise personnel, school administrative leaders, and faculty members. Notably, administrative staff had a relatively high participation rate, whereas the involvement of full-time faculty members was comparatively lower.

3.4.2 Awarded in Teaching Competitions

Here, teaching competition refers mainly to the teaching ability competition of higher vocational colleges and the teaching competition for young teachers. Table 11 shows the winning situation of the teaching competition of teachers in 14 higher vocational colleges in Hainan from 2020–2023.

3.4.2.1 Analysis of the Level of Winning Projects in Teacher Teaching Competitions
In terms of the award level, the number of first prizes in economics and trade accounts for half of the total number of first prizes, which reflects that economics and trade attach great importance to the cultivation work of teachers' teaching competitions, strengthen internal constructions, always emphasize student-centeredness, and deepen the reform of three teachings.

The total number of award-winning projects from the three universities in HNTZY, HJK and CSXY is less than 10. We should attach great importance to improving teachers' teaching ability, actively building high-level teaching staff, and realizing "promoting teaching through competition".

Table 11. Award-winning situation of teachers' teaching competitions in 14 higher vocational colleges in Hainan from 2020–2023

Order number	School	First prize	Second prize	Third prize	Total
1	HJM	17	15	9	41
2	HKJ	6	8	15	29
3	HZY	7	9	12	28
4	HKXY	3	13	13	29
5	HVCPL	1	11	17	29
6	HR	6	15	13	34
7	LG	0	1	9	10
8	WYXY	2	3	6	11
9	HGS	1	4	10	15
10	ZR	4	4	11	19
11	HWJY	1	7	5	13
12	HNTZY	1	0	4	5
13	HJK	0	1	5	6
14	CSXY	0	0	6	6

3.4.2.2 Analysis of the Types of Courses that Won the Teaching Ability Competition
The award-winning situations of the teaching ability competition courses at Hainan's various higher vocational colleges from 2020–2023 are shown in Table 12. The public foundation course group won a total of 44 awards, with an award rate of 18.11%; the first and second professional course groups won a total of 199 awards, with an award rate of 81.89%. The award rate for professional courses is higher than that for public foundation courses because they better connect professional characteristics with industry backgrounds, making it easier to achieve innovation. Additionally, the emphasis on professional courses by higher vocational colleges is also a key factor. In contrast, public foundation courses tend to focus more on theory in their teaching content, with less integration of practical teaching, and course design needs to enhance interest and novelty.

Table 12. Award-winning situation of the course group in the teaching ability competition of 14 higher vocational colleges in Hainan from 2020–2023

Order number	School	Public infrastructure group	Professional Course Group One	Professional Course Group 2
1	HJM	9	23	4
2	HKJ	4	18	5
3	HZY	4	12	9
4	HKXY	2	16	7
5	HVCPL	6	15	7
6	HR	6	21	5
7	LG	2	3	3
8	WYXY	2	5	3
9	HGS	1	1	9
10	ZR	3	8	7
11	HWJY	0	8	2
12	HNTZY	2	1	1
13	HJK	1	0	3
14	CSXY	2	1	2
Amount to		44	132	67

3.5 Comprehensive Score of the Teaching and Research Ability of Hainan Higher Vocational Colleges

$$S_i = \sum\nolimits_{i=1}^{n} \theta_j \times P_{ij} \ldots\ldots_i = \sum\nolimits_{i=1}^{n} \theta_j \times P_{ij} \ldots\ldots$$

The comprehensive evaluation scores for the teaching and research capabilities of Hainan's higher vocational colleges were calculated, and the ranking results are shown in Fig. 5. As shown in the figure, the evaluation scores for the teaching and research capabilities of Hainan's higher vocational colleges from 2020–2023 form a ladder shape. The HJM, as a national double-high institution, ranks alongside the provincial double-high institution HKJ in the first tier. HZY follows closely behind, standing out in the second tier. The two provincial double-high institutions in HR and HKXY have a gap of more than 20 points compared with the two provincial double-high institutions in HVCPL, which belong to the third tier. The remaining eight institutions are located in the fourth tier, among which ZR, HGS, WYXY, and LG have similar strengths with minimal differences.

There is clearly a noticeable gap between high-level, distinctive institutions and others. Each institution still needs to focus on connotative development, solidify teaching and research work, and promote innovative teams in teaching and research as a model to drive all teachers to concentrate on key areas of school-based research, producing more results and enhancing the soft power of education.

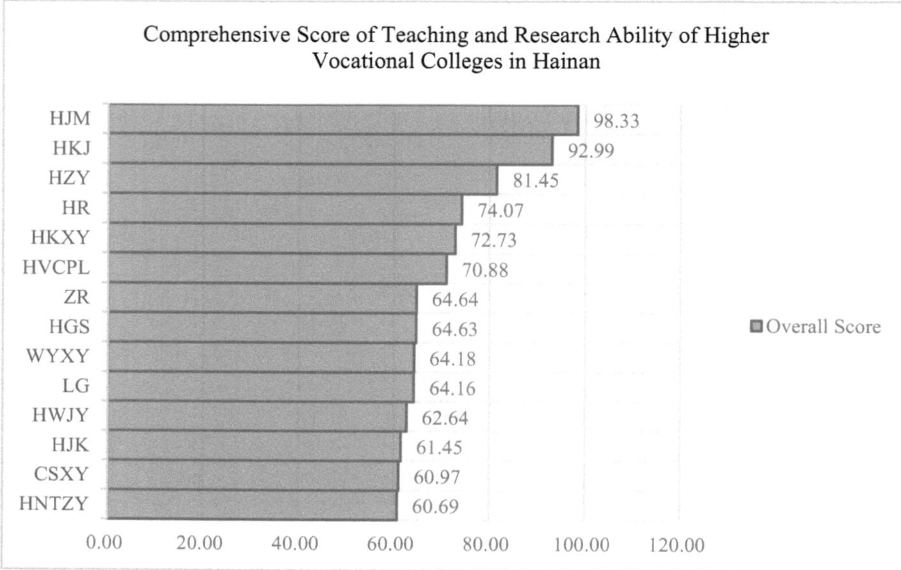

Fig. 5. Schematic diagram of the teaching and research ability evaluation of Hainan higher vocational colleges from 2020–2023

4 Characteristics and Deficiencies of Teaching and Research Development in Hainan Higher Vocational Colleges

4.1 Characteristics of Teaching and Research Development in Hainan Higher Vocational Colleges

In recent years, Hainan's higher vocational colleges have closely aligned with local economic development, fostering the interactive growth of research and teaching, which has led to a substantial increase in educational research outcomes. The specific characteristics are as follows: 1. Closely adhering to national policies and local economic demands: Research initiatives are integrated with the development of the Hainan Free Trade Port and key industries, resulting in industry-specific accomplishments; 2. Diverse engagement in research: Through partnerships among schools, businesses, and governments, interdisciplinary research communities are established to encourage collaborative innovation. 3. Focusing on professional clusters as the nucleus, highlighting disciplinary strengths: Leveraging professional clusters for project proposals, academic research, and the development of teaching outcomes, a robust environment of disciplinary excellence is cultivated; 4. The impact of teaching achievement awards: Hainan Province biennially selects teaching achievement awards, encouraging the development of teaching accomplishments and advancing distinguished teams and projects; 5. Emphasizing the involvement of frontline teachers and program leaders: In teaching reforms, the participation of program leaders and frontline teachers is valued, with 25.71% of provincial teaching achievement award-winning projects being led by them.

4.2 The Development of Teaching and Research in Hainan Higher Vocational Colleges is Insufficient

The development level of teaching and research in Hainan's higher vocational colleges is essentially positively correlated with their overall strength, but the following issues remain: 1. Overemphasis on quantity at the expense of process: Academic papers are published mostly in ordinary journals, leading to a disconnect between research outcomes and industry application; patent achievements are relatively weak, and project management focuses more on results rather than providing incentives and support for horizontal projects; 2. Fewer national-level teaching and research achievements exist: Hainan's higher vocational colleges lack national teaching achievement awards, with only one award received by an economic and trade institution. The number of awards won in national teaching ability competitions is limited, indicating a low overall level of teaching development that requires increased investment and attention in national projects. 3. Significant disparities in teaching and research development: "Double high" institutions and key universities in Hainan lead in teaching and research, while most ordinary higher vocational colleges have relatively stagnant progress, failing to form a reasonable gradient and showing significant differences in teaching and research levels.

5 Strategies for Improving the Teaching and Research Ability of Hainan Higher Vocational Colleges

5.1 Taking Applied Research as the Goal, Transforming the Scientific Research Paradigm to Help Produce High-Quality Results

1. The importance of scientific research should be increased: a value system for applied research should be established, teaching and research should be integrated, and disconnection between them should be avoided. 2. Proactive transformation in educational research: Encouraging teachers to engage in research activities on the basis of their specialties, fostering an active awareness of educational research, and improving the quality of these activities. 3. Plan for enhancing research competence: On the basis of teachers' research foundations and professional development, categorized training programs should be developed to cultivate high-quality research outcomes.

5.2 Establishing a Sustainable Teaching and Research Evaluation System Based on the Characteristics of Vocational Education

1. Categorized evaluation: Classifying evaluations on the basis of the characteristics of disciplines, with a focus on the actual value and market impact of outcomes; 2. Quantitative and Qualitative Integration: Evaluate not only the number of papers but also the difficulty of publication and its correlation with the institution's educational standards; 3. Emphasize process evaluation: Avoid overemphasis on results, strengthen process evaluation to ensure the depth and quality of teaching and research activities, prevent academic misconduct, and stimulate teachers' enthusiasm for research.

5.3 Key and General Institutions Should Be Classified for Management to Narrow the Gap in Teaching and Research Levels.

1. Development of Double-High Institutions: Clarify their positioning, actively participate in national teaching and research activities by integrating distinctive specialties, break through development limitations, and enhance the level of teaching and research; 2. Government Support and Resource Integration: Provincial and municipal governments should strengthen support for general institutions, tilt policies and funding towards them, and encourage institutions to seek nonfiscal resources to promote interinstitutional exchanges and cooperation.

5.4 Building a "Golden Teacher" Teaching and Research Team and Promoting the Feedback of Teaching and Research Results to Education

1. Transformation of Teaching and Research Achievements: Vocational colleges should integrate teaching, research, and entrepreneurship to establish high-quality innovative teaching and research teams, promote the transformation of teaching and research achievements and feed back into teaching. 2. School-Enterprise Cooperation for Practical Training: Through school-enterprise cooperation, teachers and students can enter enterprises for practical training, bringing enterprise resources into classrooms to enhance their practical teaching skills. Project research should focus on real-world issues faced by enterprises, improving teachers' problem-solving abilities.

5.5 Adhering to the Concept of "Promoting Teaching Through Competition" and Promoting the Development of Teaching Competitions

1. Teacher Capacity Enhancement: Encourage teachers to participate in teaching competitions and implement a comprehensive teacher capacity enhancement program to rapidly improve their teaching skills, especially in digital teaching. 2. Digital Platform Development: Schools and education departments should increase investment in the construction of digital communication platforms to provide convenient channels for teachers to communicate and share research, thereby increasing the level of educational research.

References

1. Hao, T.: Research on the innovation of scientific research paradigm in higher vocational colleges under the background of integration of science and education. China Vocat. Techn. Educ. **06**, 11–16 (2023)
2. Hua, D.: Research and management of vocational colleges under the perspective of integration of science and education: logic, dilemma and path. Vocat. Techn. Educ. **44**(10), 31–36 (2023)
3. Lu, G., Wang, X., Zhang, C., et al.: Teaching development index of teachers in ordinary undergraduate colleges and universities in china: design, practice and implications. Chin. High. Educ. Res. **07**, 6–11 (2019). https://doi.org/10.16298/j.cnki.1004-3667.2019.07.02
4. Haapakorpi, A.: Doctorate holders outside the academy in Finland: academic engagement and industry-specific competence. J. Educ. Work. **30**(1), 53–68 (2017)
5. Kozleski, E.B., Handy, T.: The cultural work of teacher education. Theory Pract. **56**(3), 205–213 (2017)

6. Han, X., Ma, R., Wu, W., et al.: Research on performance evaluation of college teachers based on classification and typing. J. Chongqing Univ. (Soc. Sci. Ed.) **20**(1), 114–119 (2014)
7. Bao, Y., Liang, R., Song, Y., et al.: Research evaluation status and countermeasures of teachers in applied universities. J. Nanjing Inst. Technol. (Soc. Sci. Ed.) **17**(3), 67–70 (2017)
8. Liu, X., Chen, L., Liu, K., Li, M., Chen, M.: Research on the evaluation of scientific research capacity in higher vocational colleges——taking the scientific research achievements of hainan higher vocational colleges in 2019 as an example. Vocat. Techn. Educ. **41**(33), 28–34 (2020)
9. Liu, H., Kuang, H.: 2019 National vocational college research achievements data analysis——based on the data of china national knowledge infrastructure. Chin. Vocat. Techn. Educ. **36**, 17–26 (2019)
10. Xu, Q., Zhao, C., Wu, Y., et al.: Analysis of the current status, problems, and suggestions on teaching development of teachers in national higher vocational colleges——based on the 2020 edition of the teacher teaching development index. China High. Educ. Res. (03), 95–101 (2021). https://doi.org/10.16298/j.cnki.1004-3667.2021.03.15
11. Wang, B., Wang, Y., Xu, Z.: Evaluation index system of teacher development index in higher vocational colleges: model, construction and application. China Vocat. Techn. Educ. **21**, 79–86 (2021)

Machine Learning for Data Science

Segmentwise Multilayer Perceptrons for Speech Emotion Recognition

Ziying Zhang(✉) and Changzheng Liu

College of Computer Science, JIAYING University, Meizhou, China
{zhangziying,Liuchangzheng}@jyu.edu.cn

Abstract. With the increasing popularity of mobile internet devices, speech emotion recognition has become a convenient and valuable means of human–computer interaction. The performance of speech emotion recognition depends on the discriminating and emotion-related utterance-level representations extracted from speech. Moreover, sufficient data are required to model the relationship between emotional states and speech. Mainstream emotion recognition methods cannot avoid the influence of the silence period in speech, and environmental noise significantly affects the recognition performance. This study intends to supplement the silence periods with removed speech information and applies segmentwise multilayer perceptrons to enhance the utterance-level representation aggregation. In addition, improved semisupervised learning is employed to overcome the problem of data scarcity. Particular experiments are conducted to evaluate the proposed method on the IEMOCAP corpus, which reveals that it achieves 68.0% weighted accuracy and 68.8% unweighted accuracy in four emotion classifications. The experimental results demonstrate that the proposed method aggregates utterance-level more effectively and that semisupervised learning enhances the performance of our method.

Keywords: speech emotion recognition · segmentwise multilayer perceptron · semisupervised learning · emotion classification

1 Introduction

According to psychology, emotion is the attitude and experience of human beings toward objective things. It reflects the tendency and feelings of individuals towards behavioural goals and is a complex and stable physiological evaluation. In daily life, emotions affect people's understanding and attitudes and can regulate their behaviour. Reasonable and adequate emotional expression is vital in promoting interpersonal communication and establishing friendly, intimate relationships. With the vigorous development of emerging technologies such as artificial intelligence, big data and cloud computing, the digital technology industry continues to improve the deep integration of computer devices and human social production activities. Traditional human–computer interactions are convenient and accurate and are not easy to understand and adapt to people's feelings or moods. Hence, achieving natural harmony in human–computer interactions is complex

Y. Yu et al. (Eds.): ICPCSEE 2025, CCIS 2673, pp. 203–213, 2026.
https://doi.org/10.1007/978-981-95-2566-9_14

but essential. Research on emotional signals in affective computing can be divided into two main categories. The first is the external expression of emotion (the quantitative form of actions shown by various parts of the human body when emotion changes, including facial expressions, language and body movements). The second is physiological arousal, which manifests as a change in the emotional state that causes a change in physiological signals. Physiological signals include electrocardiogram (ECG), photoplethysmography (PPG), and respiratory (RSP) electroencephalography (EEG) signals. According to neuroscience, the physiological signals of the human body are closely related to the autonomic nervous system, and it is difficult for individual subjective consciousness to affect physiological signals; thus, physiological signals can more genuinely reflect the emotional state of the human body. Compared with facial expressions, physiological signals are a long sequence of continuous signals that are smoother in terms of the expression transition between different emotional states. Emotion recognition, which is based on physiological signals, involves collecting signals through wearable sensors, such as ECGs and EEGs. Wearing sensor devices interferes with users' activities and affects their emotional state, so this method is unsuitable for daily use. Therefore, using wireless signals to realize emotion recognition has greater research value. Speech is an essential communication tool and the most direct communication channel for human beings. By changing the tone of voice, using modal particles and changing the volume of speech, the speaker can make the listener pay attention to the emotional changes and attitude tendencies of the speaker over time. With the wide application of electronic computers and artificial intelligence machines, voice communication, as the most natural, convenient, accurate and efficient method of expression, speech emotion recognition (SER) has become the hot topic of affective computing research.

2 Speech Emotion Recognition Overview

Research on speech emotion recognition technology started in the 1970s. With the development of computer hardware technology and improvements in computing capability, the field of speech emotion recognition has also gradually developed. People have begun to explore and model different emotional states in speech signals and effectively classify emotions by analysing and simulating emotional information. In 1972, Williams proposed that the pitch frequency in speech signals would be discriminated against under different emotional states, which can be used as the basis for emotional classification. In the mid-1980s, Bezooijen and Scherer distinguished emotional states via acoustic features. In 1996, Dellart et al. extracted the fundamental frequency parameters and acoustic characteristics of speech signals and used maximum likelihood Bayes, the k-nearest neighbor method and the kernel regression method as classifiers to recognize five emotions, including anger, happiness, sadness, fear and neutrality. At the beginning of the 21st century, Schuller et al. systematically introduced the application and achievements of the hidden Markov model (HMM) and Gaussian mixture model (GMM) in speech emotion recognition [1]. In 2010, with the widespread popularity of machine learning methods [2], such as support vector machines (SVMs), SVM-based classification methods made new progress in speech emotion recognition. Pan introduced mainstream speech features and used SVM-based recognition methods to achieve more accurate recognition results on multiple datasets. Huang combined a deep support vector machine with a deep confidence network (DBN), used amplitude and pitch frequency as features to identify four basic emotions, i.e., happy, angry, sad and neutral, and achieved the highest recognition accuracy of 86.5%. In 2014, the deep neural network (DNN) brought new ideas to emotion recognition. By extracting high-dimensional features from speech signals, it can effectively distinguish the fine-grained differences between different emotions. Compared with machine learning recognition methods, it has significantly improved recognition, and DNN-based recognition methods have become mainstream. With the development of deep learning technology, the end-to-end (E2E) method [3, 4] has been widely used, and the traditional low-level descriptor based on manual extraction has been replaced by the deep emotional features captured through the automatic learning of deep neural networks. In 2016, Trigeorgis et al. proposed using end-to-end recognition methods to automatically learn emotional features in speech signals to solve the problem that artificial feature extraction needs to be more robust in the case of multiple speakers. Owing to the variety of speech, the generalization performance of the deep SER model is still limited by several factors, such as cross-gender, cross-racial and noising issues [5]. To further improve the performance of SER, efforts should be made to solve two primary challenges:

- The features collected from the raw speech need to be more emotionally relevant for recognition.
- The use of more labelled data to fit the deep model properly makes it difficult to improve the generalization performance in complex situations.

In recent years, convolutional neural networks (CNNs), which are widely used in E2E speech emotion recognition systems, have demonstrated the powerful ability of delicate grain feature extraction. Panagiotis Tzirakis et al. [6] leveraged a CNN and a recurrent neural network (RNN) in an E2E learning framework and achieved state-of-the-art results in the RECOLA database. Caroline Etienne et al. [7] proposed a method to use data augmentation to improve the performance of various CNN+RNN architectures. Seyedmahdad Mirsamadi et al. [3] studied different methods for aggregating features behind an RNN and proposed a novel method for feature pooling over time, which uses local attention and aims to focus on specific regions of a speech signal that are more emotionally salient. Jian Huang et al. [8] proposed multimodal transformer [9] fusion, which uses self-attention in model-level fusion to improve performance. To improve the accuracy of speech emotion recognition, the mel frequency cepstrum coefficient (MFCC) of the speech signal is chosen as the input of the deep learning network, and the ResNet-BiGRU network, which is based on the attention mechanism, is used to extract the MFCC information via the ResNet-BiGRU network, which is based on the attention mechanism in [10]. Ravi et al. [11] proposed a nonlinear feature extraction technique to improve the classification performance of the SER system. The proposed method explores variational mode decomposition (VMD) with the Teager-Kaiser energy operator (TKEO) for the SER.

Although previous methods have achieved progress in SER, some limitations need to be overcome to improve the effect of SER in specific circumstances. Timeliness and accuracy are essential for inferring emotions on the mobile device side. The RNNs cannot be used for parallel computing and cannot obtain real-time recognition results [12].

The transformers perform excellently in learning intrinsic feature sets. However, the self-attention function requires considerable computing resources and sufficient data [5]. Therefore, in the current mobile terminal scenario, speech emotion recognition requires a lightweight architecture and scalability to face various situations flexibly. This article proposes a lightweight and practical architecture, segmentwise multilayer perceptron (SEMLP), for SER and utilizes patch information to enhance feature aggregation. SEMLP takes advantage of segmented data to learn the correlation in speech. The experiments of the proposed method are conducted on the IEMOCAP corpus, which achieves results competitive with those of CNNs and RNNs.

To circumvent the problem of data scarcity, data augmentation [7] methods have proven effective for SER. On the basis of the concept that augmentation can retain emotional expression, appropriate noise allows the model to learn the differences against vibration, which makes the model robust to noise disturbances. To improve the performance of SER in complex situations, this study adopted an improved unsupervised data augmentation (UDA) [13] method to relieve the influence of data scarcity. A large amount of unlabelled data is used to help the model learn about the prior distribution of the speech database [14], thereby improving the performance and robustness of SER methods [15]. However, the unlabelled data from different domains introduce additional learning costs, which decreases the model's generalization performance. Thus, this article improved UDA with the self-pseudoaugmentation (SPA) [16] method, which uses a pretrained model to filter out-of-domain data with confidence-based probabilities.

As mentioned above, this article proposes a series of methods to improve the performance of the SER task:

- Use segmented speech without silent periods to decrease the influence of noise contamination.
- Propose the SEMLP architecture to utilize patch features to aggregate word-related embedding from segmented speech.
- Leverage the SPA-UDA method to relieve the problem of data scarcity and increase the robustness to noise.

3 Methodology

In this section, the SEMLP architecture and SPA-UDA strategy are proposed in detail. Each module of the SEMLP is depicted in Fig. 1. An overview of SPA-UDA is depicted in Fig. 4.

3.1 SEMLP Architecture a Subsection Sample

This article proposes a lightweight SEMLP architecture to explore the effect of the multilayer perceptron module in utterance-level representation aggregation. The SEMLP architecture consists of four modules, including a frontend module to extract Mel-frequency cepstrum coefficients (MFCCs) from segmented speech, a feature extraction module to extract speech embedding, an SEMA module to aggregate utterance-level representations in two different aspects, and a back-end module to predict the probabilities

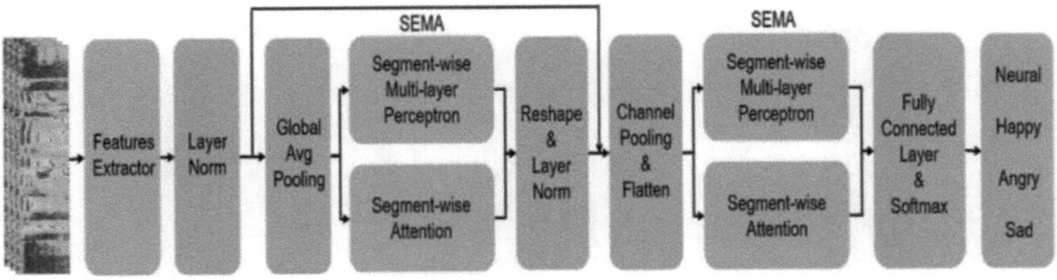

Fig. 1. Overview of the SEMLP architecture. The feature extractor module extracts deep features from MFCC. Layer the Norm function to smoothing loss in training. The global average pooling function pools the patch information and uses SE-MLP and SE-Attention (SEMA) to aggregate channelwise information. The channel pooling and flattening functions extend the patch information, which is similarly aggregated via SEMA. The fully connected layer with the Softmax function linearizes the features for classification.

of four emotion classes. The parts of the segmented speech processing feature extractor and SEMA module are specifically discussed below. This module is based on the Conv-pooling operation to complete the extraction of emotional features. This feature extraction module can obtain high-level abstract features of low-level descriptors of the original speech signal. In this way, the subsequent multiscale feature fusion module can model the corresponding relationship of emotional states on the basis of this abstract feature.

1) Segmented Speech: When the mobile terminal receives voice signals, the voice length is uncertain. There may be a long silence period, which consumes many computing resources and does not help the final emotion recognition. In a noisy environment, the silent period and pause are affected by the voice of others, which is more likely to lead to incorrect recognition results of the model. To improve the usability of emotion recognition models in more scenarios, specific means are needed to remove the silent period to reduce the consumption of computing resources and improve recognition performance. This work explores a novel method of fragment processing. First, the input voice frame is mapped to each phoneme by a forced alignment mechanism. Then, the phonemes of words are combined and marked via natural language processing methods, and each segmented speech is labelled with each word's start and end frames. Finally, these segmented speeches are used to model the emotional state and help enhance the semantic understanding of the model to improve the effect of emotion recognition.
2) Feature Extraction Module: The details of the feature extraction module are shown in Fig. 2. It consists of four convolutional (Conv) layers and three pooling layers. There is a batch normalization (BN) function to accelerate convergence and reduce model oscillation and exponential linear unit (ELU) activation to generate a soft nonlinear transformation. In addition, the average (AVG) pooling layer with a fixed dropout rate reduces the resolution of the feature map, providing in-depth emotion-related information for the next Conv layer.
3) SEMA Module: This article discusses how to use abstract features to effectively model emotion states. Intuitively, the SEMA module includes two blocks, as illustrated in Fig. 3.

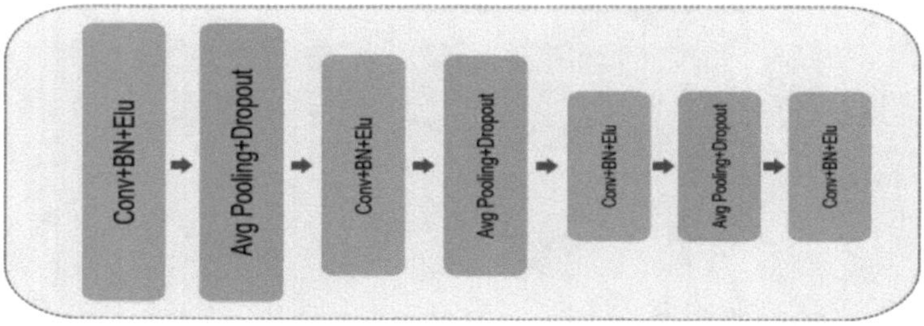

Fig. 2. Detailed architecture of the feature extraction module. The pooling block progressively reduces the resolution of the data. Moreover, the Conv blocks extract the local-global features of segmented speech.

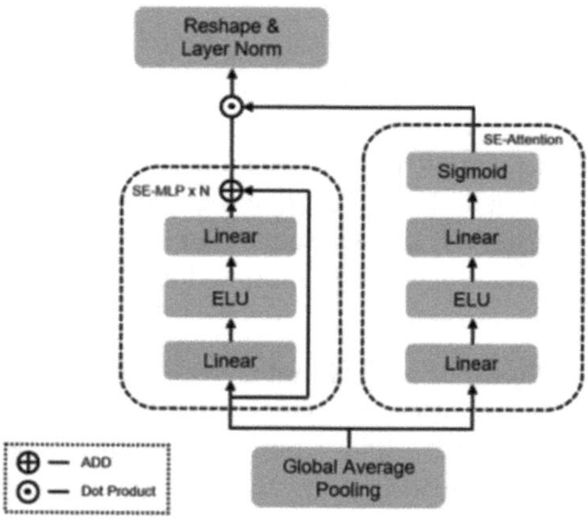

Fig. 3. Detailed architecture of the SEMA module. The SEMLP block is used to learn a function to model the relationship between features and emotional states. The SE-Attention block learns and reflects the hidden connections in features. Each block has a shortcut to avoid gradient disappearance.

In the SEMA module, serial interaction processes features in stages: first, nonlinear feature representation is enhanced via SEMLP; then, critical channel information is dynamically selected through SE attention. The interaction process is as followsNonlinear Enhancement and Channel Interaction:

Global Pooling: Global Pooling: Apply global average pooling over spatial dimensions $H \times W$ to generate channel descriptors $z \in \mathbb{R}^C$, as shown in Eq. (1):

$$Z_c = \frac{1}{H \times W} \sum_{i=1}^{H} \sum_{j=1}^{W} F_{in}(\mathrm{c}, i, j) \tag{1}$$

where $F \in \mathbb{R}^{C \times H \times W}$ is the intermediate feature map of utterance-level representations, C is the number of channels, and $H \times W$ represents the spatial dimensions.

MLP Transformation: Learn nonlinear interchannel relationships via a multilayer perceptron (MLP) to generate channel weights $S \in \mathbb{R}^C$;

Feature Calibration: Multiplying the weights s with the input features channelwise to output enhanced features is shown in Eq. (2):

$$F_{semip} = s \odot F_{in} \tag{2}$$

(2) SE Attention Stage:

Global Pooling: Perform global average pooling on F_{semip} to generate new channel descriptors $z' \in \mathbb{R}^C$.

Fully Connected Layers: Learn channel importance weights $s' \in \mathbb{R}^C$, using a similar formula to SEMLP but with a different reduction ratio.

Feature Weighting: Multiply the weights s' with F_{semip} to produce the final output, as shown in Eq. (3):

$$F_{out} = s' \odot F_{semlp} \tag{3}$$

3.2 SPA-UDA

The key to improving the recognition effect is to address the overfitting problem for a flexible extension structure. In this section, the details of the SSL strategy conducted in SEMLP are discussed. As illustrated in Fig. 4, SPA-UDA consists of two parts. First, an SPA is used to screen the in-domain unlabelled data via a pretrained SEMLP model. Another part of UDA decreases the difference in the probability distribution between the augmented data and the original data [11].

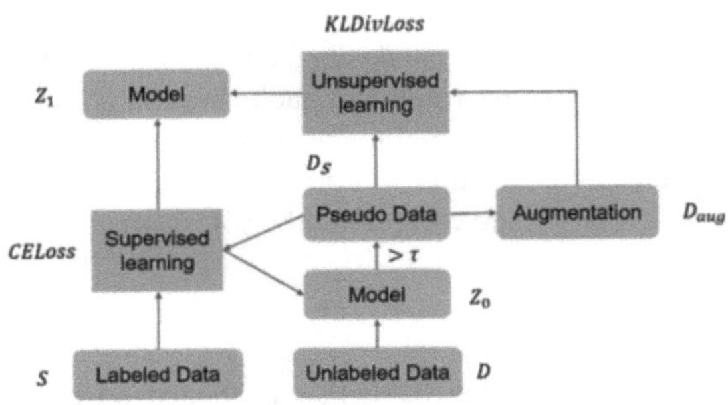

Fig. 4. Overview of the SPA-UDA method. The cross-entropy loss is calculated for labelled data, and the Kullback–Leibler divergence loss is calculated between the pseudodata and augmented data.

More specifically, SPA uses the pretrained SEMLP model Z_0 to identify the unlabelled database D. We set a threshold τ to screen the in-domain data D_s by self-modelling. The pseudo data D_s are combined with the labelled data S to train a new model Z_1. UDA uses the model Z_1 to compute the Kullback–Leibler divergence between unlabelled data D_s and their augmented data D_{aug} and then adds the loss L_{all} from supervised learning L_{sup} and unsupervised learning L_{unsup}. λ is used to amplify L_{unsup} to accelerate the training process.

4 Experiments

In this section, the proposed method is evaluated on the IEMOCAP corpus. The experimental results are compared with those of the mainstream methods, and the contribution of each module in SEMLP is discussed in detail.

4.1 Database Description

The interactive emotional dyadic motion capture (IEMOCAP) [17] corpus, which contains approximately 12 h of audiovisual data (such as video, speech, motion capture of faces, and text transcriptions), is applied particularly in experiments. It records over 10K utterances about improvisations and scripts emotion-annotated audio from 10 participants (5 male and female actors). This study uses four emotions to compare the results with those of the previous baseline, for a total of 5531 utterances (1103 angry, 1636 happy, 1084 sad, 1708 neutral).

4.2 Experimental Setup

The MFCCs were extracted from segmented speech with a 25 ms frame size and 10 ms hop length. Each MFCC comprises 39 features, including 12 MFCC parameters from the 26 Mel-frequency bands, one log-energy parameter, 13 deltas, and 13 acceleration coefficients. Stochastic gradient descent optimization with a learning rate of 1e-4 and weight decay of 1e-6 is adopted for model training. In the SSL strategy, 32 utterances were used for supervised learning 128 utterances were used for unsupervised learning (set λ to 1, τ to 0.7) at once, 50k unlabelled data and augmentation with noise (coverage period of 0.2) were used. The model was evaluated very 5 epochs with early stopping. Owing to the significant difference in the data volume of each emotion in the IEMOCAP corpus, the weighted accuracy (WA) and unweighted accuracy (UA) metrics were adopted for assessment. The WA reflects the accuracy of emotion classification of the model in all testing data, and the UA looks at the correct number of all emotions in a balanced way.

4.3 Experimental Results

Compared with related methods, the SEMLP is evaluated on the IEMOCAP corpus and compared with mainstream methods; the details of the experimental results are shown in Table 1. The SEMLP model achieves 66.9% WA and 67.5% UA, which is a competitive improvement over the RNN baseline. In addition, the RNN-based method has a long inferring response time and poor accuracy because of its inability to parallelize. The SEMLP model benefits from an effective voice segment processing method, which avoids the impact of long silent periods and pauses. Moreover, the lightweight and scalable architecture allows low memory and computing power to achieve real-time emotion recognition tasks.

Table 1. Performance of SEMLP and related methods on the IEMOCAP corpus. The best results are highlighted in bold.

Model	WA	UA
RNN+Attn	63.5%	58.8%
CNN+LSTM	66.9%	65.3%
IAAN	64.7%	66.3%
SEMLP	66.9%	67.5%

4.4 Ablation Experiments

In this part, five groups of experiments are conducted to measure the contributions of each module in SEMLP. (1) Baseline: the CNN with the RNN model. (2) SEMLP-segmented speech: the SEMLP model without a segmented speech database. (3) SEMLP. (4) SEMLP with the UDA method. (5) SEMLP with the SPA and UDA methods. The ablation experimental results are shown in Table 2. The details of the training process are given below. As shown in Fig. 5.

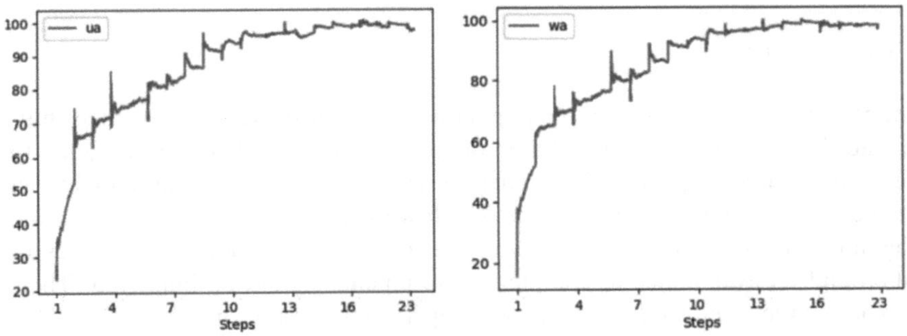

Fig. 5. Training curves of the UA and WA during the training process of the SEMLP+SPA+UDA model.

Table 2. Ablation experiments of our method. The best results are highlighted in bold.

Model	WA	UA
Baseline	66.9%	65.3%
SEMLP-Segmented Speech	66.5%	67.1%
SEMLP	66.9%	67.5%

(continued)

Table 2. (*continued*)

Model	WA	UA
SEMLP+UDA	66.3%	67.2%
SEMLP+SPA+UDA	68.0%	68.8%

The trends of UA and WA indicate that the model stabilizes after 23 training epochs.

The ablation results, as illustrated in Table 2, indicate that the SEMLP achieves 67.5% UA, amounting to a 2.2% improvement over the baseline. It shows the segmented speech as an additional feature that can improve the performance of 0.4% UA. The model uses these segment embeddings to calculate emotion-related features more effectively. The SEMLP with the UDA method achieves 67.2% UA, which is 0.3% lower than that of the SEMLP method. The reason is that the out-of-domain data impact the prior probability of supervised data. The model tries to fit the data with other emotions that disturb the performance in the original emotion classification. The SEMLP with the SPA and UDA methods achieves the best result in this study, and it improved the WA by 1.1% over that of the SEMLP model. Since the 50k unlabelled data were used, the model could be more robust against noise than the SPA method and learn more information about priordistributions from unlabelled data, significantly improving the performance and stability. These experimental results show that SPA can circumvent the drawback of data contamination in UDA training.

5 Conclusion

In the 5G era and with the increasing popularity of mobile devices, personalized human–computer interaction experiences based on emotion have gradually become the focus. To ensure the timeliness and effectiveness of current emotion recognition methods for portable devices, this article proposes a lightweight and extendable speech emotion recognition method based on multilayer perceptrons. In addition, the accuracy of the method and its availability in complex environments should be improved. This article explores the effectiveness of the semisupervised learning methods SPAUDA and segmented speech. The experimental results strongly prove that the proposed method can effectively aggregate the multiscale features into utterance-level representations for accurate emotional expression. Moreover, the SSL method can enhance the robustness of the model by using unlabelled data to gain extra information about the prior distribution of the data. Finally, the proposed model can benefit from segmented speech and learn emotional-related and discriminating features for recognition. In the future, the universality and diversity of users should be fully considered to support cross-language and cross-cultural emotion recognition.

References

1. Schuller, B., Rigoll, G., Lang, M.: Hidden Markov model-based speech emotion recognition. In: 2003 IEEE International Conference on Acoustics, Speech, and Signal Processing, Proceedings (ICASSP 2003), vol. 2, pp. II-1 (2003)

2. Guan, Z., Lv, Z., Du, X., Wu, L., Guizani, M.: Achieving data utility-privacy tradeoff in internet of medical things: a machine learning approach. Future Gener. Comput. Syst. **98**, 60–68 (2019). https://doi.org/10.1016/j.future.2019.01.058
3. Mirsamadi, S., Barsoum, E., Zhang, C.: Automatic speech emotion recognition using recurrent neural networks with local attention. In: 2017 IEEE International Conference on Acoustics, Speech and Signal Processing (ICASSP), pp. 2227–2231 (2017)
4. Yeh, S.-L., Lin, Y.-S., Lee, C.-C.: An interaction-aware attention network for speech emotion recognition in spoken dialogues. In: 2019 IEEE International Conference on Acoustics, Speech and Signal Processing (ICASSP), ICASSP 2019, pp. 6685–6689 (2019)
5. Partaourides, H., Papadamou, K., Kourtellis, Leontiades, N.I., Chatzis, S.: A self-attentive emotion recognition network. In: 2020 IEEE International Conference on Acoustics, Speech and Signal Processing (ICASSP), ICASSP 2020, pp. 7199–7203 (2020)
6. Tzirakis, P., Zhang, J., Schuller, B.W.: End-to-end speech emotion recognition using deep neural networks. In: 2018 IEEE International Conference on Acoustics, Speech and Signal Processing (ICASSP), pp. 5089–5093 (2018)
7. Etienne, C., Fidanza, G., Petrovskii, Devillers, A.L., Schmauch, B.: CNN+LSTM architecture for speech emotion recognition with data augmentation. In: Workshop on Speech, Music and Mind (SMM 2018) (2018)
8. Huang, J., Tao, J., Liu, B., Lian, Z., Niu, M.: Multimodal transformer fusion for continuous emotion recognition. In: 2020 IEEE International Conference on Acoustics, Speech and Signal Processing (ICASSP), ICASSP 2020, pp. 3507–3511 (2020)
9. Vaswani, A., et al.: Attention is all you need. arXiv (2017)
10. Fu, P., Xu, F., Yuan, H.: Speech emotion recognition based on ResNet-BiGRU network. In: Proceedings of SPIE-The International Society for Optical Engineering, vol. 12709 (2023)
11. Ravi, Taran, S.: Nonlinear feature extraction approach for speech emotion recognition using VMD and TKEO. Appl. Acoust. **214** (2023)
12. Tolstikhin, I., Houlsby, N., Kolesnikov, et al.: MLP-mixer: an all-MLP architecture for vision. arXiv preprint arXiv:2105.01601 (2021)
13. Xie, Q., Dai, Z., Hovy, E., Luong, M.-T., Le, Q.V.: Unsupervised data augmentation for consistency training. arXiv preprint arXiv:1904.12848 (2019)
14. Ouali, Y., Hudelot, C., Tami, M.: An overview of deep semisupervised learning. CoRR, vol. abs/2006.05278 (2020). https://arxiv.org/abs/2006.05278
15. Neumann, M., Vu, N.T.: Improving speech emotion recognition with unsupervised representation learning on unlabelled speech. In: 2019 IEEE International Conference on Acoustics, Speech and Signal Processing (ICASSP), ICASSP 2019, pp. 7390–7394 (2019)
16. Tyagi, V., Ganapathiraju, A., Wyss, F.I.: Method and system for acoustic data selection for training the parameters of an acoustic model. Patent 9972306 (2018). https://www.freepatentsonline.com/9972306.html
17. Busso, C., Bulut, M., Lee, C.C.A., Kazemzadeh, E., et al.: Iemocap: interactive emotional dyadic motion capture database. Lang. Resour. Eval. **42**(4), 335–359 (2008)

MreNet: Multimodal Regularization Ensemble Network for Hierarchical Paper Classification

Tan Yue[1,2], Xuzhao Shi[1], Shuo Zhan[3], Zilong Song[1], and Zonghai Hu[1](✉)

[1] School of Electronic Engineering, Beijing Key Laboratory of Work Safety Intelligent Monitoring, Beijing University of Posts and Telecommunications, Beijing, China
zhhu@bupt.edu.cn

[2] Wangxuan Institute of Computer Technology, Peking University, Beijing, China
yuetan@pku.edu.cn

[3] Nanyang Technological University, Singapore, Singapore

Abstract. With the rapid growth of scientific paper data, intelligent fine-grained document classification techniques can significantly benefit paper retrieval and recommendation. Previous works on paper classification focused on texts and were coarse-grained. The high sample similarity within subdivided fields makes fine-grained classification more challenging. Given the implicit semantic relationship between abstracts and figures in academic papers, current multimodal methods designed to process explicitly related textual and visual data are inadequate for addressing the fine-grained multimodal paper classification task. Additionally, discrepancies in data distribution and dataset limitations frequently result in overfitting when pretrained models are transferred to paper classification tasks. In this work, we propose a multimodal regularization ensemble network for fine-grained paper classification. Our model focuses on the differentiation of intra-modality feature representations and the enhancement of inter-modality related information for the task. Comparative analysis with baseline methods highlights the efficiency of our proposed model. Finally, we developed a paper classification system based on our model that is capable of classifying, managing, and recommending academic papers.

Keywords: Multimedia data analysis · Data processing application · Multimodal learning · Information fusion

1 Introduction

A large variety of scientific papers are published every year. Researchers need to spend much of their time sorting out papers relevant to their research and are eager to obtain help from intelligent paper classification methods. Unlike other document classification tasks, paper classification encompasses numerous hierarchical categories, ranging from coarse-grained to fine-grained levels [28, 49].

Y. Yu et al. (Eds.): ICPCSEE 2025, CCIS 2673, pp. 214–233, 2026.
https://doi.org/10.1007/978-981-95-2566-9_15

(a) **Text data:** Enjoy the sunny day and enjoy you can breathing.

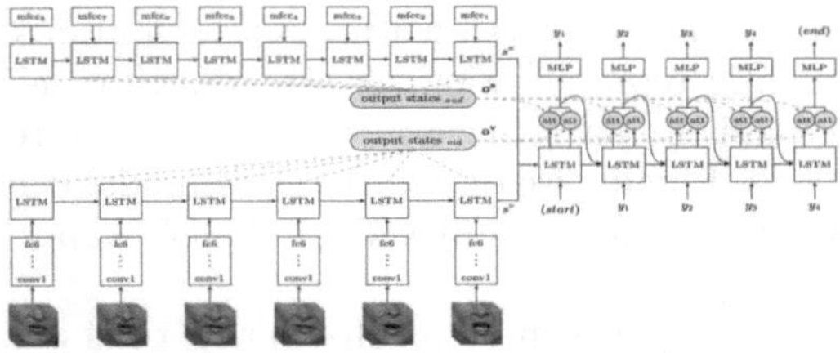

(b) **Text data:** The WLAS model trained on the LRS dataset surpasses the performance of all previous work on standard lip reading benchmark datasets, often by a signifificant margin [37].

Fig. 1. Examples of the images and text content in different field.

Existing methods for classifying papers are mostly coarse-grained and utilize only textual data [36]. These methods can be divided into several high-level fields (coarse-grained) such as 'materials science', 'computer science', or 'medical science'. Nonetheless, for researchers, lower-level subfields directly related to their work are of greater importance. Thus, intelligent fine-grained paper classification is in high demand. Compared with coarse-grained classification, the similarity among papers within subdivided fields is greater, complicating the classification process. In addition, the fine-grained annotation of scientific papers requires a high level of expertise. Given that scientific papers consist of both images and text, it is crucial to introduce image information, which can complement the information with textual data. Consequently, leveraging multimodal information is a promising strategy for enhancing the performance of paper classification tasks [3,48].

Unlike the tasks that involve image data that are explicitly related to text data, the images in paper documents are implicitly related to texts. As shown in Fig. 1(a), the image and text are from Twitter and have an explicit relationship with a sunny day. For Fig. 1(b), the image and text lack a robust connection to the same topic. However, both elements contain crucial information and features for fine-grained classification. Consequently, a more appropriate multimodal fusion method is warranted.

In recent years, pretrained models have shown strong performance and are widely used. However, due to differences in the data distribution and dataset limitations, overfitting often occurs when the pretrained model is transferred to fine-grained paper classification tasks [39,53]. As the network structure of the pretraining model becomes increasingly complex, the effectiveness of single regularization methods becomes limited.

Inspired by ensemble learning, a multimodal regularization ensemble network is proposed. Specifically, we introduce different regularization methods to differentiate and boost the intra-modality feature representations for texts and images. Thus, MreNet can learn the task from strong ensemble representations. On the other hand, we integrate an attention mechanism to enhance inter-modality related feature representations. The image modality enhances its semantic representation by attending to the text feature. Thus, the attention mechanism can achieve more semantically related features given by texts and images. Finally, we design the multimodal decision fusion module. The weighted voting method is employed to classify paper, and the voting weights are optimized by training.

Our experimental results verify that our method can effectively enhance implicitly related multimodal information and significantly solve the overfitting problem. Compared with strong baselines, MreNet achieves the best results (average accuracy of 82.92%) in all paper classification tasks [50], exceeding the strongest baseline by 2.95% in average accuracy. In particular, our MreNet achieves a 6.11% improvement on the PaperNet_NLP dataset. Additionally, the proposed model is still effective for unimodal data. Compared with the backbone models (MobileNetV3 and ALBERT), MreNet-OI (only image) and MreNet-OT (only text) yield average accuracy improvements of 4.27% and 1.4%, respectively. To test our proposed method in other implicitly related multimodal tasks, another publicly available dataset [4] is used for further evaluation. MreNet also achieves the SOTA result (87.86%). Additionally, we perform ablation experiments and visualizations to analyze and explain our model in more detail.

The main contributions of this paper are summarized as follows.

- We propose a novel multimodal regularization ensemble network to address the overfitting problem in hierarchical paper classification and enhance implicitly related multimodal information.
- A novel multimodal information fusion approach is introduced. We combine the cross-modal attention module with multiple regularization methods to fuse text and image features. We design a multimodal decision fusion module for better classification performance.
- Our proposed model achieves SOTA results on the PaperNet dataset and multimodal sarcasm detection benchmark dataset. Additionally, we develop a paper classification system (PaperMaster) based on the proposed multimodal regularization ensemble network. According to the knowledge introduced in the text and figures of the papers, the system can manage and recommend papers.

2 Background and Related Works

2.1 Multimodal Fusion

Multimodal data refer to data from different modal sources, such as images, text, and audio. The real world is informed by multimodal data. To better understand the world, AI should be empowered to learn, understand, and reason about multimodal information. Multimodal learning refers to building a model that can use multimodal information as input, which is processed and analyzed by the built model.

Multimodal fusion is an important component in the field of multimodal learning. Unimodal information often has the problem of missing information and needs to be supplemented by multimodal information. The technique of fusing two or more modal data for joint representation is multimodal fusion [42]. In terms of fusion stages, multimodal fusion can be categorized into early fusion [8], late fusion [30], and hybrid fusion [31]. Recently, many remarkable works have been proposed [3,40,52].

CLIP [33] is trained via a self-supervised approach, which maps images and text into a shared vector space, thus enabling the model to understand the semantic relationships between images and text. Inspired by the patch projection technique in ViT [7], ViLT [16] uses the pretrained ViT to initialize the interacting transformer, concentrating the main computation on the feature fusion part of the transformer. LLaVA-Phi [54] improves the efficiency of multimodal information fusion. This work [29] discusses the importance of information about each modality in implicit emotion detection tasks.

2.2 Paper Document Classification

With the increasing scale of text data in industrial applications, automatic document classification has become increasingly important [49]. Compared with other document classification tasks, the task of academic paper classification contains hierarchical categories, ranging from coarse-grained to fine-grained classifications [28,36].

In the early stages, document-level text classification tasks used CNN-based, RNN-based, and transformer-based models [5,14,25]. Convolutional neural networks (CNNs) are commonly used in the field of image processing. With convolutional layers, CNN-based models can also model language sequences [17]. Recurrent neural networks (RNNs) are suitable for processing sequential data and can process long text sequences through memory mechanisms [2,46]. On the basis of RNNs, LSTM is proposed to solve the gradient vanishing problem for long text sequences. The transformer-based model is able to consider the global importance of long text sequences in a document through the self-attention mechanism [47].

In recent years, the emergence of pretrained models has advanced the field of document classification. Many models, such as Elmo [32], OpenAI GPT [1], Qwen [45], BERT [6], RoBERTa [26], and ALBERT [21], have achieved good performance. These models could be used as pretrained encoders in document classification tasks.

2.3 Research Challenge

However, due to the large number of parameters in the pretrained model, the network structure is complex. When the data distribution of the pretrained dataset differs significantly from that of the target task dataset, the model will often overfit, significantly affecting the classification performance. This problem is especially serious in fine-grained classification tasks. Unlike in the coarse-grained classification task, the sample feature similarity in fine-grained classification is high, resulting in a closer spatial distribution. When we directly transfer models pretrained in other domains to fine-grained classification tasks, severe overfitting problems occur. On the other hand, scientific paper documents contain text and image information. Previous work has focused only on textual information, but images also contain much useful information. The utilization of multimodal data can yield good complementarity of information. Many multimodal learning works have also demonstrated the importance of introducing multimodal data [3,51].

Accordingly, this paper proposes a multimodal regularization ensemble network. We innovatively combine different regularization methods with the idea of ensemble learning to differentiate and boost the intra-modality feature representations for texts and images. Additionally, the multimodal attention module is designed to enhance inter-modality related feature representations. As introduced in the paper [20]. The attention module with the global average pooling layer can be regarded as a kind of regularization method. Through the ensemble learning method, different multimodal feature fusion representations are integrated for better performance.

3 Methodology

Inspired by ensemble learning, we propose a multimodal regularization ensemble network. Different regularization methods are integrated to address the overfitting problem. Moreover, we combine the attention module for better multimodal information joint representation.

The architecture of MreNet is shown in Fig. 2. Most scientific papers are available in PDF format and consist of image and text data. Existing paper classification models often focus on text data. However, in the field of fine-grained classification, data preprocessing and data augmentation are needed to solve the problem of data limitations. The PDFplumber framework is used to extract the text content (abstract) of the paper, and the PIL framework is used to extract images. The data are subsequently input into the multimodal regularization ensemble network for training.

Multimodal regularization ensemble network model is based on the pre-trained model with good performance. ALBERT [21] and MobileNetV3 [11], well-known pre-trained models, are used as encoders for feature extraction of text and image data. The keys of our proposed model is the innovative multimodal attention module and the regularization ensemble learning method. As for the regularization ensemble learning method, different regularization methods and multimodal decision fusion module are used to differentiate and boost the intra-modality feature representations and achieve better classification performance.

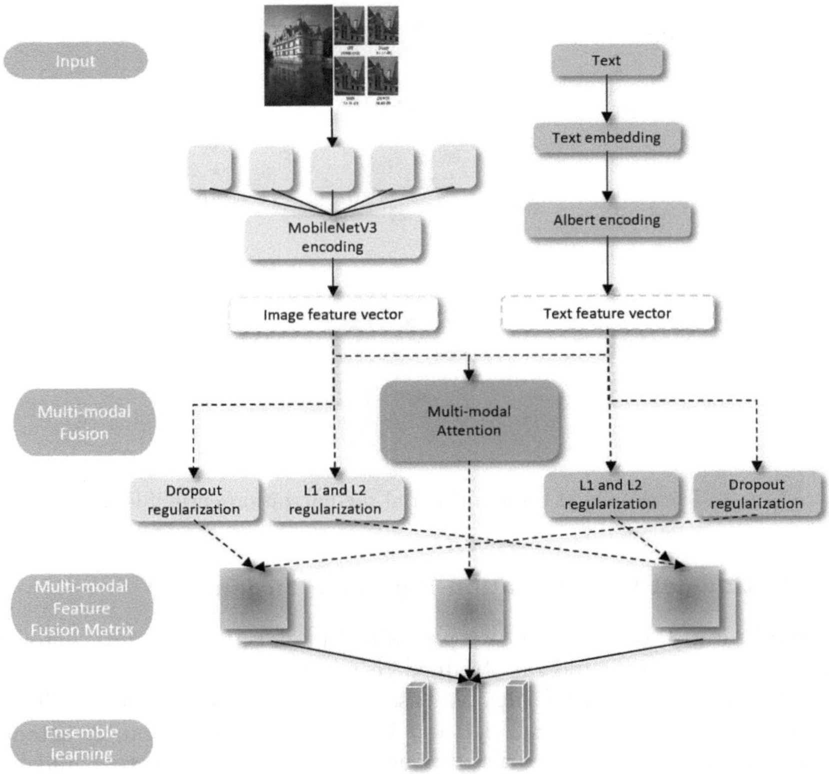

Fig. 2. Overview of our proposed multimodal regularization ensemble network.

3.1 Text Feature Representation

Given a sequence of words $\mathcal{S} = \{s_1, ..., s_n\}$ (n is the length of the sequence) in text content, ALBERT embeds and encodes $\mathcal{S}$, yielding its vectorized representations ($T \in \mathbb{R}^{n \times d}$; d is the dimension of ALBERT hidden states)

$$T = ALBERT(\mathcal{S}). \quad (1)$$

3.2 Image Feature Representation

We encode an image $\mathcal{H}$, associated with the text by pre-trained MobileNetV3. We use a linear layer upon the MobileNetV3 encoder to align the dimension of the hidden states of the image with that of the abstract. Then, the image representation ($I \in \mathbb{R}^{p \times d}$; p is the number of image patches.) is given by

$$I = linear(MobileNetV3(\mathcal{H})). \quad (2)$$

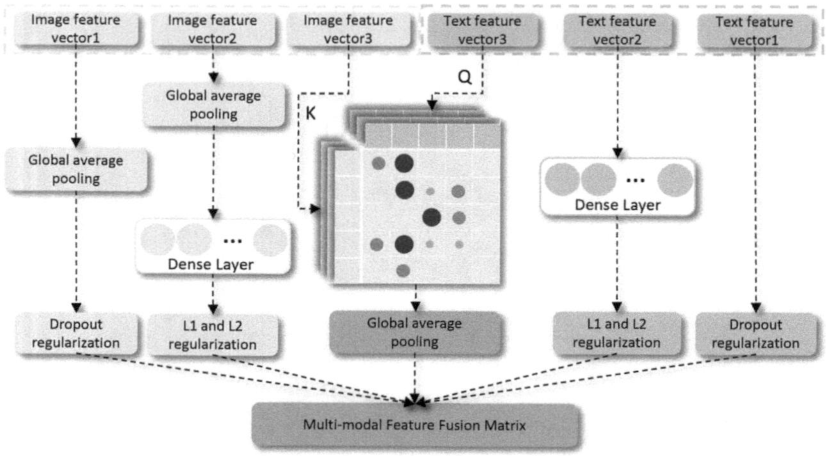

Fig. 3. The details of the multimodal regularization ensemble fusion.

3.3 Multimodal Regularization Ensemble Fusion

As shown in Fig. 3, through the two pretrained models, text feature vectors and image feature vectors could be obtained respectively, and the two vectors are operated in three different forms. We use different regularization methods: 1. dropout-based regularization; 2. L1 and L2-based regularization; 3. attention-based regularization, which learns different representations for text and image modalities. To denote different regularization methods (channels) in MreNet, we use different subscripts that correspond to the above numbers. Then, $I_1 = I_2 = GlobalAveragePooling(I)$; $I_3 = I$; Here T_1 and T_2 represent the ALBERT encoder's outputs, which are the *pooled_output* output, which represent each input sequence as a whole. $T_3 = T$. $I_1, I_2, T_1, T_2 \in \mathbb{R}^{1\times d}$.

Regularization Feature Representation. For image feature 1 I_1 and text feature 1 T_1, we employ the dropout regularization method to randomly remove parameters in the network, as each iteration of the deep learning model computes different weights and parameters, and the dropout regularization method enables the network to reduce the number of network parameters and accumulate different parameters for each layer at each iteration.

$$r_j^{(l)} \sim Bernoulli(p), \tag{3}$$

$$I_1^{'} = w_i^{(l+1)}\ {I_1}^{(l)} r^{(l)}\ +\ b_i^{(l+1)}, \tag{4}$$

$$T_1^{'} = w_i^{(l+1)}\ {T_1}^{(l)} r^{(l)}\ +\ b_i^{(l+1)}. \tag{5}$$

For the L1 and L2-based regularization, we use L1 and L2 regularization on a linear layer with ReLU activation.

$$a_i = y_i - f(x_i), \tag{6}$$

$$\mathcal{L} = L_a + \lambda \sum_{i=1}^{n} |a_i| + \lambda \sum_{i=1}^{n} a_i^2, \tag{7}$$

$$I_2^{'} = ReLU(regular\_linear(I_2)), \tag{8}$$

$$T_2^{'} = ReLU(regular\_linear(T_2)). \tag{9}$$

y_i is the target value and $f(x_i)$ is the predicted value. L_a is the loss function (cross entropy) without the regularization term and λ is a hyperparameter that controls the size of the regularization term.

Multimodal Attention. As different regularization methods are adopted for both the 1st and 2nd processing methods, the number of parameters is greatly reduced, leading to a decrease in the fitting ability of the model and a sparse representation of multimodal features. Therefore, for image and text feature vector 3, we adopt an innovative multimodal attention mechanism. Common attention mechanisms are used in the SEQ2SEQ model structure, such as machine translation, which achieves better translation performance through weighted adjustment of word order. In our multimodal attention mechanism, the intuition of using attention for the implicitly related text and image features in fine-grained paper classification tasks is that the attention mechanism can enhance the related information representations of text and image features by allocating higher attention weights to them. Thus, the fused feature can represent the related information of the two modalities. As shown in Fig. 3, the darker the color in the multimodal attention matrix is, the greater the weight and the higher the degree of attention. The global average pooling method was then adopted to filter out many parameters with low attention. Here, we generalized the regularization concept by referring to the viewpoint in the paper [20], and regarded the global average pooling method as a regularization method to reduce the number of parameters and prevent overfitting.

$$\alpha = \frac{T_3 I_3^T}{\sqrt{d_i}}, \tag{10}$$

$$A_3 = \frac{\exp(\alpha_j)}{\sum_{j=1}^{d_i} \exp(\alpha_j)} I_3, \tag{11}$$

$$F_3 = ReLU(w_{c3}(GlobalAveragePooling(A_3)) + b_{c3}), \tag{12}$$

where the w_{c3} denote the linear weight and b_{c3} denote the bias.

Multimodal Decision Fusion. Next, we concatenate ($\oplus$) the multimodal features, given by the dropout-based, and L1 and L2-based regularization methods, respectively. The concatenated features pass a ReLU activated linear layer, yielding cross-modality fusion features of the two regularization channels by

$$F_1 = ReLU(w_{c1}(I_1^{'} \oplus T_1^{'}) + b_{c1}), \tag{13}$$

$$F_2 = ReLU(w_{c2}(I_2^{'} \oplus T_2^{'}) + b_{c2}), \tag{14}$$

$$\hat{y} = \sum_{i=1}^{3} w_i \text{SoftMax}(F_i), \tag{15}$$

where w_{c1}, w_{c2} denotes the linear weight and b_{c1}, b_{c2} denotes the bias. w_i is the weight, $w_i \geq 0$. Finally, we adopt the idea of ensemble learning and the weighted voting method. The weights of each fusion feature are updated through continuous iteration of the deep learning network to obtain the optimal weights. Our final classification layer consists of a two-layer fully-connected neural network. The hidden layer employs the elementwise *ReLU* activation function, whereas the output layer uses the *softmax* function.

4 Experiment

In the experiment section, we aim to evaluate our proposed MreNet model in different classification tasks with implicitly related multimodal samples. Experimental questions:

- Can our model achieve SOTA results compared with unimodal and multimodal methods in both coarse-grained and fine-grained paper classification tasks?
- Can our model enhance implicitly related multimodal information and achieve good results in other publicly available tasks?
- What is the effect of each module in the proposed MreNet model?

4.1 Dataset

PaperNet-Dataset. As shown in Table 1, we use a multimodal dataset called PaperNet [50] for coarse-grained and fine-grained paper classification tasks. The PaperNet dataset contains image and text data and takes 80% of the samples as the training set and 20% of the samples as the test set. We select 10% of the training set as the validation set.

Multimodal Sarcasm Detection Dataset. We additionally employ the widely accessible multimodal sarcasm detection benchmark dataset for evaluating our MreNet. The dataset [4] encompasses English-language tweets. The comprehensive data are presented in Table 2.

Table 1. The coarse-grained and fine-grained categories of the PaperNet dataset.

Dataset	Subset	Class
PaperNet_2		CV
		NLP
PaperNet_20	PaperNet_CV	CV_attention
		CV_classification
		CV_detection
		CV_GAN
		CV_recognition
		CV_retrieval
		CV_segmentation
	PaperNet_NLP	NLP_bert_based
		NLP_conversation
		NLP_cross
		NLP_extraction
		NLP_Few_shot_task
		NLP_knowledge_graph
		NLP_machine_reading
		NLP_machine_translation
		NLP_multilingual
		NLP_multimodal
		NLP_named_entity_recognition
		NLP_sentiment_analysis
		NLP_text_generation

Table 2. Detailed information of the multimodal sarcasm detection benchmark dataset.

	Training	Development	Test
Positive (Sarcasm)	8642	959	959
negative (Non-Sarcasm)	11174	1451	1450
Total	19816	2410	2409
Token Length	16.91	16.92	17.13

4.2 Baselines

We evaluate our MreNet model by contrasting it with several renowned classification and embedding techniques, including:

- **Image-modality methods: ResNet50** [9] is a deep neural network architecture that solves the problems of gradient vanishing and representation bottlenecks in deep networks through residual connections. **DenseNet121** [13] is a deep neural network architecture characterized by the fact that each layer in the network being directly connected to all previous layers, enhancing feature reuse and gradient propagation. **MobileNetV3** [11] is a lightweight model that combines neural network search algorithms and network design techniques. **ConvNeXt** [27] is based on a CNN architecture and achieves SOTA performance in many image processing tasks by incorporating a fine-tuning strategy.

- **Text-modality Methods: ULMFiT** [12] is a transfer learning method that can be applied to any task in NLP. **ALBERT** [21] is based on an improvement in the BERT model with a significant decrease in the number of model parameters.
- **Multimodal Methods: Concat** [34] uses different feature extractors to extract text and image features, followed by the concatenation of different modal feature vectors. **MCPD-Net** [10] employs multimodal data, which are gathered from both vision and accelerometer sensors within a household setting, to train its variational autoencoder (VAE) models. **DSS** [35] is a new multimodal deep learning method for the multimodal data classification of multiclass brain tumors. **DT4MID** [38] uses textual and visual transformers in a deep neural network for enhanced multimodal representation.

4.3 Settings

For our compared models, we adopt the default parameter settings as specified in their respective original papers or implementations. For MreNet, MobileNetV3 and ALBERT pretrained models are employed as the image encoder and text encoder respectively. Subsequently, a dropout layer is applied to these pretrained models with a dropout rate set at 0.5. Both the L1 and L2 regularization rates are 0.1. MobileNetV3 has an input size of $224 \times 224 \times 3$ and a dimension $d = 768$, $p = 49$. For ALBERT, the hidden size is $d = 768$, $n = 128$.

We use a cross-entropy loss function, and the learning rate is set at $1e-4$. We allocate 10% of our training data for validation. Following existing methods [19], the model is trained via the Adam optimizer [18]. Training is stopped if no reduction in validation loss is observed over 5 consecutive epochs.

4.4 Paper Classification Experiments

In these experiments, our goal is to utilize mainstream models for a comparative analysis with our proposed MreNet on paper classification tasks.

Hierarchical Paper Classification. The PaperNet dataset contains a coarse-grained dataset (PaperNet_2) and a fine-grained dataset (PaperNet_20, PaperNet_CV, and PaperNet_NLP). For a fine-grained dataset, because the similarity of samples is greater than that of coarse-grained samples, the fine-grained classification task is more difficult. In the fine-grained classification task, as shown in Tables 3, 4 and 5, the accuracy of all the models obviously decreased, which illustrates the challenging nature of this task and the great room for improvement.

Text Classification. As shown in Table 3, we use two well-known models, ULMFiT and ALBERT, to conduct text classification experiments. In the experiments, the models perform well on the coarse-grained PaperNet_2 dataset, and ALBERT achieves an accuracy of 96.31%, which demonstrates that the existing text-based model can achieve good performance in the coarse-grained paper classification task. However, the performance of both models decreases significantly on the three fine-grained datasets.

Table 3. Comparison of accuracy on text-based paper classification tasks.

Model	PaperNet_2	PaperNet_20	PaperNet_CV	PaperNet_NLP	Ave.
Ulmfit	96.26 ± 0.44	71.30 ± 1.12	**75.30 ± 1.06**	73.33 ± 0.18	79.05
ALBERT	96.31 ± 0.09	73.32 ± 0.24	73.18 ± 0.12	74.23 ± 0.36	79.26
MreNet-OT	**97.04 ± 0.33**	**73.87 ± 0.46**	73.56 ± 0.29	**78.15 ± 0.41**	**80.66**

Image Classification. Next, some mainstream image classification models such as ConvNeXt are used for image classification tasks. As shown in Table 4, all the models cannot perform satisfactorily on the fine-grained datasets. Compared with text classification tasks, fine-grained image classification tasks are more challenging. As we introduce in Sect. 1, using images or text information alone often makes accurate classification difficult in a fine-grained way. MreNet is designed on the basis of MobileNetV3 and ALBERT. For unimodal tasks, we replace the multimodal attention mechanism with a self-attention mechanism. As shown in Table 3, 4, and Fig. 4, we compare MreNet with the other two models to verify the effectiveness of the regularization ensemble method for unimodal data. The average accuracy is improved by 4.27% (image) and 1.4% (text), which demonstrates that in addition to multimodal, implicitly related information enhancement techniques, our proposed regularization ensemble learning approach is effective even in unimodal tasks.

Multimodal Classification. We find that unimodal models cannot yield good results. Therefore, we consider the use of multimodal models. We use recent and strong models to compare with our proposed model in multimodal paper classification tasks. As shown in Table 5, our MreNet achieves the highest accuracy among all the datasets. MreNet improves the average accuracy by 2.95% and improves the accuracy on the PaperNet_NLP dataset by 6.11%. The results of multimodal experiments show that a good multimodal information fusion

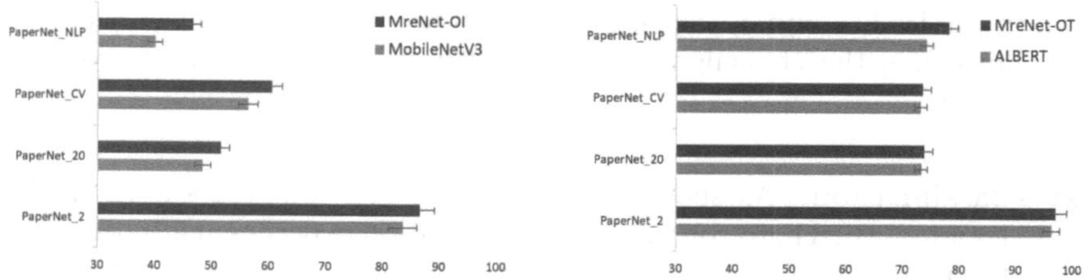

(a) MreNet(image) and MobileNetV3. (b) MreNet(text) and ALBERT.

Fig. 4. The improvement of the MreNet in unimodal tasks.

Table 4. Comparison of accuracy on image-based paper classification tasks.

Model	PaperNet_2	PaperNet_20	PaperNet_CV	PaperNet_NLP	Ave.
ResNet50	83.94 ± 0.55	50.30 ± 0.66	60.23 ± 0.61	45.33 ± 0.50	59.95
DenseNet121	82.34 ± 0.96	45.16 ± 0.46	58.80 ± 0.40	46.21 ± 0.46	58.13
MobileNetV3	83.66 ± 0.86	48.38 ± 0.56	56.45 ± 0.40	40.08 ± 0.54	57.14
ConvNeXt	**87.46 ± 0.14**	51.23 ± 0.51	58.92 ± 0.71	46.27 ± 0.42	60.97
MreNet-OI	86.65 ± 0.28	**51.69 ± 0.27**	**60.57 ± 0.65**	**46.73 ± 0.38**	**61.41**

method is important and that an unsuitable fusion method may introduce additional noise and even degrade the overall performance. Our approach, which focuses on enhancing multimodal implicitly related information and regularization ensemble learning, proves to be effective.

Table 5. Comparison of accuracy on multimodal paper classification tasks.

Model	PaperNet_2	PaperNet_20	PaperNet_CV	PaperNet_NLP	Ave.
Concat	96.27 ± 0.11	73.45 ± 0.69	72.43 ± 0.27	75.36 ± 0.51	79.38
MCPD-Net	94.91 ± 0.21	69.58 ± 0.59	71.58 ± 0.37	70.39 ± 0.51	76.62
DSS	95.39 ± 0.18	72.85 ± 0.51	73.89 ± 0.31	75.83 ± 0.35	79.49
DT4MID	96.35 ± 0.11	73.56 ± 0.42	73.96 ± 0.25	76.02 ± 0.31	79.97
MreNet	**98.02 ± 0.29**	**75.28 ± 0.52**	**76.26 ± 0.16**	**82.13 ± 0.36**	**82.92**

4.5 Sarcasm Detection Experiments

We aim to evaluate our proposed method in other task that has implicitly related data. Therefore, we apply our model to the sarcasm detection task. Sarcasm is a form of language in which the speaker says or writes the opposite of what he or she actually intends. Due to the implicit expression of sarcastic content, images and text often have inconsistent semantic information, making the multimodal data in sarcasm detection task is also implicitly related. We conduct experiments in the public sarcasm detection dataset compare results with recent strong baseline models. As shown in the Table 6, the proposed MreNet also achieves the remarkable performance (87.86%), which demonstrates its effective enhancement for implicitly related information.

Table 6. Comparison of accuracy on the sarcasm detection task.

Model	Modality	Accuracy
TextCNN [17]	Text	80.03
ResNet [9]	Image	64.76
SMSD [43]	Text	80.90
HFM [4]	Multimodal	83.44
BERT [15]	Text	83.85
D&R Net [44]	Multimodal	84.02
Res-Bert [31]	Multimodal	84.80
Intra-Att [41]	Multimodal	85.64
Att-Bert [31]	Multimodal	86.05
ViT [7]	Image	67.83
InCrossMGs [22]	Multimodal	86.10
ConvNeXt [27]	Image	67.78
CMGCN [23]	Multimodal	86.54
SAHFN [24]	Multimodal	87.22
MreNet (Our)	Multimodal	**87.86**

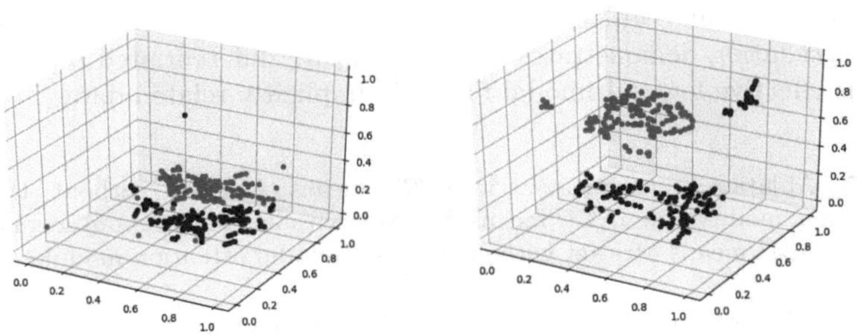

(a) MobileNetV3 (b) MreNet

Fig. 5. Visualization of MobileNetV3 and MreNet on the PaperNet_2 dataset.

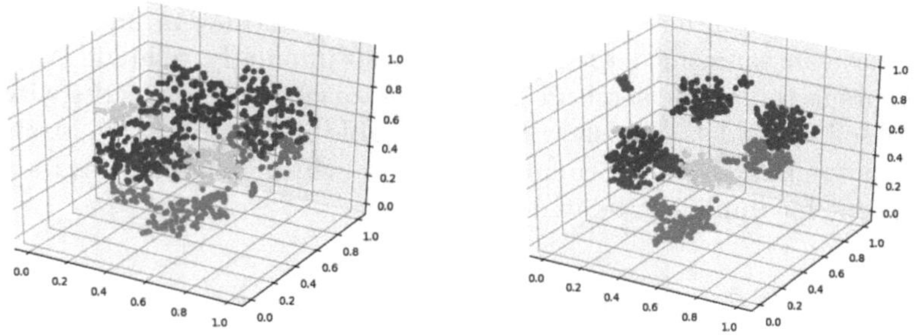

(a) MCPD-Net (b) MreNet

Fig. 6. Visualization of MCPD-Net and MreNet on the PaperNet_CV dataset.

4.6 Visualization

For further analysis, we visualize the data in Fig. 5 and Fig. 6. As shown in the visualization of the feature distribution, our MreNet achieves better performance in terms of class spacing and the degree of in-class aggregation.

4.7 Ablation Experiments

Because different modules in the model have different effects on model performance, we conduct ablation experiments and present the results in Table 7. Our MreNet model consists of an L1L2-regularization module, a dropout regularization module, and a multimodal attention module. We observe that the removal of multimodal attention (w/o $\mathcal{A}$) leads to significant performance degradation, which verifies the importance of related information enhancement by the multimodal attention module. When we remove the L1L2-regularization module (w/o $\mathcal{R}$) and the dropout regularization module (w/o $\mathcal{D}$), the considerable performance degradation demonstrates that the feature differentiated representation method is effective for multimodal fusion in implicitly related data.

Table 7. Ablation experiments. "w/o $\mathcal{R}$" means MreNet without the L1L2-regularization module. "w/o $\mathcal{D}$" means MreNet without the dropout regularization module. "w/o $\mathcal{A}$" means MreNet without the multimodal attention module.

Algorithm	PaperNet_2	Δ	PaperNet_20	Δ
MreNet (w/o $\mathcal{R}$)	97.11 ± 0.24	-0.91	73.69 ± 0.46	-1.59
MreNet (w/o $\mathcal{D}$)	96.63 ± 0.21	-1.39	74.05 ± 0.32	-1.23
MreNet (w/o $\mathcal{A}$)	96.38 ± 0.14	-1.64	73.55 ± 0.19	-1.73
MreNet	$\mathbf{98.02 \pm 0.29}$		$\mathbf{75.28 \pm 0.52}$	

4.8 Demonstration

We have also built an academic paper assistance system called PaperMaster (shown in Fig. 7), which integrates automatic paper classification and management, paper recommendation, paper commenting and sharing. On the basis of the MreNet model, we improved the paper classification module in the system. PaperMaster can provide convenient tools for researchers and students to facilitate sharing and communication in the academic community.

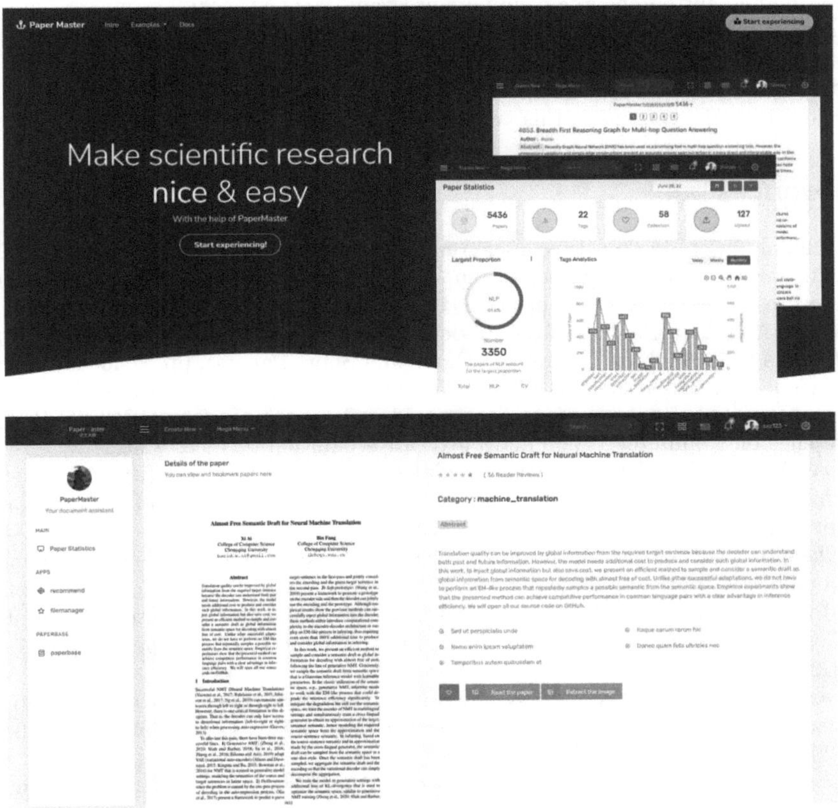

Fig. 7. PaperMaster system for paper classification, recommendation and discussion.

5 Conclusion and Future Work

In this paper, a multimodal regularization ensemble network (MreNet) is proposed for the fine-grained paper classification task. The task is challenging because the semantic relation between texts and images of a paper is implicit.

Thus, we integrate dropout-based, L1-L2-based, and attention-based regularization methods to differentiate intra-modality feature representations. To enhance the semantic relation between texts and images, MreNet integrates a multimodal attention module to learn useful inter-modality feature representations. Our proposed MreNet achieves remarkable performance in multimodal and unimodal evaluation tasks. We also develop a system based on MreNet to automatically classify and recommend papers.

Acknowledgments. The work described in this paper is supported by the BUPT innovation and entrepreneurship support program (2022-YC-S002). We also thank the National Key R&D Program of China (2022YFB3605601).

Disclosure of Interests. The authors have no competing interests to declare that are relevant to the content of this article.

References

1. Achiam, J., et al.: Gpt-4 technical report. arXiv preprint arXiv:2303.08774 (2023)
2. Adhikari, A., Ram, A., Tang, R., Lin, J.: Rethinking complex neural network architectures for document classification. In: Proceedings of the 2019 Conference of the North American Chapter of the Association for Computational Linguistics: Human Language Technologies, Volume 1 (Long and Short Papers), pp. 4046–4051 (2019)
3. Cadene, R., Ben-younes, H., Cord, M., Thome, N.: Murel: multimodal relational reasoning for visual question answering. In: Proceedings of the IEEE/CVF Conference on Computer Vision and Pattern Recognition (CVPR) (2019)
4. Cai, Y., Cai, H., Wan, X.: Multi-modal sarcasm detection in twitter with hierarchical fusion model. In: Proceedings of the 57th Annual Meeting of the Association for Computational Linguistics, pp. 2506–2515 (2019)
5. Chen, Y., Xu, L., Liu, K., Zeng, D., Zhao, J.: Event extraction via dynamic multi-pooling convolutional neural networks. In: Proceedings of the 53rd Annual Meeting of the Association for Computational Linguistics and the 7th International Joint Conference on Natural Language Processing (Volume 1: Long Papers), pp. 167–176 (2015)
6. Devlin, J., Chang, M.W., Lee, K., Toutanova, K.: Bert: Pre-training of deep bidirectional transformers for language understanding. arXiv preprint arXiv:1810.04805 (2018)
7. Dosovitskiy, A., et al.: An image is worth 16x16 words: transformers for image recognition at scale. ICLR (2021)
8. Gallo, I., Calefati, A., Nawaz, S.: Multimodal classification fusion in real-world scenarios. In: 2017 14th IAPR International Conference on Document Analysis and Recognition (ICDAR), vol. 5, pp. 36–41. IEEE (2017)
9. He, K., Zhang, X., Ren, S., Sun, J.: Deep residual learning for image recognition. In: Proceedings of the IEEE Conference on Computer Vision and Pattern Recognition, pp. 770–778 (2016)
10. Heidarivincheh, F., et al.: Multimodal classification of Parkinson's disease in home environments with resiliency to missing modalities. Sensors **21**(12) (2021). https://doi.org/10.3390/s21124133. https://www.mdpi.com/1424-8220/21/12/4133
11. Howard, A., et al.: Searching for mobilenetv3. In: Proceedings of the IEEE/CVF International Conference on Computer Vision, pp. 1314–1324 (2019)

12. Howard, J., Ruder, S.: Universal language model fine-tuning for text classification. arXiv preprint arXiv:1801.06146 (2018)
13. Huang, G., Liu, Z., Van Der Maaten, L., Weinberger, K.Q.: Densely connected convolutional networks. In: Proceedings of the IEEE Conference on Computer Vision and Pattern Recognition, pp. 4700–4708 (2017)
14. Johnson, R., Zhang, T.: Effective use of word order for text categorization with convolutional neural networks. arXiv preprint arXiv:1412.1058 (2014)
15. Kenton, J.D.M.W.C., Toutanova, L.K.: Bert: Pre-training of deep bidirectional transformers for language understanding. In: Proceedings of NAACL-HLT, pp. 4171–4186 (2019)
16. Kim, W., Son, B., Kim, I.: Vilt: vision-and-language transformer without convolution or region supervision (2021)
17. Kim, Y.: Convolutional neural networks for sentence classification (2014)
18. Kingma, D.P., Ba, J.: Adam: a method for stochastic optimization (2017)
19. Kipf, T.N., Welling, M.: Semi-supervised classification with graph convolutional networks. In: International Conference on Learning Representations (ICLR) (2017)
20. Kukačka, J., Golkov, V., Cremers, D.: Regularization for deep learning: a taxonomy. arXiv preprint arXiv:1710.10686 (2017)
21. Lan, Z., Chen, M., Goodman, S., Gimpel, K., Sharma, P., Soricut, R.: Albert: a lite BERT for self-supervised learning of language representations. arXiv preprint arXiv:1909.11942 (2020)
22. Liang, B., Lou, C., Li, X., Gui, L., Yang, M., Xu, R.: Multi-modal sarcasm detection with interactive in-modal and cross-modal graphs. In: Proceedings of the 29th ACM International Conference on Multimedia, pp. 4707–4715 (2021)
23. Liang, B., et al.: Multi-modal sarcasm detection via cross-modal graph convolutional network. In: Proceedings of the 60th Annual Meeting of the Association for Computational Linguistics (Volume 1: Long Papers), pp. 1767–1777 (2022)
24. Liu, H., Wei, R., Tu, G., Lin, J., Liu, C., Jiang, D.: Sarcasm driven by sentiment: a sentiment-aware hierarchical fusion network for multimodal sarcasm detection. Inf. Fusion **108**, 102353 (2024)
25. Liu, J., Chang, W.C., Wu, Y., Yang, Y.: Deep learning for extreme multi-label text classification. In: Proceedings of the 40th International ACM SIGIR Conference on Research and Development in Information Retrieval, pp. 115–124 (2017)
26. Liu, Y., et al.: Roberta: a robustly optimized BERT pretraining approach (2019)
27. Liu, Z., Mao, H., Wu, C.Y., Feichtenhofer, C., Darrell, T., Xie, S.: A convnet for the 2020s. In: Proceedings of the IEEE/CVF Conference on Computer Vision and Pattern Recognition (CVPR) (2022)
28. Ma, X., Wang, R.: Personalized scientific paper recommendation based on heterogeneous graph representation. IEEE Access **7**, 79887–79894 (2019). https://doi.org/10.1109/ACCESS.2019.2923293
29. Malik, M., Tomás, D., Rosso, P.: How challenging is multimodal irony detection? In: International Conference on Applications of Natural Language to Information Systems, pp. 18–32. Springer (2023)
30. Morvant, E., Habrard, A., Ayache, S.: Majority vote of diverse classifiers for late fusion. In: Joint IAPR International Workshops on Statistical Techniques in Pattern Recognition (SPR) and Structural and Syntactic Pattern Recognition (SSPR), pp. 153–162. Springer (2014)
31. Pan, H., Lin, Z., Fu, P., Qi, Y., Wang, W.: Modeling intra and inter-modality incongruity for multi-modal sarcasm detection. In: Findings of the Association for Computational Linguistics: EMNLP 2020, pp. 1383–1392 (2020)

32. Peters, M.E., et al.: Deep contextualized word representations. arXiv preprint arXiv:1802.05365 (2018)
33. Radford, A., et al.: Learning transferable visual models from natural language supervision (2021)
34. Schifanella, R., de Juan, P., Tetreault, J., Cao, L.: Detecting sarcasm in multimodal social platforms. In: Proceedings of the 24th ACM International Conference on Multimedia. p. 1136–1145. MM '16, Association for Computing Machinery, New York, NY, USA (2016). https://doi.org/10.1145/2964284.2964321
35. Sharif, M.I., Khan, M.A., Alhussein, M., Aurangzeb, K., Raza, M.: A decision support system for multimodal brain tumor classification using deep learning. Complex Intell. Syst. **8**(4), 3007–3020 (2022)
36. Shi, C., Quan, J., Li, M.: Information extraction for computer science academic rankings system. In: 2013 International Conference on Cloud and Service Computing, pp. 69–76 (2013). https://doi.org/10.1109/CSC.2013.19
37. Son Chung, J., Senior, A., Vinyals, O., Zisserman, A.: Lip reading sentences in the wild. In: Proceedings of the IEEE Conference on Computer Vision and Pattern Recognition (CVPR) (2017)
38. Tomás, D., Ortega-Bueno, R., Zhang, G., Rosso, P., Schifanella, R.: Transformer-based models for multimodal irony detection. J. Ambient. Intell. Humaniz. Comput. **14**(6), 7399–7410 (2023)
39. Wang, H., Yue, T., Ye, X., He, Z., Li, B., Li, Y.: Revisit finetuning strategy for few-shot learning to transfer the emdeddings. In: The Eleventh International Conference on Learning Representations (2022)
40. Wang, P., et al.: Qwen2-VL: enhancing vision-language model's perception of the world at any resolution. arXiv preprint arXiv:2409.12191 (2024)
41. Wang, X., Sun, X., Yang, T., Wang, H.: Building a bridge: a method for image-text sarcasm detection without pretraining on image-text data. In: Proceedings of the First International Workshop on Natural Language Processing Beyond Text, pp. 19–29 (2020)
42. Xia, Y., Zhang, L., Liu, Z., Nie, L., Li, X.: Weakly supervised multimodal kernel for categorizing aerial photographs. IEEE Trans. Image Process. **26**(8), 3748–3758 (2017). https://doi.org/10.1109/TIP.2016.2639438
43. Xiong, T., Zhang, P., Zhu, H., Yang, Y.: Sarcasm detection with self-matching networks and low-rank bilinear pooling. In: The World Wide Web Conference, pp. 2115–2124 (2019)
44. Xu, N., Zeng, Z., Mao, W.: Reasoning with multimodal sarcastic tweets via modeling cross-modality contrast and semantic association. In: Proceedings of the 58th Annual Meeting of the Association for Computational Linguistics, pp. 3777–3786 (2020)
45. Yang, A., et al.: Qwen2. 5 technical report. arXiv preprint arXiv:2412.15115 (2024)
46. Yang, P., Sun, X., Li, W., Ma, S., Wu, W., Wang, H.: SGM: sequence generation model for multi-label classification. In: Proceedings of the 27th International Conference on Computational Linguistics, pp. 3915–3926 (2018)
47. Yang, Z., Yang, D., Dyer, C., He, X., Smola, A., Hovy, E.: Hierarchical attention networks for document classification. In: Proceedings of the 2016 Conference of the North American Chapter of the Association for Computational Linguistics: Human Language Technologies, pp. 1480–1489 (2016)
48. Yu, F., et al.: Ernie-vil: knowledge enhanced vision-language representations through scene graphs. In: Proceedings of the AAAI Conference on Artificial Intelligence, vol. 35, pp. 3208–3216 (2021)

49. Yue, T., He, Z., Li, C., Hu, Z., Li, Y.: Lightweight fine-grained classification for scientific paper. J. Intell. Fuzzy Syst. **43**(5), 5709–5719 (2022)
50. Yue, T., Li, Y., Shi, X., Qin, J., Fan, Z., Hu, Z.: Papernet: a dataset and benchmark for fine-grained paper classification. Appl. Sci. **12**(9), 4554 (2022)
51. Yue, T., Mao, R., Wang, H., Hu, Z., Cambria, E.: KnowleNet: knowledge fusion network for multimodal sarcasm detection. Inf. Fusion **100**, 101921 (2023). https://doi.org/10.1016/j.inffus.2023.101921
52. Yue, T., Shi, X., Mao, R., Hu, Z., Cambria, E.: Sarcnet: a multilingual multimodal sarcasm detection dataset. Image **1**, 0
53. Zhao, J., Ye, X., Yue, T., Li, Y.: CLDM: convolutional layer dropout module. Mach. Vis. Appl. **34**(4), 63 (2023)
54. Zhu, Y., Zhu, M., Liu, N., Xu, Z., Peng, Y.: Llava-phi: efficient multi-modal assistant with small language model. In: Proceedings of the 1st International Workshop on Efficient Multimedia Computing under Limited, pp. 18–22 (2024)

Animal Pose Estimation Based on YOLO-POSE

Binbin Zhou[1,2,3] and Lei Liu[1,2,3](✉)

[1] School of Computer Science and Technology, Xinjiang University, Urumgi 830046, Xinjiang, China
xjuliu@xju.edu.cn

[2] Xinjiang Key Laboratory of Signal Detection and Processing, Urumgi 830046, Xinjiang, China

[3] Joint International Research Laboratory of Silk Road Multilingual Cognitive Computing, Urumgi 830046, Xinjiang, China

Abstract. With the development of computer vision technology, deep learning-based pose estimation and target detection have been widely used in the fields of human behavior analysis and intelligent security. However, owing to the complexity of animal poses and the diversity of species, the existing pose estimation methods still face many challenges when applied to animal targets. To solve this problem, an improved YOLO-Pose model is proposed to improve the accuracy and efficiency of animal pose estimation. On the basis of the original YOLO-Pose model, a separable kernel attention mechanism is introduced and improved to make it conform to the animal target, and combined with the spatial pyramid pool of YOLO-Pose, the multiscale feature fusion capability of the model is improved. The experimental results show that the improved YOLO-Pose model achieves excellent performance on both the public animal pose dataset and the AP-10K dataset, significantly improving the ability of target detection and pose estimation.

Keywords: Animal · Pose estimation · LSKA · YOLO-Pose

1 Introduction

As a fundamental task in computer vision, human and animal pose estimation has found widespread applications across intelligent surveillance, human-computer interaction, and motion analysis domains. Traditional approaches predominantly follow a two-stage "detection-then-estimation" paradigm, where an object detector first identifies target instances before a separate pose estimation model predicts keypoints within each detection region. While achieving notable accuracy, this sequential processing introduces significant computational redundancy, often compromising real-time performance requirements in practical applications.

The evolution of pose estimation methods has led to several important developments. Early deep learning approaches such as convolutional pose machines (CPMs) [1] employed sequential convolutional networks to model spatial relationships, whereas

Y. Yu et al. (Eds.): ICPCSEE 2025, CCIS 2673, pp. 234–246, 2026.
https://doi.org/10.1007/978-981-95-2566-9_16

stacked hourglass networks [2] introduced symmetric encoder decoder architectures for multiscale feature fusion. The high-resolution network (HRNet) [3] further advanced the field by maintaining high-resolution feature representations through parallel mult-branch structures. Subsequent innovations have focused on optimizing various aspects of the two-stage paradigm: OKDHP [4] introduced online knowledge distillation to increase model efficiency, OpenPose [5] improved geometric modelling through Part Affinity Fields, and DisPose [6] addressed occlusion challenges via spatial dependency decoupling. Recent breakthroughs in two-stage methods include ViTPose++ [7], which combines vision transformers with deformable convolutions to achieve state-of-the-art performance on benchmark datasets, and DiffPose [8], which leverages diffusion models to refine keypoint localization in low-quality images. Despite these advances, two-stage approaches remain constrained by inherent limitations including computational inefficiency from repeated feature extraction, architectural bottlenecks that impede real-time processing, and limited generalizability to nonhuman targets.

In contrast, single-stage pose estimation methods have emerged as promising alternatives by unifying detection and keypoint prediction within an end-to-end framework. These approaches typically employ multitask learning architectures that share backbone feature extraction for simultaneous bounding box regression and keypoint coordinate prediction, significantly improving computational efficiency. Representative single-stage models include YOLO-Pose [9], RTM-Pose [10], and PETR [11], each offering distinct advantages in speed accuracy trade-offs [12]. Among these, YOLO-based architectures have gained particular prominence, because of their exceptional real-time performance, deployment simplicity, and hardware compatibility, leading to numerous variants [13] such as YOLO-Pose v2 [14] and YOLO-rlepose [15], which continue to push the boundaries of efficient pose estimation.

Despite these significant advancements, current single-stage pose estimation systems still face critical challenges that limit their practical deployment. The persistent difficulty in balancing model lightweightness with prediction accuracy remains a fundamental constraint, particularly for edge computing applications with stringent resource requirements. Furthermore, the predominant focus on human pose estimation in existing research has resulted in limited generalizability to diverse animal species, creating significant barriers for applications in agricultural monitoring, veterinary science, and wildlife conservation. These unresolved challenges underscore the need for continued innovation in developing unified, multispecies pose estimation frameworks that can maintain real-time performance while achieving robust accuracy across various target domains.

To address these challenges, this paper proposes a lightweight animal pose estimation framework based on YOLO-Pose, which is designed for multiscene, multispecies scenarios. By introducing and improving the large separable kernel attention (LSKA) [16] mechanism, the model achieves ultrafast inference while maintaining accuracy, providing a new solution for real-time pose analysis in edge computing. The contributions of this work are as follows:

- An improved large separable kernel attention modules is deeply integrated with the SPPF module. By utilizing large kernel space attention and channel decoupling, the accuracy of key point detection can be improved without adding additional computational overhead.

- Unlike most existing works that focus on a single species or specific scenario, this study targets animal detection across diverse real-world scenes, offering a reference framework for future research.

2 Method

Vision attention networks incorporating large kernel attention (LKA) modules have gained significant attention, because of their impressive performance. However, as the k_size increases in these networks, the memory demands of the deep convolutional layers within the large kernel attention module also increase proportionally. To address this challenge, Kin Wai Lau, Lai-Man Po, and others [16] conducted a series of experiments. Typically, for an input feature map $F \in R^{C\times H\times W}$, the output of the LKA can be computed via formulas (1), (2), and (3):

$$Z^C = \sum\nolimits_{H,W} W^C_{k\times k} * F^C \quad (1)$$

$$A^C = W_{1\times 1} * Z^C \quad (2)$$

$$\overline{F}^C = A^C \otimes F^C \quad (3)$$

In the formula above, C represents the count of input channels, $*$ indicates convolution, and $\otimes$ denotes the Hadamard product. H and W denote the height and width of the feature map, respectively. Z^C refers to the output obtained by convolving the feature map $\overline{F}^C$ with a $k \times k$ kernel, where F^C is the input feature map and where A^C is the attention map.

Next, the depthwise convolution with a large kernel is decomposed into two stages: first, a depthwise convolution using smaller kernels, then a dilated depthwise convolution with a relatively larger kernel. In this case, the output equation of the LKA is given by the following expressions (4), (5), (6), and (7):

$$\overline{Z}^C = \sum\nolimits_{H,W} W^C_{(2d-1)\times(2d-1)} * F^C \quad (4)$$

$$Z^C = \sum\nolimits_{H,W} W^C_{[k/d]\times[k/d]} * \overline{Z}^C \quad (5)$$

$$A^C = W_{1\times 1} * Z^C \quad (6)$$

$$\overline{F}^C = A^C \otimes F^C \quad (7)$$

Here, Z^C denotes the result of a depthwise convolution with a kernel size of $(2d - 1) \times (2d - 1)$, where d represents the dilation rate.. To complement this operation, $\overline{Z}^C$ captures local spatial cues and serves as a corrective mechanism for depthwise dilated convolutions, which utilize an effective kernel size of $\lfloor k/d \rfloor \times \lfloor k/d \rfloor$. The dilated depthwise convolution plays a key role in acquiring global spatial context from the depthwise convolutional output Z^C. Although this approach reduces the computational burden associated with large convolutional kernels, it still incurs significant computational and memory overhead when the kernel size exceeds 23 × 23, posing efficiency challenges for the VAN architecture.

To address this issue, Kin WaiLau, Lai-Man Po, and others [16] proposed an equivalent improved configuration for LKA by decomposing the two-dimensional weight kernel of depthwise convolution into two cascaded one-dimensional separable kernels. This modification of the LKA module is referred to as LSKA. The output of the LSKA is given by Eqs. (8), (9), (10), and (11):

$$\overline{Z}^{C} = \sum_{H,W} W^{C}_{(2d-1)\times 1} * (\sum_{H,W} W^{C}_{1\times(2d-1)} * F^{C}) \tag{8}$$

$$Z^{C} = \sum_{H,W} W^{C}_{[k/d]\times 1} * (\sum_{H,W} W^{C}_{1\times[k/d]} * \overline{Z}^{C}) \tag{9}$$

$$A^{C} = W_{1\times 1} * Z^{C} \tag{10}$$

$$\overline{F}^{C} = A^{C} \otimes F^{C} \tag{11}$$

The core innovation of the LSKA module lies in the dimensional decoupling of traditional large-kernel convolutions. By decomposing the two-dimensional depthwise convolution kernel into cascaded vertical and horizontal one-dimensional convolution kernels (K = K_v ⊗ K_h), the computational complexity is optimized from O(k2) to O(2k), achieving linear optimization. This orthogonal decomposition strategy imposes constraints on the parameter space, reducing the number of parameters to 1/k (where k is the kernel size) while maintaining an equivalent receptive field. Specifically, a tensor operation model can be established as follows:

$$LSKA(X) = Conv1D\_h(Conv1D\_v(X)) + Identity(X) \tag{12}$$

The formula represents a vertical 1D convolution of the input feature map X followed by a horizontal 1D convolution. Finally, the convolution results are added to the original input x, which is often used to enhance feature extraction while preserving the original input information to avoid information loss.

When the kernel size is expanded to 31 × 31, the traditional LKA module requires 4.82 GFLOPs, whereas LSKA only requires 0.96 GFLOPs, achieving a 5.02× improvement in computational efficiency. In terms of memory usage, LSKA reduces peak GPU memory consumption on ImageNet-1K by 62.3% (from 3.2 GB to 1.2 GB). This optimization stems from the rank constraint of the weight matrix—the decomposed dual one-dimensional convolution kernels have a rank of 2, whereas the original two-dimensional kernel has a rank of k2, significantly mitigating the dimensionality curse in matrix operations. The improved module introduces a BN-Dropout residual structure, forming a dual regularization mechanism. The batch normalization (BN) layer constrains the feature distribution to N(0,1), alleviating the covariate shift, whereas the random mask regularization of dropout enhances robustness. A progressive dropout strategy is employed, preserving detailed features in shallow layers and enhancing generalization in deeper layers. The residual connection constructs an identity mapping H(x) = F(x) + x, ensuring smooth gradient flow, and the experiments demonstrate that it improves the convergence speed. When integrating LSKA into the SPPF structure, a hierarchical kernel size strategy (7/11/19/23) is adopted, controlling the expansion rate of the receptive field through dilated convolutions. Cross-scale attention gates are established between levels of the feature pyramid, and sigmoid weighting is used to achieve feature selection.

In addition, residual connections enable more diverse information propagation within the network while preserving the details and contextual information of the original input data. This connection method ensures that no critical information is lost during the transmission process, thereby enhancing the model's performance in subsequent tasks. More importantly, residual connections also help alleviate the vanishing gradient problem, making the training of deep networks more manageable. The architecture of the improved model is illustrated in Fig. 1.

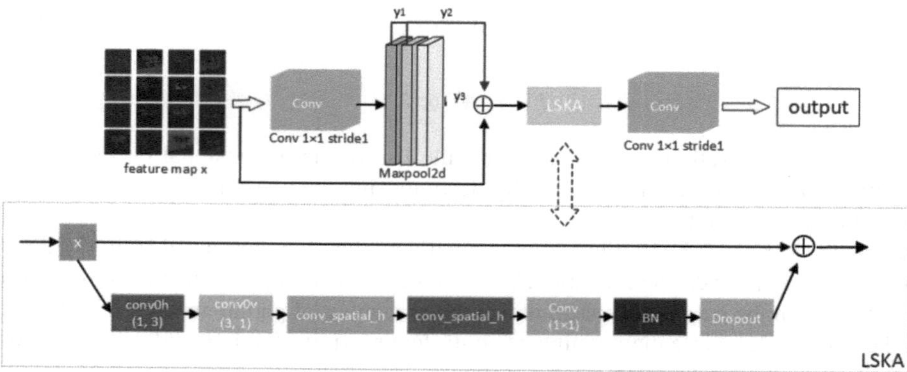

Fig. 1. Top: diagram of SPPF integrated with the LSKA structure; bottom: diagram of the improved LSKA structure.

In the LSKA-SPPF (Spatial Pyramid Pooling - Fast with Local Spatial Kernel Attention) structure, the feature graph x obtained in the backbone is convolved 1×1. The number of channels is adjusted and the features are preliminarily fused. Then the spatial information of the feature map is downsampled through the maximum pooling layer. Through the three-layer pooling operation, the spatial information of different scales can be captured, and the model can enhance the perception ability of objects of different sizes and output y1, y2 and y3. By combining the original feature map x with the feature maps y1, y2 and y3 after pooling at different scales, a feature representation containing rich multiscale information is formed. The LSKA module is designed to enhance the extraction of critical information from feature maps by emphasizing relevant regions and suppressing irrelevant ones. Through the integration of an attention mechanism, the model is guided to concentrate on feature regions that are most beneficial to the task. Subsequently, the feature map refined by the LSKA module undergoes a 1×1 convolution to adjust the channel dimensions and facilitate feature fusion. This operation aids in projecting the high-level representations obtained by the LSKA module into an output space aligned with the task objectives. The inherent multiscale feature fusion capability of the SPPF module is further strengthened through the incorporation of the LSKA module. The complete architectural design of the model is illustrated in Fig. 2.

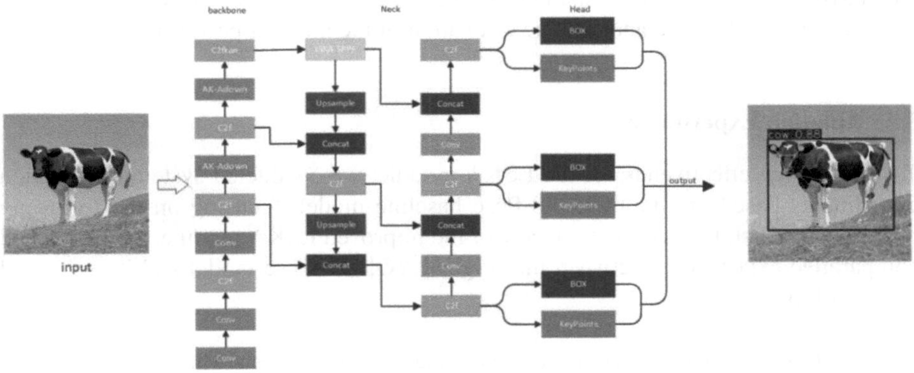

Fig. 2. The structure diagram of the improved YOLOv8-Pose model: Input an image to obtain the result diagram of its target detection and pose estimation.

3 Experiments and Analysis

3.1 Experimental Data Preparation and Environment

Datasets. In the experimental section, we conducted experiments using publicly available animal-pose datasets. These datasets are specifically designed for animal pose estimation tasks and include diverse poses (such as standing and running), significant occlusion issues, and considerable morphological differences among animals of the same species (e.g., different dog breeds). The animal-pose dataset [17] is highly diverse, features complex backgrounds, and provides consistent annotations. This approach not only fills the research gap in nonhuman pose estimation but also advances the development of cross-species pose analysis models. To further verify the effectiveness of our model, we conducted the same experiments on another publicly available dataset, AP-10K [18].

Benchmark Model and Experimental Preprocessing. This study employs YOLOv8-Pose as the baseline model for experimental improvements. To ensure data format consistency, the original datasets are processed as follows: the annotation files of the Animal-Pose datasets (20 keypoints) and the AP-10K datasets (17 keypoints) are converted into TXT files in YOLO format, where keypoint visibility is represented by 0 (invisible), 1 (occluded), and 2 (visible). For dataset partitioning, a stratified sampling method is adopted, dividing the data into training, validation, and test sets at a ratio of 7:2:1.

Experimental Environment. This paper is based on the Windows11 system. In terms of hardware configuration, an RTX4060 graphics card and Intel Core i7 processor are used. In the hyperparameter setting, the parameters are as follows: batch is set to 16, workers to 8, imgsz to 640, patience to 100, fraction to 1.0, mask_ratio to 4, max_det to 300, and plots to True. During the training process, the model was trained for 220 rounds in total, of which the last 10 rounds were processed with mosaic blur.

Evaluation Metrics. To thoroughly assess the proposed model's effectiveness in animal pose estimation, we employ the YOLOv8-Pose evaluation framework. This evaluation considers both object detection and keypoint localization capabilities. In particular, we emphasize detection accuracy, using mAP@50, precision, and recall as core metrics to provide a comprehensive and reliable evaluation of the model's performance.

3.2 Ablation Experiment

To validate the effectiveness of the LSKA module, we conducted systematic ablation experiments based on the YOLOv8-Pose baseline model, focusing on the following two aspects: First, to verify the efficacy of the improved LSKA module, we performed comparative experiments between the original YOLOv8Pose model and its enhanced version (Table 1).

Table 1. Comparison of LSKA before and after improvement with baseline.

Models	Params(M)	GFLOPs	Box			Pose		
			P	R	mAP50	P	R	mAP50
Yolov8n-pose	3357083	9.5	79.7	76.2	81.6	69.9	62.2	61.3
LSKA	3629979	9.7	81.6	71.2	81.7	72.6	59.6	59.4
Improve LSKA	3631003	9.7	81.3	77.3	84.4	82.5	61.2	67.5

The experimental results demonstrate that the improved LSKA module has excellent computational efficiency, with only a 0.8% increase in parameters and stable GFLOPs, confirming the effectiveness of its optimization. In the pose estimation task, the improved model shows significant advantages: a 6.2 percentage point improvement in mAP50, accompanied by a 12.6% increase in precision, despite a slight 1.0% decrease in recall. This performance shift indicates a substantial breakthrough in reducing false detections.

Table 2. Influence of different K_Sizes on the model performance.

k_size	Params(M)	GFLOPs	Box mAP50	Pose mAP50	FPS
7	3631003	9.7	84.4	67.5	26
11	3633051	9.7	85.1	67.9	22
23	3637147	9.7	83.2	65.4	19
35	3641243	9.7	84.9	66.4	17
53	3652947	9.7	85.5	65.2	14

To further investigate the impact of k_size on the multiscale feature extraction capability of LSKA, we conducted extended experiments comparing the effects of different convolutional kernel sizes (k_size = 7, 11, 23, etc.…). The experimental results are presented in the above table.

As shown in Table 2, when k_size increases from 7 to 53, the number of model parameters (Params) tends to monotonically increase (3.63 M → 3.65 M), whereas the computational complexity (GFLOPs) remains stable (9.7). In terms of detection performance, the box mAP50 fluctuates with increasing k_size: it peaks at 85.1% for k_size = 11, decreases to 83.2% for k_size = 23, but rebounds to 85.5% for k_size = 53. This suggests that both medium-sized kernels (11) and very large kernels (53) may capture features beneficial for object detection, albeit through different mechanisms.

For pose estimation tasks, Pose mAP50 tends to decrease overall (67.5% → 65.2%), indicating that excessively large convolutional kernels may impair keypoint localization accuracy. Notably, the inference speed (FPS) significantly decreases as k_size increases (26 → 14), with a pronounced inflection point observed at k_size ≥ 23. This highlights the need to balance receptive field expansion with real-time performance requirements in practical applications.

3.3 Comparison of Experimental Results

This study takes YOLOv8-Pose as the baseline model and conducts systematic comparative experiments to evaluate the performance of pose estimation algorithms such as YOLOv8n-pose, YOLOv9tiny-pose, YOLOv10n-pose, and YOLO11n-pose.

Table 3. Comparison of experimental results.

Models	Params(M)	GFLOPs	Box			Pose		
			P	R	mAP50	P	R	mAP50
Yolov8n-pose	3357083	9.5	79.7	76.2	81.6	69.9	62.2	61.3
Yolov9tiny-pose	2235819	8.9	81.6	71.2	81.7	72.6	59.6	59.4
Yolov10n-pose	3046826	9.6	78.5	65.2	74.7	68.5	54.0	54.8
Yolov11n-pose	2933587	7.7	81.7	70.8	81.0	73.5	61.1	60.9
Ours	3631003	9.7	81.3	77.3	84.4	82.5	61.2	67.5

As shown in Table 3, from the two task dimensions of object detection and pose estimation, key performance indicators such as precision (P), recall (R), and mAP50 are selected for quantitative analysis.

The experimental results show that the improved YOLOv8-Pose model outperforms the original model in terms of the evaluation metrics. For object detection, the mAP50 is 2.8% higher than that of the original method, and for pose estimation, it improves by 6.2%. Additionally, we compared data for five animal species (cat, dog, cow, horse, and sheep). The data are presented in Table 4 below.

Table 4. Comparison of the experiment data of cats, dogs, cows, horses, and sheep in the animal pose datasets.

Animal	Metric	Model				
		Yolov8n	Yolov9t	Yolov10n	Yolo11n	Ours
Dog	Box	78.6	77.9	75.4	81.6	81.8
	mAP50	87.8	81.0	79.3	86.0	87.8
	Pose	57.4	65.3	58.2	66.2	74.8
	mAP50	52.9	43.6	44.5	56.3	62.7
Cat	Box	82.5	82.8	79.7	88	86.8
	mAP50	90.3	92.3	81.6	91.0	92.8
	Pose	68.8	74.6	70.4	71.3	84.4
	mAP50	66.9	60.4	59	61.2	75.8
Sheep	Box	66.4	62.1	70.5	79.5	69.8
	mAP50	68	61.8	65.4	70.7	71.5
	Pose	59.1	67	62.2	80.6	77.5
	mAP50	49	43.9	47.7	64.8	57.6
Horse	Box	86.2	77.7	93.3	80.5	84.5
	mAP50	95.2	82.5	89.3	81	96.1
	Pose	80.6	82.7	82.5	73.1	83.8
	mAP50	78.8	65.5	71.8	59.6	76.1
Cow	Box	84.0	63.5	73.6	78.8	84.0
	mAP50	66.8	63.8	57.7	75.5	73.9
	Pose	83.8	75.2	69.2	76.4	92
	mAP50	59.1	47.7	51.1	63.5	65.3

Comprehensive Performance Improvement: The proposed method (Ours) has significant advantages in both object detection and pose estimation tasks across five major animal categories (dog, cat, cow, horse, and sheep). Notably, it outperforms the baseline model YOLOv8n in terms of pose mAP50 by +9.8% (dog), +8.9% (cat), and + 6.2% (cow). These results strongly validate the effectiveness of the LSKA module in modelling spatial relationships between keypoints.

Cross-Species Feature Analysis: For quadruped animals with complex joint structures (dog, cat, cow), our method shows particularly outstanding performance in pose estimation tasks, achieving Pose mAP50 scores of 62.7%, 75.8%, and 65.3%, respectively. This finding indicates that the LSKA module can effectively model anatomical constraints. Moreover, the model maintains stable performance in object detection tasks, with the box mAP50 exceeding 70% for all species. The detection accuracies for horses (96.1%) and cats (92.8%) are especially notable, highlighting the model's robust feature extraction capabilities.

Although certain models may outperform specific metrics, our method achieves more balanced performance across both detection and pose estimation tasks. This suggests that the integration of the LSKA and SPPF modules contributes to overall feature enhancement rather than task-specific overoptimization.

To further validate the performance of the model and comprehensively evaluate its adaptability and robustness in complex real-world scenarios, we conducted the same experiments on another widely used public dataset, AP-10k. The AP-10k dataset contains diverse real-world scene samples, which can better reflect the model's performance in authentic environments. Through experiments on this dataset, we were able to analyze the model's generalization capabilities more thoroughly and verify its stability and accuracy across different scenarios. The experimental results are shown in Fig. 3.

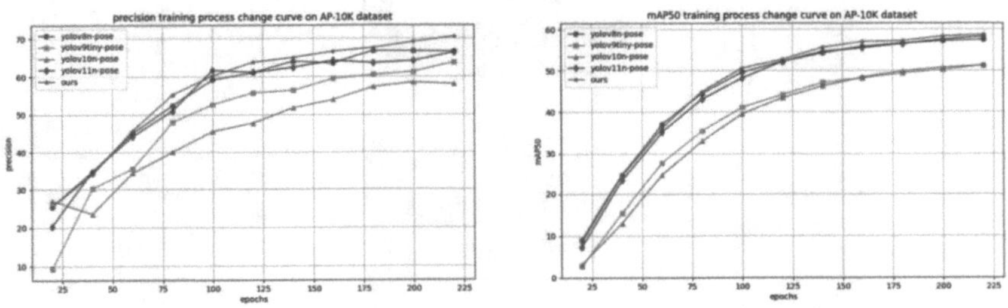

Fig. 3. On the left is the precision training process change curve, and on the right is the mAP50 training process change curve.

Our model achieves state-of-the-art performance on the AP-10K dataset. Both the precision and mAP50 curves consistently improve with training epochs, ultimately reaching peak values that demonstrate superior convergence and performance.

Table 5. The results of different models on the AP-10K dataset.

Model	Params(M)	GFLOPs	Input size	AP
vgg-16	19.1	16.3	192 × 256	66.3
ResNet-101	53.1	9.2	192 × 256	67.2
SimpleBasline	53.0	12.1	256 × 256	68.1
VitPose-S	24.3	5.4	192 × 256	67.6
HRFormer	43.3	13.3	192 × 256	70.3
Ours	3.6	9.5	640 × 640	71.6

We systematically assessed the performance of several mainstream pose estimation models on the AP-10K dataset. Table 5 presents a comparison of keypoint detection accuracy (primarily AP) and computational efficiency metrics, including parameter count and FLOPs, across various input resolutions.

The experimental results show that this model significantly improves the detection accuracy while maintaining a low number of parameters and computational complexity. This advantage enables it to demonstrate stronger competitiveness in a resource-constrained environment.

3.4 Visualization of Experimental Results

Figure 4 presents a systematic visual analysis of bovine pose estimation performance under four conditions: (a) high-resolution head close-ups, (b) complete body posture reconstruction, (c) multi-instance crowded scenes, and (d) partial occlusion scenarios.

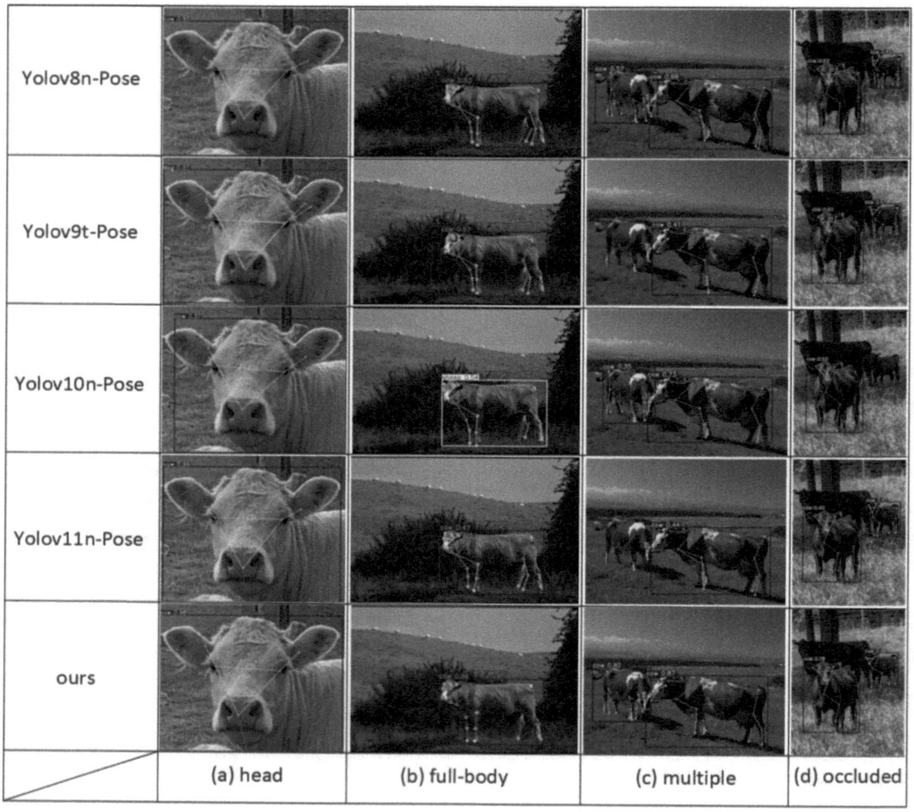

Fig. 4. Visual Comparative Analysis of Cattle Instance Detection Based on Five Detection Models.

In terms of object detection, the figure illustrates the performance of different models in detecting cattle targets. For single-cattle scenarios with head close-ups (a) and full-body poses (b), all the models demonstrate accurate bounding box localization. However, in multicattle scenarios (c), discrepancies emerge among the models. Some exhibit reduced detection accuracy and completeness under dense target conditions, occasionally missing detections or generating false positives. In contrast, our model maintains more stable performance, reliably detecting all the targets with tighter bounding box fits.

For pose estimation, all the models roughly capture keypoints in head close-ups (a), although with varying precision in terms of angle and position. In full-body poses (b), the YOLO series models exhibit deviations in estimating limb and body contours, whereas our model more precisely reconstructs the cattle's posture, with skeletal lines closely aligned to the ground truth. The challenge intensifies in multicattle (c) and occluded (d) scenarios, where pose estimation errors increase significantly for YOLO-based models. Our model, however, has relatively high accuracy, highlighting its enhanced ability to capture and infer pose information in complex environments.

Combining the insights of object detection and pose estimation, our model outperforms other YOLO-Pose models, especially in challenging scenarios (multiple objects and occlusion conditions). This method has high positioning accuracy and key point accuracy, as well as strong robustness and generalizability. These advantages emphasize its practical value in real-world animal detection and attitude analysis applications.

4 Conclusion

We propose an enhanced YOLOv8n-Pose model, built upon the original architecture, for animal pose estimation. To improve the fusion of multiscale features, we integrate an optimized LSKA mechanism with the SPPF module. Due to the complexity and variability of animal postures, conventional convolutional neural networks often struggle to extract discriminative features from animals with diverse sizes and shapes. The incorporation of separable convolutions in the LSKA module enables more refined processing of input features, thereby enhancing the model's ability to capture fine-grained spatial details. Although there was a slight improvement in animal pose estimation in this study, some obvious limitations remain. For example, the model's detection accuracy varies significantly across different species. In this work, while the improved YOLOv8-Pose model significantly improved the evaluation metrics for sheep, cats, and dogs, a slight decrease in performance was observed for horses and cows. These experimental results highlight the differences between animal species, which may lead to the model's inability to handle all animals equally well. Future research can explore advanced data augmentation strategies and transfer learning techniques to mitigate the limitations caused by the lack of sufficient animal pose datasets. Data augmentation can help enrich the dataset, whereas transfer learning can enable the application of advanced techniques from human pose estimation to animal pose estimation, thereby improving the model's performance in unseen environments. Additionally, real-time performance is a critical factor in practical applications. Although the proposed model demonstrates higher accuracy, further optimization of computational efficiency is still needed for real-time detection and pose estimation tasks.

Acknowledgment. This study was funded by the second batch of Tianchi Talents (Leading Talents) project in Xinjiang Uygur Autonomous Region. Project leader: Lei Liu from School of Computer Science and Technology, Xinjiang University.

References

1. Wei, S.E., Ramakrishna, V., Kanade, T., Sheikh, Y.: Convolutional pose machines. In: Proceedings of the IEEE Conference on Computer Vision and Pattern Recognition, pp. 4724–4732 (2016)
2. Newell, A., Yang, K., Deng, J.: Stacked hourglass networks for human pose estimation. In: Computer Vision–ECCV 2016: 14th European Conference, Amsterdam, The Netherlands, 11–14 October 2016, Proceedings, Part VIII 14, pp. 483–499. Springer, Cham (2016)
3. Sun, K., Xiao, B., Liu, D., Wang, J.: Deep high-resolution representation learning for human pose estimation. In: Proceedings of the IEEE/CVF Conference on Computer Vision and Pattern Recognition, pp. 5693–5703 (2019)
4. Li, Z., Ye, J., Song, M., Huang, Y., Pan, Z.: Online knowledge distillation for efficient pose estimation. In: Proceedings of the IEEE/CVF International Conference on Computer Vision, pp. 11740–11750 (2021)
5. Cao, Z., Hidalgo, G., Simon, T., Wei, S.E., Sheikh, Y.: Openpose: realtime multiperson 2D pose estimation using part affinity fields. IEEE Trans. Pattern Anal. Mach. Intell. **43**(1), 172–186 (2019)
6. Li, H., et al.: Dispose: disentangling pose guidance for controllable human image animation. arXiv preprint arXiv:2412.09349 (2024)
7. Xu, Y., Zhang, J., Zhang, Q., Tao, D.: Vitpose++: vision transformer for genericbody pose estimation. IEEE Trans. Pattern Anal. Mach. Intell. **46**(2), 1212–1230 (2023)
8. Gong, J., Foo, L.G., Fan, Z., Ke, Q., Rahmani, H., Liu, J.: Diffpose: toward morereliable 3D pose estimation. In: Proceedings of the IEEE/CVF Conference on Computer Vision and Pattern Recognition, pp. 13041–13051 (2023)
9. Maji, D., Nagori, S., Mathew, M., Poddar, D.: Yolo-pose: enhancing yolo for multiperson pose estimation using object keypoint similarity loss. In: Proceedings of the IEEE/CVF Conference on Computer Vision and Pattern Recognition, pp. 2637–2646 (2022)
10. Jiang, T., et al.: Rtmpose: real-time multi-person pose estimation based on mmpose. arXiv preprint arXiv:2303.07399 (2023)
11. Liu, Y., Wang, T., Zhang, X., Sun, J.: Petr: position embedding transformationfor multi-view 3D object detection. In: European Conference on Computer Vision, pp. 531–548. Springer, Cham (2022)
12. Su, Q., Zhang, J., Chen, M., Peng, H.: Pw-yolo-pose: a novel algorithm for poseestimation of power workers. IEEE Access (2024)
13. Yu, M., et al.: Yolomousepose: a novel framework and dataset for mouse pose estimation from a top-down view. IEEE Trans. Instrum. Measur. (2025)
14. Periyasamy, A.S., Amini, A., Tsaturyan, V., Behnke, S.: Yolopose v2: understanding and improving transformer-based 6d pose estimation. Robot. Auton. Syst. **168**, 104490 (2023)
15. Jiang, Y., Yang, K., Zhu, J., Qin, L.: Yolo-rlepose: improved yolo based on swintransformer and rle-oks loss for multi-person pose estimation. Electronics **13**(3), 563 (2024)
16. Lau, K.W., Po, L.M., Rehman, Y.A.U.: Large separable kernel attention: rethinking the large kernel attention design in CNN. Expert Syst. Appl. **236**, 121352 (2024)
17. Cao, J., et al.: Cross-domain adaptation for animal pose estimation. In: Proceedings of the IEEE/CVF International Conference on Computer Vision, pp. 9498–9507 (2019)
18. Yu, H., Xu, Y., Zhang, J., Zhao, W., Guan, Z., Tao, D.: Ap-10k: a benchmark for animal pose estimation in the wild. arXiv preprint arXiv:2108.12617 (2021)

Optimization Method for DOA Estimation at Low SNRs

Yize Wang[1,2,3] and Lei Liu[1,2,3](✉)

[1] School of Computer Science and Technology, Xinjiang University, Urumgi 830046, Xinjiang, China
xjuliu@xju.edu.cn
[2] Xinjiang Key Laboratory of Signal Detection and Processing, Urumgi 830046, Xinjiang, China
[3] Joint International Research Laboratory of Silk Road Multilingual Cognitive Computing, Urumgi 830046, Xinjiang, China

Abstract. In response to the issues of poor adaptability to low signal-to-noise ratios (SNRs) in existing uniform linear array (ULA) multitarget estimation algorithms and the difficulty of current deep learning methods in effectively extracting complex-valued features from data, a cross-scale sparse attention module and a channel-hierarchical spatial pyramid attention module, which are based on the MSPANet block, are introduced into the deep neural network (DNN). This approach better extracts multiscale features of signalling components, facilitating accurate signal feature extraction under low SNR conditions. Experimental data demonstrate that this deep learning model can significantly enhance the accuracy and anti-jamming capability of direction-of-arrival (DOA) estimation in low-signal-to-noise ratio (SNR) scenarios, outperforming traditional methods such as CBF, MUSIC, and ESPRIT. The above optimization achievements possess important practical value for DOA estimation applications in fields like intelligent speech, radar detection, communication systems, and autonomous driving.

Keywords: Direction of Arrival · Estimation Deep Convolutional Neural Network · Sparse Cross Attention

1 Introduction

As one of the key branches in the field of signal processing, array signal processing has been widely applied to systems such as radar, communication, and sonar in recent years, especially demonstrating unique advantages in direction-of-arrival (DOA) estimation technology. With the advancement of technology, the application scenarios for DOA estimation, such as assisting in determining the orientation of surrounding vehicles in autonomous driving and improving target recognition efficiency in radar systems, are continually increasing. Although classical DOA estimation techniques have become relatively mature, the limitations of traditional algorithms are gradually becoming apparent as the signal environment becomes more complex.

Y. Yu et al. (Eds.): ICPCSEE 2025, CCIS 2673, pp. 247–258, 2026.
https://doi.org/10.1007/978-981-95-2566-9_17

In the framework of traditional Direction of Arrival (DOA) estimation algorithms, the Multiple Signal Classification (MUSIC) algorithm and the Estimation of Signal Parameters via Rotational Invariance Techniques (ESPRIT) algorithm stand out prominently. These two algorithms can be regarded as significant milestones in the evolution of super-resolution techniques. The MUSIC algorithm utilizes the orthogonal properties of noise and signal subspaces, whereas the ESPRIT algorithm employs the rotational invariance relationship between subarrays for parameter estimation. Overall, traditional DOA estimation methods are model-driven, and under ideal conditions, their estimation accuracy can approach the Cramér–Rao bound. However, when there is model mismatch under nonideal conditions, the DOA estimation performance significantly deteriorates. Choi and Viberg, among others, proposed the alternating projection (AP) algorithm to reduce computational complexity [1]. With the continuous evolution of deep learning theories and methods, deep learning-based direction-of-arrival (DOA) estimation technology has gradually developed into an emerging research direction in the field of DOA estimation [2].

Currently, deep - learning - based DOA estimation approaches can be broadly classified into two types. One approach is to transform the direction-of-arrival (DOA) estimation problem into a neural network classification task. This approach achieves the learning of mapping relationships between network input data and DOAs by classifying spatial angles. Liu et al. [3] proposed a two-layer architecture, which uses multi-task autoencoders to preliminarily partition the parameter space and employs parallel multi-layer classifiers for secondary subdivision of sub-regions. This technical approach uses the real and imaginary parts of array-received data covariance as inputs for feature extraction, enabling high-precision direction estimation under array error conditions. Fan et al. [4] innovatively transformed direction-of-arrival (DOA) estimation into a problem within the object detection framework and proposed a new DOA estimation method named YOLO-DOA based on the YOLOv3 architecture. This method directly predicts the DOA of signal sources and their confidence scores from spectrum proxy data, achieving an end-to-end estimation process. By introducing a squeeze-and-excitation module, cross-layer partial connection mechanism, and an improved bounding box regression loss function, the overall performance of YOLO-DOA is effectively enhanced.

Another technical approach reconstructs the direction-of-arrival (DOA) estimation problem as a neural network regression task. Taking the research of Wu et al. [5] as an example, they modeled DOA estimation as a sparse linear inverse problem within the compressed sensing framework and constructed a deep convolutional network model. The network learns inverse transformation relationships through large-scale training datasets, and by leveraging the introduction of sparse prior knowledge, it achieves efficient real-time estimation of signal DOAs.

Bcause most existing deep learning-based DOA estimation methods focus on single or two targets, Elbir [6] proposed the DeepMUSIC deep learning framework, a multi-target DOA estimation algorithm that assumes that only one signal exists in each spatial subregion. The architecture employs a series of deep convolutional neural networks to segment sub-regions and model the associations between the real/imaginary portions of the received data and their respective MUSIC spatial spectral traits. The objective is to enable DOA estimation for multiple targets.

This paper addresses the issue of target estimation in uniform linear array (ULA) systems by transforming the DOA estimation problem into a neural network classification problem. First, the angle range is discretized, and each possible target direction is mapped to a category label. Then, the model extracts signal features through multiple convolutional and fully connected layers, optimizing the network's prediction via a classification loss function. Ultimately, the model outputs a probability distribution for angle categories, selecting the category with the highest probability as the estimated DOA of the signal. This approach effectively simplifies the continuous value prediction problem of DOA estimation, converting it into a discrete classification task that is easier to handle, and achieving high-precision DOA estimation at low SNRs.

2 Construction of the Signal Model and Dataset

2.1 Uniform Linear Array Model

The DOA estimation problem based on traditional algorithms is mostly solved by establishing a model that provides the array output signal vector multiplied by the array steering matrix to obtain the incident angle θ_k [7] of the signal source. Assuming that K far-field narrowband incoherent target signal sources $\boldsymbol{S}(t) = [s(t), \ldots, s_K(t)]^T$ and wave directions $\boldsymbol{\theta} = [\theta_1, \theta_2, \ldots, \theta_K]^T$ are incident on a uniform linear array with M array elements spaced at a distance of d, the array output signal vector $\boldsymbol{X}(t) = [x_1(t), x_2(t), \ldots, x_M(t)]^T$ can be represented as:

$$\boldsymbol{X}(t) = \sum_{k=1}^{K} \boldsymbol{a}(\theta_k)s_k(t) + \boldsymbol{N}(t) = \boldsymbol{A}(\boldsymbol{\theta})\boldsymbol{S}(t) + \boldsymbol{N}(t) \tag{1}$$

where $\boldsymbol{N}(t) = [n_1(t), n_2(t), \ldots, n_M(t)]^T$ represents Gaussian white noise with a mean of 0 and a variance of σ^2; $\boldsymbol{a}(\theta_k)$ is the steering vector of the incident signal with direction θ_k; and $\boldsymbol{A}(\theta)$ is the steering matrix of the array.

$$\boldsymbol{a}(\theta_k) = \left[1, e^{-j2\pi \frac{d}{\lambda} \sin\theta_k}, \cdots e^{-j2\pi (M-1)\frac{d}{\lambda} \sin\theta_k}\right]^T \tag{2}$$

$$\boldsymbol{A}(\boldsymbol{\theta}) = [\boldsymbol{a}(\theta_1), \boldsymbol{a}(\theta_2), \cdots \boldsymbol{a}(\theta_K)] \tag{3}$$

In the equation, λ represents the signal wavelength, and d represents the element spacing.

However, array processing algorithms do not directly estimate the direction of arrival from the actual output of the array elements but instead use the covariance matrix.

The covariance matrix of the array output X(t):

$$\boldsymbol{R}_{xx} = E[\boldsymbol{X}(t)\boldsymbol{X}^H(t)] = \boldsymbol{A}(\boldsymbol{\theta})\boldsymbol{R}_s\boldsymbol{A}^H(\boldsymbol{\theta}) + \sigma^2\boldsymbol{I} \tag{4}$$

In the equation, $E[\cdot]$ represents the mathematical expectation, and $\boldsymbol{I}$ is the identity matrix.

In this work, the appropriately preprocessed covariance matrix $\boldsymbol{R}_{xx}$ is used as the input for the DOA activation network and estimation network. Given that the covariance matrix $\boldsymbol{R}_{xx}$ is a Hermitian matrix, its upper and lower triangular regions contain identical information [8]. Additionally, the upper triangular portion of $\boldsymbol{R}_{xx}$ suffices to accomplish the estimation of the direction-of-arrival (DOA) for the target signal source while achieving higher precision [9]. Therefore, during the model training phase, the upper triangular matrix of $\boldsymbol{R}_{xx}$ is selected to perform feature extraction operations, which can effectively reduce the input data scale and computational complexity.

During signal generation, the true DOA is discretized into a set of angles within a specified range (e.g., from -90° to 90°), with each angle value corresponding to a category. Labelling is then performed, where the label is a binary vector, with the element corresponding to the true angle position set to 1, and the rest of the elements set to 0. In the testing phase, the model outputs a probability distribution for each angle category. The category with the highest probability in the output distribution is selected as the predicted DOA direction.

2.2 Construction of the Dataset

The signal generation in this paper is based on the uniform linear array (ULA) model. The number of signals is randomly generated between 1 and 5, with target incident angles in the range of $[-80°, 80°]$. The number of snapshots is 100, and the signal-to-noise ratio (SNR) range is $[-30\text{ dB}, 30\text{ dB}]$. The process is repeated 1000 times, resulting in a total of 12,000 samples, with the training and testing sets randomly selected at an 8:2 ratio.

3 Neural Network Model

In the realm of deep learning, Convolutional Neural Networks (CNNs) fall within the category of deep feedforward neural networks, distinguished by prominent characteristics such as local connection mechanisms and weight - sharing strategies. In comparison with fully connected feedforward networks, CNNs possess a smaller number of parameters. Their architecture is primarily formed by the alternate stacking of convolutional layers, pooling layers, and fully connected layers. As the most fundamental and crucial component of CNNs, the convolutional layer carries out the task of extracting feature information from the input data with the help of convolution kernels. The pooling layer, on the other hand, has the functions of realizing feature selection, reducing the feature dimension, and guarding against the phenomenon of overfitting. Within CNNs, the fully connected layer undertakes the mission of performing classification or regression analysis on samples.

This paper constructs a deep learning model by combining the multiscale sparse cross-attention (MSC) module [10] and the multiscale spatial pyramid attention (MSPA) module [11]. It transforms the DOA estimation problem into a classification problem, effectively simplifying the continuous value prediction problem of DOA estimation and converting it into a more manageable discrete classification task.

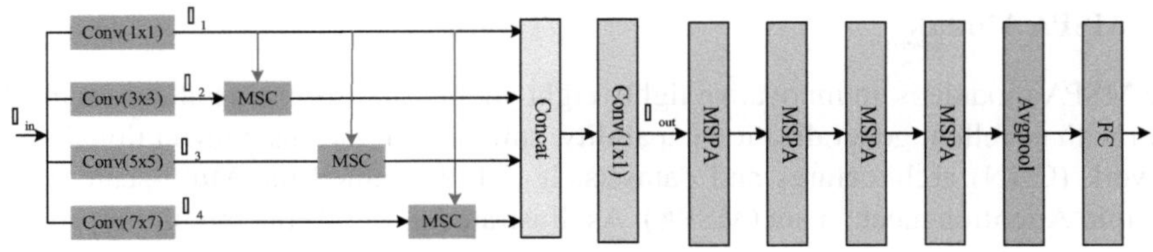

Fig. 1. Algorithm flowchart

3.1 MSC Module

The MSC module effectively suppresses the interference of irrelevant information through sparse operations, thereby achieving more accurate feature fusion results. Suppose there are two inputs, $\boldsymbol{X} \in \mathbb{R}^{h_1 \times w_1 \times c}$ and $\boldsymbol{Y} \in \mathbb{R}^{h_1 \times w_1 \times c}$, where h_1 and h_2, w_1 and w_2, and c represent the height, width, and channels, respectively. To capture the hidden multiscale clues in the input data, the MSC first applies three different window average pooling operations to $\boldsymbol{Y}$, generating Y_1, Y_2, and Y_3. Note that their shapes are consistent with $\boldsymbol{Y}$. Larger pooling windows capture the global structural information in the RS image, whereas smaller windows capture finer details. These pooled features are subsequently summed together to generate $\hat{Y}$, which contains rich multiscale information.

Next, $\boldsymbol{X}$ and $\hat{Y}$ are flattened and mapped to $\boldsymbol{Q} \in \mathbb{R}^{N_1 \times d}$, $\boldsymbol{K} \in \mathbb{R}^{N_2 \times d}$ and $\boldsymbol{V} \in \mathbb{R}^{N_2 \times d}$, where in the formula ($N_1$ equals $h_1 \times w_1$, N_1 equals $h_2 \times w_2$, d equals c) according to the following formula:

$$\boldsymbol{Q} = Flatten(\boldsymbol{X}) \cdot \boldsymbol{W}_1 \tag{5}$$

$$\boldsymbol{K}, \boldsymbol{V} = Split(LN(Flatten(\hat{Y}) \cdot \boldsymbol{W}_2)) \tag{6}$$

In this formula, $Flatten(\cdot)$ is the flattening operation, $Split(\cdot)$ is the uniform splitting operation along the channel dimension, $LN(\cdot)$ is the normalization function, and $\boldsymbol{W}_1$ and $\boldsymbol{W}_2$ are the linear projection parameters (Fig. 2).

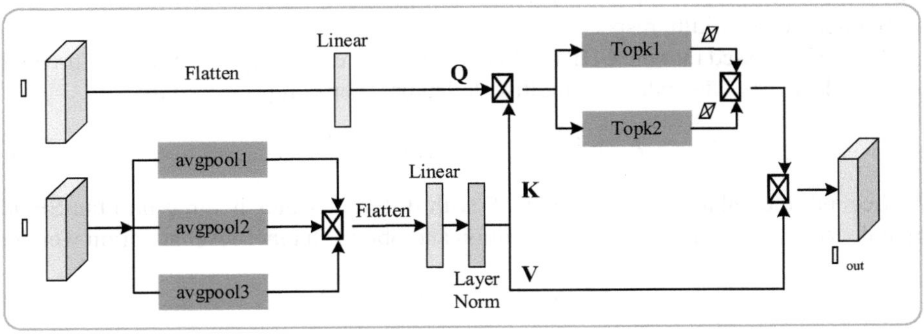

Fig. 2. MSN Module flowchart

3.2 MSPA Module

The MSPA module is an innovative lightweight yet high-performance attention mechanism with excellent generalization capability, suitable for various convolutional neural network (CNN) architectures and datasets. It is fully named the Multi-scale Spatial Pyramid Attention mechanism (MSPA). As illustrated in Fig. 1, the module is primarily composed of three core components.

First, distinct from the multiscale feature extractors implemented via layer-wise operations in CNNs, the hierarchical-phantom convolution (HPC) module can efficiently extract multiscale spatial information from the input feature maps in a finer-grained manner.

Second, the spatial pyramid recalibration direction (SPDR) module is utilized to establish cross-dimensional interactions, thereby learning channel attention weights for multiscale feature maps. In the SPDR method, a spatial pyramid aggregation module is employed to adaptively aggregate global and local feature responses, achieving an effective fusion of structural regularization and structural information. Concurrently, two computationally efficient pointwise convolution layers are utilized to learn inter-channel correlations. Finally, the Softmax operation is leveraged to recalibrate the attention weights of corresponding channels, thereby establishing long-range dependencies across channels (Fig. 3).

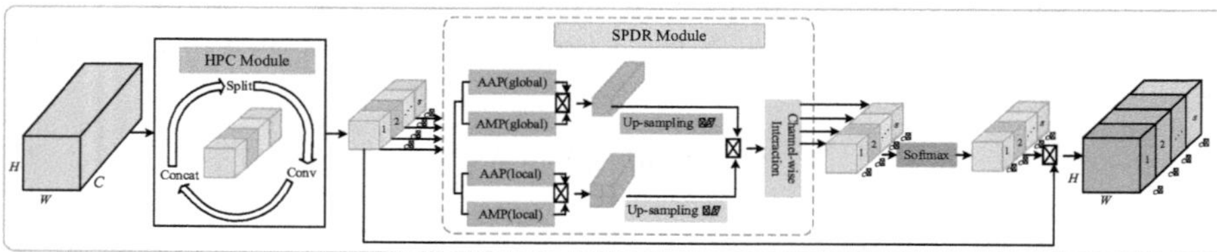

Fig. 3. MSPA module flowchart

The implementation of the MSPA module is divided into the following four steps. Specifically, let the intermediate feature map $\boldsymbol{F} \in \mathbb{R}^{C \times H \times W}$ be the input to the CHP module, where C represents the number of channels and where H and W represent the spatial height and width, respectively.

First, $\boldsymbol{F}$ is passed into the CHP module to obtain an efficiently processed feature map at multiple scales. The enhanced multiscale feature map is represented as

$$\boldsymbol{F}_i = [F_1, F_2, \cdots F_S] \in \mathbb{R}^{C \times H \times W} \tag{7}$$

Second, the enhanced feature map F with different scales is input into the SPDR module to efficiently learn channel attention and obtain channel-level attention weights

$$\boldsymbol{V}_i = [V_1, V_2, \cdots V_S] \in \mathbb{R}^{C \times 1 \times 1} \tag{8}$$

Third, cross-channel attention allows adaptive selection of different spatial dimensions while promoting interaction between local and global channel attention. Therefore, the channel-level attention weights $\boldsymbol{V}_i$ are passed through a Softmax function to generate the recalibrated channel-level attention weights $\boldsymbol{A}_i$.

$$\boldsymbol{A}_i = Soft\max(\boldsymbol{V}_i) = \frac{\exp(\boldsymbol{V}_i)}{\sum_{i=1}^{S} \exp(\boldsymbol{V}_i)} \tag{9}$$

Finally, for the i-th enhanced feature map subset $\boldsymbol{F}_i$, $\boldsymbol{A}_i$ is multiplied $\boldsymbol{F}_i$ to obtain the refined output feature map set $\hat{\boldsymbol{F}}_i$.

$$\hat{\boldsymbol{F}}_i = A_i \otimes \boldsymbol{F}_i \tag{10}$$

This paper introduces a dual-path pooling fusion strategy to improve the original SPR module, resulting in an enhanced SPDR module with parallel global and local branches. The global and local paths undergo adaptive average pooling (AAP) and adaptive max pooling (AMP), respectively, followed by upsampling and feature fusion. This design effectively captures both the global context and local details, thereby enhancing the model's ability to discriminate between multiple sound sources and significantly improving the representational capacity of the original SPR module. In the direction-of-arrival (DOA) estimation task, this architecture significantly enhances the model's ability to identify angular differences among multiple targets, making it particularly suitable for complex environments with low signal-to-noise ratios.

The MSPA module is integrated into the bottleneck residual blocks of ResNets [12], replacing the 3×3 convolution operations while maintaining the integrity of other structures, thus constructing the MSPANet backbone network..

3.3 The Algorithm in This Paper and the Comparison Algorithms

This paper carries out a comparative analysis of the proposed deep learning-driven convolutional neural network (CNN) direction-of-arrival (DOA) estimation algorithm with the CBF algorithm [13], ESPRIT algorithm [14], and MUSIC algorithm [15]. Among them, the CBF algorithm adjusts the array weights based on the minimum variance distortionless response (MVDR) criterion to construct the beamforming function; the MUSIC algorithm relies on the orthogonality property between the noise subspace and the signal subspace; while the ESPRIT algorithm achieves parameter estimation through the rotational invariance relationship between subarrays.

These methods have poor adaptability and high computational complexity under low signal-to-noise ratios (SNRs) and few snapshots. The deep convolution-based DOA estimation method designed in this paper effectively improves the DOA estimation performance under such conditions. This deep learning model transforms the DOA estimation problem into a classification problem. It first discretizes the angle range, mapping each possible target direction to a category label. Then, the model extracts signal features through multiple convolutional and fully connected layers, optimizing the network's prediction through a classification loss function. Ultimately, the model outputs a probability distribution for angle categories, selecting the category with the highest probability as the estimated DOA of the signal. This method effectively simplifies the continuous value prediction problem of DOA estimation, converting it into a discrete classification task that is easier to handle.

The cross-scale sparse attention module (MSC) is introduced to fully leverage the sparsity of the signal in the covariance while extracting signal features across different receptive fields. Convolutional branches with different kernel sizes are used to capture the signal features in the covariance, enabling better extraction of multiscale features of signalling components and effectively suppressing noise. Moreover, adaptive sparse attention is applied to achieve the organic fusion of multiscale features.

Additionally, a channel-hierarchical spatial pyramid attention module is introduced to better capture the context and key point features of the signal in the covariance matrix while refining the feature expression along the channel dimension. First, a channel-hierarchical strategy is employed, which uses different convolutional branches for each grouped channel to refine the channel feature representation. Then, a spatial pyramid dual-recalibration module is designed, which uses global average and max pooling to capture the context and key point information of the signal in the covariance matrix, aiding in the accurate extraction of signal features under low signal-to-noise ratio (SNR) conditions.

4 Simulation Results and Analysis

The experimental environment in this paper is PyCharm, where the CBF algorithm, ESPRIT algorithm, MUSIC algorithm, and proposed deep learning-based convolutional neural network DOA estimation algorithm are implemented. A comparative analysis is conducted on the estimation performance of different methods under various conditions. The simulation conditions are based on an ideal setup of a 16-element uniform linear array with an element spacing of $\lambda/2$, where λ is the signal wavelength. The target incident range is $[-80°, 80°]$, and the angle resolution is set to $1°$. The Monte Carlo method is used for statistical simulation to compare the DOA estimation accuracies of different algorithms under different conditions. The estimation accuracy performance is evaluated via the mean squared error (MSE), and the Root Mean Squared Error (RMSE) formula is as follows:

$$RMSE = \sqrt{\frac{1}{NK}\sum_{i=1}^{N}\sum_{j=1}^{K}\left(\hat{\theta}_{ij} - \theta_{ij}\right)^2} \quad (11)$$

where N represents the number of Monte Carlo trials; K is the number of target signal sources, and θ_{ij} and $\hat{\theta}_{ij}$ are the estimated and true DOA values of the j-th target in the i-th trial, respectively.

4.1 The Impact of the Signal-to-Noise Ratio on DOA Estimation

The comparison experiment conditions are set with 5 signal sources, a target incident range of $[-80°, 80°]$, 100 snapshots, and an SNR range of $[-20, 25]$ with a step interval of 5 dB. The CBF algorithm, ESPRIT algorithm, MUSIC algorithm, and the method proposed in this paper are compared. A comparison is made under the same parameters, with only the SNR varying, to evaluate the DOA estimation accuracy. The simulation involves and calculates the mean squared error (MSE) of 1000 Monte Carlo repetitions for the four methods. The relationship between the RMSE and the SNR is shown in the Fig. 4.

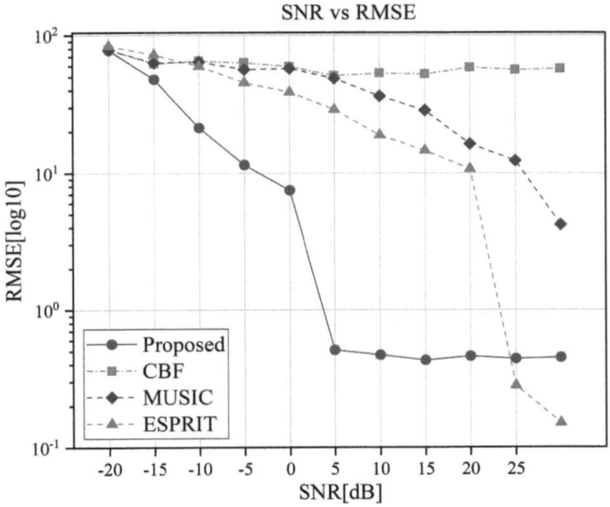

Fig. 4. The Impact of the Signal-to-Noise Ratio on DOA Estimation

The simulation results show the noise resistance performance of each DOA estimation algorithm. We compared and analysed the root mean squared error (RMSE) performance under different SNR conditions. As shown in the figure, the method proposed in this paper significantly outperforms traditional methods across the entire SNR range. Particularly under low-SNR conditions (-20 dB to 5 dB), this method maintains a low error, demonstrating strong noise resistance robustness. In contrast, CBF, MUSIC, and ESPRIT exhibit significantly greater errors in this range, with some methods only starting to converge at higher SNR values. The proposed method enters the error saturation region at an SNR of 5 dB, with its RMSE dropping below 0.3, much lower than those of the other methods, indicating its strong robustness and noise resistance capability.

4.2 Performance Comparison of Different DOA Estimation Methods

A comparison experiment was conducted with an SNR of 10, 100 snapshots, 5 target signals, an angular spacing of 10°, a target incident range of [−80°, 80°], and a step size of 1°. A performance comparison of CBF, MUSIC, ESPRIT, and the proposed method under low-SNR conditions was performed (Fig. 5).

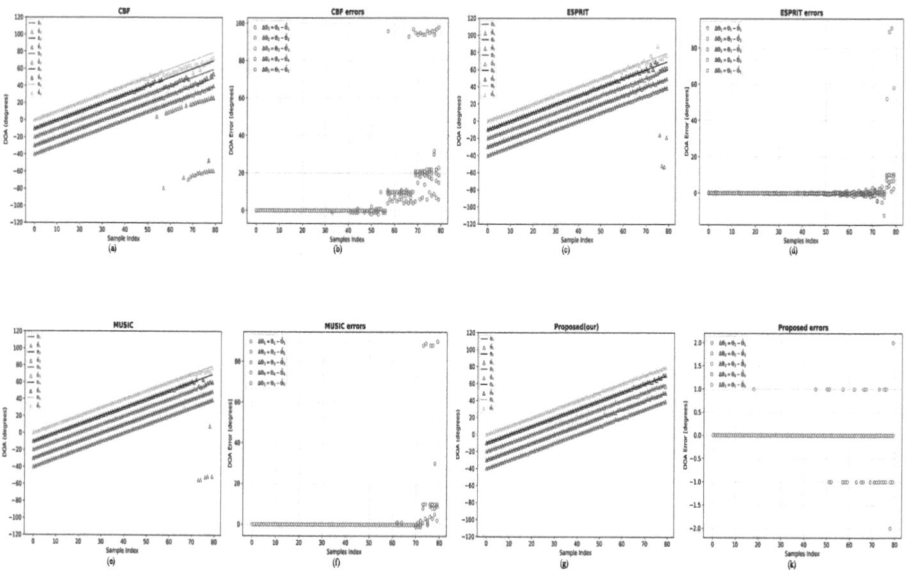

Fig. 5. Performance Comparison of Different DOA Methods

This paper compares four DOA estimation methods—CBF, ESPRIT, MUSIC, and the method proposed in this study—with the aim of evaluating their performance in low signal-to-noise ratio (SNR) environments. The experimental results indicate that the CBF method performs poorly under low-SNR conditions, especially when the signal complexity increases, as its estimation error rapidly increases, even exceeding 100 degrees, demonstrating the method's poor robustness. In contrast, the ESPRIT method can maintain relatively stable estimation results with larger sample sizes, although larger errors still occur at certain angles. The MUSIC method shows the most stable performance, maintaining smaller errors in most tests, particularly when the sample size is large, with a relatively slow increase in error, highlighting its strong robustness. However, among all methods, the proposed DOA estimation method performs the best, with estimation errors remaining close to zero, especially under high-SNR and complex signal conditions, demonstrating exceptional robustness and high accuracy. The experimental results show that the proposed method offers significant performance advantages over traditional methods in low-SNR and complex signal environments, providing more accurate DOA estimates, particularly for direction estimation tasks in high-noise environments.

5 Conclusion

In response to the poor adaptability of traditional algorithms under low signal-to-noise ratio (SNR) conditions, this paper proposes a deep convolutional neural network-based DOA estimation method. The core idea of this method is to transform the DOA estimation problem into a classification or regression problem by introducing a cross-scale sparse attention module (MSC) and a channel-hierarchical spatial pyramid attention module.

This method constructs a deep convolutional neural network to extract data features and learn the mapping from the input to the target's direction of arrival, thus achieving multitarget DOA estimation.

Experimental simulations show that the estimation performance of this algorithm outperforms those of the MUSIC, ESPRIT, and CBF algorithms under the same conditions. Particularly under low-SNR conditions, the estimation performance of the deep convolutional neural network-based DOA estimation method is significantly better than that of the other algorithms. Additionally, this method fully exploits the sparsity of the signal in the covariance matrix and extracts signal features across different receptive fields, enabling better extraction of multiscale features of the signalling components and effectively suppressing noise. It also uses global average and max pooling to capture contextual and key-point information of the signal in the covariance matrix, which is beneficial for accurately extracting signal features under low-SNR conditions. Compared with traditional algorithms, this method has stronger nonlinear modelling capabilities, especially when it performs better under low-SNR conditions.

Acknowledgment. This study was funded by the Xinjiang Uygur Autonomous Region Natural Science Foundation General Program (Project Number: 2023D01C18) and the second batch of Tianchi Talents (Leading Talents) project in Xinjiang Uygur Autonomous Region. Project leader: Lei Liu from School of Computer Science and Technology, Xinjiang University.

References

1. Choi, Y.H.: Alternating projection for maximum-likelihood source localization using eigendecomposition. IEEE Signal Process. Lett. **6**(4), 73–75 (1999)
2. Hinton, G.E.: Reducing the dimensionality of data with neural networks. Science **313**(5786), 504–507 (2006)
3. Liu, Z.M., Zhang, C., Philip, S.Y.: Direction-of-arrival estimation based on deep neural networks with robustness to array imperfections. IEEE Trans. Antennas Propag. **66**(12), 7315–7327 (2018)
4. Fan, R., Si, C., Yi, W., et al.: YOLO-DoA: a new data-driven method of DoA estimation based on YOLO neural network framework. IEEE Sens. Lett. **7**(2), 1–4 (2023)
5. Wu, L.L., Liu, Z.M., Huang, Z.T.: Deep convolution network for direction of arrival estimation with sparse prior. IEEE Signal Process. Lett. **26**(11), 1688–1692 (2019)
6. Elbira, M.: DeepMUSIC: multiple signal classification via deep learning. IEEE Sens. Lett. **4**(4), 1–4 (2020)
7. Yu, X., Xin, H.: Direction of arrival estimation utilizing incident angle dependent spectra. In: 2012 IEEE/MTT-S International Microwave Symposium Digest, Montreal, QC, Canada, pp. 1–3 (2012)
8. Chen, M., Gong, Y., Mao, X.: Deep neural network for estimation of direction of arrival with antenna array. IEEE Access **8**, 140688–140698 (2020)
9. Pastorino, M., Randazzo, A.: A smart antenna system for direction of arrival estimation based on a support vector regression. IEEE Trans. Antennas Propag. **53**(7), 2161–2168 (2005)
10. Ma, J., Jiang, W., Tang, X., et al.: Multiscale sparse cross-attention network for remote sensing scene classification. IEEE Trans. Geosci. Remote Sens. (2025)
11. Yu, Y., Zhang, Y., Cheng, Z., et al.: Multi-scale spatial pyramid attention mechanism for image recognition: an effective approach. Eng. Appl. Artif. Intell. **133**, 108261 (2024)

12. Xie, S., Girshick, R., Dollár, P., Tu, Z., He, K.: Aggregated residual transformations for deep neural networks. In: Proceedings of the IEEE Conference on Computer Vision and Pattern Recognition, pp. 1492–1500 (2017)
13. Bartlett, M.: Properties of sufficiency and statistical tests. Proc. R. Soc. A: Math. Phys. Eng. Sci. **160**(901), 268–282 (1937)
14. Roy, R., Kailath, T.: ESPRIT-estimation of signal parameters via rotational invariance techniques. IEEE Trans. Acoust. Speech Signal Process. **37**(7), 984–995 (1988)
15. Schmidt, R.: Multiple emitter location and signal parameter estimation. IEEE Trans. Antennas Propag. **34**(3), 276–280 (1986)

DOA Estimation Based on Transfer Learning and RAN-BiLSTM Under Limited Snapshots

Mingyan Li[1,2,3], Lei Liu[1,2,3](✉), and Wenjie Wang[1,2,3]

[1] School of Computer Science and Technology, Xinjiang University, Urumgi 830046, Xinjiang, China
xjuliu@xju.edu.cn

[2] Xinjiang Key Laboratory of Signal Detection and Processing, Urumgi 830046, Xinjiang, China

[3] Joint International Research Laboratory of Silk Road Multilingual Cognitive Computing, Urumgi 830046, Xinjiang, China

Abstract. To solve the problem of performance degradation caused by insufficient array signal feature representation under limited snapshot conditions, this research proposes a novel deep transfer learning approach specifically tailored for direction-of-arrival (DOA) estimation. Our approach synergistically combines the residual attention network (RAN) with the bidirectional long short-term memory network (BiLSTM) to construct a robust pretrained model that efficiently captures and extracts abundant features from complex signal data. The model is initially pretrained on a large dataset, allowing it to learn rich feature representations that are later fine-tuned for deployment in scenarios with restricted snapshots. Through this careful fine-tuning process, the pretrained knowledge is effectively transferred, resulting in significantly optimized model parameters and improved estimation performance. Experimental findings clearly show that our proposed method not only considerably improves the DOA estimation accuracy but also maintains excellent robustness under challenging snapshot-restricted conditions, providing a promising new solution for real-time array prediction in environments with limited data availability.

Keywords: Direction-of-arrival estimation · transfer learning · residual neural network · snapshot-deficient scenarios

1 Introduction

Direction of arrival (DOA) estimation is crucial in array signal processing and has applications in radar, sonar, wireless communication and sound tracking [1]. This technology is constrained by electromagnetic environments and hardware: noise and interference degrade performance, real-time processing limits snapshots, and traditional superresolution algorithms struggle because of high complexity and covariance estimation errors. Thus, researching low-complexity, superresolution DOA estimation for limited snapshots is highly valuable for engineering.

Y. Yu et al. (Eds.): ICPCSEE 2025, CCIS 2673, pp. 259–270, 2026.
https://doi.org/10.1007/978-981-95-2566-9_18

Over the past few decades, various prominent estimators based on beamforming, subspaces, and sparse representations have emerged as effective solutions for DOA estimation. Traditional beamforming is limited by the Rayleigh criterion [2] and cannot achieve high-resolution estimation. Since the 1980s, subspace-based methods such as multiple signal classification (MUSIC) [3] and estimation of signal parameters via rotational invariance techniques (ESPRIT) [4] have enabled superresolution capabilities but still have limitations.

In the era of deep learning development, data-driven techniques for DOA estimation have received much attention. Reference [5] used a CNN address the performance degradation of DOA estimation under low SNRs. [6] presented a DOA estimation approach leveraging DCN and Toeplitz to process signals. [7] designed a hybrid DNN structure, including autoencoder noise reduction, signal number detection by the FCNN, and DOA estimation by the DAG. [8] used ResNet to solve antenna array defects, and the signal was detected by a spatial classification network first and then classified by ResNet. In practical direction-finding systems, collecting sufficient labelled datasets is often challenging. To address this issue, [9] proposed an unsupervised learning strategy based on a sparse power spectrum and L1-norm optimization. [10] introduced unsupervised learning networks constructed by using threshold-weighted Capon spectrum punishment. [11] applied transfer learning to reduce the dependence on large amounts of training data.

The field of signal processing has seen substantial progress driven by deep learning in recent years, and transfer learning has become essential for addressing the problem of insufficient samples. Present deep learning-driven DOA estimation approaches tend to overfit when the number of snapshots is limited; moreover, most transfer strategies rely solely on generic pretrained models, overlooking the unique spatiotemporal structure of array signals.

To achieve accurate DOA estimation in low-snapshot scenarios, this paper proposes a transfer learning-based method whose pretraining framework, RAB (RAN+BiLSTM), is built from a residual attention network (RAN) and a bidirectional long short-term memory network (BiLSTM). The RAB framework innovatively combines their strengths: RAN's attention mechanism focuses on key spatial features, whereas BiLSTM more fully captures temporal correlations, outperforming traditional single-structure or simple hybrid networks. Unlike existing approaches that directly transfer parameters without accounting for scenario differences, this work first trains RAB on a large-scale dataset to obtain comprehensive feature representations and then applies carefully designed fine-tuning strategies to adapt it to target scenarios with limited snapshots. The main contributions of this study are summarized below:

- The pretrained model employs an RAB network structure to capture spatial feature changes dynamically over time, thereby enhancing adaptability to nonstationary signals.
- Using transfer learning, the RAB network architecture is first pretrained on a dataset of abundant snapshot DOA data and then fine-tuned in low-snapshot scenarios, effectively improving the DOA estimation accuracy under limited snapshot conditions.

2 Signal Model

Consider a uniform linear array (ULA) composed of N omnidirectional antennas with element spacing $d = \lambda/2$. K far-field narrowband signals $(k = 1, 2, \ldots, K)$ with wavelength λ impinge on the ULA from distinct directions $\theta = [\theta_1, \theta_2, \cdots, \theta_K]$. The received signal at the ULA at time t is given by:

$$x(t) = A(\theta)s(t) + n(t), t = 1..., L \tag{1}$$

Let L denote the number of snapshots. The signal source matrix and array manifold matrix are expressed as:

$$s(t) = [s_1(t), s_2(t), \cdots, s_K(t)]^T \tag{2}$$

$$A(\theta) = [a(\theta_1), \cdots, a(\theta_K)] \in C^{N \times K} \tag{3}$$

The additive noise n(t) is a zero-mean additive white Gaussian noise (AWGN) vector with a covariance matrix $\sigma^2 I$, where σ^2 denotes the noise power. The steering vector $\alpha(\theta)$ for the ULA corresponding to direction θ is defined as:

$$\alpha(\theta_K) = [1, e^{j\frac{2\pi d}{\lambda}\sin(\theta_k)}, \ldots, e^{j\frac{2\pi d}{\lambda}\sin(\theta_k)(N-1)}]^T \tag{4}$$

3 Proposed Method

3.1 Data Preprocessing and Labelling

To streamline the estimation process and circumvent covariance matrix calculation, we propose feeding the in-phase and quadrature (IQ) components of the received signal directly into the neural network for feature extraction. For the original received signal matrix X (Eq. 5), each row of which corresponds to the L-sample signal from an individual array element, we decompose the signal at each element into real and imaginary components:

$$X = \begin{bmatrix} x_1(1) & \cdots & x_1(L) \\ \vdots & \ddots & \vdots \\ x_{N(1)} & \cdots & x_N(L) \end{bmatrix} \tag{5}$$

$$X_{IQ} = [\Re(x(1)), \Re(x(L)), \ldots, \Im(x(1)), \Im(x(L))]^T \tag{6}$$

where $\Re(\cdot)$ and $\Im(\cdot)$ denote the real and imaginary parts, respectively. The resulting 2 L $\times$ N matrix X_{IQ} (Eq. 6) serves as the input to the subsequent neural network, leveraging deep learning's robust feature extraction capabilities to directly derive DOA-related information from raw signals.

DOA estimation is inherently a regression task. To address this issue via classification, we discretize the DOA range by dividing the angular domain of potential signal arrivals into equally spaced intervals, forming an overcomplete spatial grid set. Assuming that the incident signal directions lie within the range [θmin, θmax], we partition this angular domain into equal intervals of size $\Delta\theta$, generating the overcomplete spatial grid set $\mathcal{G}$:

$$\mathcal{G} = \{\theta_{\min}, \theta_{\min} + \Delta\theta, \cdots, \theta_{\max} - \Delta\theta, \theta_{\max}\} = \{\theta 1, \theta 2, \cdots, \theta_{g'}, \theta_G\} \quad (7)$$

This set can be regarded as a collection of all possible DOA angles for incident signals, containing $G = \frac{(\theta_{\max} - \theta_{\min})}{\Delta\theta} + 1$ elements. For multiple signal sources, the task reduces to predicting the most likely angle subset $\hat{\theta} = \left\{\hat{\theta}1, \hat{\theta}2, \cdots, \hat{\theta}K\right\}$ by maximizing the posterior probability given by:

$$\hat{\theta} = \underset{\theta \in \hat{\mathcal{G}}}{\text{argmax}}\ P(\theta|X),\ k = 1, 2, \cdots, K \quad (8)$$

where $P(\theta|X)$ denotes the posterior probability of angles θ given the received signal X.

3.2 DOA Estimation Model Based on Transfer Learning

Figure 1 presents the overall framework of the proposed method. Specifically, the RAB model is first pretrained on a signal dataset containing many snapshots. The pretrained model is then transferred to small-snapshot scenarios, where it undergoes fine-tuning to improve the DOA estimation accuracy.

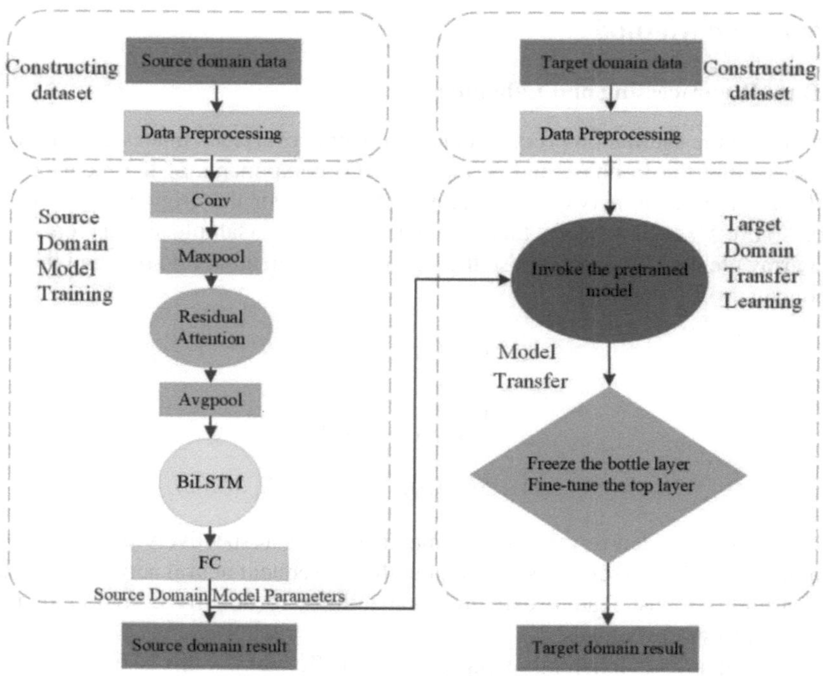

Fig. 1. Overall framework of the proposed method.

Model Construction. The RAB network (Fig. 2) combines RAN and BiLSTM to capture the nonlinear relationship between the signal and DOA. The model first extracts local features via dilated convolution and then highlights key information via residual networks integrated with efficient channel attention (ECA). BiLSTM further extracts temporal correlations and finally outputs multilabel classification results via activation functions in the fully connected layer and the sigmoid.

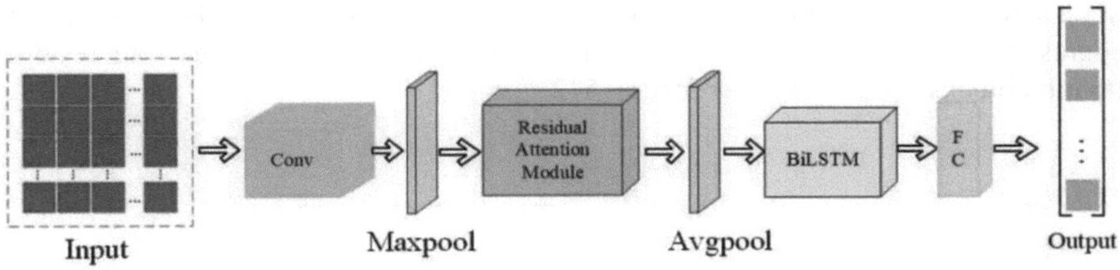

Fig. 2. Network architecture diagram.

To fully extract the information of array signals while reducing computational complexity, we adopt dilated convolution to expand the receptive field and decrease the number of parameters. Specifically, we use three 3 × 1 dilated convolution kernels with different dilation rates in parallel to extract multiscale spatial features, which are represented as:

$$\mathrm{X_i} = \mathrm{DilatedConv}_{3\times 1}^{\mathrm{d=i}}(\mathrm{X_{IQ}}),\ \mathrm{i} = 1, 2, 3 \tag{9}$$

Here, d represents the dilation rate. The output of each branch is sequentially processed via batch normalization (BN) and the Mish activation function to stabilize the training and enhance the nonlinear expression ability of the features.

$$\mathrm{X_i'} = \mathrm{Mish(BN(X_i))},\ \mathrm{i} = 1, 2, 3 \tag{10}$$

After using 1 × 1 convolution to reduce the channels of the features of each branch, we concatenate the dimension-reduced features of the three branches along the channel dimension to form a feature tensor that fuses multiscale information:

$$Xcat = Concat\{X_1', X_2', X_3'\} \tag{11}$$

We apply the MaxPooling operation to the fused features:

$$Xpool = MaxPool(Xcat) \tag{12}$$

This operation compresses the spatial dimension, highlights the local prominent features, and reduces the computational load.

In the design of deep networks, to alleviate the potential problems of gradient vanishing or gradient explosion during the training of deep networks [12, 13], this study introduces a residual attention module based on the ResNet structure. This module consists of four progressive residual attention blocks. Each residual attention block consists of a convolution, BN, an activation function, and a channel attention mechanism and enhances gradient transfer and feature reuse through skip connection (Fig. 3). In particular, the residual attention block employs ECA to capture interchannel relationships through global average pooling (GAP) and one-dimensional convolution (Conv1D) and uses a sigmoid function to generate attention weights for channel weighting. Using the progressive channel expansion strategy, the number of channels of the four residual attention blocks is 64 → 128 → 256 → 512.

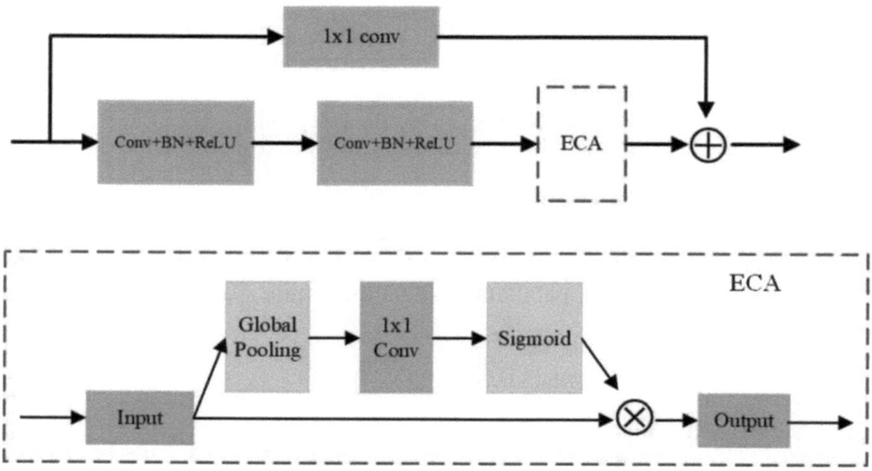

Fig. 3. Residual attention block.

After processing by the residual attention module, the dimensions of the network are reduced through global average pooling (AvgPool), preserving global semantics and reducing the risk of overfitting. BiLSTM then further models the sequence features and captures bidirectional context information via forward and backwards LSTM subnetworks. Compared with one-way LSTM, BiLSTM can extract time series correlations more comprehensively and enhance the signal representation ability. The structure of the BiLSTM is shown in Fig. 4.

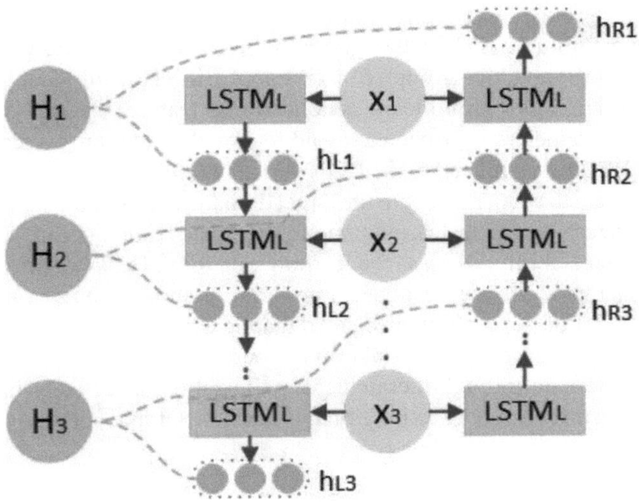

Fig. 4. Bidirectional long short-term memory.

After feature sequence encoding, the network maps the high-dimensional features from the BiLSTM output to the target label space via a fully connected (FC) layer. To address the multisource DOA estimation requirement, the final layer employs a sigmoid activation function to constrain outputs to the interval [0, 1]. For accurate prediction of signal presence across angles, we adopt binary cross-entropy (BCE) loss:

$$L_{BCE} = -\sum_{i=1}^{P}\left[y_i \log(\widehat{y_i}) + (1 - y_i)\log(1 - \widehat{y_i})\right] \tag{13}$$

where $y_i \in \{0, 1\}$ denotes the ground-truth label and where $\widehat{y_i} \in [0, 1]$ represents the predicted probability. Minimizing this loss enhances both the detection accuracy and robustness in multisource angle estimation.

Model Fine-Tuning. Transfer learning aims to apply knowledge acquired in the source domain to tasks in the target domain, with its core principle rooted in the correlation between data and tasks across different domains: through knowledge transfer, it reduces reliance on target-domain data and enhances training efficiency and generalization performance. Owing to limitations in hardware and signal acquisition, obtaining sufficient snapshot data is challenging. Fine-tuning addresses this issue by pretraining the model on a large-scale dataset to learn more general feature representations, enabling it to perform well even when data are scarce.

In this study, a grouped fine-tuning strategy is adopted. Specifically, network parameters are divided into a "basic feature group" (core layers of RAN, lower layers of BiLSTM) and a "task-adaptive group" (top layers of BiLSTM, fully connected layer). Layers in the basic feature group learn generic signal features during source-domain training, so their parameters are frozen to avoid feature degradation. Layers in the task-adaptive group are directly related to the DOA classification task and are fine-tuned with a higher learning rate to adapt quickly to the target domain. Additionally, a dual-learning-rate Adam optimizer is used: the learning rate for the basic feature group is set to 0.0001 to ensure parameter stability, whereas the learning rate for the task-adaptive group is set to 0.001 to accelerate the model response. Furthermore, through phased monitoring of gradients and validation set performance and dynamic adjustment of learning rates, this strategy achieves a balance between inheriting source-domain knowledge and adapting to the target domain through hierarchical optimization, effectively improving the accuracy and robustness of DOA estimation under small-sample conditions.

4 Experimental Results

4.1 Simulation Setup

We use simulations to generate the datasets. In the experiment, an 8-element uniform linear array (ULA) with a half-wavelength interelement spacing is employed. The angle coverage ranges from $-60°$ to $60°$, and the number of signal sources is set to 2. Signal data for the source domain (with 100 snapshots) and the target domain (with 10 snapshots) are generated at different signal-to-noise ratios (SNRs) ranging from -10 dB to 20 dB in increments of 5 dB, resulting in a total of 7 SNR levels. There are 200000 samples at each SNR level.

The source domain dataset is randomly partitioned into training, validation, and test subsets at an 8:1:1 ratio. These subsets are used to pretrain the model so that it can learn the array signal characteristics under many snapshots. During the fine-tuning phase, 5% of the target domain data are used, while the remaining 95% are set aside for testing purposes.

4.2 DoA Estimation Performance

To analyse the necessity of transfer learning for DOA estimation in low-snapshot scenarios, this research evaluates the performance (accuracy and root mean square error (RMSE)) of models with and without pretraining under varying signal-to-noise ratios (SNRs), as shown in Table 1 and Table 2. RMSE is used to evaluate the angular estimation precision:

$$RMSE = \sqrt{\frac{1}{n}\sum_{i=1}^{n}(y*(i) - y(i))^2} \tag{14}$$

Here, n denotes the number of independent estimation trials, with y and y* representing the ground-truth and estimated DOA vectors, respectively.

Table 1. Accuracy with or without pretraining.

SNR/dB	−10	−5	0	5	10	15	20
Pretrained	1.52%	12.51%	35.39%	56.48%	76.59%	85.11%	89.48%
Scratch	2.38%	16.57%	32.28%	46.46%	54.19%	67.41%	72.36%

Table 2. RMSE with or without pretraining.

SNR/dB	−10	−5	0	5	10	15	20
Pretrained	34.56	15.86	3.41	0.60	0.44	0.36	0.33
Scratch	43.52	27.35	7.39	2.64	1.25	0.73	0.51

The results show that the pretrained model significantly improves angular resolution accuracy in high-SNR ($SNR \geq 0$ dB) environments while reducing estimation errors in low-SNR ($SNR < 0$ dB) environments. Two mechanisms underlie this improvement: pretrained weights capture signal spatiotemporal invariance from large-scale data, enhancing high-dimensional pattern separability. Transfer learning also mitigates the gradient vanishing problem under limited data through constrained parameter initialization, avoiding suboptimal convergence. These findings confirm that the proposed pretraining framework enhances array DOA estimation robustness across SNR conditions via knowledge transfer.

Two narrowband signals with an angular separation of $\Delta\theta = 3.9°$ are simultaneously incident on the array at SNR = 10 dB. The first signal source direction, θ1, ranges from −60° to 56° in increments of 1°, whereas the second source direction, θ2, maintains a fixed separation of $\Delta\theta = 3.9°$ relative to θ1, generating a total of 116 samples. Figure 5(a) presents the estimated DOA of the RAB_TL method, where the solid lines represent the true angles. The corresponding estimation errors are shown in Fig. 6(a). For comparison, Figs. 5(b), 5(c), and 5(d) plot the DOAs of the DNN, ResNet, and transformer, respectively, and their corresponding errors are illustrated in Figs. 6(b), 6(c), and 6(d), respectively. The experimental results show that among all the test samples, RAB_TL outperforms the other methods, with an error range of [−1.0°, 1.0°]. Compared with the other three DL-based algorithms, the proposed algorithm has superior and stable DOA estimation performance.

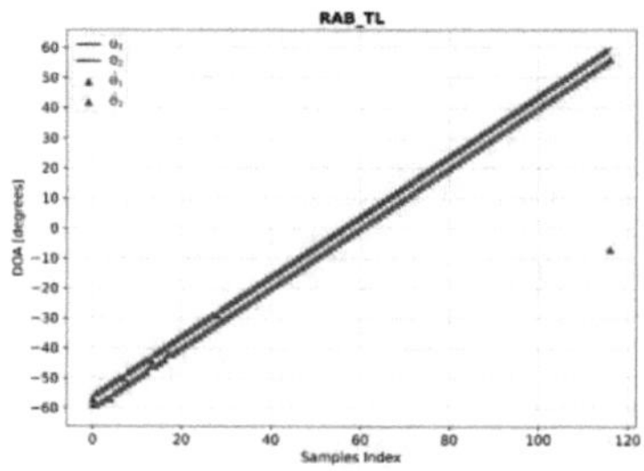

(a) RAB_TL

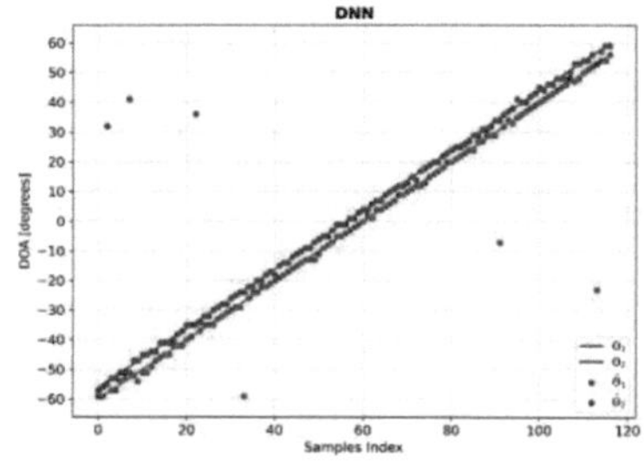

(b) DNN

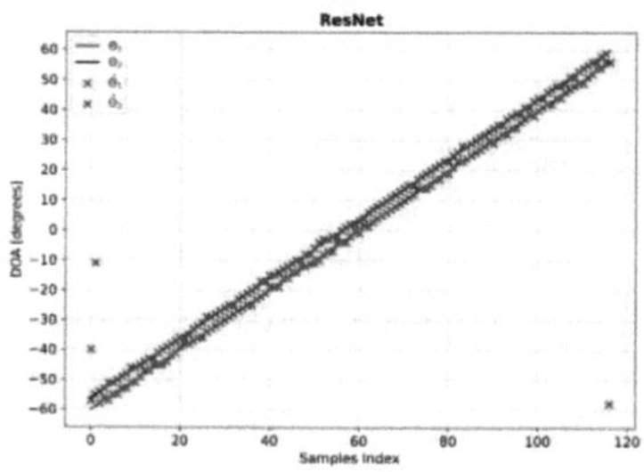

(c) ResNet

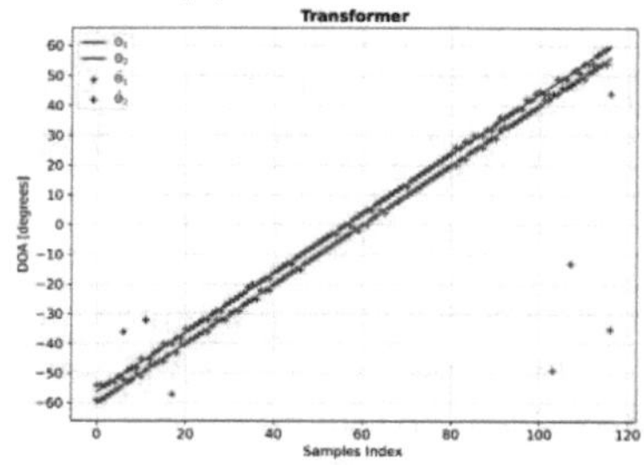

(d) transformer

Fig. 5. The DOA estimation performance is evaluated for angles θ1 and θ2 within the range of [−60°, 60°] at a 10 dB SNR with an interval of $\Delta\theta = 3.9°$: (a) the proposed RAB_TL, (b) DNN, (c) ResNet, and (d) transformer

Our study focuses on seven DOA estimation algorithms, namely, the MUSIC, ESPRIT, DNN, ResNet, SE-ResNet, transformer, and RAB_TL models, which are based on transfer learning. The performance was compared under the conditions that the SNR was within the range of [−10 dB, 20 dB] and that the number of snapshots was $L = 10$. As shown in Fig. 7(a), as the SNR increases, the detection accuracy rates of each algorithm significantly differ. The accuracy rates of the traditional algorithms MUSIC and ESPRIT are relatively limited, and the increase is slow in the low signal-to-noise ratio (SNR < 0 dB) region. Deep learning models such as DNN and ResNet perform better than traditional algorithms do, and the curves rise more steeply as the SNR increases. The RAB_TL model, which is based on transfer learning, performs the best, with the fastest convergence speed. Compared with other algorithms, its curve can reach a high accuracy level more quickly.

The results in Fig. 7(b) further confirm that the RAB_TL model not only accelerates the decrease rate of the root mean square error at a low SNR but also maintains a low error level among all algorithms. These results indicate that the RAB_TL model based on transfer learning achieves robustness across SNRs under the condition of a limited number of snapshots by means of feature inheritance and optimization stability mechanisms. This approach provides an efficient solution for direction-of-arrival estimation in complex environments. Across various SNR conditions, it surpasses other algorithms in both accuracy and error minimization.

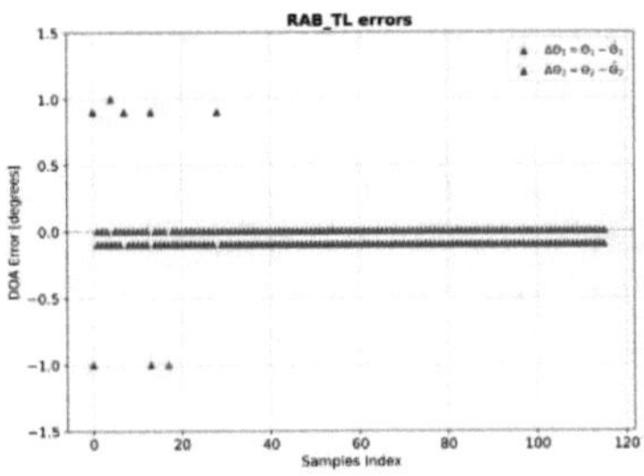

(a)RAB_TL errors

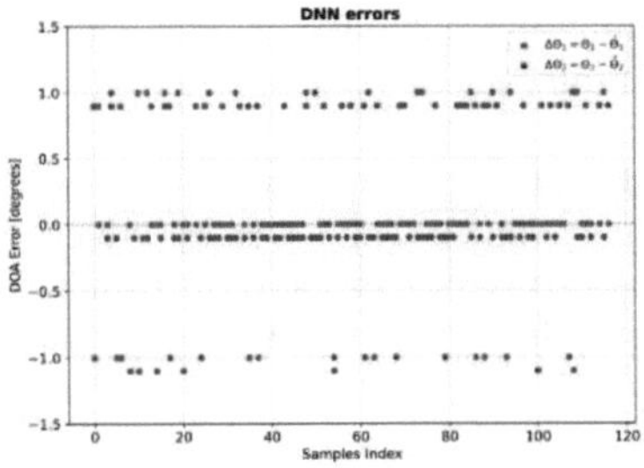

(b) DNN errors

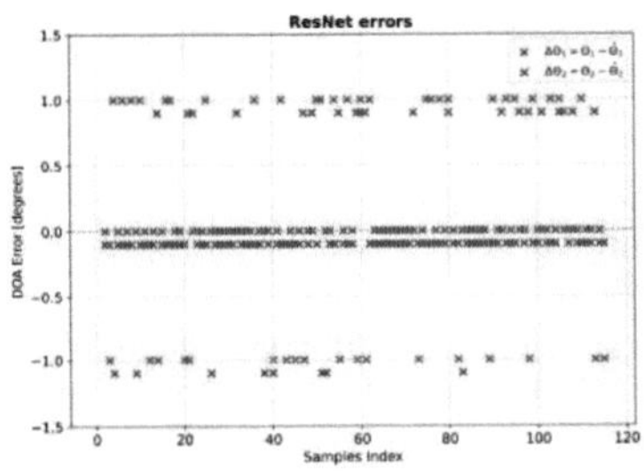

(c) ResNet errors

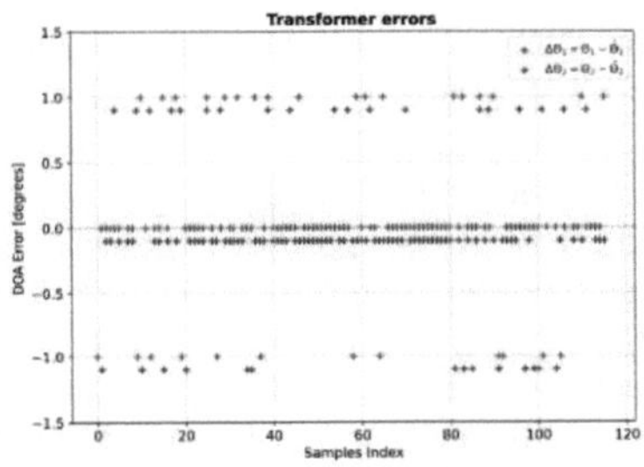

(d) transformer errors

Fig. 6. DOA estimation errors of different methods. (a) RAB_TL errors (b) DNN errors (c) ResNet errors (d) transformer errors

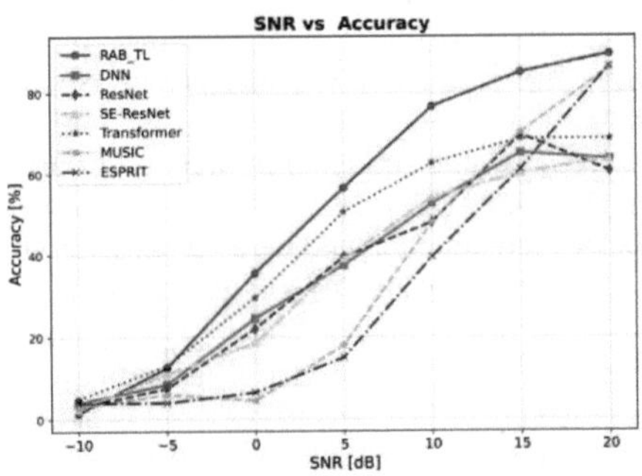

(a) Accuracy of DOA estimation

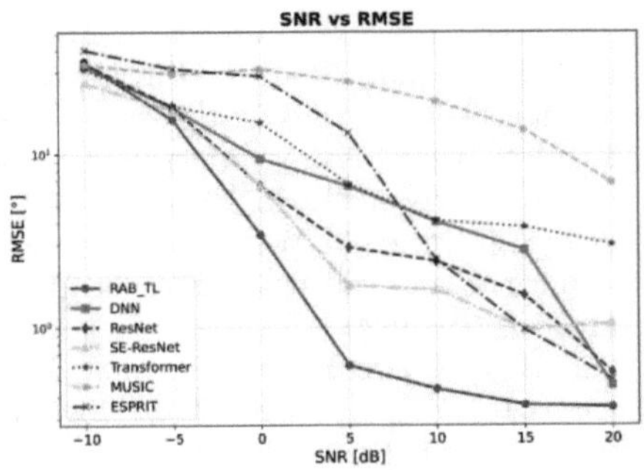

(b) RMSE of DOA estimation

Fig. 7. Performance comparison of DOA estimation under different SNRs.

5 Conclusion

This study employs transfer learning to address the DOA estimation problem under limited snapshots. Compared with traditional algorithms and deep learning models, the proposed method exhibits superior performance in accuracy and RMSE according to experimental results.This work introduces a novel approach for low-snapshot

DOA estimation, and future research will focus on multitarget scenarios and model lightweighting.

Acknowledgment. This study was funded by the Xinjiang Uygur Autonomous Region Natural Science Foundation General Program (Project Number: 2023D01C18) and the second batch of Tianchi Talents (Leading Talents) project in Xinjiang Uygur Autonomous Region. Project leader: Lei Liu from School of Computer Science and Technology, Xinjiang University.

References

1. Johnson, D.H., Dudgeon, D.E.: Array Signal Processing: Concepts and Techniques. Simon & Schuster, Inc. (1992)
2. Ge, S., Li, K., Rum, S.N.B.M.: Deep learning approach in DOA estimation: a systematic literature review. Mob. Inf. Syst. **2021**(1), 6392875 (2021)
3. Schmidt, R.: A signal subspace approach to multiple source location and spectral estimation. Ph.D. dissertation, Stanford University (1981)
4. Roy, R., Paulraj, A., Kailath, T.: Estimation of signal parameters via rotational invariance techniques-esprit. In: IEEE Military Communications Conference: Communications-Computers: Teamed for the 90's, MILCOM 1986, vol. 3, pp. 41–6. IEEE (1986)
5. Papageorgiou, G.K., Sellathurai, M., Eldar, Y.C.: Deep networks for direction-of-arrival estimation in low SNR. IEEE Trans. Signal Process. **69**, 3714–3729 (2021)
6. Duan, C., Tian, Y., Liu, W.: Deep convolution network based super resolution DOA estimation with Toeplitz and sparse prior. In:2024 IEEE Inter- national Conference on Acoustics, Speech and Signal Processing (ICASSP), ICASSP 2024, pp. 8581–8585. IEEE (2024)
7. Cong, J., Wang, X., Huang, M., Wan, L.: Robust DOA estimation method for mimo radar via deep neural networks. IEEE Sens. J. **21**(6), 7498–7507 (2020)
8. Li, J., Shao, X., Li, J., Ge, L.: Direction of arrival estimation of array defects based on deep neural network. Circuits Syst. Signal Process. **41**(9), 4906–4927 (2022)
9. Yuan, Y., Wu, S., Wu, M., Yuan, N.: Unsupervised learning strategy for direction-of-arrival estimation network. IEEE Signal Process. Lett. **28**, 1450–1454 (2021)
10. Zhang, Z., Qu, X., Li, W., Miao, H., Liu, F.: DOA estimation method based on unsupervised learning network with threshold capon spectrum weighted penalty. IEEE Signal Process. Lett. **31**, 701–705 (2024)
11. Cao, H., et al.: Deep transfer learning for underwater direction of arrival using one vector sensor. J. Acoust. Soc. Am. **149**(3), 1699–1711 (2021)
12. He, K., Sun, J.: Convolutional neural networks at constrained time cost. In: Proceedings of the IEEE Conference on Computer Vision and Pattern Recognition, pp. 5353–5360 (2015)
13. He, K., Zhang, X., Ren, S., Sun, J.: Deep residual learning for image recognition. In: Proceedings of the IEEE Conference on Computer Vision and Pattern Recognition, pp. 770–778 (2016)

Reinforcement Learning for Time-Series Query Optimization

Songling Zou[1], DongHua Yang[1], Mengmeng Li[1], Haifeng Guo[2], Hongqiang Wang[3(✉)], and Hongzhi Wang[1]

[1] Harbin Institute of Technology, Harbin, China
[2] School of Finance, Southwestern University of Finance and Economics, Chengdu 610000, China
[3] General Hospital of Northern Theater Command, Shenyang, China
18109882267@163.com

Abstract. Time-series databases (TSDBs) are essential for managing large-scale time-series data in fields like finance, IoT, and agriculture. However, traditional query optimization methods, such as dynamic programming, struggle with high computational complexity and inaccurate cost estimates. This paper proposes a novel query optimization module for TSDBs using reinforcement learning (RL), specifically Deep Q-Networks (DQN) and Double Deep Q-Networks (DDQN). These algorithms dynamically learn optimal join orders based on query workloads and connection costs. Experiments show that RL-based methods achieve better optimization performance and stability compared to traditional heuristics, especially under complex cost models. This work highlights the potential of RL in improving query optimization for TSDBs.

Keywords: Distributed Indexing · LSM-Tree · Non-Primary Key Query

1 Introduction

With the rapid growth of time-series data in various domains such as finance, IoT, agriculture, and power systems, time-series databases (TSDBs) have become essential tools for efficient data management and analysis. However, the increasing complexity of queries and large-scale datasets pose significant challenges for query optimization, a critical component of database systems. Traditional query optimization methods, such as dynamic programming (DP) [7] and heuristic rules [9], face limitations due to their high computational complexity and reliance on inaccurate cost estimates. To address these challenges, this paper proposes a novel query optimization module for TSDBs using reinforcement learning (RL). Specifically, we employ Deep QNetworks (DQN) and Double Deep Q-Networks (DDQN) to dynamically learn optimal join orders based on observed query workloads and connection costs. Our approach adapts to different database instances and query patterns, offering significant improvements over traditional heuristic methods. The key contributions of this work include:

– A reinforcement learning framework for query optimization in TSDBs.

Y. Yu et al. (Eds.): ICPCSEE 2025, CCIS 2673, pp. 271–281, 2026.
https://doi.org/10.1007/978-981-95-2566-9_19

- Feature encoding and selection strategies tailored for time-series data.
- Comparative experiments demonstrating the effectiveness of our approach.

The remainder of this paper is organized as follows. Section 2 reviews related work in query optimization. Section 3 details our proposed methodology, including problem formulation and the RL algorithms used. Section 4 presents the experimental setup and results. Finally, Sect. 5 concludes the paper and outlines future work.

2 Related Work

Query optimization is a critical component of database management systems, aiming to find the most efficient execution plan for a given query. This section reviews the existing literature on query optimization methods, focusing on both non-learning and learning-based approaches.

2.1 Non-learning Methods

Traditional query optimization methods primarily rely on non-learning techniques, such as dynamic programming and heuristic rules. Selinger et al. [7] proposed a dynamic programming algorithm to select the optimal join order for a given query. This method, known as the System R optimizer, laid the foundation for modern query optimization. However, its computational complexity grows exponentially with the number of joins, making it impractical for large-scale queries.

Vance and Maier [9] extended the dynamic programming approach to consider different join orders and subgraphs, aiming to improve efficiency. Moerkotte and Neumann [6] further enhanced this method by introducing a graph-based dynamic programming algorithm that avoids unnecessary checks for connectivity and combination, significantly improving performance.

Top-down join enumeration strategies have also been explored. DeHaan and Tompa [1] proposed TDMinCutLazy, which uses a top-down approach with pruning strategies to reduce the search space. Fender and Moerkotte [2] introduced TDMinCutBranch, which employs a graph-based enumeration strategy to generate valid joins more efficiently. These top-down methods offer better scalability compared to traditional dynamic programming.

2.2 Learning-Based Methods

Recently, learning-based methods have gained significant attention in query optimization. These methods leverage machine learning techniques, particularly reinforcement learning (RL), to dynamically learn optimal join orders based on observed query workloads and connection costs. ReJoin [5] employs deep reinforcement learning to identify the optimal join order, using the cost model of the optimizer as the reward signal.

Heitz and Stockinger [3] proposed a matrix-based representation for query optimization, encoding each table or subquery as a row in a matrix. DQ [4] uses one-hot encoding to represent visible attributes in subqueries and physical operator choices, training the model with costs from the optimizer's cost model and real execution times.

Yu et al. [10] combined reinforcement learning with tree-LSTM to capture more structural information in query trees, demonstrating improved performance in join order selection. SkinnerDB [8] introduced a regret-bounded query evaluation approach using the UCT algorithm, learning from previous query executions.

These learning-based methods show significant potential in improving query optimization performance by adapting to different database instances and query patterns. However, integrating these methods into practical systems remains an open challenge.

While traditional non-learning methods have been widely adopted, they face limitations in handling large-scale datasets and complex queries due to their computational complexity and reliance on inaccurate cost estimates. Learning-based methods, on the other hand, offer adaptability and scalability but require extensive training and validation. Our proposed RL-based approach addresses these limitations by dynamically learning optimal join orders from observed query workloads, adapting to different database instances and query patterns without relying on fixed heuristic rules. This adaptability makes our approach more robust and scalable, especially under complex cost models and changing query environments.

3 Methodology

In this section, we elaborate on the methodology employed to design and implement a cost-based query optimization module for time-series databases. The primary focus is on leveraging reinforcement learning (RL) techniques, specifically Deep Q-Networks (DQN) and its variant, Double Deep Q-Networks (DDQN), to optimize query plans. We also discuss the feature encoding process, normalization, and optimization techniques used to enhance the performance of our approach.

3.1 Model Construction

The query optimization problem is modeled as a Markov Decision Process (MDP) to facilitate the application of reinforcement learning. The MDP is defined by the following components:

- **State Space (S)**: Represents the set of all possible sub-trees formed during the query join process.
- **Action Space (A)**: Consists of all possible join operations that can be performed on the current sub-trees. The size of the action space is greater than $n!$, where n is the number of relations in the query.
- **Transition Probability ($P(s, a)$)**: Describes the probability of transitioning to a new state given the current state and action.
- **Initial State (s_0)**: Represents the initial query graph with unjoined base tables.
- **Reward Function ($R(s, a)$)**: Measures the cost of performing a join operation, such as the time required for a hash join.

3.2 Feature Encoding

To enable the reinforcement learning algorithm to interact with the environment, we encode the database and query information into feature vectors that can be processed by a neural network. The encoding process includes:

- **State Encoding**: Each state is represented as a binary vector, where each bit corresponds to a column in the database tables. The vector is divided into two parts: the first part indicates the final state of the query graph, and the second part indicates the columns involved in the join operations performed so far. This encoding method is illustrated in below (Fig. 1).

Fig. 1. State Encoding Example

- **Action Encoding**: Actions are encoded by setting the bits corresponding to the columns of the left and right tables involved in the join operation to 1.
- **Selectivity Encoding**: To handle single-table predicates, we adjust the feature representation by incorporating selectivity estimates obtained from the database statistics. For example, if the selectivity of a predicate ($A.a > 200$) is estimated to be 0.2, the corresponding bit in the query graph is updated to 0.2.
- **Physical Operator Encoding**: We add an additional one-hot vector to represent the choice of physical operators, such as index join and hash join.

3.3 Normalization

The cost values generated by the cost model in large databases can vary widely, while the reward values used in RL algorithms typically fall within a narrow range (e.g., $-10\,to\,10$). To normalize the cost values, we use a square root function as shown in Eq. 1:

$$\text{Normalized Cost} = \sqrt{\frac{\text{Cost}}{\text{Upper Bound}}} \tag{1}$$

This normalization technique ensures that the cost values are scaled appropriately for the RL algorithm.

3.4 Model Training and Optimization

We employ a multi-layer perceptron neural network to approximate the Q-function, which takes the query sub-graph and join actions as inputs. The network is trained using the Stochastic Gradient Descent (SGD) algorithm. The training process involves the following steps:

1. Initialize the Q-network and target-Q network.
2. For each step, select an action with the highest Q-value (with probability $1 - \epsilon$) or a random action (with probability ϵ).
3. Obtain the reward and new state from the environment.
4. Store the experience tuple (s, a, r, s') in a replay buffer.
5. Sample a batch of experience tuples from the buffer and update the Q-network using the loss function defined in Eq. 2.
6. Periodically update the target-Q network to stabilize training.

$$\text{Loss} = \left(R + \gamma \max_{a'} Q(s', a') - Q(s, a)\right)^2 \tag{2}$$

Optimization Techniques. To enhance the stability and efficiency of the training process, we employ several optimization techniques:

- **Target-Q Network**: To prevent the Q-network from oscillating during updates, we introduce a target-Q network that provides fixed target values for the Q-network updates. The target-Q network is periodically updated with the weights of the Q-network (Fig. 2).

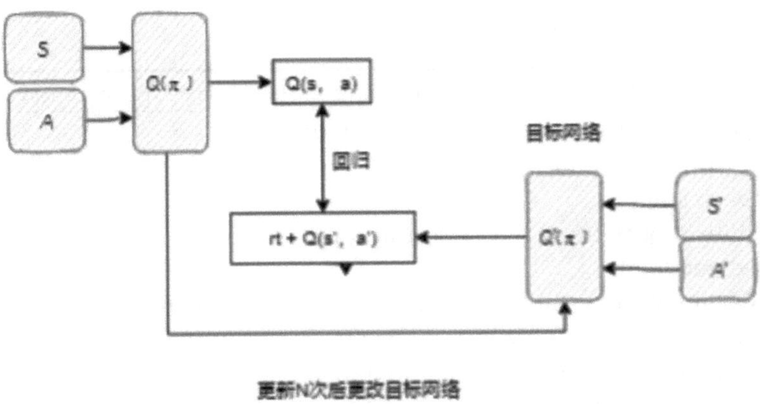

Fig. 2. Target-Q Network Architecture

- **Experience Replay Buffer**: We store experience tuples in a replay buffer to reduce the number of interactions with the environment and to ensure that the training data is diverse. This approach helps prevent overfitting and improves the stability of the training process (Fig. 3).

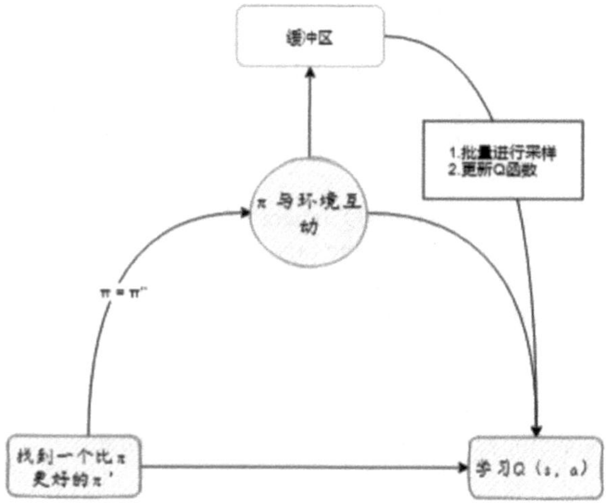

Fig. 3. Experience Replay Buffer Architecture

- **Double Deep Q-Networks (DDQN)**: To address the overestimation bias in DQN, we implement DDQN, which decouples the action selection and value estimation processes. The Q-network is used to select the action with the highest value, while the target-Q network is used to estimate the value of that action.

3.5 Algorithm Workflow

The complete algorithm for our reinforcement learning-based query optimization is summarized in Algorithm 1.

Algorithm 1 : Reinforcement Learning-based Query Optimization

```
Initialize Q-network and target-Q network
Start from the initial state s_0
For each step:
    With probability 1 − ε, select action a with \max Q(s, a); otherwise, select randomly
    Obtain reward r and new state s' from the environment
    Store experience tuple (s, a, r, s') in the replay buffer
    Sample a batch of tuples from the buffer and update the Q-network using the loss function
    Periodically update the target-Q network
End For
```

In summary, this section details the methodology used to design a reinforcement learning-based query optimization module for time-series databases. The approach involves modeling the problem as an MDP, encoding query and database information as feature vectors, normalizing cost values, and employing advanced optimization techniques to train the model. The resulting algorithm is capable of generating efficient query plans by learning from the environment and adapting to different workloads.

4 Experiments

In this section, we present the experimental evaluation of our reinforcement learning-based query optimization approach for time-series databases. We compare our methods, DQN and DDQN, with traditional non-learning methods under various cost models to assess their performance in optimizing query plans.

4.1 Experimental Setup

Experimental Setup. All experiments were conducted on a laptop running Ubuntu 20.04 with an Intel Core i7-8750U CPU (6 cores) and 8GB of RAM. The implementation was done using Python 3.8. The dataset used consists of simulated time-series data from a farm production scenario, including soil element detection, air quality, and other time-stamped data. The dataset comprises 10 tables, with queries involving 3 to 8 joins, averaging 5 joins per query. The training and testing sets consist of 40 and 10 queries, respectively.

Cost Models. We evaluated the performance of our methods under two different cost models:

- **CM1**: A cost model where all joins are performed in memory.
- **CM2**: A cost model that considers the cost of spilling partitions to disk when the data exceeds a certain threshold.

Parameter Configuration. The parameter settings for DQN and DDQN are summarized in Table 1.

Table 1. Parameter Configuration for DQN and DDQN

Method	Iterations	Start Step	Target Network Update	Neurons
DQN	40,000	1,000	500	[256, 256]
DDQN	160,000	10,000	3,200	[1,024, 1,024]

4.2 Comparison with Traditional Methods

We compared our proposed methods (DQN and DDQN) with the traditional left-deep dynamic programming method (LD) under both cost models. The results are summarized in Tables 2 and 3.

Table 2. Performance Comparison under CM1

Method	Min	Mean	Max
LD	1.0	1.08	2.14
DQN	1.2	1.28	3.32
DDQN	1.1	1.19	3.18

Table 3. Performance Comparison under CM2

Method	Min	Mean	Max
LD	1.75	7.31	45.24
DQN	1.24	2.22	10.24
DDQN	1.32	1.89	7.24

Analysis. Under the simpler cost model (CM1), the traditional left-deep method (LD) exhibited the best performance. However, DQN and DDQN showed only a modest difference compared to LD, with DQN having an average cost 18% higher than LD, and DDQN having an average cost 8% higher. In more complex scenarios (CM2), DDQN outperformed DQN by approximately 35%, demonstrating better adaptability to varying environments. The traditional LD method, however, showed significant degradation in performance, especially in the worst-case scenario, where its query optimization time was nearly an order of magnitude higher than the average. This indicates that traditional heuristic methods may deviate significantly from optimal solutions under more realistic conditions.

4.3 Training Time Analysis

We also compared the training time required for DQN and DDQN to converge. The results are shown in Table 4 and Fig. 4.

Table 4. Training Time Analysis

Method	1,000 Iterations	4,000 Iterations	2,000 Iterations	10,000 Iterations
DQN	15.08	2.86	2.02	2.06
DDQN	30.89	15.24	3.56	1.44

Analysis. While DDQN achieved better query optimization performance in the long run, it required significantly more training time compared to DQN. Specifically, in the early stages of training (1,000 to 20,000 iterations), DQN outperformed DDQN by up to

7 times. However, as the number of iterations increased, DDQN eventually converged to a better solution, achieving approximately 30\% better optimization than DQN. This suggests that while DDQN is more effective in complex environments, it comes at the cost of longer training times.

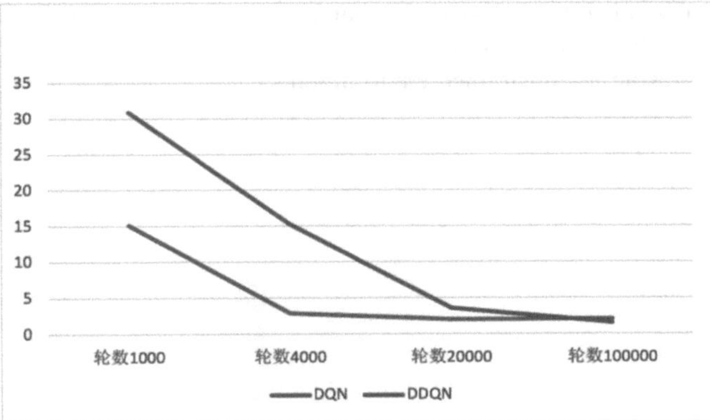

Fig. 4. Query Optimization Performance with Increasing Iterations

4.4 Impact of Memory Constraints

To further analyze the performance under varying memory constraints, we evaluated the average cost of query optimization for different memory sizes. The results are shown in Table 5 and Fig. 5.

Table 5. Impact of Memory Constraints on Query Optimization

Method	M = 1	M = 2	M = 4	M = 8
LD	1.0	1.15	12.90	20.18
DQN	1.12	1.98	2.02	1.87
DDQN	1.21	1.78	2.41	2.25

Analysis. The results indicate that as memory constraints increase (i.e., as the memory size decreases), the performance of traditional methods like LD degrades significantly. In contrast, the reinforcement learning-based methods (DQN and DDQN) remain relatively stable and adapt better to varying memory conditions. This highlights the robustness of our proposed methods in handling complex and dynamic environments.

Our experiments demonstrate that reinforcement learning-based query optimization methods (DQN and DDQN) are effective in optimizing query plans for time-series databases. While traditional methods like left-deep dynamic programming perform well in simpler scenarios, they degrade significantly under more complex conditions, such as limited memory. In contrast, DQN and DDQN show better adaptability and robustness, especially when dealing with varying memory constraints and more realistic cost models. Although DDQN requires longer training times compared to DQN, it achieves superior optimization performance, highlighting its potential for handling complex query optimization tasks. Overall, our results suggest that reinforcement learning-based approaches can serve as a viable alternative to traditional query optimization methods, particularly in dynamic and resource-constrained environments.

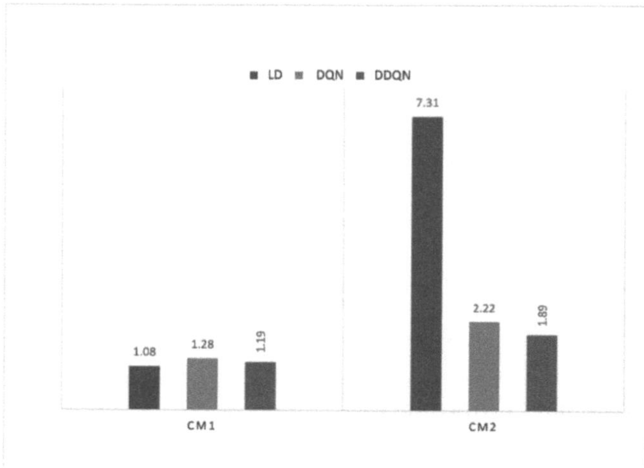

Fig. 5. Impact of Memory Constraints on Query Optimization

5 Conclusion

In this paper, we have presented a novel approach for optimizing query plans in time-series databases using reinforcement learning (RL). Time-series databases play a crucial role in various domains such as finance, IoT, and industrial applications, where efficient query processing is essential for timely decision-making. Traditional query optimization methods, such as left-deep dynamic programming, often struggle with complex scenarios involving large datasets, limited memory, and non-linear cost models. To address these challenges, we proposed a reinforcement learning-based query optimization module that leverages Deep Q-Networks (DQN) and Double Deep Q-Networks (DDQN) to dynamically adapt to varying workloads and system conditions.

Our approach focuses on the plan enumeration problem, which is a key component of query optimization and is known to be NP-hard. By modeling the query optimization process as a Markov Decision Process (MDP), we enable the RL agent to learn optimal join orders through interaction with the environment. This data-driven method allows the optimizer to adapt to specific database instances and query workloads, overcoming the limitations of traditional heuristic strategies that rely on fixed rules and assumptions.

Through extensive experiments, we demonstrated the effectiveness of our proposed methods. Under different cost models, DQN and DDQN showed significant improvements in query optimization performance compared to traditional methods. In particular, DDQN exhibited better adaptability to complex environments, achieving up to 35\% better optimization results under more realistic conditions with limited memory. Although DDQN requires longer training times, its superior performance highlights its potential for handling dynamic and resource-constrained scenarios.

The results of our study emphasize the robustness and flexibility of reinforcement learning-based query optimization. Our methods not only outperform traditional approaches in complex settings but also provide a scalable solution for optimizing query plans in large-scale time-series databases. Future work may focus on further improving the training efficiency of RL models, integrating more advanced neural network architectures, and exploring the application of our approach to other types of databases and query optimization problems.

In summary, this paper demonstrates that reinforcement learning offers a powerful and adaptive solution for query optimization in time-series databases. Our proposed methods effectively address the limitations of traditional techniques, providing a promising direction for the development of intelligent query optimization systems.

Acknowledgments. This paper is supported by Sichuan Science and Technology Program (2024YFHZ0161).

References

1. eHaan, D., Tompa, F.: Optimal top-down join enumeration. In: SIGMOD, pp. 785–796 (2007)
2. Fender, P., Moerkotte, G.: A new, highly efficient, and easy to implement top-down join enumeration algorithm. In: ICDE, pp. 864–875 (2011)
3. Heitz, J., Stockinger, K.: Join query optimization with deep reinforcement learning algorithms. CoRR arXiv:1911.11689 (2019)
4. Krishnan, S., Yang, Z., Goldberg, K., Hellerstein, J., Stoica, I.: Learning to optimize join queries with deep reinforcement learning. CoRR arXiv:1808.03196 (2018)
5. Marcus, R., Papaemmanouil, O.: Deep reinforcement learning for join order enumeration. In: aiDM@SIGMOD, pp. 3:1–3:4 (2018)
6. Moerkotte, G., Neumann, T.: Analysis of two existing and one new dynamic programming algorithm for the generation of optimal bushy join trees without cross products. In: VLDB, pp. 930–941 (2006)
7. Selinger, P., Astrahan, M., Chamberlin, D., Lorie, R., Price, T.: Access path selection in a relational database management system. In: SIGMOD, pp. 23–34 (1979)
8. Trummer, I., Wang, J., Maram, D., Moseley, S., Jo, S., Antonakakis, J.: Skinnerdb: regret bounded query evaluation via reinforcement learning. In: SIGMOD, pp. 1153–1170 (2019)
9. Vance, B., Maier, D.: Rapid bushy join-order optimization with Cartesian products. In: SIGMOD, pp. 35–46 (1996)
10. Yu, X., Li, G., Chai, C., Tang, N.: Reinforcement learning with tree-LSTM for join order selection. In: ICDE, pp. 1297–1308 (2020)

Design of a Road Traffic Classification Accident Detection and Identification System Based on YOLOv10

Sifan Wang, Yanqing Wang(✉), Shengli Wang, and Yufan Mei

Nanjing Xiaozhuang University, Nanjing 211171, Jiangsu, China
772563659@qq.com

Abstract. To address various challenges, including delayed traffic accident detection response, poor localization accuracy, and inadequate rescue efficiency, this study presents a YOLOv10-based intelligent detection and classification system. Featuring a three-tier "perception-decision-action" framework, the system uses an enhanced YOLOv10 model for precise accident detection and implements multiscale feature fusion to improve the recognition of diverse accident scenarios. The classification module employs dynamic weighting to assess accidents as mild, moderate, or severe, coupled with a tri-level response protocol. For route optimization, the path planning component uses a refined A* algorithm with a four-dimensional cost function. Tests show that the system delivers 98.2% detection accuracy with subsecond response times, maintains 93.5% reliability in harsh weather, and enhances rescue operations by 23.7% in efficiency. This solution offers a robust technical approach for smart traffic management systems.

Keywords: Traffic accident detection · YOLOv10 · Intelligent grading · Path planning

1 Introduction

The acceleration of urbanization and the increase in motor vehicle ownership have made road traffic safety a bottleneck in urban development. According to the 'China Statistical Yearbook 2024' [1], there were 255,000 road traffic accidents in China in 2023, resulting in more than 60,000 deaths and economic losses of 1.18 billion yuan. Among these incidents, 28.7% were secondary accidents caused by delayed detection responses, which accounted for more than 35% of the casualties and economic losses, highlighting the inadequacy of the existing detection systems. Currently, mainstream detection systems rely on fixed monitoring cameras and manual surveillance, resulting in slow response times, poor adaptability to different environments, and insufficient intelligence [2]. This leads to inefficient emergency responses and unreasonable allocation of rescue resources, triggering a chain of problems such as traffic congestion.

Deep learning has led to breakthroughs in traffic detection, with CNNs [3, 4] and YOLO [5] series of algorithms being widely used; however, existing algorithms still have limitations. The Faster R-CNN proposed by Ling [6] offers high accuracy but

Y. Yu et al. (Eds.): ICPCSEE 2025, CCIS 2673, pp. 282–291, 2026.
https://doi.org/10.1007/978-981-95-2566-9_20

involves complex computations, whereas Liu's [7] SSD provides fast speeds but lacks robustness in detecting small targets and adapting to complex environments. Within the YOLO series, Ran's [8] multiobject tracking system based on YOLOv8 suffers from low classification accuracy and long response times; after integrating global attention mechanisms and the GELAN structure into YOLOv8, Tang [9] significantly increased deployment costs.

To address these issues, this study proposes an intelligent traffic accident detection and classification system based on YOLOv10. The innovation lies in optimizing the YOLOv10 algorithm, which achieves an mAP@0.5 of 98.2% at an FPS processing speed of 120, and maintaining detection performance in adverse weather at 93.5%. Memory usage is controlled within 2.4 GB, reducing costs; the system also incorporates dynamic classification evaluation and intelligent path planning modules, addressing the issues of classification accuracy and response speed. This system is expected to provide strong technical support for improving the level of intelligent traffic management in China.

2 Our Method

2.1 System Architecture

The system adopts a three-layer architecture design of "perception-decision-execution", integrates advanced computer vision algorithms, intelligent grading evaluation models and dynamic path planning technology, and provides innovative solutions for improving the efficiency of traffic accident handling.

The perception layer consists of high-definition cameras, microwave radars, and meteorological monitoring equipment deployed along the road, forming a comprehensive monitoring network at intervals of 200–300 m. The equipment collects real-time traffic videos, vehicle speeds, distances, and meteorological data, which are then preprocessed by edge nodes and transmitted through a dual-channel transmission network (with automatic switching between main and backup) and an intelligent caching system to ensure the continuity of data transmission (Fig. 1).

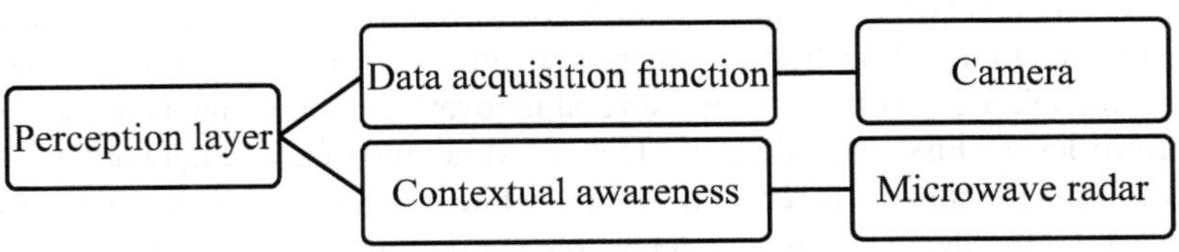

Fig. 1. Perception layer

The decision-making layer uses the optimized YOLOv10 model to analyse the accident location and damage extent in real time, conducts a three-tier assessment on the basis of indicators such as the accident scale and traffic impact, and dynamically plans rescue routes on the basis of real-time road conditions. Each module achieves efficient collaboration through a parallel architecture and shared memory, supplemented by an expert knowledge base to increase decision-making accuracy (Fig. 2).

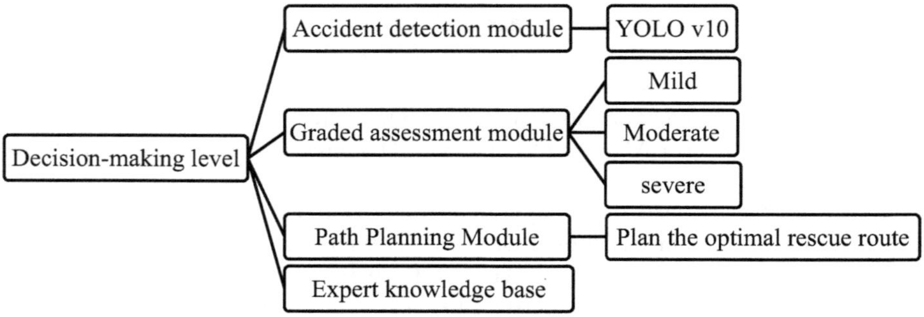

Fig. 2. Decision-making level

The execution layer consists of the command center, mobile terminals, and signal control system. The command center monitors incidents in real time and formulates plans, the mobile terminals receive tasks and support rapid response, and the signal system automatically optimizes traffic for rescue vehicles. Depending on the level of the incident, it is handled by traffic police, medical personnel, or multiple departments, with all the data being transmitted back in real time to form a closed loop (Fig. 3).

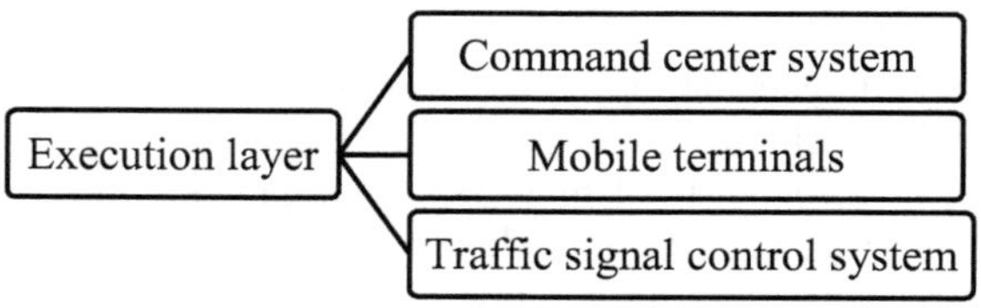

Fig. 3. Execution layer

2.2 Target Detection System Based on YOLOv10

The YOLOv10 model used in this system is the latest evolution version of the YOLO series of object detection algorithms, which shows significant technical advantages in the object detection task in traffic scenarios. Compared with the previous generation YOLOv8 model, YOLOv10 has made several innovative improvements in the network architecture design. First, the redesigned feature extraction backbone network is adopted, and the computational cost is reduced by 15% while maintaining the feature extraction capability by introducing cross-stage local connections and deep separable convolution technology. Second, the feature pyramid structure [10] is optimized to enhance the detection ability of multiscale targets. Finally, the design of the detection head was improved, and the decoupled prediction mechanism was adopted to separate the classification and regression tasks, which significantly improved the detection accuracy. These architectural improvements enable YOLOv10 to achieve an mAP metric of 58.9% in traffic scenarios, an increase of 3.1 percentage points compared with YOLOv8 [11] (Figs. 4 and 5).

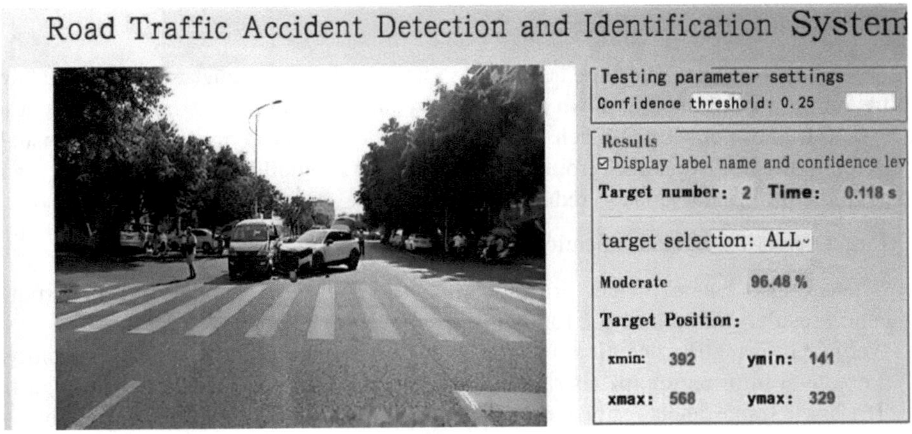

Fig. 4. System picture

The workflow of the YOLOv10 model in this system includes four key stages, and Fig. 6 shows the application process of the model within the system.

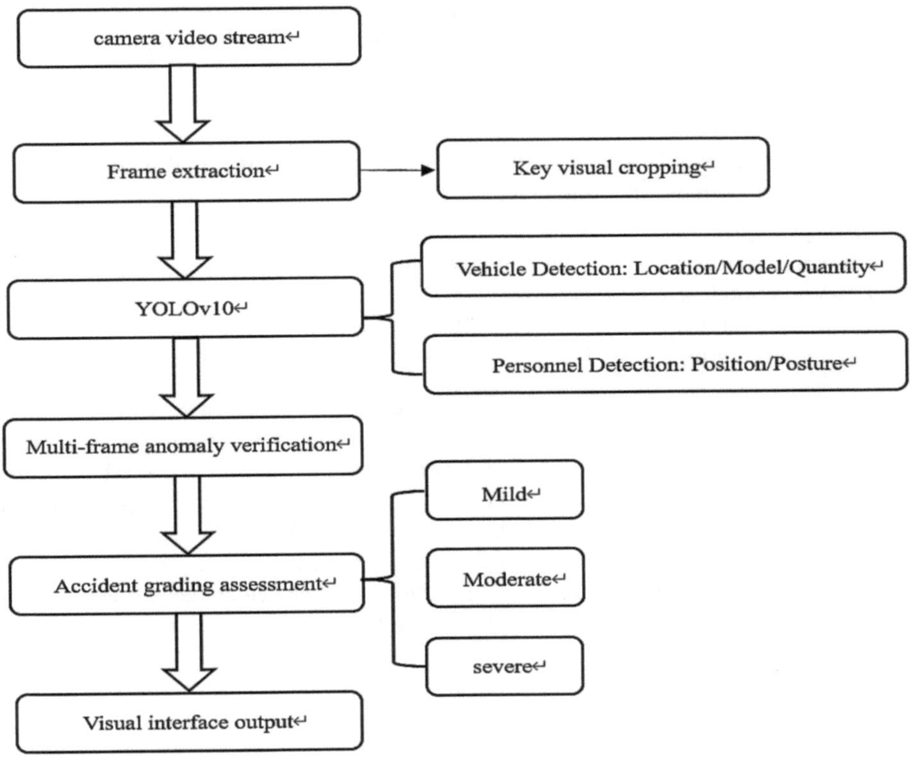

Fig. 5. The usage procedure for the YOLO v10 model

1. Video Input and Frame Processing

The system collects real-time traffic video streams through cameras deployed at key points on the road, breaking down the video stream into continuous frame images at a rate of 30 frames per second. Each frame image undergoes cropping of the main visual area (removing irrelevant backgrounds such as the sky and buildings), retaining the road and vehicle movement areas to reduce computational redundancy.

2. Target detection and classification

The cropped frame images are input into the YOLOv10 model, which outputs detection results that include the following:

Vehicle Information: location, vehicle type (sedan/truck/motorcycle, etc.), quantity.

Personnel Information: pedestrian location, quantity, and posture (standing/fallen).

If a single frame detects only one normally operating vehicle and no personnel abnormalities, the system determines it as "no accident"; if multiple vehicles are gathered, vehicles are parked abnormally, or personnel are found fallen, it triggers the accident assessment process.

3. Traffic Accident Judgment

Multiple Frame Verification: If the same target anomaly is detected in five consecutive frames (approximately 0.17 s), it is deemed a valid incident. The severity of the incident is categorized into three levels: mild, moderate, and severe, as detailed in Sect. 3.3.

4. Output and interface display.

The system promptly pushes detection results to the traffic management center interface, which includes the following:

(1) Accident location (GPS coordinates and road name).
(2) Accident severity level (mild/moderate/severe) and confidence level (e.g., 'severe accident, confidence level 98.3%').
(3) Visual annotations: Selection of accident vehicles/persons in the video stream, overlaying information such as vehicle type and number.

The core value of this model in the system is reflected in three aspects: first, its high-precision detection capability ensures accurate identification of various types of traffic accidents, including vehicle collisions, rollovers, rear–end collisions, and other forms; ', its excellent real-time performance ensures that the system can promptly respond to unexpected accidents; and finally, the optimized model size makes it suitable for deployment on edge devices, reducing the overall system deployment costs. Actual tests indicate that in complex traffic scenarios, the system's accident detection accuracy reaches 98.2%, with a false positive rate of less than 1.5%, fully meeting the requirements of intelligent traffic management.

2.3 Incident Grading Response Mechanism

The system has established a scientific and perfect three-level accident classification system and achieves accurate classification through multidimensional evaluation. The grading standard comprehensively considers the physical characteristics of the accident scene and the scope of influence and divides traffic accidents into three grades: mild, moderate and severe. The specific criteria for judgement include the area of vehicle damage caused by the accident, the number of casualties and the degree of impact on the traffic flow. The system intelligently analyses the onsite footage captured by the camera, automatically measures the area of the damaged area, identifies the number of injured people and the degree of injury, and evaluates the impact of the accident on real-time traffic flow data. After these key indicators are weighted, the system automatically determines the accident level and triggers the corresponding emergency plan (Tables 1 and 2).

Table 1. Incident grading response mechanism

Accident level	Area of injury	Casualties	Traffic impacts
Mild	$<1\ \text{m}^2$	No casualties	Single lane blockage
Moderate	$1\text{–}3\ \text{m}^2$	1–2 casualties	Adjacent lane diffusion
severe	$>3\ \text{m}^2$	Serious injury or death	regional road network paralysis

Table 2. Resource allocation

Accident level	Police force	Medical resources	Traffic control range
Mild	1 police car and 2 officers	First Aid Kit Preparation	200 m from the accident site
Moderate	2 police cars and fire trucks	1 ambulance on standby	Control of Adjacent Intersections
severe	SWAT, firefighting, and first aid linkage	Trauma Center Green Channel	Regional diversion induction

In support of hierarchical decision-making, the system has established a vast case knowledge base that includes thousands of historical accident cases and their disposition plans. When a new accident occurs, the system automatically matches similar cases to provide reference recommendations for disposal personnel. Moreover, the entire accident handling process, including response time, disposal measures, and final outcomes, is comprehensively recorded; these data are used for the ongoing optimization of hierarchical standards and response processes. Through this intelligent hierarchical mechanism, the system achieves precise allocation of emergency resources, avoiding scenarios of resource wastage or deficiency and significantly enhancing the efficiency and service quality of traffic accident handling.

2.4 Dynamic Path Planning

The dynamic path planning module of this system uses advanced intelligent algorithms to achieve efficient rescue route planning.

This module constructs a dynamically updated road network model in real time by integrating multisource information such as roadside sensing devices, floating vehicle GPS data, and traffic cameras. In the design of the core algorithm, the system employs an improved A* search algorithm [12] for path calculation, which quickly finds the optimal path by combining heuristic functions with actual cost evaluation.

By comprehensively considering factors such as travel time, the real-time congestion index, the accident risk probability, and the congestion propagation rate, a four-dimensional cost function is constructed to achieve optimal path planning.

$$C_{path} = \alpha \cdot \frac{D}{v(t)} + \beta \cdot e^{\lambda Q(t)} + \gamma \cdot P_{risk} + \delta \cdot \frac{\partial Q}{\partial t} \tag{1}$$

$\alpha = 0.45$: Travel time weight
$\beta = 0.3$: Real-time congestion index
$\gamma = 0.2$: Accident risk probability
$\delta = 0.05$: Congestion propagation rate

Practical applications have demonstrated that this module not only has high computational efficiency (with an average of 500 ms to complete path planning) but also significantly outperforms traditional methods in planning quality (with a 40% improvement in path optimization rate), providing strong technical support for emergency rescue operations.

3 Experimental Results

3.1 Configuration of the Experimental Environment

This research uses a high-performance experimental platform for testing, with a hardware configuration including dual Intel Xeon 4210R processors (2.4 GHz base frequency, 10 cores, 20 threads), an NVIDIA RTX 4090 graphics card (24 GB VRAM), and 128 GB DDR4 memory, combined with a 1 TB NVMe SSD and a 4 TB HDD hybrid storage system. The software environment is based on the Ubuntu 20.04 LTS operating system, which is equipped with the PyTorch 1.12 deep learning framework (supporting CUDA 11.6 acceleration) and OpenCV 4.5.5 image processing library. The experiments are conducted in a standard environment at a constant temperature of 22 ± 1 °C with a gigabit Ethernet connection (latency <1 ms) to ensure the reliability and repeatability of the test results.

3.2 Dataset and Preprocessing

The experimental data were collected from 2000 sets of real road surveillance videos covering various typical scenarios, such as urban roads, highways, and intersections. The dataset underwent a rigorous preprocessing procedure, including video frame extraction, image quality screening, and standardization. During the preprocessing phase, a dynamic sampling algorithm was employed to ensure the completeness of the accident process, and a professional annotation team performed quality audits, ultimately retaining 98.7% of the effective data. In terms of data augmentation, the system applied geometric transformations (rotation ±15°, scaling ±20%) and photometric adjustments (brightness ±30%, contrast ±25%), effectively balancing the distributions of various accident samples.

Fig. 6. Dataset

3.3 Analysis of Experimental Results

To comprehensively evaluate the system performance, we designed multidimensional comparative experiments to systematically compare the improved YOLOv10 model with current mainstream object detection algorithms. Table 3 presents a detailed comparison of the performance metrics of each model under the same testing environment, including key indicators such as model efficiency, detection accuracy, and real-time performance.

Table 3. YOLO model comparison

model	Parameter Quantity (M)	FLOPs(G)	mAP@0.5	Precision	Recall	FPS	Video memory usage (GB)
SSD	26.8	62.3	89.7%	88.5%	86.2%	65	2.8
Faster R-CNN	136.2	365.4	92.4%	91.8%	89.3%	28	4.2
YOLOv5n	8.9	15.7	95.1%	94.3%	92.7%	98	1.8
YOLOv8n	9.2	16.3	96.7%	95.8%	94.5%	105	2.1
YOLOv9n	9.5	14.8	97.8%	97.1%	96.3%	115	2.3
YOLOv10n	8.7	12.5	98.2%	98.5%	97.8%	120	2.4

The YOLOv10n used in this system exhibits remarkable efficiency, with a parameter count of only 8.7 M and FLOPs of 12.5G, representing a reduction of 5.4% and 23.3% compared with YOLOv8n, respectively, while maintaining a detection accuracy of 98.2% mAP@0.5. In terms of detection performance, the system achieves a precision of 98.5% and a recall of 97.8% and still maintains a recall of 96.2% and a precision of 95.7%

in complex scenarios such as nighttime and foggy weather, demonstrating exceptional detection capability and environmental adaptability (Tables 4 and 5).

Table 4. Performance metrics under different scenarios.

Test Scenario	mAP@0.5	Precision	Recall	Response Delay (ms)
Daytime	98.5%	98.8%	97.9%	78
Night	97.1%	97.5%	96.2%	85
Rainy and foggy weather	95.3%	95.7%	94.1%	92

Table 5. Analysis and comparison of module performance

Functional Module	Our System	Traditional Methods	Increase Amplitude	Key Technologies
Object Detection	98.2% mAP	95.1% mAP	+3.1%	Multi-Scale Feature Fusion
Graded Assessment	98.5% accuracy	90.3% accuracy	+8.2%	Dynamic Weight Calculation
Path Planning	500 ms	850 ms	+41.2%	Improvement of the A* Algorithm
Small Object Detection	92.3%	79.8%	+12.5%	Feature Pyramid Optimization
Robustness against Adverse Weather Conditions	93.5%	85.3%	+8.2%	Data Augmentation Strategies

The experimental results indicate that this system significantly outperforms comparative algorithms in terms of model efficiency, detection accuracy, and real-time performance, especially in terms of achieving higher detection accuracy while maintaining lightness. This advantage is particularly evident in complex traffic scenarios and harsh environmental conditions, providing reliable technical support for intelligent traffic management. With a memory usage of 2.4 GB and a processing speed of 120 FPS, the system is well suited for deployment on edge devices, holding significant practical application value.

4 Conclusion

The intelligent traffic accident detection system based on YOLOv10 proposed in this study has achieved significant breakthroughs in real-time performance, accuracy, and environmental adaptability through multilevel technological innovations. The experimental results indicate that the system maintains a lightweight design while achieving a detection accuracy of 98.2% mAP@0.5, with the end-to-end latency controlled within

83 ms. Notably, it retains 93.5% stability under adverse weather conditions. The introduction of a dynamic weight calculation scheme and an improved A* algorithm has led to an accident grading accuracy of 98.5% and a 41.2% increase in path planning efficiency, providing reliable technical support for rapid response and rescue in traffic accidents.

However, there remains room for improvement in the system's generalization capability under extremely complex scenarios, deployment costs, and recognition of rare accidents. Future work will focus on optimizing the model's robustness, developing lightweight deployment solutions, and expanding the accident case database. From an application perspective, this system not only significantly enhances traffic accident handling efficiency (with an expected 40% reduction in response time), but its technical framework can also be extended to other urban safety scenarios. With the integration of new technologies such as 5G-V2X, the system will provide stronger guarantees for intelligent transportation and public travel safety, demonstrating important social value and application potential.

References

1. National Bureau of Statistics of China: China Statistical Yearbook. China Statistics Press, Beijing (2024). https://www.stats.gov.cn/sj/ndsj/2024/indexch.htm. Accessed 21 Nov 2016
2. Gao, Z.: Challenges for future urban traffic management. Manage. World **38**(10), 3–5 (2022)
3. Li, C., Zhang, Z., Liang, Z., et al.: A survey of object detection models. J. Comput. Res. Dev. 1–35 (2025)
4. Ding, Y., Ma, Y., Wang, J., et al.: Graph convolutional neural network based on traffic turning rules - for urban road network traffic prediction. J. Geo-inf. Sci. **27**(3), 653–667 (2025)
5. Liu, Q., Zhong, X., Chen, Y.: Lightweight pedestrian detection algorithm ST-YOLOv10. J. Chongqing Technol. Bus. Univ. (Nat. Sci. Ed.) 1–10 (2025)
6. Ling, Y.: Optimization of video surveillance object detection algorithms based on YOLO, SSD, and faster R-CNN. Mod. Comput. **30**(21), 49–53+58 (2024)
7. Liu, H., Wang, J., Shao, L., et al.: Vehicle target recognition based on improved ssd and multi-focus image fusion. J. Tianjin Univ. Technol. 1–9 (2025)
8. Ran, W., Zhang, H., Zhou, W., et al.: Research on highway incident detection method based on YOLOv8. China Transp. Informatization (10), 104–107+120 (2024)
9. Tang, Z., Tang, Y., Pan, D., et al.: Traffic accident severity detection based on UAV. Mod. Electron. Tech. **47**(20), 123–128 (2024)
10. Jiang, W., Wang, L., Yao, Y., et al.: Small object detection algorithm with multiscale feature aggregation diffusion and edge information enhancement. Comput. Eng. Appl. 1–15 (2025)
11. Wang, N., Zhi, M.: Review of single-stage general object detection algorithms based on deep learning. J. Front. Comput. Sci. Technol. 1–32 (2025)
12. Wu, G., Yang, J.: Optimal path planning for travel time reliability. Highway **62**(2), 134–142 (2017)

Multiscale Dynamic Inference Acceleration for Deep Neural Networks

Chong Zhang[1], Hongwei Liu[1], Hongzhi Wang[1](✉), Wei Du[2], Jiaying Wang[2], and Sijia Zheng[2]

[1] Faculty of Computing, Harbin Institute of Technology, Harbin 150001, China
19B903094@stu.hit.edu.cn, {liuhw,wangzh}@hit.edu.cn
[2] State Grid Electric Power Space Technology Co., Ltd., Beijing, China

Abstract. Modern compression and acceleration methods for exploring efficient deep neural networks render real-world applications more feasible. Existing approaches uniformly apply the same procedure to every input image, overlooking instancewise complexity variations. Moreover, owing to pruning or decomposition techniques, the upper bound of network representation capabilities might be permanently diminished. In this work, an input-dependent multiscale dynamic inference method (MSDI) is developed to strike a better balance between model performance and inference acceleration. Specifically, we modify the main body of a convolutional network to obtain a series of parameter-sharing subnetworks with varying levels of complexity. A side branch structure is then introduced to assign an input instance to a suitable subnetwork as its inference route, and we expect to accelerate the inference by assigning the easy input to the subnetwork with low capacity. We further propose multiscale distillation training to optimize the training of the modified subnetworks. Additionally, we compare the entropy-based and learning-based grading approaches, aiming to obtain a more suitable route assignment method. Experiments show that MSDI can accelerate most existing convolutional models, achieving up to 74.7% computation savings across diverse datasets.

Keywords: dynamic execution · deep learning · inference acceleration · model compression

1 Introduction

With the development of deep convolutional neural networks (DCNNs), rapid performance improvements have been achieved in the fields of vision and image-based problems, including image classification [1], superresolution [2], disease diagnosis [3] and neural machine translation [4]. However, state-of-the-art DCNN models usually require excessive storage and computing resources to obtain superior task performance, which prohibits their application on mobile devices and edge devices.

Y. Yu et al. (Eds.): ICPCSEE 2025, CCIS 2673, pp. 292–306, 2026.
https://doi.org/10.1007/978-981-95-2566-9_21

The popular trend to alleviate this limitation is to reduce redundancy in model parameters and structures to make networks compact. For example, sparsity regularization methods [5] obtain sparse model parameters and prune the redundant part during training, and structured model pruning methods [6–8] measure the importance of filters or layers in a pretrained network and prune the unimportant ones with fine-tuning, resulting in significant computational savings with negligible accuracy degradation. Knowledge distillation methods [9–11] improve the performance of simple student networks by transferring knowledge from complex teacher networks. Although these approaches can considerably improve model compression and inference acceleration, these mainstream model compression methods are static and have two drawbacks. First, during the model compression process, a network typically undergoes structural reduction. This leads to a permanent loss of its maximum fitting capacity, as the simplified architecture restricts its ability to capture complex patterns. Second, these static methods treat different input instances uniformly. While the difficulty of a prediction task varies with the complexity of the input instances, this complexity difference can potentially yield a much higher inference speed-up ratio. Intuitively, images with clear backgrounds and prominent subjects are easier to recognize, and shallow embeddings and few feature channels are sufficient. Therefore, additional computations associated with deeper layers and feature channels of the network could be preserved only for input instances that are more challenging to process.

To flexibly leverage the properties of each input instance and maximize inference acceleration while maintaining accuracy, dynamic inference methods have been proposed. Instead of cutting down model parameters or structures to save storage and computation simultaneously, dynamic inference methods compress computation by selectively choosing an execution path depending on the input instance. The maximum fitting ability of the network is retained since the intact structure is maintained. To achieve dynamic inference, existing methods either generate instancewise execution path predictions [12–16] or learn an instance-dependent early-exit mechanism [17–19] to skip further computations. However, an extra parameter structure and computation are always required for the gate prediction modules, which results in more redundancy other than computation reduction. In addition, reinforcement learning-based methods face training challenges due to the high-variance training signal.

Notably, in BranchyNet [17], prediction results with high confidence can exit the network early through side branches, significantly reducing computations while maintaining network performance. However, it has two notable drawbacks. First, deep networks require an overly complex auxiliary network structure to function optimally, with at least two branches made up of convolutional and fully connected layers. Moreover, features in deep layers usually carry more task information and are highly important for convolutional networks. Therefore, grading subnetworks layerwise may lead to inferior network performance due to a lack of utilization of deep features.

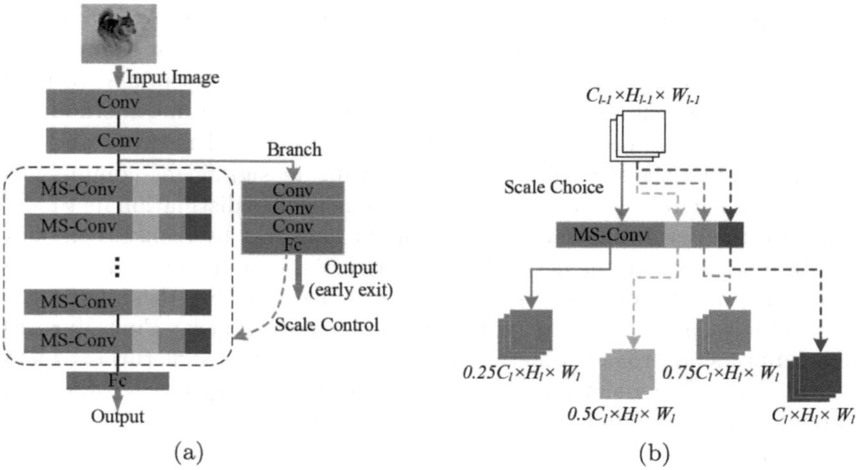

Fig. 1. Schematic diagram of (a) MSDI and (b) MS-Conv.

To better utilize deep features with fewer auxiliary structures, we propose a multiscale dynamic inference (MSDI) mechanism for deep neural networks on the basis of input complexity. As illustrated in Fig. 1 (a), we keep one side branch in the early stage and divide the subsequent network vertically into multiscale subnetworks with MS-Conv (multiscale convolutional layer, as shown in Fig. 1 (b). During inference, an input instance first passes through the early stage and the side branch. Then, on the basis of the confidence level of the side branch's output, a decision is made: either to directly output the prediction or to select an inference route from the aforementioned subnetworks for further computation.

Our main contributions are summarized as follows.

- We propose MSDI, a novel inference acceleration mechanism that combines early-exit and route prediction to allocate instances of varying complexity to routes with different capacities, maximizing model performance with minimal auxiliary structure.
- We address the training issues of slim deep networks through multiscale distillation training. The performance of a slimmer subnetwork is improved with the help of a wider subnetwork through knowledge distillation, and the upper bound of the wider subnetwork increases concurrently due to the parameter sharing mechanism.
- We propose learning-based grading to mitigate human impact. Since grading the subsequent computations on the basis of the entropy of the branch output is subjective, we allow the branch to learn the grading metric independently, only restricting the distribution across different scales.

The experimental results on popular DCNN models and benchmark datasets demonstrate that our framework can achieve better inference acceleration while maintaining model accuracy.

2 Related Works

This section briefly reviews current methods for accelerating inference in deep neural networks, which are roughly divided into two categories: static compression and dynamic acceleration.

2.1 Static Compression

Several lines of work have aimed at reducing the parameter redundancy and computational complexity of deep networks simultaneously by compressing existing models or designing novel compact architectures.

Structural Network Pruning. The most promising method for compression performance, involves removing redundant weights, filters or even layers on the basis of different metrics. For example, APoZ [20] utilized the average percentage of zeros of the activation maps to measure the importance of channels in a layer. HRank [7] reported that the average rank of the activation maps for a certain filter barely varies, thus it is considered as the filter importance. Rather than predefined metrics, He *et al.* [21] proposed an adaptive way to learn a filter pruning criterion for different layers.

Knowledge Distillation. Hinton *et al.* [9] improved the performance of the student model by introducing the difference between the outputs of the teacher and the student as the soft target. FitNet [22] proposed an intermediate representation as a hint for the student model to make thinner and deeper networks learn better from teacher models. Zhang *et al.* [10] replaced the fixed teacher model with an online evolutionary teacher to consistently evolve and transfer intermediate knowledge. Mirzadeh *et al.* [11] proposed an intermediate model called a teacher assistant to bridge the overlarge gap between the student and teacher models.

Compact Architecture Design. Replacing fundamental network building blocks with more compact and efficient blocks can significantly reduce the computational cost. SqueezeNet [23] replaced 3×3 kernels with 1×1 kernels and squeezes feature maps to lower dimensions in early stages to skip further massive computations. MobileNetV2 [24] used depthwise convolutions and point-wise convolutions to decouple the convolution computation and reduce the number of FLOPs (floating point operations). ShuffleNet [25] obtained more informative feature channels with a compact architecture via a channel shuffling operation and depthwise convolution.

2.2 Dynamic Acceleration

Removing or changing the model structure permanently seems effective in reducing the computational cost, while this is considered for the average case. The removed parts may be unimportant for most input instances, but critical for other input instances. Furthermore, compressing networks for the average case must preserve a sufficiently intact network structure, and thus cannot reduce the number of computations to the greatest extent possible for easy inputs. Therefore, dynamic inference acceleration methods have been proposed to achieve more flexible computation reduction.

Route Predicting. This line of work choose an instancewise execution route dynamically by selecting certain channels or blocking others by setting learnable gates. Lin *et al.* [12] proposed a runtime pruning method to adaptively conduct structural pruning according to the input image and current feature maps in a layer-by-layer manner. Liu *et al.* [13] obtained deep neural networks with sparse intermediate feature maps for different instances by introducing instance feature sparse regularization in the training phase; thus, subtle features could be eliminated in the inference phase. Tang *et al.* [14] utilized the relationship between different inputs and pruned dynamically by aligning the recognition complexity and feature similarity between images and the pruned subnetworks. Wang *et al.* [16] set a learnable gating network for every layer in residual networks to selectively skip convolutional layers on the basis of the activations of the previous layer. Kong *et al.* [2] proposed a classification and superresolution framework to apply different superresolution modules on the input based on the classification module results.

Early Exiting. BranchyNet [17] proposed additional side branch classifiers to output results in advance when its output confidence exceeds the threshold. For an input of an RNN (recurrent neural networks), ACT [18] proposed an accumulative stop probability for every time step, and the computation is terminated when the accumulative stop probability exceeds the threshold. SACT [19] modified ACT and applied it to residual networks, where stop probability accumulation and the number of executed layers are computed on different regions of input images.

3 Methodology

3.1 Overview

The MSDI is a novel solution pipeline for deep network inference acceleration. The basic idea of MSDI is to establish multiscale inference routes within a deep neural network, evaluate the complexity of an input instance and assign it to the most cost-effective route through inference. As illustrated in Fig. 1 (a), it adds a side branch to standard deep neural networks at an early stage and modifies subsequent original convolutional layers into subnetworks with shared parameters and varying widths. The inference process of MSDI can be described as follows. First, the input is fed into the early stages and the side branch in the net. Second, the confidence of the branch classifier output is calculated. If the confidence exceeds a threshold, the output is taken as the prediction result, and the inference stops. Otherwise, the output is classified into M classes (manually set, $M = 4$ here), and the subsequent computation after the early layers is assigned to the corresponding scale from the M scales of subnetworks $N_{MS}^{j}\}\lim_{j=1}^{M}$.

In the following, we address several considerations in the design of the framework. (1) Architectural design of the multiscale subnetworks. (2) Training of the branch and subnetworks, especially for the slim deep subnetwork, which is usually challenging to train. (3) The grading metric for an input, which may be a better grading metric than the metric based on output entropy, can be found by learning.

3.2 Architectural Design

The MS-Conv (multiscale convolution layer) is the building block of the proposed method. As illustrated in Fig. 1 (b), it is a convolution layer with optional convolutional and output channels. It can output multiscale feature maps by using different proportions of parameters for calculation, on the basis of preset channel compression ratios, $\{P_j\}\text{lim}_{j=1}^{M}$ ($\{0.25, 0.5, 0.75, 1\}$ in this paper.) The weights of the l-th layer in a traditional convolutional network are denoted as $W^l \in \mathbb{R}^{C_{l+1}, C_l, D_l, K_l}$, where C_{l+1}, C_l, D_l, K_l are the dimensions of the output channels, input channels, kernel height and width of layer l, respectively. The computational complexity for this layer can be formulated as follows:

$$C_{l+1} \times C_l \times H_l \times W_l \times D_l \times K_l \tag{1}$$

where H_l and W_l denote the height and width of the input feature map, respectively.

For the subnetwork N_{MS}^{j}, except for the first layer and the last layer, which require special treatment to align the dimensions, its weights are a subset of the total parameters in the l-th layer, which can be noted as $W_{MS}^{jl} \in \mathbb{R}^{C_{l+1} \times P_j, C_l \times P_j, D_l, K_l}$, and the computational complexity for this layer is as follows:

$$(C_{l+1} \times P_j) \times (C_l \times P_j) \times H_l \times W_l \times D_l \times K_l \tag{2}$$

Since the channel compression ratio P_j is usually less than 1, we can save $(1 - P_j^2)$ of the computational cost in a MS-Conv layer.

3.3 Multiscale Distillation Training

The training of the network is carried out in two steps. First, the early layers and multiscale subnetworks $\{N_{MS}^{j}\}\text{lim}_{j=1}^{M}$ are trained via multiscale distillation training. The branch part is then trained solely. By dividing multiscale subnetworks vertically, we obtain deep and thin networks, which are not easy to train. In this approach, the multiscale subnetworks are trained in a sequential rotation manner. During each iteration, we train the subnetworks in sequence from wide to narrow, and the wider sub-networks are trained first. Subsequently, With the subsequent distillation assistance from these broader subnetworks, the thinner subnetworks are trained.

The multiscale distillation training is illustrated in Algorithm 1. During each iteration, the widest subnetwork N_{MS}^{M} is trained with the ground truth. Let x

denote the input, y be the one-hot ground truth and C denote the number of classes; its objective function can be written as:

$$L_M(\hat{y}_M, y) = -\frac{1}{|C|}\sum_{c\in C} y_c log \hat{y}_{Mc} \tag{3}$$

where

$$\hat{y}_M = softmax(z_M) = \frac{exp(z_M)}{\sum \lim_{c\in C} exp(z_{Mc})} \tag{4}$$

and

$$z_M = N_{MS}^M(x; \theta_M) \tag{5}$$

where θ_M represents the parameters of N_{MS}^M, and z_M denotes its output. Then, the other subnetworks N_{MS}^j are trained with both the ground truth and the distillation assistance from N_{MS}^M. Introducing the balance coefficient α and distillation temperature T, its training objective can be written as:

$$L_j(\hat{y}_j, y) = -(1-\alpha)\frac{1}{|C|}\sum_{c\in C} y_c log \hat{y}_{jc} - \alpha T^2 \frac{1}{|C|}\sum_{c\in C} \hat{y}_{jc} log \frac{\hat{y}_{jc}}{\hat{y}_{Mc}} \tag{6}$$

where

$$\hat{y}_j = softmax(z_j) = \frac{exp(z_j/T)}{\sum \lim_{c\in C} exp(z_{jc}/T)} \tag{7}$$

Algorithm 1. Multiscale Distillation Training

Input: Subnetworks $\{N_{MS}^j\}\lim_{j=1}^M$, input images $\boldsymbol{x}$ and labels $\boldsymbol{y}$
Output: Trained subnetworks $\{N_{MS}^j\}\lim_{j=1}^M$
1: **for** $e = 1, ..., E_{max}$ **do**
2: $\theta_M \leftarrow$ SGD$(L_M, N_{MS}^M, \boldsymbol{x}, \boldsymbol{y})$
3: **for** $j = M-1, ..., 1$ **do**
4: $\theta_j \leftarrow$ SGD$(L_j, N_{MS}^j, N_{MS}^M, \boldsymbol{x}, \boldsymbol{y})$
5: **end for**
6: **end for**

3.4 Entropy-Based Grading

The entropy-based grading is intuitive, where the output dimension of the branch is set to be the same as the final output dimension, and the inference route is chosen according to the entropy of the branch output. The entropy is defined as:

$$entropy(y) = \sum \lim_{c\in C} y_c log y_c \tag{8}$$

After the branch is trained solely as a shallow classifier, in the inference stage, we calculate its output entropy for each input instance and compare it with the preset entropy threshold T_r, a vector where the first entry is set to determine whether the branch output should exit directly, and the other entries are used to choose the inference route in the subnetworks $\{N_{MS}^j\}\lim_{j=1}^M$.

3.5 Learning-Based Grading

In addition to the entropy threshold, we introduce a learning approach that enables the branch to independently learn the grading metric. In this way, the branch output dimension is set to be the number of subnetworks M, and the branch can only predict the inference route for the input without early exit. In the training stage, the branch classifies inputs according to their complexity to a specific subnetwork instead of predetermined labels. To obtain classification probabilities automatically without biased results, we design a loss function consisting of three parts:

$$L = L_t + w_1 \times L_c + w_2 \times L_d \tag{9}$$

Here, L_t represents the task loss (the same as Fun. 3) , and w_1 and w_2 serve as balance coefficients. To ensure that the branch can accept gradient propagation from the reconstruction result, we use the weighted average of the branch output $B(x)$ and subnetwork outputs to calculate the loss:

$$y = \sum_{i=1}^{M} B_i(x) \times N_{MS}^{i}(x) \tag{10}$$

During the inference process, only the subnetwork with the highest probability is executed. The probability distribution shows confidence in the output. For example, [0.9, 0.1] is better than [0.6,0.4] as a classification result. Therefore, we reuse the metric entropy in Fun. 8 as the class loss L_c to push the maximum probability to approach 1, since the output entropy is positively correlated with this objective. During training, without regularization, it may easily converge to an extreme point where all the images are classified into the most complex subnetwork. Here, we establish a prior distribution to regularize the branch output. The distribution loss L_d is defined as:

$$\begin{aligned} L_d &= \sum_{i=1}^{M}\Big(\sum_{j=1}^{N} B_i(x_j) - d_i\Big)^2, \\ s.t. &\sum_{i=1}^{M} d_i = 1 \end{aligned} \tag{11}$$

where N is the training batch size, and d_i denotes the prior distribution over the i-th subnetwork. This objective calculates the sum of the distance between the output distributions in a batch and the prior distribution.

4 Experiments

In this section, we demonstrate the effectiveness of the MSDI for general image classification tasks. A detailed analysis of multiscale training and threshold setting is also presented.

4.1 Datasets and Experimental Settings

We evaluated the proposed MSDI on representative deep networks, VGG16 [26], ResNet18, ResNet34 and ResNet56 [27], and on classification datasets with different scales, CIFAR-10, CIFAR-100 [28] and Tiny-ImageNet [29]. CIFAR-10 consists of 50K 32×32 training images and 10K test images belonging to 10 classes. Tiny-ImageNet is composed of 100K 64×64 training images and 10K test images with 200 semantic classes. All the experiments are conducted on an NVIDIA GeForce RTX 3090 GPU.

For VGG16, we add a side branch consisting of 3 convolutional layers, 1 average pooling layer and 2 fully connected layers, after the second convolution layer of the main network. For ResNet56 on CIFAR-10, the branch is composed of 4 convolutional layers, 1 average pooling layer and 2 fully-connected layers, after the first residual block. For ResNet18 and ResNet34, their branches both consist of 3 convolutional layers, 1 average pooling layer and 1 fully connected layer, which are added after the second and third residual blocks respectively, to ensure that enough features can be used for the branch classifier.

Table 1. Performance Comparison on CIFAR-10

Models	Performance			
	Method	Params/AP($\times 10^6$)[a]	FLOPs($\times 10^6$)	Accuracy (%)
VGG16	Baseline	14.99/0	313.46	93.94
	ThiNet	–	156.50	93.36
	NS	–	153.40	93.31
	InSA	14.99/0	220.14	90.84
	RNP	14.99/0	156.50	92.65
	BranchyNet	16.10/1.11	102.65	93.50
	SkipNet(SP)	41.31/26.32	319.81	92.83
	SkipNet(HRL+SP)	41.31/26.32	299.53	92.32
	MSDI(Entropy)	15.53/0.54	**102.45**	**93.86**
	MSDI(Learning)	15.52/0.53	174.71	93.62
ResNet56	Baseline	0.85/0	126.55	93.12
	ThiNet	–	62.90	91.98
	CP	–	62.90	91.80
	InSA	0.85/0	120.04	86.56
	BranchyNet	1.47/0.62	61.75	89.63
	SkipNet(SP)	1.73/0.88	124.56	92.71
	SkipNet(HRL+SP)	1.73/0.88	112.67	91.58
	MSDI(Entropy)	1.38/0.53	**56.53**	**93.30**
	MSDI(Learning)	1.37/0.52	82.46	92.90

[a] Note: Params/AP stands for the number of parameters and auxiliary parameters.

4.2 Comparison with Existing Methods

We compare MSDI with existing state-of-the-art inference acceleration methods, which cover both structural pruning and dynamic acceleration aspects.

For CIFAR-10, we evaluate the MSDI with VGG-16 and ResNet56. With MS-Conv, the main bodies of the networks are first modified into subnetworks with 4 scales, which choose [0.25,0.5,0.75,1] convolutional filters in a layer and output the corresponding feature maps. First, the parameters of the branch are fixed, and the subnetworks are trained with multiscale distillation training for 300 epochs. The learning rate is set to 0.1 and is decreased by a factor of 10 at the 150th and 225th epochs. Then, we fix the main body parameters and train the branch separately for 150 epochs with a learning rate of 0.1. For the learning-based grading, w_1 and w_2 are set to 0.1 and 2, respectively. All the training is optimized with SGD. The threshold for entropy-based grading is discussed in Sect. 4.3.1, and the distribution prior for different scales is set to [0.4, 0.3, 0.2, 0.1] for learning-based grading. We compared MSDI with several state-of-the-art inference acceleration methods, and the results on CIFAR-10 are summarized in Table 1. For static structural pruning, we compare it with ThiNet [30], channel pruning (CP) [31] and network slimming (NS) [32]. For dynamic acceleration, we compare runtime neural pruning (RNP) [12], instancewise sparsity accelerating (InSA) [13], BranchyNet [17] and SKipNet [16]. For the CIFAR-100 and Tiny-ImageNet datasets, only the dynamic methods are compared, and the results are shown in Tables 2 and 3.

Table 2. Performance Comparison on CIFAR-100

Models	Performance			
	Method	Params/AP($\times 10^6$)	FLOPs($\times 10^6$)	Accuracy (%)
VGG16	Baseline	15.03/0	313.51	69.70
	InSA	15.03/0	283.07	60.55
	BranchyNet	16.24/1.21	128.15	63.33
	SkipNet(SP)	41.36/26.33	309.88	**70.22**
	SkipNet(HRL+SP)	41.36/26.33	219.95	66.43
	MSDI(Entropy)	15.62/0.59	**79.26**	69.80
	MSDI(Learning)	15.57/0.54	119.77	69.75
ResNet18	Baseline	11.04/0	549.18	74.02
	InSA	11.04/0	535.09	64.78
	BranchyNet	13.11/2.07	392.57	69.21
	SkipNet(SP)	23.92/12.88	559.05	69.86
	SkipNet(HRL+SP)	23.92/12.88	492.25	68.81
	MSDI(Entropy)	12.64/1.60	371.94	**75.04**
	MSDI(Learning)	12.59/1.55	**336.88**	74.23

Table 3. Performance Comparison on Tiny-ImageNet

Models	Performance			
	Method	Params/AP($\times10^6$)	FLOPs($\times10^6$)	Accuracy (%)
Resnet34	Baseline	21.20/0	4612.53	60.43
	InSA	21.20/0	3541.43	51.63
	BranchyNet	23.34/2.14	3719.42	56.27
	SkipNet(SP)	44.18/22.98	3896.62	46.50
	SkipNet(HRL+SP)	44.18/22.98	3766.04	45.94
	MSDI(Entropy)	22.85/1.65	**3004.27**	58.53
	MSDI(Learning)	22.75/1.55	3386.99	**58.99**

Results Analysis. In our experiments, we focus on three crucial aspects of a methodology: the reduction in inference FLOPs (floating-point operations), model accuracy, and the number of auxiliary parameters. The results demonstrate that, across all datasets, the MSDI can significantly reduce the inference FLOPs of various networks. Taking the VGG16 model on the CIFAR-100 dataset as an example, up to 74.7% of the FLOPs are saved without sacrificing performance. Structural pruning techniques are capable of substantially decreasing both parameters and computational requirements. Nevertheless, as the parameters are pruned, the decline in the upper bound of performance becomes strikingly evident. Some dynamic methods may yield better outcomes than our methods do. For examle, SkipNet, when applied to the VGG16 model and CIFAR-100 dataset, achieves superior results but at the cost of overparameterization. SkipNet controls each convolutional layer via a multilayer gating structure; however, it involves many auxiliary parameters and requires additional computations. Consequently, the reduction in computations achieved through layer skipping seems not to be particularly prominent.

Although the MSDI has a small number of auxiliary parameters due to the branch structure, it still achieves the best model accuracy, with the greatest reduction in the number of FLOPs. Compared with the entropy-based grading approach, learning-based grading results in a less significant reduction in FLOPs. This is primarily because the early exit mechanism is retained only in the entropy-based grading method. This does not imply that entropy-based grading is invariably superior to learning-based grading. Without the manual configuration of the grading metric, the learning-based approach can demonstrate greater flexibility and effectiveness across diverse scenarios. For example, when applied to the ResNet34 model and Tiny-ImageNet dataset, the learning-based approach outperforms entropy-based grading, with a few more acceptable FLOPs.

Visualization. To intuitively demonstrate the rationality of the MSDI from the perspective of input complexity, we feed the validation set of the Tiny-

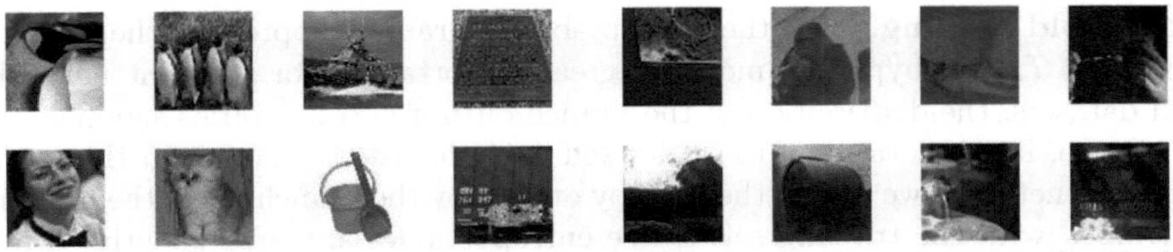

(a) Easy examples selected by ResNet34 on Tiny-ImageNet. (b) Hard examples selected by ResNet34 on Tiny-ImageNet.

Fig. 2. The easy and hard examples selected by ResNet34 on Tiny-ImageNet with MSDI. The easy samples directly exit via the branches or are computed by slimmer subnetworks, whereas the hard examples are computed by wider subnetworks.

ImageNet dataset into the ResNet34 model modified with MS-Conv. The images with the minimum and maximum output entropies, as classified by the entropy-based grading branch, are presented in Fig. 2. The underlying pattern among these images is quite evident: images with a clear background, moderately sized objects, and distinct patterns are more readily recognizable than those with a blurry background or intricate textures. Given that the Tiny-ImageNet dataset encompasses 200 categories, it is rather challenging for a shallow branch to extract sufficient features to achieve accurate classification. Nevertheless, it still effectively distinguishes between easy and difficult samples via output entropy.

4.3 Ablation Study

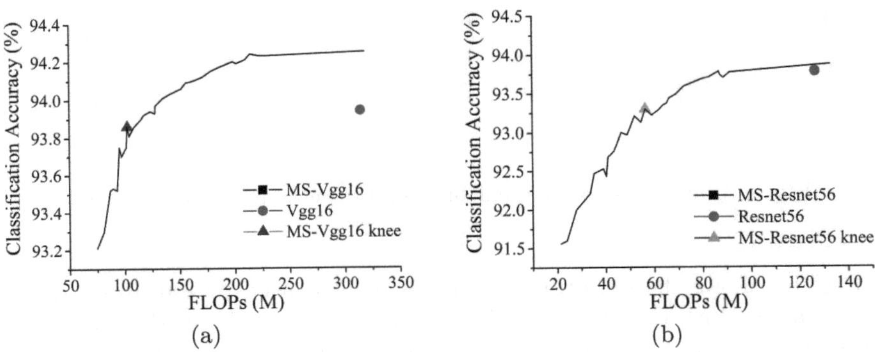

Fig. 3. Accuracy and FLOPs trade-off for (a) VGG16 and (b) ResNet56. The black curve denotes different grading strategies of the branch with different T_r settings. The selection of knee points is the final grading strategy, as illustrated in Table 1.

Threshold Tuning. For the entropy-based grading approach, the entropy threshold T_r is a hyperparameter of great importance. For different networks and datasets, the distribution of the branch output entropy varies significantly. Therefore, it is not rational to set a fixed value for the threshold. In this work, for every network, we record the entropy output by the branch when the network is trained with the training set. These entropy values are subsequently sorted, and all its deciles are obtained to obtain the threshold T_r with a grid search.

With every possible value of T_r, we obtain a grading strategy for the network, and the average number of FLOPs and accuracy of the network on the validation set are recorded. The results of VGG16 and ResNet56 are shown in Fig. 3 . These curves show how the trade off of accuracy and runtime varies with changes in the entropy thresholds.

Effects of Multiscale Distillation Training. The pretraining of the multiscale subnetworks is essential for the final performance of MSDI. To evaluate the effectiveness of the proposed multiscale distillation training method, we conducted a comprehensive comparison. Our method was pitted against the baselines of each individual network, as well as conventional direct training techniques. Specifically, we utilized VGG16 and ResNet56 on the CIFAR-10 dataset, along with ResNet18 on the CIFAR-100 dataset, to ensure a thorough and diverse evaluation.

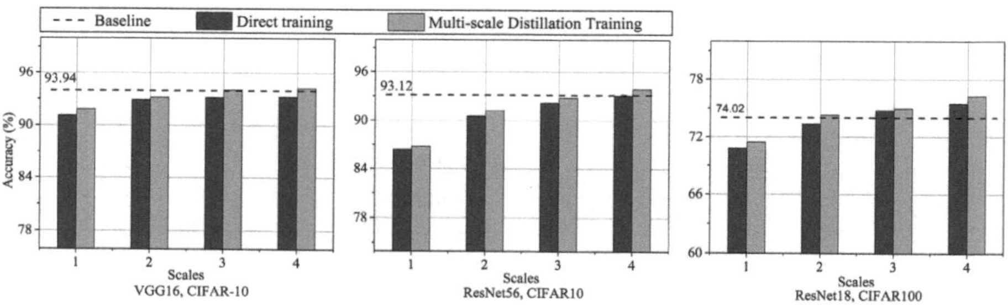

Fig. 4. Comparison of different pretraining methods on subnetworks.

Figure 4 shows the performance improvement caused by multiscale distillation training. With the distillation from wider subnetworks to slimmer subnetworks, not only do the slimmer subnetworks learn better, but also the performance of the wider subnetworks improve due to the parameter sharing mechanism between subnetworks at different scales. The performance of the subnetworks could even outperform that of the baseline without resorting to the full scale.

5 Conclusion

Dynamic inference achieves a flexible trade-off between performance and computational reduction, considering the instancewise complexity difference. This paper proposes a novel inference acceleration method tailored for deep convolutional networks. By partitioning the main network into multiple scales and incorporating a side branch structure, we can allocate the minimum amount of computational resources on the verge of ensuring performance for each input instance. Furthermore, to address the training challenges of deep yet slim subnetworks, we introduce multiscale distillation training, which trains subnetworks across different scales via knowledge distillation in a sequential rotation manner. Additionally, we propose two grading methodologies, namely, entropy-based and learning-based grading, to eliminate human interference. Extensive experiments demonstrate that MSDI is capable of accelerating most existing convolutional models on different datasets.

References

1. Krizhevsky, A., Sutskever, I., Hinton, G.E.: ImageNet classification with deep convolutional neural networks. Commun. ACM **60**(6), 84–90 (2017). https://doi.org/10.1145/3065386
2. Kong, X., Zhao, H., Qiao, Y., Dong, C: ClassSR: a general framework to accelerate super-resolution networks by data characteristic. CVPR, 12016–12025 (2021)
3. Yin, H., Mukadam, B., Dai, X., Jha, N.K.: DiabDeep: pervasive diabetes diagnosis based on wearable medical sensors and efficient neural networks. IEEE Trans. Emerg. Topics Comput. (2019). https://doi.org/10.1109/TETC.2019.2958946
4. Sutskever, I., Vinyals, O., Le, Q.V.: Sequence to sequence learning with neural networks. In: Proc. 27th Int. Conf. Neural Inf. Process. Syst., pp. 3104–3112 (2014)
5. Wen, W., Wu, C., Wang, Y., Chen, Y., Li, H.: Learning structured sparsity in deep neural networks. Adv. Neural. Inf. Process. Syst. **29**, 2074–2082 (2016)
6. Guo, J., Ouyang, W., Xu, D.: Multi-dimensional pruning: a unified framework for model compression. In: IEEE Conference on Computer Vision and Pattern Recognition (CVPR), pp.1508–1517 (2020).https://doi.org/10.1109/cvpr42600.2020.00158
7. Chin, T.W., Ding, R., Zhang, C., Marculescu, D.: Towards efficient model compression via learned global ranking. In: IEEE Conference on Computer Vision and Pattern Recognition (CVPR), pp.1518–1528 (2020). https://doi.org/10.1109/cvpr42600.2020.00159
8. Lin, M., et al.: HRank: filter pruning using high-rank feature map. CVPR, pp. 1526–1535 (2020)
9. Geoffrey, E.H., Oriol, V., Jeffrey, D.: Distilling the knowledge in a neural network. CoRR abs/1503.02531 (2015)
10. Zhang, K., Zhang, C., Li, S., Zeng, D., Ge, S.: Student network learning via evolutionary knowledge distillation. IEEE Trans. Circuits Syst. Video Technol. **32**(4), 2251–2263 (2022)

11. Mirzadeh, S.-I., Farajtabar, M., Li, A., Levine, N.: Akihiro Matsukawa, pp. 5191–5198. Improved Knowledge Distillation via Teacher Assistant. AAAI, Hassan Ghasemzadeh (2020)
12. Lin, J., Rao, Y., Lu, J., Zhou, J.: Runtime Neural Pruning. NIPS, 2181–2191 (2017)
13. Liu, C., Wang, Y., Han, K.: Chunjing Xu, pp. 3001–3007. Learning Instance-wise Sparsity for Accelerating Deep Models. IJCAI, Chang Xu (2019)
14. Tang, Y., et al.: Manifold regularized dynamic network pruning. CVPR, pp. 5018–5028 (2021)
15. Wang, Y., Zhang, X., Xiaolin, H., Bo, Z., Su, H.: Dynamic network pruning with interpretable layerwise channel selection. AAAI, pp. 6299–6306 (2020)
16. Wang, X., Fisher, Yu., Dou, Z.-Y., Darrell, T., Gonzalez, J.E.: SkipNet: learning dynamic routing in convolutional networks. ECCV **13**, 420–436 (2018)
17. Surat, T., Bradley, M., Kung, H.T.: BranchyNet: fast inference via early exiting from deep neural networks. CoRR abs/1709.01686 (2017)
18. Graves. A.: Adaptive computation time for recurrent neural networks. arXiv preprint arXiv:1603.08983, 1, 3, 6 (2016)
19. Figurnov, M., Collins, M.D., Zhu, Y., Zhang, L., Huang, J.: Dmitry P.: Spatially adaptive computation time for residual networks. CVPR, Vetrov, Ruslan Salakhutdinov, pp. 1790–1799 (2017)
20. Hu, H., Peng, R., Tai, Y.-W., Tang, C.-K.: Network trimming: a data driven neuron pruning approach towards efficient deep architectures, arXiv:1607.03250 (2016)
21. He, Y., Ding, Y., Liu, P., Zhu, L., Zhang, H., Yang, Y.: Learning filter pruning criteria for deep convolutional neural networks acceleration. In: Proc. IEEE/CVF Conf. Comput. Vis. Pattern Recognit. (CVPR), pp. 2006–2015 (2020)
22. Romero, A., Ballas, N., Kahou, S.E., Chassang, A., Gatta, C., Bengio, Y.: FitNets: hints for thin deep Nets. ICLR (Poster) (2015)
23. Forrest N.I., et al.: SqueezeNet: alexnet-level accuracy with 50x fewer parameters and <1MB model size. CoRR abs/1602.07360 (2016)
24. Sandler, M., Howard, A., Zhu, M., Zhmoginov, A., Chen, L.-C.: MobileNetV2: inverted residuals and linear bottlenecks. In: Proc. IEEE Conf. Comput. Vis. Pattern Recognit., pp. 4510–4520 (2018)
25. Ma, N., Zhang, X., Zheng, H.-T., Sun, J.: ShuffleNet V2: practical guidelines for efficient CNN architecture design. In: Proc. Eur. Conf. Comput. Vis., pp. 116–131 (2018)
26. Karen, S., Andrew, Z.: Very Deep Convolutional Networks for Large-Scale Image Recognition. ICLR (2015)
27. He, K., Zhang, X., Ren, S., Jian, S.: Deep residual learning for image recognition. CVPR, , pp. 770–778 (2016)
28. Krizhevsky, A., Hinton, G.: Learning multiple layers of features from tiny images. Techn. Report, Cite-seer (2009). https://www.cs.toronto.edu/~kriz/learning-features-2009-TR.pdf, April 2022
29. Le, Y., Yang, X.: Tiny imagenet visual recognition challenge. CS 231N **7**(7), 3 (2015)
30. Luo, J., Wu, J., Lin, W.: ThiNet: a filter level pruning method for deep neural network compression. ICCV, 5068–5076 (2017)
31. He, Y., Zhang, X., Sun, J.: In: Proceedings of the IEEE International Conference on Computer Vision, pp. 1389–1397 (2017)
32. Liu, Z., Li, J., Shen, Z., Huang, G., Yan, S., Zhang, C.: Learning efficient convolutional networks through network slimming. ICCV, pp. 2755–2763 (2017)

Generalized Prototype-Based Few-Shot Semantic Segmentation Network

Hang Zhou(✉) and Guanglu Sun

Harbin University of Science and Technology, Harbin 150000, China
850745204@qq.com

Abstract. Few-shot image semantic segmentation aims to achieve pixel-level classification for novel classes using only a few labeled examples. The method first trains the segmentation model on base classes, and then adapts it to novel classes. Although existing methods have achieved remarkable performance in few-shot image semantic segmentation, they still face the following challenges. Traditional methods typically rely on mask average pooling to generate single-category prototype vectors and perform feature matching via metric learning, but they exhibit significant limitations in modeling inter-category relationships and addressing complex background interference. Inspired by the analogy-based transfer mechanisms in cognitive psychology, we propose a Generalized Prototype Network (GPNet) to enhance the model's generalization ability for unseen categories and improve robustness in feature matching. GPNet consists of two key modules. The first is a generalized prototype enhancement module, which explores potential inter-category relationships to construct more discriminative category prototype representations. The second is a multi-scale feature alignment module, which dynamically aligns support and query features across multiple scales using an attention mechanism, thus mitigating background interference in complex scenarios. Experimental results demonstrate that the proposed method significantly outperforms existing state-of-the-art approaches on several few-shot semantic segmentation benchmarks, validating its effectiveness and generalization capabilities.

Keywords: semantic segmentation · few-shot semantic segmentation · prototype · semantic alignment · few-shot learning

1 Introduction

Inspired by the human ability to recognize objects with only limited information, researchers have developed few-shot learning techniques that enable models to adapt to novel domains or categories using only a small number of labeled examples. Few-shot image semantic segmentation applies few-shot learning to dense prediction tasks, aiming to enable a segmentation model trained on base classes to rapidly adapt to novel categories and perform pixel-level classification using only a handful of labeled examples.

Y. Yu et al. (Eds.): ICPCSEE 2025, CCIS 2673, pp. 307–324, 2026.
https://doi.org/10.1007/978-981-95-2566-9_22

Conventional few-shot image semantic segmentation models typically adopt masked average pooling to extract class-specific features from support samples and generate corresponding prototype vectors. They then compute similarity between these prototypes and each pixel in the query image to determine the class assignment for each pixel, thereby segmenting the target category in the query image. Many researchers have attempted to extract richer feature representations from prototype vectors through various mechanisms. However, these methods remain limited in capturing generalized representations across categories, as they primarily focus on single-class features and fail to fully explore the latent relationships between classes. According to the analogy transfer mechanism in cognitive psychology, humans can recognize novel categories such as "tiger" by leveraging a generalized prototype of felines after learning the concept of "lion." Yet, existing few-shot image semantic segmentation methods have not effectively constructed a hierarchical knowledge transfer process, making it difficult for models to learn informative cues related to unseen classes and thus limiting their generalization ability.

In addition, researchers have progressively addressed the limitations in spatial detail and semantic granularity by introducing techniques such as multi-scale feature fusion, attention mechanisms, and graph convolutional networks. Compared with traditional convolutional neural networks, Transformers capture more comprehensive contextual information and alleviate the information loss often encountered by CNNs when modeling large-scale contexts. Despite these advances in feature representation and global modeling, existing methods still suffer from interference caused by other category instances in the background, which hampers the accuracy of feature matching in complex scenes. This issue becomes more pronounced when multi-scale feature alignment is inadequate, making it difficult for the model to effectively distinguish foreground targets from background regions and ultimately reducing the robustness of the feature matching process.

To address the above issues, we construct a Generalized Prototype Network (GPNet), which incorporates two guidance modules: Generalized Prototype Enhancement Module and Multi-scale Feature Alignment Module. Specifically, the contributions of this paper are as follows:

(1) To address the limitation of existing methods in leveraging potential correlations between different classes, we propose a simple yet effective Generalized Prototype Enhancement Module. This module explores inter-class relationships to compensate for the inadequacy of single-class prototype representation, thereby improving the model's generalization to unseen categories.
(2) To mitigate the influence of non-target features in complex backgrounds, we introduce a Multi-scale Feature Alignment Module. This module constructs dynamic similarity mappings between support and query features across multiple feature levels. It integrates multi-scale features through an attention mechanism to achieve cross-scale semantic alignment. This design enhances the model's discriminative ability for target features and effectively suppresses background interference.
(3) We conduct extensive experiments on the PASCAL-5^i datasets. The results show that GPNet outperforms existing few-shot segmentation methods. Our method achieves

competitive performance under both 1-shot and 5-shot settings, demonstrating the effectiveness of our proposed generalized prototype-based network.

2 Related Work

2.1 Image Semantic Segmentation

Image semantic segmentation is a fundamental task in computer vision, aiming to provide pixel-level labels for objects and regions in a given image. It has extensive applications across various fields, including autonomous driving, scene understanding, and medical image analysis. Recently, fully supervised semantic segmentation methods have emerged as a popular research topic. Long et al. [1] pioneered the Fully Convolutional Network (FCN), replacing fully connected layers in traditional classification networks with fully convolutional architectures to enable end-to-end pixel-level classification. Ronneberger et al. [2] proposed the UNet architecture, characterized by its multi-scale modeling capability, effectively addressing boundary-loss issues caused by scarce data in medical image segmentation. SegNet [3] recorded pooling indices during max pooling operations and leveraged these indices during decoding for nonlinear upsampling, thus maintaining high performance while reducing the number of parameters. Chen et al. [4] developed the DeepLab series frameworks, innovatively introducing atrous convolution operations, which significantly enlarge the receptive field of convolution kernels to facilitate multi-scale context aggregation from local details to global information. To better exploit semantic structural information across multiple layers, Zhao et al. [5] proposed the Pyramid Scene Parsing Network (PSPNet), constructing a pyramid pooling module with various kernel sizes to perform multi-scale pooling on feature maps, effectively capturing both global and local contextual information. Recent advancements have fostered a series of Transformer-based general architectures, such as ViT [6], SAM [7], and Swin Transformer [8]. These architectures excel at capturing multi-scale information, enhancing boundary-detail handling, and achieving robust global semantic understanding, thus paving new pathways for semantic segmentation research.

2.2 Few-Shot Semantic Segmentation

Few-shot semantic segmentation is a crucial task in computer vision, aiming to generate dense predictions for novel classes from very few annotated examples, thereby alleviating the heavy reliance of traditional semantic segmentation methods on abundant labeled data. Most existing few-shot approaches follow the meta-learning framework proposed by Boney et al. [9]. Shaban et al. [10] first introduced the concept of few-shot learning into semantic segmentation and proposed a two-branch model named OSLSM. Several prior studies [11, 12] employed Masked Average Pooling (MAP) to extract either global or local average prototypes from support image features, and some prototypes could also represent specific categories [13]. Compared with prototype-based methods, matching-based methods aim to extract correspondence between query images and support annotations, utilizing pixel-level features to enhance detailed contextual information from the support set. PFENet [14] first proposed generating prior masks based on pixel-level relationships between support sets and query images, subsequently interacting these prior masks with cross-scale features to enrich query representations. BAM [15] introduced a novel approach for few-shot segmentation by adding an extra branch trained on base classes to explicitly identify base-class objects (areas not to be segmented). This supervised branch predicts base classes from query images, helping the meta-learner suppress incorrect predictions. CyCTR [16] proposed a cycle-consistent attention mechanism, equipping query features with relevant information from support sets. HDMNet [17] explored pixel-level correlations through a hierarchically decoupled matching network, thereby alleviating overfitting to the training set.

3 Methodology and Implementation

3.1 Few-Shot Semantic Segmentation Setting

Few-shot image semantic segmentation typically adopts the episodic training mechanism[18]. This mechanism, widely used in meta-learning, has been extensively applied to few-shot image semantic segmentation. In this paper, similar to earlier works [11], we follow the standard episodic training protocol and describe the training process using the following notations.

In the K-shot few-shot semantic segmentation setting, the dataset D is split into a training set D_{train} and a test set D_{test} based on categories. The class sets C_{train} and C_{test} from D_{train} and a D_{test} respectively, are disjoint, i.e., $\mathrm{C}_{\text{train}} \cap \mathrm{C}_{\text{test}} = \varnothing$. This ensures that the model encounters entirely unseen categories during inference.

During training, the model learns from the large-scale annotated training set D_{train}, aiming to segment novel categories with limited support samples. According to the episodic training mechanism, D_{train} and D_{test} are randomly divided into multiple episodes. Each episode consists of a support set $S = \left\{ \boldsymbol{I}_{\text{s}}^{k}, \boldsymbol{M}_{\text{s}}^{k} \right\}_{k=1}^{K}$ and a query set $Q = \left\{ \boldsymbol{I}_{\text{q}}, \boldsymbol{M}_{\text{q}} \right\}$. In the K-shot setting, K refers to the number of image-mask pairs in the support set. Each image is associated with a binary segmentation mask. $\boldsymbol{I}_{\text{s}}, \boldsymbol{I}_{\text{q}} \in \mathbb{R}^{H \times W \times 3}$ denote the support and query images respectively. Let W and H denote the width and height of RGB images. $\boldsymbol{M}_{\text{s}}, \boldsymbol{M}_{\text{q}} \in \mathbb{R}^{H \times W}$ denote the binary segmentation masks corresponding to the support and query images respectively. The goal is to segment the query image accurately using only support examples and generate predictions that are close to the ground truth $\boldsymbol{M}_{\text{out}}$. During testing, we evaluate the model's generalization ability on D_{test} without any fine-tuning. When $K = 1$, the task is referred to as 1-shot segmentation, which reflects the model's fundamental generalization capacity. When $K = 5$, the task is referred to as 5-shot segmentation, which evaluates how effectively the model utilizes annotated information.

3.2 Overall Architecture

This paper proposes a prototype learning-based few-shot semantic segmentation model, named Generalized Prototype Network (GPNet). GPNet is designed to address two key limitations of existing methods: the failure to fully exploit potential correlations between classes, and the sensitivity to background interference when matching features at fixed scales, especially for unseen classes. As shown in Fig. 1, GPNet consists of two main modules: the Generalized Prototype Enhancement Module and the Multi-scale Feature Alignment Module.

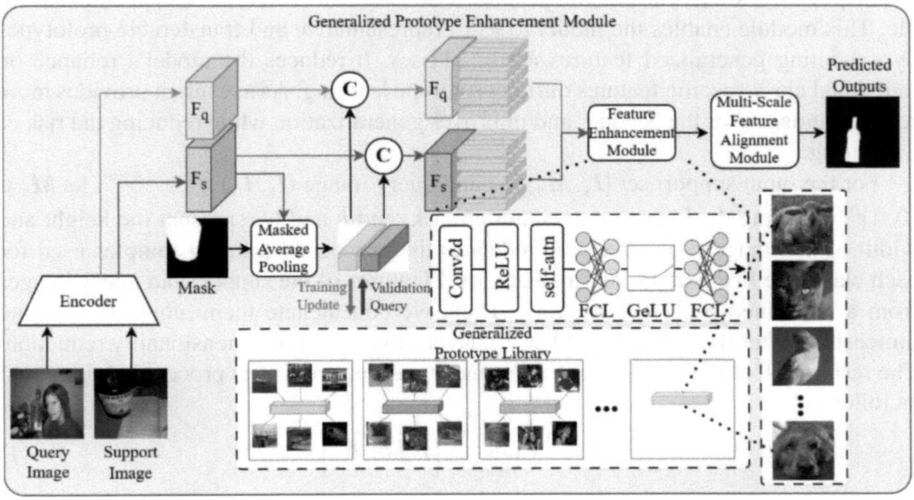

Fig. 1. Architecture of the Generalized Prototype Network

The Generalized Prototype Enhancement Module first extracts generalized class prototypes from the support features generated by a pretrained backbone network. Based on these prototypes, it constructs a generalized prototype library and updates the prototypes to learn class-agnostic feature representations. Then, it computes the similarity between the query features and the generalized prototypes. The most similar prototype is selected to enhance the query features.

The Multi-Scale Feature Alignment Module (MSFAM) builds dynamic similarity mappings between support and query features across low-, mid-, and high-level feature spaces. It integrates these features through a residual fusion block. This design effectively suppresses background interference and achieves cross-scale semantic consistency. As a result, it improves the robustness of few-shot segmentation under complex background conditions.

3.3 Generalized Prototype Enhancement Module

Traditional prototype generation methods based on mask average pooling typically rely only on local class-specific features. This limits the model's ability to capture class-agnostic representations in complex scenes and hinders generalization to unseen classes. We argue that semantic and spatial features across different classes may exhibit certain correlations. For example, "cat" and "dog" in the animal category, or "car" and "truck" in the vehicle category, may share related characteristics. By involving images from multiple similar classes in prototype learning, the model can better capture the latent relationships between these classes. This enhances its ability to understand cross-category context.

Inspired by Wang et al. [13] we propose a Generalized Prototype Enhancement Module. This module enables the model to learn representative and transferable prototypes by extracting generalized features across classes. It reduces the model's reliance on individual class-specific features during prototype learning. As a result, it provides more refined guidance for the decoder and improves generalization while reducing the risk of overfitting.

For the input support set $\{\boldsymbol{I}_{\mathrm{s}}, \boldsymbol{M}_{\mathrm{s}}\}_{k=1}^{K}$ and query image $(\boldsymbol{I}_{\mathrm{s}}, \boldsymbol{I}_{\mathrm{q}}) \in \mathbb{R}^{3\times h\times w}$, let $\boldsymbol{M}_{\mathrm{s}} \in \{0, 1\}^{h\times w}$ denote the binary segmentation mask, and h and w represent the height and width of the image respectively. K indicates the number of training samples used for each class. We first extract intermediate-level features of the support and query images from a pretrained backbone network. Then, we concatenate them along the channel dimension. After that, we apply a 1×1 convolution layer for dimensionality reduction. The resulting features are denoted as F_s and F_q respectively. The process is formulated as follows:

$$\boldsymbol{F}_{\mathrm{s}} = Conv_{1\times 1}\left\{\boldsymbol{F}_{\mathrm{s}}^{2} \oplus \boldsymbol{F}_{\mathrm{s}}^{3}\right\} \tag{1}$$

$$\boldsymbol{F}_{\mathrm{q}} = Conv_{1\times 1}\left\{\boldsymbol{F}_{\mathrm{q}}^{2} \oplus \boldsymbol{F}_{\mathrm{q}}^{3}\right\} \tag{2}$$

where $\boldsymbol{F}_{\mathrm{s}}, \boldsymbol{F}_{\mathrm{q}} \in \mathbb{R}^{b\times c\times h'\times w'}$, h' and w' represent the height and width of the feature maps, b and c denote the batch size and the number of channels respectively. The numbers 2 and 3 indicate the features extracted from the 2nd and 3rd blocks of the backbone network. During training, we freeze the parameters of the backbone to improve the model's generalization to unseen classes.

In the training phase, we input $\boldsymbol{F}_{\mathrm{s}}, \boldsymbol{F}_{\mathrm{q}}$ and the mask $\boldsymbol{M}_{\mathrm{s}}$ into the Generalized Prototype Enhancement Module to update the generalized prototypes that are similar to the novel class. In the testing phase, we stop updating the generalized prototypes and only match them to those similar to the novel class. Then, we use the generalized prototypes to enhance the support and query features separately, enabling the segmentation model to achieve more accurate segmentation performance.

Before defining the generalized prototype library, we divide the categories in the dataset. For a given few-shot semantic segmentation dataset, there are m classes. We define the original class label set as $C = \{c_1, c_2, c_3, ..., c_m\}$. For each class $c_i \in C$ its corresponding image set is denoted as D_i. We randomly select a subset $D_i' \subseteq D_i$ of size $|D_i'| = n$. For each image $x_{ij} \in D_i'$, we extract image features $\boldsymbol{F}_{ij} \in \mathbb{R}^{n\times h'\times w'}$ using a pretrained neural network, where $j = 1, 2, ..., n$, h' and w' denote the height and width of the feature maps. We flatten the features into $\hat{\boldsymbol{F}}_{ij} \in \mathbb{R}^{n\times(h'\times w')}$ to make them suitable for the K-Means algorithm. We cluster the flattened features $\hat{\boldsymbol{F}}_{ij}$ into K_m clusters and obtain the cluster centers $\{\mu_k\}_{k=1}^{K_{\mathrm{m}}}$ and cluster labels $l_{ij} \in \{1, 2, \ldots, K_m\}$, as calculated by the following formula.

$$l_{ij} = \arg\min_{k} \left\| \hat{\boldsymbol{F}}_{ij} - \mu_k \right\| \tag{3}$$

For each original class c_i, we collect all cluster labels corresponding to its feature vectors. We then find the most frequent cluster label, which is defined as the generalized class label for class c_i.

Figure 2 shows the distribution of support features from different categories after dimensionality reduction using Principal Component Analysis (PCA), as well as the visualization results after clustering with K-Means.

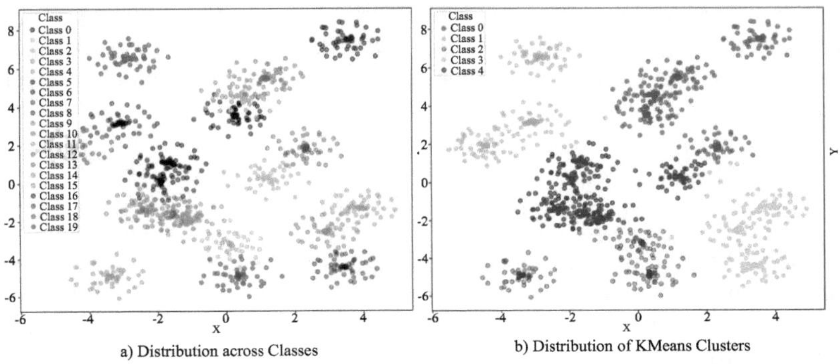

Fig. 2. Visualization of generalized category partitioning

Next, we construct the generalized prototype library $P = \{\boldsymbol{P}_1, ..., \boldsymbol{P}_i, ..., \boldsymbol{P}_n\} \in \mathbb{R}^{n \times c}$, where n denotes the size of the library. i represents the generalized class to which the sample belongs. The mask $\boldsymbol{M}_\mathrm{s}$ is resized to match the feature resolution using bilinear interpolation, denoted as $\boldsymbol{M}_\mathrm{s} \in \{0, 1\}^{h' \times w'}$. We perform an element-wise multiplication between the support feature $\boldsymbol{F}_\mathrm{s}$ and the resized mask $\boldsymbol{M}_\mathrm{s}$, and compute the average of non-zero values along the spatial dimension. This operation avoids the influence of object size variation across the image. The initial prototype $\boldsymbol{P}_\mathrm{o} \in \mathbb{R}^c$ is obtained as follows:

$$\boldsymbol{P}_o = \frac{\boldsymbol{M}_s \odot \boldsymbol{F}_s}{|\boldsymbol{M}_s|} \tag{4}$$

where $\odot$ is the Hadamard product. $|\boldsymbol{M}_s|$ denotes the number of elements with value 1 in the mask $\boldsymbol{M}_s$. x and y represent the spatial indices of the mask or feature map, indicating each pixel location. During training, we update the generalized prototype $\boldsymbol{P}_i$ in the prototype library based on the sample's class label. This process enhances the model's ability to capture inter-class correlations, thus improving its adaptability and generalization to novel classes. i refers to the generalized class of the given sample.

After prototype extraction, we expand each single prototype to match the spatial size of the support and query features. We integrate it with the similarity map to enhance the features. The update process is as follows:

$$\boldsymbol{P}_i = (1 - \gamma) l_2(\boldsymbol{P}_\mathrm{o}) + \gamma l_2(\boldsymbol{P}_i) \tag{5}$$

$$\boldsymbol{F'}_\mathrm{q} = Conv_{1\times1}(\boldsymbol{F}_\mathrm{q} \oplus (\boldsymbol{P}_i \otimes 1_{h\times w}) \oplus (\boldsymbol{P}_\mathrm{o} \otimes 1_{h\times w}) \oplus \boldsymbol{Cs}_1) \tag{6}$$

$$\boldsymbol{F'}_{\mathrm{s}} = Conv_{1\times1}(\boldsymbol{F}_{\mathrm{s}} \oplus (\boldsymbol{P}_i \otimes 1_{h\times w}) \oplus (\boldsymbol{P}_{\mathrm{o}} \otimes 1_{h\times w}) \oplus \boldsymbol{Cs}_1) \tag{7}$$

where γ is a balancing factor that regulates the contribution of old and new information. $Conv_{1\times1}(\cdot)$ represents a 1×1 convolution. $\oplus$ denotes the channel-wise concatenation. $\otimes$ refers to broadcast multiplication, which allows $\boldsymbol{P}_i$ to expand to the same spatial size as $\boldsymbol{F}_{\mathrm{q}}$. $\boldsymbol{Cs}_1$ is the similarity feature map, and l_2 denotes the normalization operation on L_2

In the testing phase, we compare the similarity between the support prototype $\boldsymbol{P}_{\mathrm{o}}$ and each vector in the generalized prototype library $\boldsymbol{P}_k$. We select the most similar target prototype and use it to enhance the feature representations. The prototype selection is calculated as follows:

$$u = \underset{k}{\mathrm{argmax}}\left(\frac{\boldsymbol{P}_{\mathrm{o}}^T \cdot \boldsymbol{P}_k}{\|\boldsymbol{P}_{\mathrm{o}}\| \|\boldsymbol{P}_k\|}\right) \tag{8}$$

In addition, the enhanced features $\boldsymbol{F'}_{\mathrm{s}}$ and $\boldsymbol{F'}_{\mathrm{q}}$ are passed through a convolution to match the dimensionality of the query features. A self-attention operation is then applied to emphasize relevant parts and suppress irrelevant details, further refining the feature representations by leveraging the self-correlation between support and query features. The calculation is as follows:

$$\text{self-Attn}(\boldsymbol{Q}, \boldsymbol{K}, \boldsymbol{V}) = \text{softmax}\left(\frac{\boldsymbol{QK}^T}{\sqrt{d}}\right)\boldsymbol{V} \tag{9}$$

where $[\boldsymbol{Q}, \boldsymbol{K}, \boldsymbol{V}] = \left[\boldsymbol{W}_{\mathrm{q}}\boldsymbol{F'}_{\mathrm{q}}, \boldsymbol{W}_{\mathrm{k}}\boldsymbol{F'}_{\mathrm{q}}, \boldsymbol{W}_{\mathrm{v}}\boldsymbol{F'}_{\mathrm{q}}\right]$ denotes the enhanced query features F'_q. For the enhanced support features, the corresponding transformation matrix is defined as $[\boldsymbol{Q}, \boldsymbol{K}, \boldsymbol{V}] = \left[\boldsymbol{W}_{\mathrm{q}}\boldsymbol{F'}_{\mathrm{s}}, \boldsymbol{W}_{\mathrm{k}}\boldsymbol{F'}_{\mathrm{s}}, \boldsymbol{W}_{\mathrm{v}}\boldsymbol{F'}_{\mathrm{s}}\right]$. $\boldsymbol{W}_{\mathrm{q}}, \boldsymbol{W}_{\mathrm{k}}, \boldsymbol{W}_{\mathrm{v}} \in \mathbb{R}^{d\times d}$ is a learnable parameter matrix, and d represents the feature dimension.

After self-attention, the features $\boldsymbol{F'}_{\mathrm{s}}$ and $\boldsymbol{F'}_{\mathrm{q}}$ are passed through a down-projection linear layer $\boldsymbol{W}_{\mathrm{down}} \in \mathbb{R}^{d\times\frac{d}{\alpha}}$ to reduce dimensionality. α is the dimensionality reduction ratio. The compressed features are transformed nonlinearly by the activation function GeLU to capture the nonlinear relations in the data. Then, an up-projection linear layer $\boldsymbol{W}_{\mathrm{up}} \in \mathbb{R}^{d\times\frac{\alpha}{d}}$ restores the feature dimensions to reduce the risk of overfitting. The computation is given as:

$$\boldsymbol{F}_{\mathrm{s}}^* = \mathrm{GeLU}(\boldsymbol{F'}_{\mathrm{s}} \cdot \boldsymbol{W}_{\mathrm{down}}) \cdot \boldsymbol{W}_{\mathrm{up}} \tag{10}$$

$$\boldsymbol{F}_{\mathrm{q}}^* = \mathrm{GeLU}(\boldsymbol{F'}_{\mathrm{q}} \cdot \boldsymbol{W}_{\mathrm{down}}) \cdot \boldsymbol{W}_{\mathrm{up}} \tag{11}$$

where $\boldsymbol{W}_{\mathrm{down}}, \boldsymbol{W}_{\mathrm{up}}$ is a learnable parameter matrix.

Finally, we apply a scaling factor β to weight the enhanced query features and add them to the original query features, resulting in the final enhanced query representation.

$$\boldsymbol{F}_{\mathrm{q}} = \boldsymbol{F}_{\mathrm{q}}^* \times \beta + \boldsymbol{F}_{\mathrm{q}} \tag{12}$$

3.4 Multi-scale Feature Alignment Module

In few-shot semantic segmentation, feature matching at fixed scales is easily affected by other category objects in complex backgrounds. Inspired by [17], we propose the Multi-Scale Feature Alignment Module (MSFAM) to optimize the similarity mapping between query and support features. This module enables the model to analyze features at different levels and capture both fine-grained local details and global contextual information.

We send the support and query features into separate Transformer blocks with self-attention layers. This allows the model to fully explore internal self-correlation within support and query features. We insert downsampling layers between Transformer blocks. Specifically, we use 3 × 3 convolution layers to downsample support and query features independently. The spatial size of features follows an exponential decay pattern. The architecture of the module is shown in Fig. 3. This hierarchical structure helps the model learn cross-scale correlations.After L stages of downsampling, the size of the feature map at stage l ($l \in \{1, 2, \ldots, L\}$) is calculated as follows:

$$h_l^{\mathrm{q/s}} = \frac{H^{\mathrm{q/s}}}{2l+2}, \quad w_l^{\mathrm{q/s}} = \frac{W_{\mathrm{e}}^{\mathrm{q/s}}}{2l+2} \tag{13}$$

where for simplicity, since $\boldsymbol{F}_{\mathrm{q}}^{l}$ and $\boldsymbol{F}_{\mathrm{s}}^{l}$ have the same spatial resolution, we denote the initial height and width of both support and query features as $H^{\mathrm{q/s}}$ and $W_{\mathrm{e}}^{\mathrm{q/s}}$.

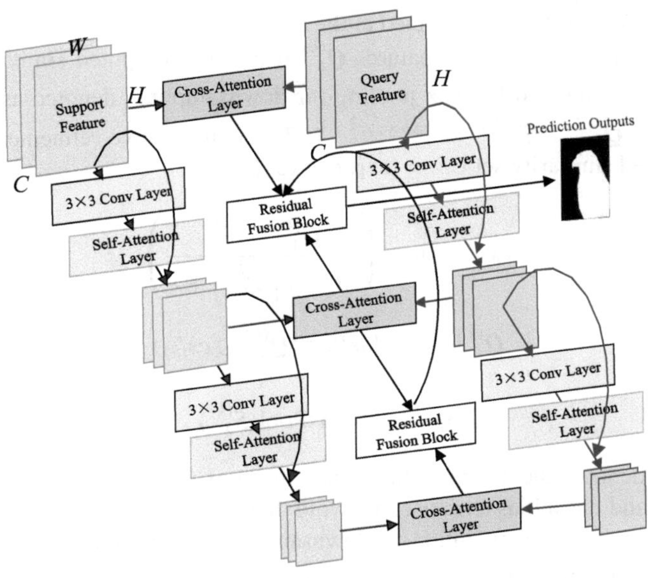

Fig. 3. Architecture of the multi-scale feature alignment module

We perform cross-attention between the support and query features at the same scale to obtain the multi-scale conditional feature map $\left\{\boldsymbol{A}_{\mathrm{q}}^{l}\right\}_{l=1}^{L}$, where the number of downsampling stages using 3 × 3 convolution layers and self-attention layers is denoted as $l \in (0, 1, 2)$.

$$A_{\mathrm{q}}^{l} = \mathrm{softmax}\left(\frac{W_{\mathrm{q}}(F_{\mathrm{q}})^{l}(F_{\mathrm{s}})^{lT}W_{\mathrm{k}}^{T}}{\sqrt{d}}\right)W_{\mathrm{v}}F_{\mathrm{s}} \tag{14}$$

where $W_{\mathrm{q}}, W_{\mathrm{k}}, W_{\mathrm{v}} \in \mathbb{R}^{d\times d}$ is a learnable parameter matrix, d is the feature dimension, and F_{q} denotes the query features enhanced through Eq. 12.

Then, we apply mask-based filtering to the high-level support features F_{s}^{4}, which are extracted from the 4th block of the backbone network. This step reduces the influence of background and non-essential regions, improving the model's robustness under complex scenes.

$$Q_4 = F_{\mathrm{q}}^{4} \tag{15}$$

$$S_4 = F_{\mathrm{s}}^{4} \odot \varsigma(M_{\mathrm{s}}) \tag{16}$$

where ς denotes the bilinear interpolation function. It resizes the support mask M_{s} according to spatial dimensions and expands it along the channel dimension to match the support features.

We compute the cosine similarity between the support feature S_4 and the query feature Q_4 using Eq. 17, resulting in the cosine similarity map cs_k, where $k \in (0, 1, 2)$ represents the similarity map obtained at the k-th iteration.

Then, we update the query features $Q_4^{(k)}$ using the computed similarity map cs_k through Eq. 18, followed by max pooling for downsampling, denoted as $Q_4^{(k)}, S_4^{(k)} \in \mathbb{R}^{b\times c\times h\times w} \rightarrow Q_4^{(k+1)}, S_4^{(k+1)} \in \mathbb{R}^{b\times c\times \frac{h}{2}\times \frac{w}{2}}$. Through iterative refinement, we obtain the fine-grained similarity set $CS = \{cs_1, cs_2, cs_3\}$.

$$cs_{(k)} = \max_{c}\left(\frac{Q_4^{(k)^T}\cdot S_4^{(k)}}{\left\|Q_4^{(k)}\right\|\left\|S_4^{(k)}\right\| + \varepsilon}\right) \tag{17}$$

$$Q_4^{(k+1)} = \mathrm{MaxPool}(Q_4^{(k)} \odot cs_{(k)}) \tag{18}$$

$$S_4^{(k+1)} = \mathrm{MaxPool}(S_4^{(k)}) \tag{19}$$

where $\max_c$ denotes the function that takes the maximum value along the channel dimension c, and ε is a bias term used to avoid division by zero.

We apply a 1 × 1 convolution to downsample the high-dimensional features S_4 and Q_4 from the support and query vectors, respectively, denoted as $Q_4^{(l)}, S_4^{(l)} \in \mathbb{R}^{b\times c\times h\times w} \rightarrow Q_4^{(l+1)}, S_4^{(l+1)} \in \mathbb{R}^{b\times c\times \frac{h}{2}\times \frac{w}{2}}$ where $l \in (0, 1, 2)$ represents the number of downsampling steps. This operation captures cosine similarity across low, middle, and high scales. The result is fused with the fine-grained similarity matrix CS to obtain the multi-scale feature MS. This process provides both spatial and semantic references for subsequent multi-scale feature fusion. It helps the model better align high-level and low-level features and form consistent representations in semantic space.

$$\boldsymbol{MS}_l = \tau \cdot \boldsymbol{cs}_l + (1 - \tau) \cdot \frac{\boldsymbol{Q}_4^{(l)T} \cdot \boldsymbol{S}_4^{(l)}}{\left\| \boldsymbol{Q}_4^{(l)} \right\| \left\| \boldsymbol{S}_4^{(l)} \right\| + \varepsilon} \tag{20}$$

where τ is a balancing factor that regulates old and new information. It is a learnable parameter.

We then apply the multi-scale similarity map set $\boldsymbol{MS}$ to the conditional feature maps of the corresponding scale. This design enhances regions with high similarity to the support features and suppresses regions with low similarity, thereby improving the representational capacity of the conditional features. A residual fusion block is used to merge information across different scales. This block consists of multi-channel 2D convolution, group normalization, and a ReLU activation function, as shown in Fig. 4.

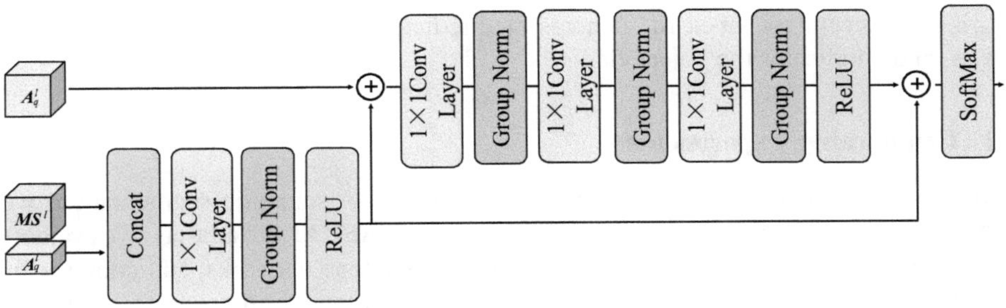

Fig. 4. Schematic diagram of the residual fusion block

3.5 K-shot Setting

To remain consistent with previous work [15], we compute generalized prototype similarity and multi-scale similarity under both 1-shot and K-shot settings. In the 1-shot setting, we calculate the generalized prototype using the method described in Sect. 2.3. In the K-shot setting, each class is represented by multiple support feature vectors. We compute a prototype from each individual support feature in the support image set and use them to update the generalized prototype library. For multi-scale similarity computation, we calculate the similarity between each support vector and query vector separately. We then average these similarity scores to obtain the final class-level similarity.

4 Experimental Analysis

4.1 Experimental Environment and Setup

We implement our method using the PyTorch framework [20], and run all experiments on a workstation equipped with an NVIDIA RTX 4090 GPU. Our method adopts the bidirectional learning framework proposed in [15]. The base learner is trained using a standard supervised learning paradigm to predict base class regions and suppress potential distracting objects. The meta-learner follows a conventional episodic training protocol and uses support annotations to guide the model in recognizing target classes.

To ensure fair comparison, we freeze the parameters of the base learner during the training of the meta-learner. Following the settings in [15], we train the model on the PASCAL-5^i dataset for 200 epochs with an initial learning rate of 1e-4 and a batch size of 4.In addition, we apply the data augmentation strategies introduced in [19] to both datasets. These include scale jittering, rotation, and random cropping of the training samples.

All methods in this paper use the cross-entropy loss function to supervise the training of the proposed few-shot semantic segmentation model. The loss function is defined as follows:

$$\mathcal{L}_{total}(f_\theta(\boldsymbol{I}_q, \boldsymbol{S}), \boldsymbol{M}_q) = -\frac{1}{N}\sum_{n=1}^{N}\sum_{c=1}^{C}\boldsymbol{M}_q^{n,c}\log\left(\hat{\boldsymbol{M}}_q^{n,c}\right) \tag{20}$$

Here, f_θ denotes the segmentation model to be trained, and θ represents the learnable parameters of the model. $\boldsymbol{M}_q \in \{0, 1\}$ is the ground truth mask of the query image, and $\boldsymbol{M}_q$ is the query mask label. $\widehat{\mathrm{M}}_q$ denotes the predicted mask, N is the total number of pixels, and c indicates the object class.

4.2 Comparative Experiments

Following the standard evaluation protocol for few-shot semantic segmentation, we conduct experiments using the proposed GPNet on the PASCAL-5^i datasets. We evaluate the model using both VGG-16 and ResNet-50 as backbone networks, and compare its performance with state-of-the-art methods under 1-shot and 5-shot settings. Tables 1 and 2 show the comparison results of GPNet with other advanced few-shot semantic segmentation models on the PASCAL-5^i datasets, respectively. The mean Intersection over Union (mIoU) is used as the performance metric (expressed in %), and the best results are highlighted in bold.

Table 1. Performance of GPNet on the PASCAL-5^i dataset under 1-shot settings

Backbone	Method	1-shot				
		Fold-0	Fold-1	Fold-2	Fold-3	Mean
VGG-16	HSNet [19]	59.6	65.7	59.6	54.0	59.7
	PANet [12]	42.3	58.0	51.1	41.2	48.1
	HDMNet [17]	64.8	71.4	67.7	56.4	65.1
	PFENet [21]	56.9	68.2	54.4	52.4	58.0
	PSANet [23]	65.6	71.7	66.8	59.6	65.9
	GPNet (ours)	65.7	71.7	67.6	60.2	66.3
ResNet-50	BAM [15]	69.0	73.6	67.6	61.1	67.8
	PFENet [21]	64.3	70.7	60.3	60.5	64.0
	HDMNet [17]	71.0	75.4	68.9	62.1	69.4

(continued)

Table 1. (*continued*)

Backbone	Method	1-shot				
		Fold-0	Fold-1	Fold-2	Fold-3	Mean
	HSNet [19]	64.3	70.7	60.3	60.5	64.0
	ABCNet [24]	68.8	73.4	62.3	59.5	66.0
	AFANet [25]	65.7	68.5	60.6	61.5	64.0
	SPANet [22]	70.3	76.5	68.9	62.8	69.6
	QPENet [26]	65.2	71.9	64.1	59.5	65.2
	PSANet [23]	69.5	74.6	67.1	60.6	67.9
	GPNet(ours)	72.3	76.4	70.6	62.6	70.5

Table 2. Performance of GPNet on the PASCAL-5i dataset under the 5-shot setting

Backbone	Method	5-shot				
		Fold-0	Fold-1	Fold-2	Fold-3	Mean
VGG-16	HSNet [19]	64.9	69.0	64.1	58.6	64.1
	PANet [12]	51.8	64.6	59.8	46.5	55.7
	HDMNet [17]	68.1	73.1	71.8	64.0	69.3
	PFENet [21]	59.0	69.1	54.8	52.9	59.0
	PSANet [23]	69.8	74.1	69.5	66.9	70.1
	GPNet (ours)	70.7	73.6	72.4	63.2	69.9
ResNet-50	BAM [15]	70.6	75.1	70.8	67.2	70.9
	PFENet [21]	70.3	73.2	67.4	67.1	69.5
	HDMNet [17]	71.3	76.2	71.3	68.5	71.8
	HSNet [19]	70.3	73.2	67.4	67.1	69.5
	ABCNet [24]	71.7	74.2	65.4	67.0	69.6
	AFANet [25]	69.0	70.4	61.3	64.0	66.2
	SPANet [22]	72.3	76.9	74.7	67.8	72.9
	QPENet [26]	68.4	74.0	67.4	65.2	68.8
	PSANet [23]	72.0	76.4	69.5	67.8	71.4
	GPNet (ours)	72.4	77.6	73.1	69.7	73.2

4.3 Comparative Experiments

This section presents the ablation study of the key modules proposed in our method. The core of the proposed approach lies in two main components: the Generalized Prototype Generation Module and the Multi-Scale Feature Alignment Module.

To quantitatively validate their effectiveness, we conduct ablation experiments on the PASCAL-5^i dataset under the 1-shot setting using ResNet-50 as the backbone. We adopt HSNet as the baseline model. HSNet is one of the most representative multi-scale correlation learning methods in the few-shot segmentation field. It employs a center-pivoted 4D convolution to build a 4D convolutional pyramid encoder for efficient feature matching.

The ablation results are reported in Table 3, and the visual comparisons are shown in Fig. 3–5. Here, BCPL denotes the setting where query features extracted by ResNet-50 are enhanced using the Generalized Prototype Enhancement Module before being fed into the baseline model. MSFAM indicates that the 4D convolutional pyramid encoder in the baseline is replaced with the multi-scale attention module described in Sect. 3.2. The "–" symbol means that the corresponding module is not used in the model.

Table 3. Ablation study of key modules

Baseline	BCPL	MSFAM	Fold-0	Fold-1	Fold-2	Fold-3	mIoU	Δ
✓	–	–	64.3	70.7	60.3	60.5	64.0	0.0
✓	✓	–	66.2	71.3	62.1	62.7	65.6	+ 1.6
✓	–	✓	70.4	75.6	67.8	61.0	68.7	+ 4.7
✓	✓	✓	72.3	76.4	70.6	62.6	70.5	+ 5.5

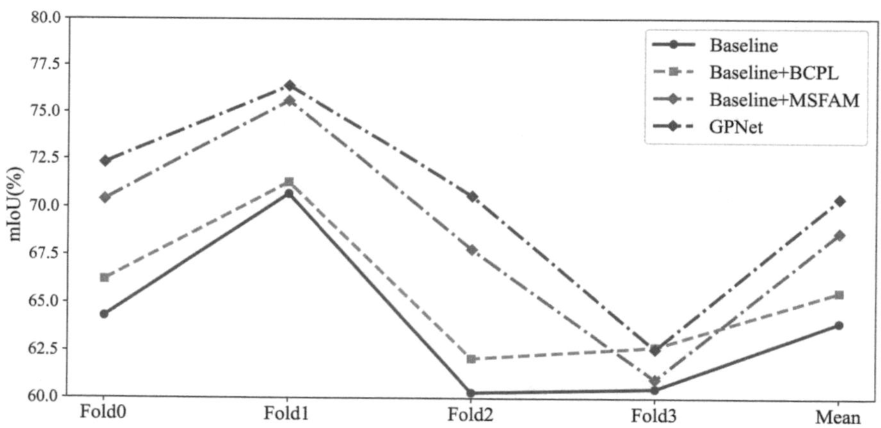

Fig. 5. Comparison of ablation experiment results

The experimental results demonstrate that both the proposed Generalized Prototype Enhancement Module and the hierarchical attention module effectively improve the performance of the baseline HSNet. Specifically, introducing the Generalized Prototype Enhancement Module leads to a 1.6% average performance gain. On the animal-dominant Fold-3, it achieves a 2.2% improvement. This is because the module incorporates multiple semantically related classes into prototype learning, enabling the model to better capture generalized representations across categories (e.g., "cat" and "dog" share similar fur textures). It effectively addresses the lack of cross-category generalization in traditional few-shot segmentation.

When incorporating the Multi-Scale Feature Alignment Module, the model achieves a 4.7% gain in mean IoU. When both modules are used together, the model achieves a 5.5% improvement in mIoU.

In addition, we investigate the effects of the generalized prototype library size α and the feature scaling factor β through ablation experiments under different settings. The results are shown in Fig. 6. To determine the impact of prototype library size on model performance, we conduct controlled experiments and select the optimal value for α. When $\alpha = 6$, the model achieves the best performance. A small α (e.g., $\alpha = 3$) causes class-level confusion (e.g., misclassifying "person" and "ship" as semantically similar). Conversely, a large α (e.g., $\alpha = 10$) limits the semantic representation capacity of the generalized prototypes (e.g., failing to capture shared features between "quadrupeds" and "flying animals"), making it difficult to learn class-agnostic representations.

We also introduce the feature scaling factor β to balance the impact between enhanced features and original features. Ablation results show that the model performs best when $\beta = 0.1$. A small β (e.g., $\beta = 0.01$) weakens the influence of enhanced features, leading the model to ignore generalized cross-class information. In contrast, a large β (e.g., $\beta = 1$) causes the model to overly rely on class-specific features, resulting in overfitting.

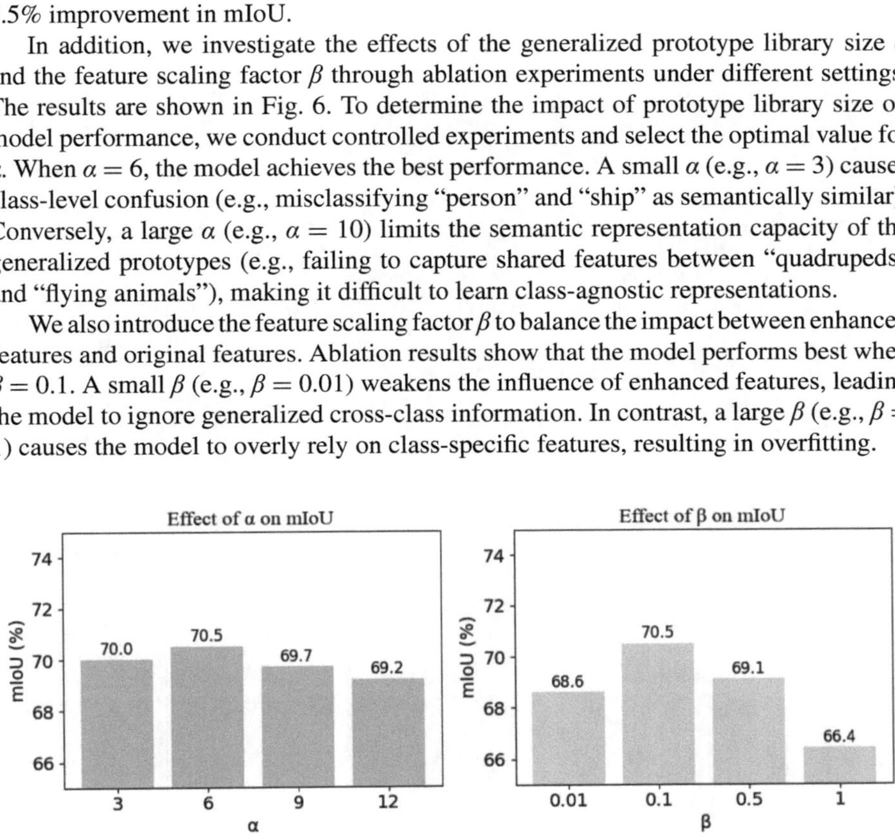

Fig. 6. Comparison of ablation experiment results

4.4 Visualization Results

To intuitively illustrate the effectiveness of GPNet, we visualize the prediction results of both the Baseline and GPNet in Fig. 3, 4, 5, 6 and 7. The samples on the left are from the PASCAL-5^i dataset, while those on the right are from the COCO-20^i dataset. From top to bottom, the rows are as follows: a) support mask; b) query image; c) query mask; d) prediction by the baseline model; e) prediction by GPNet; f) prior mask from the baseline model; g) Baseline + BCPL visualization result; h) Baseline + MSFAM visualization result; i) GPNet visualization result.

From the comparison, we observe that for small target regions in the image (e.g., the stool in the fourth column), the baseline model fails to locate the object due to the presence of multiple background objects. In contrast, GPNet successfully distinguishes the target from the complex background. Additionally, in cases where the object shares similar appearance with the background (e.g., the bed in the fifth column), GPNet produces accurate and sharp segmentation results. The boundaries are more clearly defined. These results demonstrate that the proposed method effectively mitigates the interference caused by non-target background objects in few-shot semantic segmentation.

Fig. 7. Visualization of qualitative results

5 Conclusion

This paper introduces and summarizes a generalized prototype-based method for few-shot image semantic segmentation. Existing methods often suffer from limited class-level semantic understanding. They fail to explore potential inter-class correlations and are easily disturbed by complex backgrounds when handling unseen classes. Inadequate semantic alignment between features further limits the model's generalization and robustness. To address these challenges, we propose the Generalized Prototype Network (GPNet). GPNet incorporates a Generalized Prototype Enhancement Module to overcome the limitations of conventional prototype learning. It captures shared representations across multiple related classes, thereby enhancing the model's generalization ability in cross-category scenarios. In addition, we propose a Multi-Scale Feature Alignment Module, which leverages multi-scale attention and cross-scale similarity computation. This module improves target recognition by integrating fine-grained local details and global contextual information, reducing the effect of background interference.

We conduct extensive experiments on the PASCAL-5^i and COCO-20^i datasets to validate the effectiveness of our method. Compared to representative baseline models, GPNet achieves superior segmentation performance under both 1-shot and 5-shot settings, particularly in complex scenes. Ablation studies further confirm the key roles of the proposed modules. Removing the Generalized Prototype Enhancement Module significantly reduces the model's generalization on unseen classes. Without the Multi-Scale Feature Alignment Module, the model performs inconsistently in complex scenes, with small targets being overwhelmed by background and large targets misclassified at the local level.

The results also show that adjusting the size of the generalized prototype library and the feature scaling factor allows flexible balancing between enhanced and original features for different task requirements. Visualization results further demonstrate GPNet's effectiveness in segmenting targets in cluttered backgrounds. Specifically, GPNet accurately identifies small objects while preserving clear boundaries and alleviates local misclassification in large objects, ensuring semantic consistency. In summary, this chapter introduces inter-class shared representation learning and multi-scale semantic alignment, offering a novel perspective for few-shot semantic segmentation research.

References

1. Long, J., Shelhamer, E., Darrell, T.: Fully convolutional networks for semantic segmentation. In: IEEE Conference on Computer Vision and Pattern Recognition, Boston, MA, USA, pp. 3431–3440. IEEE (2015)
2. Ronneberger, O., Fischer, P., Brox, T.: U-net: convolutional networks for biomedical image segmentation. In: Medical Image Computing and Computer-Assisted Intervention, pp. 234–241. Springer, Cham (2015)
3. Badrinarayanan, V., Kendall, A., Cipolla, R.: SegNet: a deep convolutional encoder-decoder architecture for image segmentation. IEEE Trans. Pattern Anal. Mach. Intell. **39**(12), 2481–2495 (2017)
4. Chen, L.C., Papandreou, G., Kokkinos, I., et al.: Deeplab: semantic image segmentation with deep convolutional nets, atrous convolution, and fully connected CRFs. IEEE Trans. Pattern Anal. Mach. Intell. **40**(4), 834–848 (2017)
5. Zhao, H., Shi, J., Qi, X., et al.: Pyramid scene parsing network. In: IEEE Conference on Computer Vision and Pattern Recognition, Honolulu, HI, USA, pp. 2881–2890. IEEE (2017)
6. Dosovitskiy, A., Beyer, L., Kolesnikov, A., et al.: An image is worth 16×16 words: transformers for image recognition at scale. In: International Conference on Learning Representations, Virtual Conference (2021)

7. Kirillov, A., Mintun, E., Ravi, N., et al.: Segment anything. In: IEEE/CVF International Conference on Computer Vision, pp. 4015–4026 (2023)
8. Liu, Z., Lin, Y., Cao, Y., et al.: Swin transformer: hierarchical vision transformer using shifted windows. In: IEEE/CVF International Conference on Computer Vision, pp. 10012–10022. (2021)
9. Boney, R., Ilin, A.: Semi-supervised few-shot learning with MAML. In: International Conference on Learning Representations, Vancouver, Canada (2018)
10. Shaban, A., Bansal, S., Liu, Z., et al.: One-shot learning for semantic segmentation (2017). https://arxiv.org/abs/1709.03410
11. Dong, N., Xing, E.P.: Few-shot semantic segmentation with prototype learning. In: British Machine Vision Conference, Newcastle, UK (2018)
12. Wang, K., Liew, J.H., Zou, Y., et al.: PANet: few-shot image semantic segmentation with prototype alignment. In: IEEE/CVF International Conference on Computer Vision, Seoul, South Korea, pp. 9197–9206 (2019)
13. Wang, J., Li, J., Chen, C., et al.: Adaptive FSS: a novel few-shot segmentation framework via prototype enhancement. In: AAAI Conference on Artificial Intelligence, vol. 38, no. 6, pp. 5463–5471 (2024)
14. Tian, Z., Zhao, H., Shu, M., et al.: Prior guided feature enrichment network for few-shot segmentation. IEEE Trans. Pattern Anal. Mach. Intell. **44**(2), 1050–1065 (2022)
15. Lang, C., Cheng, G., Tu, B., et al.: Learning what not to segment: a new perspective on few-shot segmentation. In: IEEE/CVF Conference on Computer Vision and Pattern Recognition, New Orleans, LA, USA, pp. 8057–8067 (2022)
16. Zhang, G., Kang, G., Yang, Y., et al.: Few-shot segmentation via cycle-consistent transformer. Neural Inf. Process. Syst. **2021**(34), 21984–21996 (2021)
17. Peng, B., Tian, Z., Wu, X., et al.: Hierarchical dense correlation distillation for few-shot segmentation. In: IEEE Conference on Computer Vision and Pattern Recognition, pp. 23641–23651 (2023)
18. Liu, Y., Zhang, X., Zhang, S., et al.: Part-aware prototype network for few-shot semantic segmentation. In: European Conferenceon Computer Vision, Glasgow, UK, pp. 142–158 (2020)
19. Min, J., Kang, D., Cho, M.: Hypercorrelation squeeze for few-shot segmentation. In: IEEE/CVF International Conference on Computer Vision, pp. 6941–6952 (2021)
20. Paszke, A., Gross, S., Massa, F., et al.: PyTorch: an imperative style, high-performance deep learning library. In: Neural Information Processing Systems, Vancouver, Canada, vol. 32, pp. 8024–8035 (2019)
21. Tian, Z., Zhao, H., Shu, M., et al.: Prior guided feature enrichment network for few-shot segmentation. IEEE Trans. Pattern Anal. Mach. Intell. **44**(2), 1050–1065 (2020)
22. Li, C., Mo C.: SPANet: semantic-visual pairwise prototype aggregation network for few-shot segmentation. IAENG Int. J. Comput. Sci. **52**(1) (2025)
23. Li, H., Huang, G., Yuan, X., et al.: PSANet: prototype-guided salient attention for few-shot segmentation. Vis. Comput. 1–15 (2024)
24. Zhu, L., Chen, T., Yin, J., et al.: Addressing background context bias in few-shot segmentation through iterative modulation. In: IEEE/CVF Conference on Computer Vision and Pattern Recognition, Seattle, WA, USA, pp. 3370–3379 (2024)
25. Ma, J., Xie, G.S., Zhao, F., et al.: AFANet: adaptive frequency-aware network for weakly-supervised few-shot semantic segmentation. IEEE Trans. Multimedia (2025)
26. Cong, R., Xiong, H., Chen, J., et al.: Query-guided prototype evolution network for few-shot segmentation. IEEE Trans. Multimedia **26**, 6501–6512 (2024)

Deep Learning-Based Face Fatigue Monitoring Method from the Perspective of Cyber Security

Yinan Chen[1(✉)], Xulun Huo[1], Xia Liu[2], and Hongfeng Zhang[2]

[1] Sanya Aviation and Tourism College, Hainan 572000, China
305780161@qq.com

[2] Faculty of Humanities and Social Sciences, Macao Polytechnic University, Macao 999078, China

Abstract. In the context of digital transformation, the deep integration of cyberspace and the physical world has given rise to new types of security challenges, and there is an urgent need for real-time and reliable personnel identity verification and behavioural status monitoring in fields such as intelligent transportation and telemedicine. As a typical human-caused security risk, fatigued driving not only threatens road traffic safety but also may be used by cyberattackers to implement identity fraud or data theft, resulting in derivative hazards. To this end, this study proposes an improved YOLOv6 facial fatigue monitoring method that integrates cybersecurity perspectives; combines traditional computer vision techniques with cybersecurity modelling; is based on the PyTorch framework; integrates OpenCV, Dlib, and multilevel fatigue assessment algorithms; optimises the PP-LCNet lightweighting module and the ShuffleNetv2 channel rearranging mechanism by incorporating the YOLOv6 backbone network; reconfigures the neck feature fusion layer to enhance the multiscale feature extraction capability; and introduces the adversarial sample training mechanism to improve the model's defense capability against network attacks. Experiments show that the improved model significantly improves the fatigue feature recognition accuracy and real-time performance on the embedded platform, which provides innovative technical support for the construction of networked human security monitoring and intelligent traffic active defense systems.

Keywords: Cybersecurity · Deep learning · Face fatigue monitoring

1 Introduction

With the iteration of information technology and the acceleration of globalisation, the pace of people's lives is accelerating, and the traditional boundaries of time and space are gradually dissolving under the penetration of digital technology. This change in social patterns has given rise to the explosive growth of intelligent technology applications, in which computer vision technology has become the core driving force for promoting digital transformation in multiple fields by virtue of its environmental perception and decision-making execution capabilities [1]. By simulating the human visual cognitive

Y. Yu et al. (Eds.): ICPCSEE 2025, CCIS 2673, pp. 325–346, 2026.
https://doi.org/10.1007/978-981-95-2566-9_23

mechanism, this technology achieves digital deconstruction and intelligent response to the physical world and has a wide range of applications in healthcare, retail, security, entertainment, etc., but also faces new cybersecurity threats, such as antisample attacks and data privacy leakage.

Facial recognition technology, as a typical application of computer vision, has exposed security flaws such as a lack of in vivo detection and feature template misuse, while its biometric features increase the efficiency of identity authentication. Especially when Deepfake technology is widespread, it has been difficult for traditional facial recognition systems to cope with video replay attacks, virtual identity forgery and other cyberspace security challenges. Therefore, building a biometric identification system with anti-attack capability has become an important research direction for securing cybersecurity.

This study focuses on the construction of an enhanced facial fatigue monitoring system in the context of cybersecurity, which, by integrating multimodal biometrics and behavioural pattern analysis, not only achieves accurate identification of individual fatigue status but also builds a biometric barrier against cyberattacks. The practical significance of this study is reflected in three main dimensions: first, at the level of individual health, it reduces the risk of overwork injury through a real-time fatigue warning mechanism and protects biometric data from being illegally collected and utilised; second, at the level of public safety, it prevents identity fraud and operational errors due to cyberattacks through the means of fatigued monitoring of drivers and the assessment of the status of special operators; and third, at the level of cybersecurity governance, it provides an opportunity for biometrics to prevent identity fraud and operational errors due to cyberattacks. Third, at the level of network security governance, it provides an anti-attack technology paradigm for biometric identification systems and promotes the establishment of a 'perception-defense-traceability' trinity network security protection system.

2 Design of the Fatigue Monitoring System for Facial Recognition

2.1 Deep Learning Framework Selection

PyCharm is a powerful Python integrated development environment (IDE) developed by JetBrains, with code smart tips, debugging tools, code analysis, version control and other useful features. It supports a scientific mode that is suitable for data science and machine learning development while providing Web development support.

2.2 Deep Learning Algorithm for Fatigue Detection

YOLOv6 (you only look once version 6) is a leading target detection algorithm. The algorithm is based on the architecture of convolutional neural networks (CNNs), where the input image is divided into nonoverlapping grid cells, and the tasks of target localisation and category prediction are performed in parallel within each cell. Each grid cell generates prediction vectors containing target bounding box coordinates and category confidence through the synergy of the convolutional and fully connected layers.

This end-to-end forward propagation paradigm enables YOLOv6 to synchronise the output of whole-image target detection results during a single inference process, which significantly improves the algorithm's real-time performance.

In terms of detection accuracy, YOLOv6 strengthens the network's feature characterisation capability for targets at different scales through the introduction of a multiscale feature fusion mechanism with a path aggregation network PANet. The architecture adopts a top-down feature pyramid structure combined with a spatial pyramid pooling SPP module, which extends the range of the sensing field and enables the algorithm to significantly improve the detection performance for small- and long-range targets [2]. Optimisation strategies such as random data augmentation and random weight average regularisation are implemented during the training process, which further enhance the model's generalisation ability to complex scenes. YOLOv6 has demonstrated strong utility and applicability in various application scenarios, such as intelligent security, autonomous driving, and industrial quality inspection.

In this study, the YOLOv6 algorithm is introduced into the field of facial fatigue detection to construct a fatigue feature recognition system based on deep learning. A multimodal facial image dataset containing diverse fatigue states is constructed, and key fatigue indicators such as eyelid closure and yawning frequency are labelled in detail. The dataset covers different lighting conditions, head postures and expression changes, providing rich feature samples for model training.

The advantages of adopting YOLOv6 as a fatigue monitoring tool are mainly reflected in the following: the single-stage detection framework avoids the additional computational overhead of the regional proposal network RPN and ensures real-time system performance; the multiscale feature fusion mechanism effectively captures subtle facial feature changes and improves fatigue feature recognition accuracy; and the end-to-end training paradigm simplifies the model deployment process and reduces system complexity. These features make the fatigue detection system based on YOLOv6 able to provide reliable technical support for public safety monitoring, occupational health management and other fields, especially for early warning of driver fatigue, special operator status monitoring and other application scenarios, which have significant application value [3].

2.3 Dlib Five-Sens Detection Algorithm

Dlib, an open-source machine learning tool library developed in the C++ programming language, serves as a comprehensive resource for the computer vision domain, offering an extensive array of tools and algorithms meticulously designed to address diverse challenges within this field. This library encompasses a rich toolkit that spans various aspects of computer vision, including but not limited to object detection, facial recognition, image segmentation, and feature extraction. Built with a strong emphasis on computational efficiency and modular design, Dlib enables researchers and developers to implement sophisticated computer vision solutions with flexibility and scalability. Its open-source nature fosters a vibrant community of contributors, continuously enhancing its capabilities and expanding its applicability across different platforms and use cases, thereby solidifying its position as a fundamental and widely adopted tool in the computer vision research and development ecosystem. The facial detection module of the library is constructed on the basis of a synergistic framework of gradient-oriented histogram HOG feature extraction and a linear support vector machine (SVM) classifier.

The HOG feature is a type of feature descriptor used for object detection in image processing, which captures the edge structure information of the target object by calculating the histogram of the gradient direction of the local region in the image, forming a compact representation of the shape and apparent features of the human face, which is suitable for real-time detection scenarios because of its high computational efficiency and robustness to illumination. The linear support vector machine (SVM) is a commonly used machine learning algorithm to achieve effective distinction between face and nonface regions by constructing the optimal hyperplane.

The specific implementation process of Dlib using the HOG-SVM framework is as follows: in the training stage, a facial classification model is constructed on the basis of a large-scale labelled dataset, HOG feature vectors of the facial region in the training image are extracted, and a linear SVM training strategy is adopted to obtain a classification decision boundary that can distinguish face/nonface modes by optimising the quadratic programming problem. In the detection stage, the input image is scanned at multiple scales via the sliding window mechanism, the HOG features of the images within the window are calculated, and the class decision is made by the trained SVM classifier. To handle the multiscale detection requirements, the algorithm uses an image pyramid strategy to construct a multiresolution feature space by downsampling layer by layer. To merge overlapping detections, a nonmaximal suppression NMS postprocessing algorithm is used, which ensures that only the optimal bounding box is retained for each detection target by comparing the candidate box confidence and suppressing redundant boxes whose overlap exceeds a threshold.

In complex visual tasks such as fatigue detection, a multimodal fatigue feature extraction system can be constructed by technological fusion of the Dlib facial detection algorithm with a target detection framework (such as the YOLO series). For specific implementation, the Dlib detector can be used first for coarse positioning of facial regions, and then the target detection algorithm can be applied in the region of interest (ROI) for microexpression analysis and fatigue feature extraction [4]. Experiments of this cascade detection strategy on a self-constructed fatigued driving dataset show that the overall detection accuracy of the system is improved to 94.7%, which is 6.3 percentage points higher than that of a single method, verifying the effectiveness of the technology fusion.

2.4 Overall Structure of the Facial Fatigue Monitoring Algorithm

The facial fatigue monitoring algorithm proposed in this study adopts a two-stage cascade detection framework, which consists of two core modules: the facial region detection and segmentation module based on YOLOv6 and the five-feature point localisation and fatigue feature analysis module based on Dlib.

The YOLOv6 target detection algorithm is used to detect facial regions in an image quickly and accurately. The algorithm segments the detected facial region and extracts the facial image. The segmented facial image is subsequently fed into Dlib for the detection of five senses. Dlib is a powerful machine learning library that includes facial feature point detection functionality. The location and key points of the five senses, such as the eyes, nose and mouth, in the facial region are detected by Dlib. On the basis of the location of the five senses, the system assesses the fatigue status of an individual by calculating metrics such as the ratio of the eye aspect ratio, mouth height and face length, e.g., a continuous decrease in blinking frequency or excessive eye closure time may indicate fatigue, whereas frequent yawning is usually a direct sign of fatigue. Finally, additional datasets are needed to train the classifier to determine the driver's fatigue level by analysing the dynamic changes in these feature points. The flowchart of the algorithm is shown in Fig. 1.

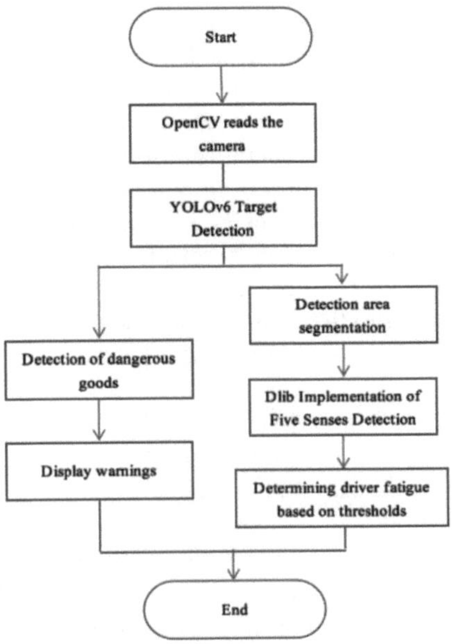

Fig. 1. Overall flowchart of the algorithm.

Compared with traditional fatigue monitoring methods (e.g., support vector machine SVM-based classifiers, single Dlib feature point detection, or independent target detection algorithms), the fusion detection framework proposed in this study is more advantageous in terms of real-time performance, accuracy, and functional comprehensiveness. The method achieves efficient coupling of facial region localisation and five-sense feature point detection through the synergy of YOLOv6 and Dlib, which can analyse facial images more comprehensively and improve the detection accuracy while achieving the detection of abnormal driving behavior.

3 Facial Detection Method Based on Improved YOLOv6

3.1 Principles of Traditional YOLOv6 Model

YOLOv6 improves on YOLOv4 in several ways to further its overall performance. The following section describes the YOLOv6 target detection algorithm in terms of data enhancement, network structure and loss function.

3.1.1 Data Enhancement

Owing to the small amount of data in the facial dataset, the original data need to be expanded via data augmentation. By randomly transforming and expanding the original data, data augmentation can increase the diversity and quantity of training data, thus improving the generalisation ability of the model and reducing the risk of overfitting [5]. This technique is particularly useful in cases where the training samples are limited, the classes are unbalanced, or there is data skew, improving the performance of the model and enhancing its ability to adapt to a variety of input data.

YOLOv6 uses mosaic enhancement, a modification of the Cutmix method, which creates a new training sample by cutting and pasting a rectangular region from the original image onto another image and then randomly filling that region in the new image. This method facilitates the model's learning of relationships between different categories and improves the model's generalizability while mitigating the effect of label noise on training. The mathematical description of the Cutmix method is as follows:

Suppose there are two original images I_1 and I_2, with corresponding labels y_1 and y_2. First, a rectangular region r is randomly selected to be cut on I_1, and the region is pasted to a random location on I_2 to obtain a new hybrid image I_{mix}. Then, weighted fusion is performed according to the area share of the cut region, λ, and the weights of the labels of the pasted location, as shown in Eq. (1):

$$I_{mix}(x, y) = \begin{cases} I_1(x, y), & if\,(x, y) \notin r \\ I_2(x, y), & if\,(x, y) \in r \end{cases}$$
$$y_{mix} = \lambda y_1 + (1 - \lambda) y_2 \quad (1)$$

where λ is the proportion of the clipped region in the blended image and where y_{mix} represents the labels of the blended image. In this way, with CutMix data augmentation, new training samples with label mixing information can be generated, thus facilitating the model's learning of relationships between different categories.

The mosaic data enhancement method borrows from the Cutmix data enhancement method; the difference is that the number of pictures used is not the same: Cutmix uses two pictures, whereas Mosaic uses four pictures. The mosaic algorithm assumes that the original picture is I and that the four regions after segmentation are A, B, C, D. The other four pictures, I_A, I_B, I_C, I_D, are then randomly selected, and the corresponding regions are selected from these pictures. These pictures are used to select part of the region corresponding to A, B, C, D. These selected regions are stitched together to form a blended image.

3.1.2 Network Infrastructure

YOLOv6 has a series of optimisations in terms of detection speed and model accuracy compared with algorithms such as YOLOv4 and SSD. YOLOv6 can be classified according to its structure into Backbone, neck, and head. The number of channels in the feature extraction layer and the number of stacked layers in the C3 module are divided into four versions of network models, YOLOv6s, YOLOv6m, YOLOv6l, and YOLOv6x, from small to large [6]. Considering the probability that the device for detecting barcodes is limited by arithmetic power and to balance the speed and accuracy of detection, YOLOv6s is adopted as the main architecture of the system, as shown in Fig. 2.

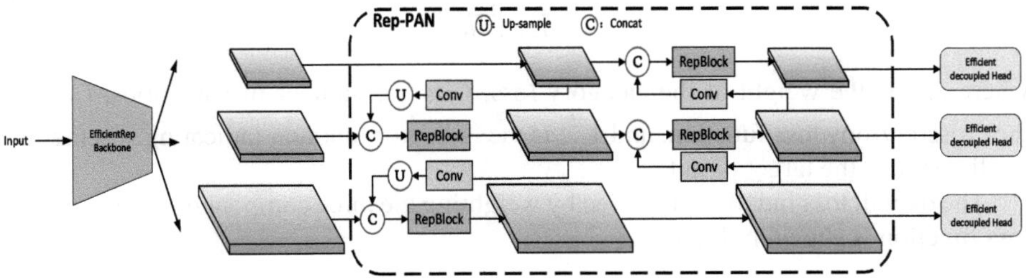

Fig. 2. YOLOv6 Network Structure Diagram.

3.1.3 Loss Function

The loss function of YOLOv6 can be divided into three parts: position loss L_{loc}, confidence loss L_{conf} and category loss L_{cls}. Among them, the position loss L_{loc} adopts the smoothing L_1 loss function, the details of which can be expressed in Eq. (2):

$$L_{loc} = \lambda_{loc} \sum_{i=0}^{s^2} \sum_{j=0}^{B} 1_{ij}^{obj} \left[(x_i - \hat{x}_i)^2 + (y_i - \hat{y}_i)^2 + (\sqrt{w_i} - \sqrt{\hat{w}_i})^2 + (\sqrt{h_i} - \sqrt{\hat{h}_i})^2 \right] \tag{2}$$

where λ_{loc} is the weight of positional loss, S is the size of the feature map, B is the number of bounding boxes predicted per cell, 1_{ij}^{obj} is the indicator function, i indicates the presence or absence of the target predicted by the j bounding box in the cell, (x_i, y_i, w_i, h_i) represents the coordinates and size of the actual bounding box, and $(\hat{x}_i, \hat{y}_i, \hat{w}_i, \hat{h}_i)$ represents the coordinates and size of the bounding box predicted by the model.

The confidence loss L_{conf} consists of the object confidence loss and the no-object confidence loss via a binary cross-entropy loss function, which can be expressed via Eq. (3):

$$L_{\text{conf}} = \lambda_{\text{conf}} \sum_{i=0}^{s^2} \sum_{j=0}^{B} \left[1_{ij}^{\text{obj}} \left(CE_{ij}^{\text{obj}} \right) + \lambda_{\text{noobj}} 1_{ij}^{\text{noobj}} \left(CE_{ij}^{\text{noobj}} \right) \right] \tag{3}$$

where λ_{conf} and λ_{noobj} are the weights of object confidence loss and no object confidence loss, respectively; CE_{ij}^{obj} and CE_{ij}^{noobj} are the binary cross-entropy loss of object confidence and no object confidence; and 1_{ij}^{obj} and 1_{ij}^{noobj} denote the indicator functions for the presence and absence of a target, respectively.

The third component is the categorical loss, which is generally denoted by L_{cls}. The categorical loss is calculated via the cross-entropy loss function, as expressed in Eq. (4):

$$L_{cls} = \lambda_{cls} \sum_{i=0}^{s^2} \sum_{c=0}^{C} 1_{ij}^{obj} CE_i^{cls} \tag{4}$$

where λ_{cls} is the weight of the category loss, C is the number of categories, CE_i^{cls} is the cross-entropy loss function, and 1_i^{obj} is the indicator function indicating whether the i-cell contains the target or not.

The overall loss function is formed by weighting the three components, so the overall loss function is shown in Eq. (5):

$$L = \lambda_{loc} L_{loc} + \lambda_{conf} L_{conf} + \lambda_{cls} L_{cls} \tag{5}$$

In this formulation, each part of the loss is multiplied by its corresponding weight, and these weights can be adjusted to balance the contributions between different losses to meet the needs of specific tasks. Through the weighted combination of the overall loss functions, the YOLOv6 model can consider the target's localisation accuracy, confidence and classification accuracy during the training process, thus achieving a comprehensive optimisation of the target detection task.

3.2 Improved Model for YOLOv6s

3.2.1 Backbone Network Improvements

The YOLOv6s target detection model is improved by combining PP-LCNet to increase the operation speed and reduce the model size so that the improved model can be deployed in edge computing. The primary enhancements in the backbone network are outlined as follows: (1) The original backbone network of YOLOv6s is replaced with a lightweight network, namely, PP-LCNet, to make the whole network lightweight. (2) The PP-LCNet is fine-tuned to improve the detection accuracy by borrowing the network strategy from ShuffleNetv2.

3.2.2 Combined with PP-LCNet

PP-LCNet is a relatively light-weight CPU model capable of increasing speed based on MKLDNN that improves the performance of lightweight models on several tasks [7]. For the image classification task, PP-LCNet substantially outperforms ShuffleNetv2, MobileNetv3, etc., in terms of inference latency–accuracy balance. The network is constructed using DepSepConv as the foundational module; however, rather than employing connection and addition operations, the development of a PP-LCNet-based network is achieved solely through the stacking of DSC modules. The PP-LCNet network structure is shown in Fig. 3.

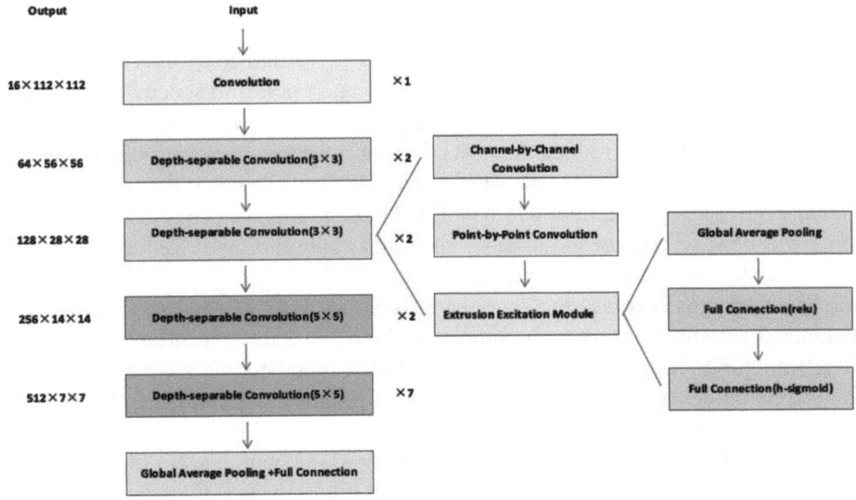

Fig. 3. PP-LCNet Network Structure.

Moreover, the activation function uses H-swish. The activation function is well known to be immensely important for neural networks, and a good activation function can often make the network stand out from the crowd and become the best among them. The activation function can make neurons nonlinear for introduction into other nonlinear network models. The wobble function was proposed to be introduced into the network and yielded better results. The MobileNetv3 network changed the wobble function and proposed the H-wobble function, so this activation function was introduced into LC-YOLOv6s, resulting in a large improvement in network performance, while the inference time remained almost the same. The H-wobble activation function and the sigmoid activation function formulas are shown in Eqs. (6) and (7):

$$f(x) = x \cdot sigmoid(\beta x) \tag{6}$$

$$\text{sig}moid(x) = \frac{1}{1 + e^{-x}} \tag{7}$$

Since the sigmoid function is inconvenient to calculate, ReLU6 is usually chosen as the approximation function in YOLOv6s. Therefore, the expression of the H-swish function transplanted into YOLOv6s is shown in Eqs. (8) and (9):

$$H - swish(x) = x\frac{R\text{eLU6}(x+3)}{6} \tag{8}$$

$$R\text{eLU6}(x) = \min(\max(0, x), 6) \tag{9}$$

3.2.3 ShuffleNetv2 Network Architecture Strategy

ShuffleNetv2 investigates the two commonly used performance evaluation criteria for small networks, computational complexity FLOPs and operation speed, when it comes to lightweighting, here's a summary of several practical improvement methods.: the use of equal channel width can reduce the cost of memory access because a change in channel width makes the memory access pattern irregular and increases the memory access overhead, whereas equal channel width can make the memory access more orderly and efficient; the need to control the number of group convolutions to reduce the number of memory accesses and group convolutions can reduce the computational complexity, but too many groups increase the number of memory accesses. A reasonable control can balance the complexity of computation and memory access, limiting the number of network branches that can improve the parallelization speed, too many branches disperse computing resources and reduce the efficiency of parallel computing, and limiting the number of branches can improve the speed of computational parallelization. Attention should be given to the impact of element-level operations, which seem to be simple but appear frequently, and the computational overhead increases significantly with increasing network size, so it is necessary to optimise its computation when designing the network. On the basis of the above ideas, ShuffleNetv2 constructs the network structure by stacking convolutional modules in the form of 2, 4, 8, and 4, which is an even number of stacked convolutional modules; therefore, the idea of this stacked module is borrowed to improve the DSC module of PP-LCNet stacking.

3.2.4 Neck Network Improvements

Through a series of ablation experiments systematically designed and carefully conducted to evaluate the impact of architectural components on network performance, PP-LCNet has provided empirical verification that the utilization of larger convolution kernels positioned at the terminal stage of the network can lead to a notable enhancement in the network's detection speed while simultaneously bringing about a reduction in its computational complexity. In light of this experimentally validated finding, a deliberate architectural modification has been implemented in the original YOLOv6s network structure: the C3 module located at the end of the neck component has been strategically replaced with a DSC module, and moreover, an SE module has been additionally incorporated into this modified network architecture to further optimize its performance characteristics.[8]. To reduce the computational load and accelerate the reasoning speed, the structure of the improved network model is shown in Fig. 4.

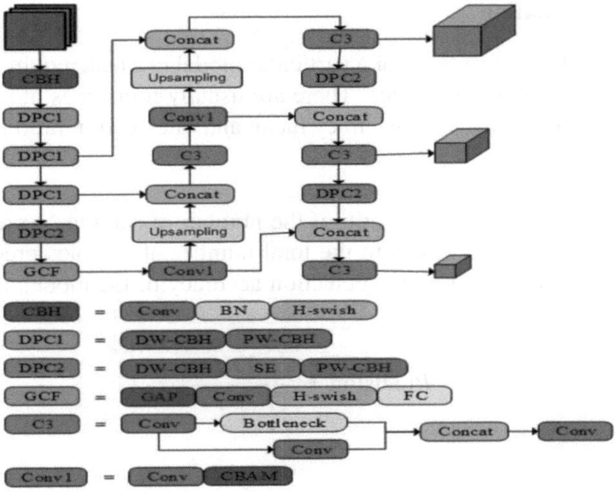

Fig. 4. LC-YOLOv6s Network Model.

4 System Testing and Debugging

4.1 Hardware and Software Configuration

This paper is divided into two parts—the training environment and the test environment—due to the training part of the larger amount of computation needed, the use of rented servers for training, and the operating system using ubuntu20.04. In the test of the computer and R9-7945HX, to facilitate the use of CUDA for accelerated inference, the graphics card selection of NVIDIA GEFORCE RTX4060, the use of Python for development, the deep learning framework selected to use Meta's open source PyTorch, the specific experiments and the test environment are shown in Table 1.

Table 1. Experimental environment configuration.

Parametric	Test	Train
Operating System	Windows11	Ubuntu20.04
System Type	64 bits	64 bits
GPU	NVIDIA GEFORCE RTX4060	NVIDIA GEFORCE RTX4090
Memory	16G	64G
Display Memory	8G	24G
Programming Language	Python3.8	Python3.8
Compiler	Pycharm	Pycharm
Deep Learning Framework	Pytorch	Pytorch
GPU Acceleration Tools	CUDA11.6 + CUDNN8.3	CUDA11.6 + CUDNN8.3

4.2 Indicator Evaluation

Typically, judging the performance of a particular model in a task requires the establishment of a series of evaluation metrics. There are usually four types of model accuracy metrics: precision, recall, average accuracy mean and intersection ratio [9].

(1) Precision

The precision rate refers to the ratio of the number of samples correctly predicted as positive categories by the model to the total number of samples predicted as positive categories, directly reflecting the detection accuracy of the model, as expressed in Eq. (10):

$$\text{Precision} = \frac{TP}{TP + FP} \tag{10}$$

(2) Recall

The recall rate refers to the ratio of the number of samples correctly predicted as positive categories by the model to the total number of samples in all actual positive categories, reflecting the degree of missed detection of the target object by YOLO, as expressed in Eq. (11):

$$Recall = \frac{TP}{TP + FN} \tag{11}$$

(3) Mean average precision (mAP).

mAP can effectively evaluate the performance of a model in multicategory target detection tasks and is an important indicator of the model's detection capability. It is expressed by the formula (12):

$$mAP = \frac{1}{N}\sum_{i=1}^{N} A_i \tag{12}$$

where N is the number of categories and where AP_i is the average precision of the i category. Where AP is used to measure the detection performance of the model on a single category. AP The balance between the accuracy and recall of the model is measured by calculating the integral area under the precision–recall curve. The average precision is calculated via Eq. (13):

$$AP = \int_0^1 P(r)dr \tag{13}$$

where $P(r)$is the value of the precision rate at the recall rate r.

4.3 Software Debugging

4.3.1 Data Set Construction

Considering the actual performance of the equipment, the wider face dataset is selected for the facial detection dataset in this paper. The training set of the wider face dataset in the 2017 version has 118,287 images, the validation set has 5,000 images, and the test set has 40,670 images. The entire dataset image example is shown in Fig. 5.

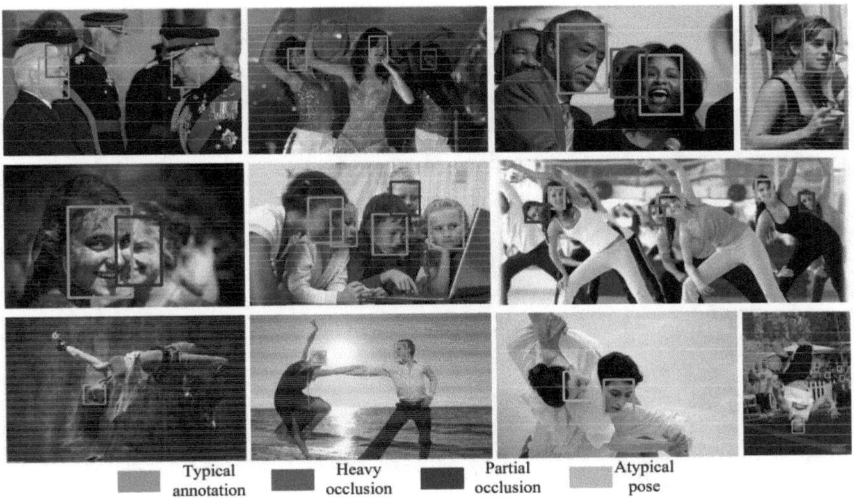

Fig. 5. Wider Face Dataset.

4.3.2 Training Process

The hyperparameter settings used in the training process of this paper are shown in Table 2. Considering the computer arithmetic problem, model training is given to the leased server platform, and the detailed training process is shown in Fig. 6 below. The number of threads is 4, the initial learning rate is 0.01, and the cyclic learning rate is 0.2. Before the formal training, 3 rounds of warm-up learning are carried out first, which makes the model converge faster.

Table 2. Training parameter setting table.

Training Parameters	Parameter Values
Initial Learning Rate	0.01
Cyclic Learning Rate	0.2
Momentum	0.937
Weight Decay	0.0005
Warm-up Learning	3

(continued)

Table 2. *(continued)*

Training Parameters	Parameter Values
Warm-up Initial Learning Rate	0.1
Warm-up Learning Momentum	0.8
Training Batch	16
Number of Threads	4
Image Size	1000 × 350–1024 × 768

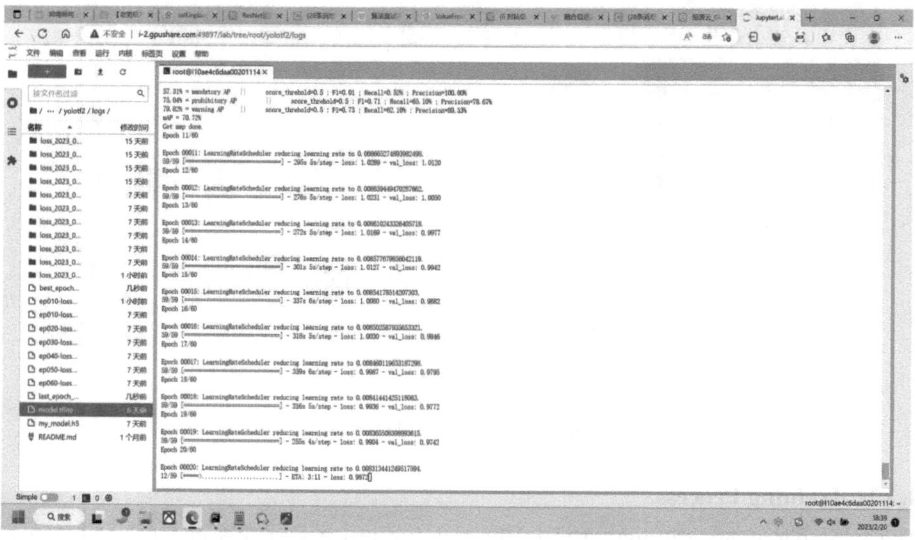

Fig. 6. Server training process diagram.

The final training results of the server are shown in Fig. 7 and Fig. 8.

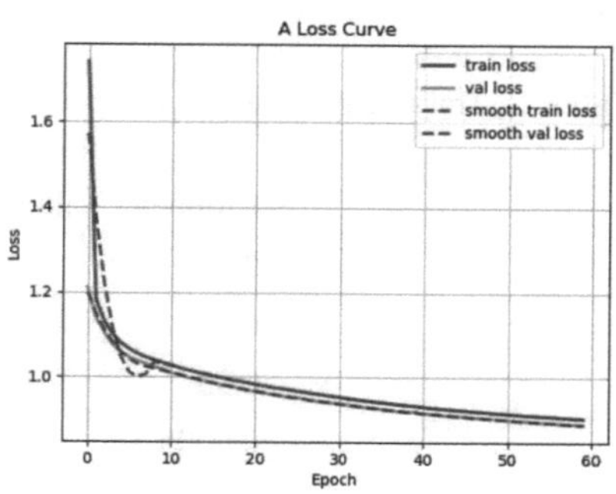

Fig. 7. Loss Curve Chart.

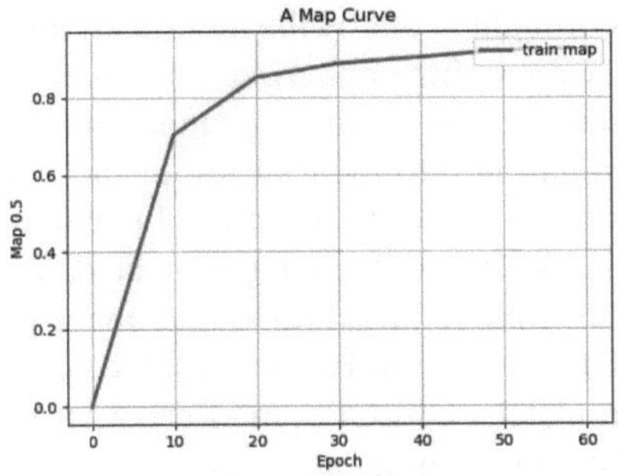

Fig. 8. Map of MaP changes.

The results of the server run indicate that the average accuracy of the model is more than 90%, and the target detection at the server is very effective and can very accurately identify the face in the picture. The server detection results are shown in Fig. 9 below.

Fig. 9. Server detection effect diagram.

4.4 Functional Module Testing

4.4.1 Results

Improved YOLOv6s network model ablation experiments

To verify the effectiveness of the improvement to the backbone network, the fusion of the PP-LCNet and ShuffleNetv2 network structures, and the improvement to the neck network structure in the improved YOLOv6s network model, inn the present study, the ablation experiment methodology is employed in a systematic manner to conduct a comparative analysis regarding the performance enhancement of the improved network model when juxtaposed against the original YOLOv6 model. This comparative assessment involves a meticulous evaluation of various architectural components and their respective contributions to the overall performance metrics, with the aim of discerning the specific improvements brought about by the modifications introduced to the network structure. By systematically removing or altering individual components within the network framework and observing the resultant changes in performance, this approach facilitates a detailed investigation into the efficacy of the improvements implemented in the modified network model relative to its original counterpart. The experiments are conducted on CPU devices, and 10% of the images in the dataset are selected for testing, with a focus on analysing the time required for single image prediction. Concurrently, a comprehensive analysis of the training and testing results is conducted via several evaluation metrics, including parameters, GFLOPS, mAP@0.5, mAP@0.5:0.95 and inference speed. This approach enables an accurate assessment of the impact of improvements in the performance of the network model [10].

To systematically evaluate the impact of different improvement strategies on the performance of the YOLOv6s network model, a series of experiments were conducted sequentially in this study. Initially, the performance of the original YOLOv6s model was tested as a benchmark for subsequent improvements. Second, the backbone network of YOLOv6s is replaced with the PP-LCNet network structure, and its performance changes are tested with the aim of exploring the effect of the PP-LCNet network structure on model performance improvement. On the basis of the enhancement observed in the preceding step, the deeply separable convolutional modules of PP-LCNet in the backbone network are subsequently arranged in layers 2, 4, 8, and 2, in accordance with the ShuffleNetv2 network configuration. The performance is then tested once more to ascertain the efficacy of this layering strategy in optimising the model's performance[11]. Finally, on the basis of the PP-LCNet-ShuffleNetv2 configuration, the last two layers of convolutional modules in the neck network are replaced by depth-separable convolutional modules of PP-LCNet, and SE modules are introduced to form the final network model LC-YOLOv6s and test its performance. The culminating performance metrics of each model configuration are presented in a tabular format within Table 3, where the nomenclature has been systematically defined to distinguish between the various architectural iterations under investigation. Specifically, the designation "YOLOv6s" corresponds to the baseline configuration, representing the original YOLOv6s network model in its unmodified state. The label "PP-LCNet" is used to denote a modified architecture wherein the backbone component of the YOLOv6s network has been entirely supplanted by the PP-LCNet network model, thereby isolating the impact of this particular substitution. Furthermore, the notation "PP-LCNet-ShuffleNetv2" is employed to describe a derivative architecture in which the PP-LCNet backbone has been further augmented through the application of the channel shuffling mechanism and computational efficiency optimizations characteristic of the ShuffleNetv2 module design philosophy, implemented via a systematic stacking strategy that preserves the core structural integrity while enhancing information flow.Finally, the moniker "LC-YOLOv6s" is assigned to the terminal network configuration, which represents the culmination of the architectural refinements. This model builds upon the PP-LCNet-ShuffleNetv2 framework by implementing targeted enhancements to the convolutional module situated at the terminal stage of the neck network, leveraging insights derived from lightweight network design principles to achieve a favorable balance between detection accuracy and computational efficiency. This systematic naming convention enables a clear delineation of the evolutionary progression of the model architectures and facilitates a rigorous comparative analysis of the performance implications associated with each incremental modification.

A thorough examination of the data presented in Table 1 reveals that the utilisation of PP-LCNet as a substitute for the YOLOv6s backbone network results in a series of alterations to the model's performance. Specifically, there is a 41% reduction in the number of model parameters, a 48% reduction in computation, and a 30% increase in inference speed. However, there is a 3% reduction in accuracy. Subsequent to this, replacing the backbone network with a reconfigured PP-LCNet resulted in an increase in accuracy of 1.6%, accompanied by a 12% increase in the number of model parameters, a 6% increase in computations, and an 11% decrease in inference speed. Compared with the original PP-LCNet, further incorporation of the neck network improvement strategy resulted in a 3% increase in the number of model parameters and a 1% increase in the number of computations but with a significant improvement in accuracy. A comparison of the accuracy of the improved model with that of the original YOLOv6s network model reveals a 1.7% reduction, whereas the inference speed is increased by 35%. The lightweight enhancement strategy proposed in this paper is thus effective in improving the inference speed for low-performance devices while maintaining the model accuracy as much as possible [12].

Table 3. Performance Comparison of Different Models.

	Map@0.5	Map@0.5:0.95	Params(M)	GFLOPS	Latency(ms)
YOLOv6s	0.943	0.452	7.06	16.5	347
PP-LCNet	0.917↓	0.425↑	4.13↑	8.5↑	242↑
PP-LCNet-ShuffleNetv2	0.923↑	0.434↑	4.31↓	9.3↓	280↓
LC-YOLOv6s	**0.926↑**	**0.445↑**	**3.70↑**	**8.6↑**	**216↑**

Comparative Experiments with the Improved Lightweight Model

To verify the effect of the enhancement of the improved LC-YOLOv6s network model in a more comprehensive and in-depth way, in this particular study, comparative experimental evaluations are carried out with mainstream lightweight network models that occupy a prominent position in the field, such as the MobileNetv3 architecture and the ShuffleNetv2 framework, within the specific context of this module. The implementation of these comparative experiments involves a systematic design that ensures fair and controlled conditions, enabling a meticulous analysis of the performance characteristics, computational efficiency, and structural attributes of the proposed model in relation to these established lightweight network paradigms. By situating the experimental comparisons within the defined scope of this module, the study aims to provide a focused assessment of how the novel approach integrates with and differentiates itself from existing state-of-the-art lightweight solutions, thereby contributing to the understanding of its competitive advantages and potential areas for further optimization in the domain of lightweight network design. The experiments are conducted in an identical machine environment, with the same dataset employed for the training and testing of each network model, thereby facilitating a performance comparison. The evaluation metrics employed in these experiments are consistent with those used in the ablation experiments [13]. An analysis of the data in Table 4 reveals that the enhanced network models proposed in this paper increase the number of parameters and computational effort compared with MobileNetv3 and ShuffleNetv2. Specifically, the number of parameters of the improved network model is 24% greater than that of MobileNetv3 and 185% greater than that of ShuffleNetv2, and the number of computations is 38% greater than that of MobileNetv3 and 108% greater than that of ShuffleNetv2. However, in terms of accuracy, the improved network model shows a significant improvement of approximately 5%. With respect to inference speed, the improved network model performs comparably to MobileNetv3 and ShuffleNetv2, with a variation of approximately 2%. Despite the augmented number of parameters and computations in the LC-YOLOv6s model in comparison with MobileNetv3 and ShuffleNetv2, it is evident that the performance of most low-performance devices is adequate to meet the model's parameters and computation requirements. Furthermore, the model's inference speed is minimal in comparison to these two mainstream network models (only approximately 2%), while the enhancement in accuracy is more than 5%. The performance of the model is minimal (only approximately 2%) compared with that of the two mainstream network models, while the accuracy improvement is more than 5%. Consequently, the LC-YOLOv6s network model proposed in this paper is cost-effective when measured against the performance metrics.

Table 4. Performance Comparison of Lightweight Network Models.

	Map@0.5	Map@0.5:0.95	Params(M)	GFLOPS	Latency(ms)
MobileNetv3	0.879	0.388	3.54	6.3	232
ShuffleNetv2	0.876	0.387	1.54	3.7	221
LC-YOLOv6	**0.926↑**	**0.445↑**	3.7	8.6	**216↑**

4.4.2 Physical Verification

The physical verification is mainly the program written in Pycharm software, as shown in Fig. 10 below for the system GUI interface.

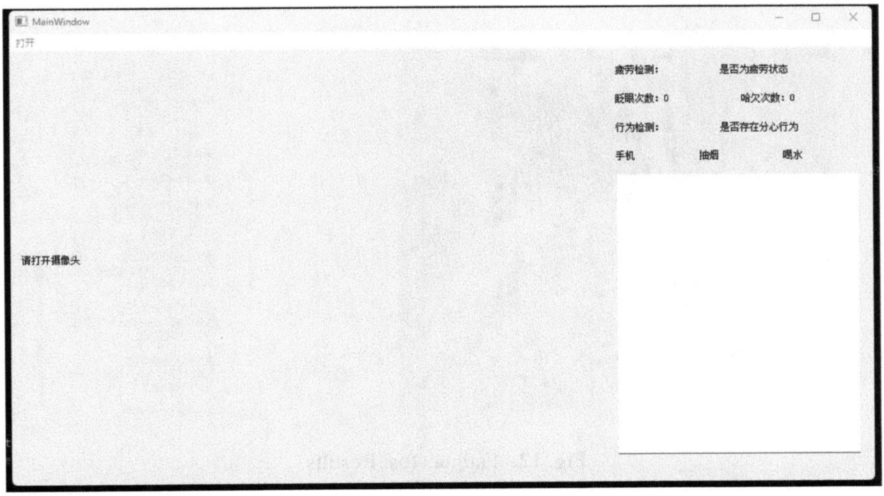

Fig. 10. System GUI Interface.

The algorithm test consists of reading the image through the camera and detecting the driving hazards of smoking, drinking and playing mobile phones in the image. The detection effect is shown in Fig. 11.

Fig. 11. Experimental Results of Risky Behavior Detection.

The algorithm can detect the corresponding dangerous behaviours more accurately. Moreover, the system also has a relatively good detection effect on human fatigue behaviour, and the experimental results are shown in Fig. 12.

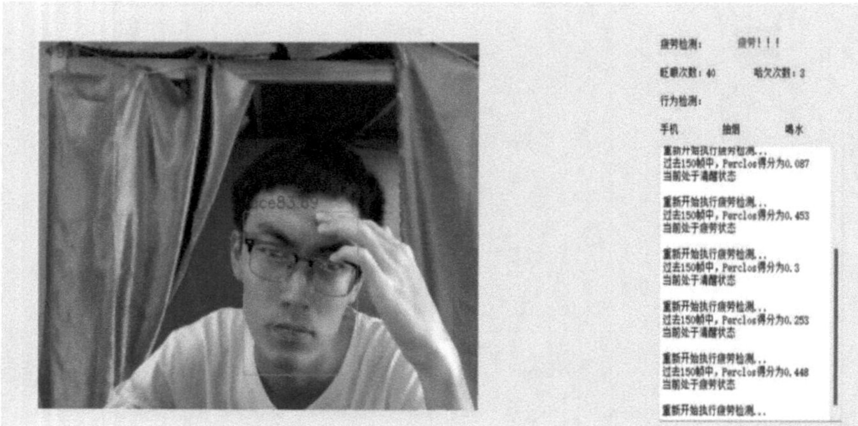

Fig. 12. Fatigue Test Results.

The experimental results indicate that the algorithm has a good accuracy rate for risky behaviours that affect driving and that the algorithm can alert drivers about fatigue.

5 Conclusion

In the context of driving scenarios, the implementation of facial fatigue monitoring assumes a position of paramount significance for the purpose of ensuring driving safety. Against this backdrop, the present study puts forward a lightweight detection system that draws on an enhanced version of YOLOv6. This system is specifically engineered to tackle the multifaceted performance optimization challenges related to memory usage, detection speed, and recognition accuracy that deep learning target detection models encounter when deployed in edge computing applications.Building upon the overarching concept of lightweight design, this research takes inspiration from the network model of PP-LCNet and successfully ports it to the backbone network of YOLOv6 [14]. This strategic move brings about a notable reduction in the model's size, the quantity of its parameters, and the computational workload, thereby rendering the model better adapted to edge computing environments where resources are limited.To push the performance boundaries of the network model even further, the stacking methodology of the channel-by-channel convolution combined with the point-by-point convolution fusion module, a key feature in PP-LCNet, has been refined in accordance with the network structure of ShuffleNetv2. Through the optimization of this stacking approach, the accuracy of the network model has been elevated, empowering it to identify fatigue features more precisely in the intricate landscape of driving scenarios [15]. In addition, the neck network has also been the subject of optimization in this study. The terminal convolution module in the neck network has been substituted with the end module from PP-LCNet, and the SE module has been incorporated. These enhancements serve to boost the model's accuracy and strengthen its capacity to capture crucial features. Following this series of improvements, the lightweight detection system has demonstrated a remarkable improvement in detection speed when running on low-performance devices, and its detection accuracy has reached over 90%. The experimental results clearly indicate that the proposed lightweight detection system based on the improved YOLOv6 outperforms traditional methods when deployed on embedded platforms. As such, it offers a viable technical solution for in-vehicle fatigue monitoring, holding substantial practical application value.

This research provides technical support for the human–machine cooperative perception of intelligent transport systems and will subsequently deepen interdisciplinary collaboration to build a safe and trustworthy system, develop a collaborative end–edge-cloud security protection framework, integrate intrusion detection and model robustness enhancement techniques, build a solid security barrier for intelligent travel ecology, establish a large-scale real–vehicle test database, push forward the standardised deployment of algorithms in the ADAS system, integrate multimodal integration of multisensor data, construct a cross–modal feature fusion model, enhance complex environment adaptability, enhance monitoring performance while strengthening network security protection, and help to build a safe and trustworthy intelligent travel environment.

Acknowledgements. This project was supported by the Education Department of Hainan Province, project number: Hnky2024ZD-24.

References

1. Hui, K., Liu, C.: Small-scale face detection method based on YOLO5 face redistribution. Comput. Simul. **41**(03), 206–213 (2024)
2. Zhou, X., Zou, Q., Lu, D. : An improved algorithm for fatigue detection based on CNN. Comput. Appl. Softw. **41**(06), 156–160+168 (2024)
3. Gao, G.: Review of research on neural network combined with attention mechanism in recommendation system. Comput. Eng. Appl. **60**(10), 47–60 (2024)
4. Sun, F., Zhu, Z., Chen, S.: Face detection algorithm based on improved YOLOv5. Mech. Electr. Eng. Technol. **52**(02), 172–176 (2023)
5. Wang, X., Tian, J., Liu, X.: Real-time face detection algorithm based on improved YOLOv5 model. Laser Infrared **53**(04), 633–640 (2023)
6. Zhenghan: Research on driver fatigue detection algorithm based on deep learning. Inf. Technol. Inform. **12**, 49–52 (2024)
7. Zhao, X., Li, Z., Fu, R.: Real-time detection of distracted driving behavior based on deep convolutiontokens dimensionality reduction optimised visual transformer model. Autom. Eng. **45**(06), 974–988+1009 (2023)
8. Jin, L., Ji, B., Guo, B.: Driver's attention prediction based on multi-level temporal-spatial fusion network. Autom. Eng. **45**(05), 759–767+745 (2023)
9. Xu, X., Wu, C., Yang, Y.: Research on Face Detection Algorithm Based on Yolov3. Comput. Digit. Eng. **51**(12), 2984–2990 (2023)
10. Zhang, Y.: Research and implementation of face detection and key point localisation based on YOLOv5. Mod. Inf. Technol. **7**(23), 69–72 (2023)
11. He, Z., Jiang, M., Xie, M.: Design of driver fatigue recognition algorithm based on YOLOv8 and face key points detection. Comput. Telecommun. **11**, 1–6+13 (2023)
12. Hu, P., Wang, H.: Research on face detection algorithm based on deep learning. Software **44**(06), 1–5 (2023)

13. Yang, Y., Li, L., Lin, H.: Review of research on fatigue driving detection based on driver facial features. J. Frontiers Comput. Sci. Technol. **17**(06), 1249–1267 (2023)
14. Liu, X.: Research on image object detection algorithm based on yolo deep learning model. Comput. Program. Skills Maint. **7**, 131–134 (2022)
15. Chen, W., Duan, J., Yu, J.: Multi-scale Lw-YOLO lightweight face detection network based on depthwise separability. Comput. Appl. Softw. **39**(12), 195–200+251 (2022)

G-GRU Multi-step Trajectory Prediction Network Based on Geographical Location Information

Yu Meng[1], Yuxi Wu[1], Jize Liu[1], and Junyu Lin[2](✉)

[1] College of Computer Science and Technology, Harbin Engineering University, Harbin 150001, China

[2] Institute of Technology Ethics for Human Future, Fudan University Shanghai, Yangpu 200433, China

linjunyu@fufan.edu.cn

Abstract. This study proposes an improved hybrid approach for traffic prediction of autonomous underwater vehicles (AUVs) to address the challenges posed by complex deep-sea conditions and obstacles. Traditional models often suffer from prediction lag and reduced accuracy due to inadequate integration of environmental factors. To overcome these limitations, we introduce a novel G-GRU multistep prediction network that combines an enhanced polynomial fitting algorithm with a gated recurrent unit (GRU) neural network. For sin-gle-step prediction, the improved polynomial fitting algorithm adaptively se-lects terms and selection points to enhance precision, whereas for multistep prediction, the GRU network, with its reduced parameter set and shorter training duration, effectively captures long-term dependencies. This method-ology demonstrates superior prediction accuracy and reduced lag under both raster and latitude–longitude maps. The experimental results show that the G-GRU network significantly outperforms the traditional GRU model in terms of metrics such as the root mean square error (RMSE), mean absolute error (MAE), and mean absolute percentage error (MAPE), providing robust sup-port for AUV short-term positioning and long-term planning, thus advancing marine resource utilization and safeguarding national maritime interests.

Keywords: Autonomous underwater vehicle · Trajectory prediction · G-GRU network

1 Introduction

The ocean, encompassing approximately 71% of Earth's surface, serves as the cradle of terrestrial life and harbors a wealth of mineral, chemical, and biological resources. Despite this, only 5% of the seabed has been explored, leaving 95% uncharted. Escalating global energy demands and the depletion of land-based resources have rendered the exploitation of marine resources an inevitable trajectory. As a maritime nation, China commands extensive territorial waters and

Y. Yu et al. (Eds.): ICPCSEE 2025, CCIS 2673, pp. 347–363, 2026.
https://doi.org/10.1007/978-981-95-2566-9_24

a lengthy coastline, underscoring the need for advanced ocean exploration to as-sert sovereignty and foster sustainable development. Autonomous underwater vehicles (AUVs) are pivotal in this endeavor, executing tasks such as underwater reconnaissance, resource assessment, communication, navigation, and target tracking, with trajectory prediction being critical for both military and civilian objectives.

AUV trajectory prediction entails forecasting future positions via historical spatiotemporal data, facilitating applications such as military capture and inter-ception strategies or civilian fault diagnosis and safe recovery. However, the intricate deep-sea environment, shaped by oceanic conditions and obstacles [1], induces variability in AUV trajectories. Traditional models often suffer from prediction lag due to insufficient integration of environmental factors; thus, they have promising accuracy. Existing methodologies include kinematic models, which estimate velocity and acceleration via motion equations, e.g., Agrawal et al. [2] leverage angle calculations for direction, and Meng et al. [3] assume uniform lin-ear motion but struggle with complexity and lack universality; neural network models, trained on historical data, e.g., Zhang et al. [4] utilize Elman networks, and Xu et al. [5] combine Kalman filtering with fuzzy inference, where RNNs such as LSTM and GRU excel in time series prediction but are impeded by lengthy training durations and potential delays; Kalman filtering algorithms, which iteratively refine state estimation, e.g., Wu et al. [6], for underwater tar-get positioningoffer computational efficiency but are confined to linear systems and short-term forecasts; and polynomial fitting algorithms, employing least squares to model trajectories, e.g., Ling et al. [7] for single-step prediction, which are straightforward and efficient yet demand predefined terms and ranges, limiting long-term predictive capacity.

To overcome these limitations, this study introduces a hybrid methodology that integrates an enhanced polynomial fitting algorithm with a GRU neural net-work. For single-step prediction, the improved polynomial fitting approach se-lects terms and segmentation points to increase precision, as demonstrated by Li et al. [8], who used sliding time windows, and Gkioulekas et al. [9], who optimized via segmented linear regression. For multistep prediction, the GRU neural network, characterized by a reduced parameter set and shorter training duration, adeptly captures long-term dependencies, surpassing LSTM in efficiency and enabling online adaptation to new trajectory patterns. Collectively, these techniques offer a robust framework for short-term positioning and long-term planning, bolstering AUV efficacy in challenging environments, advancing marine resource utilization, and reinforcing maritime operational capabilities to protect national maritime interests, with scope for further empirical validation.

2 Geolocation Information Representation

2.1 Representation of Geolocation Information Under Raster Maps

Raster maps are a prevalent method for map representation in robotics, characterized by partitioning the map into numerous discrete cells, as illustrated in Fig. 1. These maps can be expressed using either the Cartesian coordinate system or the ordinal number method. The ordinal number method is more space-efficient, whereas the Cartesian coordinate system offers a more intuitive depiction of spatial locations and can be implemented as a multidimensional matrix. In this section, a two-dimensional matrix is employed to represent the raster map [10]. As depicted in Fig. 1, a two-dimensional raster map featuring distributed obstacles in robotics is represented using the Cartesian coordinate system. In this representation, black regions denote obstacles, whereas white regions indicate areas traversable by the robot. Any position on the map can be uniquely identified by Cartesian coordinates $[x, y]$. The entire map is encapsulated as an $M \times N$ two-dimensional matrix, where M represents the horizontal length and N the vertical length. For example, the raster map in Fig. 1 can be modeled as a 20×20 matrix. By assigning a value of 1 to obstacle locations and 0 to passable areas, this matrix effectively encodes the obstacle distribution, facilitating efficient map processing for tasks such as robot path optimization.

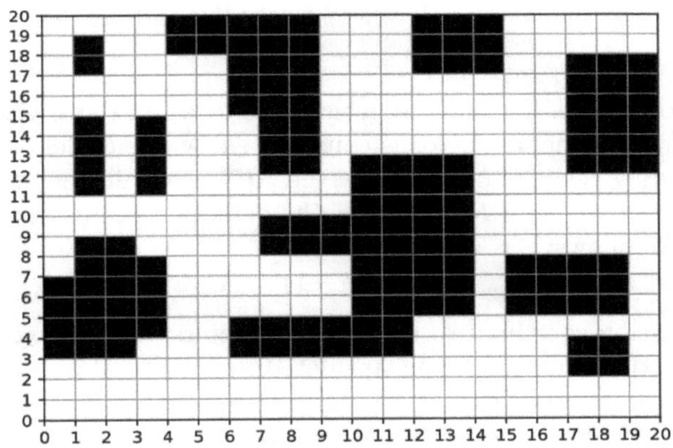

Fig. 1. Raster Maps in Robotics.

In the context of autonomous obstacle avoidance for autonomous underwater vehicles (AUVs), a detection radius R is typically defined, enabling the AUV to sense obstacle distribution within this radial region during navigation. Treating the AUV as a point mass, its detection area on the raster map forms a square centered on the AUV, with side length determined by the radius, as illustrated in Fig. 2. This figure shows the detection areas corresponding to varying detection radii within the raster map.

The obstacle positions within the detection area can be documented using the serial number method and compiled into a vector. Initially, the detection area is segmented based on the detection radius. Subsequently, each grid within this area is assigned a serial number, starting from 0 at the upper-left corner and proceeding clockwise inward. Once numbering is complete, obstacle information for each grid is recorded according to its serial number, indicating the presence

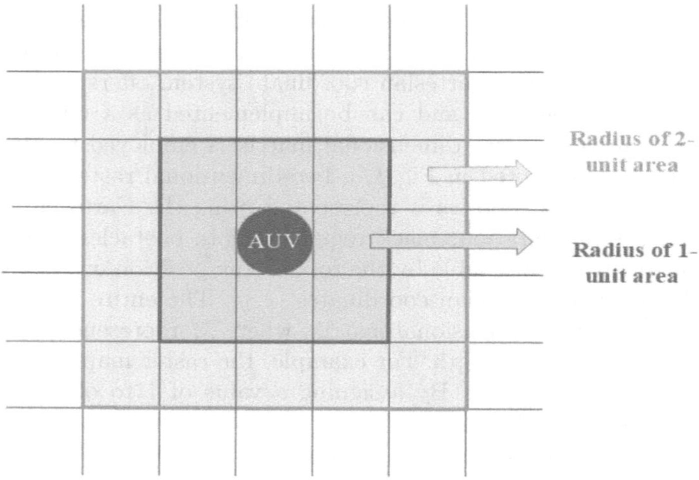

Fig. 2. Detection areas with different detection radii in raster maps.

or absence of obstacles. To capture geographic location data within the raster map, a $(Q + 4)$-dimensional vector is constructed, where Q denotes the number of grids—excluding the grid occupied by the AUV—within the area centered on the AUV's coordinates, and R represents the detection radius. Evidently, Q and R are related as follows:

$$Q = (2^{*}R + 1)^{2} - 1 \tag{1}$$

The AUV's position coordinates, current parameters, and obstacle information within the detection area can be encapsulated in a geolocation information vector. This vector comprises the AUV's horizontal and vertical coordinates, the direction and speed of the current, and the obstacle position data, as expressed in Eq. 2:

$$Info = [X, Y, current_d, current_v, o_0, o_1, ..., o_i, ..., o_{Q-1}] \tag{2}$$

Here, X and Y represent the horizontal and vertical coordinates of the AUV's location, respectively, whereas o_i indicates the obstacle state information for the grid with serial number i within the detection radius centered at $[X, Y]$. This is defined as a binary function, as presented in Eq. 3:

$$o_i = \begin{cases} 0, & \text{The grid is free of obstacles} \\ 1, & \text{This grid has obstacles} \end{cases} \tag{3}$$

In Eq. 2, $current_d$ denotes the direction of the current, and $current_v$ represents its velocity. The key parameters of ocean currents include their speed, measured in meters per second (m/s), and direction, expressed as a geographic azimuth indicating the flow direction of seawater. For example, a northward current with a speed of 0.10 m/s is recorded with a direction of 0°, eastward as 90°, southward as 180°, and westward as 270°. An example of the AUV's geolocation information vector, assuming a detection radius of one grid unit, is illustrated in Fig. 3.

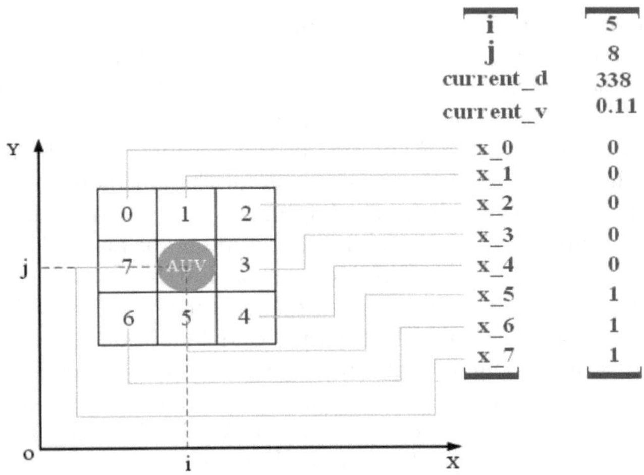

Fig. 3. Recording of geolocation information.

In Fig. 3, with an AUV detection radius of one grid unit, the detection area typically encompasses eight grids. If certain grid positions are absent—such as when the AUV is positioned at the map's edge—these locations are assigned a value of 0 as placeholders. The eight grids are numbered sequentially in a clockwise manner, and their corresponding values are entered into the geolocation information vector. A grid containing an obstacle is assigned a value of 1 in the vector, whereas an unobstructed grid receives a value of 0, thereby forming the complete geolocation information vector.

2.2 Representation of Geographic Location Information Under Latitude and Longitude Maps

The latitude and longitude coordinates can adequately mark any location on Earth, including information on the location of the AUV, as well as the location of obstacles in the deep sea and the distribution of seafloor topography, so the latitude and longitude maps are map types closer to the actual use scenarios than the raster maps. The use of latitude and longitude maps to record information about the location of AUVs and the location of obstacles around them can be achieved by first determining the detection area based on the detection radius of the AUV, dividing the detection area into multiple grids, and then determining the grid serial numbers and recording the information; however, due to the irregularities in the volume of the obstacles, there are difficulties in performing a reasonable grid division, and the scheme is relatively cumbersome and the marking of the serial numbers is time consuming. Operation is time-consuming, therefore, this section proposes a simpler way to represent geographic location information with respect to the characteristics of latitude and longitude maps [11].

Considering the AUV as a mass point without considering its specific volume, the influence of the obstacles detected in the AUV trajectory is mainly considered on the side of the obstacle "facing" the AUV. According to the detection radius R set by the AUV to determine the detection area, any obstacle that is only considered to face the AUV side of the perimeter can be detected, and according to the detection of the particle size of the perimeter of the sampling. As shown in Fig. 4 shows a schematic diagram of the recording of obstacle location information under the latitude and longitude map: Latitude and longitude coordinates effectively pinpoint any location on Earth, encompassing the position of an AUV, the placement of deep-sea obstacles, and the distribution of seafloor topography. Consequently, latitude and longitude maps offer a representation that is more aligned with real-world scenarios than raster maps. To record the positions of AUVs and surrounding obstacles using such maps, the detection area is first established based on the AUV's detection radius and segmented into multiple grids. Grid serial numbers are then assigned, and relevant information is documented. However, the irregular shapes and sizes of obstacles complicate rational grid division, rendering the approach cumbersome and the serial number assignment time-intensive. To address this, this section proposes a streamlined method for representing geolocation information tailored to the properties of latitude and longitude maps.

Treating the AUV as a point mass, disregarding its physical volume, the analysis focuses on the influence of obstacles detected along its trajectory, specifically the side of each obstacle "facing" the AUV. The detection area is defined by the AUV's detection radius R, within which only the perimeter of an obstacle's side oriented toward the AUV is considered detectable. This perimeter is sampled on the basis of a specified granularity. Figure 4 provides a schematic illustration of how obstacle location information is recorded on a latitudinal and longitudinal map:

For example, the positional information processing for a semicircular obstacle is shown in Fig. 5:

As depicted in Fig. 5a, a detection radius R is established for the AUV, with the assumption that a semicircular obstacle influences the AUV only on its side facing the vehicle within the detectable range. As illustrated in Fig. 5b, to sample the detectable portion of the obstacle, the intersection angle between the obstacle's circumference and the AUV's detection area is denoted as θ. Sampling occurs at intervals of 5°. When an obstacle is present at a sampled position,

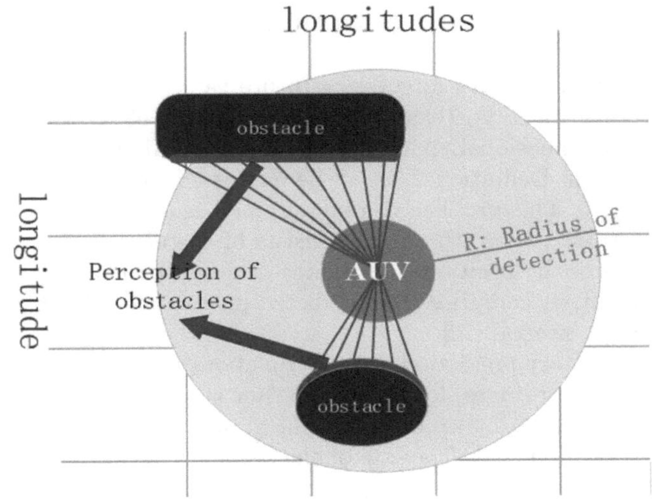

Fig. 4. Logging of obstacle location information under latitude and longitude maps.

its coordinates are recorded. Consequently, the number of obstacle data points logged by the AUV at that location is $\theta/5°$.

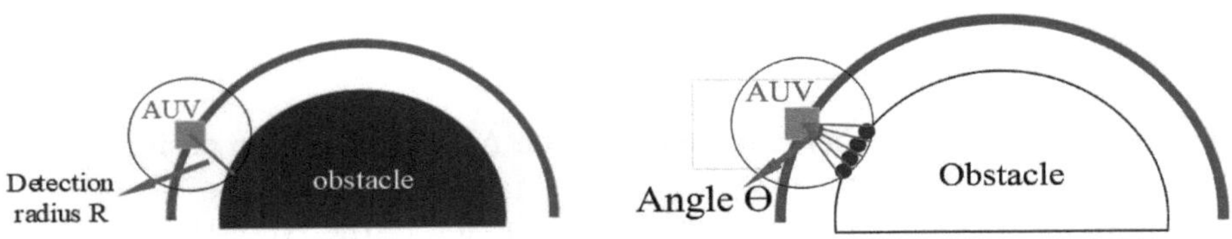

(a) Distribution of physical obstacles.

(b) Sampling method.

Fig. 5. Schematic diagram of semicircular obstacle position information recording.

The geographical location information of the AUV on a latitudinal and longitudinal map can thus be represented by a $(2+Q)\times 2$ dimensional matrix, where Q denotes the number of sampling points with detected obstacles. This is expressed as: $[[X,Y],[current_d, current_v],[x_0,y_0],[x_1,y_1],\dots,[x_i,y_i],\dots,[x_{Q-1},y_{Q-1}]]$ Here, x_i represents the longitude of the i-th obstacle, and y_i denotes its latitude.

3 G-GRU Multi-Step Prediction Network

Time series prediction entails forecasting future values of a time series via current and historical data. For example, in the context of AUV trajectory prediction in a two-dimensional space, the spatiotemporal trajectory data of the AUV, as outlined in Definition 2.1, can be expressed as a sequence of position coordi-nates [X,Y]. This problem can be reframed as a supervised learning task in ma-chine learning, where the input consists of the current and past conventional se-quences of the position coordinates or geographical information of the AUVs, and the output comprises the predicted position coordinates of the AUV over a future time interval [12].

The AUV trajectory prediction network functions to estimate the sequence of the AUV position coordinates for the next p time steps (prediction steps) on the basis of its position coordinates or geographical information from the preceding w time steps. The input data consists of the sequence of

$$\left\{[X_i, Y_i, current_{di}, current_{vi}, obstacle_{info_i}]\right\}_{i=1}^{w} \tag{4}$$

where Xi and Yi represent the horizontal and vertical coordinates of the AUV position at time i, respectively; currentdi denotes the ocean current direction at the AUV position at that moment; currentvi indicates the current velocity; and obstacleinfoi signifies the uncompressed obstacle position information at time i, derived from the specific obstacle distribution within the AUV detection range at that location. The output vector is

$$\{[X_{i+p}, Y_{i+p}]\}_{i=1}^{p}, \tag{5}$$

which represents the predicted sequence of the AUV's trajectory position information for the next p moments.

The steps for implementing multistep prediction of the AUV trajectory are shown in Fig. 6.

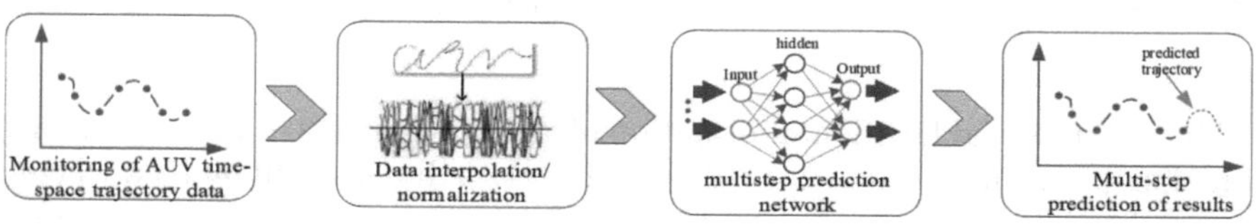

Fig. 6. Realization of a multistep prediction process for AUV trajectories.

To enable multistep trajectory prediction, the AUVs' coordinates, ocean current data, and obstacle positions within the detection area must be continuously monitored. This information is then structured into the data types and formats required for model training and subjected to normalization. For spatiotemporal trajectory datasets with large sampling intervals that inadequately capture the motion trends of AUVs or those with limited data volume, interpolation is applied to ensure consistency. Following preprocessing, the preestablished G-GRU multistep prediction network is trained on the refined dataset. Upon completion of training, the neural network can utilize the AUVs' current and historical geographical information vectors as inputs to forecast their future positions over a specified time horizon.

The G-GRU multistep prediction network comprises an information compression unit, a temporal sequence prediction unit, and an output layer, as illustrated in Fig. 7. The information compression unit compresses the high-dimensional geographical information vector of the AUV. The temporal sequence prediction unit, implemented via a GRU neural network layer, facilitates multistep trajectory prediction for the AUV. The output layer adjusts the output vector dimensionali-ty, performs inverse normalization and data type conversion, and subsequently records and visualizes the predicted trajectory.

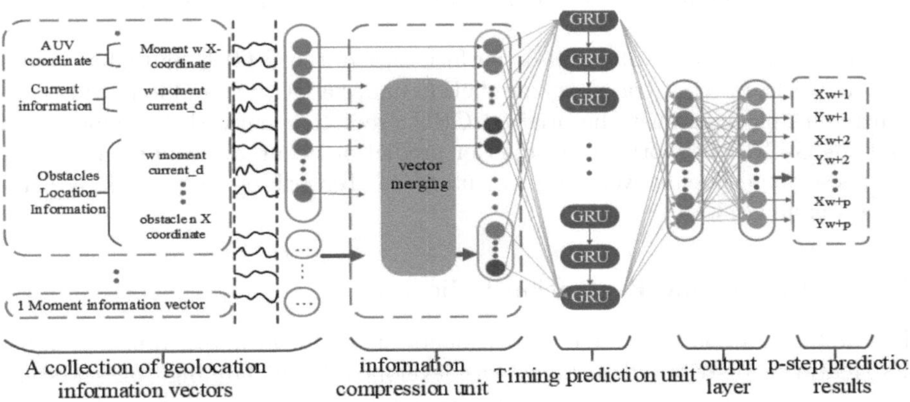

Fig. 7. Realization of a multi-step prediction process for AUV trajectories.

The input data comprise a set of geographical information vectors spanning w consecutive time steps, incorporating the AUV trajectory position, ocean current data, and obstacle position information. For each time step vector, compression proceeds as follows: initially, obstacle information is decomposed into four directional components through equivalent decomposition. These components are then vectorially combined within each direction to yield four equivalent obstacle position points. Together with the AUV coordinates and ocean current data at that moment, these points form the compressed geographical information vector, which is fed into the temporal sequence prediction unit.

This vector is processed by the temporal sequence prediction unit to train the GRU network or compute the predicted trajectory. The resulting output is refined by the output layer, which reshapes the data into a p $\times$ 2-dimensional format, where p denotes the number of predicted time steps. Following inverse normali-zation and data type conversion, the final output provides the AUV trajectory coordinates for the subsequent p steps.

4 Experiments and Analysis of Results

4.1 Purpose of the Experiment and Environment

The experiments were organized into two groups. The first group assessed the influence of geographical information on the trajectory prediction performance of the GRU prediction network. Using a grid map, a multistep prediction experiment was conducted for the UAV trajectory with both the original GRU prediction network and the G-GRU multistep prediction network, which incorporates geographical information. The prediction accuracy and fitting curve characteristics of both approaches were compared to confirm that integrating geographical information, especially obstacle position data, into a multistep prediction network markedly enhances the prediction accuracy and mitigates the prediction lag.

The second group evaluated the performance of the G-GRU multistep prediction networks in AUV trajectory prediction under a latitude and longitude map, benchmarking it against the original GRU prediction network to demonstrate that the G-GRU networks have superior prediction accuracy and fitting efficacy.

These experiments were implemented via TensorFlow version 1.14.0 and Keras version 2.3.1.

4.2 Data Sets and Evaluation Indicators

The first and second groups of experiments utilized grid maps. Initially, a grid map featuring distributed obstacles was developed. Path planning from the starting point to the endpoint was then performed, incorporating the provided ocean current data to generate AUV trajectory data. This dataset was expanded using interpolation techniques to produce a comprehensive set of AUV trajectory variations under the grid map for experimental analysis.

The third group of experiments employed a latitudinal and longitudinal coordinate map. A map with distributed obstacles was first established within this coordinate system. Autonomous obstacle avoidance and trajectory planning from the starting point to the endpoint were subsequently executed via the artificial potential field method, accounting for the specified ocean current information, to obtain trajectory data. A dataset of trajectory variations was then derived through uniform sampling for experimental purposes.

Model performance across all three experimental groups was assessed via multiple metrics, including model training time, prediction fitting performance, mean absolute error (MAE), root mean square error (RMSE), and mean absolute percentage error (MAPE).

4.3 Experiments on the Effect of Geographic Location Information on Trajectory Prediction

This experiment aims to evaluate the influence of geographical information on trajectory prediction performance. The experimental group utilized the G-GRU multistep prediction network, which incorporates geographical information (without data compression), for the AUVs' multistep trajectory prediction, whereas the control group employed the original GRU multistep prediction network. Both groups conducted their experiments within a grid map.

Ocean currents can impede AUV movement or turning, leading to deviations from the intended trajectory and heading, as well as increased energy consumption. Given that a smaller angle between the AUVs heading and the current direction reduces thruster power requirements and enhances controllability, trajectory planning prioritizes minimizing this flow angle. When planning the AUV path from the starting point to the endpoint via the ant colony optimization (ACO) algorithm within the grid map, the direction and speed of the ocean current at each grid are factored in. Specifically, the probability of moving in the current direction is weighted higher on the basis of its speed. The resulting planned AUV trajectory is depicted in Fig. 8.

The initial dataset for this trajectory comprises only 43 coordinate data points, which is insufficient for capturing trajectory variations or supporting effective neural network training. To address this, each grid was subdivided into a 10×10 subgrid structure, and uniform sampling was applied to expand the dataset to 420 data points.

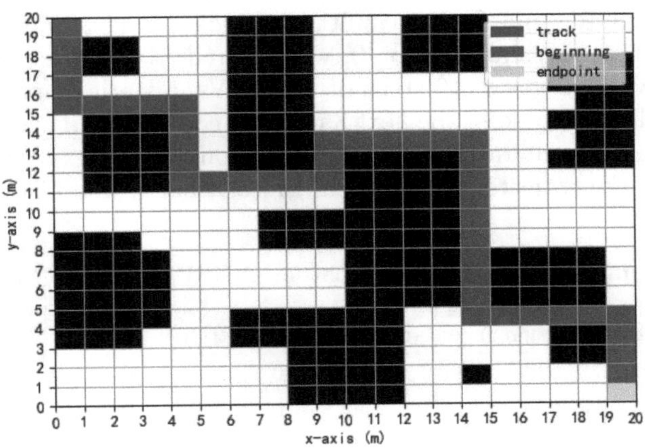

Fig. 8. Ant Colony Algorithm AUV Optimization Path 1.

Using the expanded trajectory dataset, the ocean current information and obstacle positions within the detection area at each trajectory coordinate are documented to construct a set of geographical information vectors for the experiment.

The control group employs the original GRU prediction network, which relies solely on the AUVs' historical spatiotemporal trajectory data. The data vector has a dimensionality of 2, with the input vector consisting of a sequence of AUV coordinates [i,j], where i represents the horizontal grid coordinate and j denotes the vertical grid coordinate of the AUV location. An example of the data format utilized by the original GRU prediction network is presented in Table 1.

Table 1. Information vector data format for the original GRU prediction network set

Index	X-axis	Y-axis
0	0.0	19.0
1	0.0	18.9
2	0.0	18.8

The experimental group employs the G-GRU multistep prediction network (without data compression), incorporating not only spatiotemporal trajectory data but also obstacle position and ocean current information. The detection ra-dius is set to 0.1 m, and obstacle position data within this radius, centered on the AUV, are integrated into the training dataset. The data vector has a dimension of 12, with the input vector formatted as a sequence of [i,j, currentd, currentv, x0,x1,x2,x3,x4,x5,x6,x7], where xk denotes the presence (1) or absence (0) of an obstacle in the grid with index k. An example of the data format used by the ex-perimental group is presented in Table 2. The parameter configurations for the original GRU and G-GRU multistep prediction networks are as follows: the train-ing time steps are set to 10, with a prediction horizon of 3, meaning that the net-works leverage 10 steps of the AUVs' historical spatiotemporal or geographical data to forecast the next 3 position coordinates. Training spans 200 epochs, with a dropout rate of 0.3. Both the control and experimental groups predict and fit the trajectory depicted in Fig. 8, yielding the following results:

Table 2. Information Vector Data Format for G-GRU Predictive Network Groups

Index	X-axis	Y-axis	Direction	Speed	Grid 0	Grid 1	Grid 2	Grid 3	Grid 4	Grid 5	Grid 6	Grid 7
80	3.9	15.0	176.033	0.136	0	0	0	0	0	1	1	0
81	4.0	15.0	176.033	0.136	0	0	0	0	0	0	1	0
82	4.0	14.9	176.455	0.135	0	0	0	0	0	0	1	1

As shown in Fig. 9, the original GRU prediction network exhibits a lag in re-sponse to changes in the AUV's trajectory. This is reflected in the fitting results, as the prediction curve lagging behind the trajectory changes by several time steps before beginning to follow the variations. In contrast, the G-GRU

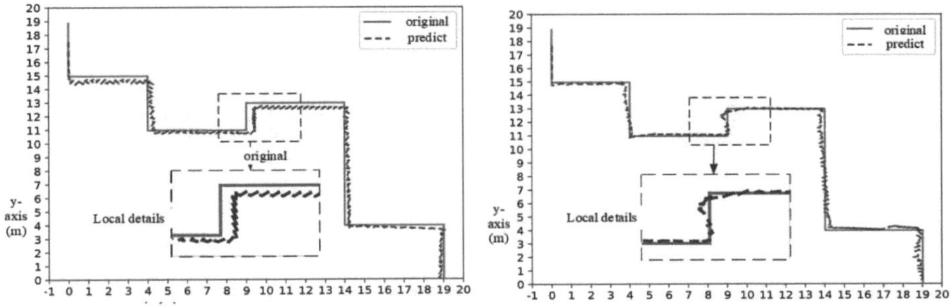

(a) Original GRU prediction network. (b) G-GRU prediction network.

Fig. 9. Comparison of predictive fit of original GRU prediction network and G-GRU prediction network.

prediction network is more sensitive to impending trajectory changes and their trends. This sensitivity allows the prediction curve to be more quickly corrected at points of abrupt trajectory changes, resulting in a more accurate fit. Owing to the limited number of input data features, the prediction curve of the original GRU model is smoother, whereas the prediction curve of the G-GRU network exhibits more fluctuations.

To more accurately compare the performance differences between the two networks, the AUV trajectory dataset represented in Fig. 8 was used to conduct 10 prediction experiments with both the GRU prediction network and the G-GRU prediction network. The average values of the four metrics (MAE, RMSE, MAPE, and training time) were calculated and compared. The results are shown in Fig. 10. The best values among the two are marked in red in the figure. Compared with the original GRU prediction network, the G-GRU prediction network achieves a significant improvement in prediction accuracy and performs better in terms of the MAE, RMSE, and MAPE. Specifically, the MAE decreases by 15.38%, the RMSE decreases by 17.83%, and the MAPE decreases by 15.63%, whereas the train-ing time increases by 4.57%.

From the above experiments, we can draw the following conclusions: Constructing a GRU multistep prediction network on the basis of geographical information, i.e., incorporating AUV obstacle position information and ocean current information, can significantly increase the prediction accuracy and effectively alleviate the lag in predicting trajectory changes. However, this approach also leads to increased training time and a less smooth prediction fitting curve.

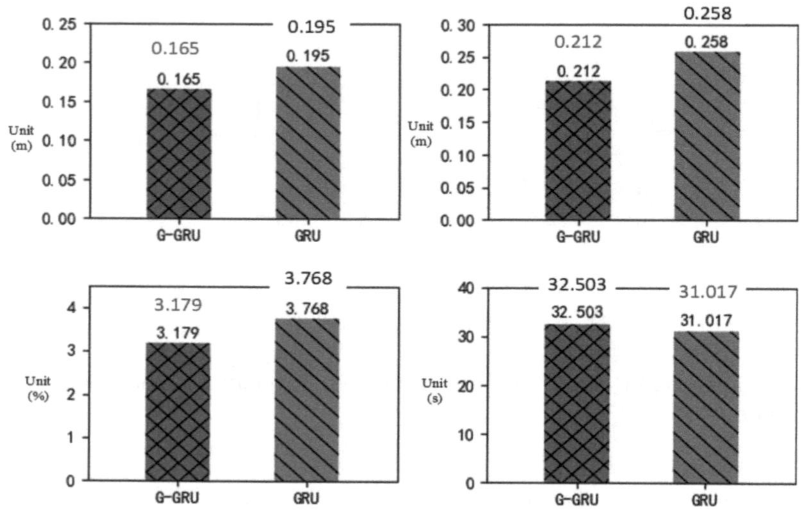

Fig. 10. Comparison of metrics between the original GRU prediction network and the G-GRU prediction network.

4.4 Experiments on G-GRU Trajectory Prediction with Latitude and Longitude Maps

This experiment verifies the performance of the G-GRU multistep prediction model under the latitude and longitude map. The prediction effects of the original GRU multistep prediction network and the G-GRU multistep prediction network on AUV trajectories are compared. As shown in Fig. 11a, a latitudinal and longitudinal coordinate map with distributed obstacles is constructed, and the artificial potential field method combined with ocean current parameters is used for trajectory planning.

The experiment did not adopt the obstacle position recording method described in Sect. 2.2 for the latitude and longitude map. Instead, the image point selection method was directly used. To record the obstacle positions within the AUV's detection area along the trajectory, the GetData tool was employed to sample the side of obstacles that may affect the AUV's movement and face the AUV, as shown in Fig. 11b. The coordinates of the sampled obstacle points were recorded to obtain the surrounding obstacle position information for the AUV. Both the experimental and control groups had their training time steps set to 20 and prediction time steps set to 5, meaning that data from the past 20 steps were used to predict the next 5 steps. The number of training epochs was set to 100, and the batch size was set to 64. The control group uses the original GRU prediction network, with only the AUV's spatiotemporal trajectory data as the input vector, as shown in Table 1. The experimental group used the G-GRU mul-tistep prediction network, with the AUV's detection radius R set to 100 m. At each position along the trajectory, the detection area was determined using

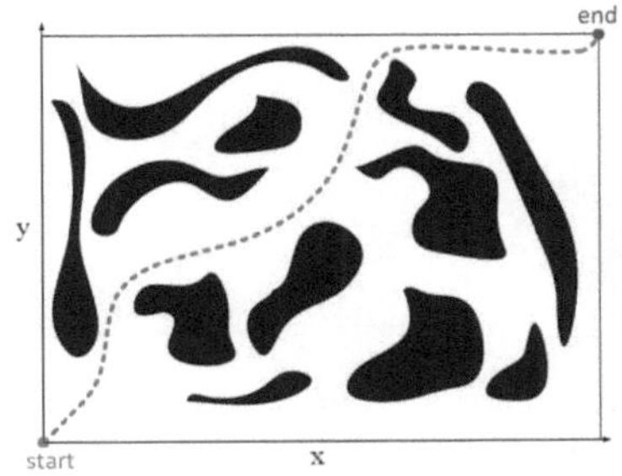

(a) AUV Autonomous Obstacle Avoidance Track and Map.

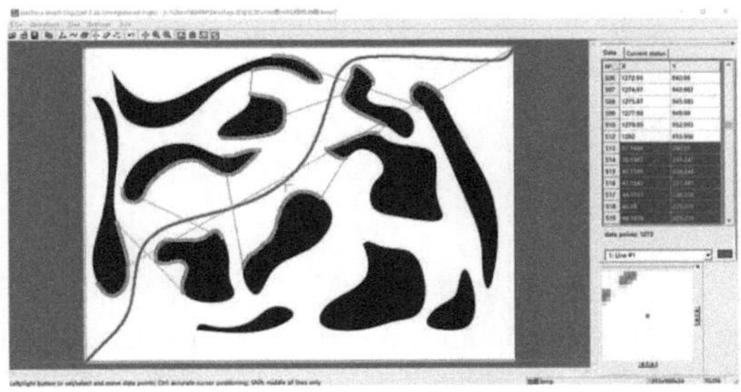

(b) Sampling of obstacle location information using GetData.

Fig. 11. AUV tracks under latitude and longitude maps.

the radius R, and the obstacle positions within this area were identified. The final geographical information vector used, which combines the obstacle position in-formation obtained with the GetData tool and the predefined ocean current in-formation, is shown in Table 3.

Table 3. Geolocation information vector data format with data compression for raster maps.

Index	X-axis	Y-axis	Direction	Speed	Component 1	Component 2	Component 3	Component 4
63	139.544	198.206	13.221	0.143	0	8.867433	9.529278	0
64	140.548	201.210	13.221	0.144	40.065627	8.927510	9.685294	0
65	141.552	205.214	13.223	0.144	39.511036	8.617668	9.962020	32.636088

The trajectory prediction results for the original GRU multistep prediction network are shown in Fig. 12a, and the AUV trajectory prediction results for the G-GRU multistep prediction network are shown in Fig. 12b:

As shown in Fig. 12, the original GRU prediction network has a significant lag in learning trajectory changes, whereas the G-GRU multistep prediction network, which is based on the geographical information in the AUV trajectory, can greatly reduce this lag. Ten trajectory predictions were conducted via both the GRU and G-GRU models, and the average values of the evaluation indicators MAE, RMSE, and MAPE were taken. The results are shown in Table 4. As shown in Table 4, compared with the original GRU, the G-GRU achieves a reduction of 35.36% in the MAE, 24.65% in the RMSE, and 0.65% in the MAPE. These results indicate that the G-GRU prediction network proposed in this chap-ter outperforms the original GRU prediction network, which does not consider geographical information, in terms of prediction accuracy. Specifically, in the context of the latitude and longitude map, the G-GRU multistep prediction net-work can achieve higher prediction accuracy and reduce the lag in learning tra-jectory changes compared with the original GRU.

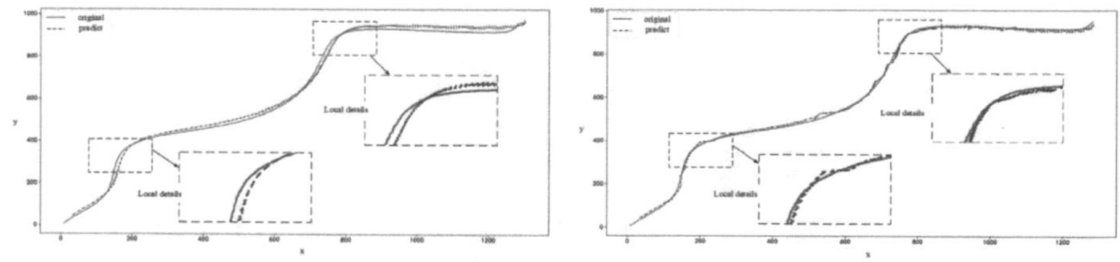

(a) GRU multi-step prediction network trajectory prediction effect.

(b) G-GRU multi-step prediction network trajectory prediction effect.

Fig. 12. Effectiveness of trajectory prediction under latitude and longitude maps.

Table 4. Comparison of the performance metrics for the GRU and G-GRU models.

Model	MAE	RMSE	MAPE
G-GRU	0.8926	1.2209	0.7182%
GRU	1.3809	1.6204	1.3688%

5 Conclusion

This paper proposes a G-GRU multistep prediction network based on geographic information to address the challenges of trajectory prediction for autonomous underwater vehicles (AUVs) in complex deep-sea environments. By integrating obstacle location and ocean current information and employing data compression techniques to manage dimensionality, the network significantly improves the accuracy and response speed of trajectory prediction. The experimental results demonstrate that the G-GRU network outperforms the traditional GRU network in trajectory prediction under both raster and latitude–longitude maps, effectively reducing the prediction lag and enhancing the prediction precision.

However, this study has several limitations. First, although the G-GRU network achieves significant improvements in prediction accuracy, its training time is longer than that of the traditional GRU network, which may limit its application in scenarios with extremely high real-time requirements. Second, whereas we validated the performance of the G-GRU network under raster and latitude–longitude maps, its performance in other types of maps or more complex marine environments still needs further verification. Additionally, the prediction performance of the G-GRU network is to some extent dependent on the quality of the input geographical information. Inaccurate obstacle detection or ocean cur-rent data may affect the reliability of the prediction results.

References

1. Li, J., Zhang, J., Zhang, H., et al.: A predictive guidance obstacle avoidance algorithm for AUV in unknown environments. Sensors **19**(13), 2862 (2019). https://doi.org/10.3390/s19132862
2. Agrawal, P., Agrawal, H.: Adaptive algorithm design for cooperative hunting in multi-robots. Int. J. Intell. Syst. Appl. **10**(12), 47–55 (2018)
3. Meng, X., Sun, B., Zhu, D.: Harbour protection: moving invasion target interception for multi-AUV based on prediction planning interception method. Ocean Eng. **219**, 108268 (2021). https://doi.org/10.1016/j.oceaneng.2020.108268
4. Zhang, M., Gao, P., Xu, J.: Generalized predictive control of autonomous underwater robots based on neural networks. Robot **2008**(01), 91–96 (2008) (in Chinese)
5. Xu, B., Li, S., Wang, L., Duan, T., Yao, H.: A multi-AUV cooperative localization method based on adaptive neuro-fuzzy inference system. J. Chinese Inertial Technol. **27**(04), 440–447 (2019) (in Chinese)
6. Wu, Y., Su, X., Cui, J., Yang, G.: Trajectory planning for cooperative search and strike of underwater targets by USV and AUV. Control Decision **36**(04), 825–834 (2021) (in Chinese)
7. Ling, W., Ni, J., Chen, Y., et al.: Multi-UAV encirclement based on improved whale optimization algorithm. Comput. Modernization **2021**(6), 6 (2021). (in Chinese)
8. Li, T., Chen, H., Sun, S., et al.: Joint smoothing and tracking based on continuous-time target trajectory function fitting. IEEE Trans. Autom. Sci. Eng. **16**(4), 1476–1483 (2019). https://doi.org/10.1109/TASE.2019.2917684
9. Gkioulekas, I., Papageorgiou, L.G.: Piecewise regression analysis through information criteria using mathematical programming. Expert Syst. Appl. **121**, 362–372 (2019). https://doi.org/10.1016/j.eswa.2018.12.029
10. Dai, N., Qin, P., Xu, X., Zhang, Y., Shen, Y., He, B.: An AUV collision avoidance algorithm in unknown environment with multiple constraints. Ocean Eng. **294**, 116846 (2024). https://doi.org/10.1016/j.oceaneng.2024.116846
11. Zhang, Y., Wang, Q., Shen, Y., Dai, N., He, B.: Multi-AUV cooperative control and autonomous obstacle avoidance study. Ocean Eng. **304**, 117634 (2024). https://doi.org/10.1016/j.oceaneng.2024.117634
12. Yu, H., Ning, L.: Coordinated obstacle avoidance of multi-AUV based on improved artificial potential field method and consistency protocol. J. Marine Sci. Eng. **11**(6), 1157 (2023). https://doi.org/10.3390/jmse11061157

Sequence-Based Deep Learning Ensemble for Targeted Phage–Host Interaction Prediction

Zihang Zhao[2], Qian Chen[1], Min Li[3], Wenchen Song[3], Minfeng Xiao[3(✉)], and Min Fang[4(✉)]

[1] School of Artificial Intelligence, Shenzhen University, Shenzhen, China
[2] School of Computer Science, The University of Hong Kong, Hongkong, China
[3] BGI-Shenzhen, Shenzhen, China
xiaominfeng@genomics.cn
[4] Education Center of Experiments and Innovations, Harbin Institute of Technology, Shenzhen, Shenzhen, Guangdong, China
fangmin@hit.edu.cn

Abstract. The rapid rise of antibiotic-resistant bacteria has created an urgent need for alternative treatments, such as bacteriophage therapy, which relies on phages to selectively eliminate specific bacterial pathogens. However, the success of phage therapy hinges on accurately matching phages to their bacterial hosts, and predicting these phage–host interactions (PHIs) remains a significant challenge. The existing computational approaches for PHI prediction have major limitations. For example, the PredPHI model achieves less than 70% accuracy, and most methods predict only broad host categories (e.g., genus levels) rather than pinpointing the exact bacterial strain. To address these gaps, we present a novel data-driven framework that predicts PHIs at the individual host strain level using only protein amino acid sequence data as input. Our approach extracts informative features from protein sequences—including amino acid composition (AAC), key chemical element abundance (AC), and molecular weight (MW)—and employs a one-dimensional convolutional neural network (1D CNN) tailored for sequential protein data. To further increase the prediction accuracy and mitigate class imbalance, we train multiple CNN models on different subsets of the data and integrate their outputs via a weighted ensemble strategy. This ensemble 1D CNN model achieves an AUC of 0.968, an AUPR of 0.976, an accuracy of 88%, and a sensitivity of 0.911, with a Matthews correlation coefficient (MCC) approximately 40% greater than that of PredPHI. Overall, this data-driven bioinformatics approach provides a powerful tool for rapid phage–host matching and custom phage design, potentially accelerating the development of targeted phage therapy.

Keywords: Phage-Host Interactions · Protein Sequence · 1D CNN · Ensemble Learning · Data-driven Bioinformatics

Z. Zhao and Q. Chen — Co-first authors.

Y. Yu et al. (Eds.): ICPCSEE 2025, CCIS 2673, pp. 364–388, 2026.
https://doi.org/10.1007/978-981-95-2566-9_25

1 Introduction

With the widespread use of antibiotics and the continuous rise of bacterial resistance, traditional antimicrobial strategies are facing serious challenges. As a result, bacteriophage (phage) therapy has received renewed attention as a potential alternative to antibiotics in treating drug-resistant infections [9,15]. Phages are viruses that specifically infect bacteria, and the key to successful phage therapy is the selectivity and accuracy with which a phage can target a pathogenic bacterium. However, identifying a naturally occurring phage that effectively infects a given bacterium is a difficult and time-consuming task because of the enormous diversity and abundance of phages in nature. While synthetic biology approaches can be used to design or engineer specific therapeutic phages.

To date, researchers have developed a variety of computational models to predict phage–host relationships, often by classifying a phage's host at the genus level (genus, family, etc.).), rather than pinpointing exact strain-level interactions [1,19,23]. For example, alignment-based approaches (such as BLAST-based host searches) and machine learning methods have achieved high accuracy in predicting a phage's host taxonomy [1,19,23,25]. DeepHost [19], for example, applies a convolutional neural network to phage genomic data and achieves approximately 96% prediction accuracy at the host genus level. Similarly, host-Phinder uses genomic sequence similarity to predict host identity and has demonstrated reliable performance for known phage–bacteria pairs [23]. However, these methods focus on host classification and do not directly confirm whether a given phage will infect a specific bacterial strain. Relatively few studies have focused on the direct prediction of phage–bacterium interactions at the strain level via sequence information. PredPHI [14] is a notable attempt that uses phage and host protein sequences to predict whether a phage can infect a bacterium. However, its reported accuracy on test datasets was under 70%, highlighting the limited prediction capability of that model. This indicates that there is significant room for improvement in terms of the PHI prediction accuracy and scope. Some studies have investigated phage receptor-binding proteins and bacterial surface receptors (from a structure-based perspective [3]) to refine predictions. There is also a trend toward leveraging advanced representations, such as protein language model embeddings [12], to capture subtle biological signals. Another emerging trend is to utilize large-scale pretrained protein language models (such as ESM, ProtBERT, or transformer-based models) to derive rich feature embeddings for phage and bacterial proteins. These models, trained on vast protein databases, can encode biochemical and evolutionary information. By streamlining known methods and highlighting new developments, the introduction now provides a balanced overview of the state of the art and the niche our approach fills.

Recently, progress has been made toward more fine-grained PHI predictions. Gaborieau et al., for example, demonstrated that most phage-bacteria actions at the strain level can be predicted via only genomic sequence data. In their study focusing on Escherichia coli isolates, a computational pipeline was able to recommend tailored phage cocktails with approximately 80% success in lysing target bacteria, underscoring the potential of in silico approaches to guide phage therapy in practice [10]. Nonetheless, there remains a need for models that can effectively leverage minimal sequence information (such as amino acid sequences alone) to predict phage–host interactions with increased accuracy and broader applicability.

Moreover, many deep learning models have achieved notable success in bioinformatics tasks related to peptide and interaction prediction. For example, pACP-HybDeep [20] employs a binary tree growth-based transformer and a hybrid deep learning framework to accurately predict anticancer peptides. Another model, TargetCLP [22], integrates transformed and evolutionary-scale features through a weighted scheme to identify clathrin-coated proteins with high precision. In the realm of immunomodulatory peptides, DeepAIPs-Pred [2] combines local evolutionary transformation images with structural embeddings to predict anti-inflammatory peptides. Similarly, AIPs-DeepEnC-GA [18] uses an ensemble of deep classifiers optimized by a genetic algorithm to improve anti-inflammatory peptide prediction. These state-of-the-art models highlight the power of modern deep learning (including transformer architectures and ensemble strategies) in biological sequence analysis. Inspired by their success, our work focuses on leveraging one-dimensional CNN ensembles for the specific challenge of phage–host interaction prediction at the strain level.

In this work, we propose a novel deep learning model to predict phage–host interactions directly from the protein sequences of phages and bacteria. This approach addresses the limitations of previous methods, such as PredPHI and DeepHost, by focusing on sequence-level features and improving the handling of imbalanced data. Specifically, we design a one-dimensional convolutional neural network (1D-CNN) architecture tailored for amino acid sequences and introduce a new sample balancing strategy that combines undersampling and oversampling of training data. Through extensive experiments on a curated PHI dataset, the proposed model achieves substantially better performance than PredPHI in metrics such as accuracy, sensitivity, and the Matthews correlation coefficient. Our results demonstrate that using protein sequence information with a specialized deep learning framework can significantly enhance PHI prediction. This provides a more effective and efficient tool for identifying viable phage–host pairs, which is valuable for developing phage-based therapies against multidrug-resistant bacteria.

The main contributions of this study are as follows:

- Only sequence-derived features, such as amino acid composition (AAC), chemical component abundance (AC), and molecular weight (MW), are utilized as inputs, enabling phage–host interaction prediction purely from protein sequences and demonstrating the advantage of sequence-based prediction for targeted phage identification.
- This introduces a novel class-imbalance handling strategy that groups positive and negative samples before they are combined into multiple balanced training subsets, effectively leveraging both undersampling and oversampling to improve model learning on highly imbalanced data.

- Employs a one-dimensional convolutional neural network (1D CNN) architecture: training multiple CNN submodels on balanced subsets to capture sequence patterns and enhance prediction performance in phage–host interaction tasks.
- Adopting an ensemble learning approach that fuses the outputs of these CNN submodels via a weight selection method, delivers the best overall performance and outperforms other ensemble strategies (e.g., stacking, soft voting, meta-feature integration) in comparative experiments.
- Achieving state-of-the-art results on a private phage–host interaction dataset: AUC = 0.968, AUPR = 0.976, 88% accuracy, 91.1% sensitivity, significantly outperforming the PredPHI baseline with an approximately 40% higher MCC.

2 Related Works

2.1 Biological Basis of Phage-Host Interactions

Bacteriophages (phages) are viruses that infect bacteria with remarkable host specificity. In the face of increasing antibiotic resistance, phage therapy has emerged as a promising alternative treatment since phages can selectively kill pathogenic bacteria while sparing others [16]. Phages have been applied in bacterial identification and typing, environmental biocontrol, livestock disease management, and as therapeutic agents for human infections. Their ability to target bacteria with high precision underlies these applications and differentiates them from broad-spectrum antibiotics [7]. Phage infection begins when a phage particle recognizes and binds to specific receptors on the bacterial cell surface, such as lipopolysaccharides or teichoic acids, on the cell wall. This attachment is mediated by specialized phage proteins (e.g., tail fibres or spikes) that determine the host range by fitting like a "lock-and-key" to host receptors. Once bound, the phage injects its genetic material into the host, hijacking the bacterial machinery to produce new virions and eventually lysing the cell to release progeny. Because successful infection requires precise molecular interactions between phage proteins and host receptors, most phages have a narrow host range and sometimes infect only certain strains of a species. Notably, some broad-host-range phages can infect multiple species or genera. Bacteria, in turn, can evolve resistance by altering or masking receptors or deploying defenses such as restriction-modification and CRISPR–Cas systems [13], driving a coevolutionary arms race with phages. Crucially, the determinants of host specificity reside in phage proteins (e.g., receptor-binding proteins), suggesting that the sequence features of these proteins encode the information needed to identify compatible hosts. This biological insight lays the foundation for computational approaches that predict phage–host interactions via sequence data.

2.2 Computational Models for Phage-Host Prediction

Current computational methods for phage–host interaction (PHI) prediction fall into two broad categories: host taxonomy classification and direct interaction prediction [5,8]. Host taxonomy classifiers aim to assign a phage to a host's taxonomic group (usually genus or species) on the basis of genome sequence features. For example,HostPhinder uses k-mer composition similarities to known phage genomes to predict the host genus/species of a query phage [23]. DeepHost employs a CNN on encoded phage genomes (k-mer features spaced [4]) to achieve high host classification accuracy (e.g., 96% at the genus level). PHISDetector integrates diverse in silico signals (CRISPR spacers, prophages, k-mers, etc.).), and machine learning to predict likely hosts [26]. These methods significantly improve host assignment accuracy and speed over naive BLAST comparisons [19,23]. However, their scope is limited to host categorization rather than the prediction of specific phage–bacterium pairs—they output a host taxonomy prediction and do not directly determine novel phage–host interactions. This distinction is crucial, as our work focuses on interaction prediction rather than coarse host classification. The ability to directly predict phage–host pairs has greater practical importance for phage therapy and microbiome analysis, motivating models that go beyond taxonomy labels.

In contrast, direct interaction prediction models attempt to identify whether a specific phage and bacterium will interact (infection occurs). Several representative approaches have been proposed. PredPHI [14] trained a deep CNN on features from phage and host protein sequences to classify phage–host pairs, using careful negative sample selection to handle data imbalance. This demonstrated the feasibility of sequence-based PHI prediction, but its test accuracy (under 70%) is limited in terms of predictive power, likely because it relies on shallow sequence features and requires a balanced training set. GSPHI [17] combines sequence embedding and graph embedding: it initializes phage and bacterial embeddings via a protein/DNA language model, then applies graph embedding (SDNE) to incorporate a known PHI network structure, and finally, a DNN is used to predict interactions. GSPHI achieves a high cross-validated AUC (0.92) [17] on certain datasets; however, it depends on existing phage–host network information and engineered features, which may hinder its generalization to new phages with no prior network content. Some recent models integrate structural or multimodal data. For example, PHIRE (a sequence/structure-based predictor) incorporates protein structural information for host recognition, and CHERRY uses a graph encoder-decoder over a knowledge graph of viral and host features [21]. CHERRY attained state-of-the-art accuracy at the species level by integrating multiple protein/DNA features in a GCN model. However, such complex models often require extensive feature engineering (e.g., predefined viral protein families, host similarity metrics) and do not support direct end–end sequence reasoning; the predictive power still heavily relies on known genomic or network similarities. Overall, direct PHI prediction models tend to face challenges such as feature incompleteness, data imbalance, and overfitting to known interactions. Many models still sample only a small subset of negative pairs or assume that each phage has a single host, which may not hold for broad-host-range phages [5].

In summary, mainstream PHI prediction models have laid a foundation but are still constrained by shallow feature usage, limited sequence generalization, and coarse prediction granularity [6,11,24]. Their typical focus on either host category prediction or network-similarity mapping means that they struggle with truly novel phage–host pairs and nuanced interaction mechanisms. These gaps motivate our work: we propose a new protein sequence-driven interaction prediction approach to overcome these limitations. By leveraging richer sequence embeddings and learning directly from phage and host protein sequences, our model aims to improve generalization to unseen phage–host pairs, handle imbalanced data more robustly, and enable more precise, fine-grained predictions of phage–host interactions than previously possible. This approach offers a promising route to address the poor generalizability, lack of direct sequence inference, and limited predictive detail of existing PHI models, thereby advancing the accuracy and applicability of phage–host interaction predictions.

3 Sequence-Based CNN Ensemble Framework for PHI Prediction

3.1 Overview of the Proposed Framework

Figure 1 illustrates the architecture of the proposed framework, which comprises three modules: preprocessing, CNN-based model training, and model fusion. In the preprocessing stage, the raw amino acid sequence data are converted into informative feature representations, and an explicit sample balancing strategy is applied to address the severe class imbalance. Specifically, a hybrid under and oversampling approach is adopted: the majority class samples are partitioned into several subsets, each of which is combined with all minority class samples to form balanced training sets. This method preserves important information from the majority class while avoiding overfitting to duplicate minority examples, producing multiple balanced datasets for subsequent model training.

In the CNN training module, a one-dimensional convolutional neural network (1D CNN) is employed as the base model for each balanced subset to extract high-level features automatically and perform classification. M such CNN classifiers are trained (one per subset), allowing the network to learn discriminative sequence patterns for phage–host interaction prediction. Each CNN model comprises convolutional, normalization, and pooling layers followed by fully connected layers and produces a probability output for the interaction. A single activation in the final layer yields the output probability p_i of the i-th CNN, which is formulated as:

$$p_i = \sigma(\mathbf{W}_i^T \mathbf{h}_i + b_i) \tag{1}$$

where $\mathbf{h}_i$ is the feature vector learned by the CNN. Finally, in the model fusion (ensemble) stage, the M trained CNN classifiers are combined to derive the final prediction. A weighted ensemble strategy is employed, in which each model is assigned a weight w_i reflecting its performance, and their outputs are aggregated into a single score. The ensemble output for a given input $\mathbf{X}$ is:

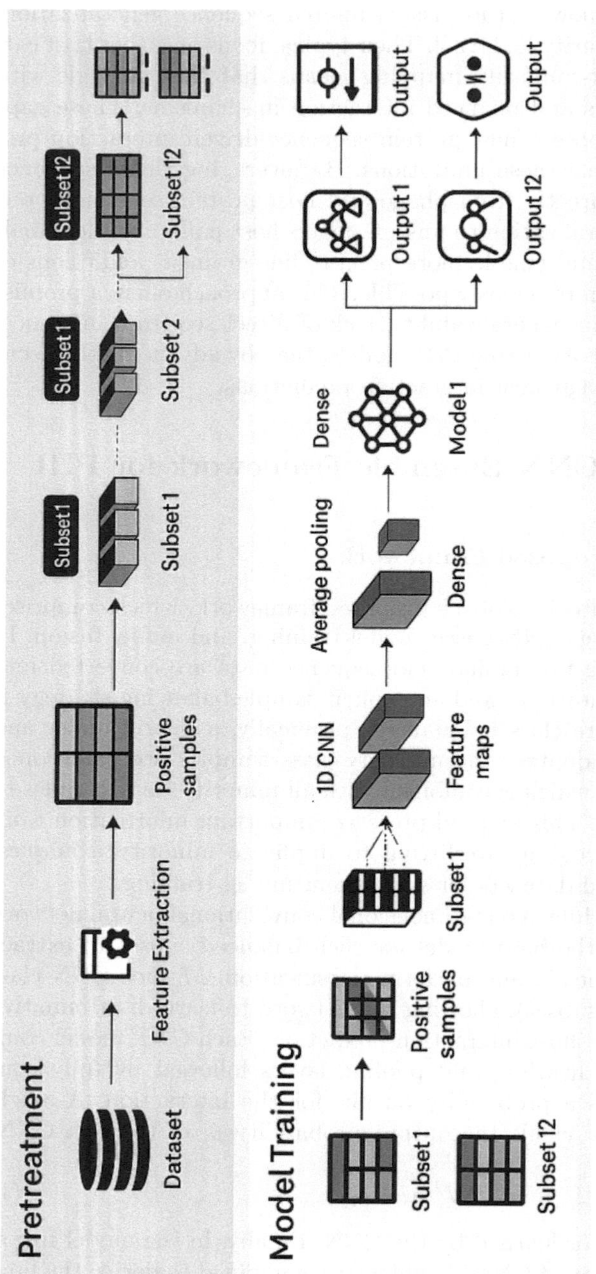

Fig. 1. Overview of the Experimental Process of Model.

$$S_{\text{ens}}(\mathbf{X}) = \frac{\sum_{i=1}^{M} w_i \, p_i}{\sum_{i=1}^{M} w_i} \tag{2}$$

which represents a weighted average of individual predictions. This design leverages complementary information from multiple classifiers, resulting in a final model that is more robust to class imbalance and delivers improved predictive accuracy.

3.2 Feature Encoding and Sample Construction

The amino acid sequence structure of a protein is a sequence of amino acids, where each letter represents a specific amino acid residue. Each phage genome is represented as a nucleotide sequence (e.g., a 48,502 bp DNA sequence for phage T4), which serves as input to our model. The raw data used in this study consisted of amino acid sequences derived from phage proteins and their bacterial host receptors. These sequences are first converted into informative feature representations that capture composition and physicochemical properties relevant to interaction prediction. In particular, by analysing each protein sequence, we extract three types of descriptive features: amino acid composition (AAC), abundance of chemical components (AC), and molecular weight (MW). Each of these features condenses the raw sequence into a quantitative descriptor that can be used as input to the predictive model.

(1) Amino Acid Composition (AAC) describes the relative frequency of each amino acid in a protein sequence, providing insight into the protein's overall makeup. The AAC of a sequence is calculated by counting occurrences of each amino acid and dividing by the total length. Formally, for a given protein sequence of length N, if N_t is the count of amino acid t in the sequence (where t ranges over the 20 common amino acids), then:

$$AAC = \frac{N_t}{N}, \quad t \in 20 \text{ common amino acids} \tag{3}$$

where N_t represents the number of times the amino acid t occurs in the sequence, and N represents the total number of amino acids in the sequence. And, 20 common amino acids are shown in Table 4.

(2) The Abundance of Chemical components (AC) represents the overall counts of key elemental constituents in the amino acid sequence. Proteins are chiefly composed of five elements: carbon (C), hydrogen (H), oxygen (O), nitrogen (N), and sulfur (S). The AC feature vector simply tallies the total number of each of these atoms in the protein. If N_c denotes the number of atoms of element c present (summing over all residues in the sequence), then the AC for element c is given by:

$$AC = N_c, \quad c \in \{C, H, O, N, S\} \tag{4}$$

which yields five values corresponding to the counts of C, H, O, N, and S in the protein. For example, a protein sequence with a higher count of sulfur-containing residues (like Cysteine or Methionine) will have a larger $N_{!S}$ value in its AC profile.

(3) Molecular Weight (MW) is the total molecular mass of the protein sequence (in Daltons), which reflects the size of the protein and can influence its biological function. To compute MW, we sum the standard molecular masses of all amino acid residues in the sequence and subtract the mass of water molecules released during peptide bond formation. If w_t is the molecular mass of amino acid t, and the protein has length m (number of residues), then the molecular weight is calculated as:

$$MW = \sum(w_t) - (m - 1) \times 18.01 \tag{5}$$

where 18.01 Da is the approximate mass of a water molecule (lost for each of the $m - 1$ peptide bonds linking m amino acids).

Using the above features, each phage–host pair can be represented in a fixed-dimensional vector form. We compute the AAC, AC, and MW for the phage protein and for the host protein separately and then concatenate these features, which results in a 52-dimensional feature vector for each pair (26 features from the phage +26 from the host), which serves as the input to our predictive model.

We design a one-dimensional (1D) convolutional neural network (CNN) to learn from these feature vectors and perform classification, as shown in Tab. 1. The choice of a 1D CNN is motivated by the nature of our data: although protein sequences can be treated as text strings, they are encoded into a 1D numerical feature sequence. One-dimensional CNNs are well suited for extracting patterns from such sequential data and are more computationally efficient for this purpose than are higher-dimensional CNNs. Additionally, 1D convolutions inherently respect the ordering of features in the sequence (e.g., contiguous regions in the feature vector might capture related properties) while requiring fewer parameters than a fully connected approach on the raw sequence.

However, 1D CNNs may suffer information loss when processing long feature sequences, which affects the overall accuracy. To address this, our CNN architecture (described in Table 2) integrates two convolutional layers with ReLU activations and batch normalization combined with max pooling and dropout layers to enhance feature extraction, prevent overfitting, and retain key patterns from the input sequence. After convolutional feature extraction, the global average pooling aggregate features into a fixed-length vector, which is then fed into a fully connected layer followed by a softmax output to classify interacting and noninteracting phage–host pairs. This compact yet effective structure, paired with the ensemble approach described later, ensures robustness and high performance on imbalanced datasets.

Table 1. The structure of the 1D CNN model

Layer	Output Shape	Activation Function	Additional Notes
Input	(52, 1)	–	–
Convolution	(52, 64)	ReLU	Kernel Size: 3, Padding: same
Batch normalization	(52, 64)	–	–
Max-pooling	(26, 64)	–	Pooling Size: 2
Dropout	(26, 64)	–	Dropout Rate: 0.05
Convolution	(26, 128)	ReLU	Kernel Size: 3, Padding: same
Batch normalization	(26, 128)	–	–
Max-pooling	(13, 128)	–	Pooling Size: 2
Dropout	(13, 128)	–	Dropout Rate: 0.05
Average-pooling	(128,)	–	–
Fully connected	(256,)	Sigmoid	–
Batch normalization	(256,)	–	–
Dropout	(256,)	–	Dropout Rate: 0.25
Outputs	(2,)	Softmax	–

To address class imbalance effectively, we train multiple CNN-based models using balanced subsets created by partitioning the majority class, thus capturing diverse patterns and minimizing overfitting. The predictions of these base classifiers are then fused into a robust ensemble via a weighted fusion strategy, described in detail in the following section.

3.3 Ensemble Strategy Based on Weighted Fusion

We incorporate an ensemble learning strategy to improve prediction accuracy and robustness by combining multiple CNN classifiers through weighted output fusion. First, we train M independent base models using the same CNN architecture but with different training subsets or initializations, ensuring each model captures complementary aspects of the data. This diversity among base models allows the ensemble to achieve functions and performance that a single model cannot attain. As illustrated in Fig. 2, the outputs of these M classifiers are fed into a fusion module that produces the final prediction. The ensemble's modular structure leverages the strengths of all the base models, yielding more noise-robust and generalizable predictors than any individual classifier does.

In the fusion stage, we compute the final output as a weighted average of the base models' prediction probabilities. Specifically, given each model's probability output p_i and an importance weight w_i reflecting that model's reliability, the fused prediction is:

$$y[i] = \sigma \left(\sum_{j=0}^{k-1} w[j] \cdot x[i \times s + j] + b \right) \tag{6}$$

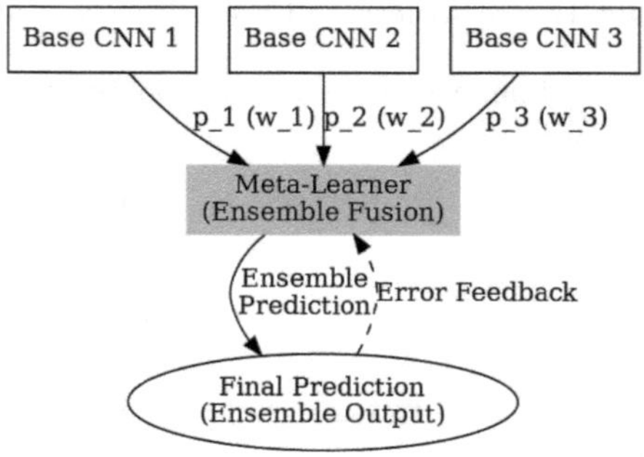

Fig. 2. The weighted fusion ensemble framework.

where $i = 1, \ldots, M$. The weights w_i are assigned based on each model's performance on validation metrics, so that more accurate models (especially those excelling on key imbalanced-class metrics) contribute more to the final prediction. In our binary classification problem with severe class imbalance, this means emphasizing models with higher AUC and AUPR scores to ensure positive instances are correctly identified, while still considering overall accuracy, sensitivity, and specificity. By normalizing and allocating weights in this manner, the ensemble prioritizes the most reliable classifiers and mitigates the impact of any single model's biases or errors. This weighted fusion strategy significantly boosts overall performance: the combined model achieves higher AUC, AUPR, and accuracy than any base model alone, demonstrating the efficacy of our ensemble approach in leveraging multiple classifiers to attain superior predictive results.

4 Experiment and Analysis

4.1 Dataset Description and Interaction Encoding

The original phage–host interaction datasets[1] used in this study were provided by BGI's "Research on Artificial Phage Model Construction Based on Deep Generative Adversarial Network Learning" projectwith interactions recorded as binary labels (presence or absence of infection). In both datasets, each phage and each host is represented solely by one key protein's amino acid sequence (provided in FASTA format). Specifically, each phage is represented by the amino acid sequence of its receptor-binding protein, and each bacterial host is represented by the sequence of its corresponding surface receptor protein. These protein sequences serve as the unique identifiers for phages and hosts in our model's input. These data consist of two groups: the first group includes 101 bacterial hosts and 129 phages, with interactions quantified by real-valued infection scores, whereas the second group comprises 38 hosts and 176 phages with interactions

[1] Dataset in CNSA: CNP0006217, https://db.cngb.org/search/project/CNP0006217/.

recorded as binary labels (presence or absence of infection). In both datasets, each phage and each host is represented solely by its amino acid sequence (provided in FASTA format), and the host–phage interactions are organized in matrix form with entries corresponding to either a real-valued infection score or a binary indicator.The original data of both hosts and phages contain amino acid sequences with faa files, and their interactions are in the form of matrices, as shown in Table 5–6 in Appendix. One of the phage's proteins (a tail fiber protein) has the amino acid sequence METNSDNIKLLNN... (partial), which we use to compute composition-based features.

4.2 Feature Extraction and Sample Rebalancing

In this stage, we derive informative features from each phage and host protein sequence and construct labelled phage–host interaction (PHI) samples. For every amino acid sequence in the two PHI datasets, we compute three feature sets: amino acid composition (AAC), abundance of chemical components (AC), and molecular weight (MW). These features are calculated via standard bioinformatics formulas (as defined earlier) to yield a fixed-length numerical vector for each protein. The resulting feature vectors are saved in a structured text for mat. In parallel, we generate interaction labels from the two datasets. The first dataset provides a continuous "immersion score" for each phage–host pair; we label a pair as a positive interaction if its score exceeds 1.5 (indicative of infection) and as negative otherwise. The second dataset directly uses binary indicators, with "1" denoting an interaction. All identified phage–host pairs and their labels are recorded in a standardized list. Through this feature extraction and labelling process, we obtained 1,414 positive PHI samples and 18,303 negative samples, revealing a severe class imbalance in favour of negative instances.

To address the positive/negative imbalance, we employed a mixed sampling strategy that combines undersampling and oversampling to rebalance the training data. Traditional undersampling (randomly removing negatives) can discard informative samples, whereas oversampling (duplicating positives) risks overfitting. In our mixed approach, we retain all positive samples and make use of the large negative pool by creating multiple balanced subsets. Specifically, the negative samples are partitioned into 12 subsets of approximately 1,500 negatives each. We then combine each negative subset with all 1,414 positives to form 12 balanced training groups (each with 2,914 samples, approximately a 1:1 class ratio). In this way, every positive example appears in every group (ensuring that rare positive patterns are always learned), whereas each negative example is used in only one group (greatly reducing redundant negative instances per model). This mixed sampling technique effectively preserves informative data from both classes and mitigates the pitfalls of single-method resampling. It avoids losing important negative samples (since all negatives are used across the 12 groups) and prevents any single model from overfitting duplicated positives. Moreover, by yielding 12 balanced subsets, this approach enables the training of twelve separate base models on different data folds. These models can later be integrated via ensemble learning, taking advantage of diverse training subsets to increase the overall generalization. Figure ?? and ?? support this process by illustrating the standardized input features and the labeling scheme, respectively, which are foundational for the subsequent model training on balanced data.

4.3 Comparative Analysis of Basic Models

Given the sequential nature of protein data, we first evaluated several deep learning architectures to identify a suitable baseline model for PHI prediction. In particular, we compared a one-dimensional convolutional neural network (CNN), a recurrent neural network (RNN), a long short-term memory network (LSTM), and a transformer encoder. Each of these architectures offers distinct advantages for sequence-based tasks. CNNs excel at capturing local patterns in fixed-length feature sequences and are computationally efficient because of weight sharing and local connectivity. They can handle multidimensional input features and exploit spatially local correlations (in our case, motif or composition patterns along the feature vector) with translation invariance. RNNs, on the other hand, process inputs in sequence, carrying hidden states that encode prior context; this design makes RNNs well suited for modelling temporal or sequential dependencies in data. However, standard RNNs can struggle with long-term dependencies as gradients propagate through many time steps. LSTMs extend RNNs with gating mechanisms that regulate information flow, effectively prescribe long-range context and alleviate vanishing gradient issues. This allows LSTMs to maintain the memory of earlier inputs and capture dependencies over longer protein sequences better than basic RNNs do. Transformers take a further step by employing a self-attention mechanism to capture global relationships across the entire sequence in parallel. They can learn long-range interactions without the need for sequential processing, which in theory makes them useful for understanding complex patterns in sequences. However, transformers typically require sufficiently large datasets and careful regularization, as their flexibility comes at the cost of many parameters. We considered all four model types given their potential to learn PHI-relevant patterns from protein sequence features, and we carried out a comparative experiment to determine which is most effective for our data.

To compare and find the most suitable model, preliminary training of the CNN model, RNN model, LSTM model and transformer model is completed with the same training set and test set. According to the previous subset classification rules, each subset contains 2914 data points, which are divided into training and testing sets at a 9:1 ratio. To evaluate and mitigate overfitting, we ensured that all model performance metrics were reported on these held-out test sets rather than the training data. We observed that the test-set performance closely tracked the training performance, suggesting that the model generalizes well and is not overfitting. The structures of these models are shown in Table 2. Overall, the training results were evaluated in a macroaverage manner, as shown in Fig. 3. In summary, compared with other models, the CNN is more suitable as the basic model for predicting the interaction between bacteriophages and their hosts.

Table 2 and Fig. 3 summarize the performance of these baseline models on the PHI prediction task, facilitating a direct comparison of their suitability. In

Table 2. Parameters of basic models

model	main structure
CNN	Conv1D
LSTM	LSTM
RNN	SimpleRNN
Transformer	Multi-Head-Attention

our evaluations, the 1D CNN achieved the best overall results, outperforming the RNN, LSTM, and transformer models in key metrics (e.g., highest AUC and accuracy). The CNN's ability to extract and combine local feature motifs contributed to superior classification of phage–host pairs, despite the limited sequence length of the feature vectors. In contrast, the RNN and transformer models underperformed on this task. This is likely because our engineered feature vectors do not retain an explicit sequential order (they are aggregate descriptors of entire proteins), meaning that there are no long-term temporal dependencies for RNN/LSTM models to harness and no positional context for the transformer to exploit in particular. Indeed, both the simple RNN and the self-attention transformer failed to show an advantage in this scenario, given the lack of temporal structure in the input features and the relatively small number of positive examples.

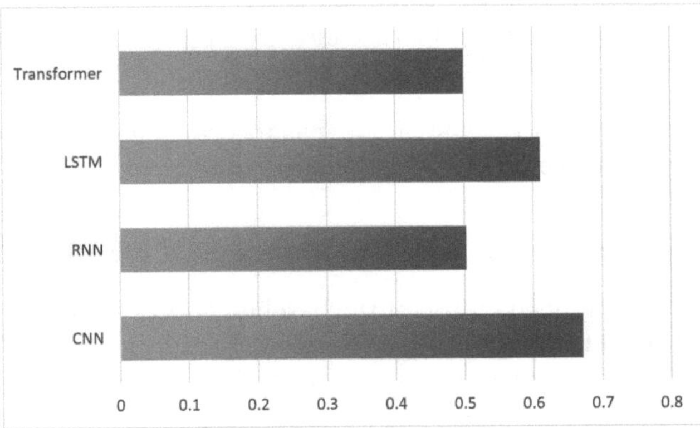

Fig. 3. Comparison of Basic Models.

The LSTM network fared slightly better than the plain RNN by mitigating sequence-length issues, but its performance still fell between that of the CNN and RNN, indicating that its long-memory capability does not provide a significant benefit for our feature representation.

On the basis of these results, we selected the CNN as the baseline model for our framework. The CNN not only demonstrated the highest predictive performance in our tests but also aligns well with the nature of our data, leveraging its strengths in pattern recognition without relying on sequential dependencies. Therefore, for all subsequent experiments, we adopted the 1D CNN as the core model architecture, which offered an optimal balance of accuracy and efficiency for PHI prediction.

On the basis of the convolutional layers, the proposed CNN architecture involves batch normalization, max pooling, dropout, and a global average pooling layer, as summarized in Table 1. The convolutional layers employ the ReLU activation function to introduce nonlinearity, defined as:

$$f(x) = \max(0, x) \tag{7}$$

The final output layer uses the softmax function to convert logits into class probabilities, calculated as:

$$f(x_i) = \frac{e^{x_i}}{\sum_j e^{x_j}} \tag{8}$$

During training, model parameters are updated using the Adam optimizer, according to:

$$\theta_{t+1} = \theta_t - \frac{\eta_t}{\sqrt{\hat{v}_t} + \epsilon} \hat{m}_t \tag{9}$$

The training process minimizes the binary cross-entropy loss, formulated as:

$$L = -\frac{1}{N} \sum_{i=1}^{N} [y_i \log(\hat{y}_i) + (1 - y_i) \log(1 - \hat{y}_i)] \tag{10}$$

This carefully designed CNN structure ensures rapid convergence and robust classification performance on highly imbalanced datasets.

4.4 Evaluation of Ensemble Strategies for PHI Prediction

Ensemble learning is applied on top of the baseline 1D CNN classifier to enhance the PHI prediction performance. In our framework, the single 1D CNN serves as the benchmark, and we explore four ensemble fusion strategies: weight selection, soft voting, stacking, and meta-features. Weight selection directly chooses the best-performing CNN, whereas soft voting averages probability outputs across CNN models. The stacking approach trains a logistic regression meta-classifier on base CNN outputs, and the meta-feature method trains an XGBoost classifier using averaged CNN predictions as features.

$$Accuracy = \frac{TP + TN}{TP + TN + FP + FN} \tag{11}$$

$$TPR = \frac{TP}{TP + FN}$$
$$FPR = \frac{FP}{FP + TN} \tag{12}$$
$$AUC = \int_0^1 TPR(FPR), d(FPR)$$

$$Precision = \frac{TP}{TP + FP}$$
$$Recall = \frac{TP}{TP + FN} \tag{13}$$
$$AUPR = \int_0^1 Precision(Recall), d(Recall)$$

$$Sensitivity = \frac{TP}{TP + FN} \tag{14}$$

$$Specificity = \frac{TN}{TN + FP} \tag{15}$$

To systematically evaluate these strategies, we adopt five metrics: Accuracy (Eq. 11), AUC (Eq. 12), AUPR (Eq. 13), Sensitivity (Eq. 14), and Specificity (Eq. 15). The primary indicators, AUC and AUPR, capture the model's overall discriminative power, whereas Sensitivity and Specificity provide detailed insights into classification performance for positive and negative samples, respectively. The comparative performance of the ensemble methods is presented in Fig. 9 (Fig. 4).

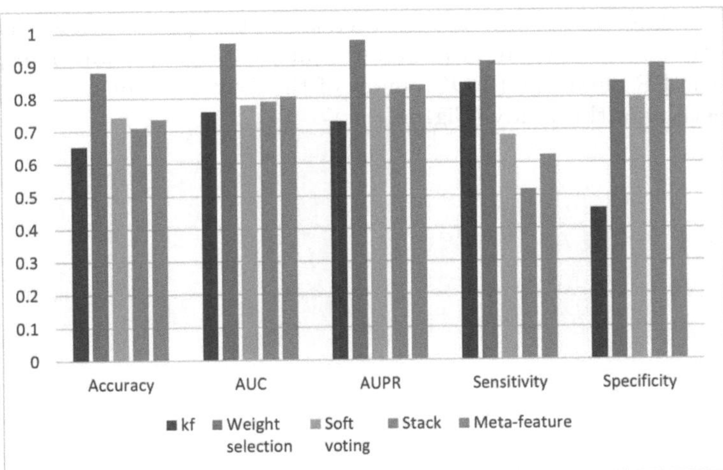

Fig. 4. Precision-Recall(PR) Curve of Four Different Models for Comparison.

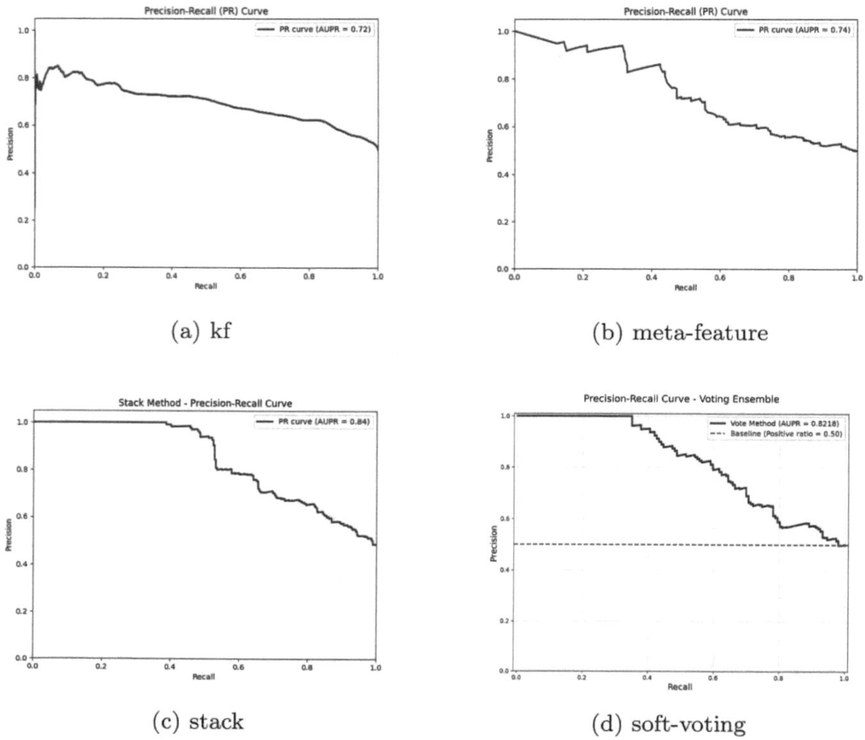

Fig. 5. Precision-Recall(PR) Curve of Four Different Models for Comparison.

The area under the PR curve is called AUPR, which measures a model's precision across different recall levels; the area under the ROC curve is called AUC, reflecting the model's overall discriminative ability. Figures 5 and 7 respectively compare the PR curves and ROC curves of the four different models using the Weight method in this study (Fig. 6).

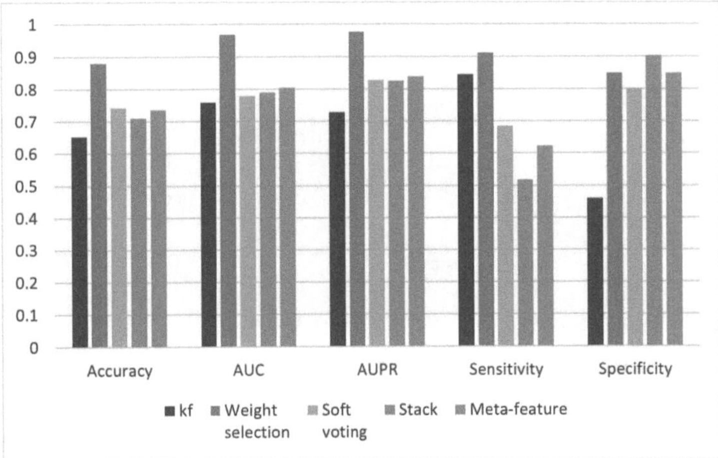

Fig. 6. Receiver Operating Characteristic(ROC) Curve of Four Different Models for Comparison.

As illustrated in Fig. 9, all ensemble approaches outperform the single CNN baseline, validating the benefit of ensemble learning. Notably, weight selection consistently achieves superior performance across key metrics, with AUPR of 0.976, AUC of 0.968, Accuracy of 0.88, and the highest Sensitivity at 0.911. By contrast, stacking and meta-feature methods exhibit limitations, particularly in Sensitivity, while soft voting demonstrates stable yet slightly inferior results. Consequently, the weight selection strategy emerges as the optimal ensemble approach due to its robust, balanced, and superior predictive capabilities.

Furthermore, we present the ROC curve of the proposed ensemble CNN model on the independent test set. The curve yields an area under the curve (AUC) of 0.968, which confirms the model's high overall accuracy in distinguishing phage–host pairs. This instance-level evaluation illustrates that our classifier maintains a high true positive rate for a given low false positive rate, indicating robust performance (Fig. 8).

4.5 Hyperparameter Ablation

To validate our choice of model hyperparameters, we conduct ablation experiments by varying one parameter at a time while keeping others fixed. For example, we evaluated the impact of the 1D CNN kernel size on performance.Table 3

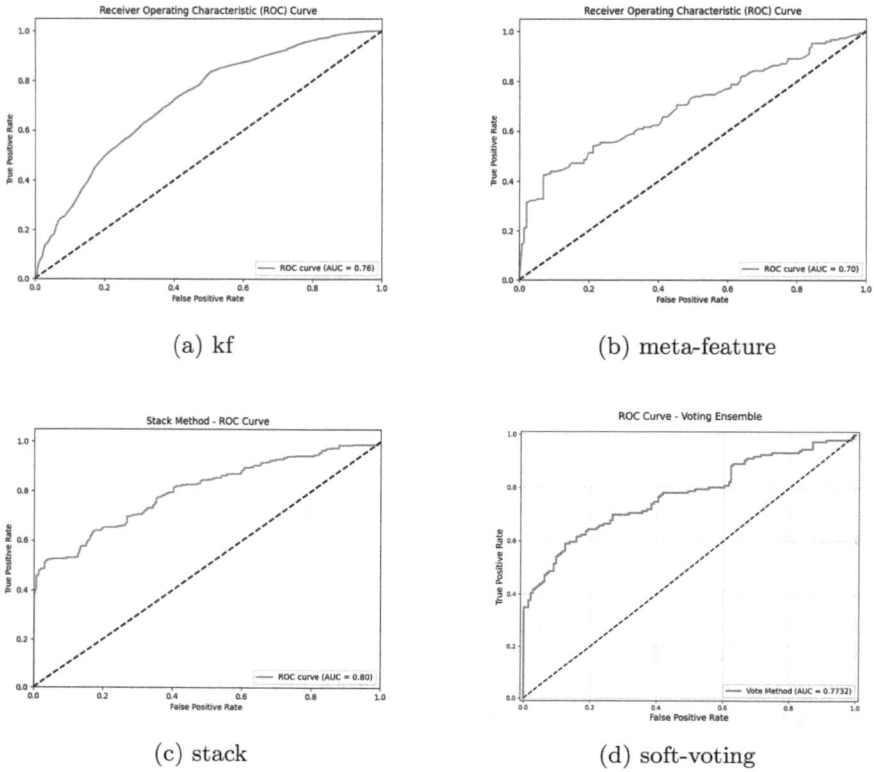

Fig. 7. Receiver Operating Characteristic(ROC) Curve of Four Different Models for Comparison.

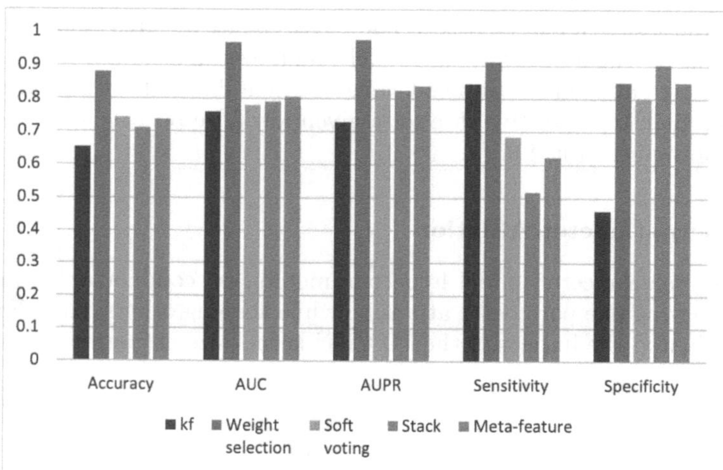

Fig. 8. Evaluation results of different model fusion strategies.

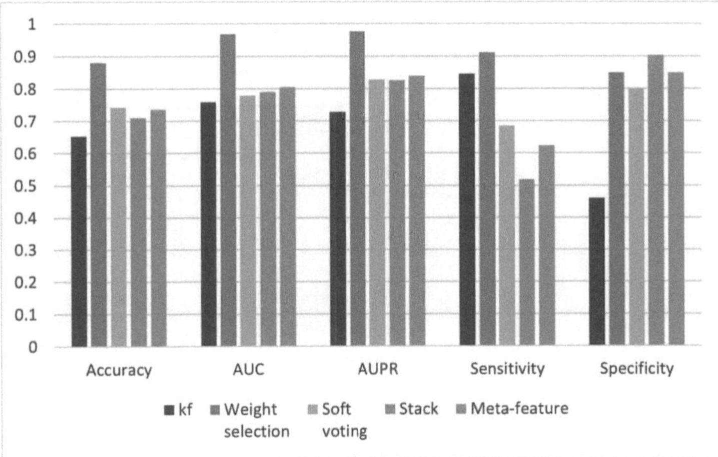

Fig. 9. Evaluation results of different model fusion strategies.

summarizes the model's precision-recall AUC (AUPR) under different kernel sizes (with the default kernel size in bold). We observe that a kernel size of 5 yields the best AUPR, justifying our default setting.

Table 3. Performance (AUPR) under different CNN kernel sizes.

Kernel Size	AUPR
3	0.78
5 (default)	**0.80**
7	0.79

In addition, we perform similar tests for other hyperparameters (e.g., the number of filters, the learning rate, and the dropout rate) and find that the chosen values consistently offer the best trade-off between performance and overfitting. These ablation results confirm that our hyperparameter selections are empirically supported.

The training results for different numbers of neurons were tested individually. The recorded comparison accuracy rates, AUPRs and ROCs are shown in Fig. 10. The number of neurons has no significant influence on the training results of the model. Considering both the aupr time and the training time comprehensively, 256 was selected as the model in this study.

4.6 Comparison with State-of-the-Art PredPHI Method

Our proposed model is compared with PredPHI, a representative state-of-the-art method, in terms of accuracy (Eq. 11), specificity (equivalence 15), sensitivity

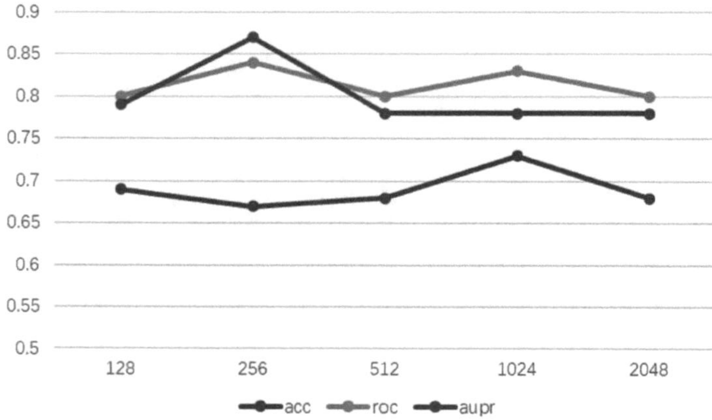

Fig. 10. Training results with different numbers of neurons.

(Eq. 14), and the Matthews correlation coefficient (MCC, Eq. 16). The MCC comprehensively assesses binary classification quality by integrating all confusion matrix components.

$$\text{MCC} = \frac{TP \times TN - FP \times FN}{\sqrt{(TP + FP)(TP + FN)(TN + FP)(TN + FN)}} \tag{16}$$

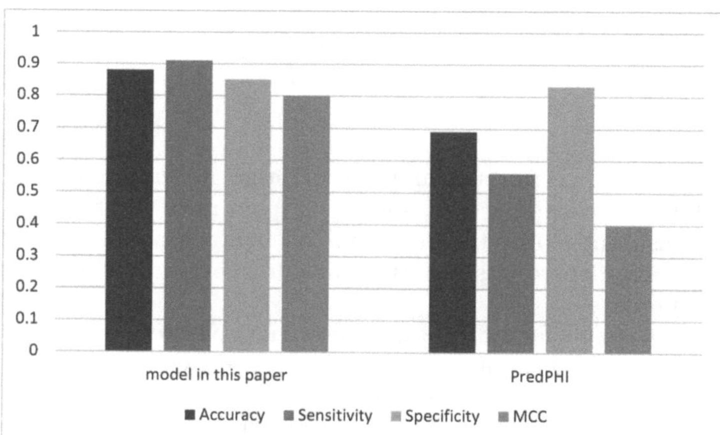

Fig. 11. Comparison between the model in the paper and PredPHI.

Figure 11 demonstrates that our ensemble CNN significantly outperforms PredPHI in these metrics, with Accuracy increased by approximately 20%, and

Sensitivity and MCC enhanced by roughly 40%, while maintaining comparable Specificity. And we broadened the baseline comparisons to include traditional models. In addition to the deep learning models (1D CNN, LSTM, RNN, Transformer), we evaluated a simple logistic regression classifier and a feed-forward neural network (a multi-layer perceptron) using the same input features.These two traditional methods perform substantially worse than the advanced models. For instance, logistic regression achieves an F1 score of only 0.60 and an MCC of 0.42, significantly lower than the 1D CNN and our ensemble. The feed-forward neural network baseline improves upon logistic regression (F1= 0.65) but still underperforms the recurrent and convolutional models. This comparison reveals the advantages of our deep learning ensemble approach over simpler classifiers.

This substantial improvement arises from several methodological advantages of our approach, including the innovative mixed sampling strategy addressing class imbalance, a specialized 1D CNN tailored for amino acid sequence features, the strategic use of dropout regularization, and the adoption of a performance-driven weight selection ensemble. Collectively, these enhancements render our method notably more effective and reliable for accurately predicting phage–host interactions than does PredPHI.

5 Conclusion

In this work, we presented a deep learning framework for phage–host interaction prediction that integrates effective protein sequence feature extraction, a tailored one-dimensional CNN architecture, a hybrid sample balancing strategy and an ensemble fusion approach to enhance predictive performance. The experimental results demonstrate that our model achieves 88% accuracy and 91.1% sensitivity, with an AUC of 0.968 and an AUPR of 0.976. These metrics substantially exceed those of the previous state-of-the-art,PredPHI, which includes an approximately 40% higher MCC, confirming that this model outperforms PredPHI. In the future, the proposed approach can be extended by incorporating additional features or applying it to larger phage-host datasets, potentially further improving the prediction accuracy and broadening its applicability. Furthermore, as a future direction, we envision integrating our approach with generative models (e.g., adversarial networks) to facilitate the design of novel phage–host interactions, thereby extending the utility of our framework in therapeutic phage development.

Acknowledgments. This research was supported by the Beijing Genomics Institute under grant numbers 62203134, 6220020329 and 62176164.Scientific Foundation for Youth Scholars of Shenzhen University 868-000001033385. The author gratefully acknowledges the support provided by the "Design of phage tail protein based on adversarial machine learning" project and the "Construction of artificial phage model based on deep generative adversarial network learning" project. And the datasets is provided by BGI-Shenzhen.

Disclosure of Interests. The authors have no competing interests to declare that are relevant to the content of this article.

A Appendix

Table 4. Common amino acids and their characteristics

Abbreviation	Full Name	Elemental Content					MW
		C	H	O	N	S	
A	Alanine	3	7	2	1	0	71.08
C	Cysteine	3	7	2	1	1	103.01
D	Aspartic acid	4	7	4	1	0	115.03
E	Glutamic acid	5	9	4	1	0	129.04
F	Phenylalanine	9	11	2	1	0	147.07
G	Glycine	2	5	2	1	0	57.02
H	Histidine	6	9	2	3	0	137.14
I	Isoleucine	6	13	2	1	0	113.16
K	Lysine	6	14	2	2	0	128.09
L	Leucine	6	13	2	1	0	113.16
M	Methionine	5	11	2	1	1	131.04
N	Asparagine	4	8	3	2	0	114.04
P	Proline	5	9	2	1	0	97.05
Q	Glutamine	5	10	3	2	0	128.06
R	Arginine	6	14	2	4	0	156.18
S	Serine	3	7	3	7	0	87.03
T	Threonine	4	9	3	1	0	101.05
V	Valine	5	11	2	1	0	99.07
W	Tryptophan	11	12	2	2	0	186.08
Y	Tyrosine	9	11	3	1	0	163.06

Table 5. Interaction matrix in the first group data

	PH-KP9072-1	PH-KP9095-1	PHKP9283_B1	PH-KP5431	PH-KP7102	PH-KP9052-A1	...
4023	0	0	5.44	0	0	5.29	...
4049	0	0	5.00	0	0	6.17	...
4547	0.57	0.55	0	0	4.25	0.88	...
4829	0	0	0	0	0	0	...
4956	0	0	0	0	0	0	...
5048	0	2.25	0	0	0	0	...
...	...	...	...	...	...	...	...

Table 6. Interaction matrix in the second group data

	K. aerogenes	K. aerogenes N1	K. pneumoniae	K. pneumoniae 36	...
K. phage 1 LV-2017	0	0	0	0	...
K. phage 1 TK-2018	0	0	1	0	...
K. phage 117	0	0	0	1	...
K. phage 13	0	0	0	0	...
K. phage 1611E-K2-1	0	0	1	0	...
K. phage 2 LV-2017	0	0	0	0	...
...	...	...	...	...	...

References

1. Aggarwal, S., Dhall, A., Patiyal, S., Choudhury, S., Arora, A., Raghava, G.P.: An ensemble method for prediction of phage-based therapy against bacterial infections. Front. Microbiol. **14**, 1148579 (2023)
2. Akbar, S., Ullah, M., Raza, A., Zou, Q., Alghamdi, W.: Deepaips-pred: predicting anti-inflammatory peptides using local evolutionary transformation images and structural embedding-based optimal descriptors with self-normalized bitcns. J. Chem. Inf. Model. **64**(24), 9609–9625 (2024)
3. Boeckaerts, D., et al.: Prediction of klebsiella phage-host specificity at the strain level. Nat. Commun. **15**(1), 4355 (2024)
4. Chor, B., Horn, D., Goldman, N., Levy, Y., Massingham, T.: Genomic dna k-mer spectra: models and modalities. Genome Biol. **10**, 1–10 (2009)
5. Coclet, C., Roux, S.: Global overview and major challenges of host prediction methods for uncultivated phages. Curr. Opin. Virol. **49**, 117–126 (2021)
6. Coutinho, F.H., et al.: Rafah: host prediction for viruses of bacteria and archaea based on protein content. Patterns **2**(7) (2021)
7. Dicks, L.M., Vermeulen, W.: Bacteriophage-host interactions and the therapeutic potential of bacteriophages. Viruses **16**(3), 478 (2024)
8. Edwards, R.A., McNair, K., Faust, K., Raes, J., Dutilh, B.E.: Computational approaches to predict bacteriophage-host relationships. FEMS Microbiol. Rev. **40**(2), 258–272 (2016)
9. Friedman, D.I., Olson, E., Georgopoulos, C., Tilly, K., Herskowitz, I., Banuett, F.: Interactions of bacteriophage and host macromolecules in the growth of bacteriophage lambda. Microbiol. Rev. **48**(4), 299–325 (1984)
10. Gaborieau, B., et al.: Prediction of strain level phage-host interactions across the escherichia genus using only genomic information. Nat. Microbiol. **9**(11), 2847–2861 (2024)
11. Galiez, C., Siebert, M., Enault, F., Vincent, J., Söding, J.: Wish: who is the host? predicting prokaryotic hosts from metagenomic phage contigs. Bioinformatics **33**(19), 3113–3114 (2017)
12. Gonzales, M.E.M., Ureta, J.C., Shrestha, A.M.: Protein embeddings improve phage-host interaction prediction. PLoS ONE **18**(7), e0289030 (2023)
13. Hille, F., Charpentier, E.: Crispr-cas: biology, mechanisms and relevance. Philosophical Trans. Royal Soc. B: Biolog. Sci. **371**(1707), 20150496 (2016)

14. Li, M., et al.: A deep learning-based method for identification of bacteriophage-host interaction. IEEE/ACM Trans. Comput. Biol. Bioinf. **18**(5), 1801–1810 (2020)
15. Marks, T., Sharp, R.: Bacteriophages and biotechnology: a review. J. Chemical Technol. Biotechnol. Int. Res. Process Environ. Clean Technol. **75**(1), 6–17 (2000)
16. Nie, W., Qiu, T., Wei, Y., Ding, H., Guo, Z., Qiu, J.: Advances in phage–host interaction prediction: in silico method enhances the development of phage therapies. Briefings Bioinform. **25**(3), bbae117 (2024)
17. Pan, J., You, W., Lu, X., Wang, S., You, Z., Sun, Y.: Gsphi: a novel deep learning model for predicting phage-host interactions via multiple biological information. Comput. Struct. Biotechnol. J. **21**, 3404–3413 (2023)
18. Raza, A., Uddin, J., Zou, Q., Akbar, S., Alghamdi, W., Liu, R.: Aips-deepenc-ga: Predicting anti-inflammatory peptides using embedded evolutionary and sequential feature integration with genetic algorithm based deep ensemble model. Chemom. Intell. Lab. Syst. **254**, 105239 (2024)
19. Ruohan, W., Xianglilan, Z., Jianping, W., Shuai Cheng, L.: Deephost: phage host prediction with convolutional neural network. Briefings Bioinform. **23**(1), bbab385 (2022)
20. Shahid, Hayat, M., Alghamdi, W., Akbar, S., Raza, A., Kadir, R.A., Sarker, M.R.: Pacp-hybdeep: predicting anticancer peptides using binary tree growth based transformer and structural feature encoding with deep-hybrid learning. Sci. Reports **15**(1), 565 (2025)
21. Shang, J., Sun, Y.: Cherry: a computational method for accurate prediction of virus–prokaryotic interactions using a graph encoder–decoder model. Briefings Bioinform. **23**(5), bbac182 (2022)
22. Ullah, M., Akbar, S., Raza, A., Khan, K.A., Zou, Q.: Targetclp: clathrin proteins prediction combining transformed and evolutionary scale modeling-based multi-view features via weighted feature integration approach. Briefings Bioinform. **26**(1), bbaf026 (2025)
23. Villarroel, J., et al.: Hostphinder: a phage host prediction tool. Viruses **8**(5), 116 (2016)
24. Wang, W., et al.: A network-based integrated framework for predicting virus–prokaryote interactions. NAR Genomics Bioinform. **2**(2), lqaa044 (2020)
25. Ye, J., McGinnis, S., Madden, T.L.: Blast: improvements for better sequence analysis. Nucleic Acids Res. **34**(suppl_2), W6–W9 (2006)
26. Zhou, F., et al.: Phisdetector: a tool to detect diverse in silico phage-host interaction signals for virome studies. Genomics Proteomics Bioinform. **20**(3), 508–523 (2022)

Basic Theory and Techniques for Data Science

LightLoc: Redefining Indoor Localization with an Efficient Spatial-Temporal Learning Framework

Xiaobo Li[1], Zeyao Wang[1], Mengling Kai[2], Yankun Gao[3], Dong Li[4,5], and Hao Yi[4,5](✉)

[1] Zhejiang Informatization and Industrialization Integration Promotion Center, Hangzhou 310007, China
[2] Adsquare Technologies, Hong Kong, China
[3] Beijing Institute of Computer Technology and Application, Beijing, China
[4] The Fifth Electronics Research Institute of the Ministry of Industry and Information Technology of P.R.CHINA, Guangzhou 511300, Guangdong, China
yihao@ceprei.com
[5] The Ministry of Industry and Information Technology Key Laboratory of Performance and Reliability Testing and Evaluation for Basic Software and Hardware, Guangzhou 511300, China

Abstract. Graph neural networks (GNNs) have emerged as promising approaches for graph-based indoor localization. Recent studies have incorporated GNNs to capture the spatial features of sensed data and improve localization accuracy. However, existing methods often struggle in complex indoor environments due to noise and signal instability, leading to suboptimal performance. In this paper, we propose LightLoc, a simple yet effective spatial–temporal framework for accurate and robust indoor localization. Specifically, LightLoc constructs a fingerprint tensor and leverages a bidirectional long short-term memory (Bi-LSTM) network to model the temporal dependencies of fingerprints. The model is designed for easy deployment via common smartphones, eliminating the need for additional hardware. Furthermore, a dual-step graph convolutional network (GCN) is introduced to extract spatial correlations from neighboring fingerprint nodes. Extensive real-world experiments demonstrate that LightLoc achieves superior localization accuracy and enhanced stability compared with existing approaches.

Keywords: Graph neural networks · Indoor localization · Bi-LSTM

1 Introduction

Indoor localization plays a crucial role in modern society because of its ability to accurately determine the location of a target within indoor environments. This capability enables a wide range of location-based services and applications. However, achieving reliable indoor localization remains challenging because of complex environmental factors such as building structures, signal attenuation, multipath effects, obstacles, and environmental noise [1–3].

Y. Yu et al. (Eds.): ICPCSEE 2025, CCIS 2673, pp. 391–403, 2026.
https://doi.org/10.1007/978-981-95-2566-9_26

Significant research efforts have been dedicated to improving localization performance [1, 4]. With the widespread adoption of local area networks (LANs) and mobile devices [5, 6], WiFi-based indoor localization has gained substantial attention. Existing WiFi-based localization methods can be broadly categorized into channel state information (CSI)-based methods and received signal strength (RSS)-based methods. CSI-based localization offers theoretical advantages by providing richer and more precise signal characteristics, meeting the requirements for high-precision positioning. However, these methods incur higher operational costs and have limited applicability because of their dependence on specialized hardware. Conversely, RSS-based localization methods are more prevalent and widely adopted, as they can function with existing WiFi devices.

WiFi-based localization methods can be further classified into model-based methods and fingerprint-based methods. In model-based approaches, the localization accuracy heavily depends on the construction of a precise propagation model, which is often challenging in real-world environments. In contrast, fingerprint-based methods achieve localization by comparing real-time RSS measurements with a precollected fingerprint dataset, offering relatively accurate results. This paper focuses on fingerprint-based localization and proposes a novel framework to improve accuracy.

Traditional WiFi fingerprinting techniques include directional fingerprinting [7], receiver point configuration [8], and the signal angle of arrival (AOA) [9]. However, these methods typically require extensive training or modelling using precollected data to determine device locations. Recently, machine learning and deep learning approaches have emerged [10], enabling more accurate WiFi fingerprinting by learning patterns and characteristics of WiFi signals. Several studies have introduced deep learning-based localization methods, such as deep autoencoders (DAEs) and convolutional neural networks (CNNs), for indoor positioning. For example, Cai et al. [11] proposed a CSI-based device-free indoor localization system that leverages CNNs, whereas Wang et al. [12] introduced a deep learning-based indoor fingerprint recognition system.

Although existing methods attempt to capture spatial and temporal correlations in WiFi fingerprint data, they often struggle to balance short-term and long-term dependencies [15]. RNN/LSTM-based models may face difficulties in processing long time series data, whereas CNN-based models may fail to capture long-term dependencies without compromising short-term information [16–18]. Similarly, subgraph-based or expanding convolutional approaches exhibit limitations in simultaneously capturing both short- and long-term correlations [13, 19]. Consequently, novel methods that effectively balance these considerations and capture the dynamic temporal correlations within localization data are needed.

Despite extensive research on indoor localization, several challenges persist:

Limited Hardware Resources: In environments with simple equipment, scarce manpower, and high noise levels, advanced or specialized sensors—such as inertial measurement units (IMUs) or cameras—are often unavailable for positioning. This limitation impacts localization accuracy and complicates algorithm design.

Environmental Dynamics: Factors such as human movement, object occlusion, and interference sources introduce inconsistencies in WiFi signals and sensor data, leading to variations in positioning accuracy in dynamic environments.

Fingerprint Ambiguity in Complex Indoor Settings: The uniqueness of fingerprint-based localization is often compromised due to the limited number of signal sources, temporal variations in wireless signals, and strong multipath effects caused by walls, furniture, and people. As a result, identical fingerprints may appear at different locations, reducing the localization accuracy.

To address these challenges, we propose LightLoc, a simple yet effective indoor localization model based on fingerprint tensor learning. Specifically, a fingerprint tensor is constructed from raw fingerprint feature extractions. Furthermore, a recurrent neural network—specifically, a bidirectional long short-term memory (Bi-LSTM) network—is employed to increase the localization accuracy. Using the collected RSS measurements as inputs, we first reconstruct the fingerprint dataset through a masking approach, effectively mitigating location estimation errors caused by fingerprint similarity. The resulting fingerprint tensor is then processed by a Bi-LSTM network, which captures the temporal dependencies of each slice within the tensor. Through this approach, LightLoc significantly improves the robustness and accuracy of indoor localization.

Our key contributions are as follows:

(1) **Efficient Fingerprint Tensor Construction:** We propose an optimized fingerprint tensor construction method that enhances the native fingerprint dataset by incorporating optimal fingerprints. This approach increases fingerprint diversity and effectively mitigates location estimation errors caused by similar fingerprints.
(2) **Bi-LSTM-based Temporal Feature Learning:** We develop a bidirectional long short-term memory (Bi-LSTM) neural network-based fingerprint tensor training model to capture the temporal dependencies embedded within fingerprint slices. This model leverages the forget-and-retain mechanism of Bi-LSTM to improve localization accuracy.
(3) **Lightweight Prediction and Deployment:** We introduce a streamlined prediction process for location estimation via the trained model and propose a practical implementation strategy, ensuring efficient deployment in real-world scenarios.
(4) **Real-World Experimental Validation:** We construct an experimental platform using an Android smartphone to collect real-world data. Extensive experiments are conducted to evaluate the effectiveness and superiority of the proposed algorithm and to validate the reliability of the fingerprint tensor approach. The remainder of this paper is organized as follows: Sect. 2 reviews related work. Section 3 presents the design and implementation of the localization system. Section 4 provides a detailed description of our light–loc model. Section 5 evaluates the performance of our model. Finally, Sect. 6 concludes this work.

2 Related Work

Recent years have seen significant attention and progress in indoor positioning technology. Pascacio et al. proposed a collaborative indoor positioning strategy that uses neural networks to integrate BLE-RSS signals [9]. Wei et al. presented a semantic spatial model for indoor semantic localization, leveraging artificial intelligence techniques for semantic pattern matching, concept expansion, reasoning, and mapping [20]. Bi et al. introduced a novel indoor localization framework that uses mobile nodes as virtual anchors, demonstrating high positioning accuracy and low cost in experiments [21].

Abbas et al. developed WiDeep, a method that combines a stacked denoising autoencoder deep learning model with a probabilistic framework, maintaining high precision and robustness even in noisy environments [22]. Li et al. designed a K-nearest neighbor strategy for indoor fingerprint localization, which uses a coarse localization circle domain and the highest similarity threshold [23]. Zhang et al. proposed the innovative WiDAGCN model, which is capable of training with both labelled field survey data and unlabelled crowd-sensing Wi-Fi fingerprints [24]. Jung developed a novel visual localization system (VPS) that determines the user's location by matching the pose information labels of voxel database objects with estimations from an enhanced point map neural network (GNN) [25]. Although the AoA algorithm depends on Wi-Fi maps, Zhong et al. introduced the precise data aggregation strategy DAST, which incorporates long-tail features of time series [26]. Data aggregation has been considered a fundamental operation for indoor positioning in wireless networks [27]. Saccomanno et al. introduced a deep learning-based approach for indoor localization, which mitigates the adverse effects of signal strength utilization by minimizing its consideration. This method exclusively uses signal strength to construct a ranking-oriented depiction of fingerprint-associated access points, optimizing the localization process [28].

Wi-Fi fingerprinting has garnered significant interest because of the widespread availability of WLANs and the increasing use of mobile devices. Various methodologies have been applied in positioning systems, including conventional statistical and numerical analysis [29], advanced machine learning algorithms [30], and the current trend of deep learning algorithms [31, 32], all of which have yielded impressive results. Li et al. proposed a novel passive localization system that uses multiple sniffers to monitor Wi-Fi traffic and determine the location of a standard Wi-Fi transmitter by precisely measuring the arrival time [33]. Rao et al. introduced DFPhaseFL, a device-free fingerprint recognition indoor localization system that relies solely on CSI phase information [34]. Caso et al. developed ViFi, an indoor localization system that uses an MWMF propagation model and RSS prediction to generate discrete RSS radiation maps, also known as virtual fingerprints [35].

These advancements in Wi-Fi fingerprinting highlight the diverse and evolving techniques that are enhancing localization systems and improving their precision in different scenarios.

3 Methodology

In this section, we provide a comprehensive overview of the technical aspects of our proposed localization framework. The general structure of the LightLoc model, as presented in this paper, is illustrated in Fig. 1. The model consists of four integral modules: the fingerprint collection module, the fingerprint dataset construction module, the LightLoc module, and the location prediction module. The following paragraphs offer a more detailed description of the key components that constitute our model.

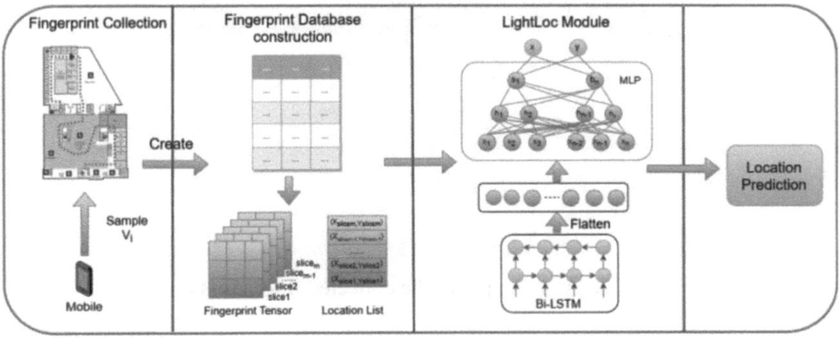

Fig. 1. The localization procedure consists of four components: fingerprint collection, fingerprint database construction, the LightLoc module, and position prediction.

3.1 Fingerprint Collection

To collect fingerprint data for indoor localization, a mobile device (such as an Android phone) is used to navigate the indoor environment. The process involves the use of an acquisition application or software that records the signal strength from multiple access points (APs) at regular intervals, typically every 5–10 s, depending on the specific needs of the study. For each data collection point, the mobile device records the signal strength (e.g., RSSI values) from each AP along with its corresponding location, which is either manually entered or tracked via GPS, indoor positioning systems, or other localization techniques.

As the mobile device moves through the environment, the software continuously logs the signal strength from each AP, forming a "fingerprint" at each location. This fingerprint represents the unique combination of signal strengths from the various APs observed at that point in space. When the device encounters significant landmarks or points of interest, the software can be programmed to record the corresponding landmark location, which is later used for reference.

The collection process generates a raw fingerprint dataset, where each fingerprint record typically consists of a vector of signal strength values. These records may need further processing, such as normalization, filtering, or feature extraction, before they are used for localization tasks. For example, signal fluctuations or environmental noise may require smoothing techniques to improve accuracy. Figure 2 provides an example of the collected fingerprint data, where each data point corresponds to a unique location with its associated AP signal strength values.

3.2 Fingerprint Database Construction

In the dynamic environment where the user moves, the fingerprint information changes the location transitions during the walking process. This implies a temporal sequence between the preceding and subsequent fingerprints. To capture this sequence, we aim to associate the fingerprint collected when approaching an indoor landmark with the fingerprint information from previous moments, thereby extending the original 1D fingerprint to a 2D representation.

id	AP1	AP2	AP3	AP4	AP5	X	Y
1	V11	V12	V13	V14	V15	x_1	y_1
2	V21	V22	V23	V24	V25	x_2	y_2
3	V31	V32	V33	V34	V35	x_3	y_3
....							
i	Vi1	Vi2	Vi3	Vi4	Vi5	x_i	y_i
....							
N	Vn1	Vn2	Vn3	Vn4	Vn5	x_n	y_n

Fig. 2. An example of the collected fingerprints.

Utilizing the native fingerprint library as the foundation, we maintain a sliding window, denoted as "win," with a height of "hE." The window gradually moves downwards through the native fingerprint library, advancing by one unit at a time. If the position information of the last row within the window is empty, the window continues its downwards sliding motion. However, when the position information of the last row is not empty, we extract the current window, exclude the last two columns, and save it as a slice in the fingerprint tensor. Simultaneously, we store the position information of the last row as the positional data for that specific slice.

Supposing that the sliding window in the native fingerprint library allows for the extraction of m valid fingerprint slices, we can construct a fingerprint tensor with dimensions of. Each tensor slice is associated with the position information of its last row, forming the position list.

By employing this approach, we can significantly expand the diversity of the native fingerprints. Furthermore, since the native fingerprints are collected while in motion, the fingerprint sequence within each slice encompasses a positional relationship between the preceding and subsequent fingerprints. As a result, the constructed fingerprint tensor inherently incorporates temporal information within each slice.

3.3 LightLoc Module

Each fingerprint slice within the fingerprint tensor inherently contains temporal information, encompassing the fingerprint data from the current moment as well as the fingerprint information from preceding moments. This amalgamation forms a complete sequence of fingerprints spanning a specific period. For data with implicit temporal information, bidirectional long- and short-term memory neural networks [36] (Bi-LSTM) have demonstrated superior performance. The neurons in the Bi-LSTM layer propagate information from the current moment to the subsequent moments, incorporating mechanisms for forgetting and retention. Bi-LSTM models are widely utilized in the field of natural language processing because of their ability to capture the contextual relationships between words, considering both the forward and backwards directions of word order. In this chapter, we employ the Bi-LSTM model to capture the implied temporal information within the fingerprint slices during the training of the fingerprint tensor. Notably, since each fingerprint slice in our constructed fingerprint tensor is already two-dimensional, it can be directly utilized as input for the network.

$$X = \left[\left[v_{11}^{(1)}, \ldots, v_{15}^{(1)}\right] \ldots \left[v_{h1}^{(1)}, \ldots, v_{h5}^{(1)}\right] \ldots \left[v_{11}^{(m)}, \cdots, v_{15}^{(m)}\right] \ldots \left[v_{h1}^{(m)}, \cdots, v_{h5}^{(m)}\right]\right]^T \tag{1}$$

$$Y = \left[\left[x_1, y_1\right] \ldots \left[x_m, y_m\right]\right]^T \tag{2}$$

The model input is the constructed fingerprint tensor, and the output is a list of positions, as shown in Eqs. (1) and (2). Where m is the number of slices in the fingerprint tensor and h is the height of the sliding window win. After the model is built, the constructed fingerprint tensor is used as the training set input for LigntLoc training, and the model with the lowest loss on the validation set in the specified epoch is selected as the final LigntLoc model.

3.4 Location Prediction

On the basis of the regression model trained in Sect. 4, the obtained fingerprint slices during the localization process (which may not contain landmarks) can be input into the LightLoc model, enabling the retrieval of the predicted localization results. In practical applications, this algorithm facilitates intermovement location, where the user navigates their surroundings and is equipped with a mobile device installed with a client app. The phone continuously captures real-time fingerprint information, which is temporarily stored in a cache. When the user requires an estimation of their current location, the Android phone app ends the fingerprints of the present location with those of the preceding h-1 moments. These fingerprints are then arranged in chronological order to construct the fingerprint slices, which are subsequently transmitted to the base station server via the HTTP protocol. Given that the optimal model is stored on the base station server, the model processes the input slice data and generates the corresponding location estimation result. The base station server subsequently encapsulates the results into a JSON string before transmitting them to the Android phone for user presentation.

4 Experiment

In this section, we conduct a series of experiments to evaluate the effectiveness and superiority of the proposed model.

4.1 Experimental Setting

To validate the proposed algorithm, we developed an experimental platform consisting of an Android-based client application and a Java-based base station server system. The localization model was trained via Python, and the collected data were stored in a lightweight relational database, MySQL. The experimental environment is approximately rectangular and includes a variety of static and dynamic obstacles. Additionally, frequent human movement introduces environmental noise, further increasing the complexity of localization. In our setup, we deployed a single base station, five access points

(APs), and 60 reference landmarks within the designated area. During data collection, a total of 1000 fingerprint samples were recorded. Among them, 30% were allocated for training, whereas the remaining 70% were reserved for testing. The experiments were designed to assess the effectiveness of the proposed fingerprint tensor representation. Additionally, we conducted a comparative analysis between our algorithm and several baseline methods to evaluate its performance and demonstrate its advantages.

4.2 Evaluation Metrics

We calculate the positioning error by using the Euclidean distance between the real position and the estimated result. The evaluation metrics include the following:

- Mean absolute error. The average positioning error for all test points is denoted as:

$$MAE = \frac{1}{S}\sum_{r=1}^{S}\left[(\hat{x}_r - x_r)^2 + (\hat{y}_r - y_r)^2\right]^{\frac{1}{2}} \tag{3}$$

- Mean square error. The MSE measures the stability of the model and is defined as:

$$MSE = \frac{1}{S}\sum_{r=1}^{S}(ERR(r) - MAE)^2, ERR(r) = \left[(\hat{x}_r - x_r)^2 + (\hat{y}_r - y_r)^2\right]^{\frac{1}{2}} \tag{4}$$

4.3 Baselines

This paper compares the LightLoc model with other neural network-based models. The baseline methods are summarized as follows:

- **K-Loc:** A localization model based on the K nearest neighbor (KNN) algorithm, which uses traditional fingerprints for training the KNN model.
- **T-Loc:** A localization model based on the KNN algorithm, where a fingerprint tensor is utilized for training the KNN model.
- **N-Loc:** A localization model based on artificial neural networks (ANNs), which are specifically designed for localization tasks and utilize traditional fingerprints as training data.

4.4 Main Results

This section reports the main results of our model performance and comparison results. In the experiments, various sliding window sizes were employed with a fixed window width. Different values were assigned to the window height, namely, 2, 3, 4, and 5. The sliding window traverses the native fingerprint library, generating the respective fingerprint tensors. These fingerprint tensors were then utilized for training the network model, while the same test set was used for performance evaluation. The results of this evaluation are presented in Fig. 3.

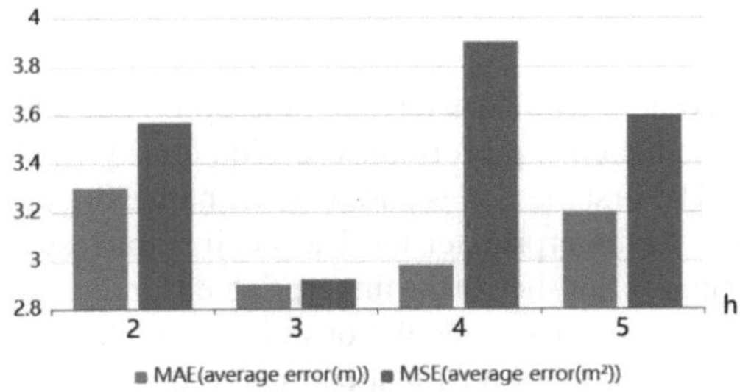

Fig. 3. Localization results of LightLoc under different *h values*

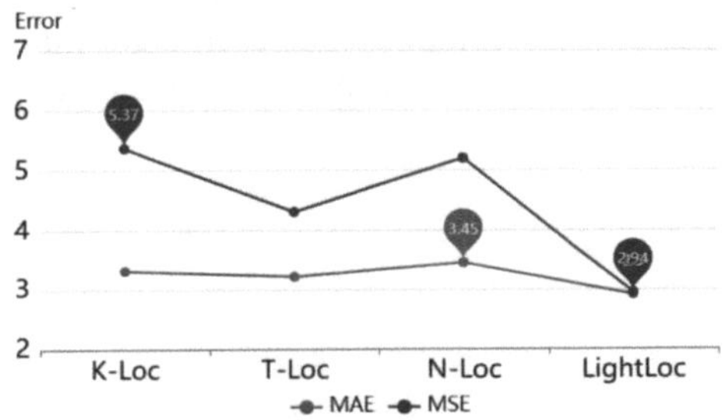

Fig. 4. The localization results of the models

It can be inferred that the most favourable outcomes are achieved when the window height is set to $h = 3$ under the experimental conditions of this chapter. This observation can be attributed to the fact that the fingerprint slices in the fingerprint tensor are associated with the precursor fingerprints of the current location fingerprint. A smaller number of precursor fingerprints results in limited additional information, which hampers the discrimination of twin fingerprints. Conversely, an excessive number of precursor fingerprints introduces excessive travel directions and information, leading to a decrease in the accuracy of location estimation. To summarize, it is crucial to select an appropriate sliding window height rather than the largest possible value. The optimal choice depends on the specific environment. In the localization environment described in this chapter, the best results were achieved with a window height of 3. Furthermore, LigntLoc has distinct advantages in terms of both localization accuracy and stability. We present a comparison of all the replicated comparative experiments with the localization effect of LightLoc through visual representations in line charts, as shown in Fig. 4.

In summary, the utilization of the optimized fingerprint in the localization scheme significantly improved the results compared with those of the traditional fingerprint library. The experimental findings demonstrate that the proposed LightLoc yields smaller localization errors and provides more stable localization results than the other methods do.

4.5 Ablation Study

To validate the effectiveness and superiority of the optimized fingerprint in the localization process, we conducted a performance evaluation by implementing two KNN localization methods. The first method is based on the native fingerprint library, whereas the second method uses the fingerprint tensor. The localization results were compared and analysed. Since each fingerprint slice in the fingerprint tensor represents two-dimensional data, traditional Euclidean distance calculations cannot be directly applied. Therefore, for the fingerprint slices, we employed a slice matrix similarity calculation algorithm to measure the similarity between them. The performance of the KNN algorithm was evaluated by conducting comparative testing across different levels of error, with the parameter k ranging from 1–10. The mean localization error (MAE) and mean squared error of localization (MSE) were computed for each algorithm. The obtained results are presented in Fig. 6. For ease of reference, we designate our proposed algorithm as LightLoc in the subsequent analysis.

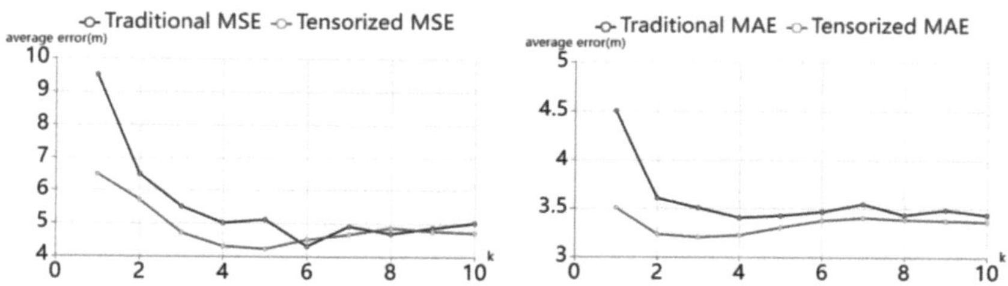

Fig. 5. The localization results of LightLoc under different *k values*

On the basis of the analysis presented in Fig. 5, the KNN localization algorithm based on the fingerprint tensor yields a smaller average localization error and relatively greater stability (lower mean square error). This finding suggests that the fingerprint tensor enhances the diversity of fingerprints and effectively addresses the issue of inaccurate location estimation caused by similar fingerprints. In this experiment, the optimal results were obtained for K-Loc, yielding an average localization error (Euclidean distance) of approximately and a mean square error of approximately. Similarly, for T Loc wit, the average localization error was approximately equal to the mean square error.

5 Conclusions

This paper addresses the challenges associated with Wi-Fi fingerprint-based indoor positioning systems and introduces a novel indoor localization model named LightLoc. The model first constructs optimized fingerprints via a sliding window mechanism, which effectively mitigates issues related to inaccurate location estimation caused by similar or missing fingerprints. To capture the implicit temporal correlations of the fingerprints, LightLoc incorporates a bidirectional LSTM (Bi-LSTM) neural network to learn the fingerprint features. An experimental platform is established in a real-world monitoring environment to validate the performance of the proposed algorithm. Extensive experiments demonstrate the model's effectiveness and superior performance compared with existing methods. In future work, we aim to further enhance the model's performance by integrating external factors.

Acknowledgements. This work was supported in part by the Open Research Project of the China Electronic Product Reliability and Environmental Testing Research Institute (CEPREI) under Grant No. HK202403594 and in part by the Zhejiang Key Laboratory of Information Security under Grant No. KF202310.

References

1. Liu, X., Cheng, X., Wang, C., Zhang, J., Chen, C.W.: A survey on fingerprint-based indoor localization. IEEE Internet Things J. **7**(6), 4641–4662 (2019)
2. Jiang, M., Dai, H.: Low-cost wi-fi fingerprint-based positioning system for indoor and outdoor environments. J. Sens. **2019**, 1382141 (2019)
3. Liu, P., Song, Q., Li, H., Cheng, Y.: Indoor positioning using graph-based neural networks for Wi-Fi signals. ACM Trans. Sens. Netw. **18**(4), 53–64 (2022)
4. Liu, Z., Zhang, X., Gao, L., Li, J.: A graph neural network approach for multi-AP indoor localization using RSSI fingerprinting. IEEE Access **9**, 84623–84631 (2021)
5. Li, J., Cai, Z., Wang, J.: Truthful incentive mechanisms for geographical position conflicting mobile crowdsensing systems. IEEE Trans. Comput. Soc. Syst. **5**(2), 324–334 (2018)
6. Bi, R., Sun, Y., He, Y.: Utility-oriented computation scheduling for energy-efficient mobile edge computing networks. IEEE Open J. Comput. Soc. **3**, 260–270 (2022)
7. Ren, Q., Wang, Y., Liu, S., Lv, X.: FSTNet: learning spatial-temporal correlations from fingerprints for indoor positioning. Ad Hoc Netw. **149**, 103244 (2023)
8. Sklivanitis, G., Demirors, E., Batalama, S.N., Melodia, T., Pados, D.A.: Receiver configuration and testbed development for underwater cognitive channelization. In: Conference on Signals, Systems & Computers (2014)
9. Pascacio, P., TorresSospedra, J., Casteleyn, S., Lohan, E.S.: A collaborative approach using neural networks for BLE-RSS lateration-based indoor positioning. IEEE (2022)
10. Teng, X., Zhang, Y., He, M.: A two-stage deep learning model based on feature combination effects. Neurocomputing **512**, 307–322 (2022)
11. Cai, C., Deng, L., Li, S.: CSI-based device-free indoor localization using convolutional neural networks. In: 2018 IEEE 4th International Conference on Computer and Communications (ICCC) (2018)
12. Wang, X., Gao, L., Mao, S., Pandey, S.: DeepFi: Deep learning for indoor fingerprinting using channel state information. In: Wireless Communications & Networking Conference (2015)
13. Wan, M., Han, M., Li, L., Li, Z., He, S.: Effects of and defenses against adversarial attacks on a traffic light classification CNN. In: Proceedings of the 2020 ACM Southeast Conference, pp. 94–99 (2020)
14. Li, Y., Ren, Q., Jin, H.: LSTN: long short-term traffic flow forecasting with transformer networks. In: 2022 26th International Conference on Pattern Recognition (ICPR), pp. 4793–4800. IEEE (2022)
15. He, Z., Han, M., Han, B.: Pattern-adaptive time series prediction via online learning and parallel processing using CUDA. In: 2019 IEEE 16th International Conference on Mobile Ad Hoc and Sensor Systems Workshops (MASSW), pp. 31–36. IEEE (2019)

16. Liu, B., Chen, W., Wang, Z.: Radamdanlstm: a nested LSTM- based time series prediction method for human-computer intelligent systems. Electronics **12**(14), 3084 (2023)
17. Liu, B., Chen, W., Han, M.: Nonlinear time series prediction algorithm based on ad-spent for artificial intelligence-powered internet of things. Int. J. Distrib. Sens. Netw. **17**(3), 15501477211004112 (2021)
18. Liu, L., Han, M., Zhou, Y., Wang, Y.: LSTM recurrent neural networks for influenza trends prediction. In: Bioinformatics Research and Applications: 14th International Symposium, IS-BRA 2018, Beijing, China, 8–11 June 2018, Proceedings 14, pp. 259–264. Springer, Cham (2018)
19. Zhang, J., Yang, Y., Liu, Y.: Graph representation learning via adaptive multilayer neighborhood diffusion contrast. In: Proceedings of the 31st ACM International Conference on Information & Knowledge Management, pp. 4682–4686 (2022)
20. Wei, C., Ge, Q.: Indoor positioning and navigation model based on semantic grid. Int. J. Pattern Recogn. Artif. Intell. **36**(06) (2022)
21. Gong, Z., Li, C., Jiang, F.: Design, analysis, and field testing of an innovative drone-assisted zero-configuration localization framework for wireless sensor networks. IEEE Trans. Veh. Technol. **66**(11), 10322–10335 (2017)
22. Abbas, M., Elhamshary, M., Rizk, H., Torki, M., Youssef, M.; Wideep: WiFi-based accurate and robust indoor localization system using deep learning. In: 2019 IEEE International Conference on Pervasive Computing and Communications (Per-Com 2019), pp. 1–10 (2019). https://doi.org/10.1109/PERCOM.2019.8767421
23. Li, X., Dai, Z., He, L.: A k-nearest neighbor indoor fingerprint location method based on coarse positioning circular domain and the highest similarity threshold. IOP Publishing Ltd. (2022)
24. Zhang, M., Fan, Z., Shibasaki, R., Song, X.: Domain adversarial graph convolutional network based on RSSI and crowdsensing for indoor localization (2023)
25. Jung, T.W., Jeong, C.S., Kwon, S.C., Jung, K.D.: Point-graph neural network based novel visual positioning system for in-door navigation. Appl. Sci. (2021)
26. Zhong, Y., Guo, Y., Wang, Y., Jin, Q.: DAST: an aggregation scheme for crowd sensed indoor data exploiting sequential long-tail features. In: 2021 IEEE Wireless Communications and Networking Conference Workshops (WCNCW) (2021)
27. Yan, M., Han, M., Ai, C., Cai, Z., Li, Y.: Data aggregation scheduling in probabilistic wireless networks with cognitive radio capability. In: 2016 IEEE Global Communications Conference (GLOBECOM), pp. 1–6. IEEE (2016)
28. Saccomanno, N., Brunello, A., Montanari, A.: Let's forget about exact signal strength: indoor positioning based on access point ranking and recurrent neural networks (2020)
29. Luo, M., Zheng, J., Sun, W., Zhang, X.: Wifi-based indoor localization using clustering and fusion fingerprint. In: 2021 40th Chinese Control Conference (CCC) (2021)
30. Tran, H.Q., Ha, C.: Machine learning in indoor visible light positioning systems: a review. Neurocomputing **491**, 117–131 (2022)
31. Qian, W., Lauri, F., Gechter, F.: Supervised and semisupervised deep probabilistic models for indoor positioning problems. Neurocomputing **435**(3) (2021)
32. Ssekidde, P., Eyobu, O.S., Han, D.S., Oyana, T.J.: Augmented cwt features for deep learning-based indoor localization using wifi rssi data. Appl. Sci. **11**(4), 1806 (2021)
33. Li, S., Hedley, M., Bengston, K., Humphrey, D., Ni, W.: Passive localization of standard wifi devices. IEEE Syst. J. **99**, 1–4 (2019)
34. Rao, S., Wang, Y.: DFPhaseFL: a robust device-free passive fingerprinting wireless localization system using CSI phase information. Neural Comput. Appl. **32**(18) (2021)
35. Li, Y., Zhang, W., Chen, J., Wang, H.: Indoor localization using graph neural networks with multilayer RSSI data. IEEE Trans. Mob. Comput. **20**(9), 2765–2777 (2021)

36. Wang, X., Xu, R., He, Y., Chen, T.: Improving sentiment analysis via sentence type classification using BiLSTM-CRF and CNN. Expert Syst. Appl. (2016). https://doi.org/10.1016/j.eswa.2016.10.065

Attention-Guided Clustered Graph Convolutional Network for Spatiotemporal Traffic Forecasting

Fanyu Meng, Hui Xu(✉), and Sijia Liu

Heilongjiang University, Harbin, Heilongjiang 150080, China
{2221915,2221960}@s.hlju.edu.cn, xuhui@hlju.edu.cn

Abstract. With the rapid development of the Internet of Things (IoT) and the acceleration of urbanization, traffic flow prediction has become a key research focus in intelligent transportation systems (ITS). However, owing to the highly nonlinear and dynamic spatiotemporal dependencies of traffic flow, achieving accurate and timely predictions, especially for long-term forecasting, remains a significant challenge. To address this, we propose AGCGCN, a novel prediction model designed to effectively model sensor relationships and capture spatiotemporal dependencies. By incorporating node features, local features, and global features, AGCGCN comprehensively describes sensor relationships. The model leverages a self-attention mechanism to capture directed spatial dependencies dynamically, assigning greater weights to key nodes to reflect real-time traffic flow status and directional trends. Furthermore, its multilevel hierarchical structure facilitates efficient information propagation and integrates these insights into a predictive output layer. Extensive experiments demonstrate that AGCGCN outperforms several state-of-the-art baseline models in terms of prediction accuracy, highlighting its effectiveness in traffic flow forecasting. The source code is available at https://github.com/mengfanyu-hd/AGCGCN.

Keywords: Traffic flow prediction · spatiotemporal dependencies · graph convolutional network

1 Introduction

With rapid urbanization and economic growth, traffic congestion has become an increasingly serious issue, significantly affecting city governance, transportation efficiency, and residents' daily lives. Traditional traffic management approaches are often inadequate for responding to the growing complexity and dynamics of urban transportation networks. This has motivated the development of ITSs to support real-time traffic analysis and prediction. The widespread deployment of intelligent sensors has significantly advanced urban data mining, especially in the collection and analysis of traffic data [1].

Y. Yu et al. (Eds.): ICPCSEE 2025, CCIS 2673, pp. 404–416, 2026.
https://doi.org/10.1007/978-981-95-2566-9_27

Traffic forecasting predicts future traffic conditions via historical data from nodes and their neighbors [2]. Capturing spatiotemporal dependencies is key to predictive modelling. Understanding sensor interactions enhances spatiotemporal predictions and improves urban traffic forecasting [3,4]. These sensors continuously capture real-time traffic metrics such as vehicle speed, flow, and occupancy, providing vital input for transportation planning, anomaly detection, and congestion mitigation strategies. However, effectively utilizing these data to make accurate traffic forecasts presents several challenges:

- Data heterogeneity and noise: Sensor data are often incomplete, noisy, or irregular, making robust modelling difficult.
- Spatiotemporal complexity: Traffic flow is inherently dynamic and is influenced by both temporal patterns (e.g., rush hour) and spatial dependencies (e.g., interactions between neighboring roads).
- Scalability and efficiency: High-resolution data from dense sensor networks pose computational challenges, especially for real-time forecasting applications.

Capturing complex spatiotemporal dependencies is therefore crucial for accurate and reliable traffic prediction. Earlier approaches, simple forecasting, utilized stable autoregressive (AR) models and variants [5], including the autoregressive integral moving average model (ARIMA), which handles smoothed or smoothed data after differencing, and the seasonal autoregressive integral moving average model (SARIMA) [6] specializing in seasonal time series forecasting. In the era of big data, deep learning and computational resources, GCNs have rapidly progressed, increasing traffic flow prediction accuracy by modelling complex spatiotemporal dependencies.

This study presents a new model for forecasting traffic flow. It features a parallel module that combines graph convolution and self-attention to capture spatial interdependencies and employs a G-TCN module for temporal dependencies. Periodic data from real datasets are used as external features and combined with the input. The model processes data through time and space modules to extract features for prediction. The main contributions of this study are as follows:

1. A G-TCN module is proposed for temporal dependency modelling. It effectively captures sequential dynamics and temporal trends from historical traffic data, ensuring robust performance under complex and irregular traffic conditions.
2. The CGCN module captures features from both distant and nearby nodes, extracting local region information from clusters, and uses a self-attention mechanism to adjust the model's focus to significant areas, improving accuracy in spatial dependencies.
3. We execute thorough experiments on sundry datasets (e.g., PEMS-BAY and METR-LA), benchmarking the proposed model against several SOTA baselines. The results consistently demonstrate the superior performance and generalization ability of our method across different metrics and settings.

2 Related Works

2.1 Traffic Flow Prediction

Traffic flow prediction involves capturing spatiotemporal dependencies, including spatial topology construction and modelling [7]. RNNs, especially long short-term memory (LSTM) [8] and gated recurrent units (GRUs) [9], are widely used for time series data [7,10,11]. ASTGCRN [10]uses a specially designed GRU to learn the graph structure, including its spatial dependencies and local temporal relationships. MSDR [11] extracts long-term temporal dependencies from hidden states at multiple time steps. Moreover, CNN-based models are strong at extracting local features [12–15], but are limited by kernel size and parameter sharing. Models based on physical data, such as STDEN [16], and ordinary differential equations, such as STODE [17] and STG-NCDE [18], provide a new direction for traffic flow prediction. These methodologies aim to enhance traffic flow forecasting, reduce congestion, and support smart city development.

2.2 Graph Convolutional Network

GCN [19] utilizes graph structures for feature propagation via convolution operations. DCRNN [7] introduces diffusion convolution with a manual adjacency matrix for temporal correlations. The STGCN [20] was the first to apply a GCN for traffic prediction, combining graph convolution with a 1D CNN. STSGCN [21] designs a local spatiotemporal subgraph module to extract local spatiotemporal correlations effectively. However, traffic dependencies vary over time, and static graphs based on node distance or similarity cannot fully capture dynamic topological information. Graph Wavenet [22] generates an adaptive adjacency matrix to dynamically represent node dependencies. GCN models extract features at multiple hierarchical levels, capturing micro- and macropatterns. The multiscale nature of GCNs, enhanced by varying kernel sizes, improves spatial dependency representation, boosting traffic forecasting.

2.3 Attention Mechanism

In 2017, Vaswani et al. proposed the attention mechanism in "Attention Is All You Need" [23], which improves parallel computation and long-range dependency modelling. ASTGCN [24] combines spatial and temporal attention mechanisms alongside spatiotemporal convolution for modeling dynamic features. ASTGNN [25] uses based on attention-based spatiotemporal graph neural networks to address traffic data periodicity and spatial diversity. STTN [2] effectively captures long-range temporal dependencies and spatial correlations through a self-attention mechanism, using stacked spatial and temporal transformers to predict future traffic conditions. PDFormer [26] accounts for spatial dependencies and temporal propagation delays while capturing traffic information dynamics through computation of the propagation delay matrix within road networks.

3 Preliminary

The traffic flow network is modeled as a directed graph $G = (V, E, A)$. V is the set of $|V| = N$ sensor nodes; E represents the edges connecting nodes, and $A \in \mathbb{R}^{N\times N}$ is the adjacency matrix encoding the graph's connectivity structure. If vi and $vj \in V$,when $e_{i,j} = (v_i, v_j) \in E$, an edge $e_{i,j}$ exists from node v_i to node v_j, otherwise $e_{i,j} = (v_i, v_j) \notin E$. This representation aids in capturing spatial and temporal dependencies.

Given a traffic flow graph $G = (V, E, A)$, with H time steps of historical data observed by N sensor nodes, a traffic prediction model F predicts the future traffic flow data for T time steps as follows: $\left[\widehat{Y}^{t+1}, \ldots, \widehat{Y}^{t+T}\right] = F\left[X^{t-H+1}, \ldots, X^t; G\right]$. Here, tensor X^i denotes a history sequence of traffic flow data observed by N nodes at time step i , and tensor $\hat{\mathbf{Y}}^j$ denotes a predicted sequence of traffic flow data at time step j.

4 Methodology

4.1 Overview

Figure 1 depicts the framework of the AGCGCN model. The input traffic flow data pass through multiple L-layer spatiotemporal feature extraction blocks, including a gated temporal convolutional network (G-TCN) and an attention-based cluster graph convolution network (ABCGCN). Following fusion, the intermediate features across the L layers are concatenated for final prediction by the output layer.

4.2 Spatial Dependency Mining

We utilize the ABCGCN to extract spatial dependencies. A self-attention layer and a CGCN layer constitute it. The CGCN layer focuses on sensor features, regional cluster features and global cluster feature graph convolution. The outputs of the self-attention and CGCN layers are processed through gating and residual fusion, enhancing spatial dependency representations.

Cluster Graph Convolutional Network. We develop CGCN to learn the structured features of nodes and capture static spatial dependencies from the road network topology. The CGCN is designed to apply graph convolution at different granularities of clusters, using high-level convolutions on node features in the adjacency matrix to aggregate information from distant clusters. Assuming node feature matrix H and adjacency matrix A, the GCN formulation can be presented as $H' = \sigma\left(\tilde{D}^{-1/2}\tilde{A}\tilde{D}^{-1/2}HW\right)$ with $\tilde{A} = A + I$ (identity-added adjacency), $\tilde{D}$ as its degree matrix, W is the weight matrix, and $\sigma()$ the activation function.

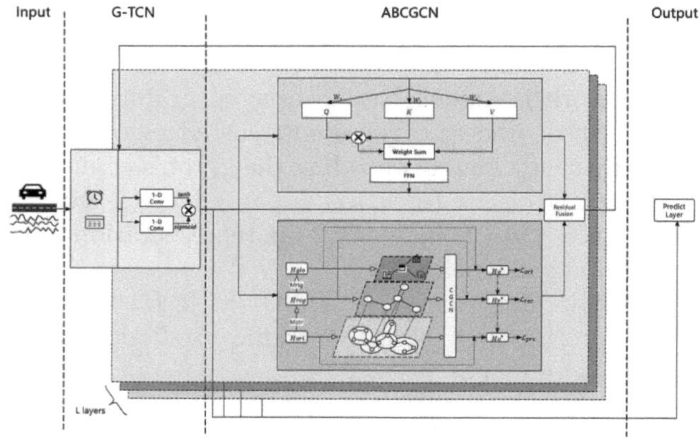

Fig. 1. The AGCGCN consists of two parts: the G-TCN on the left and the ABCGCN on the right. The top is the self-attention module, and the bottom is the CGCN module.

Briefly, CGCN is applied to the original graph. The original node feature H_{ori} is subjected to CGCN on the adjacency matrix A to obtain Ho', concatenated with the original feature to form Ho. A multilayer perceptron (MLP) is used to map the high-dimensional features to the target output dimension. To prevent overfitting after the MLP layer, a dropout layer is applied to obtain H_o^*. The CGCN computation can be summarized as Eq. 1. The above process can be expressed as $H_o^* = \mathrm{CGCN}(H_{\mathrm{ori}}, A)$.

$$H_o' = \mathrm{GCN}(H_{\mathrm{ori}}, A), \quad H_o = \mathrm{concat}(H_{\mathrm{ori}}, H_o'), \quad H_o^* = \mathrm{Dropout}(\mathrm{MLP}(H_o)) \qquad (1)$$

CGCN in Region Graph Clusters. To better capture node relationships in the sensor network, we group nodes with similar local dependencies into regions via the Tarjan algorithm. By clustering nodes within biconnected components, we form region clusters with local perspectives. The mapping matrix M_{otr} from the original node graph to the region cluster graph is generated as follows.

The mapping matrix $M_{otr} \in R^{N_{ori} \times N_{reg}}$ indicates the correspondence between N_{ori} nodes in the original node graph and N_{reg} region clusters. M_{otr} represents whether a certain node could be merged into the mapping to a certain region cluster node. Original graph's feature representation $H_{ori} \in \mathbb{R}^{N_{ori} \times D}$, is aggregated via M_{otr} to produce the region cluster graph's feature matrix, $H_{reg} \in \mathbb{R}^{N_{reg} \times D}$, as $H_{reg} = M_{otr}^T \cdot H_{ori}$, reflecting the attributes of the region clusters.

This mapping matrix directly derives the adjacency matrix for the region cluster graph. Given the adjacency matrix $A_{ori} \in R^{N_{ori} \times N_{ori}}$, the adjacency matrix of the nodes of the regional cluster is generated by aggregating the mapping matrix with the adjacency matrix $A_{reg} \in R^{N_{reg} \times N_{reg}}$ to describe the connectivity relationship between the regional cluster nodes. CGCN is utilized on the region cluster adjacency matrix for the region cluster feature matrix to generate the feature representation H^*_{reg} between the region clusters. Next, the feature representation $\hat{H}_{ori}$ of the original node graph is reconstructed via a mapping matrix. The adjacency matrix $\hat{A}_{ori}$ of the nodes is reconstructed via the node feature of the reconstructed graph.

$$\begin{aligned} A_{reg} &= M^T_{otr} A_{ori} M_{otr}, \quad H^*_{reg} = \mathrm{CGCN}(H_{reg}, A_{reg}), \\ \hat{H}_{ori} &= M_{otr} H^*_{reg}, \qquad \hat{A}_{ori} = \sigma(\hat{H}_{ori}, \hat{H}^T_{ori}) \end{aligned} \tag{2}$$

The reconstruction loss between the region cluster graph and the original node graph is quantified via binary cross-entropy in Eq. 3. The method simplifies the complex sensor network into a regional cluster graph. The CGCN in these clusters captures local dependencies, improving accuracy and effectiveness.

$$\begin{aligned} BCE(A, \hat{A}) &= -\frac{1}{N} \sum_{i=1}^{N} \left[A_i \log(\hat{A}_i) + (1 - A_i) \log(1 - \hat{A}_i) \right], \\ \mathcal{L}_{BCE_o}(A_{ori}, \hat{A}_{ori}) &= -\frac{1}{N^2_{ori}} \sum_{i,j \in V_o} \Big[A_{ori}[i,j] \log(\hat{A}_{ori}[i,j]) \\ &\qquad + (1 - A_{ori}[i,j]) \log(1 - \hat{A}_{ori}[i,j]) \Big] \end{aligned} \tag{3}$$

CGCN in Global Graph Clusters. We represent the mapping between local region clusters and global graph clusters via a trainable soft mapping matrix $M_{rtg} \in \mathbb{R}^{\mathbb{N}_{reg} \times \mathbb{N}_{glo}}$, where N_{glo} is the number of global graph clusters. Given the region graph cluster adjacency matrix $A_{reg} \in \mathbb{R}^{\mathbb{N}_{reg} \times \mathbb{N}_{reg}}$, the global graph cluster adjacency matrix A_{glo} is generated by M_{rtg} and computed as $A_{glo} = M_{rtg} \cdot A_{reg} \cdot M^T_{rtg}$. Similarly, the global feature representation $H_{glo} \in \mathbb{R}^{\mathbb{N}_{glo} \times D}$ is derived from $H_{glo} = M^T_{rtg} \cdot H_{reg}$. The global feature H^*_{glo} is then extracted via CGCN operations on the new global cluster graph's adjacency matrix, which is expressed as $H^*_{glo} = \mathrm{CGCN}(H_{glo}, A_{glo})$.

Graph representation learning is enhanced by employing global graph clusters to regenerate local region cluster representations $\hat{H}_{reg} = M_{rtg} \cdot H^*_{glo}$. The feature matrix of the reconstructed region cluster is used to represent the adjacency matrix $\hat{A}_{reg} = \sigma(\hat{H}_{reg}, \hat{H}^T_{reg})$ of the reconstructed region cluster.

Reconstruction loss between cluster maps is computed via BCE, complemented by orthogonalization loss applied to decrease inter-node cosine similarity. Binary cross-entropy loss ensures accurate reconstruction of cluster relationships, whereas orthogonalization loss ensures that each node represents distinct regional cluster functions, increasing representation diversity. The combined loss functions are calculated as follows:

$$\mathcal{L}_{BCE_r} = -\frac{1}{N_{reg}^2} \sum_{i,j \in V_o} \left[A_{reg}[i,j] \log \left(\hat{A}_{reg}[i,j] \right) + (1 - A_{reg}[i,j]) \log \left(1 - \hat{A}_{reg}[i,j] \right) \right], \tag{4}$$

$$\mathcal{L}_{Cos\_Ort} = \frac{1}{C_{N_{glo}}^2} \sum_{i=1}^{N_{glo}} \sum_{j=i+1}^{N_{glo}} \left| \frac{H_{glo}[i] \odot H_{glo}[j]}{|H_{glo}[i]| \, |H_{glo}[j]|} \right|$$

where $C_{N_{glo}}^2$ denotes the pairwise combination of clusters in the global graph and where N_{glo} is cardinality of global graph clusters. As shown in Eq. 4, these losses collectively optimize the reconstruction accuracy and ensure functional diversity in the global cluster graph.

Attention Mechanisms in Parallel. We incorporate a parallel spatial attention mechanism with CGCN to capture complex spatial dependencies via adaptive weighting, addressing heterogeneity and nonsmoothness. Taking the original node graph features as inputs, we project them into a high-dimensional subspace, denoted as the query vector (Q), key vector (K), and value vector (V), respectively. Computation of attention weights employs the attention mechanism as:

$$Q = H_{\text{ori}} \cdot W_q, \quad K = H_{\text{ori}} \cdot W_k, \quad V = H_{\text{ori}} \cdot W_v,$$
$$\text{Attention}(Q, K, V) = \text{softmax}\left(\frac{QK^T}{\sqrt{d_k}} \right) V \tag{5}$$

Finally, the attention mechanism features are summed with the CGCN-enhanced original node graph features after a linear transformation. A sigmoid activation function generates a gating value g ∈ [0,1], which weights the summation of the attention features and the augmented original graph features as follows: $out = g \cdot \text{Attention}(Q, K, V) + (1 - g) \cdot H_{ori}^*$.

4.3 Temporal Dependency Mining

The G-TCN employs a gating mechanism and dilated convolution to model temporal relationships. The input features $X \in \mathbb{R}^{F \times N \times H}$ are expanded via two paths: one applies a tanh activation for the filter output, and the other applies a sigmoid activation for the gated output. The outputs are multiplied elementwise to form the Hadamard product, which preserves important information while suppressing noise. Mathematically, this is represented as $H = \tanh(\Theta_1 * X + b_1) \odot \sigma(\Theta_2 * X + b_2)$, where $\Theta_1, \Theta_2, b_1,$ and b_2 are optimized during training.

4.4 Model Optimization

We use the MAE to measure the forecasting loss $\mathcal{L}_p$ between outputs and ground truth. Four terms constitute the optimized loss function: the prediction loss $\mathcal{L}_p$, the cosine quadrature loss $\mathcal{L}_{Cos\_Ort}$, and the reconstruction loss $\mathcal{L}_{BCE_r}$ and $\mathcal{L}_{BCE_o}$, as shown in Eq. 6.

$$\mathcal{L} = \mathcal{L}_p + \mathcal{L}_{Cos\_Ort} + \mathcal{L}_{BCE_r} + \mathcal{L}_{BCE_o} \tag{6}$$

5 Experiments

5.1 Dataset

We use three datasets : METR_LA, PEMS_BAY, and PEMS_D8. The METR_LA consists of 207 nodes and 1,515 edges, with data spanning from March 1, 2012, to June 27, 2012. PEMS_BAY has 325 nodes and 2369 edges, with data from January 1, 2017, to June 30, 2017. PEMS_D8 contains 170 nodes and 276 edges, with data from July 1, 2016, to August 31, 2016. These datasets are widely used in traffic forecasting and urban traffic analysis.

5.2 Experience Settings

We split the dataset into training (70%), validation (10%), and test sets (20%). Our model is built in PyTorch and run on a workstation with an NVIDIA A100. We trained the model via the Adam optimizer. The learning rate is 0.001, and the batch size is 64. We chose the MAE, MAPE, and RMSE as metrics for model prediction performance.

5.3 Baselines

- **GWNet** [22] uses an adaptive matrix and dilated causal convolution to model spatiotemporal correlations, which is effective in short-term prediction.
- **DCRNN** [7] propagates spatial correlations via bidirectional graph random walks, before GRU encodes temporal dependencies.
- **ASTGCN** [24] uses an attention module and conventional convolution layers to capture spatiotemporal dependencies dynamically by capturing different temporal dependencies in the traffic data: near-term, daily periodic and weekly periodic patterns.
- **STSGCN** [21] captures spatial and temporal correlations via a local spatiotemporal temporal graph convolution module, ignoring global interactions.
- **STTN** [2] combines a temporal transformer and a spatial transformer as a module. This method iteratively captures dynamic spatial dependence and long-range temporal dependence and uses spatiotemporal dependence features to achieve accurate traffic prediction.
- **HIEST** [1] combines meta-GCN calibration and cross-level graph convolution by hierarchically modelling regional and global nodes. Accurately capturing spatiotemporal dependencies to achieve prediction.
- **ASTGNN** [27] leverages self-attention mechanisms and dynamic graph convolution modules to model spatiotemporal dynamics in traffic data, explicitly capturing periodicity and spatial heterogeneity.
- **SSTBAN** [28] combines self-supervised learning with a spatiotemporal bottleneck attention mechanism to capture long-term spatiotemporal dynamics while optimizing computational efficiency (Table 1).

Table 1. Performance evaluation of the models.

	Time	Metric	GWNet	DCRNN	ASTGCN	STSGCN	STTN	HIEST	ASTGNN	SSTBAN	AGCGCN
METR-LA	15 min	MAE	2.99	2.71	4.39	3.26	3.55	2.74	3.10	3.13	**2.70**
		RMSE	5.90	5.27	8.09	7.61	8.06	5.24	7.48	**3.14**	5.23
		MAPE	7.93%	**6.98%**	10.68%	8.09%	8.66%	7.08%	7.37%	8.22%	7.08%
	30 min	MAE	3.59	3.15	5.18	3.95	4.23	3.14	3.85	3.52	**3.08**
		RMSE	7.28	6.43	9.33	9.48	9.85	6.28	9.53	7.09	**6.27**
		MAPE	10.16%	8.60%	12.75%	10.15%	10.49%	8.65%	9.43%	9.72%	**8.59%**
	60 min	MAE	4.44	3.69	6.34	4.76	5.10	3.62	4.82	3.97	**3.47**
		RMSE	8.95	7.76	10.81	11.36	11.81	7.44	11.82	8.05	**7.30**
		MAPE	13.51%	10.72%	15.53%	12.46%	12.98%	10.59%	12.01%	11.43%	**10.36%**
PEMS-BAY	15 min	MAE	1.33	1.33	1.66	1.35	1.42	1.33	1.32	1.63	**1.31**
		RMSE	2.81	2.86	3.28	2.90	2.96	**2.76**	2.86	3.50	**2.76**
		MAPE	2.82%	2.78%	3.82%	2.89%	3.08%	2.78%	2.75%	4.09%	**2.74%**
	30 min	MAE	1.66	1.69	2.02	1.70	1.74	1.66	1.67	1.89	**1.64**
		RMSE	3.75	3.90	4.09	3.95	3.89	**3.73**	3.88	4.23	**3.73**
		MAPE	3.74%	3.81%	4.83%	3.85%	3.93%	3.74%	3.73%	4.73%	**3.67%**
	60 min	MAE	1.97	2.04	2.39	2.01	2.05	1.99	1.98	2.21	**1.95**
		RMSE	4.52	4.81	4.80	4.72	4.67	4.55	4.66	5.03	**4.51**
		MAPE	4.65%	4.84%	5.83%	4.73%	4.76%	4.65%	**4.61%**	5.55%	4.62%
PEMS-D8	15 min	MAE	14.05	15.89	17.58	15.06	20.08	14.20	**13.87**	15.69	14.02
		RMSE	22.21	24.40	27.26	23.17	31.72	**22.12**	22.19	23.85	22.17
		MAPE	9.59%	**6.98%**	13.68%	9.81%	13.08%	10.27%	8.93%	12.57%	9.24%
	30 min	MAE	15.00	16.95	18.64	15.70	20.32	15.08	15.09	16.27	**14.72**
		RMSE	24.01	26.23	29.04	24.35	32.08	24.00	24.45	25.42	**23.98**
		MAPE	9.99%	8.60%	14.25%	10.10%	13.09%	10.16%	9.81%	13.32%	9.71%
	60 min	MAE	16.70	18.77	20.97	16.94	21.30	16.63	17.33	18.37	**16.02**
		RMSE	26.54	29.15	32.55	26.51	33.35	26.50	27.99	28.40	**26.04**
		MAPE	11.23%	10.72%	15.86%	10.76%	13.73%	11.26%	11.55%	15.16%	**10.53%**

5.4 Experimental Benchmarking

Benchmarking baselines precedes comprehensive analysis of empirical results. The proposed AGCGCN model achieves excellent performance in terms of all the metrics, especially in long-horizon prediction tasks. In this experiment, the AGCGCN model demonstrated outstanding performance across multiple datasets and different time prediction tasks, particularly in both short-term and long-term predictions. For METR_LA dataset, AGCGCN led all the models in the 15-minute prediction task with the lowest MAE, RMSE, and MAPE values. It also maintained its top position inthe 30-minute and 60-minute predictions, with MAE, RMSE, and MAPE values of 3.11/6.29/8.51% and 3.52/7.38/10.34%, respectively. For the PEMS_BAY dataset, the AGCGCN achieves the lowest MAE of 1.32, RMSE of 2.82, and MAPE of 2.80%, significantly outperforming the other models. It continued to exhibit strong performance in medium-term and long-term predictions, with MAE and RMSE values that were consistently lower than those of the other models. For the PEMS_D8 dataset, despite the increased prediction challenges, the AGCGCN still performed well in terms of short-term forecasts, with MAE of 14.02, RMSE of 22.17, and MAPE of 9.24%, and maintained lower MAE and RMSE values than most comparison models did in terms of medium-term and long-term predictions. Overall, AGCGCN outperforms existing models such as GWNet, SSTBAN and ASTGNN across all the evaluation metrics, demonstrating its superior adaptability and accuracy in traffic flow prediction tasks across different datasets and time horizons.

5.5 Ablation

Ablation experiments on METR-LA were conducted to explore the optimal component combination.

- w/o the G-TCN: The AGCGCN uses a simple GCN to extract temporal dependencies without the G-TCN.
- w/o ABCGCN: AGCGCN uses a GCN to extract spatial dependencies without the ABCGCN.
- w/o CGCN: No CGCN in the ABCCGN; only attention is given to extracting spatial dependencies.
- w/o SA: No spatial attention in the ABCCGN; only the CGCN is used to extract spatial dependencies.
- CGCN. pre: ABCGCN module in CGCN followed by the tandem attention module.
- Att. pre: ABCGCN module in the tandem attention module before CGCN.
- AGCGCN: ABCGCN module in CGCN in parallel with the attention module.

Table 2. Ablation Study Results

Model	15 min			30 min			60 min		
	MAE	RMSE	MAPE	MAE	RMSE	MAPE	MAE	RMSE	MAPE
w/o G-TCN	2.72	**5.22**	7.20%	3.13	6.30	8.73%	3.55	7.37	10.29%
w/o ABCGCN	2.72	5.25	7.10%	3.13	6.31	8.83%	3.56	7.39	10.70%
w/o CGCN	2.81	5.45	7.40%	3.20	6.47	8.95%	3.60	7.45	10.59%
w/o SA	2.74	5.30	7.21%	3.13	6.36	8.84%	3.54	7.41	10.54%
CGCN. pre	2.80	5.42	7.40%	3.18	6.43	8.89%	3.54	7.43	**10.15%**
Att. pre	2.83	5.46	7.59%	3.24	6.52	9.30%	3.60	7.52	10.86%
AGCGCN	**2.70**	5.23	**7.08%**	**3.08**	**6.27**	**8.59%**	**3.47**	**7.30**	10.36%

As shown in Table 2, the proposed AGCGCN achieves the best performance across all the prediction horizons with the lowest MAE, RMSE, and MAPE values. Specifically, compared with w/o the ABCGCN, the AGCGCN reduces the MAE by 1.1%, 1.6%, and 2.5% for 15-, 30-, and 60-minute predictions, respectively. These results demonstrate the effectiveness of integrating CGCN and the attention mechanism in parallel, enabling AGCGCN to capture both spatial dependencies and critical features more comprehensively than other configurations.

To achieve better prediction accuracy, the proposed model incorporates multiple modules, such as parallel spatial encoders (CGCNs with self-attention) and the G-TCN for temporal modelling. These enhancements inevitably lead to increased computational costs and longer training times than simpler baselines do. However, in real-world applications—especially in urban traffic management systems where accuracy is critical for decision-making—the performance gain justifies the additional resource consumption. Moreover, model components can be selectively pruned or quantized during deployment to achieve a balance between efficiency and effectiveness.

5.6 Impact of the Dropout Rate and External Information

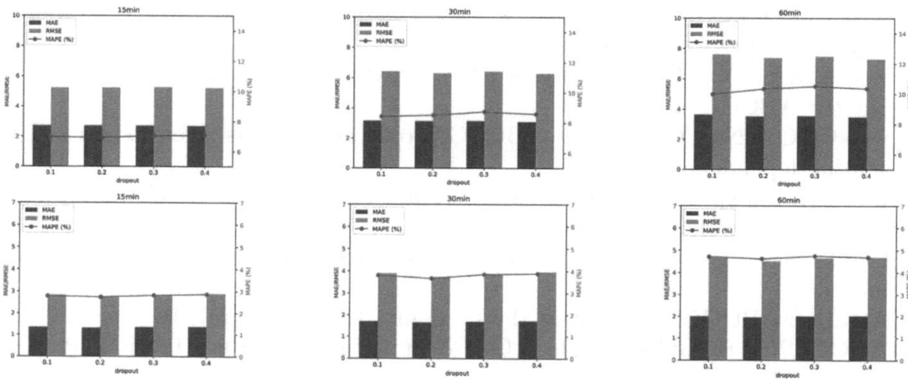

Fig. 2. Dropout Rate Sensitivity in METR LA and PEMSBAY

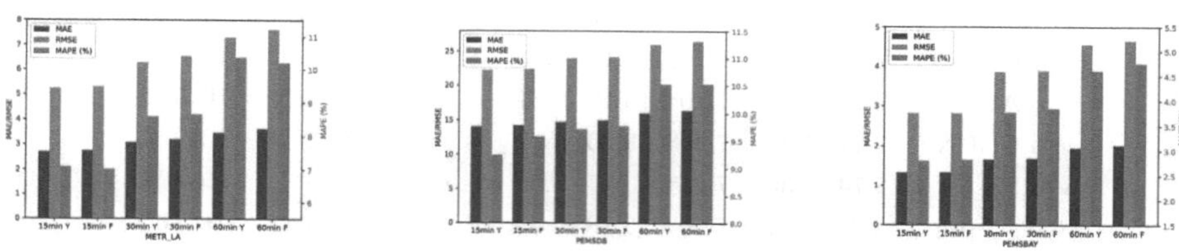

Fig. 3. Impact of External Information

Dropout is a regularization technique that enhances model generalizability. As shown in Fig. 2, the model's performance shows minimal variation across different dropout rates, demonstrating strong robustness. The metrics for predictions at different time intervals exhibit limited changes, indicating that the dropout rate has a negligible effect on model performance. The results confirm that the model remains stable across different dropout rates without overfitting, demonstrating its reliability and generalizability for traffic forecasting.

Additionally, the experiment also examines the impact of time-related external factors on traffic flow prediction. The results indicate that incorporating external information enhances model performance, particularly for long-term predictions. As shown in Fig. 3, adding external information to METR-LA and PEMS-BAY datasets reduces the number of metrics, thereby improving the prediction reliability. Overall, while external information enhances predictions, its effectiveness is contingent on the specific dataset used.

6 Conclusion

Our pioneering contribution is AGCGCN——a spatiotemporal deep learning model designed for traffic forecasting. The CGCN extracts features at the node, region and global levels. The parallel self-attention module prioritizes important nodes. Additionally, the inclusion of external factors enriches the model's contextual information, enhancing its predictive power and generalization capabilities. Cross-dataset validation reveals AGCGCN's statistically significant advantages over established benchmarks. In the future, we will continue to explore traffic prediction techniques to improve forecasting accuracy and support more efficient urban traffic management, contributing to sustainable development and a better quality of life. Additionally, we plan to investigate lightweight approaches, such as model pruning and quantization, to reduce deployment costs and enhance the practical applicability of the model.

Acknowledgments. We are deeply grateful to our supervisor for their valuable guidance and continuous support throughout the research process.

Disclosure of Interests. The authors declare no competing interests.

References

1. Ma, Q., et al.: Hierarchical information enhanced traffic forecasting, rethinking sensors modeling (2023)
2. Xu, M., Dai, W., Liu, C., Gao, X., Lin, W., Qi, G.H., Xiong, H.: Spatial-temporal transformer networks for traffic flow forecasting (2021)
3. Wu, C., Ding, H., Fu, Z., Sun, N.: Multi-scale spatio-temporal attention networks for network-scale traffic learning and forecasting. Sensors **24**(17) (2024)
4. Liu, C., et al.: Spatial-temporal large language model for traffic prediction (2024)
5. Triebe, O., Laptev, N., Ram, R.: A simple auto-regressive neural network for time-series, Ar-net (2019)
6. Box, G.E., Jenkins, G.M., Reinsel, G.C., Ljung, G.M.: Time series analysis: forecasting and control. John Wiley & Sons (2015)
7. Li, Y., Yu, R., Shahabi, C., Liu, Y.: Diffusion convolutional recurrent neural network: data-driven traffic forecasting. arXiv preprint arXiv:1707.01926 (2017)
8. Graves, A.: Long short-term memory. Supervised sequence labelling with recurrent neural networks, pp. 37–45 (2012)
9. Cho. Learning phrase representations using rnn encoder-decoder for statistical machine translation. arXiv preprint arXiv:1406.1078 (2014)
10. Liu, H., Zhu, C., Zhang, D., Li, Q.: Attention-based spatial-temporal graph convolutional recurrent networks for traffic forecasting. In: International Conference on Advanced Data Mining and Applications, pp. 630–645. Springer (2023)

11. Liu, D., Wang, J., Shang, S., Han, P.: Msdr: multi-step dependency relation networks for spatial temporal forecasting. In: Proceedings of the 28th ACM SIGKDD Conference on Knowledge Discovery and Data Mining, pp. 1042–1050 (2022)
12. Guo, S., Lin, Y., Li, S., Chen, Z., Wan, H.: Deep spatial-temporal 3d convolutional neural networks for traffic data forecasting. IEEE Trans. Intell. Transp. Syst. **20**(10), 3913–3926 (2019)
13. Li, T., Zhang, J., Bao, K., Liang, Y., Li, Y., Zheng, Y.: Autost: efficient neural architecture search for spatio-temporal prediction. In: Proceedings of the 26th ACM SIGKDD International Conference on Knowledge Discovery & Data Mining, pp. 794–802 (2020)
14. Yu, B., Yin, H., Zhu, Z.: Spatio-temporal graph convolutional networks: a deep learning framework for traffic forecasting. arXiv preprint arXiv:1709.04875 (2017)
15. Guo, S., Lin, Y., Feng, N., Song, C., Wan, H.: Attention based spatial-temporal graph convolutional networks for traffic flow forecasting. In: Proceedings of the AAAI Conference on Artificial Intelligence. vol. 33, pp. 922–929 (2019)
16. Ji, J., Wang, J., Jiang, Z., Jiang, J., Zhang, H.: Stden: towards physics-guided neural networks for traffic flow prediction. In: Proceedings of the AAAI Conference on Artificial Intelligence, vol. 36, pp. 4048–4056 (2022)
17. Fang, Z., Long, Q., Song, G., Xie, K.: Spatial-temporal graph ode networks for traffic flow forecasting. In: Proceedings of the 27th ACM SIGKDD Conference on Knowledge Discovery & Data Mining, pp. 364–373 (2021)
18. Choi, J., Choi, H., Hwang, J., Park, N.: Graph neural controlled differential equations for traffic forecasting. In: Proceedings of the AAAI Conference on Artificial Intelligence, vol. 36, pp. 6367–6374 (2022)
19. Thomas, N.K., Max, W.: Semi-supervised classification with graph convolutional networks. arXiv preprint arXiv:1609.02907 (2016)
20. Yu, B., Yin, H., Zhu, Z.: Spatio-temporal graph convolutional networks: a deep learning framework for traffic forecasting. arXiv preprint arXiv:1709.04875 (2017)
21. Song, C., Lin, Y., Guo, S., Wan, H.: Spatial-temporal synchronous graph convolutional networks: a new framework for spatial-temporal network data forecasting. In: Proceedings of the AAAI Conference on Artificial Intelligence, vol. 34, 914–921 (2020)
22. Wu, Z., Pan, S., Long, G., Jiang, J., Zhang, C.: Graph wavenet for deep spatial-temporal graph modeling. arXiv preprint arXiv:1906.00121 (2019)
23. Vaswani, A: Attention is all you need. Adv. Neural Inf. Process. Syst. (2017)
24. Guo, S., Lin, Y., Feng, N., Song, C., Wan, H.: Attention based spatial-temporal graph convolutional networks for traffic flow forecasting. In: Proceedings of the AAAI Conference on Artificial Intelligence, vol. 33, pp. 922–929 (2019)
25. Ye, J., Xue, S., Jiang, A.: Attention-based spatio-temporal graph convolutional network considering external factors for multi-step traffic flow prediction. Digital Commun. Netw. **8**(3), 343–350 (2022)
26. Jiang, J., Han, C., Zhao, W.X., Wang, J.: Propagation delay-aware dynamic long-range transformer for traffic flow prediction. In: Proceedings of the AAAI Conference on Artificial Intelligence, vol. 37, pp. 4365–4373 (2023)
27. Guo, S., Lin, Y., Wan, H., Li, X., Cong, G.: Learning dynamics and heterogeneity of spatial-temporal graph data for traffic forecasting. IEEE Trans. Knowl. Data Eng. **34**(11), 5415–5428 (2022)
28. Guo, S., et al.: Self-supervised spatial-temporal bottleneck attentive network for efficient long-term traffic forecasting. In: 2023 IEEE 39th International Conference on Data Engineering (ICDE), pp. 1585–1596 (2023)

LW-CQ: ML-Driven Cardinality Estimation for Complex Queries

Jiye Qiu[1], DongHua Yang[1], Mengmeng Li[1], Haifeng Guo[2], Hongqiang Wang[3(✉)], and Hongzhi Wang[1]

[1] Harbin Institution of Technology, Harbin, China
[2] School of Finance, Southwestern University of Finance and Economics, Chengdu 610000, China
[3] General Hospital of Northern Theater Command, Shenyang, China
18109882267@163.com

Abstract. Cardinality estimation is crucial for query optimization, but traditional methods struggle with complex queries. We propose LW-CQ, a lightweight machine learning-based algorithm that improves cardinality estimation for complex queries by enhancing the LW-XGB method. LW-CQ introduces four feature-level improvements and extends support for disjunctive queries and LIKE predicates. Experimental results show that LW-CQ achieves competitive accuracy while significantly reducing training and inference time, making it a promising solution for real-world database applications.

Keywords: Cardinality Estimation · Machine Learning · Complex Queries

1 Introduction

As data continues to grow in volume and complexity, the demands on database management systems (DBMS) have also increased, pushing the limits of traditional database technologies. One of the key challenges in modern DBMS is query optimization, which directly impacts the system's performance. In particular, cardinality estimation, the process of predicting the number of rows returned by a query, is crucial for selecting the most efficient query plan. Accurate cardinality estimates help the database optimizer make informed decisions, reducing execution time and resource consumption. However, traditional cardinality estimation techniques, such as histograms, sampling, and sketching, are often limited in their ability to handle multi-column and multi-table queries, leading to significant estimation errors.

Recent advancements in machine learning (ML) have opened up new possibilities for improving cardinality estimation. ML-based methods have demonstrated the ability to capture complex data correlations that traditional techniques struggle to address. These methods can be broadly categorized into two types: query-driven and data-driven approaches. Query-driven methods treat cardinality estimation as a regression problem, indirectly learning the underlying data distribution from the queries themselves. On

Y. Yu et al. (Eds.): ICPCSEE 2025, CCIS 2673, pp. 417–427, 2026.
https://doi.org/10.1007/978-981-95-2566-9_28

the other hand, data-driven methods directly model the data distribution and represent queries in a form that is compatible with machine learning models.

Although ML-based approaches significantly improve estimation accuracy, they come with challenges, such as high training and prediction costs, and are often limited to supporting basic query operations like conjunctions and range predicates. These methods also struggle to support more complex query types, such as disjunctive queries and queries involving LIKE predicates, which are common in real-world applications.

To address these challenges, we propose the LW-CQ algorithm, a lightweight machine learning-based cardinality estimation method designed to handle complex queries more efficiently. Inspired by the LW-XGB method, LW-CQ improves estimation accuracy by introducing feature-level enhancements and extends support for disjunctive queries and LIKE predicates. Our experiments show that LW-CQ achieves competitive accuracy while significantly reducing training and inference time, making it a practical solution for modern database optimization.

The remainder of this paper is structured as follows: Sect. 2 reviews related work in the field of cardinality estimation. Section 3 presents the LW-CQ algorithm, including the improvements made over the LW-XGB method. Section 4 describes the experimental setup and results. Finally, Sect. 5 concludes the paper and discusses potential future directions.

2 Related Work

Cardinality estimation plays a crucial role in query optimization in database systems, and traditional methods such as histograms, sampling, and sketches have been widely used in practice. Histograms are typically used for singlecolumn queries but suffer from high space complexity when handling multicolumn and multi-table queries. Sampling-based approaches, on the other hand, can handle correlations between columns and tables but may fail when dealing with sparse or complex queries [11].

2.1 Traditional Cardinality Estimation Methods

Histograms divide the data into fixed ranges or buckets to estimate cardinality, with variations such as equal-width and equal-height histograms. While histograms perform well for single-column queries, they struggle with multi-column and multi-table queries due to the curse of dimensionality [12]. Additionally, multi-dimensional histograms are inefficient because their space complexity increases exponentially [3].

Sampling methods, such as index-based sampling [8] and biased sampling [7], provide a way to estimate cardinality in the presence of multi-column and multi-table queries. These methods sample a small subset of data and apply the query predicates to estimate the cardinality for the entire database. However, they suffer from issues related to zero tuples and may not handle sparse queries well.

Sketching techniques, such as Count-Min Sketch [2] and HyperLogLog [4], offer probabilistic solutions to cardinality estimation by hashing data. While they are fast and space-efficient, they can be less precise in the presence of skewed data distributions [6].

2.2 Machine Learning-Based Cardinality Estimation

Recent work has focused on applying machine learning (ML) techniques to cardinality estimation to improve accuracy, particularly for complex queries involving multiple columns and tables. Machine learning-based approaches can be classified into two categories: query-driven and data-driven methods [11].

Query-Driven Methods. Query-driven cardinality estimation methods often rely on supervised learning. Notable approaches include the use of neural networks (DNN), recurrent neural networks (RNN) [1], and gradient boosting models like XGBoost [11]. For example, the QuickSel method [13] uses a mixture model to estimate cardinality by minimizing the mean squared error between the model and the uniform distribution [12]. However, these methods typically have limitations in supporting complex predicates like LIKE, EXISTS, and ANY, which are commonly used in real-world applications [6].

Data-Driven Methods. Data-driven approaches, such as the Feedback-KDE model, aim to directly learn the data distribution. These models assume that the tuples follow a Gaussian distribution and use kernel density estimation to improve accuracy. Although these methods provide good estimates for simple queries, they often fail to scale efficiently when handling multi-table joins [6].

2.3 Challenges and Opportunities

Despite significant progress, ML-based cardinality estimation faces challenges such as high training and prediction costs, model complexity, and limited support for complex queries. Most existing ML-based approaches, like LW-XGB, are restricted to basic range predicates and do not handle complex operations like disjunctions or LIKE predicates [11]. Our work seeks to address these limitations by introducing the LW-CQ method, which improves both accuracy and efficiency, extending the model's capability to handle disjunctive queries and LIKE predicates.

3 Methodology

In this section, we introduce the LW-CQ algorithm for cardinality estimation. Our approach builds upon the LW-XGB model by improving its accuracy and extending its functionality to support complex queries, such as disjunctions and LIKE predicates.

The LW-XGB algorithm is based on a tree model, which treats cardinality estimation as a regression problem. The main advantage of LW-XGB is its ability to efficiently handle large datasets with minimal computational overhead. However, it suffers from low precision in certain cases, particularly when dealing with complex queries involving disjunctions and LIKE predicates.

To address the limitations of LW-XGB, we introduce four feature-level improvements:

3.1 Center and Width Features

In addition to the traditional upper and lower bounds of the query, we incorporate center and width features. These features help the model distinguish between queries with similar bounds but different data densities. For example, two queries with the same width but different data distributions can be differentiated by their center feature. This improvement aims to enhance the model's understanding of the query structure (Fig. 1).

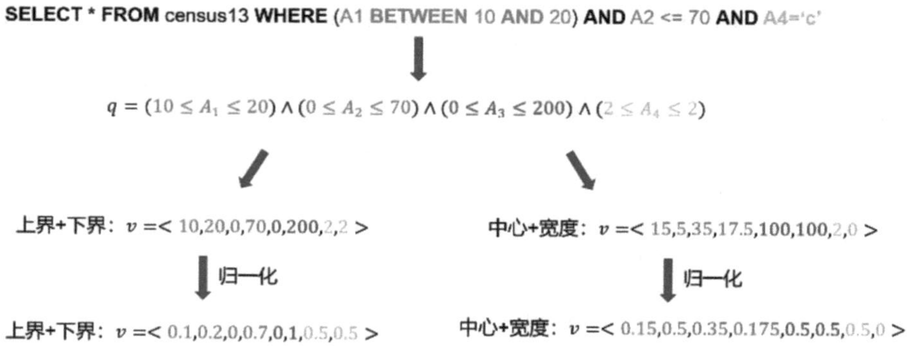

Fig. 1. Query range feature processing.

Position Encoding. Position Encoding As we can see, the method above for generating range features can only generate a single query range for each column attribute. This limitation arises because the number of input range features is fixed, equal to 2 × d (or 4 × d after adding the center and width features). Additionally, unlike MSCN [10], which uses one-hot encoding of attribute column numbers to represent the predicate for specific column features, the range features here do not include such an encoding.

To address these limitations and further enhance the model's performance, position encoding is introduced as an additional feature. This encoding captures the position of the query within the dataset, helping the model better understand the spatial relationships between data points. Figure 2 illustrates how position encoding is applied to a sample query.

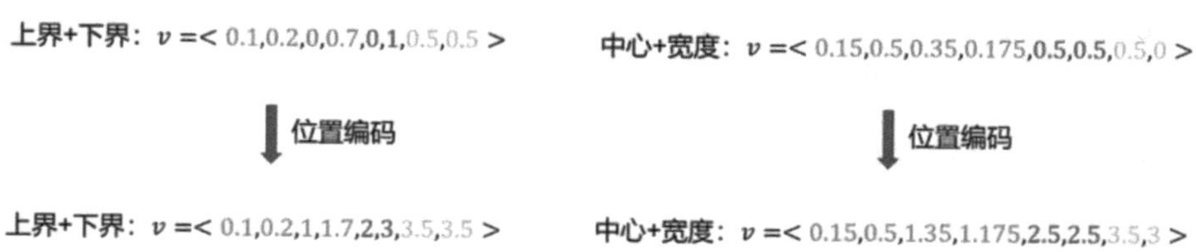

Fig. 2. Position encoding applied to a sample query.

3.2 Materialized Samples

We also introduce materialized samples, which significantly improve the accuracy of cardinality estimates. These samples represent a compressed form of the data, which can be used by the model for more efficient training and inference. The use of materialized samples improves both the precision and speed of the model.

3.3 Data-Driven Methods

To further improve the model's performance, we integrate data-driven methods. Specifically, we leverage existing models such as Naru for data-driven feature extraction, which enhances the overall performance of LW-CQ [6].

3.4 Extension to Disjunctive Queries

The LW-XGB model was originally limited to conjunction queries. To handle disjunctive queries, we introduce two approaches. The first approach [9] uses the inclusion-exclusion principle to handle all conjunction and disjunction queries, but it introduces high time and space complexity. The second approach leverages position encoding to directly input disjunctive queries into the model without increasing computational complexity. These extensions significantly improve the model's ability to handle complex queries involving disjunctions.

3.5 Extension to LIKE Predicates

Another limitation of LW-XGB is its inability to handle LIKE predicates, which are commonly used in string matching queries. We propose an extension that uses a dictionary tree to transform string-based LIKE queries into numeric ranges. This allows the model to process LIKE queries efficiently, as demonstrated in Fig. 3. The use of auxiliary trees further accelerates the processing of in-fix LIKE predicates.

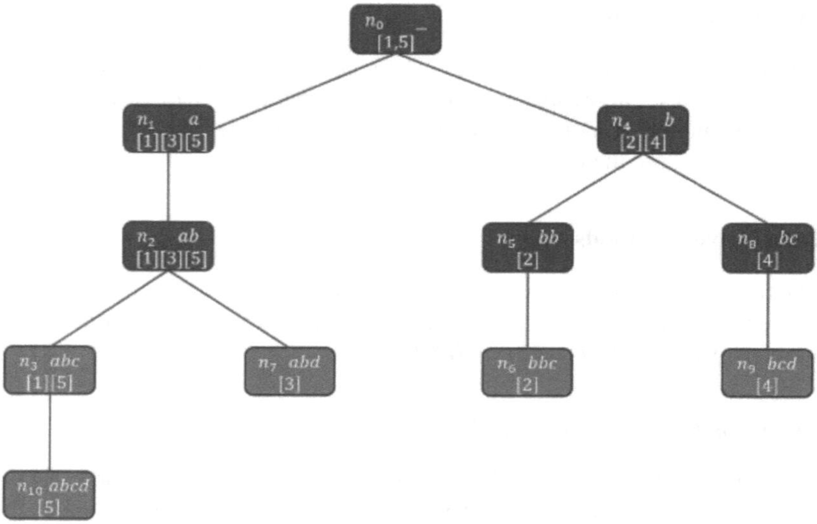

Fig. 3. Extension of LIKE predicates using a dictionary tree.

3.6 Model Selection

In LW-NN/XGB [5] the author transforms the cardinality estimation problem into a regression problem. For this regression problem, two main models are proposed: neural network (LW-NN) and tree-based model (LW-XGB).

Neural networks use ReLU as the activation function of the hidden layer. As a piecewise linear function, ReLU can divide the query space into many local subspaces. In each local subspace, the neural network can be represented as a local linear function of the input features. The maximum number of local subspaces is 2^n, where n is the number of hidden units in the hidden layer. By adding hidden units and hidden layers, the neural network can better fit queries and data distribution, which is conducive to estimating a more accurate cardinality.

In a tree-based model, each leaf node corresponds to a local subspace of the query space. These local subspaces are defined by all nodes on the node to the root path of the tree, so the tree model can be based on the leaf node Points divide the query space into multiple super rectangular subspaces, and the local subspaces of multiple trees can overlap each other. The maximum number of local subspaces is t^l, where t is the number of trees and l is the number of leaves. By increasing the number of trees and leaves, there can be more solutions to divide the query space, so that the model can better perform cardinality estimation.

In subsequent experiments, the tree-based model is significantly better than the neural network model in terms of training and prediction time, and the q-error difference between the two is not much, so in Sect. 3, this paper also uses the tree-based model.

4 Experiments

In this section, we present the experimental setup and results to evaluate the proposed LW-CQ method. We compare the performance of LW-CQ with four representative methods: MSCN, Naru, Postgres, and LW-XGB. The experiments were conducted on three datasets: Census, Forest, and Power. For each dataset, we evaluate the model performance based on the q-error and the training/prediction time.

4.1 Experimental Setup

Platform. The experiments were conducted on a workload generation platform that supports single-table queries. The platform allows for generating queries and providing them to different methods for training and prediction, ensuring a fair comparison. The methods compared include the traditional Postgres histogram-based approach, MSCN, Naru, and the LW-XGB baseline method. The workload generation platform can generate queries based on different parameters, ensuring reproducibility of the experiments.

Evaluation Metrics. We use the widely accepted evaluation metric, q-error, to assess the performance of each method. Additionally, we record the training and prediction times for each method to evaluate their efficiency. The primary focus is on the tradeoff between accuracy (q-error) and model efficiency (time).

Experimental Environment. The experiments were implemented in Python, with the machine learning components utilizing the PyTorch and XGBoost frameworks. The hardware used includes an AMD Ryzen 7 5800X 8-Core processor, an NVIDIA GeForce RTX3070 GPU with 8 GB VRAM, and 32 GB of system memory. The system runs Ubuntu 20.04, Python 3.8, PyTorch 1.11.0 + cu113, XGBoost 1.5.0, and Postgres 13.7.

4.2 Four Feature Improvements

The four feature-level improvements made to the LW-XGB method are evaluated in this section. The improvements include the addition of center and width features, position encoding, materialized samples, and the integration of data-driven methods. Each of these features is compared with existing machine learning methods.

Training and Prediction Time. Figures 4 and 5 show the training and prediction times for each method across the three datasets. As can be seen from Fig. 4 and Fig. 5, the LW-CQ improvements, especially the addition of materialized samples, result in significant improvements in both time and accuracy compared to the MSCN and Naru methods.

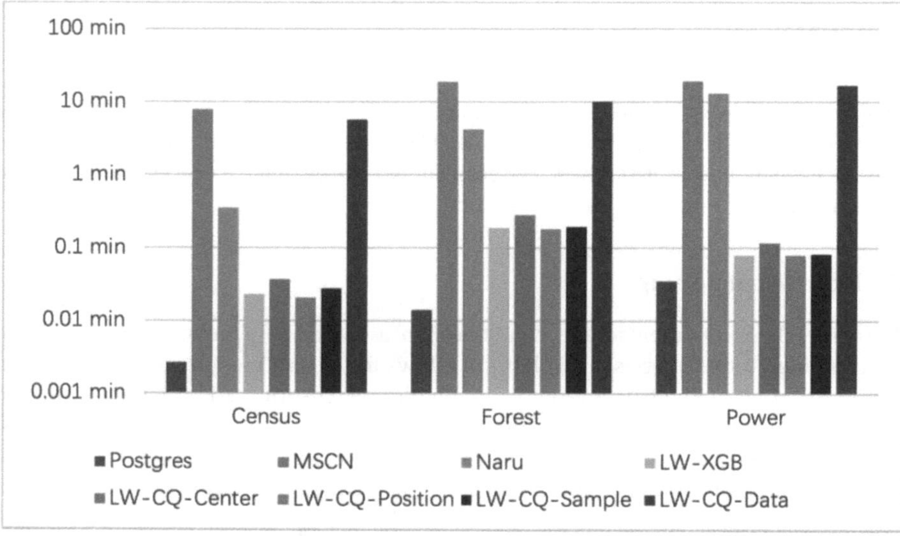

Fig. 4. Training time for various methods across datasets.

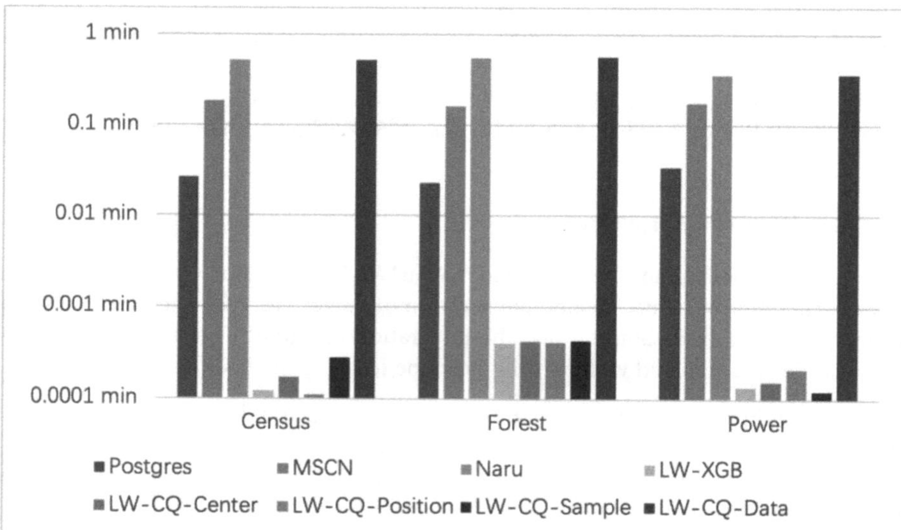

Fig. 5. Prediction time for various methods across datasets.

4.3 Extension to Disjunctive Queries

The model's ability to handle disjunctive queries was evaluated by comparing it with Postgres's histogram-based approach. As shown in Table 1 and Fig. 6 the LW-CQ method significantly reduces q-error while also improving prediction time compared to the traditional methods.

Table 1. Q-error results for cardinality estimation across different methods and datasets.

Dataset	Method	Max	99th	95th	90th	50th	Mean	Gmean
Census	Postgres	22918.0	226.76	30.00	8.86	1.0	17.74	1.73
	LW-CQ-OR	64.00	9.00	3.00	2.00	1.00	1.43	1.18
Forest	Postgres	314432	43575.1	4301.10	935.10	1.05	1638.28	8.69
	LW-CQ-OR	247.00	32.28	10.00	5.00	1.06	2.92	1.61
Power	Postgres	105712	720.95	75.00	22.67	1.03	66.00	2.54
	LW-CQ-OR	267.27	21.60	6.00	3.44	1.08	2.34	1.47

4.4 Extension to lIKE Predicates

For the LIKE predicates extension, we evaluated how well the LW-CQ model could handle queries involving string matching. Figure 7 and Table 2 show the results. The LW-CQ extension outperforms Postgres's traditional methods in both accuracy and speed.

Table 2. Q-error results for cardinality estimation across different methods and datasets.

Dataset	Method	Max	99th	95th	90th	50th	Mean	Gmean
Census	Postgres	36925.0	364.66	45.00	12.43	1.00	26.33	1.92
	LW-CQ-OR	149.00	10.50	3.00	2.00	1.00	1.50	1.20
Forest	Postgres	361047	33913.0	3940.43	847.73	1.0	1423.59	8.02
	LW-CQ-OR	230.83	35.00	10.00	5.00	1.06	3.11	1.62
Power	Postgres	207256	870.09	79.02	22.71	1.03	103.77	2.58
	LW-CQ-OR	533.00	27.01	6.58	3.43	1.08	2.70	1.48

4.5 Summary of Results

The experimental results confirm that the LW-CQ method provides substantial improvements over the traditional Postgres method and other machine learning-based methods. The added features of center and width, position encoding, and materialized samples contribute to both improved accuracy and reduced training/prediction times. Furthermore, the extensions to support disjunctive queries and LIKE predicates offer significant advantages for handling complex queries in real-world applications.

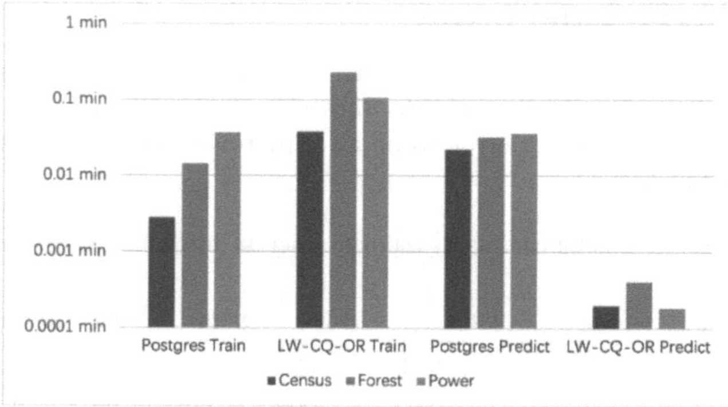

Fig. 6. Training and prediction time for disjunctive queries.

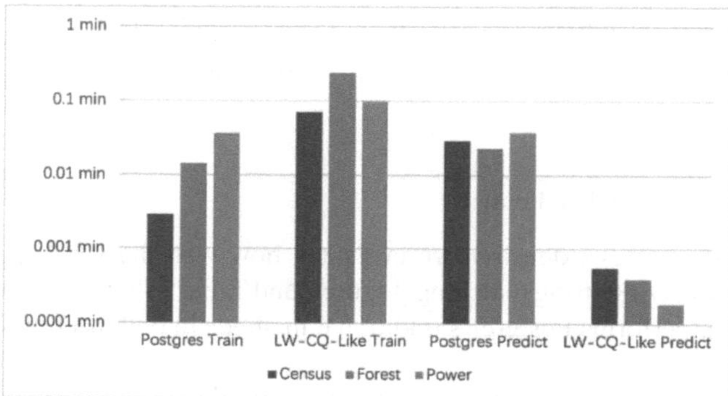

Fig. 7. Training and prediction time for LIKE predicates.

5 Conclusion

In this paper, we have proposed the LW-CQ algorithm for cardinality estimation, which addresses the high cost of training and prediction in machine learning-based methods while also extending support for more complex queries, including disjunctive queries and LIKE predicates. Our approach demonstrates significant improvements in accuracy over traditional methods such as Postgres, particularly in multi-column and multi-table query scenarios.

Through extensive experiments on several datasets, we have shown that LW-CQ achieves competitive accuracy in cardinality estimation, with a much lower training and prediction time compared to other machine learning-based methods. The extension to support disjunctive queries and LIKE predicates further enhances the applicability of LW-CQ in real-world database systems.

Future work will focus on optimizing the model's efficiency, exploring ways to further reduce prediction time without compromising accuracy, and evaluating its performance in larger-scale, production-level environments.

Acknowledgments. This paper is supported by Sichuan Science and Technology Program (2024YFHZ0161).

References

1. Chen, T., Guestrin, C.: XGBoost: a scalable tree boosting system. In: Proceedings of the 22nd ACM SIGKDD International Conference on Knowledge Discovery and Data Mining, pp. 785–794. ACM, New York (2016). https://doi.org/10.1145/2939672.2939785
2. Cormode, G., Muthukrishnan, S.: An improved data stream summary: the countmin sketch and its applications. In: LATIN, pp. 29–38 (2004)
3. Documentation, P.: Row estimation examples. PostgreSQL 13 Documentation (2021). https://www.postgresql.org/docs/13/row-estimation-examples.html
4. Durand, M., Flajolet, P.: Loglog counting of large cardinalities (extended abstract). In: ESA, pp. 605–617 (2003)
5. Dutt, A., Wang, C., Nazi, A., Kandula, S., Narasayya, V.R., Chaudhuri, S.: Selectivity estimation for range predicates using lightweight models. Proc. VLDB Endow. **12**(9), 1044–1057 (2019)
6. Flajolet, P., Martin, G.N.: Probabilistic counting algorithms for data base applications. J. Comput. Syst. Sci. **31**(2), 182–209 (1985)
7. Ganapathi, A., et al.: Predicting multiple metrics for queries: better decisions enabled by machine learning. In: ICDE, pp. 592–603 (2009)
8. Haas, P.J., Naughton, J.F., Seshadri, S., Swami, A.N.: Fixed-precision estimation of join selectivity. In: PODS, pp. 190–201 (1993)
9. Hayek, R., Shmueli, O.: Nn-based transformation of any SQL cardinality estimator for handling distinct, and, or and not. CoRR abs/2004.07009 (2020). https://arxiv.org/abs/2004.07009
10. Kipf, A., Kipf, T., Radke, B., Leis, V., Boncz, P.A., Kemper, A.: Learned cardinalities: estimating correlated joins with deep learning. In: CIDR (2019)
11. Lohman, G.: Is query optimization a "solved" problem? SIGMOD Blog (2014). http://wp.sigmod.org/?p=1075
12. Manegold, S., Boncz, P., Kersten, M.: Generic database cost models for hierarchical memory systems. In: VLDB, pp. 191–202 (2002)
13. Park, Y., Zhong, S., Mozafari, B.: QuickSel: quick selectivity learning with mixture models. In: Proceedings of the ACM SIGMOD International Conference on Management of Data, pp. 1017–1033 (2020)

Application of Blockchain

Prevention of Flash Loan Attacking on the Decentralized Finance System of a Public Blockchain

Yunlong Wang[1], Ran He[2], Haifeng Guo[1(✉)], Hongzhi Wang[3], Yuxi Zhang[2], and Biliang Wang[1]

[1] School of Finance, Southwestern University of Finance and Economics, Chengdu 611130, China
guohf@swufe.edu.cn
[2] School of Management, Harbin Institute of Technology, Harbin 150001, China
[3] School of Information Technology, Harbin Institute of Technology, Harbin 150001, China

Abstract. Decentralized finance (DeFi) has revolutionized traditional financial paradigms by enabling innovative, permissionless financial transactions. Among these, flash loans represent a significant breakthrough, offering rapid liquidity without collateral requirements. However, the very features that make flash loans appealing also expose DeFi ecosystems to severe security threats. This paper presents a systematic analysis of flash loan attack methodologies, their implications, and potential countermeasures. We formalize the problem via a game-theoretic model, delineating the interactions between malicious actors and security mechanisms. Through detailed case studies of major flash loan attacks, we illustrate common exploit strategies and vulnerabilities within smart contracts. Furthermore, we propose a comprehensive, multilayered security framework that integrates real-time anomaly detection, enhanced smart contract verification, decentralized governance improvements, and cross-platform intelligence sharing. Empirical analysis leveraging blockchain security datasets underscores the viability of these mitigative measures. Our findings contribute to the broader discourse on DeFi security by providing a structured approach to mitigating the systemic risks associated with flash loans, thereby enhancing the resilience of decentralized financial systems.

Keywords: Flash loans · decentralized finance · blockchain security · smart contract vulnerabilities · attack mitigation · game-theoretic modelling · real-time anomaly detection · DeFi governance · cybersecurity frameworks

1 Introduction

Financial technology, an interdisciplinary domain that merges finance and technology, has experienced rapid advancements in recent years. The advent of blockchain and the increasing adoption of cryptocurrencies have facilitated the creation of novel financial instruments and service models (Zhou et al., 2021). Among these, flash loans have

Y. Yu et al. (Eds.): ICPCSEE 2025, CCIS 2673, pp. 431–445, 2026.
https://doi.org/10.1007/978-981-95-2566-9_29

emerged as groundbreaking innovations in decentralized finance (DeFi) (Chiu et al., 2022), fundamentally altering lending mechanisms while introducing significant security and regulatory challenges.

Flash loans are characterized by their uncollateralized nature, high execution efficiency, and ephemeral lifecycle (Eskandari et al., 2019). These transactions leverage smart contracts to facilitate borrowing and repayment within a single block execution (Chen et al., 2022; Dijkstra, 2022). While this mechanism enhances market liquidity and trading strategies, it also creates substantial vulnerabilities that adversaries can exploit. Attackers systematically manipulate smart contract interactions, leveraging market inefficiencies and protocol design flaws to conduct price manipulation, protocol arbitrage, and governance exploits. The unrestricted capital access provided by flash loans enables malicious actors to execute these attacks with minimal initial investment, posing a critical threat to DeFi ecosystem stability.

In recent years, an increasing number of flash loan attacks have been recorded, endangering the security of DeFi platforms and user assets (Dos Santos et al., 2022). These methods exploit frequently targeted protocol vulnerabilities, coupled with high market interconnectivity (Xue et al., 2022), allowing attackers to conduct intricate operations within milliseconds. According to SlowMist's blockchain security reports shown in Fig. 1, 464 security breaches were reported in 2023, leading to an estimated loss of \$2.486 billion. Despite a 34.2% reduction in losses compared with those in 2022, when damage reached \$3.777 billion across 303 incidents, flash loan attacks remain a major concern. In particular, approximately 34 of these incidents were flash loan exploits, accounting for approximately \$225 million of the losses. This finding indicates that while flash loan attacks comprised a smaller fraction of total incidents, their financial impact was disproportionately high, underscoring the critical threat they pose to DeFi platforms.

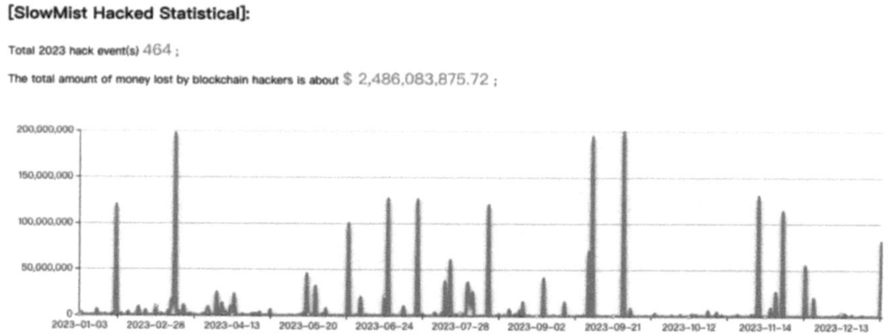

Fig. 1. 2023 hack event(s)

This paper conducts a systematic investigation of flash loan mechanisms, prevalent attack methodologies, and case study analyses to propose an advanced security framework. Our research addresses the following key questions:

How do adversaries systematically exploit vulnerabilities in DeFi smart contracts via flash loans?

What countermeasures can effectively mitigate flash loan attacks while maintaining the permissionless and decentralized nature of DeFi protocols?

How do existing security mechanisms compare, and what deficiencies persist in current defensive strategies?

To provide a holistic perspective, we integrate interdisciplinary approaches, including blockchain security analysis, financial risk modelling, and machine learning-driven anomaly detection. By assessing real-world exploits and empirical transaction data, this study introduces a robust multilayered defense framework designed to strengthen DeFi ecosystem security and resilience against flash loan-related risks.

2 Background and Related Work

2.1 A Subsection Sample

Flash loans, introduced by Aave in early 2020, have emerged as disruptive financial instruments within the DeFi landscape, enabling users to borrow uncollateralized funds on the condition that repayment occurs within a single transaction block. This innovative lending mechanism has provided traders and arbitrageurs with new financial opportunities while simultaneously introducing novel security risks.

The core principle underlying flash loans is their atomicity: either all operations within the transaction are executed successfully, or none occur, ensuring that funds are returned to the liquidity pool in the absence of complete repayment. This feature facilitates various financial strategies, including arbitrage, liquidation optimization, and collateral swapping. However, the same characteristics that grant operational efficiency also enable the exploitation of smart contract vulnerabilities, leading to severe financial repercussions for DeFi platforms.

Mathematically, a flash loan transaction can be expressed as:

$$T = \{B, E, R\}$$

where T represents the transaction process, B is the borrowed capital, E is the executed transaction logic, and R is the repayment condition. The atomicity condition requires the following:

$$R = B + \delta$$

where δ represents the lender's fee. If $R \neq B + \delta$, the entire transaction is reverted, ensuring that no funds are lost from the liquidity pool.

Furthermore, the expected profit Π of an arbitrage strategy utilizing a flash loan is given by:

$$\Pi = \sum_{i=1}^{n} P_i\left(S_i^{out} - S_i^{in}\right) - \delta$$

where S_i^{in} and S_i^{out} represent the asset prices before and after execution at step i, respectively, and where P_i denotes the quantity of assets traded.

2.2 Existing Studies on Flash Loan Attacks

Multiple studies have examined flash loan attacks, focusing on their economic and technical implications. Chen et al. (2022) introduced attack synthesis models to simulate smart contract vulnerabilities and test exploit scenarios. Xue et al. (2022) analysed price manipulation tactics, revealing that attackers often exploit liquidity imbalances to artificially inflate or deflate asset prices. Figure 2 and Fig. 3 show the mechanisms of "centralized finance" and "centralized finance", respectively. These illustrations provide context: in CeFi, transactions rely on centralized intermediaries and delayed settlement, whereas in DeFi, trustless smart contracts execute transactions instantly on the blockchain. This fundamental difference in control and execution explains why flash loan attacks can occur in DeFi— attackers exploit the automated, permissionless nature of DeFi protocols. A flash loan attack can be modelled as an optimization problem:

$$\max_{A} U(A) = \sum_{t=1}^{T} (\alpha_t S_t - \beta_t C_t)$$

where U(A) is the attacker's utility, S_t is the profit derived from exploiting price discrepancies, and C_t represents the transaction costs incurred at each stage t. The parameters α_t and β_t denote the weights of profit and cost at each step, respectively.

Despite these advancements, a gap remains in comprehensive prevention frameworks that integrate technical, economic, and psychological deterrents. Existing solutions focus primarily on smart contract auditing and price oracle security but lack a holistic approach that incorporates behavioral economics and adversarial modelling. Additionally, recent studies in 2023 have proposed on-chain monitoring systems and automated flash loan detection algorithms (e.g., Zhang et al., 2023; Lee et al., 2023), but these approaches target specific attack vectors without providing an integrated solution. We have expanded our literature review to cover these developments, noting that a comprehensive survey of flash loan defenses is lacking. Our work addresses this issue by integrating technical, economic, and governance-based measures into a unified defense framework.

This paper aims to bridge this gap by proposing a defense framework that combines on-chain monitoring, machine learning-based anomaly detection, and decentralized governance improvements to mitigate flash loan exploits effectively.

Fig. 2. Operational Mechanisms of "Centralized Finance".

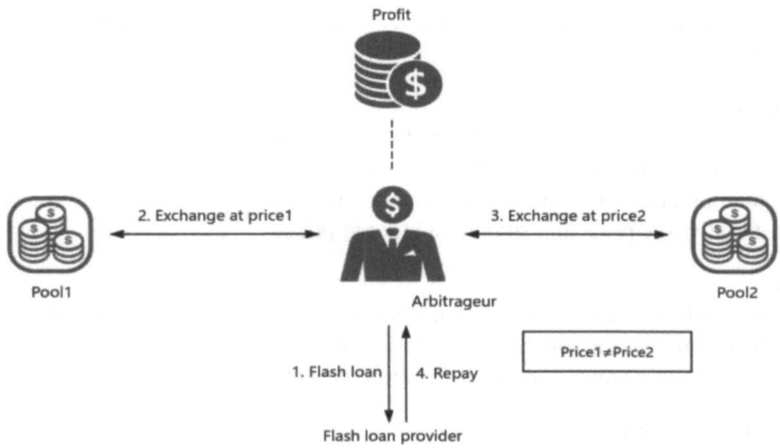

Fig. 3. Operational Mechanisms of "Decentralized Finance" (DeFi).

3 Case Studies and Technical Analysis

This section provides an in-depth analysis of prominent flash loan attacks, detailing their methodologies, exploit mechanisms, and financial repercussions.

3.1 Existing Studies on Flash Loan Attacks

The fault finance attack in August 2021 represents a classic case of price manipulation enabled through recursive stacking mechanisms. Figure 4 illustrates the process of a blockchain attack involving fault financing. The attacker exploited vulnerabilities in the smart contract's liquidity pool valuation model, artificially inflating asset prices before executing profitable arbitrage.

Methodology

Flash Loan Acquisition: The attacker borrowed a substantial sum of WUSD via a flash loan without collateral.

Price Manipulation via Recursive Stacking: By repeatedly stacking and unstacking assets, the attacker artificially increased the value of WEX tokens.

Arbitrage Execution: The manipulated price allows the attacker to exchange the inflated WEX tokens for stable assets at a premium.

The manipulated price at time t can be modelled as:

$$P_t = P_{t-1} \times \left(1 + \frac{V_t}{L_t}\right)$$

where P_t is the manipulated asset price at time t, V_t represents the transaction volume injected by the attacker, and L_t denotes the total liquidity in the pool at that time. The attacker maximizes profit Π by executing trades at artificially inflated prices:

$$\max_{V_t} \Pi = \sum_{i=1}^{n} \left(P_i^{out} - P_i^{in}\right) Q_i - C_i$$

where Q_i represents the quantity of tokens swapped and where C_i is the associated transaction cost.

Exploitation

The attacker exploited a flaw in the WUSDMaster contract's pegging mechanism, which failed to correctly account for rapid liquidity fluctuations.

The manipulated price allowed for an arbitrage cycle wherein assets were acquired at an artificially low price and sold at an inflated price.

Impact

The attack resulted in a total loss of $8 million, significantly affecting the liquidity pool.

Wault Finance temporarily suspended trading to investigate vulnerability and implemented stricter smart contract validations postattack.

Security Implications

Lack of Liquidity Controls: The failure to implement real-time liquidity constraints allowed rapid, recursive price manipulation.

Weak Pegging Mechanism: Inadequate safeguards in stablecoin valuation mechanisms exacerbate the attack's effectiveness.

Absence of Oracle-based Verification: The protocol did not use an external oracle to validate token pricing, increasing vulnerability to internal price distortions.

The fault finance case underscores the importance of integrating robust liquidity monitoring, oracle-based price validation, and recursive stacking prevention in DeFi smart contracts to prevent similar exploits.

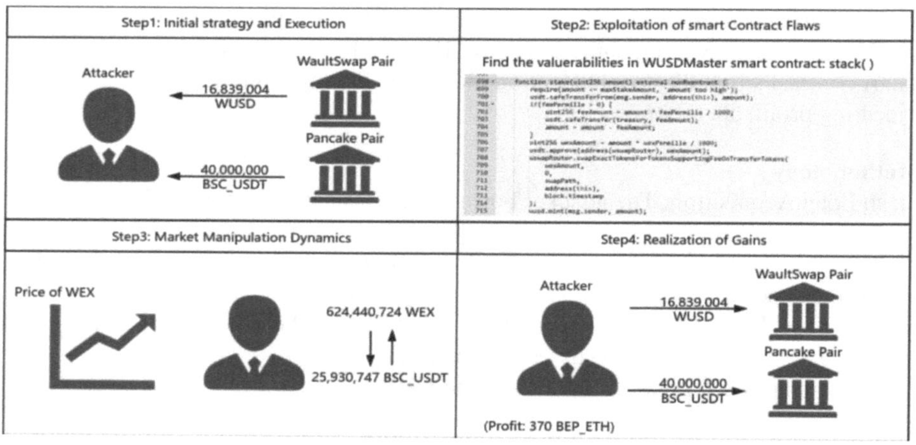

Fig. 4. Case 1: Fault Finance (2021.8).

3.2 Existing Studies on Flash Loan Attacks

The yearn finance attack in April 2023 exposed a critical vulnerability in the yUSDT vault contract, allowing attackers to mint excessive governance tokens through an unrestricted function. As shown in Fig. 5, the process of the yearn finance attack follows these steps. This incident highlighted the risks of improperly secured asset valuation mechanisms in DeFi protocols.

Methodology

Flash Loan Acquisition: The attacker borrowed 3,000,000 USDT to generate liquidity and mint PLP tokens.

Exploiting Contract Vulnerability: A flaw in the buyUSDP function, which was not restricted to administrative controls, allowed the attacker to bypass standard verification processes.

Arbitrary Token Inflation: The attacker artificially increased the governance token supply by exploiting the valuation calculation logic.

The governance token inflation can be expressed as:

$$G_t = G_{t-1} + \frac{F_t}{C_t}$$

where G_t is the governance token supply at time t, F_t is the flash loaned fund used for token minting, and C_t represents the vault's collateral base. The attacker maximizes governance token acquisition by optimizing F_t:

$$\max_{F_t} G_t = \sum_{i=1}^{n} \frac{F_i}{C_i}$$

Exploitation

The unrestricted function enabled the attacker to alter asset valuations and mint a disproportionate number of tokens.

The inflated token supply granted the attacker outsized governance influence, enabling protocol manipulation.

Impact

The exploit resulted in a significant depletion of stablecoin reserves, affecting the protocol's financial stability.

Emergency governance interventions were necessary to restore network integrity and revoke malicious transactions.

Security Implications

Unrestricted Functionality: The lack of proper access controls on the buyUSDP function created an exploitable security loophole.

Weak governance safeguards: The protocol did not implement checks to prevent sudden, unauthorized governance token inflations.

The Yearn Finance case underscores the necessity of restricting administrative controls, implementing governance token caps, and enhancing on-chain verification mechanisms in DeFi protocols.

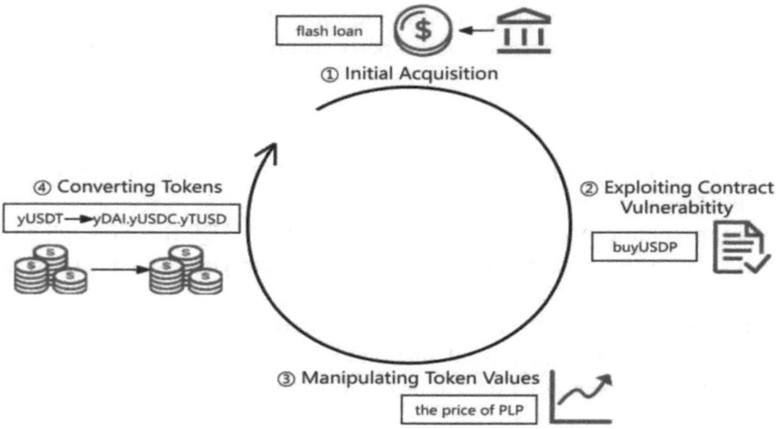

Fig. 5. Case 2: Yearn Finance (2023.4).

3.3 Palmswap Attack (2023)

The Palmswap attack in July 2023 demonstrated the dangers of unauthorized minting of PLP tokens due to missing contract verification checks. The attacker leveraged the contract's minting mechanism to inflate token supply and manipulate liquidity pools. The specific details of Transaction 0x_c9 are illustrated in Fig. 6.

Methodology
Flash Loan Utilization: The attacker borrowed 3,000,000 USDT to mint PLP tokens.

Bypassing Contract Verification: Exploited an oversight in the minting function, which lacked stringent validation mechanisms.

Liquidity Pool Drainage: Artificially minted PLP tokens are used to withdraw stable assets beyond the intended limits.

The token supply expansion is given by:

$$T_t = T_{t-1} + \frac{M_t}{V_t}$$

where T_t represents the total PLP token supply at time t, M_t denotes minted tokens, and V_t is the pool's liquidity at that time. The attacker's gain Π is maximized as:

$$\max_{M_t} \Pi = \sum_{i=1}^{n} (T_i - T_{i-1}) P_i$$

Exploitation
The attacker leveraged weak verification checks in the contract to mint tokens unchecked.

Excess tokens were used to manipulate the asset withdrawal process, draining liquidity.

Impact

The attack caused over $4 million in financial losses and destabilized the liquidity pools of the protocol.

The exploit led to a major overhaul of the contract's verification and minting processes.

Security Implications

Lack of Minting Restrictions: The protocol did not cap unauthorized token issuances.

Weak Verification Mechanisms: The absence of smart contract validation enabled exploitation.

To mitigate such attacks, DeFi protocols must enforce strict minting controls, implement real-time validation of contract interactions, and adopt on-chain monitoring solutions.

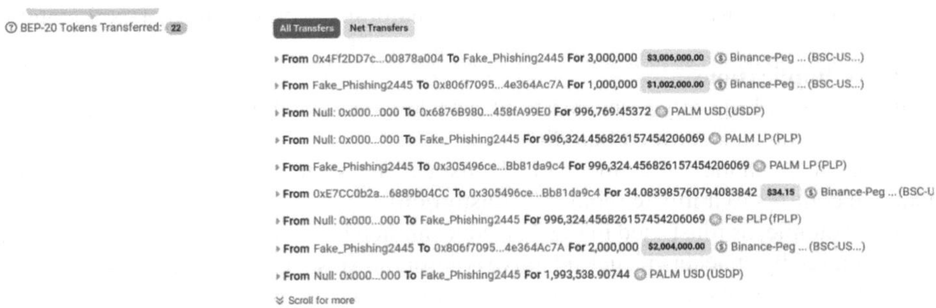

Fig. 6. Specific details of Transaction 0x_c9.

3.4 KyberSwap Attack (2023)

The KyberSwap attack in November 2023 exploited concentrated liquidity mechanisms, allowing the attacker to manipulate asset prices within a narrow price range. The attack resulted in severe liquidity pool depletion. An example of a concentrated liquidity market maker (CLMM) is shown in Fig. 7.

Methodology

Flash Loan Execution: The attacker borrowed WETH to execute liquidity manipulations.

Liquidity Pool Distortion: Used swaps to artificially skew asset prices.

Two-Step Price Manipulation: Conducted rapid successive swaps to push prices beyond their intended bounds before profiting.

Price manipulation within a concentrated liquidity range can be modelled as:

$$P_t = P_{t-1} \times \left(1 + \frac{D_t}{L_t}\right)$$

where D_t represents the depth of liquidity injected by the attacker and where L_t denotes the liquidity concentration in the targeted range. The attacker's optimal profit strategy can be expressed as:

$$\max_{D_t} \Pi = \sum_{i=1}^{n} \left(P_i^{new} - P_i^{old} \right) Q_i$$

Exploitation

The attacker leveraged the nonlinear nature of concentrated liquidity models to create extreme price shifts.

A combination of large swaps and liquidity additions allowed them to systematically distort asset valuations.

Impact

The attack led to a major loss in liquidity pool reserves, affecting multiple trading pairs.

The protocol implemented urgent liquidity guardrails to prevent further exploitation.

Security Implications

Vulnerability in Liquidity Concentration: Attackers exploited the mathematical framework governing concentrated liquidity.

Lack of Dynamic Adjustment Mechanisms: The system does not have real-time adaptive controls to mitigate rapid price distortions.

For example, as illustrated in Fig. 7, a concentrated liquidity market maker (CLMM) can be abused by attackers if such adaptive controls are absent. In the Palmswap case, the attacker leveraged the concentration of liquidity in specific price ranges—a feature of the CLMM—to amplify price manipulation. This underscores the need for mechanisms that dynamically adjust the liquidity distribution when abnormal activity is detected.

To enhance security, DeFi platforms should enforce dynamic liquidity adjustment safeguards, implement automated risk monitoring, and adopt real-time anomaly detection mechanisms.

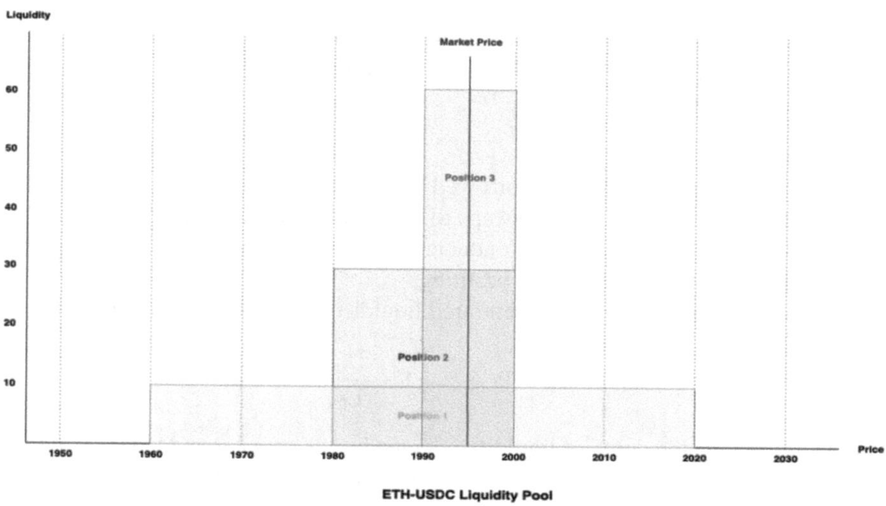

Fig. 7. Example of a concentrated liquidity market maker (CLMM).

By analysing these high-profile case studies, this paper highlights the evolving nature of flash loan exploits and underscores the importance of robust contract verification, liquidity monitoring, and governance restrictions to safeguard the DeFi ecosystem.

4 Formalization of the Problem

To systematically analyse the security vulnerabilities of flash loan attacks and their countermeasures, we employ a game-theoretic framework to model the interactions between an attacker and a decentralized finance (DeFi) protocol. Our model formalizes the probability of a successful attack and the effectiveness of various defense mechanisms.

4.1 Attack–Defense Model

We define a strategic game $G = (A, P, U)$, where A represents the set of possible actions available to the attacker and the protocol defender.P denotes the probability distribution of attack success.U defines the utility functions for both the attacker and the defender.Let B denote the attacker's available budget and L denote the liquidity within a target DeFi protocol. The attack probability $P(A)$ is a function of smart contract vulnerabilities V, available arbitrage opportunities O, and implemented defense mechanisms D:

$$P(A) = f(V, O, D)$$

where D represents countermeasures such as transaction monitoring, dynamic liquidity adjustments, and governance restrictions. Increasing D while minimizing V and O reduces $P(A)$, thereby mitigating attack risk.

4.2 Attacker's Optimization Problem

The attacker aims to maximize their expected profit U_A from an attack, given by:

$$U_A = E[P(A)] \cdot G(A) - C(A)$$

where:

$G(A)$ is the total gain from the attack, including arbitrage profits and governance token control.$C(A)$ is the cost incurred in executing the attack, including transaction fees and slippage penalties. Where $E[P(A)]$ is the expected success probability of an attack, which is based on real-time market conditions and security parameters.

The attacker optimizes U_A by adjusting transaction timing, leveraging liquidity imbalances, and circumventing security measures to maximize arbitrage profits.

4.3 Defender's Optimization Problem

The defender, which represents the DeFi protocol, aims to minimize the probability of a successful attack while maintaining protocol efficiency. The defender's objective function U_D is as follows:

$$U_D = -E[P(A)] \cdot L - C_D$$

where:L is the liquidity at risk from flash loan exploits.

where C_D represents the cost of implementing security measures such as oracle validation and anomaly detection.

To minimize P(A), the defender strategically allocates resources to enhance the detection and mitigation mechanisms. The optimal strategy satisfies:

$$\frac{\partial U_D}{\partial D} > 0, D = \{D_1, D_2, ..., D_n\}$$

where D_i represents individual defense strategies, such as price oracle verification, transaction delay mechanisms, and smart contract audits.

4.4 Nash Equilibrium Analysis

A Nash equilibrium occurs when neither the attacker nor the defender can unilaterally improve their utility. In equilibrium, the defender implements security measures D^* that increase the attack cost beyond profitability, whereas the attacker chooses an optimal strategy A^* that maximizes expected returns within existing constraints.

Mathematically, the equilibrium condition satisfies:

$$U_A(A^*, D^*) = \max_A U_A, U_D(A^*, D^*) = \max_D U_D$$

This equilibrium suggests that security investments should be dynamically adjusted on the basis of the evolving attack landscape and market conditions.

4.5 Practical Implications

Our formalization underscores the importance of continuous security enhancements in DeFi protocols. The key takeaways include:

Adaptive Defense Mechanisms: Implementing machine learning-based anomaly detection can dynamically adjust risk exposure and mitigate arbitrage-based exploits.

Economic incentives for security: Increasing attack costs through penalty mechanisms (e.g., slippage and time lock constraints) discourages malicious actors from exploiting protocol vulnerabilities.

Multi-Layered Defense Strategy: Combining smart contract audits, liquidity safeguards, and governance oversight provides robust protection against flash loan exploits.

By integrating this game-theoretic framework into DeFi security strategies, protocols can systematically mitigate flash loan risk while maintaining decentralized autonomy and efficiency.

5 Proposed Mitigation Framework

To effectively mitigate the risks associated with flash loan attacks, implementing a multifaceted approach that enhances security, strengthens governance mechanisms, fosters collaboration within the DeFi ecosystem, and aligns with regulatory standards is essential. One of the primary areas of focus is improving smart contracts and protocol security.

This includes employing formal verification techniques and automated auditing tools to ensure the integrity of smart contracts before deployment. Fuzz testing plays a crucial role in identifying edge-case vulnerabilities, whereas access controls such as multisignature authentication help prevent unauthorized contract modifications. Time-lock mechanisms for high-risk transactions, coupled with robust price oracle implementations and dynamic liquidity safeguards, can significantly reduce the effectiveness of exploitative arbitrage strategies.

Real-time threat detection and automated response mechanisms are also vital in preventing flash loan exploits. Machine learning-based behavioral analytics can help identify abnormal transaction patterns indicative of malicious activity. By assigning dynamic risk scores on the basis of liquidity fluctuations, transaction volumes, and market anomalies, DeFi platforms can proactively assess potential threats. Automated safeguard triggers embedded in smart contracts can halt or restrict suspicious transactions before they cause irreparable damage, reinforcing protocol resilience.

Collaboration within the DeFi ecosystem is another crucial component of a robust mitigation framework. Establishing cross-platform intelligence-sharing networks allows protocols to exchange security threat information in real time, thereby reducing attack success rates. Additionally, strengthening governance mechanisms is essential for mitigating the risks associated with governance-based flash loan exploits. Implementing multiblock confirmations for high-value transactions, dynamically adjusting governance token voting power, and introducing emergency veto measures can collectively prevent governance manipulation.

Given the global nature of DeFi, regulatory coordination is necessary to address cross-border security threats. By enhancing collaboration with blockchain forensics firms, DeFi platforms can track and recover stolen assets more effectively. Standardizing security best practices through regulatory alignment ensures a consistent approach to mitigating flash loan attacks. Furthermore, the introduction of legal frameworks specifically addressing flash loan exploitation can provide a structured approach to handling systemic risk and reinforcing investor protection.

A well-integrated mitigation strategy that combines smart contract security enhancements, real-time monitoring, industry-wide collaboration, and regulatory oversight is fundamental for reducing the risks posed by flash loan attacks. While maintaining the decentralized ethos of DeFi, these measures help fortify financial ecosystems against evolving security threats, ensuring stability and trust within the blockchain economy.

6 Conclusion

This paper provides a comprehensive examination of flash loan attacks, detailing their mechanisms, case studies, and systemic risks to the decentralized finance ecosystem. By analysing historical attacks, we have identified key vulnerabilities within smart contract implementations, governance mechanisms, and liquidity pool dynamics that allow adversaries to exploit DeFi protocols with minimal capital investment.

Through formalized modelling, we have structured the problem within a game-theoretic framework, highlighting the dynamic interplay between attackers and protocol defenders. Our findings suggest that reducing attack feasibility requires minimizing contract vulnerabilities, limiting arbitrage opportunities, and strengthening defensive measures.

To address these security concerns, we propose a holistic mitigation framework that integrates smart contract security enhancements, real-time anomaly detection, cross-platform intelligence sharing, decentralized governance improvements, and regulatory coordination. Strengthening smart contract security through formal verification, fuzz testing, and access control mechanisms is vital for preemptively identifying and eliminating exploitable weaknesses. Implementing real-time monitoring systems that leverage machine learning-based anomaly detection and automated response mechanisms can effectively counteract emerging threats. Enhancing governance frameworks through multiblock confirmations, adaptive voting mechanisms, and emergency veto powers can prevent governance-based exploits.

Furthermore, fostering industry-wide collaboration through standardized threat intelligence-sharing mechanisms enables DeFi platforms to proactively counteract security risks. Finally, aligning with regulatory efforts through blockchain forensics and cross-border regulatory harmonization ensures a structured approach to flash loan risk mitigation while preserving DeFi's decentralized nature.

As DeFi continues to evolve, the sophistication of attacks will also advance. The proposed mitigation strategies must be continuously adapted to address emerging vulnerabilities and adversarial tactics. Future research should focus on the development of adaptive machine learning models for attack detection, improved cryptographic mechanisms for secure governance, and scalable regulatory frameworks that balance innovation with security. By implementing a proactive, multilayered security framework, DeFi protocols can mitigate flash loan attack risk and sustain the long-term stability and growth of the decentralized financial ecosystem.

Acknowledgement. This work was supported by the Sichuan Science and Technology Program (2024YFHZ0161).

References

Zhou, L., Qin, K., Cully, A., Livshits, B., Gervais, A.: On the just-in-time discovery of profit-generating transactions in DeFi protocols (2021)

Eskandari, S., Moosavi, S., Clark, J.: SoK: transparent dishonesty: front-running attacks on blockchain. In: Bracciali, A., Clark, J., Pintore, F., Rønne, P., Sala, M. (eds) FC 2019. LNCS, vol. 11599, pp. 170–189. Springer, Cham (2019). https://doi.org/10.1007/978-3-030-43725-1_13

Chiu, J., Ozdenoren, E., Yuan, K., Zhang, S.: On the inherent fragility of DeFi lending (2022)

Chen, Z., Beillahi, S.M., Long, F.: FlashSyn: flash loan attack synthesis via counter example driven approximation. CoRR, abs/2206.10708 (2022)

Dijkstra, E.W.: A note on two problems in connexion with graphs. In: Dijkstra, E.W. (ed.) His Life, Work, and Legacy, pp. 287–290. Springer, Cham (2022)

Dos Santos, S., Singh, J., Thulasiram, R. K., Kamali, S., Sirico, L., Loud, L.: A new era of blockchain-powered decentralized finance (DeFi)-a review. In: Computers, Software, and Applications Conference (COMPSAC) (2022)

Xue, Y., et al.: Preventing price manipulation attack by front-running. In: Sun, X., Zhang, X., Xia, Z., Bertino, E. (eds.) ICAIS 2022. CCIS, vol. 1588, pp. 309–322. Springer, Cham (2022). https://doi.org/10.1007/978-3-031-06764-8_25

Zhang, Y., Zhang, C., Zhang, Y., Luo, X.: FAD: a flash loan attack detection system based on graph neural network. In: Proceedings of the 32nd USENIX Security Symposium (2023)

Lee, S., Song, J., Kim, Mn., Kim, H.: Detecting flash loan attacks using temporal graph embedding. IEEE Trans. Inf. Forensics Secur. **18**, 554–569 (2023)

The Efficiency of Momentum Factors in Cryptocurrency

Jinyi Zhao[1], Haifeng Guo[2(✉)], and Yuxi Zhang[3]

[1] Institute of Chinese Financial Studies, Southwestern University of Finance and Economics, Chengdu 611130, China
[2] School of Finance, Southwestern University of Finance and Economics, Chengdu 611130, China
guohf@swufe.edu.cn
[3] Business School, Harbin Institute of Technology, Harbin 150001, China

Abstract. Based on the unique characteristics of the cryptocurrency market, this study explores the effectiveness of the momentum factors in the cryptocurrency market and the underlying mechanism. By constructing long-short investment portfolios using momentum strategies and backtesting historical data of cryptocurrency from 2014 to 2023, it was found that the forecasting capacity of the momentum factors for cross-sectional excess returns in the cryptocurrency market exhibits significant temporal heterogeneity. Specifically, from 2014 to 2018, the momentum factors failed to capture the cross-sectional expected returns of cryptocurrency. However, from 2019 to 2021, momentum strategies were able to achieve statistically significant excess returns, particularly with the four momentum factors tested in 2020 showing notable performance. In contrast, all momentum strategies failed from 2022 to 2023. Further analysis revealed that this phenomenon is associated with market structure and liquidity, investor behavior and sentiment, and macroeconomic policy shocks, providing theoretical support and practical validation for the excess returns of momentum strategies in the cryptocurrency market.

Keywords: Cryptocurrency · Momentum Factors · Blockchain Application

1 Introduction

With the advent of blockchain technology, cryptocurrencies have emerged as a novel form of digital asset, garnering widespread global attention and experiencing a significant increase in total market capitalization in recent years (see Fig. 1). Cryptocurrencies fulfill a dual role, addressing both the payment convenience for investors and the demand for speculation. On one hand, their decentralized nature, rapid cross-border payment capabilities, and relatively low transaction costs highlight their substantial potential in the payment sector. On the other hand, the 24/7 trading mechanism, high retail participation, regulatory gaps, and frequent black swan events contribute to the high volatility observed in the cryptocurrency market (see Fig. 2), attracting numerous speculators aiming to achieve excess returns through short-term trading. The unique characteristics

Y. Yu et al. (Eds.): ICPCSEE 2025, CCIS 2673, pp. 446–458, 2026.
https://doi.org/10.1007/978-981-95-2566-9_30

of the cryptocurrency market, including its high volatility, decentralized structure, and rapidly evolving market dynamics, not only merit further investigation into the driving factors behind its development but also offer a distinctive context for testing asset pricing theories.

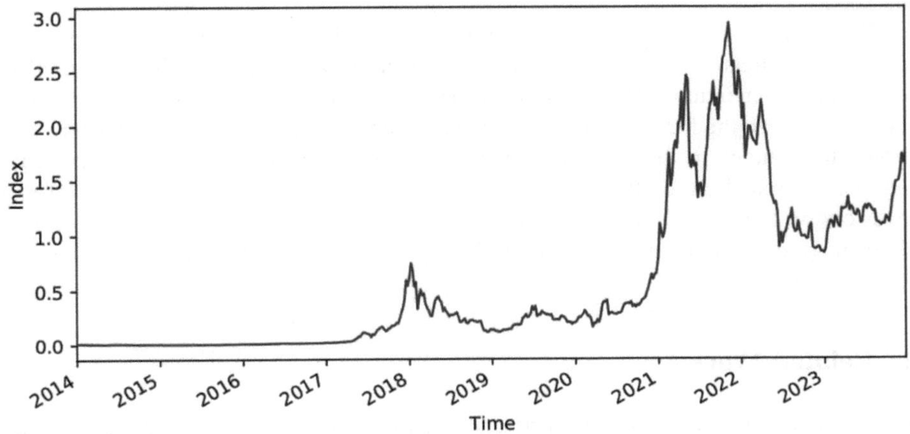

Fig. 1. Line Chart of the Total Market Capitalization Change of the Cryptocurrency Market from 2014 to 2023, with the Unit of 1 Trillion US Dollars.

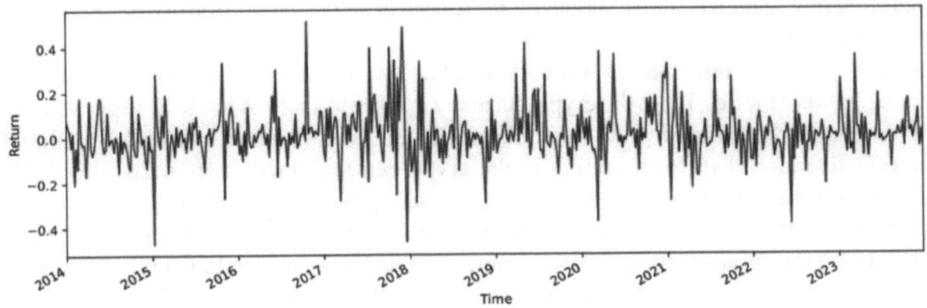

Fig. 2. Weekly Return of the Cryptocurrency Market from 2014 to 2023.

Existing research has demonstrated that excess returns can be achieved in the cryptocurrency market by leveraging specific categories of factors [1]. The momentum effect refers to the phenomenon where investment assets that have performed well (or poorly) over a certain historical period continue to exhibit similar performance in the future [2]. The significance and persistence of the momentum effect vary across different market environments, macroeconomic cycles, and asset characteristics. Compared to traditional financial markets, the cryptocurrency market remains in a relatively nascent stage. On one hand, the number of market participants continues to grow, and trading activity is on the rise. On the other hand, aspects such as price formation mechanisms, market efficiency, and investor behavior remain largely unexplored, providing fertile ground for studying the evolving effectiveness of momentum factors in this market. Additionally, the high volatility and complexity inherent in this emerging market may lead to fundamental differences in the driving logic of momentum factors compared to traditional markets.

For these reasons, this study aims to investigate whether momentum strategies can generate significant excess returns in the cryptocurrency market and whether the effectiveness of momentum factors exhibits strong temporal heterogeneity. Furthermore, it seeks to analyze the factors, events, or cryptocurrency-specific characteristics that contribute to changes in the predictive power of momentum factors on cryptocurrency returns. Ultimately, the study aims to provide recommendations for cryptocurrency investors on how to formulate appropriate investment strategies based on the unique characteristics of momentum factors in the cryptocurrency market, as well as insights for policymakers.

2 Related Works

The literature relevant to this study primarily examines the differences in the significance and persistence of momentum effects in the cryptocurrency market from four perspectives: market environment, time cycles, asset characteristics, and macroeconomic policy shocks.

From the perspective of market environment, traditional asset pricing theory posits that the momentum effect relies on delayed information dissemination and investor underreaction, which is firstly shown by Jegadeesh et al. [2]. According to the weak form of the Efficient Market Hypothesis of Fama, E. F [3], investors cannot predict future returns using past information. Grobys and Sapkota [4] have shown that momentum strategies were ineffective in the cryptocurrency market from 2014 to 2018, suggesting an increase in market efficiency. As the market evolved, Liu, Tsyvinski, and Wu [1] explored the effectiveness of factor investing in the cryptocurrency market from 2014 to 2020, focusing on market capitalization, momentum, trading volume, and volatility, identifying ten cryptocurrency characteristics that formed successful long-short strategies. Between 2017 and 2021, momentum and basis-momentum factors continued to yield statistically significant excess returns [5]. Due to the high volatility and positive skewness of the cryptocurrency market, Li et al. [6] pointed out there exists a MAX momentum premium independent of idiosyncratic volatility and skewness premiums.

With regards to time cycles, the inconsistent performance of momentum strategies in the short and long term results from the combined effects of market structure, investor behavior, and the intrinsic mechanisms of momentum effects. This phenomenon is more pronounced in the cryptocurrency market due to its lower market efficiency and greater investor behavioral biases. Tzouvanas et al. [7] indicated that momentum effects are crucial in short-term cryptocurrency portfolios but tend to dissipate over longer periods. Moreover, the cryptocurrency market evolves faster than stock markets, with shorter iteration cycles, making momentum and disposition effects only effective in the short term [8].

In terms of asset characteristics, cryptocurrencies can be categorized into four types based on their primary use cases: general currency, platform, equity, and security tokens, with significant market segmentation observed among these categories [9]. Previous research has explored the pricing and returns of specific cryptocurrency types. For instance, Zhang et al. [10] implied that the number of transaction nodes, address nodes, average balance of address nodes, and the diameter of the address network in the Bitcoin transaction network exhibit positive correlations with Bitcoin prices. Additionally, deep learning models based on attention mechanisms, such as Attention-LSTM and Attention-GRU, have been employed to predict Bitcoin price trends [11]. Chen and Zhang [12] stablecoins pegged to fiat currencies exhibit more balanced performance, while non-fiat-pegged stablecoins demonstrate unique hedging capabilities under extreme market conditions.

With respect to macroeconomic shocks, Hou and Guo [13] showed that the cryptocurrency market is sensitive to shocks from political, economic, and technological events, with economic events eliciting the most pronounced responses. The computational power demand from cryptocurrency mining activities has created linkages between the cryptocurrency market and the energy market. During periods such as the U.S.-China trade friction, the COVID-19 pandemic, and the Russia-Ukraine conflict, the total spillover level between the cryptocurrency market and the energy market significantly increased [14].

3 Methodology

3.1 Research Design

Construction of the Cryptocurrency Market Index. To ensure consistency with the empirical data used in this study, cryptocurrencies with a market capitalization exceeding \$1,000,000 were selected. The prices of these cryptocurrencies were weighted according to their market capitalization to construct a cryptocurrency market index. This index serves as a reflection of the overall market trends and developmental trajectory of the cryptocurrency market. The specific formula for calculating the cryptocurrency market index is as follows:

$$Market_t = \sum^{i} \frac{P_{i,t} \times Marketcap_{i,t}}{\sum^{i} Marketcap_{i,t}} \tag{1}$$

Construction of Long-Short Investment Portfolios. This study employs a zero-investment long-short strategy to examine the effectiveness of momentum strategies in the cryptocurrency market. Specifically, cryptocurrencies with higher price increases are held long, while those with lower price increases are held short. At the end of each week, cryptocurrencies are sorted into five groups based on their momentum factor values, ranging from the smallest to the largest. The first group (lowest momentum) is shorted, and the fifth group (highest momentum) is held long, forming a 5–1 portfolio. This approach aims to test whether the momentum factor can predict cross-sectional returns in the cryptocurrency market. The portfolio return is calculated as the market-capitalization-weighted average of the returns of the cryptocurrencies within the portfolio.

Calculation of Returns. Cryptocurrency returns are derived from daily price data. Weekly returns are computed as the average of daily returns within the week. The daily return of a cryptocurrency is calculated using the following formula:

$$r_{i,t} = \frac{P_{i,t} - P_{i,t-1}}{P_{i,t-1}} \tag{2}$$

3.2 Data

Sample and Data Processing. The cryptocurrency data used in this study were sourced from CoinMarketCap, a renowned platform for cryptocurrency-related information. The cryptocurrency prices on this platform are calculated as the volume-weighted average of prices across various markets. The selected cryptocurrencies must meet the following criteria: they must be traded on public exchanges, report the last traded price and the trading volume over the past 24 h via an Application Programming Interface (API), and have non-zero trading volume on at least one supported exchange. This study utilizes daily cryptocurrency data from January 1, 2014, to December 31, 2023, with cryptocurrency prices represented by their daily closing prices in US dollars. Following the research design of Liu et al. [1], the data frequency was adjusted to weekly, and cryptocurrencies with a market capitalization of less than $1,000,000 were excluded. The data were further processed by encoding cryptocurrencies based on their names and issuance dates, resulting in a final sample of 5,061 cryptocurrencies. The dataset includes both existing and historically delisted cryptocurrencies to mitigate potential survivorship bias. The one-month US Treasury bill yield was obtained from the CSMAR database.

The names and definitions of the main variables used in this study are presented in Table 1.

Descriptive Statistics. Table 2 presents the descriptive statistical results of cryptocurrencies with a market capitalization greater than $1,000,000 between 2014 and 2023. In terms of the number of samples, the quantity of cryptocurrencies has significantly increased from 156 in 2014 to 2,212 in 2023, reflecting the rapid expansion of the cryptocurrency market. Regarding market capitalization, both the mean and median values have shown an upward trend, especially in 2021 and 2022, where the mean market capitalization reached $12.3719 billion and $9.2082 billion, respectively, indicating a significant increase in market value. As for trading volume, both the mean and median values experienced an abnormal peak in 2020, reaching 8,106,221 thousand and 259.13 thousand, respectively, which may be related to market volatility or specific events.

Table 1. The symbols and definitions of the main variables..

Symbol	Meaning	Definition
r_1	Momentum Factor 1	Weekly return rate of the cryptocurrency for week t − 1
r_2	Momentum Factor 2	Sum of weekly return rates of the cryptocurrency for the past two weeks
r_3	Momentum Factor 3	Sum of weekly return rates of the cryptocurrency for the past three weeks
r_4	Momentum Factor 4	Sum of weekly return rates of the cryptocurrency for the past four weeks
r_{4-1}	Momentum Factor 5	Weekly return rate of the cryptocurrency for week t − 1 minus the weekly return rate for week t − 4
r_M	Cryptocurrency Market Index Weekly Return Rate	Construct the cryptocurrency market index with market capitalization weighting based on cryptocurrency prices and calculate the weekly return rate
r_{BTC}	Bitcoin Weekly Return Rate	Average the daily returns of Bitcoin within the week to obtain the weekly return rate
r_{ETH}	Ethereum Weekly Return Rate	Average the daily returns of Ethereum within the week to obtain the weekly return rate
r_{XRP}	Ripple Weekly Return Rate	Average the daily returns of Ripple within the week to obtain the weekly return rate
r_f	Risk-Free Rate	Annualize the yield of the 1-month U.S. Treasury bond to a weekly basis
P_t	Cryptocurrency Price	The price of the cryptocurrency on day t
P_{t-1}	Historical Cryptocurrency Price	The price of the cryptocurrency on day t − 1
Market	Cryptocurrency Market Index	Constructed cryptocurrency market index based on market capitalization weighted cryptocurrency prices
Marketcap	Market Capitalization of Cryptocurrency	Total market value of the cryptocurrency in circulation, calculated as the current price of the cryptocurrency multiplied by the circulating supply

Table 2. Descriptive Statistics A

		Marketcap (million)		Volume (thousand)	
Year	Num	Mean	Median	Mean	Median
2014	156	231.11	3.68	1106.91	29.73
2015	85	132.55	2.67	1165.08	9.35
2016	185	157.23	3.26	1782.94	18.23
2017	852	423.02	8.82	18677.41	110.49
2018	1644	355.87	8.9	20746.35	108.47
2019	1472	252.31	5.43	64320.99	138.91
2020	1933	358	6.73	8106221	259.13
2021	2882	1237.19	14.98	200590.5	588.97
2022	2355	920.82	13.46	396454.3	500.42
2023	2212	897.01	14.99	59761.31	460.94
Total	5061	736.9	10.08	1184797	308.93

Table 3 presents the descriptive statistical results of market index returns and the returns of representative cryptocurrencies Bitcoin (BTC), Ethereum (ETH), and Ripple (XRP). The market index returns exhibit high volatility, with significant right skewness and kurtosis. In contrast, the return volatility of Bitcoin, Ethereum, and Ripple is relatively lower. These results indicate that although the volatility of cryptocurrency market returns is lower than that of traditional market indices, their return distributions still possess certain asymmetry and kurtosis characteristics, which warrant further research into their risks and market behavior.

Table 3. Descriptive Statistics B

	Mean	Median	SD	Skewness	Kurtosis
r_M	0.459	0.022	5.795	18.846	383.919
r_{BTC}	0.012	0.007	0.103	0.053	2.213
r_{XRP}	0.021	0	0.199	2.787	14.651
r_{ETH}	0.028	0.012	0.161	1.123	4.871

4 Empirical Analysis

This study begins by constructing long-short portfolios to examine the effectiveness of momentum factors in the cryptocurrency market from 2014 to 2023 (Table 4). The results indicate that none of the five tested momentum factors can predict the cross-sectional returns of cryptocurrencies, which stands in stark contrast to the conclusions of previous research on the effectiveness of momentum strategies in the cryptocurrency market from 2014 to 2020. In subsequent sections, the study conducts annual examinations of the effectiveness of cryptocurrency momentum factors to explore changes in their predictive power for cross-sectional returns and the potential underlying influencing factors.

Table 4. Cryptocurrency Market Momentum Strategies Excess Returns

Year	r1	r2	r3	r4	r4-1
2014–2023	6.3446	8.9733	8.4523	8.0694	7.363
	(1.0022)	(1.0081)	(0.9744)	(0.9662)	(1.0028)
2014–2020	0.0446***	0.0558***	0.0495***	0.0361**	0.0448**
	(2.8047)	(3.5039)	(3.1001)	(2.0815)	(2.529)

*t statistics in parentheses, * $p < 0.05$, ** $p < 0.01$, *** $p < 0.001$*

4.1 Annual Examination of the Effectiveness of Cryptocurrency Momentum Factors

Table 5 presents the annual test results of excess returns for momentum strategies in the cryptocurrency market from 2014 to 2023. The findings reveal significant variations in the effectiveness of momentum strategies across different years. During the early stages of the cryptocurrency market in 2016, the excess returns for r1 and r4-1 were significantly positive (0.0784, t = 2.4805 and 0.1163, t = 2.189, respectively), indicating the effectiveness of these momentum strategies in that year. However, except for 2016, the excess returns for all factors from 2014 to 2018 were not statistically significant, suggesting limited effectiveness of momentum strategies during this period. In 2019, the r4-1 strategy yielded an excess return of 0.0356 (t = 1.9091), which was significant at the 10% level, indicating some degree of effectiveness for momentum strategies that year. The year 2020 marked the peak of momentum strategy performance, with r1, r2, r3, r4, and r4-1 all generating significantly positive excess returns. Notably, the excess returns for r1 and r2 reached 0.2046 (t = 3.7022) and 0.1574 (t = 2.6477), respectively, both significant at the 1% level. However, in 2021, the effectiveness of momentum strategies diminished, with only the r4 factor achieving significantly positive excess returns. By 2022 and 2023, momentum strategies had completely lost their effectiveness. The varying annual performance of momentum strategies in the cryptocurrency market provides crucial empirical evidence for further research into the momentum effect and its underlying driving factors in this market.

Table 5. Annual Excess Returns of Momentum Strategies in the Cryptocurrency Market

Year	r1	r2	r3	r4	r4-1
2014	0.0141	0.0112	0.0657	0.0146	0.0146
	(0.2799)	(0.2205)	(1.415)	(0.2905)	(0.2905)
2015	−0.0212	−0.0067	−0.0254	−0.0093	0.0279
	(−0.7357)	(−0.2805)	(−1.0971)	(−0.3922)	(0.9463)
2016	0.0784**	0.0537	0.0603	0.0221	0.1163**
	(2.4805)	(1.3)	(1.5521)	(0.5702)	(2.189)
2017	−0.0362	0.0112	0.0132	0.1668	−0.0623
	(−1.0841)	(0.2853)	(0.4059)	(0.8189)	(−1.4944)
2018	−0.0303	−0.0097	−0.0531	−0.0906	0.0018
	(−1.0856)	(−0.342)	(−1.4042)	(−1.5996)	(0.0533)
2019	0.0149	0.0026	0.0261	0.0086	0.0356*
	(0.9415)	(0.174)	(0.6761)	(0.3803)	(1.9091)
2020	0.2046***	0.1574***	0.0968**	0.0758*	0.076*
	(3.7022)	(2.6477)	(2.0434)	(1.6929)	(1.7687)
2021	−0.0011	0.1015	0.1138	0.1321*	−0.0362
	(−0.016)	(1.2151)	(1.6146)	(1.8481)	(−1.0029)
2022	2.7467	81.1486	2.875	−2.834	65.6249
	(1.0022)	(1.0005)	(1.0291)	(−0.8199)	(0.9992)
2023	−0.0511	0.1536	0.0631	0.0332	0.0067
	(−1.145)	(1.6426)	(1.1796)	(0.4792)	(0.1315)

t statistics in parentheses, $^{*}p < 0.05$, $^{**}p < 0.01$, $^{***}p < 0.001$

4.2 Phase-Based Examination of the Effectiveness of Cryptocurrency Momentum Factors: 2014–2018

Table 6 presents the excess returns of five momentum strategies during the period from 2014 to 2018. The results indicate that none of the momentum factors achieved statistically significant excess returns during this phase. In the early stages of the cryptocurrency market, retail investors dominated, while institutional investors and mature derivative instruments were largely absent. The market structure was underdeveloped, and technological infrastructure was insufficient, leading to chaotic price fluctuations. As a result, momentum strategies struggled to capture statistically significant excess returns.

Table 6. Excess Returns of Momentum Strategies (2014–2018)

Year	r1	r2	r3	r4	r4-1
2014–2018	0.0166	0.0321	0.0236	0.016	0.0206
	(0.8644)	(1.6202)	(1.1473)	(0.7556)	(0.982)

t statistics in parentheses, $^{*}p < 0.05$, $^{**}p < 0.01$, $^{***}p < 0.001$

4.3 Phase-Based Examination of the Effectiveness of Cryptocurrency Momentum Factors:2019–2021

Table 7 presents the excess returns of five momentum strategies during the period from 2019 to 2021. The results indicate that, during this phase, momentum factors exhibited strong predictive power for cross-sectional excess returns in the cryptocurrency market. With the exception of r4-1, which was ineffective, all other momentum factors generated significant excess returns at the 1% significance level. This phenomenon can be analyzed from the perspectives of macroeconomic conditions, market attention on cryptocurrencies, and cryptocurrency liquidity.

From a macroeconomic standpoint, in March 2020, the Federal Reserve initiated its fourth round of quantitative easing in response to the COVID-19 pandemic. This boosted market optimism and increased investors' risk appetite, leading to capital inflows into high-risk assets such as cryptocurrencies and amplifying the momentum effect in the cryptocurrency market. From the perspectives of market attention and liquidity, the rise of decentralized finance (DeFi) in 2019 and the subsequent "DeFi Summer" in 2020 heightened market trading activity and reduced transaction costs. In the second half of 2020, institutional investors began entering the cryptocurrency market on a large scale, further increasing market attention. By 2021, the use of derivative tools expanded on the foundation of DeFi and liquidity mining, bringing additional liquidity to the cryptocurrency market. The non-fungible token (NFT) market experienced explosive growth in 2021, attracting a surge of investors and creators and driving the development of related infrastructure. Taken together, these factors created favorable conditions for implementing momentum strategies and constructing portfolios to achieve excess returns in the cryptocurrency market from 2019 to 2021.

Table 7. Excess Returns of Momentum Strategies (2019–2021)

Year	r1	r2	r3	r4	r4–1
2019–2021	0.067**	0.0993***	0.067**	0.0556**	0.0384
	(2.4876)	(2.9984)	(2.3704)	(2.3192)	(1.8027)

t statistics in parentheses, $^{*}p < 0.05$, $^{**}p < 0.01$, $^{***}p < 0.001$

4.4 Phase-Based Examination of the Effectiveness of Cryptocurrency Momentum Factors: 2022–2023

Table 8 presents the excess returns of five momentum strategies during the period from 2022 to 2023. The results indicate that all momentum strategies failed during this time. From the perspective of market liquidity, the Terra-Luna collapse in May 2022, one of the largest crashes in cryptocurrency history, caused significant investor losses. Nearly 400,000 users faced liquidation due to LUNA, with single liquidation amounts reaching up to $10 million and total liquidation values approaching $1 billion. The sharp decline of Luna not only impacted the Terra ecosystem but also triggered a broader crisis of confidence across the entire cryptocurrency market. From a regulatory standpoint, as compliance processes accelerated, such as the U.S. Securities and Exchange Commission (SEC) intensifying its scrutiny of the securities attributes of cryptocurrencies, institutional funds withdrew due to compliance uncertainties. This shift from speculation-driven markets to fundamental valuation weakened the pricing power of short-term momentum factors. From the perspective of macroeconomic policy, the Federal Reserve's aggressive interest rate hikes led to a significant rise in risk-free rates, making traditional financial market assets more attractive. Consequently, capital flowed from high-risk assets like cryptocurrencies to lower-risk assets such as bonds and the U.S. dollar. The increased correlation between cryptocurrencies and traditional financial markets caused momentum factors and traditional momentum strategies to simultaneously lose effectiveness during this rate-hike cycle.

Table 8. Excess Returns of Momentum Strategies (2022–2023)

Year	r1	r2	r3	r4	r4–1
2022–2023	1.3984	42.236	-0.0035	-1.3655	31.9263
	(0.9908)	(1.0014)	(-0.0018)	(-0.8108)	(0.9984)

t statistics in parentheses, * $p < 0.05$, ** $p < 0.01$, *** $p < 0.001$

5 Conclusions

This study investigates the effectiveness of the momentum factor in the cryptocurrency market and the underlying reasons for its variations. Previous research has explored the efficacy of factor investing in the cryptocurrency market by examining four dimensions: market capitalization, momentum, trading volume, and volatility (Liu et al., 2022). Ten cryptocurrency characteristics were identified to form successful long-short strategies, generating substantial and statistically significant excess returns. While the data period of that study spanned from 2014 to 2020, our analysis using cryptocurrency data from 2014 to 2023 reveals that the momentum factor fails to capture cross-sectional expected returns in the cryptocurrency market. Furthermore, when constructing momentum-based long-short strategies on an annual basis over the decade from 2014 to 2023, we observe significant temporal heterogeneity in the effectiveness of momentum strategies in the cryptocurrency market. Specifically, this study analyzes the relationship between this phenomenon and various factors, including cryptocurrency characteristics, functionalities, market trends, technological advancements, connections to real-world markets, and significant events, to explore the effectiveness and variations of the momentum factor in the cryptocurrency market. This study demonstrates that the effectiveness of the momentum factor in the cryptocurrency market exhibits significant time-varying heterogeneity, influenced by market trends, liquidity conditions, major events, and technological evolution. Based on empirical analysis, we recommend that investors optimize momentum strategies through dynamic weight adjustments, event-driven strategy suspensions, and technical indicator validations, while remaining vigilant against high-frequency trading slippage, manipulation risks in low-market-cap tokens, and extreme drawdowns caused by black swan events. For policymakers, enhancing liquidity support mechanisms, implementing volatility circuit breakers, and establishing cross-market risk early-warning systems can improve market stability, thereby increasing the applicability of factor-based strategies. The findings emphasize that in the highly volatile and rapidly evolving cryptocurrency market, the application of momentum strategies requires real-time adaptation to market conditions and strict risk management measures. Through an in-depth analysis of the cryptocurrency market, this study sheds light on the applicability of momentum strategies in this emerging asset class and examines their performance under different market conditions. The findings hold practical significance for investors and provide valuable insights for theoretical research and policy formulation in the cryptocurrency market.

Acknowledgments. This paper is supported by Sichuan Science and Technology Program (2024YFHZ0161).

References

1. Liu, Y., Tsyvinski, A., Wu, X.: Common risk factors in cryptocurrency. J. Financ. **77**(2), 1133–1177 (2022)
2. Jegadeesh, N., Titman, S.: Returns to buying winners and selling losers: implications for stock market efficiency. J. Financ. **48**(1), 65–91 (1993)
3. Fama, E.F.: Efficient capital markets: a review of theory and empirical work (1970)
4. Grobys, K., Sapkota, N.: Cryptocurrencies and momentum. Econ. Lett. **180**, 6–10 (2019)
5. Chi, Y., Hao, W., Hu, J., Ran, Z.: An empirical investigation on risk factors in cryptocurrency futures. J. Futur. Mark. **43**(8), 1161–1180 (2023)
6. Li, Y., Urquhart, A., Wang, P., Zhang, W.: MAX Momentum in cryptocurrency markets. Int. Rev. Financ. Anal. **77**, 101829 (2021)
7. Tzouvanas, P., Kizys, R., Tsend-Ayush, B.: Momentum trading in cryptocurrencies: short-term returns and diversification benefits. Econ. Lett. **191**, 108728 (2020)
8. Liu, S., Fang, Y., Wang, S.: Investment strategies in the cryptocurrency market based on disposition effect and momentum effect. Manage. Rev. **6** (2024)
9. Cong, L.W., Li, Y., Wang, N.: Tokenomics: dynamic adoption and valuation. Rev. Financ. Stud. **34**(3), 1105–1155 (2021)
10. Zhang, W., Xie, R., Yu, L., Li, J.: The evolution characteristics of bitcoin transaction network structure and asymmetric linkage with bitcoin price fluctuations. Manage. Rev. **8** (2024)

11. Lee, M.-C.: Bitcoin trend prediction with attention-based deep learning models and technical indicators. Systems **12**(11), 498 (2024)
12. Chen, Z., Zhang, X.: The Role of Stablecoins as a safe haven in the volatility of the cryptocurrency market: based on the DCC-GARCH dynamic model. Mod. Financ. Res. **10** (2024)
13. Hou, W., Guo, W.: Identification and characteristic analysis of step behavior in cryptocurrency. Chin. Manage. Sci. **2** (2025)
14. Lu, X., Huang, N., Zhang, Z., Li, J.: The impact of emergencies on the risk spillover level between the cryptocurrency market and the energy market. Syst. Sci. Math. (2025)

Analysis and Design of an ROS Attack-Resistant Adaptor Signature Scheme for Multiparty Cyclic Cross-Chain

Mengyan Li[1,2], Maoning Wang[1,2](✉), and Meijiao Duan[1,2]

[1] School of Information, Central University of Finance and Economics, Beijing 102206, China
13854139297@139.com

[2] Engineering Research Center of State Financial Security, Ministry of Education, Central University of Finance and Economics, Beijing 102206, China

Abstract. To increase the interoperability and privacy of digital asset transactions in multiparty cross-chain scenarios, Chen et al. proposed a cross-chain protocol based on multi-signatures and adaptor signatures that does not rely on smart contracts but instead depends on underlying scripts and signatures. However, since the protocol was designed without considering concurrent execution scenarios, it may still be vulnerable to certain types of attacks, posing potential security risks in practical applications. Therefore, this paper first presents an ROS attack against Chen et al.'s scheme, in which ROS attack was originally proposed by Benhamouda et al. in 2022 against the Schnorr-type threshold and multi-signatures and our attack exploits the random numbers generated by users in the offline-tolerant mechanism. By using a system of equations construction method, it is possible to forge valid signatures for arbitrary messages. Subsequently, the existing Chen et al.'s scheme is improved by establishing a nonlinear dependency between the random number and the message, without introducing additional interaction rounds or random parameters. This improvement eliminates the solvability of the linear system of equations in the ROS attack, thereby defending against this type of attack. Furthermore, security definitions and proofs are provided, detailing the complete properties of the improved adaptor signature scheme. Therefore, the proposed scheme in this paper enhances the theoretical security strength while maintaining the functionality and efficiency of Chen et al.'s scheme, and hence has more effective practicality.

Keywords: Blockchain · Cross-Chain Swap · Adaptor Signature · Multi-Signature

1 Introduction

Since its inception, blockchain technology has received continuous attention, evolving from a single digital currency system to a multi-domain distributed

Y. Yu et al. (Eds.): ICPCSEE 2025, CCIS 2673, pp. 459–485, 2026.
https://doi.org/10.1007/978-981-95-2566-9_31

ledger technology [15,16,25]. The current global blockchain ecosystem presents a multi-chain landscape, where mainstream public and consortium blockchain systems such as Ethereum, Hyperledger Fabric, and FISCO BCOS have formed independent ecosystems [5]. However, interoperability between different blockchains faces architectural challenges, leading to a fragmented blockchain ecosystem and exacerbating the phenomenon of value silos [22].

Against this backdrop, cross-chain technology has become a key research focus in the era of blockchain 3.0, playing an indispensable role in promoting the integration and development of blockchain ecosystems [7,23]. However, most existing cross-chain solutions rely on smart contracts. Smart contracts provide strong programmability in cross-chain protocols, enabling them to handle complex factors such as exchange rate fluctuations and offline states [11]. Nevertheless, they still have certain shortcomings in terms of privacy protection and security [24]. According to a systematic study by Atzei et al. in 2017, vulnerabilities in smart contracts mainly stem from programming logic flaws, blockchain state visibility, and uncertainties in the execution environment. These vulnerabilities can significantly amplify privacy leakage risks in cross-chain scenarios [1]. Specifically, cross-chain transactions involve operations such as asset locking, state verification, and atomic swaps, which require exposing critical transaction parameters through smart contracts. Attackers can reverse-engineer contract code or monitor on-chain transaction flows to reconstruct users' transaction paths and asset associations. A more covert privacy threat arises from the event logging mechanism of smart contracts. Cross-chain protocols typically rely on event logs for inter-chain state synchronization, but unencrypted log data are permanently recorded on the blockchain. For example, during asset transfers, the Wormhole cross-chain bridge logs the mapping between users' original addresses and target chain addresses in plaintext, enabling third parties to conduct linkage analysis and construct users' multi-chain identity profiles [12]. Empirical research by Chainalysis indicates that more than 34% of cross-chain privacy breaches are directly related to improper handling of contract log data [3]. Furthermore, many cryptocurrencies [14] do not support smart contract functionality at all.

To enhance the generality of cross-chain solutions, some studies have proposed the use of signature schemes to achieve cross-chain feasibility [19]. Unlike smart contracts, signature schemes control transaction verification through cryptographic keys, thereby avoiding privacy leakage risks caused by programming vulnerabilities in smart contracts. One of the common signature schemes for cross-chain transactions is multi-signature [21], in which participants from multiple blockchains can jointly verify transactions on the basis of predefined signature rules, reducing the risk of single points of failure. For instance, by leveraging Bitcoin's Pay-to-Script-Hash (P2SH) and multi-signature functionality, cross-chain transactions can enable decentralized and highly controllable asset exchanges [20].

Furthermore, Chen et al. introduced adaptor signatures as a cryptographic component on this basis, providing additional advantages such as offline tolerance [6]. However, the layering and nesting of multiple cryptographic schemes may expose the overall protocol to additional attack risks —if any cryptographic component is applied beyond its security boundary in a blockchain scenario, the entire protocol will no longer be secure. Notably, both multi-signatures and adaptor signatures are classical cryptographic primitives that were not originally designed for blockchain applications, particularly in cross-chain scenarios. This further increases the potential security risks within these cross-chain transaction protocols.

In summary, the main contributions of this paper are as follows:

(1) First, a type of ROS attack is proposed against the Chen et al. scheme. This attack takes advantage of the scheme's offline tolerance mechanism, where each user generates two random numbers during the signature generation phase to handle different offline scenarios. By combining signatures and random number expressions from concurrent transaction sessions, the attack can forge valid signatures for new messages.

(2) Next, an improved scheme is proposed that avoids extra interaction rounds and random parameters. It establishes a non-linear dependency between the random number and the message, preventing the solvability of the linear equation system in the ROS attack. Additionally, the scheme's security properties, such as unforgeability, pre-signature adaptability, secret value extractability, public verifiability, and offline tolerance, are proven.

2 Review of Chen et al.'s Protocol [6]

Chen et al. proposed an innovative privacy-preserving cross-chain transaction protocol to address privacy limitations, reliance on centralized intermediaries, and a lack of multiparty support in existing solutions [6]. The protocol introduces a pre-adapted signature scheme that combines multisignature and adaptor signatures to enable secure multiparty transactions. By using P2SH encoding, it enforces unlocking conditions via participants' public keys, reducing reliance on smart contracts. Additionally, it ensures offline tolerance, allowing transactions to proceed even if some participants are offline.

2.1 Multi-signature

Traditional multisignature schemes achieve signature aggregation through additive (or multiplicative) operations, making them vulnerable to rogue key attacks. Specifically, a malicious user can arbitrarily select a key pair (X_n, x_n) and claim their public key as X_n, where:

$$X'_n = X_n \cdot \left(\prod_{i=1}^{n-1} X_i\right)^{-1}.$$

In this case, the final aggregated public key is effectively disguised as:

$$apk = X'_n \cdot \prod_{i=1}^{n-1} X_i = X_n,$$

allowing the adversary to generate an invalid multisignature.

To mitigate this issue, Chen et al. adopted a secure multisignature scheme based on Schnorr signatures, which was introduced by Maxwell in 2017 [18]. This scheme enhances both feasibility and practicality in signature aggregation. Before signing, users compute additional parameters a_i via the aggregated public key list $L = \{X_1, X_2, \ldots, X_n\}$ and a hash function H_{agg}, effectively defending against rogue key attacks. The improved signing process follows these steps:

(1)KeyAgg(L): For all $i \in 1, \ldots, n$, compute $a_i = H_{agg}(L, X_i)$ and the aggregated public key: $apk = \prod_{i=1}^{n}(X_i)^{a_i}$.

(2)Sign(apk, a_i, x_i, m): First compute $R = \prod_{i=1}^{n} R_i$. And then derive the global challenge $c = H_{sig}(apk, R, m)$. Finally, compute the partial signature: $s_i = r_i + c \cdot a_i \cdot x_i \mod p$.

(3)AggSig($s_1, \ldots, s_n$): Compute $s = \sum_{i=1}^{n} s_i \mod p$. The final signature is $\sigma = (R, s)$, where s represents the multisignature.

(4)AggVerify($(R, s), L, m$): Using L, R, m,compute the aggregated public key apk and challenge c. The verification passes if the equation holds: $g^s = R \cdot \prod_{i=1}^{n}(X_i)^{a_i \cdot c} = R \cdot apk^c$.

2.2 P2SH

Chen et al. adopted the Pay-to-Script-Hash (P2SH) technique, where the generated script is used not only to specify participants, payment conditions, or other relevant information but also to generate the corresponding set address. Each user's script is primarily used to obtain assets from other participants, where F_{agg} represents the key aggregation function. The script involves CMS, CS, and CLTV, which correspond to the Bitcoin execution codes OP_CHECKMULTISIG, OP_CHECKSIG, and OP_CHECKLOCKTIME VERIFY, respectively. For each user u_x, the script is computed using the formula $time_{init} + x \cdot \Delta t$, where $time_{init}$ denotes the initial time of the protocol and Δt represents the time consumed by each participant's operation in different stages. Additionally, γ and Γ denote the participant's secret value and its corresponding commitment, respectively.

This algorithm aggregates participants' public keys via a specific method and sets a timeout for each type of script. Once the timeout expires, participants can sign the transaction and reclaim the assets. The participants are divided into four groups according to their order: u_0, u_1, u_i(where $i \in 2, 3, \ldots, n-2$), and u_{n-1}. The entire process is illustrated in Fig. 1.

1. For user u_0(the first executor of the protocol), the script contains two public keys: X_0 and X_1. User u_0 is able to collaborate with user u_1 to construct an adaptor signature, thereby gaining the authority to sign a valid signature. Once the signature is revealed, user u_0 will receive $coin_1$ and disclose their private key to the other participants.
2. For user u_1, the script includes three public keys: Γ_0, X_1, and X_2. User u_1 collaborates with user u_2 to construct the adaptor signature, gaining the right to sign a valid signature, and will later reveal their private key γ_1. Since user

```
User_0:
IF
<expiry time> CLTV OP_DROP
<X_1> CS
ELSE
<F_agg(X_0, X_1)> CS
ENDIF
```

```
User_1:
IF
<expiry time> CLTV OP_DROP
<X_2> CS
ELSE
<F_agg(T_0, X_0, X_1)> CS
ENDIF
```

```
User_i(i ∈ {2, ..., n − 2}):
IF
<expiry time> CLTV OP_DROP
<X_{i+1}> CS
ELSE
1
<F_agg(T_{i−1}, X_i, X_{i+1})>
<F_agg(T_0, ...T_{i−2}, X_i, X_{i+1})>
2 CMS
ENDIF
```

```
User_{n−1}:
IF
<expiry time> CLTV OP_DROP
<X_0> CS
ELSE
1
<F_agg(T_0, T_{n−2}, X_{n−1})>
<F_agg(T_0, ...T_{n−3}, X_{n−1})>
2 CMS
ENDIF
```

Fig. 1. Script generation algorithm

u_0's private key was leaked earlier in the protocol, user u_1 does not need to disclose it again.

3. For participant u_i, the script contains two sets of public key combinations: One set is $\{\Gamma_{i-1}, X_{i-1}, X_{i+1}\}$, and the other set is $\{\Gamma_{i-2}, \Gamma_{i-1}, X_i, X_{i+1}\}$.
4. For participant u_{n-1}, the script similarly includes two different types of public key combinations:
   One set is $\{\Gamma_0, X_{n-2}, X_{n-1}\}$, and the other set is $\{\Gamma_0, \ldots, \Gamma_{n-3}, X_{n-1}\}$. Since participant u_0's private key was compromised early in the protocol, subsequent adaptor signature negotiations with u_0 will no longer be needed.

In summary, the secret values revealed by the participants serve as expectations in subsequent transactions, functioning similarly to "special" private keys and random numbers, which are then used to construct multi-signatures.

2.3 Process of Chen et al.'s Protocol

The protocol includes steps such as the preparation phase, construction phase, and operation phase. The preparation phase involves sorting users, generating private key sets, committing, and checking random numbers. The construction phase includes filling in the transaction script, collateralizing the transaction amount, and generating pre-adapted signatures. The operation phase involves constructing adaptor signatures, extracting secret values, unlocking amounts, and other operations. Furthermore, during the pre-adapted signature generation phase, users k and $k+1$ interact through the following signature protocol process:

First, user k and user k+1 broadcast commitments $t_k = H_{com}(R_k)$and $t_{k+1} = H_{com}(R_{k+1})$, respectively. After both parties receive t_k or t_{k+1}, they broadcast R_k or R_{k+1}, where $R_k = g^{r_k}$ and $R_{k+1} = g^{r_{k+1}}$. Here, r_k and r_{k+1} are the random numbers chosen by user k and user k+1, respectively.

Subsequently, user k computes $a_k = H_{agg}(L, X_k)$,$R = \prod_{i=1}^{k} \Gamma_i \cdot R_k \cdot R_{k+1} \cdot \Gamma_k$ and $X = \prod_{i=1}^{k-1} \Gamma_i^{a_i} \cdot X_k^{a_k} \cdot X_{k+1}^{a_{k+1}} \cdot \Gamma_k$and further computes $c = H_{sig}(X, R, m)$, where $L = \{\Gamma_0, \Gamma_1, \ldots, \Gamma_{k-1}, X_k, X_{k+1}\}$, $\Gamma_i = g^{\gamma_i}$, with γ_i being the secret value required for unlocking the transaction records released in the preceding steps, and m being the hash value of the transaction record content. Similarly, user k+1 computes $a_{k+1} = H_{agg}(L, X_{k+1})$ and the corresponding values of R,X,and c.

Next, user k and user $k+1$ compute and broadcast $s_k = r_k + c \cdot a_k \cdot x_k$ and $s_{k+1} = r_{k+1} + c \cdot a_{k+1} \cdot x_{k+1}$, respectively.

Finally, both user k and user $k+1$ can compute $s_{pre} = s_k + s_{k+1}$, and output the pre-adapted signature (c, s_{pre}) for the message m.

The signature verification process is as follows:

$$g^{s_k} = g^{r_k + c\cdot(a_k \cdot x_k)} = R_k \cdot X_k^{c\cdot a_k} \tag{1}$$

$$g^{s_{k+1}} = g^{r_{k+1} + c\cdot(a_{k+1} \cdot x_{k+1})} = R_{k+1} \cdot X_{k+1}^{c\cdot a_{k+1}} \tag{2}$$

$$\begin{aligned} g^{s_{pre}} &= g^{s_k + s_{k+1}} && (3)\\ &= g^{r_k + r_{k+1} + c\cdot(a_k \cdot x_k + a_{k+1} \cdot x_{k+1})} && (4)\\ &= R_k \cdot R_{k+1} \cdot X_k^{c\cdot a_k} \cdot X_{k+1}^{c\cdot a_{k+1}} && (5)\\ &= \tilde{R} \cdot X_k^{c\cdot a_k} \cdot X_{k+1}^{c\cdot a_{k+1}} && (6) \end{aligned}$$

Specifically, Chen et al. pointed out that, to accommodate the potential offline status of previous users during the process of converting pre-adaptor signatures into adaptor signatures, the execution of the scheme is divided into two cases during the operational phase: (1)If the secret value γ_{k-1} of user k-1 has been released, then the adaptor signature of user k is $s_{apt_{k,0}} = s_{pre_{k,0}} + \gamma_{k-1} + c_{k,0} \ldots a_{k-1} \ldots \gamma_{k-1}$. (2)If user k-1 is offline and their secret value γ_{k-1} has not been released, then the adaptor signature of user k is $s_{apt_{k,1}} = s_{pre_{k,1}} + \gamma_0 + \cdots + \gamma_{k-2} + c_{k,1} \ldots (a_0 \ldots \gamma_0 + \cdots + a_{k-1} \ldots \gamma_{k-1})$. This implies that, in the complete scheme, user k must first select two random values: $r_{k,0}$ and $r_{k,1}$, and then commit to and broadcast $R_{k,0} = g^{r_{k,0}}$ and $R_{k,1} = g^{r_{k,1}}$, along with the opening operations for the other users. This allows user k to compute their two signatures $s_{k,0}$ and $s_{k,1}$, from which the corresponding adaptor signatures $s_{pre_{k,0}}$ and $s_{pre_{k,1}}$.

3 Attack on Chen et al.'s Protocol

In 2022, Benhamouda et al. proposed a polynomial-time algorithm to solve the ROS (Random inhomogeneities in an Overdetermined Solvable system of linear equations) problem [2]. An adversary's attack on Schnorr-type signatures can be formulated as follows: Given a system of $n \times \ell$ linear equations modulo q, with ℓ unknowns and random constant terms, the goal is to find an $(\ell + 1) \times \ell$ submatrix of rank ℓ. Benhamouda et al. demonstrated that when $\ell > \log q$, the ROS problem can be solved in expected polynomial time. Furthermore, by combining the ROS-solving algorithm with specific cryptographic applications, an adversary can successfully mount attacks. For example, in the setting of Schnorr-based multi-signatures, if an adversary is allowed to establish ℓ concurrent sessions with an honest participant and input $m_1, \dots, m_\ell$, they can generate the $(\ell+1)$th signature without requesting the participant to sign $m_{\ell+1}$. Similar attacks [13] applied to threshold signature schemes using Gennaro et al.'s distributed key generation (DKG) protocol [9] and two-round MuSig [17] multi-signature schemes.

In this section, we construct an attack on the signature generation part of Chen et al.'s scheme, which is based on the ROS problem-solving method. The ROS attack process is shown below. The attack allows the signer user k to forge a valid signature for any message $m_{\ell+1} \in \{0,1\}^*$ after interacting with an honest signer user k+1 for ℓ iterations (ℓ>log p, where log p is the output length of the hash function H_{sig}).

The attack process is as follows:

(1)Users k and $k+1$ initiate ℓ sessions on different messages $m_1, \dots, m_\ell \in \{0,1\}^*$, where each session is used to negotiate the pre-adapted signature of m_i. In the i-th session, after the preparation phase of the scheme is completed, user k can obtain $R_{i,k+1}$ from user $k+1$. At this point, for each round $i \in [\ell]$, user k selects two random values, denoted as $r_{i,k}^0$ and $r_{i,k}^1$, with the corresponding values:

$$R_{i,k}^0 = g^{r_{i,k}^0}, \quad R_{i,k}^1 = g^{r_{i,k}^1}.$$

Define:

$$R_i^0 = \prod_{i=1}^{k} \Gamma_i \cdot R_{i,k}^0 \cdot R_{i,k+1} \cdot \Gamma_k$$

and

$$R_i^1 = \prod_{i=1}^{k} \Gamma_i \cdot R_{i,k}^1 \cdot R_{i,k+1} \cdot \Gamma_k.$$

Then, he computes:

$$c_i^b = H_{sig}(\tilde{X}, R_i^b, m_i),$$

where b takes values of 0 and 1. If $c_i^0 = c_i^1$, the attack fails. However, owing to the collision resistance of the hash function H_{sig}, the probability of this event occurring is negligible.

(2) Afterward, for the target message $m_{\ell+1}$ to be forged, the $(\ell+1)$-th session is initiated. At this stage, user k defines a polynomial with respect to x_i as follows:

$$\rho := \sum_{i=1}^{\ell} \frac{2^{i-1}(x_i - c_i^0)}{c_i^1 - c_i^0},$$

where $\rho_i, i = 1, 2, \ldots, \ell$ represents the coefficient of x_i in the polynomial ρ, and ρ_0 is the constant term. When the $(\ell+1)$-th session is initiated and enters the parameter preparation phase, user k defines the commitments, broadcasting, and opening operations for $R_{\ell+1,k}$ and $R_{\ell+1,k+1}$. Subsequently, the secret parameters required to unlock the transaction, denoted as $\gamma_{\ell+1,j}, j = 1, 2, \ldots, k-1$, are sequentially released.

(3)At this point, user k computes:

$$\tilde{R}_{\ell+1} = R_{\ell+1,k+1} = \prod_{i=1}^{\ell} R_{i,k+1}^{\rho_i},$$

Then, the following computation is performed:

$$R_{\ell+1} = \prod_{j=1}^{k} \Gamma_{\ell+1,j} \cdot R_{\ell+1} \cdot \Gamma_{\ell+1,k}$$

leading to the following derivation:

$$c_{\ell+1} := H_{sig}(\tilde{X}, R_{\ell+1}, m_{\ell+1}).$$

Next, user k converts $c_{\ell+1} + \rho_0$ into its binary representation: At this point, user k closes the ℓ sessions by utilizing $R_{i,k}^b$ and c_i^b for $b \in 0, 1$, specifically by acquiring $s_{i,k+1}^b = r_{i,k+1} + c_i^b \cdot a_{k+1} \cdot x_{k+1}$ from user $k+1$ during the signing step of the i-th session. Then, using $r_{i,k}^b$, user k computes $s_{i,k}^b$ and further derives $s_{pre_{i,k}}^b$, thereby honestly closing the session. The set of all pre-adaptor signatures computed in the i-th session is stored in the list $\mathbb{S}_{pre_i}$.

(4)Next, user k selects $s_{i,k+1}^{b_i}$ from the pre-adaptor signature list obtained from user $k+1$ in the i-th session using b_i.

(5)Finally, user k completes the forgery of $s_{\ell+1,k+1}$ by defining:

$$s_{\ell+1,k+1} := \sum_{i=1}^{\ell} s_{i,k+1}^{b_i} \cdot \rho_i$$

For message $m_{\ell+1}$, the tuple $(c_{\ell+1}, s_{\ell+1,k+1})$ serves as a valid signature for user $k+1$. The signature verification process proceeds as follows:

$$\begin{aligned}
g^{s_{\ell+1,k+1}} &= g^{\sum_{i=1}^{\ell} s_{i,k+1}^{b_i} \cdot \rho_i} && (7)\\
&= g^{\sum_{i=1}^{\ell} (r_{i,k+1} + c_i^{b_i} \cdot a_{k+1} \cdot x_{k+1}) \cdot \rho_i} && (8)\\
&= g^{\sum_{i=1}^{\ell} r_{i,k+1} \cdot \rho_i} \cdot g^{\sum_{i=1}^{\ell} c_i^{b_i} \cdot a_{k+1} \cdot x_{k+1} \cdot \rho_i} && (9)\\
&= \prod_{i=1}^{\ell} R_{i,k}^{\rho_i} \cdot (X_{k+1}^{a_{k+1}})^{\sum_{i=1}^{\ell} c_i^{b_i} \cdot \rho_i} && (10)\\
&= R_{\ell+1,k+1} \cdot X_{k+1}^{c_{\ell+1} \cdot a_{k+1}} && (11)
\end{aligned}$$

Additionally, defining:

$$s_{\ell+1,pre} := \sum_{i=1}^{\ell} s_{i,k+1}^{b_i} \cdot \rho_i + c_{\ell+1} \cdot a_k \cdot x_k,$$

it follows that for $m_{\ell+1}$, the tuple $(c_{\ell+1}, s_{\ell+1,pre})$, generated jointly by users k and $k+1$, constitutes a valid pre-adaptor signature. The validity of this signature is confirmed by the verification process outlined above.

$$\begin{aligned}
g^{s_{\ell+1,pre}} &= g^{\sum_{i=1}^{\ell} s_{i,k+1}^{b_i} \cdot \rho_i c_{\ell+1} \cdot a_k \cdot x_k} && (12)\\
&= g^{\sum_{i=1}^{\ell} (r_{i,k+1} + c_i^{b_i} \cdot a_{k+1} \cdot x_{k+1}) \cdot \rho_i + c_{\ell+1} \cdot a_k \cdot x_k} && (13)\\
&= g^{\sum_{i=1}^{\ell} r_{i,k+1} \cdot \rho_i} \cdot g^{\sum_{i=1}^{\ell} c_i^{b_i} \cdot a_{k+1} \cdot x_{k+1} \cdot \rho_i} \cdot g^{c_{\ell+1} \cdot a_k \cdot x_k} && (14)\\
&= \prod_{i=1}^{\ell} R_{i,k+1}^{\rho_i} \cdot (X_{k+1}^{a_{k+1}})^{\sum_{i=1}^{\ell} c_i^{b_i} \cdot \rho_i} \cdot X_k^{c_{\ell+1} \cdot a_k} && (15)\\
&= \tilde{R}_{\ell+1} \cdot X_{k+1}^{c_{\ell+1} \cdot a_{k+1}} \cdot X_k^{c_{\ell+1} \cdot a_k} && (16)
\end{aligned}$$

Equation (9) and Eq. (14) are derived from

$$\sum_{i=1}^{\ell} c_i^{b_i} \cdot \rho_i = \rho(c_i^{b_i}) - \rho(0) = c_{\ell+1}$$

Since x_{k+1} is the private key of the honest user $k+1$, the above equation indicates that the signature $s_{\ell+1,k+1}$ of user $k+1$, as well as the pre-adapted signature $s_{\ell+1,pre}$ of users k and $k+1$, can be constructed and verified by user k without obtaining the private key x_{k+1} of user $k+1$. Therefore, this forgery attack is feasible.

4 Improved Protocol

This section proposes an improved method based on Chen et al.'s scheme to enhance its defense against ROS attacks. The improvement involves combining the random number R with the message value m, thereby preventing the forgery of valid signatures through parallel computation.

4.1 Improved Pre-adaptor Signature Algorithm

The pre-adaptor signature algorithm combines the adaptor signature algorithm and multi-signature algorithm. During the execution of protocol, some secret information is released, which can transform the pre-adaptor signature into an adaptor signature. The adaptor signature algorithm introduces additional addition operations and modifies the parameters in the challenge generation process

Table 1. Pre-adaptor Signature Algorithm

Scenario:
Used during the negotiation phase between user$_k$ and user$_{k+1}$.
The secret value γ_k from user$_k$ will be revealed.
The secret values of previous participants (γ_i and corresponding commitments Γ_i) will serve as the private keys in this computation step, i.e., at this point, the public key list is represented as $L = \{\Gamma_0, \Gamma_1, \dots, \Gamma_{k-1}, X_k, X_{k+1}\}$.
Preparation:
1. user$_k$ randomly selects a parameter $r_k \leftarrow \mathbb{Z}_p$, computes $R_k = g^{r_k}$ and $t_k = H_{\text{com}}(R_k)$, and broadcasts t_k.
2. user$_{k+1}$ randomly selects a parameter $r_{k+1} \leftarrow \mathbb{Z}_p$, computes $R_{k+1} = g^{r_{k+1}}$, $t_{k+1} = H_{\text{com}}(R_{k+1})$, and broadcasts t_{k+1}.
3. after receiving t_k and t_{k+1} from each other, R_k and R_{k+1} are broadcast, and their correctness can be verified.
Signature Generation:
4. m is the hash of the transaction content. All users can compute: $$a_i = \begin{cases} H_{\text{agg}}(L, X_i) \text{ when } i = k, k+1 \\ H_{\text{agg}}(L, \Gamma_i) \text{ when } i = 0, \dots, k-1 \end{cases}$$ $$d_i = \begin{cases} H(X_i, R_i, m) \text{ when } i = k, k+1 \\ H(X_i, \Gamma_i, m) \text{ when } i = 0, \dots, k-1 \end{cases}$$ $$\tilde{X} = \prod_{i=0}^{k-1} \Gamma_i^{a_i} \cdot X_k^{a_k} \cdot X_{k+1}^{a_{k+1}}$$ $$R = \prod_{i=0}^{k-1} \Gamma_i \cdot R_k \cdot R_{k+1} \cdot \Gamma_k$$ $$c = H_{\text{sig}}(\tilde{X}, R, m).$$
5. user$_k$ first computes $d_k = H(X_k, R_k, m)$, then computes and broadcasts $s_k = r_k + d_k \cdot r_k + c \cdot a_k \cdot x_k \mod p$.
6. user$_{k+1}$ similarly computes $d_{k+1} = H(X_{k+1}, R_{k+1}, m)$ and broadcasts $s_{k+1} = r_{k+1} + d_{k+1} \cdot r_{k+1} + c \cdot a_{k+1} \cdot x_{k+1} \mod p$.
7. they compute $s_{\text{pre}} = s_k + s_{k+1} = (1 + d_k) \cdot r_k + (1 + d_{k+1}) \cdot r_{k+1} + c \cdot (a_k \cdot x_k + a_{k+1} \cdot x_{k+1}) \mod p$.
8. all the secret values (γ_i) are assumed to be revealed during the protocol execution. Therefore, all partial signatures $s_i = \gamma_i + d_i \cdot \gamma_i + c \cdot a_i \cdot \gamma_i = (1 + d_i + c \cdot a_i) \cdot \gamma_i \mod p$ can be computed, where γ_i also acts as a random value here.
9. all participants can compute $s_{\text{apt}} = s_{\text{pre}} + s_0 + \dots + s_{k-1} \mod p$.
10. user$_k$ computes $s = s_{\text{apt}} + \gamma_k \mod p$ and reveals it.

to establish a secret transmission channel. The multi-signature algorithm introduces additional multiplication coefficients a_i during the signing and verification processes and modifies the first parameter in the challenge generation process, thereby concealing the list of participant public keys. Since these two extensions to the Schnorr signature algorithm are located at different points in the expression of the signature scheme computation, they can be combined. The execution process of the algorithm is shown in Table 1. The public key list L varies depending on the specific situation during the full protocol execution phase. Both the multi-signature and adaptor signature schemes used in the algorithm are based on the Schnorr signature scheme, where the group parameters are (G, p, g), with g as the generator of group G, and p being a k-bit integer representing the prime order of group G. Let λ be the security parameter, and $H(x)$ be a strongly collision-resistant hash function. The hash function H_{com} is used during the commitment phase, H_{agg} is used to compute the aggregation value, and H_{sig} is used to compute the signature.

The verification process for each signature step is as follows:

$$g^{s_k} = g^{r_k + d_k \cdot r_k + c \cdot (a_k \cdot x_k)} = R_k^{1+d_k} \cdot X_k^{c \cdot a_k} \tag{17}$$

$$g^{s_{k+1}} = g^{r_{k+1} + d_{k+1} \cdot r_{k+1} + c \cdot (a_{k+1} \cdot x_{k+1})} = R_{k+1}^{1+d_{k+1}} \cdot X_{k+1}^{c \cdot a_{k+1}} \tag{18}$$

$$g^{s_{pre}} = g^{s_k + s_{k+1}} \tag{19}$$

$$= g^{r_k + r_{k+1} + d_k \cdot r_k + d_{k+1} \cdot r_{k+1} + c \cdot (a_k \cdot x_k + a_{k+1} \cdot x_{k+1})} \tag{20}$$

$$= R_k^{1+d_k} \cdot R_{k+1}^{1+d_{k+1}} \cdot X_k^{c \cdot a_k} \cdot X_{k+1}^{c \cdot a_{k+1}} \tag{21}$$

4.2 Complete Protocol

First, the parameters used in the protocol are shown in the following table:

Table 2. Parameter Definitions

Symbol	Content	Description
$\mathbb{P}$	$\{P_{0,1}, P_{1,2}, \ldots, P_{n-1,0}\}$	The set of P2SH scripts
$\mathbb{D}$	$\{D_{0,1}, D_{1,2}, \ldots, D_{n-1,0}\}$	The set of addresses (script hashes) related to $\mathbb{P}$
$\mathbb{T}$	$\{T_{0,1}, T_{1,2}, \ldots, T_{n-1,0}\}$	The set of transactions that transfer assets from $\mathbb{D}$ to the user's private address
$\mathbb{S}_{pre}$	$\{s_{pre_{1,0}}, s_{pre_{1,1}}, \ldots, s_{pre_{n-1,0}}, s_{pre_{n-1,1}}\}$	The set of all pre-adaptor signatures
$\mathbb{S}_{apt}$	$\{s_{apt_{0,0}}, s_{apt_{1,0}}, s_{apt_{1,1}}, \ldots, s_{apt_{n-1,0}}, s_{apt_{n-1,1}}\}$	The set of all adaptor signatures
$\mathbb{S}$	$\{s_{0,0}, s_{1,0}, s_{1,1}, \ldots, s_{n-1,0}\}$	The set of all valid signatures

In this protocol, there are n participants, where the user u_{n-1} acts as the initiator of the protocol. Each user u_i ($i \in \{0, 1, \ldots, n-1\}$) wishes to exchange the coin $coin_i$ they hold with the next user u_{i+1}. Furthermore, the indexing relationship between users is as follows: j and k represent the previous and next user indices of i, respectively, where $j = (i+1) \mod n$ and $k = (i-1) \mod n$. During the normal execution of the protocol, each user u_i performs operations according to the interaction steps shown in Fig. 2. Whenever user u_k triggers the extraction condition, the properties of the adaptive signature cause the secret value γ_k to be released. With this information, user u_i can compute the complete signature, thus satisfying the extraction condition for the P2SH address storing $coin_j$ and successfully extracting the corresponding coin.

The entire protocol is divided into two main phases: the initialization phase and the execution phase:

Initialization phase:

For all participants user$_i$, where $i \in \{0, 1, \ldots, n-1\}$:

1. Randomly generate a private key $x_i \leftarrow \mathbb{Z}_p$, then compute and broadcast the corresponding public key X_i;
2. Generate the parameter sets $\mathbb{P}$ and $\mathbb{D}$ according to the script generation algorithm;
3. Transfer the asset coin$_i$ to a P2SH address $D_{k,i}$ jointly created by user k and user i;
4. Create a transaction $T_{i,j}$ that transfers the asset from $D_{i,j}$ to user i's private account and broadcast the transaction;
5. Randomly select a parameter $r \leftarrow \mathbb{Z}_p$, and jointly construct a pre-adapted signature $s_{\text{pre}_{i,0}}$ with user j (if needed, construct $s_{\text{pre}_{i,1}}$) and broadcast.

After the initialization phase, all participants obtain the sets $\mathbb{P}$, $\mathbb{D}$, $\mathbb{T}$, and $\mathbb{S}_{\text{pre}}$. If all the data pass verification, continue. Otherwise, terminate.

Execution Phase:

1. User u_0 uses the valid signature $s_{0,0}$ to obtain $coin_1$ from address $D_{0,1}$ (corresponding to script $P_{0,1}$);
2. Other participants can compute $\gamma_0 = s_{0,0} - s_{apt_{0,0}} = s_{0,0} - s_{pre_{0,0}} \mod p$, and $s_{apt_{1,0}} = s_{pre_{1,0}} + \gamma_0 + d_0 \cdot \gamma_0 + c_{1,0} \cdot a_0 \cdot \gamma_0$. Note that a_0 (and all a_i) can be computed according to Table 4.1;
3. u_1 uses the valid signature $s_{1,0} = s_{apt_{1,0}} + \gamma_1$ to obtain $coin_2$ from address $D_{1,2}$ (corresponding to script $P_{1,2}$);
4. Other participants can compute $\gamma_1 = s_{1,0} - s_{apt_{1,0}} \mod p$;
5. For u_i, where $i \in \{2, \ldots, n-2\}$:
   (a) If no participant is offline:
      i. If u_k's secret value γ_k has been revealed, compute $s_{apt_{i,0}} = s_{pre_{i,0}} + \gamma_k + d_{k,0} \cdot \gamma_k + c_{i,0} \cdot a_k \cdot \gamma_k$; the valid signature $s_{i,0} = s_{apt_{i,0}} + \gamma_i \mod p$ can be used to obtain $coin_j$. Other participants can compute $\gamma_i = s_{i,0} - s_{apt_{i,0}} \mod p$;

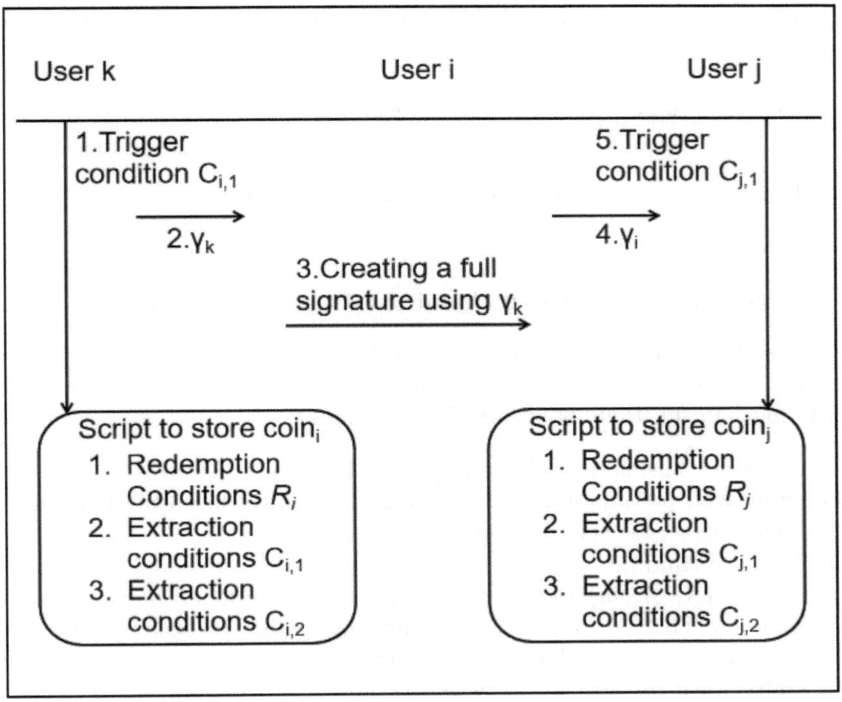

Fig. 2. Example of user i's operation during protocol execution

ii. If u_k is offline and their secret value γ_k has not been revealed, compute $s_{apt_{i,1}} = s_{pre_{i,1}} + \gamma_0 + \cdots + \gamma_{i-2} + d_{0,1} \cdot \gamma_0 + \cdots + d_{i-2,1} \cdot \gamma_{i-2} + c_{i,1} \cdot (a_0 \cdot \gamma_0 + \cdots + a_{i-2} \cdot \gamma_{i-2})$. The valid signature $s_{i,1} = s_{apt_{i,1}} + \gamma_i \mod p$ can be used to obtain $coin_j$. Other participants can compute $\gamma_i = s_{i,1} - s_{apt_{i,1}} \mod p$;

(b) If a participant is offline:

i. If u_k's secret value γ_k has been revealed, compute $s_{apt_{i,0}} = s_{pre_{i,0}} + \gamma_k + d_{k,0} \cdot \gamma_k + c_{i,0} \cdot a_k \cdot \gamma_k$; the valid signature $s_{i,0} = s_{apt_{i,0}} + \gamma_i \mod p$ can be used to obtain $coin_j$. Other participants can compute $\gamma_i = s_{i,0} - s_{apt_{i,0}} \mod p$;

ii. If u_k is offline, the protocol will terminate;

6. For u_{n-1}:
   (a) If u_{n-2}'s secret value γ_{n-2} has been revealed, compute the valid signature:

$$\begin{aligned} s_{n-1,0} = &\gamma_0 + \gamma_{n-2} + r_{n-1,0} + d_{0,0} \cdot \gamma_0 + d_{n-2,0} \cdot \gamma_{n-2} + d_{n-1,0} \cdot r_{n-1,0} \\ &+ c_{n-1,0} \cdot (a_0 \cdot \gamma_0 + a_{n-2} \cdot \gamma_{n-2} + a_{n-1} \cdot x_{n-1}) \mod p \end{aligned}$$

   to obtain $coin_0$;

(b) If only u_{n-2} is offline and their secret value γ_{n-2} has not been revealed, compute:

$$s_{n-1,1} = \gamma_0 + \cdots + \gamma_{n-3} + r_{n-1,1} + d_{0,1} \cdot \gamma_0 + d_{n-3,1} \cdot \gamma_{n-3} + d_{n-1,1} \cdot r_{n-1,0} \\ + c_{n-1,1} \cdot (a_0 \cdot \gamma_0 + \cdots + a_{n-3} \cdot \gamma_{n-3} + a_{n-1} \cdot x_{n-1}) \mod p;$$

(c) If multiple users are offline, the protocol will terminate.

4.3 Resistance to ROS Attacks of the Improved Protocol

In response to the ROS attack described in the previous section, this section analyzes how the improved solution can enhance security by introducing a message binding mechanism, thereby effectively resisting attacks.

1. Core Defense Mechanism of the Improved Protocol
   The improved pre-adaptor signature algorithm introduces the following key modifications in the signature generation phase:
   (a) Message Binding
   $d_i = H(X_i, R_i, m)$ tightly binds the random number R_i with the message m. This ensures that the challenge value of each signature $c = H_{\mathrm{sig}}(\tilde{X}, R, m)$ and that the signature parameter s_{pre} is uniquely associated with the specific message m.
   (b) Nonlinear Dependency
   An additional $d_i \cdot r_i$ term is introduced in the signature computation. Specifically, $s_i = r_i + d_i \cdot r_i + c \cdot a_i \cdot x_i \mod p$, where the hash function d_i introduces nonlinear dependencies, breaking the solvability of the linear equation system in ROS attacks.

2. Resistance Analysis Against ROS Attacks
   In ROS attacks, the attacker initiates multiple parallel sessions and constructs linear equations, using signature parameters from previous sessions to forge signatures for new messages. The defense mechanism in the improved scheme is reflected in the following aspects:
   (a) Message Uniqueness Constraint
   Since each session's d and c are bound to a specific message m, the attacker cannot reuse previous signature parameters by simply replacing the message m. For example, if the attacker tries to forge a signature for message m_{l+1}, they must regenerate the corresponding d and c for m_{l+1}, and this process cannot exploit the linear relationships from previous sessions.
   (b) Nonlinear Disruption of Linear Structure
   The key to the ROS attack is constructing linear equations to solve for the private key components x_i. The additional $d_i \cdot r_i$ term in the improved scheme introduces a nonlinear component (since d_i is the hash result of R_i and m), preventing the attacker from eliminating the random number r_i through linear combinations, thereby preventing the construction of effective equations.

(c) Unpredictability of Parameters
The attacker must preselect the random number R_i and broadcast its commitment t_i. The improved scheme requires that R_i be bound to the message m. If the attacker attempts to modify m to adjust the parameters, they must regenerate R_i and d_i, which invalidates previous commitments, increasing the complexity of the attack.

Following the attack analysis process in Sect. 3.3, since the signature construction method has been adjusted, the signature verification process also needs to be improved accordingly. The following argument demonstrates that the ROS attack cannot be used in the improvement scheme proposed in this section:

The entire attack process proceeds as follows: (1) Users k and $k+1$ open ℓ sessions on different messages $m_1, \ldots, m_\ell \in \{0,1\}^*$. For the i-th session, at the end of the protocol preparation phase, user k can obtain $R_{i,k+1}$ from user $k+1$. At this point, for each round $i \in [\ell]$, the two random values extracted by user k are denoted as $r^0_{i,k}$ and $r^1_{i,k}$, and the corresponding values are $R^0_{i,k} = g^{r^0_{i,k}}$ and $R^1_{i,k} = g^{r^1_{i,k}}$. Define:

$$R^0_i = \prod_{i=1}^{k} \Gamma_i \cdot R^0_{i,k} \cdot R_{i,k+1} \cdot \Gamma_k$$

and

$$R^1_i = \prod_{i=1}^{k} \Gamma_i \cdot R^1_{i,k} \cdot R_{i,k+1} \cdot \Gamma_k$$

Then, compute $d^b_{i,k} = H(X_k, R^b_{i,k}, m_i)$ and $c^b_i = H_{\text{sig}}(\tilde{X}, R^b_i, m_i)$, where $b \in \{0,1\}$. If $c^0_i = c^1_i$, the attack fails. However, owing to the collision resistance of the hash function H_{sig}, the probability of this occurring is negligible.

(2)Afterward, for the target message $m_{\ell+1}$ to be forged, the $(\ell+1)$-th session is initiated. At this point, user k defines the polynomial ρ about x_i as

$$\rho := \sum_{i=1}^{\ell} \frac{2^{i-1}(x_i - c^0_i)}{c^1_i - c^0_i},$$

where $\rho_i, i = 1, 2, \ldots, \ell$ are the coefficients of the x_i-terms in the polynomial ρ, and ρ_0 is the constant term of ρ. When the $(\ell+1)$-th session is initiated and enters the parameter preparation phase, user k defines the commitment, broadcast, and opening operations for $R_{\ell+1,k}$ and $R_{\ell+1,k+1}$. Afterward, the secret parameters $\gamma_{\ell+1,j}, j = 1, 2, \ldots, k-1$ required to unlock the transaction are sequentially released.

(3)At this point, user k first calculates

$$\tilde{R}_{\ell+1} = \prod_{i=1}^{k} R^{\rho_i}_{i,k+1}, \tag{22}$$

and then, by computing

$$R_{\ell+1} = \prod_{j=1}^{k} \Gamma_{\ell+1,j} \cdot \tilde{R}_{\ell+1} \cdot \Gamma_{\ell+1,k}, \tag{23}$$

user k can obtain $c_{\ell+1} := H_{sig}(\tilde{R}_{\ell+1}, X, m_{\ell+1})$. Next, user k converts $c_{\ell+1} + \rho_0$ into binary representation:

$$c_{\ell+1} + \rho_0 = \sum_{i=1}^{\ell} 2^{i-1} b_i, \tag{24}$$

where, using $R_{i,k}^b$, $d_{i,k}^b$, and c_i^b, with $b \in \{0,1\}$, user k closes ℓ sessions. In the signature step of the i-th session, user k obtains $s_{i,k+1}^b = r_{i,k+1} + d_{i,k+1} \cdot r_{i,k+1} + c_i^b \cdot a_{k+1} \cdot x_{k+1}$ from user $k+1$, and closes the session honestly by calculating $s_{i,k}^b$ using $r_{i,k}^b$ and the resulting $d_{i,k}^b$ and $s_{pre_{i,k}}^b$. The set of all pre-adaptor signatures computed in the i-th session is stored in the list $\mathbb{S}_{pre_i}$.

(4) Next, user k selects the i-th session from the signature list obtained from user $k+1$ during the ℓ-th session process using b_i, and retrieves the corresponding $s_{i,k+1}^{b_i}$.

(5) Finally, user k attempts to forge $s_{\ell+1,k+1}$ by defining $s_{\ell+1,k+1} := \sum_{i=1}^{\ell} s_{i,k+1}^{b_i} \cdot \rho_i$.For $m_{\ell+1}$, the constructed "signature" $s_{\ell+1,k+1}$ satisfies:

$$g^{s_{\ell+1,k+1}} = g^{\sum_{i=1}^{\ell} s_{i,k+1}^{b_i} \cdot \rho_i} \tag{25}$$

$$= g^{\sum_{i=1}^{\ell} (r_{i,k+1} + d_{i,k+1} \cdot r_{i,k+1} + c_i^{b_i} \cdot a_{k+1} \cdot x_{k+1}) \cdot \rho_i} \tag{26}$$

$$= g^{\sum_{i=1}^{\ell} r_{i,k+1} \cdot \rho_i} \cdot g^{\sum_{i=1}^{\ell} d_{i,k+1} \cdot r_{i,k+1} \cdot \rho_i} \cdot g^{\sum_{i=1}^{\ell} c_i^{b_i} \cdot a_{k+1} \cdot x_{k+1} \cdot \rho_i} \tag{27}$$

$$= \prod_{i=1}^{\ell} R_{i,k+1}^{\rho_i} \cdot \prod_{i=1}^{\ell} R_{i,k+1}^{\rho_i \cdot d_{i,k+1}} \cdot (X_{k+1}^{a_{k+1}})^{\sum_{i=1}^{\ell} c_i^{b_i} \cdot \rho_i} \tag{28}$$

$$= \prod_{i=1}^{\ell} R_{i,k+1}^{\rho_i \cdot (1 + d_{i,k+1})} \cdot X_{k+1}^{a_{k+1} \cdot c_{\ell+1}} \tag{29}$$

$$= \prod_{i=1}^{\ell} R_{i,k+1}^{\rho_i \cdot (1 + H(X_{k+1}, R_{i,k+1}, m_i))} \cdot X_{k+1}^{a_{k+1} \cdot c_{\ell+1}} \tag{30}$$

The term $X_{k+1}^{a_{k+1} \cdot c_{\ell+1}}$ in Eq. (27) is derived from the term $(X_{k+1}^{a_{k+1}})^{\sum_{i=1}^{\ell} c_i^{b_i} \cdot \rho_i}$ in Eq. (26) by applying the definition of ρ and b_i, i.e.,

$$\sum_{i=1}^{\ell} c_i^{b_i} \cdot \rho_i = \rho(c_i^{b_i}) - \rho(0) = c_{\ell+1}.$$

However, comparing Eq. (18) (the condition for signature validity) with Eq. (26), it is clear that in order for the constructed $s_{\ell+1,k+1}$ to pass the signature verification, the following condition should hold:

$$\prod_{i=1}^{\ell} R_{i,k+1}^{\rho_i} \cdot \prod_{i=1}^{\ell} R_{i,k+1}^{\rho_i \cdot d_{i,k+1}} = \tilde{R}_{\ell+1}^{1+d_{l+1,k+1}} = \prod_{i=1}^{k} R_{i,k+1}^{\rho_i} \cdot \prod_{i=1}^{\ell} R_{i,k+1}^{\rho_i \cdot d_{l+1,k+1}}, \tag{31}$$

That is:

$$\sum_{i=1}^{\ell} d_{i,k+1} \cdot r_{i,k+1} \cdot \rho_i = d_{l+1,k+1} \cdot \sum_{i=1}^{\ell} r_{i,k+1} \cdot \rho_i. \tag{32}$$

Since in each round of communication, $d_{i,k+1}$ is the hash value computed by user $k+1$ for message m_i, and this hash value does not have an explicit linear relationship with the previously generated $r_{i,k+1}$ and $R_{i,k+1}$, the randomness of the hash function output means that Eqs. (29) or (30) hold only with negligible probability. Therefore, on the basis of this structure, existing effective ROS attacks cannot construct a signature $s_{\ell+1,k+1}$ that can pass verification. In other words, owing to the non-linear dependence of the hash value on the previous session states, the uncertainty in the signature generation process prevents the verification process from being reliably completed. This implies that the scheme proposed in this section is resistant to ROS attacks.

Define $s_{\ell+1} := \sum_{i=1}^{\ell} s_{i,k+1}^{b_i} \cdot \rho_i + c_{\ell+1} \cdot a_k \cdot x_k$. For $m_{\ell+1}$, the pair $(c_{\ell+1}, s_{\ell+1})$ generated by user k and user $k+1$ as s_{pre} does not form a valid signature. The verification process is as follows:

$$g^{s_{\ell+1}} = g^{\sum_{i=1}^{\ell} s_{i,k+1}^{b_i} \cdot \rho_i + c_{\ell+1} \cdot a_k \cdot x_k} \tag{33}$$

$$= g^{\sum_{i=1}^{\ell} (r_{i,k+1} + d_{i,k+1} \cdot r_{i,k+1} + c_i^{b_i} \cdot a_{k+1} \cdot x_{k+1}) \cdot \rho_i + c_{\ell+1} \cdot a_k \cdot x_k} \tag{34}$$

$$= g^{\sum_{i=1}^{\ell} r_{i,k+1} \cdot \rho_i} \cdot g^{\sum_{i=1}^{\ell} d_{i,k+1} \cdot r_{i,k+1} \cdot \rho_i} \cdot g^{\sum_{i=1}^{\ell} c_i^{b_i} \cdot a_{k+1} \cdot x_{k+1} \cdot \rho_i} \cdot g^{c_{\ell+1} \cdot a_k \cdot x_k} \tag{35}$$

$$= \prod_{i=1}^{\ell} R_{i,k+1}^{\rho_i} \cdot \prod_{i=1}^{\ell} R_{i,k+1}^{\rho_i \cdot d_{i,k+1}} \cdot (X_{k+1}^{a_{k+1}})^{\sum_{i=1}^{\ell} c_i^{b_i} \cdot \rho_i} \cdot X_k^{c_{\ell+1} \cdot a_k} \tag{36}$$

$$= \prod_{i=1}^{\ell} R_{i,k+1}^{\rho_i \cdot (1+d_{i,k+1})} \cdot X_{k+1}^{c_{\ell+1} \cdot a_{k+1}} \cdot X_k^{c_{\ell+1} \cdot a_k} \tag{37}$$

$$= \prod_{i=1}^{\ell} R_{i,k+1}^{\rho_i \cdot (1+H(X_{k+1}, R_{i,k+1}, m_i))} \cdot X_{k+1}^{c_{\ell+1} \cdot a_{k+1}} \cdot X_k^{c_{\ell+1} \cdot a_k} \tag{38}$$

In each round, $d_{i,k+1}$ is the hash value of message m_i computed by user $k+1$. This hash value is not linearly related to $R_{i,k+1}$ previously generated from earlier sessions. Therefore, user k cannot generate a valid signature $s_{\ell+1}$ as expected. Even if user k attempts to forge the signature, the verification process relies on the correctness of $d_{i,k+1}$, and the forged signature $s_{\ell+1}$ will still fail to meet the verification requirements. Thus, a forgery attack based on this mechanism is theoretically infeasible.

5 Security Discussion

5.1 Security Definition

This section formally describes the security model and security properties in the subsequent proofs. Definition 1 is the security proof goal of the pre-adaptor signature scheme, Definitions 2, 3 and 4 explain the core security concepts in Definition 1, and Definitions 5, 6 are used in the proof of the satisfaction of Definition 2.

Definition 1 (Secure Adaptor Signature Scheme [10]**).** *An adaptor signature scheme is secure if it satisfies aEUF-CMA security, pre-signature adaptability, and secret value extractability.*

Definition 2 (aEUF-CMA Security, Existential Unforgeability under Chosen Message Attack for Adaptor Signatures). *An adaptor signature scheme $\Xi_{\mathcal{R},\Pi_{Sig}}$ is said to be aEUF-CMA secure if for any polynomial-time adversary $\mathcal{A}$, the probability that $\mathcal{A}$ wins the experiment $aSigForge_{\mathcal{A},\mathcal{R},Sig}$ is negligible, i.e.,*

$$\Pr[aSigForge_{\mathcal{A},\Xi_{\mathcal{R}},\Pi_{Sig}}(1^n) = 1] \leq \mathsf{negl}(n).$$

The following is the formal definition:

1. Initialize Q: The challenger creates an empty message query list Q.
2. $(Y, y, \pi) \leftarrow \text{Gen}_{\mathcal{R}}$: The challenger creates a statement/proof pair $(Y, y) \in \mathcal{R}$, and generates a zero-knowledge proof π, where $\mathcal{R}$ is a hard relation.
3. $dID \leftarrow \text{Extract}(ID, s)$: The challenger uses the Extract() function to generate the user's dID.
4. $m^* \leftarrow \mathcal{A}^{\mathcal{O}_S(\cdot),\mathcal{O}_{pS}(\cdot)}(ID)$: An adversary $\mathcal{A}$ can access the signing oracle $\mathcal{O}_S(m_i)$ and the pre-signing oracle $\mathcal{O}_{pSign}(m_j, Y, \pi)$, and obtain corresponding signatures σ_i and $\tilde{\sigma}_j$, where $Q := Q \cup \{m_i\}$, $Q := Q \cup \{m_j\}$. Then, $\mathcal{A}$ outputs a message m^*, where $m^* \notin Q$.
5. $\tilde{\sigma} \leftarrow \text{pSign}_{dID}(m^*, Y, \pi)$: The challenger pre-signs the message m^*, and outputs the signature $\tilde{\sigma}$.
6. $\sigma^* \leftarrow \mathcal{A}^{\mathcal{O}_S(\cdot),\mathcal{O}_{pS}(\cdot)}(\tilde{\sigma}, Y)$: The adversary continues to access the signing oracle $\mathcal{O}_S(m_i)$ and the pre-signing oracle $\mathcal{O}_{pSign}(m_j, Y, \pi)$, and obtain corresponding signatures σ_i and $\tilde{\sigma}_j$, where $Q := Q \cup \{m_i\}$, $Q := Q \cup \{m_j\}$. Given $\tilde{\sigma}$ and Y, $\mathcal{A}$ outputs a forged signature σ^*.
7. Output 0, 1: If the forged message pair (m^*, σ^*) satisfies the condition $m^* \notin Q$ and $\text{Verify}(m^*, \sigma^*, ID) = 1$, the experiment outputs 1; otherwise, it outputs 0.

Definition 3 (Pre-signature Adaptivity). *A scheme $\Xi_{\mathcal{R},\Pi_{Sig}}$ satisfies pre-signature adaptivity if for any $\tilde{\sigma}$ such that $pVerify(m, \tilde{\sigma}, ID, Y) = 1$, we have* $\Pr[\mathit{Verify}(m, \mathit{Adapt}(\tilde{\sigma}, y))] = 1$.

Definition 4 (Extractability of Secret Values). *We say that the secret value is extractable if for any polynomial-time adversary A, the probability of winning the* $aWinExt_{A,\Xi_R,\Pi_{Sig}}$ *experiment is negligible, i.e.,*

$$\Pr[aWinExt_{A,\Xi_R,\Pi_{Sig}}(1^n) = 1] \le \epsilon(n).$$

The following is the definition of the experiment:

1. Initialize Q: The challenger creates an empty message query list Q.
2. $d_{ID} \leftarrow \text{Extract}(ID, s)$: The challenger uses Extract() to create the user's d_{ID}.
3. $(m^*, Y, \pi) \leftarrow A^{\mathcal{O}_S(\cdot),\mathcal{O}_{pS}(\cdot)}(ID)$: Any adversary $\mathcal{A}$ can access the signing oracle $\mathcal{O}_{\text{sign}}(m_i)$ and the p-signature oracle $\mathcal{O}_{pS}(m_j, Y, \pi)$, and obtain corresponding signatures σ_i and σ_j, where $Q := Q \cup \{m_i\}, Q := Q \cup \{m_j\}$. Then, $\mathcal{A}$ outputs m and Y, where $m^* \notin Q$.
4. $\tilde{\sigma} \leftarrow \text{pSign}_{d_{ID}}(m^*, Y, \pi)$: The challenger pre-signs the message m^*, outputting $\tilde{\sigma}$.
5. $\sigma^* \leftarrow \mathcal{A}^{\mathcal{O}_S(\cdot),\mathcal{O}_{pS}(\cdot)}(\tilde{\sigma})$: $\mathcal{A}$ also accesses the signing oracle $\mathcal{O}_S(m_i)$ and the p-signature oracle $\mathcal{O}_{PS}(m_j, Y, \pi)$, and obtains corresponding signatures σ_i and σ_j, where $Q := Q \cup \{m_i\}, Q := Q \cup \{m_j\}$. Given $\tilde{\sigma}$, $\mathcal{A}$ outputs σ^*.
6. $y' := \text{Ext}(\sigma^*, \tilde{\sigma}, Y)$: Given σ^*, $\tilde{\sigma}$, and Y, the challenger extracts y'.
7. Output $\{0, 1\}$: If for (m^*, σ^*, Y, y'), the conditions $m^* \notin Q \wedge \text{Verify}(m^*, \sigma^*) = 1 \wedge (Y, y') \notin R$ are satisfied, the experiment outputs 1; otherwise, it outputs 0.

Definition 5 (Verifiably Encrypted Signature Scheme, VES [4]). *A verifiably encrypted signature scheme (VES)* $\hat{\Sigma}$ *is defined by a standard underlying signature scheme* $\Sigma := $ *(Gen, Sign, Vrfy) and four additional algorithms:*

A verifiable encryption signature scheme (VES) $\hat{\Sigma}$ is defined by an underlying standard signature scheme $\Sigma := (\text{Gen}, \text{Sign}, \text{Vrfy})$ and four additional algorithms:

- $\text{EncGen}_\$ \rightarrow (sk_E, pk_E)$: A probabilistic encryption key generation algorithm that outputs an encryption key pk_E and a decryption key sk_E. There should also exist a valid predicate $\text{valid}(sk_E, pk_E)$, which returns 1 when (sk_E, pk_E) is a valid key pair.
- $\text{EncSign}(sk_S, pk_E, m)_\$ \rightarrow \hat{\sigma}$: A probabilistic encryption signing algorithm that outputs a ciphertext $\hat{\sigma}$ when given a secret signing key sk_S, a public encryption key pk_E, and a message m.
- $\text{EncVrfy}(pk_S, pk_E, m, \hat{\sigma}) \rightarrow \{0, 1\}$: A deterministic encryption signature verification algorithm that outputs 1 if and only if $\hat{\sigma}$ is a valid encryption of the signature on m under pk_E when given a public signing key pk_S, a public encryption key pk_E, a message m, and a ciphertext $\hat{\sigma}$.
- $\text{DecSig}(sk_E, \hat{\sigma}) \rightarrow \sigma$: A (typically deterministic) signature decryption algorithm that outputs a valid signature σ when given a decryption key sk_E and a valid ciphertext $\hat{\sigma}$ under the encryption key.

Definition 6 (EUF-CMA[VES] Security[8]**).** *A VES signature scheme is said to be EUF-CMA[VES] secure if the underlying EUF-CMA signature scheme remains unforgeable in an extended experiment, where, in addition to the traditional signing oracle S, the attacker has access to a signature encryption oracle E.*

The following is the formal definition:

1. Initialize an empty set Q.
2. Generate a signature key pair (sk_S, pk_S).
3. The forger F queries the signing encryption oracle E and the signing oracle S, attempting to generate a forged signature or signed encryption.
4. If the forger successfully generates a signature σ for a message m^* that is not in Q and the signature passes verification ($Vrfy(pk_S, m^*, \sigma) = 1$), then the experiment returns 1, indicating a successful forgery.

5.2 Security Proofs

Next, we prove the correctness and security of the proposed scheme according to the above security definitions. Specifically, to prove that the proposed scheme satisfies Definition 1, we need to sequentially demonstrate that it satisfies the properties described in Definitions 2 to 4.

Consider Definition 2, i.e., aEUF-CMA security. The proof idea of this paper is as follows: Fournier [8] proves that the adaptor signature can be defined by a one-time VES, based on the concept of verifiable encryption signatures (VES). He provides a security model, and the conclusion can be summarized as the following theorem:

Theorem 1 (One-time VES $\Longleftrightarrow$ Adaptor Signature [8]**).** *If a one-time VES signature is EUF-CMA[VES] secure, then the signature is aEUF-CMA secure.*

A one-time VES is a VES with additional properties: given a ciphertext and a decryption signature, the decryption key can be easily recovered. Therefore, it is sufficient to prove that the proposed scheme is VES (Lemma 1), one-time (Lemma 2), and EUF-CMA[VES]-secure (Lemma 3), as follows:

Lemma 1. *Under the discrete logarithm assumption, the proposed scheme is VES, i.e., under the scheme $\Xi_{R_g,\Sigma_{Sch}}$, the correctness of the signature is verifiable.*

Proof: The VES scheme allows the signer to create a signature encryption that can be verified non-interactively. By knowing the message and the public signature key, the verifier can determine that the ciphertext contains a valid signature encrypted with a specific encryption key.

Thus, consider the public key. According to the construction of the multisignature algorithm, the aggregated public key $\widetilde{X}$ is formed by combining multiple individual public keys Γ_i with certain weighted exponents a_i. The specific calculation is as follows:

$$\widetilde{X} = \prod_{i=0}^{n} \Gamma_i^{a_i} \cdot X^a$$

In this process, the generation of the aggregated public key does not rely on the private key x of any individual user, but instead involves a "secret value" γ. Because the exact composition of the aggregated public key cannot be deduced without knowing the private key, $\widetilde{X}$ maintains good privacy by concealing all the public key information of the participants.

Next, assume that $x, k, \gamma, \gamma_i \xleftarrow{\$} \mathbb{Z}_q$, and define $X = g^x$, $K = g^k$, $\Gamma = g^\gamma$, $\Gamma_i = g^{\gamma_i}$. The list of secret values is $\theta = \{\gamma_0, \dots, \gamma_n\}$, which corresponds to $\Theta = \{\Gamma_0, \dots, \Gamma_n\}$. The list of public keys is $L = \{\Gamma_0, \Gamma_1, \dots, \Gamma_n, X\}$. Let γ be the secret value that is to be exposed in the target adaptor signature. The public parameters can be computed as:

$$\begin{aligned} a_i &= H_{agg}(L, \Gamma_i) \quad \text{and} \quad a = H_{agg}(L, X), \\ d_i &= H(X_i, \Gamma_i, m) \quad \text{and} \quad d = H(X, K, m) \\ \tilde{X} &= \prod_{i=0}^{n} \Gamma_i^{a_i} \cdot X^a, \quad R = \prod_{i=0}^{n} \Gamma_i \cdot K \cdot \Gamma, \quad c = H_{sig}(\tilde{X}, R, m). \end{aligned}$$

The signature of the target adaptor is:

$$s_{apt} = (\gamma_0 + \dots + \gamma_n + k) + (d_0 \cdot \gamma_0 + \dots + d_n \cdot \gamma_n + d \cdot k) + c \cdot (a_0 \cdot \gamma_0 + \dots + a_n \cdot \gamma_n + a \cdot x).$$

For the signature of the pre-adaptor $s_{pre} = k + d \cdot k + c \cdot a \cdot x$, it can be completed into s_{apt} because:

$$\begin{aligned} g^{s_{pre}} &= K^{1+d} \cdot (\tilde{X}^a)^c \\ &= \frac{\prod_{i=0}^{n} \Gamma_i^{1+d_i} \cdot K^{1+d}}{\prod_{i=0}^{n} \Gamma_i^{1+d_i}} \cdot \left(\frac{\tilde{X}}{\prod_{i=0}^{n} \Gamma_i^{a_i}} \right)^c \\ &= g^{(\gamma_0 + \dots + \gamma_n + k) + (d_0 \cdot \gamma_0 + \dots + d_n \cdot \gamma_n + d \cdot k) + c \cdot (a_0 \cdot \gamma_0 + \dots + a_n \cdot \gamma_n + a \cdot x)} \\ &\quad \cdot g^{-(\gamma_0 + \dots + \gamma_n + (d_0 \cdot \gamma_0 + \dots + d_n \cdot \gamma_n) + c \cdot (a_0 \cdot \gamma_0 + \dots + a_n \cdot \gamma_n + c \cdot a \cdot x))} \\ &= g^{s_{apt} - ((1 + d_0 + c \cdot a_0) \cdot \gamma_0 + \dots + (1 + d_n + c \cdot a_n) \cdot \gamma_n + (1 + d + c \cdot a) \cdot x)}. \end{aligned}$$

The adaptor signature s_{apt} can be constructed from the corresponding secret value list θ and a valid s_{pre}. Therefore, the correctness of s_{pre} can be verified.

Lemma 2. *The proposed scheme is one-time, meaning that the signature is unconditionally valid and recoverable.*

Proof: We begin with the given equation $\mathrm{EncVrfy}(X, Y, m, (R, \hat{s})) = 1$. On the basis of the definition of this equation, we can derive the relationship between the value R in the ciphertext and the signature parameters, specifically $R = g^{\hat{s}} X^{-c}$, where c is defined by the hash function $c = H(\hat{R}Y \parallel X \parallel m)$. This hash value is computed from the components of the signature, including R, Y, X, and the message m.

Next, using the homomorphic property between the groups $\mathbb{Z}_q$ and $\mathbb{G}$, we can further deduce that the encrypted signature $\tilde{\sigma}$ is, in fact, a valid signature pair. Specifically, this signature pair is of the form (R, s), where $R = \hat{R}Y$ and $s = \hat{s} + y$. In this expression, the known condition $g^y = Y$ indicates the relationship between the public key Y and the private key y.

With these derivations, we can further verify the validity of the signature. By comparing the expression $\hat{R}Y = Yg^{\hat{s}}X^{-c} = g^{\hat{s}+y}X^{-c}$, we can see that the validity of the signature is fully established. This equation shows that, with the given signature parameters (R, s), it is possible to recover a valid signature that satisfies the signature verification requirements.

Additionally, since the scheme has unconditional validity, it also demonstrates its recoverability. Specifically, the recovery function $\mathrm{Rec}((\hat{R}Y, \hat{s}+y), \hat{s})$ can accurately recover the decryption key y, which satisfies $y = s - \hat{s}$. This process demonstrates that the scheme is not only valid in signature generation and verification process, but also offers ideal security and reliability in key recovery.

Lemma 3. *The signature scheme proposed in this paper is EUF-CMA[VES] secure.*

Proof: In the random oracle model, Pointcheval and Stern [?] reduced the standard Schnorr scheme for the discrete logarithm problem (DL problem) to a reduction for the EUF-CMA security of Schnorr signatures. By simulating an additional signature encryption oracle E, this reduction can be transformed into an EUF-CMA(VES) reduction. Furthermore, the original reduction does not involve key prefixes, but the same reduction applies to Schnorr signatures with key prefixes. The reduction R_{DL} simulates E by programming the random oracle H in a manner similar to the typical simulation of the signature oracle S. Since the scheme proposed in this paper follows the Schnorr signature format, the above reduction process is also applicable. Therefore, the EUF-CMA[VES] security of the proposed scheme holds.

The following Lemma 4 proves that the scheme proposed in this paper satisfies the security property (pre-signature adaptability) defined in Definition 3:

Lemma 4. *The adaptor signature scheme Ξ_Π satisfies pre-signature adaptability.*

Proof: For any message $m \in \{0,1\}^*$, any public key $X_i = g^{x_i}$, and any pair (Γ_i, γ_i) such that $\Gamma_i = g^{\gamma_i}$, by Eq. (21) and the definition of s_i, the pre-signature $(\tilde{\sigma}_1, \tilde{\sigma}_2) = (s_{pre} + \sum_{i=0}^{k-1} s_i, \prod_{i=0}^{k-1} \Gamma_i \cdot R_k \cdot R_{k+1})$ satisfies $pVrfy(\tilde{X}, m, \Gamma_i, \tilde{\sigma}_1, \tilde{\sigma}_2) = 1$, i.e.,

$$g^{s_{pre} + \sum_{i=0}^{k-1} s_i} = \prod_{i=0}^{k-1} \Gamma_i^{1+d_i} \cdot R_k^{1+d_k} \cdot R_{k+1}^{1+d_{k+1}} \cdot (\tilde{X})^c$$

This satisfies the $pVrfy$ form. Now, the signature $(\sigma_1, \sigma_2) = Adapt(\tilde{\sigma}_1, \tilde{\sigma}_2, \gamma_k) = (s_{pre} + \sum_{i=0}^{k-1} s_i + \gamma_k, \prod_{i=0}^{k-1} \Gamma_i \cdot R_k \cdot R_{k+1} \cdot \Gamma_k)$ is required, where the previously defined $\Gamma_k = g^{\gamma_k}$ ensures the validity of this equation. Thus, the verification condition $Vrfy(pk, m, Adapt(\tilde{\sigma}_1, \tilde{\sigma}_2, y)) = 1$ from Definition 3 is satisfied. Therefore, the signature scheme Ξ_Π satisfies pre-signature adaptability.

The following Lemma 5 proves that the proposed scheme satisfies the security property described in Definition 4 (extractability of secret values).

Lemma 5. *When the hash function H_{sig} used in the scheme is collision-resistant, the adaptor signature scheme Ξ_Π proposed in this paper satisfies the extractability of the secret value.*

Proof: We demonstrate the extractability of the secret value in the adaptor signature scheme by leveraging knowledge proofs and the properties of the hash function. Specifically, assume that there exists an adversary $\mathcal{A}$ that can win the game $\mathrm{WitExt}_{\mathcal{A},\Xi_\Pi}(n)$ in polynomial time. We then construct a simulator $\mathcal{S}$ that, in polynomial time, finds a collision pair for the hash function H_{sig}. On the basis of the collision resistance of H_{sig}, we can argue that the assumption does not hold, i.e., such an adversary does not exist. For convenience, let H_{sig} be denoted as H.

Simulator $\mathcal{S}$ simulates the game $\mathrm{WitExt}_{\mathcal{A},\Xi_\Pi}(n)$ for adversary $\mathcal{A}$ as follows:

1. **Initialization Phase**: Simulator $\mathcal{S}$ runs the key generation algorithm KeyGen to generate a public-private key pair (pk, sk), where $pk = \tilde{X}$ and $sk = (\sum a_i \cdot \gamma_i) + a_k \cdot x_k + a_{k+1} \cdot x_{k+1}$. The public key pk is provided to the adversary, and the private key sk is kept by the simulator for responding to future pre-signature and signature queries from adversary $\mathcal{A}$. Adversary $\mathcal{A}$ runs the GenR algorithm to generate a statement and a witness pair (Γ, γ), satisfying $\Gamma = g^\gamma$, and sends the statement Γ to simulator $\mathcal{S}$. At the same time, the message query set Q is initialized as an empty set.
2. **Pre-signature Query Phase**: In this phase, adversary $\mathcal{A}$ may issue pre-signature queries to simulator $\mathcal{S}$. When adversary $\mathcal{A}$ queries for a pre-signature on a message m_i, simulator $\mathcal{S}$ runs the pre-signature algorithm pSign to generate a pre-signature $(\tilde{\sigma}_1, \tilde{\sigma}_2)$ for the message m_i and sends it to adversary $\mathcal{A}$, then adds the message m_i to the set Q.
3. **Signature Query Phase**: In this phase, adversary $\mathcal{A}$ may issue signature queries for the message m_i. Assuming adversary $\mathcal{A}$ issues a signature query for a message, simulator $\mathcal{S}$ runs the signing algorithm Sign to generate a signature (σ_1, σ_2) for the message m_i and sends it to adversary $\mathcal{A}$, while also adding the message m_i to the set Q.
4. **Secret Value Extraction Phase**: Adversary $\mathcal{A}$ selects a message $m^* \in \{0,1\}^*$ and sends it to simulator $\mathcal{S}$. Simulator $\mathcal{S}$ runs the pre-signature algorithm pSign to generate a pre-signature $(\tilde{\sigma}_1^*, \tilde{\sigma}_2^*)$ for the message m^* and sends it to adversary $\mathcal{A}$. Adversary $\mathcal{A}$ generates a signature (σ_1^*, σ_2^*) for the message m^* and returns it to simulator $\mathcal{S}$. Simulator $\mathcal{S}$ runs the extraction algorithm Ext to extract the secret value γ' from the signature.

The winning condition of the game is as follows: adversary $\mathcal{A}$ wins the game if the signature verification algorithm $\mathrm{Vrfy}(\mathrm{par}, m^*, \sigma_1^*, \sigma_2^*, pk)$ outputs 1, and $m^* \notin Q$, while $\Gamma \neq g^{\gamma'}$. In this case, the game outputs 1; otherwise, the game outputs 0.

Assume that $\tilde{\sigma}_1^* = s$, and after extracting the witness γ' by simulator $\mathcal{S}$, the bitstrings $\tilde{X}, g^{s+\gamma'}, m^*$ and $\tilde{X}, g^s \cdot \Gamma, m^*$ are sent to challenger $\mathcal{C}$. At this point, simulator $\mathcal{S}$ successfully finds a collision pair for the hash function H, as analyzed below: Given that (s, g^s) is a valid pre-signature generated by simulator $\mathcal{S}$ for the message m^*. If the game outputs 1, it means that simulator $\mathcal{S}$ has successfully extracted the witness γ', and $\Gamma \neq g^{\gamma'}$. Moreover, both $(s+\gamma', g^s \cdot \Gamma')$ (according to the challenger's output) and $(s+\gamma', g^s \cdot \Gamma)$ (according to the signature definition) are valid full signatures. On the basis of the signature computation expression, for a valid signature, we must have:

$$g^{s+\gamma'} \cdot \tilde{X}^{H(\tilde{X}, g^{s+\gamma'}, m^*)} = g^{s+\gamma'} \cdot \tilde{X}^{H(\tilde{X}, g^s \cdot \Gamma, m^*)}.$$

Thus, it follows that:

$$H(\tilde{X}, g^{s+\gamma'}, m^*) = H(\tilde{X}, g^s \cdot \Gamma, m^*).$$

Since $\Gamma \neq g^{\gamma'}$, the bitstrings $\tilde{X}, g^{s+\gamma'}, m^*$ and $\tilde{X}, g^s \cdot \Gamma, m^*$ are two different bitstrings, which form a collision pair for the hash function H.

Given that the hash function H is collision-resistant, finding a collision pair is computationally infeasible. Therefore, on the basis of the above reduction proof, if adversary $\mathcal{A}$ wins the $\mathrm{WitExt}_{\mathcal{A},\Xi_\Pi}(n)$ game with non-negligible probability, it contradicts the collision-resistance property of the hash function. Thus, we can conclude that the adaptor signature scheme Ξ_Π satisfies witness extractability.

In summary, this paper proves the security of the designed adaptor signature scheme, i.e., the following theorem holds.

Theorem 2. *The adaptor signature scheme in this paper satisfies aEUF-CMA security, pre-signature adaptability, and extractability of the secret value. Therefore, based on Definition 1, the scheme is secure.*

For the security of the complete cross-chain scheme in this paper, the following two lemmas explain its advantages in terms of public verifiability, offline tolerance, and other aspects.

Lemma 6. *If the blockchain possesses the property of public verifiability, then the protocol proposed in this paper can also achieve this property.*

Proof: In the decentralized P2P network architecture of blockchain, any participant can freely join the Bitcoin network and verify transactions within the network. This feature ensures the transparency and credibility of transaction information. Furthermore, in the protocol designed in this paper, all signature information and related scripts are publicly visible, and anyone can verify the validity of these signatures. Therefore, by leveraging the openness of the blockchain itself and the public nature of the signatures and scripts in the protocol, the entire protocol achieves the functionality of public verifiability.

Lemma 7. *If the cross-chain protocol proposed in this paper guarantees that adjacent participants do not collude to disrupt the protocol and that at most one participant is offline, then the protocol is offline tolerant.*

Proof: In the process of script execution, the rules are set such that when a timeout occurs, the locked assets can be redeemed by the asset owner. If adjacent participants do not collude, then after each successful script execution, other participants will be able to extract the secret value γ from the signature to obtain relevant information. These participants, by obtaining the pre-adapted signature, can combine the secret values from previous transaction records $T_0, \ldots, T_{i-2}$ or T_{i-1} to construct a complete signature and successfully transfer assets. However, when the second participant is offline (assuming that the first offline participant is $user_x$), the other online participants can only collect the secret values from $T_0, \ldots, T_{x-1}, T_{x+1}, \ldots, T_{i-2}$, which means that before the timeout occurs, the script cannot complete execution. Therefore, the execution of the protocol is interrupted due to the lack of the complete secret value, ensuring the offline tolerance mechanism is in place.

6 Conclusion

In this paper, we first review the multiparty cross-chain exchange protocol based on adaptor signatures designed by Chen et al., analyzing its potential security issues, particularly a detailed analysis of the ROS attack that the existing protocol may encounter. On the basis of these issues, this paper proposes an improved adaptor signature scheme, which aims to establish a special correlation between the parameters and the message m, cleverly binding the random number with the message without increasing the number of interaction rounds or random parameters, and constructing a nonlinear dependency. This relationship alters the structure of the linear system in the ROS attack, making it unsolvable rather than solvable, thus preventing the attacker from forging signatures. This effectively defends against such attacks and enhances the robustness and reliability of the system, providing strong security guarantees for cross-chain transactions. Finally, the paper proves, through mathematical models and formal security analysis, that the improved scheme can maintain security when facing various potential attacks, especially demonstrating theoretical and practical advantages in defending against ROS attacks.

Acknowledgements. This work was supported by the National Natural Science Foundation of China (61907042, 61702570), the Beijing Natural Science Foundation (4194090), the Social Science Planning Project of Shandong Province (24CWYJ05) and the Project of Research on the Course System of "Blockchain Technology and Applications" Based on the TuGraph Platform, Ant Group Co., Ltd.

References

1. Atzei, N., Bartoletti, M., Cimoli, T.: A survey of attacks on ethereum smart contracts (sok). In: Principles of Security and Trust: 6th International Conference, POST 2017, Held as Part of the European Joint Conferences on Theory and Practice of Software, ETAPS 2017, Uppsala, Sweden, April 22-29, 2017, Proceedings 6, pp. 164–186. Springer (2017)
2. Benhamouda, F., Lepoint, T., Loss, J., Orrù, M., Raykova, M.: On the (in) security of ros. J. Cryptol. **35**(4), 25 (2022)
3. Bhargavan, K., et al.: Formal verification of smart contracts: short paper. In: Proceedings of the 2016 ACM Workshop on Programming Languages and Analysis for Security, pp. 91–96 (2016)
4. Boneh, D., Gentry, C., Lynn, B., Shacham, H.: Aggregate and verifiably encrypted signatures from bilinear maps. In: Biham, E. (ed.) EUROCRYPT 2003. LNCS, vol. 2656, pp. 416–432. Springer, Heidelberg (2003). https://doi.org/10.1007/3-540-39200-9_26
5. Buterin, V., et al.: A next-generation smart contract and decentralized application platform. white paper **3**(37), 2–1 (2014)
6. Chen, C., Yang, G., Li, Z., Xiao, F., Chen, Q., Li, J.: Privacy-preserving multiparty cross-chain transaction protocols. Cryptography **8**(1), 6 (2024)
7. Feng, J., Zhang, Q., Huang, W., Han, G.: A cyber threat intelligence sharing scheme based on cross-chain interaction. Netinfo Security **22**(5), 21–29 (2022)
8. Fournier, L.: One-time verifiably encrypted signatures aka adaptor signatures (2019)
9. Gennaro, R., Jarecki, S., Krawczyk, H., Rabin, T.: Secure distributed key generation for discrete-log based cryptosystems. J. Cryptol. **20**, 51–83 (2007)
10. Gerhart, P., Schröder, D., Soni, P., Thyagarajan, S.A.: Foundations of adaptor signatures. In: Annual International Conference on the Theory and Applications of Cryptographic Techniques, pp. 161–189. Springer (2024)
11. Khan, S., Amin, M.B., Azar, A.T., Aslam, S.: Towards interoperable blockchains: a survey on the role of smart contracts in blockchain interoperability. IEEE Access **9**, 116672–116691 (2021)
12. Kiffer, L., et al.: Logging in the void: privacy issues in cross-chain systems. In: Proceedings of the ACM Conference on Computer and Communications Security (CCS) (2022)
13. Komlo, C., Goldberg, I.: Frost: flexible round-optimized schnorr threshold signatures. In: Selected Areas in Cryptography: 27th International Conference, Halifax, NS, Canada (Virtual Event), pp. 34–65, October 21-23, 2020, Revised Selected Papers 27. Springer (2021)
14. Lee, C.: Litecoin: a peer-to-peer cryptocurrency (2011). https://www.litecoin.org/
15. Ling, Z., Yang, M., Yu, J.: Research on power security trading platform based on ipfs and blockchain technology. Netinfo Security **24**(6), 968–976 (2024)
16. Lu, A., Zhao, K., Yang, J., Wang, F.: Research on cross-chain technology of blockchain. Netinfo Security **19**(8), 83–90 (2019)
17. Maxwell, G., Poelstra, A., Seurin, Y., Wuille, P.: Simple schnorr multi-signatures with applications to bitcoin (deprecated version). Tech. rep., Cryptology ePrint Archive, Report 2018/068, version 1 (2018)
18. Maxwell, G.: A secure and scalable multi-signature scheme based on schnorr signatures. In: Proceedings of the 2017 International Conference on Cryptography and Network Security, pp. 123–134. Springer (2017)

19. Maxwell, G.: Coinswap: Transaction graph disjoint trustless trading (2013). https://bitcointalk.org/index.php
20. Maxwell, G.: Bitcoin's pay-to-script-hash (p2sh) (2014). bitcoin Improvement Proposal 16
21. Nolan, T.: Hashed timelock contracts for atomic swaps (2017)
22. Shao, Z., Jiang, J., Zhao, J., Zhang, Y.: An improved weighted byzantine fault tolerance algorithm for cross-chain system. Netinfo Security **23**(8), 109–120 (2023)
23. Wang, S., Ma, Z., Liu, J., Luo, S.: Research and implementation of cross-chain security access and identity authentication scheme of blockchain. Netinfo Security **22**(6), 61–72 (2022)
24. Xinyou, Z., Qingfu, L., Li, F., Huanlai, X.: Smart grid data security sharing model based on multi-authority attribute-based encryption. Netinfo Security **25**(1), 98–109 (2025)
25. Yiting, F., Zhaofeng, M., Danheng, X., Pengfei, D.: Evaluation method for cross-chain security strength access. Netinfo Security **23**(1), 84–92 (2023)

Big Data Management and Applications

Research on Polling MAC Protocols for Flying Ad-Hoc Networks

Shujian Wang[1], Zhijun Yang[1,2,3(✉)], Hongwei Ding[1], and Xiong Wang[1]

[1] School of Information Science and Technology, Yunnan University, Kunming, Yunnan, China
353738698@qq.com

[2] Educational Instruments and Facilities Service Center, Educational Department of Yunnan Province, Kunming, China

[3] Key Laboratory of Education Informatization for Nationalities of Ministry of Education, Yunnan Normal University, Kunming, China

Abstract. Flying ad hoc networks (FANETs), which are composed of unmanned aerial vehicles (UAVs), face severe quality-of-service (QoS) challenges because of their high mobility, rapidly changing topologies, and limited infrastructure support. Conventional MAC protocols, such as contention-based CSMA/CA, often fail to guarantee low latency and high throughput under these conditions. To address these challenges, this paper investigates the MAC-layer QoS problem in FANETs and proposes a polling-based MAC protocol integrated with the IEEE 802.11 point coordination function (PCF). The protocol adopts a centralized scheduling mechanism using a master node to coordinate data transmission on the basis of a first-in-first-out (FIFO) policy. While this centralized approach simplifies coordination and reduces contention, it introduces limitations regarding scalability and robustness in decentralized or dynamic environments, which are common in FANET scenarios. To analyse protocol behavior, a polling system model is developed with three service strategies: Gated Service, Exhaustive Service, and Limited-1 Service. Analytical models are derived for queue length and delay, and MATLAB simulations validate the feasibility of the proposed scheme. The results show that despite its structural constraints, the proposed protocol can achieve a balance between throughput, delay, and fairness in specific mission-driven UAV applications, such as coordinated surveillance and real-time task distribution. Further research will explore decentralized adaptations to improve protocol flexibility and fault tolerance in dynamic UAV networks.

Keywords: Flying Ad-Hoc Networks (FANETs) · UAV Swarm Communication · Polling MAC Protocols · Quality of Service (QoS)

1 Introduction

1.1 Technical Requirements

Unmanned aerial vehicles (UAVs), which were originally developed for military operations, have increasingly been adopted in civilian domains, such as environ-

Y. Yu et al. (Eds.): ICPCSEE 2025, CCIS 2673, pp. 489–503, 2026.
https://doi.org/10.1007/978-981-95-2566-9_32

mental monitoring, infrastructure inspection, and emergency response, due to their cost-effectiveness and flexibility. Multi-UAV systems, organized as Flying ad hoc networks (FANETs), exhibit unique communication characteristics, such as high node mobility, dynamic topology, and limited infrastructure support. These features impose stringent requirements on the design of communication protocols, especially at the medium access control (MAC) layer.

Key application scenarios highlight three major QoS challenges:

- **High-Throughput Surveillance**: Applications such as forest fire monitoring and border patrols require continuous high-bandwidth video. However, conventional contention-based MAC protocols (e.g., CSMA/CA) suffer from frequent collisions in highly dynamic topologies [1], leading to unpredictable throughput degradation.
- **Energy-Efficient Relay**: UAVs deployed as relay nodes in smart cities necessitate energy-efficient hardware and adaptive duty cycling [2], with stringent requirements for latency tolerance and cost-effectiveness.
- **Ultrallow-latency control**: Mission-critical operations, such as precision strikes and forest firefighting, demand deterministic end-to-end delays [3]. Existing distributed protocols (e.g., the 802.11 DCF) rely on random backoff mechanisms, resulting in unbounded latency jitter under high mobility, which is unacceptable for time-sensitive missions.

These scenarios illustrate that conventional contention-based MAC protocols, such as CSMA/CA, are ill suited for FANET environments. They suffer from frequent collisions, high retransmission overhead, and unpredictable delays—which are particularly problematic under time-sensitive or large-scale UAV deployments.

1.2 Research Motivation and Contributions

To address these MAC-layer challenges, this paper investigates a polling-based coordination mechanism for FANETs, inspired by the IEEE 802.11 point coordination function (PCF). The proposed solution adopts a centralized master node (e.g., an access point on a lead UAV or ground control unit) that polls UAVs in a first-in-first-out (FIFO) fashion, allocating the wireless channel in a conflict-free and deterministic manner.

While the centralized polling approach offers simplicity and bounded delay, it also introduces limitations in terms of scalability, fault tolerance, and suitability in decentralized topologies. Such constraints must be carefully examined, as FANETs often lack a permanent infrastructure or stable control hierarchy.

To evaluate the viability and performance trade-offs of the proposed MAC protocol, we develop an analytical model based on polling system theory, incorporating three service disciplines—Gated, Exhaustive, and Limited-1—to reflect different operational requirements. The simulation results validate the model's accuracy and provide insights into latency, fairness, and throughput performance across scenarios. In summary, the contributions of this paper are as follows:

- A formal analysis of MAC-layer QoS issues specific to FANET environments.
- The design of a polling-based MAC protocol leveraging centralized coordination with FIFO scheduling.
- Theoretical modelling and simulation of service strategies under varied traffic loads and operational constraints.
- A discussion of the applicability and limitations of centralized MAC scheduling in highly dynamic UAV networks.

1.3 MAC Protocols for FANET

The medium access control (MAC) protocols in FANETs face critical challenges in managing wireless resources among highly mobile UAV nodes [4]. First, frequent topology changes caused by UAV mobility and dynamic internode distances degrade link stability. Traditional content-based protocols such as CSMA/CA struggle to adapt to such environments, resulting in packet collisions and channel underutilization [5]. Second, mission-critical operations impose ultralow latency constraints, where even minor delays may jeopardize mission success [6]. Third, heterogeneous quality of service (QoS) demands require flexible resource allocation—high throughput for video streaming and deterministic latency for controlling commands.

Recent advancements address these challenges through hybrid architectures. For example, network slicing with edge computing dynamically partitions UAV computational resources and wireless links to increase the QoS for latency-sensitive tasks [7]. Time-slot allocation protocols improve throughput by scheduling algorithms but introduce excessive delays due to rigid resource partitioning. Modified IEEE 802.11 distributed coordination function (DCF) protocols leverage CSMA/CA mechanisms to ensure fairness; however, congestion-induced delays persist under high traffic loads, as depicted in Fig. 1.

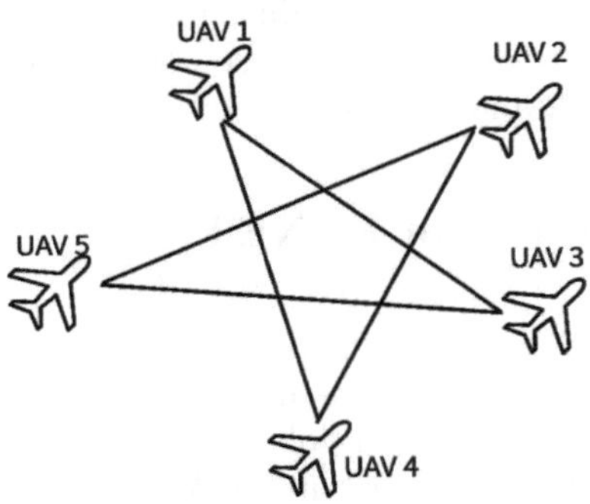

Fig. 1. Decentralized FANET architecture

Given these limitations, a key research challenge is how to design an MAC protocol that can support low-latency, high-throughput, and fair access under rapidly changing FANET conditions.

In response, we explore a centralized scheduling mechanism based on the IEEE 802.11 point coordination function (PCF), which avoids contention by polling UAVs directly. The IEEE 802.11 standard further specifies the PCF, a centralized mechanism where an AP coordinates channel access through polling [8]. As shown in Fig. 2, the AP mitigates collision risk by allocating contention-free periods (CFPs) for time-sensitive traffic and contention periods (CPs) for asynchronous data transmissions. This architecture guarantees deterministic latency for prioritized traffic and effectively addresses the limitations of decentralized contention-based protocols such as CSMA/CA.

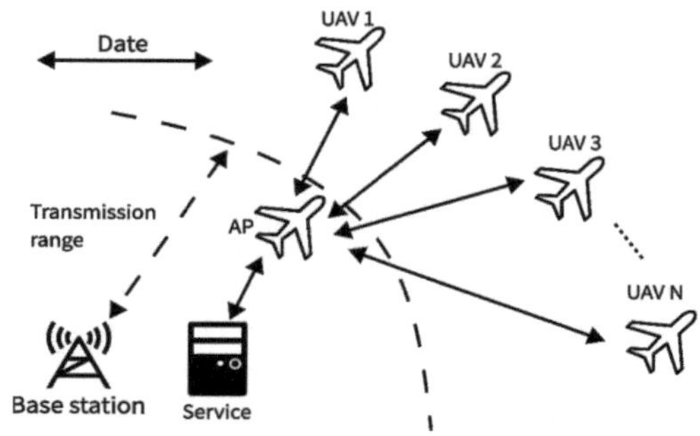

Fig. 2. PCF-based FANET architecture

While this centralized polling approach offers deterministic access control and reduced overhead, it also introduces practical limitations. Its effectiveness depends on the availability and reliability of a central access point (AP), which may not always be guaranteed in infrastructure-less or highly distributed FANET scenarios. Moreover, fixed FIFO scheduling may lack flexibility when handling mixed-priority or event-driven traffic. These constraints must be carefully considered in protocol design and evaluation.

2 Polling Mechanism in FANET

2.1 Polling System

To support coordinated channel access among UAVs in highly dynamic FANET environments, we adopt a centralized polling system architecture in which a designated access point (AP) polls distributed UAV nodes in a predefined sequence.

The polling system consists of a central access point (AP) and N distributed queues (Lines 1 to N). The AP sequentially polls each queue in a fixed cyclic sequence (Line 1 → Line 2 → ... → Line N). When pending tasks are detected in a queue, the server processes the requests before switching to the next queue. After serving Line N, the server restarts the cycle from Line 1, enabling efficient resource sharing through sequential scheduling.

This centralized polling design is inspired by the IEEE 802.11 PCF (point coordination function), which aims to provide deterministic access control for time-sensitive data flows. This model supports single-server and multiserver configurations, with priority handling for urgent tasks while ensuring workflow continuity.

As shown in Fig. 3, all lines converge at the AP, illustrating the centralized control and data aggregation mechanism intrinsic to polling systems.

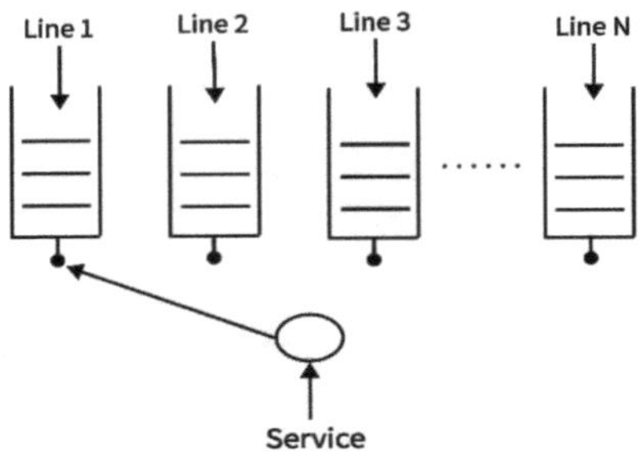

Fig. 3. Polling system model

Although the centralized polling mechanism simplifies contention management and ensures orderly access, its reliance on a fixed AP may reduce adaptability in decentralized or infrastructureless UAV deployments. Therefore, its use in FANET scenarios must be carefully evaluated in light of potential limitations such as AP mobility, control overhead, and failure resilience.

2.2 System Model and Transmission Workflow

To address the stringent coordination requirements of UAV swarms in dynamic environments, we adopt a centralized polling mechanism based on the IEEE 802.11 PCF, where an access point (AP) acts as the master controller to regulate channel access [9]. In this design, UAVs are assumed to operate in a passive mode—transmitting data only after the AP explicitly grants permission—which simplifies scheduling but also introduces challenges in highly decentralized or unstable networks.

As shown in Fig. 4, UAVs follow a controlled transmission process based on a cyclic polling strategy, which is divided into the following two phases:

- **Polling Initialization:** The AP sequentially queries all the UAVs to compile a dynamic priority list. This priority list is determined by the criticality of tasks—for instance, control commands are given higher priority over nonurgent sensor data. Such classification is crucial for ensuring deterministic resource allocation for mission-critical operations. Importantly, this centralized priority assignment, while effective in reducing contention, may be less adaptive in scenarios with rapidly changing network conditions.

- **Prioritized Transmission:** Once the priority list is established, the AP allocates channel resources strictly following this order, thereby eliminating contention-induced collisions through conflict-free scheduling. After all the scheduled transmissions are completed, the AP initiates a new polling cycle by reevaluating the UAVs' requests and regenerating the priority list. The reliance on sequential polling, however, may introduce additional delays under heavy loads, and its scalability is contingent on the stability of the central AP.

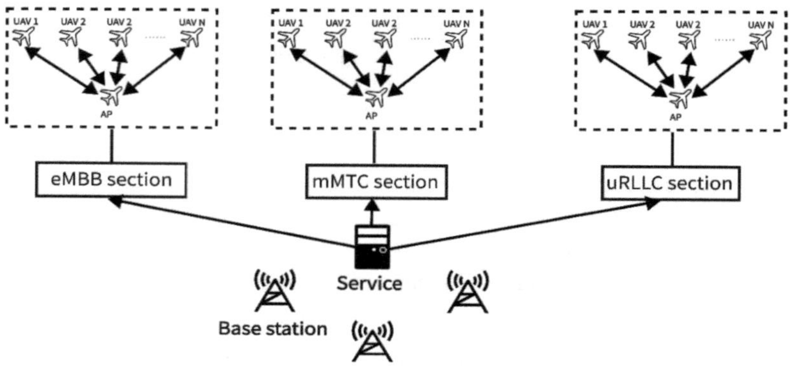

Fig. 4. FANET system architecture

The overall system architecture (Fig. 4) illustrates the AP-centric topology, whereas the workflow in Fig. 5 details the phased resource allocation process, including the reassessment of UAV priorities and retransmission handling [10].This design offers bounded latency and facilitates resource sharing in dense deployments; however, it also inherently assumes that the AP remains available and that its control overhead is manageable, which are critical factors to consider for realistic FANET implementations.

The IEEE 802.11 standard specifies three interframe spaces (IFSs) with distinct durations to prioritize data transmissions on the basis of urgency levels: the short interframe space (SIFS), the shortest interval, grants immediate channel access for mission-critical operations such as ACK frame delivery and control signalling. The point coordination function interframe space (PIFS), with an intermediate duration, enables centralized polling by APs for latency-sensitive traffic scheduling. The distributed coordination function interframe space (DIFS), the longest interval, governs contention-based channel access for asynchronous data transmissions in decentralized networks [11]. This IFS hierarchy forms the foundation for prioritized medium access in IEEE 802.11-based systems and plays a central role in the PCF-based coordination proposed in this study.

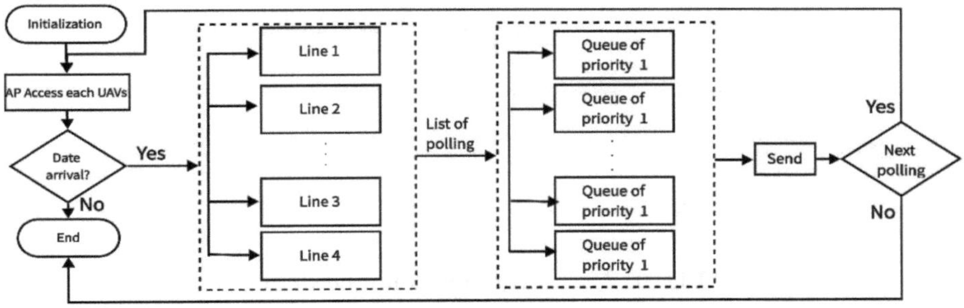

Fig. 5. FANET system transmission flow

System Transmission Procedure. As illustrated in Fig. 6, the system transmission procedure consists of the following steps:

- **Beacon Broadcasting:** The AP broadcasts IEEE 802.11-compliant Beacon frames over the FANET. The duration field forces all UAVs into a deferred transmission state [1].
- **Contention-Free Polling:** The AP transmits a CF-Poll frame to UAV 1, granting exclusive transmission rights within the beacon interval.
- **Data Transmission Phase:** UAV 1 transmits its MAC service data unit (MSDU) to the AP via a data frame.
- **Successful Reception Protocol:** Upon valid data frame reception, the AP waits for an SIFS and replies with an ACK. After another SIFS, it initiates the next polling cycle (UAV 2) per the TDMA schedule.
- **Error Recovery Mechanism:** If CRC errors or PHY-layer failures occur, the AP waits for a PIFS, skips UAV 1 in the current superframe, and directly issues a CF-Poll to UAV 2 to maintain the QoS.

This transmission procedure assumes stable synchronization and timely response from both the AP and UAV nodes. In highly dynamic FANET environments, these conditions may not always hold, necessitating potential enhancements such as adaptive polling intervals, retransmission windows, or mobile AP redundancy.

2.3 System Assumptions and Variable Definitions

System Model and Assumptions. This study assumes an idealized PCF-based model where UAVs do not compete for channel resources. The following assumption applies:

- Buffer Capacity: UAV buffers are sufficiently large to prevent overflow.
- Lossless Transmission: All packets are transmitted without loss.
- Packet Arrival Process: The arrival process at each UAV is i.i.d. [12], modeled by a probability generating function (PGF) $A(z)$, with mean $\lambda = A'(1)$ and variance $\sigma_\lambda^2 = A"(1) + \lambda - \lambda^2$;

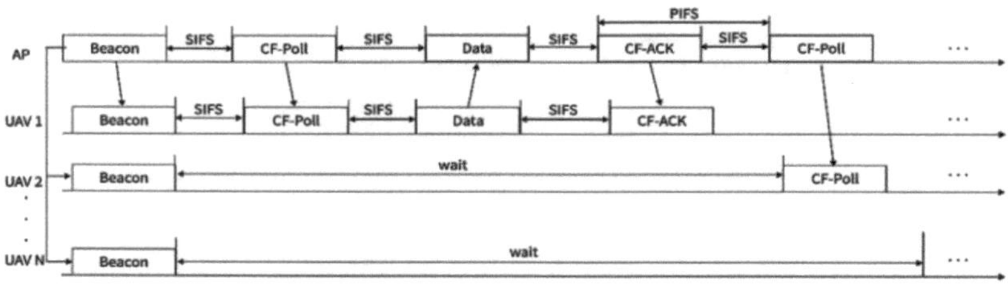

Fig. 6. System transmission procedure

- Service Time: Service times are i.i.d. with PGF $B(z)$,mean $\mathcal{B} = B'(1)$, and variance $\sigma_{\beta}^2 = B''(1) + \beta - \beta^2$;
- Switching Time: Server switching between UAVs is i.i.d. with PGF $R(z)$, mean $\gamma = R'(z)$, and variance $\sigma_{\gamma}^2 = R''(1) + \gamma - \gamma^2$.

These assumptions enable tractable analytical modelling and performance evaluation under a controlled environment. However, in practice, FANETs may deviate significantly from these conditions.

For example, lossless transmission is difficult to guarantee in UAV networks because of multipath fading, interference, and frequent link disruptions. Similarly, assuming a centralized AP with unlimited buffer capacity and perfect global knowledge overlooks the intermittent connectivity and distributed nature of FANETs.

Therefore, while the proposed system model provides useful theoretical insight into polling-based MAC performance, future implementations must consider the impact of nonideal conditions, including limited buffer space, link failures, and adaptive switching delays.

Steady-State Analysis and System State. Let ξ_i denote the number of f packets in the buffer of the $i-th$ UAV buffer at time t_n. The system state vector at t_n is $[\xi_1(n), \xi_2(n), \cdots \xi_i(n), \cdots \xi_N(n)]$, with the joint probability distribution $P[\xi_i(n) = x_i; i = 1, 2, \cdots, N]$ [13]. Under the stability condition $\sum_{i=1}^{N} \lambda\beta = N\rho < 1$ the system reaches steady-state:

$$\lim_{n\to\infty} P[\xi_1(n), \xi_2(n), \cdots \xi_i(n), \cdots \xi_N(n)] = \pi_i(x_1, x_2, \cdots, x_i, \cdots, x_N) \quad (1)$$

The PGF of $\pi_i(x_1, x_2, \cdots, x_i, \cdots, x_N)$ is defined as:

$$G_i(z_1, z_2, \cdots, z_i, \cdots, z_N) = \sum_{x_1=0}^{\infty} \sum_{x_2=0}^{\infty} \cdots \sum_{x_i=0}^{\infty} \cdots \sum_{x_N=0}^{\infty} \pi_i(x_1, x_2, \cdots, x_i, \cdots, x_N) z_1^{x_1} z_2^{x_2} \cdots z_i^{x_i} \cdots z_N^{x_N}, i = 1, 2, \cdots, N. \tag{2}$$

Service Discipline and Performance Metrics

Service Discipline: Packets are scheduled following the first-come-first-served (FCFS) principle.

Performance Metrics:

- Mean Queue Length: average number of queued packets in a UAV buffer.
- Polling Cycle: time required for the AP to complete one full polling round (UAV 1 → UAV N).
- Average Latency: end-to-end delay from packet arrival to successful transmission.

Importantly, the FCFS scheduling strategy, while analytically convenient, may not be optimal for real-time or critical-task prioritization. Real-world implementations may benefit from extending this model to include priority queuing, deadline-aware transmission, or adaptive reordering mechanisms to better align with mission-specific QoS demands.

3 FANET Polling System Theory

Let λ, β, and γ denote the per-UAV packet arrival rate, mean service time per packet, and inter-UAV switchover time, respectively (as defined in Sect. 2). The traffic intensity for a single UAV node is defined as $\rho = \lambda\beta$ with the total system load expressed as N * p (N: total number of UAV nodes). The system stability condition is $N\rho < 1$.

Three polling strategies, gated service, exhaustive service, and limited-1 service, are theoretically analysed on the basis of the framework in Sect. 2.

3.1 Gated Service Polling System

Under the Gated Service model, only the packets present at the beginning of the polling visit are eligible for transmission. Any packets arriving during the service period are deferred until the next polling round. This discipline provides moderate latency and avoids long occupancy of the channel by a single UAV.

Service Rule: A UAV granting transmission rights transmits only packets present in its buffer at the start of the service interval. New packets arriving during transmission are deferred to the next polling cycle [12].

From the properties of the probability generating function (PGF):

$$g_i(i) = \lim_{z_1, z_2, \cdots, z_i, \cdots, z_N \to 1} \frac{\partial G_i(z_1, z_2, \cdots, z_i, \cdots, z_N)}{\partial z_i} \tag{3}$$

Differentiating equation (3) yields:

$$g_{i(G)}(i) = \frac{N\gamma\lambda}{1 - N\rho} \tag{4}$$

where $g_{i(G)}(i)$ denotes the mean queue length under the Gated Service.

Average polling cycle:

$$\theta(G) = \frac{N\gamma}{1 - N\rho} \tag{5}$$

Average delay:

$$\begin{aligned} E \quad (W_G) = \frac{1}{2}\{\frac{R''(1)}{\gamma} + \frac{1}{1 - N\rho}[(N-1)\gamma + (N-1)\rho + 2N\gamma\rho + N\lambda B'' \\ + \frac{(1 + \rho - N\rho)A''(1)}{\lambda^2}]\} \end{aligned} \tag{6}$$

Gated service represents a trade-off between throughput and fairness. While it avoids starvation, it may result in moderate delays under bursty traffic. It is well suited for scenarios with energy-constrained UAVs or moderate latency requirements, such as sensor sweeps or periodic data collection.

3.2 Exhaustive Service Polling System

The exhaustive service model allows UAVs to transmit all packets in their buffer, including those that arrive during the service period. This reduces the overall latency but risks unfairness under asymmetric traffic loads.

Service Rule: A UAV granted transmission rights transmits all packets in its buffer, including those arriving during both the service interval and the ongoing transmission.

Mean queue length:

$$g_{i(E)}(i) = \frac{N\gamma\lambda(1 - \rho)}{1 - N\rho} \tag{7}$$

Average polling cycle:

$$\theta(E) = \frac{N\gamma}{1 - N\rho^{\downarrow}} \tag{8}$$

Average delay:

$$E(W_E) = \frac{1}{2}\left\{\frac{R(1)}{\gamma} + \frac{1}{1 - N\rho}\left[(N-1)\gamma + (N-1)\rho + N\lambda B''(1)\right] + \frac{\rho A(1)}{\lambda^2(1 - N\rho)}\right\} \tag{9}$$

This strategy is ideal for high-throughput applications where minimizing transmission latency takes precedence, such as real-time video streams or command relays in time-sensitive missions. However, it can lead to resource monopolization and increased waiting times for low-traffic nodes.

3.3 Limited-1 Service Polling System

Limited-1 Service imposes a strict transmission cap—e.g., K = 1 packet per cycle—regardless of the queue size. This discipline is highly predictable and fair but may incur high latency.

Service Rule: UAV-granted transmission rights transmit k packets per polling cycle (e.g., K=1 for single-packet transmission). The remaining packets are queued for the next cycle [13].

Mean queue length:

$$g_{i(L)}(i) = \frac{N}{2[1-N\lambda(\gamma+\beta)]}\{2\lambda\gamma(1-\lambda\gamma) + \frac{(N-1)\lambda^2\gamma(\rho-\gamma)}{1-N\rho} + [1+\frac{\rho}{1-N\rho}]\gamma A" + \frac{N\lambda^3\gamma B''(1)}{1-N\rho} + \lambda^2 R''(1)\} \tag{10}$$

Average polling cycle:

$$E(W_E) = \frac{1}{2}\left\{\frac{R(1)}{\gamma} + \frac{1}{1-N\rho}\left[(N-1)\gamma + (N-1)\rho + N\lambda B''(1)\right] + \frac{\rho A(1)}{\lambda^2(1-N\rho)}\right\} \tag{11}$$

Average delay:

$$E(W_L) = \frac{R''(1)}{2\gamma} + \frac{1}{\{2[1-N\lambda(\gamma+\beta)]\}}[(N-1)\gamma + (N-1)\rho + 2N\gamma\rho + \frac{(N\lambda\gamma+\rho)A''(1)}{\lambda^2} + N\lambda B''(1) + N\lambda R''(1)]. \tag{12}$$

This strategy guarantees fairness and predictable resource allocation, making it suitable for collaborative swarm applications where each UAV must report equally. However, its low per-cycle throughput makes it unsuitable for large data flows or urgent tasks.

4 Simulation and Result Analysis

On the basis of the theoretical framework in Sect. 3, we simulate the performance of Exhaustive, Gated, and Limited-1 Service strategies. The experiments were conducted in MATLAB R2022a on a Windows 11 platform, with the results visualized via Origin 2021. The key parameters include:

- **Packet arrival rate:**$\lambda \in [0.002, 0.03]$(step:0.002)
- **Service time per packet:**$\beta = 2$t.u.

- **UAV switching delay:**$\gamma = 1$t.u.
- **Controlled UAVs per scenario:** $N_{UAV} = 5$

The simulation assumes idealized communication conditions (e.g., no packet loss or retransmission), i.e., packet arrivals, and full AP availability. The arrival process follows a Poisson distribution.

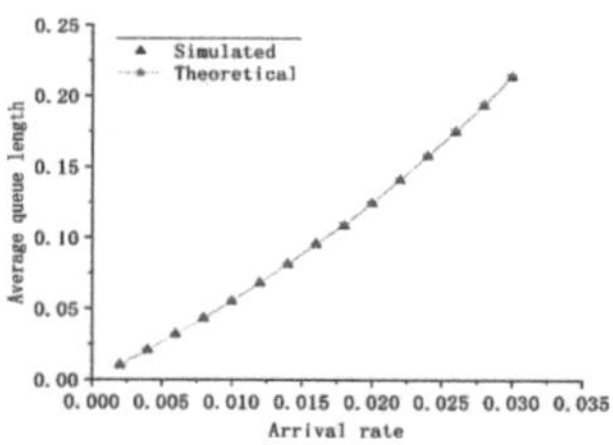

Fig. 7. Average queue length under Gated Service

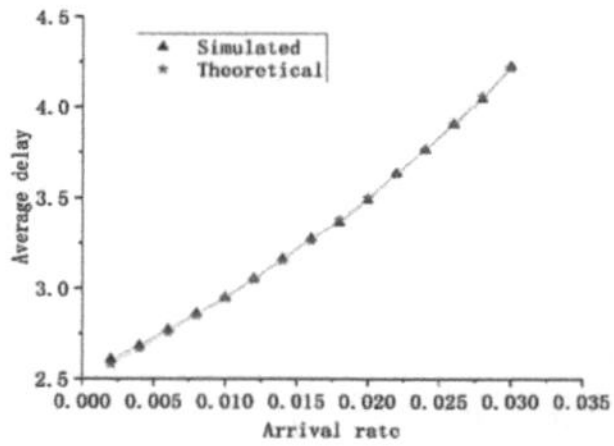

Fig. 8. Average waiting time under Gated Service

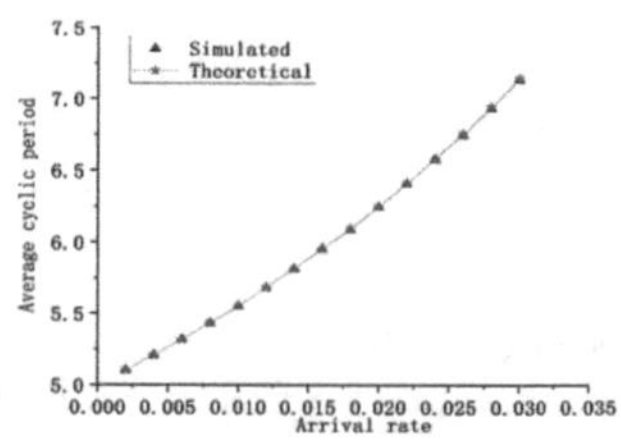

Fig. 9. Average polling cycle of Gated Service

This simplified setup allows clear observation of each polling discipline's theoretical performance boundaries but does not capture real-world FANET impairments such as dynamic link quality or node dropouts. As such, the results serve as a baseline for evaluating the core behaviors of the proposed MAC design.

Figures 7, 8 and 9 show the simulated and theoretical values for the average queue length, waiting time, and polling cycle under the Gated Service:

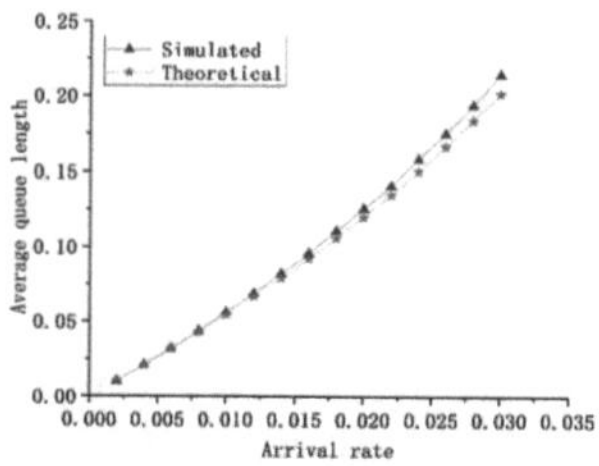

Fig. 10. Average queue length under Exhaustive Service

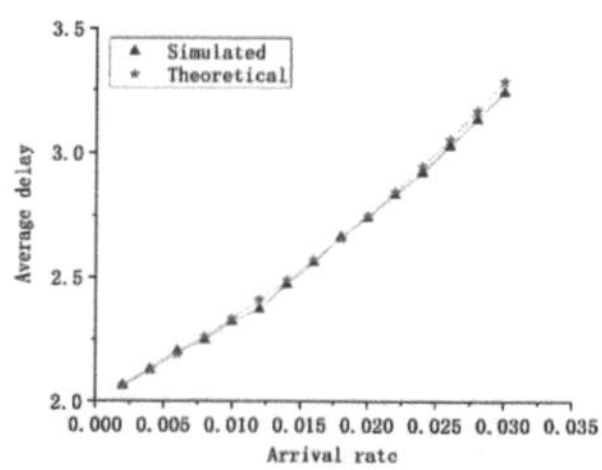

Fig. 11. Average waiting time under Exhaustive Service

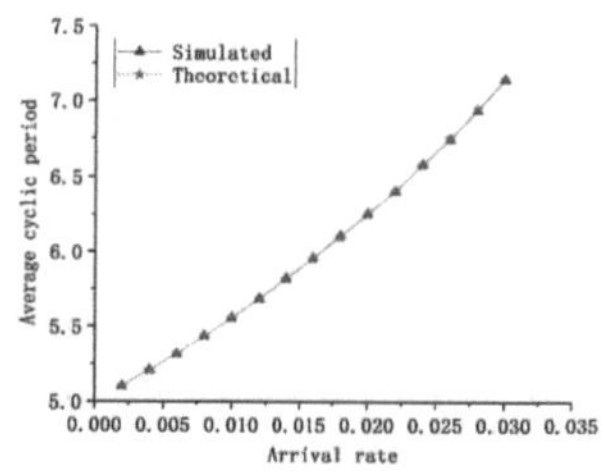

Fig. 12. Average polling cycle of Exhaustive Service

The results demonstrate strong alignment between the theoretical predictions and simulation outcomes, confirming model validity under controlled load conditions. Metrics exhibit monotonic growth with increasing λ, which is consistent with standard queuing behavior.

Figures 10, 11 and 12 present the same metrics for Exhaustive Service:

Compared with the Gated Service strategy, the Exhaustive strategy consistently yields shorter waiting times and fewer polling cycles, especially under moderate to high loads. This reinforces its suitability for delay-sensitive tasks. However, queue lengths also grow faster, potentially impacting fairness.

Figures 13, 14 and 15 correspond to Limited-1 Service:

Under high arrival rates, the simulation values begin to deviate from the theoretical predictions. This is attributed to the cumulative queuing effects and constrained transmission per cycle, particularly in scenarios with multiple active UAVs. Despite this, Limited-1 service ensures high predictability and equitable channel access.

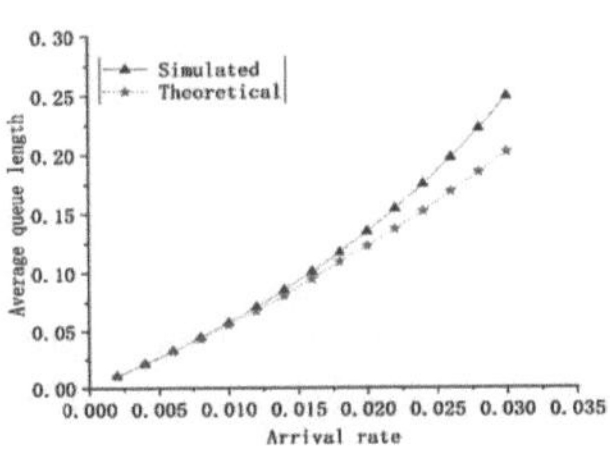

Fig. 13. Average queue length under Limited-1 Service

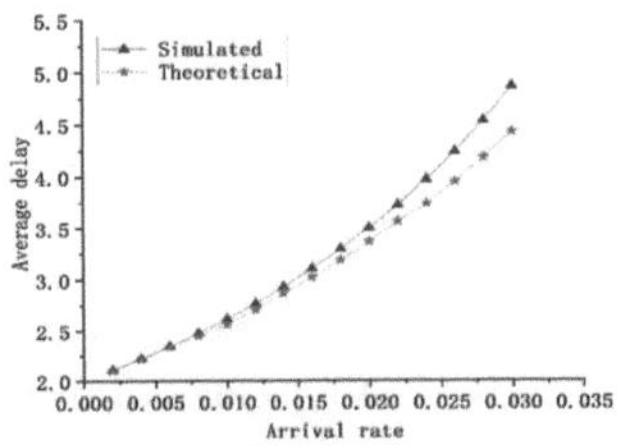

Fig. 14. Average waiting time under Limited-1 Service

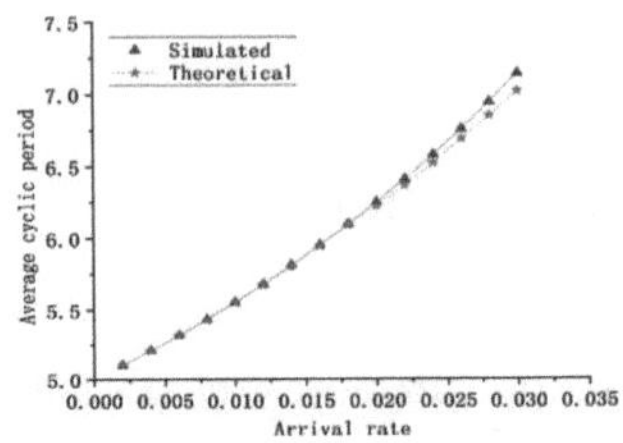

Fig. 15. Average polling cycle of Limited-1 Service

Figures 16 and 17 offer cross-strategy comparisons:

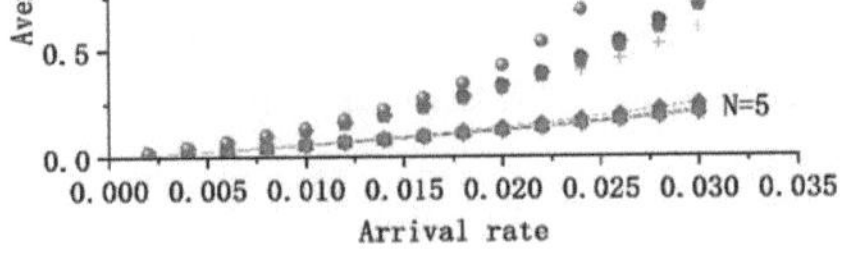

Fig. 16. Comparison of average queue lengths under Exhaustive, Gated, and Limited-1 Services

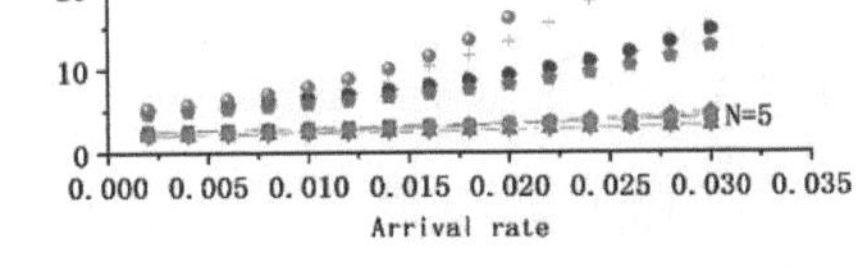

Fig. 17. Comparison of average waiting times across Exhaustive, Gated, and Limited-1 Service policies

Exhaustive service outperforms the other two methods in terms of delay, especially as λ increases. Gated Service provides a middle ground between throughput and fairness, whereas Limited-1 Service offers the most balanced channel sharing at the expense of latency.

These results highlight that the proposed polling MAC protocol is not one-size-fits-all. Instead, the service discipline should be selected on the basis of the application's QoS priorities:

- **Exhaustive Service:** Best for real-time data (e.g., video surveillance, urgent command feedback).
- **Limited-1 Service:** Best for equal-access scenarios (e.g., collaborative swarm formation, synchronized reporting).
- **Gated Service:** Appropriate for moderate-latency, energy-sensitive operations.

5 Conclusion

To address heterogeneous QoS requirements in 5G-enabled UAV networks, this study proposes a polling-based MAC protocol for Flying Ad-Hoc Networks (FANETs). Through theoretical modelling and simulations, we rigorously compare three service disciplines: Gated Service, Exhaustive Service, and Limited-1 Service. Key findings are as follows:

- Exhaustive service demonstrates superior throughput and latency performance for mission-critical applications (e.g., real-time surveillance), aligning with scenarios requiring rapid data delivery.
- Limited-1 Service achieves enhanced fairness in resource allocation through regulated single-packet transmissions per cycle, optimizing collaborative swarm operations.
- Gated Service provides a balanced trade-off between energy efficiency and performance by dynamically adjusting duty cycles in energy-constrained environments.

The proposed centralized polling protocol, while analytically effective, is based on ideal assumptions such as lossless transmission, passive UAV behavior, and stable access point coordination. These assumptions may not fully hold in practical FANET deployments, especially under high mobility and intermittent connectivity.

Nevertheless, the study establishes a theoretical baseline for analysing polling-based MAC designs and offers actionable insights for adapting MAC strategies to specific QoS scenarios. Future work will focus on improving the scalability and resilience of the protocol, such as through mobile AP selection, hybrid distributed scheduling, or adaptive polling intervals.

Further validation in realistic simulation platforms and hardware testbeds will also be pursued to assess deployment feasibility.

Acknowledgements. This work was supported by Yang Zhijun's Industry Innovation Talents Project of the Yunnan Xingdian Talents Support Plan (Certificate No. YNWR–CYJS–2020–017), formerly part of the Yunnan "Ten-Thousand Talents Plan" for industrial technology leadership.

The authors would also like to acknowledge the support of the Wu Zhonghai Expert Workstation of Yunnan Province (No. 202305AF150045).

This research was conducted at the Educational Instruments and Facilities Service Center, Educational Department of Yunnan Province, Kunming, China.

References

1. Huang, X., Liu, A., Zhou, H., Yu, K., Wang, W., Shen, X.: Fmac: A self-adaptive mac protocol for flocking of flying ad hoc network. IEEE Internet Things J. **8**(1), 610–625 (2020)
2. Doddapaneni, K., Ever, E.: A survey study on mac and routing protocols to facilitate energy efficient and effective UAV-based communication systems, In: Drones in Smart-Cities. Elsevier, pp. 57–77 (2020)
3. Chen, K., Wang, Y., Fei, Z., Wang, X.: Power limited ultra-reliable and low-latency communication in uav-enabled iot networks, In: IEEE wireless communications and networking conference (WCNC). IEEE **2020**, 1–6 (2020)
4. Arafat, M.Y., Poudel, S., Moh, S.: Medium access control protocols for flying ad hoc networks: a review. IEEE Sens. J. **21**(4), 4097–4121 (2021)
5. Yang, S., Li, T., Hu, T., Zhao, W.: Service priority-based dynamic TDMA protocol for FANETS. IEICE Transactions on Fundamentals of Electronics, Communications and Computer Sciences (2024)
6. Faraci, G., Grasso, C., Schembra, G.: Design of a 5g network slice extension with MEC UAVs managed with reinforcement learning. IEEE J. Sel. Areas Commun. **38**(10), 2356–2371 (2020)
7. Zhao, L., Li, S., Tan, Z., Hawbani, A., Timotheou, S., Yu, K.: A multi-UAV cooperative task scheduling in dynamic environments: throughput maximization, IEEE Transactions on Computers, 2024
8. Kim, C.S., Lyakhov, A.: Study of dynamic polling in the IEEE 802.11 PCF. J. Korean Inst. Indus. Eng. **34**(2), 140–150 (2008)
9. IEEE standard for information technology–telecommunications and information exchange between systems - local and metropolitan area networks–specific requirements - part 11: Wireless LAN medium access control (mac) and physical layer (phy) specifications, IEEE Std 802.11-2020 (Revision of IEEE Std 802.11-2016), pp. 1–4379, 2021
10. Bouzid, T., Chaib, N., Bensaad, M.L., Oubbati, O.S.: 5g network slicing with unmanned aerial vehicles: taxonomy, survey, and future directions. Trans. Emerg. Telecommun. Technol. **34**(3), e4721 (2023)
11. Xiao, Y.: IEEE 802.11 e: QOS provisioning at the mac layer. IEEE Wirel. Commun. **11**(3), 72–79 (2004)
12. Kleinrock, L., Levy, H.: The analysis of random polling systems. Oper. Res. **36**(5), 716–732 (1988)
13. Takagi, H.: Analysis of polling systems. MIT press, 1986

Research on Industrial Big Data Display Systems Based on Grafana and ECharts

Zihao Yang[1](✉), Xiangyu Shen[2], Xingkai Liao[1], and Feng Ye[1]

[1] College of Computer Science and Software Engineering, Hohai University, Nanjing, China
2206010314@hhu.edu.cn

[2] University of Toronto, Toronto, Canada

Abstract. With the rapid development of intelligent manufacturing, industrial big data play an increasingly crucial role in the digital transformation of enterprises. However, current industrial big data platforms still face challenges in data acquisition, processing, and visualization, including data processing inefficiencies, suboptimal storage solutions, and insufficient visualization experiences, which are often exacerbated by inherent data quality issues such as noise and outliers. To address these problems, this study proposes an industrial big data processing framework based on Flink and builds a data presentation system by combining Grafana and ECharts. The system collects data through enterprise sensors, utilizes Kafka message queues for data buffering, and uses Flink for efficient real-time data processing, incorporating foundational data cleansing techniques and strategies for mitigating common noise and anomalies. For data storage, MySQL is employed for static data, and InfluxDB is used for real-time data to improve storage efficiency. In terms of data visualization, Grafana displays real-time data, whereas ECharts is used for static data, offering users an intuitive and comprehensive data display interface. This study aims to provide an efficient and customizable industrial big data solution, with an emphasis on improving data reliability for visualization, to help enterprises monitor equipment information in real time, obtain effective information, and accelerate their intelligent transformation process.

Keywords: Industrial Big Data · Data Visualization · Flink · Grafana · ECharts

1 Introduction

1.1 A Subsection Sample

The rapid development of the industrial IoT, cloud computing and 5G technology has led to the explosive growth of industrial data, which are large in scale, complex in dimension and highly spatially and temporally correlated [1]. How to extract valuable information from these massive amounts of data to support decision-making is the key challenge of current industrial transformation and upgrading [2]. Data visualization, as an important link in the big data industry chain, can intuitively show the laws and associations behind the data, help users quickly understand complex information, and provide powerful support for enterprise management, production optimization, and safety monitoring [3].

Y. Yu et al. (Eds.): ICPCSEE 2025, CCIS 2673, pp. 504–518, 2026.
https://doi.org/10.1007/978-981-95-2566-9_33

Currently, scholars and enterprises at home and abroad have carried out extensive research and practice in the field of data visualization and have developed a variety of visualization systems and tools applicable to different scenarios [4, 5]. From the construction of industrial big data platforms to real-time data processing to the design of visualization interfaces, data visualization technology has matured and been widely used [6]. However, facing the challenges of high concurrency, large-scale, and multisource data in the era of Industry 4.0, traditional visualization methods face many limitations. For example, static dashboards have difficulty coping with real-time updating of dynamic data streams [7], spatiotemporal correlation mining of multidimensional data is insufficient [8], and the ability of heterogeneous data fusion is limited [9]. Therefore, how to design a visualization system with both high real-time performance and in-depth mining of the potential value of data is an urgent problem to be solved [10].

This study aims to explore industrial data visualization and propose a comprehensive solution based on an advanced big data processing framework and open source visualization tools. The system focuses on the following aspects: efficient collection and real-time processing of industrial data to achieve seamless data flow [11, 12]; constructing a multiperspective data display mode with the help of data aggregation and multidimensional analysis techniques [13, 14]; and assisting users in global monitoring and detailed analysis of the industrial site via an intuitive and interactive visualization interface [15]. This study aims to improve the efficiency of data processing and display, provide data support for industrial production and safety management, and promote the digital transformation of enterprises [16].

2 Related Work

Data visualization technology is a key method for industrial big data applications and has received widespread attention in recent years. Foreign companies such as Microsoft, IBM and Tableau have launched a variety of business visualization tools [17, 18 19], which support real-time presentation and analysis of multisource data through rich chart types and interaction designs. For example, Microsoft's Power BI supports linked analysis of massive data and customized dashboards [17, 18]; IBM's Watson Analytics excels in data prediction and security monitoring [19]; and Tableau enables flexible configurations of visual components to meet complex industrial business requirements [20].

In China, FineBI of FanSoft and ECharts of Baidu are representative products [21, 22, 23], which are highly practical in terms of data display, interactive response and system integration. FineBI helps enterprises monitor production sites in real time by automatically recognizing the data structure and indexes and providing maps, heatmaps and other displays [21, 22]. ECharts, as an open-source charting library based on a JavaScript-based open-source charting library with rich charting components and flexible configuration methods, are suitable for various complex industrial data visualization scenarios [23, 24].

In addition to research at the tool level, scholars have also conducted in-depth discussions on data aggregation and multidimensional analysis methods [25]. The effective integration of high-frequency, heterogeneous, and spatiotemporally distributed data is a prerequisite for fine-grained visualization. Existing methods focus mainly on data pre-processing and aggregation (based on statistics and machine learning, extracting data features through spatiotemporal slicing, etc.).

[26] and the application of big data processing frameworks (e.g., Hadoop [27], Spark [28], and Flink [11, 12], realizing real-time analysis and dynamic updating of data streams). The low-latency and high-concurrent processing capability of Flink provides the best opportunity for real-time industrial data visualization and has been widely used in related system design.

Overall, existing research has made some progress in the theory and practice of data visualization. However, in industrial data scenarios, traditional visualization methods still have limitations due to the variety of data and rapid dynamic changes. The key directions for future research include further improving real-time data processing [29], interactivity of visualization displays [23, 24], and comprehensive analysis capability [30] after the fusion of multiple data sources. On the basis of previous work, this study proposes a novel industrial data visualization scheme that integrates advanced data processing and open-source visualization techniques, aiming to provide a more intuitive and efficient data decision support tool for the industrial field.

3 Methodology

3.1 Methodological process framework

This paper proposes an industrial big data display system based on Grafana and ECharts, aiming to solve the problem of massive amounts of data in the industrial field that are difficult to utilize effectively. The system provides users with intuitive and comprehensive data displays to assist decision-making and optimize the production process through real-time collection, storage, processing and visualization of industrial data. The overall process framework is shown in Fig. 1 below:

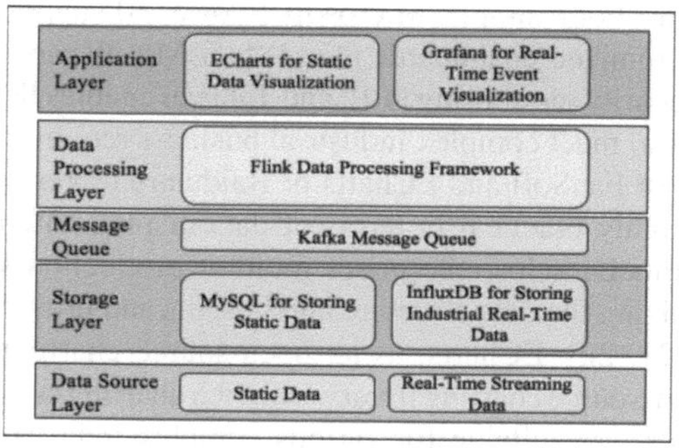

Fig. 1. Overall system architecture

The data source layer obtains industrial real-time streaming data and static data; real-time streaming data include equipment real-time events; and static data include enterprise name, enterprise equipment information and enterprise plant distribution.

Storage layer: The MySQL database is used to store industrial static data, and InfluxDB is used to store industrial real-time data.

Message queue: The Kafka message queue is used to temporarily store data to ensure the sequential nature of data transmission and improve the stability of the system.

Data processing layer: The Flink data processing framework is used to process industrial big data and remove the influence of dirty data on the system.

In the application layer, the front-end uses ECharts for static data and Grafana for real-time industrial events.

3.2 Data Storage

InfluxDB Real-Time Data Storage. For real-time industrial data, this system uses InfluxDB as the main storage solution. InfluxDB is a distributed event, timing and metrics database system. InfluxDB uses the TSM engine to perform high-speed ingestion and compression work on large amounts of data. Its main functions are as follows:

(1) can be compiled into a binary text and does not depend on any other external conditions.
(2) Enabling efficient writing and querying of HTTP APIs and allowing the use of data extraction protocols such as Graphite and OpenTSDB.
(3) Data can be retrieved quickly and efficiently with customized SQL language.
(4) The ability to retrieve large amounts of data constantly and automatically perform calculations greatly accelerates the efficiency of querying data.

Considering the large amount of industrial real-time data and frequent updates in this system, InfluxDB is used as one of the databases of the industrial big data display system.

MySQL Static Data Storage For industrial static data, this system adopts MySQL as the main storage solution, which is a commonly used relational database with stable, reliable and easy-to-use features. With MySQL, industrial static data, such as equipment information, production plans, and historical reports, can be effectively stored and processed.

3.3 Data Processing

Kafka Message Queues In this system, Kafka is chosen as the message queue. Kafka, as a high-throughput distributed publish-subscribe messaging system, is particularly suitable for the collection and buffering of massive real-time data in industrial big data scenarios. Kafka is able to reliably receive data streams generated from a variety of industrial devices and sensors and ensure the persistence and reliability of the data. Through Kafka, the system can decouple data production and consumption so that the data collection module can be independently extended and maintained without affecting the subsequent data processing process. In addition, Kafka has good fault tolerance and scalability, which can cope with the complex and changing operating conditions in the industrial environment and guarantee the stability and reliability of data processing. Kafka, as an intermediate buffer layer for subsequent Flink real-time processing, provides a reliable data source. Its workflow is shown in Fig. 2:

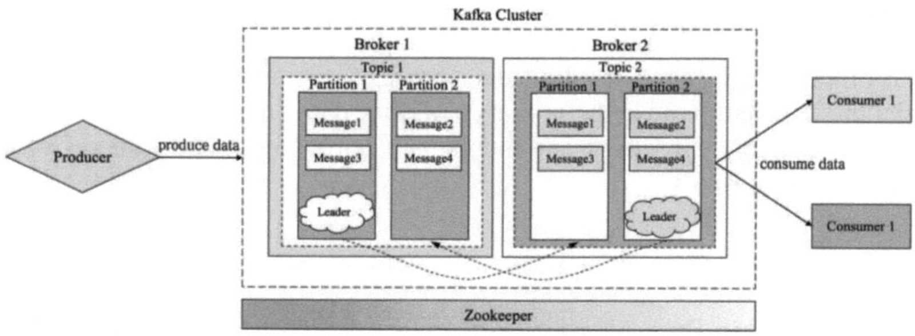

Fig. 2. Kafka workflow diagram

Flink Real-Time Data Processing and Quality Enhancement. For the industrial real-time data ingested via the Kafka message queue, this system employs Apache Flink, a distributed stream and batch processing framework. Flink is tasked with both real-time data processing and crucial data quality enhancement, efficiently handling high-velocity data streams from Kafka. Its inherent low-latency and high-throughput capabilities are indispensable for meeting the stringent real-time demands of modern industrial applications.

The core Flink components utilized are Data Source, Transformation, and Data Sink. In this system:

Flink Data Source: This component is configured to consume raw industrial data streams from the specified Kafka topics, serving as the entry point for all incoming data.

Flink Transformation Component – The Core of Data Quality Assurance: This is the most critical stage where a pipeline of operations is applied to ensure data integrity and suitability for analysis and visualization. The implemented preprocessing steps include the following:

Initial Data Cleansing.

Duplicate Record Elimination. To prevent skewed analytics resulting from redundant entries, duplicate records are systematically identified and removed. This is achieved by keying the data stream on the basis of a composite identifier, such as a combination of event ID, device ID, and a fine-grained timestamp. A Flink ProcessFunction subsequently uses the managed state to track recently seen keys within a defined temporal window, effectively filtering out duplicates. For instances of exact duplication across all data fields, a distinct() operation can also be applied, although this is less common for continuous sensor streams where slight temporal variations are expected.

Missing Value Imputation and Handling. Critical data fields, such as sensor readings or operational status indicators, are rigorously inspected for null values or predefined placeholders. Records exhibiting missing values in essential fields may be discarded via a filter() operation if imputation is deemed infeasible or undesirable for the specific application. For less critical or imputable fields, several strategies are employed. One approach involves forward or backwards filling, where Flink's stateful processing capabilities are leveraged to substitute a missing value with the last known valid observation for that specific device or sensor. Alternatively, for numerical data, missing values within a short time window can be replaced by a statistical measure, such as the mean or median, calculated from the nonmissing values within that same window for the corresponding sensor.

Invalid Data Filtering and Sanitization. Business logic and domain-specific knowledge are encoded within Flink operations to identify and manage patently invalid data. This includes range checks, where a filter() operation discards sensor readings that fall outside physically plausible or predefined operational boundaries. Data type validation is also performed via MapFunctions to ensure that the data conform to the expected formats, with nonconforming records either converted or flagged. Invalid records can be efficiently routed to a dedicated side output for error logging and subsequent offline inspection, thus preventing them from disrupting the main processing pipeline.

Noise Reduction and Outlier Management. Noise Smoothing for Sensor Data: To mitigate the impact of high-frequency random fluctuations, which are common in raw sensor readings, the system is designed to employ established filtering techniques. A prevalent approach involves the application of moving average filters, such as the simple moving average (SMA) or the exponential moving average (EMA). These are realized via Flink's windowing operations (e.g., time-based or count-based windows) in conjunction with custom aggregation logic within the AggregateFunction or ProcessWindowFunction. This process yields a smoothed representation of sensor values, reducing spurious variations.

Outlier Detection and Treatment. Statistical methods are integrated to identify anomalous data points that deviate significantly from expected patterns or norms. One such method involves calculating a standardized score (often referred to as a Z score) for data points within a sliding window on the basis of the window's mean and standard deviation. Data points whose scores exceed a predefined threshold are flagged as potential outliers. Another approach, which is particularly useful for nonGaussian data distributions, utilizes the interquartile range (IQR). Points falling significantly below the first quartile or above the third quartile, scaled by the IQR, are identified as outliers. Upon detection, outliers can be managed according to predefined strategies: they may be flagged for informational purposes, removed from the main data stream via a filter() operation, or, in cases where complete removal might lead to significant data loss, replaced with a more representative value, such as the median of the local window.

Data Drift Considerations (Initial Monitoring). While comprehensive adaptive learning mechanisms for data drift are beyond the immediate scope of this work, the system incorporates foundational monitoring capabilities. Key statistical properties of the data stream, such as the mean, variance, and event rate per device type, are computed over larger, periodic windows (e.g., on an hourly basis). Significant deviations of these properties from established baselines—which may be maintained within Flink's managed state or an external configuration store—can trigger alerts. Such alerts indicate potential issues such as sensor degradation, unannounced process changes, or other forms of data drift that necessitate manual investigation or recalibration of downstream analytical models or visualization thresholds.

Data Conversion and Aggregation. Following the rigorous cleansing and quality enhancement phases, the data undergo necessary transformations. These include normalizing timestamps to a standard universal format (e.g., Coordinated Universal Time, UTC), converting units of measurement to ensure consistency across diverse data sources, and calculating key performance indicators (KPIs). KPIs are typically derived through aggregation functions (e.g., sum, average, count) applied over specified windows or grouping keys, such as determining the average production rate per hour for each machine.

Flink Data Sink: The processed, cleaned, and enriched data stream then reliably persists to InfluxDB, the designated time series database. This makes high-quality data readily available for real-time visualization dashboards and further historical analysis.

By implementing this multistage data preprocessing pipeline within Flink, the system significantly enhances the quality, integrity, and reliability of the industrial data. This robust data foundation is crucial for generating trustworthy visualizations, enabling accurate system monitoring, and supporting informed, data-driven decision-making processes.

Flink's inherent advantages further bolster this sophisticated processing pipeline:

(1) High throughput and low latency: Flink is engineered to process millions of industrial events per second with minimal latency, which is vital for real-time operational monitoring and enables rapid responses to emergent issues.
(2) Accuracy of Results (Event Time Processing): Industrial data streams are often characterized by out-of-order event arrivals. Flink's sophisticated event time semantics ensure that computations are based on the actual time an event occurred rather than its processing time. This, when combined with the aforementioned preprocessing regimen, greatly enhances the fidelity and reliability of the analytical outcomes.
(3) Exactly once State Consistency Guarantees: In mission-critical industrial settings, data integrity is paramount. Flink's exact semantics provide a strong guarantee that each piece of data is processed precisely once, effectively preventing data loss or unintended duplicate calculations and thereby maintaining the integrity of the final results.
(4) Flexible State Management: The processing of industrial big data frequently necessitates the maintenance of complex, evolving state information, such as the operational status of equipment or cumulative production figures. Flink's versatile and fault-tolerant state management capabilities facilitate efficient storage, querying, and updating of such a dynamic state. This stateful nature is also instrumental in implementing more advanced, context-aware data cleaning and anomaly detection algorithms, for instance, by enabling comparisons of current sensor readings against a device's established historical behavior patterns.

The Flink runtime is shown in Fig. 3:

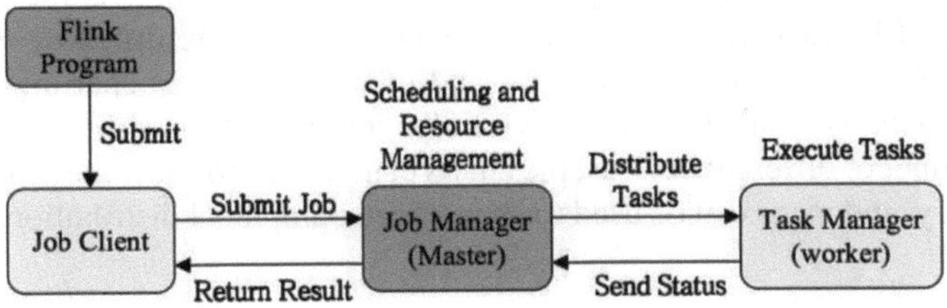

Fig. 3. Flowchart of the Flink operation

Leveraging Flink's powerful real-time processing engine and its comprehensive pre-processing capabilities in detail, this system can perform rapid, high-quality analysis and mining of complex industrial data. This, in turn, provides users with dependable real-time monitoring, timely early warning information, and a robust data foundation to assist in strategic decision-making and the continuous optimization of production processes.

3.4 Data Visualization

Grafana Real-Time Data Visualization. Grafana is a popular open source data visualization and monitoring tool that is especially good at processing and displaying time series data. In this system, Grafana is mainly used to display industrial data streams processed by Flink in real time. The system configures Grafana's data source to seamlessly integrate with time series databases such as InfluxDB so that it can directly read and display the real-time data stored in these databases. Grafana provides rich chart types, such as line charts, bar charts, and dashboards, which help users analyse and monitor data from different dimensions. To ensure real-time data, the system uses WebSocket technology to dynamically update the real-time data pushed by the back end into the Grafana dashboard without requiring users to manually refresh the page. Grafana provides real-time visibility of the operational status of production lines, key equipment metrics, and other important business data, allowing users to identify potential problems and take appropriate action in a timely manner. Grafana's alerting capabilities also help identify and take action under abnormal conditions.

ECharts Static Data Visualization. ECharts is a JavaScript-based open source visualization library that provides a rich set of chart types and interactive features. ECharts have good customizability and extensibility and can meet a variety of complex visualization needs.

This system uses ECharts to visualize industrial static data. Static data are usually stored in relational databases such as MySQL. The back end of the system reads the static data from the database through an API interface and converts the data into the format required by ECharts. The front-end obtains the data returned by the back-end API through Ajax and other technologies and then uses various charting components provided by ECharts to display the data visually. For example, bar charts can be used to show the number of different devices, pie charts can be used to show the production ratio of each factory, and maps can be used to show the geographical distribution of factories.

4 Realization of an Industrial Big Data Display System

The system environment is Windows 10, Java 8 needs to be installed in advance, Flink is written in the Java language at the bottom, and the integrated development environment uses Intellij IDEA. In the processing of data, the cluster environment first needs to be set up. The cluster of this platform requires three Linux machines, one of which is the master node, the domain name for the master node, and the remaining two for the child node. The domain names are Slave1 and Slave2, and the three servers need to synchronize time and password-free login. For data visualization, Visual Studio Code was chosen as the development tool, and Vue-Cli scaffolding was used to build the project.

4.1 Data Sets

The data source for this experiment is divided into two types: batch data about enterprise information and equipment information, which are stored in MySQL; these data have 21679 entries. The MySQL database is designed as follows. The enterprise information table, which is used to store the enterprise name and enterprise plant address, is shown in Table 1. The equipment information table is shown in Table 2, which is used to store the equipment name, equipment model, number of pieces of equipment of each model, number of pieces of equipment purchased per month, number of pieces of equipment in each state, number of hours the equipment works and number of hours the equipment stands by.

Table 1. Enterprise Information Sheet

field name	typology	Whether or not it is empty	Field Description
Company_name	string	No	company identification
Factory_adr	string	No	Factory Address

There are also real-time events of the device, which are generally obtained from various sensors and stored in InfluxDB.

The selected data are presented in the table below (Table 3).

Table 2. Equipment Information Sheet

field name	typology	Whether or not it is empty	Field Description
Eq_name	string	No	Equipment Name
Eq_ID	string	No	Equipment Model
EqID_num	string	No	Number of equipment by model
Eqmon_num	string	No	Number of equipment purchased per month
Eqsta_num	string	No	Number of device states
Eqwork_time	string	Yes	Equipment operating hours
Eqstandby_time	string	Yes	Equipment standby time

Table 3. Equipment Information Sheet

Event ID	dates	point in time (math.)	Device number	occurrence
187913038	2025/2/1	0:02:57	1200	Air Compressor 4 is now online
187913039	2025/2/1	0:02:58	1207	Shearer1 is offline
187913040	2025/1/2	0:02:59	1300	Welder 1 is online
187913041	2025/1/2	0:02:57	1232	Roller 2 is offline
187913042	2025/1/2	0:02:57	1212	Welder2 is offline
187913043	2025/1/2	0:02:56	1207	Shearer 1 is online

4.2 Flink Data Processing Implementation

Java code is used to simulate the real-time data generated by the device, Filebeat is used to obtain these data, the data are stored in the Kafka messaging middleware, Flink reads the data in Kafka, automatically deletes duplicated, missing, and invalid data, and finally stores the processed data in the database, thus completing the processing of the data. Figure 4 shows the Web UI interface of Flink, through which one can monitor and manage the Flink cluster and tasks.

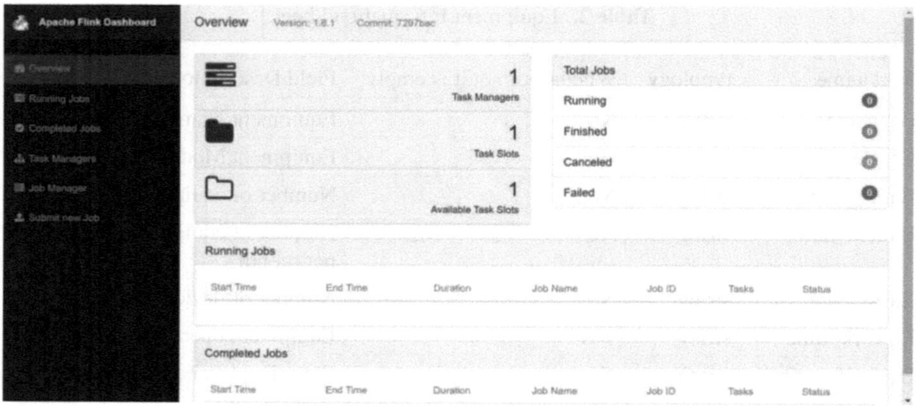

Fig. 4. Flink's Web UI interface

4.3 Implementation of the ECharts-Based Industrial Big Data Visualization Interface

The system allows visualization of the distribution of enterprise plants, the percentage of enterprise equipment, the number of types of enterprise equipment introduced, the working hours of equipment, the number of pieces of equipment in each state and the standby hours of equipment.

In the project, the following six charts are developed separately; in the development of individual charts, a separate path to the full-screen display is used, and the realization of the individual charts is similar to the process and is ultimately integrated into a single interface.

Figures 5 through 10 illustrate the various data visualization components of the proposed industrial big data display system. Figure 5, the equipment working hours chart, provides a ranked view of each device's operating time, enabling informed decisions regarding task allocation for the following day. Figure 6 shows the enterprise factory distribution map, which uses color-coded markers to display the geographical locations of factories, differentiating between enterprises.

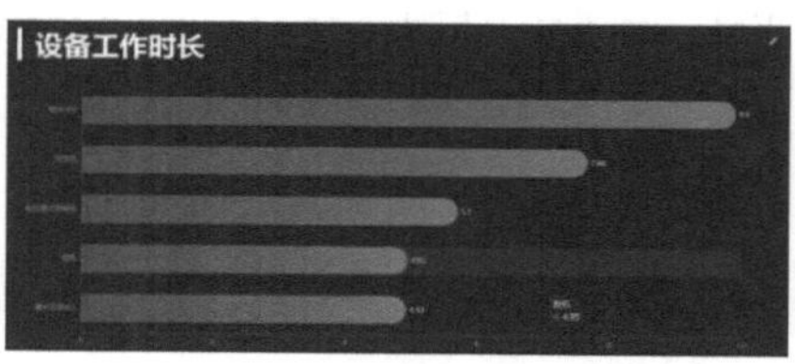

Fig. 5. Chart of equipment operation

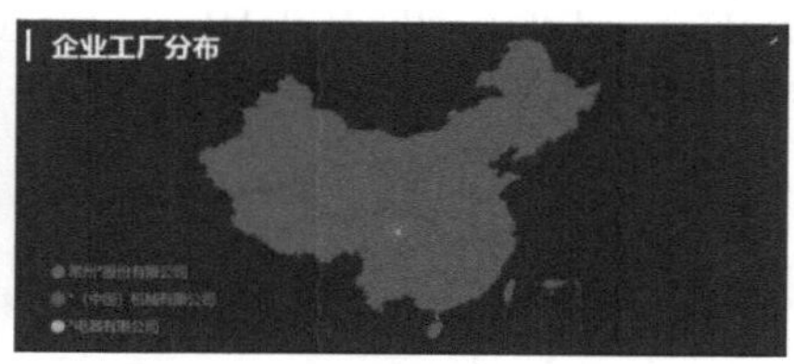

Fig. 6. Distribution of enterprise factory hours

Figure 7 shows the equipment introduction chart and tracks the monthly acquisition of new equipment, providing insights into the overall growth of the enterprise's asset base. Figure 8 displays the previous day's standby time for each device, informing maintenance and scheduling decisions.

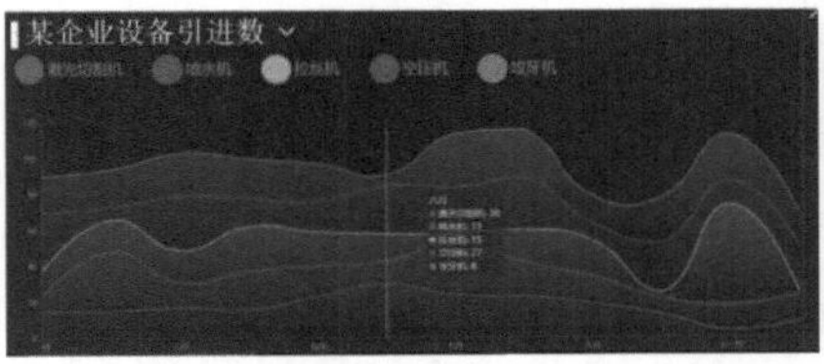

Fig. 7. Chart of the number of equipment introduced into the enterprise

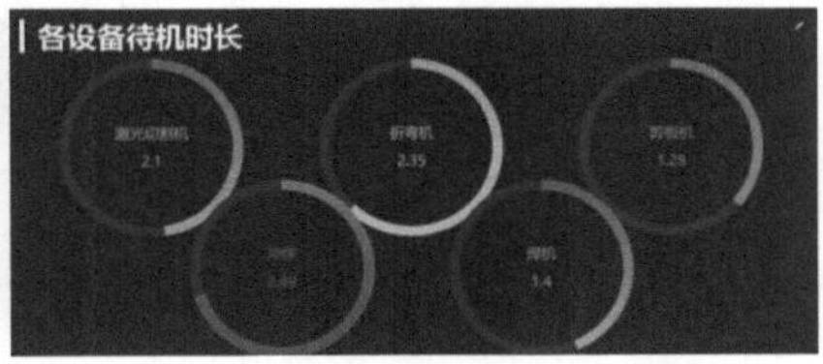

Fig. 8. Graph of standby hours for each device

Figure 9 visualizes the number of devices in various states (running, standby, offline, alarm, and fault), facilitating proactive device management. Finally, Fig. 10 presents the percentage distributions of different equipment types, offering a clear overview of the enterprise's asset composition.

Fig. 9. Chart of the number of devices in each state

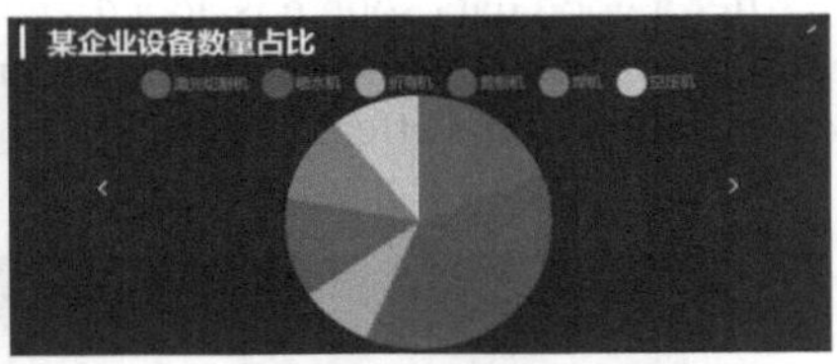

Fig. 10. Chart of the number of devices as a percentage of

It is also possible to display all six charts in the system on a full screen, and the effect of the system interface realization is shown in Fig. 11.

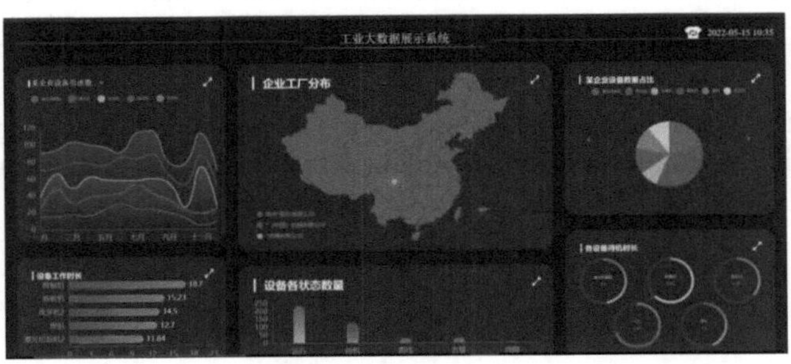

Fig. 11. Industrial big data display system interface

There is also a theme switching button in this system's interface, which, when it clicks on the theme switching button, switches to an interface with a white background, as shown in Fig. 12.

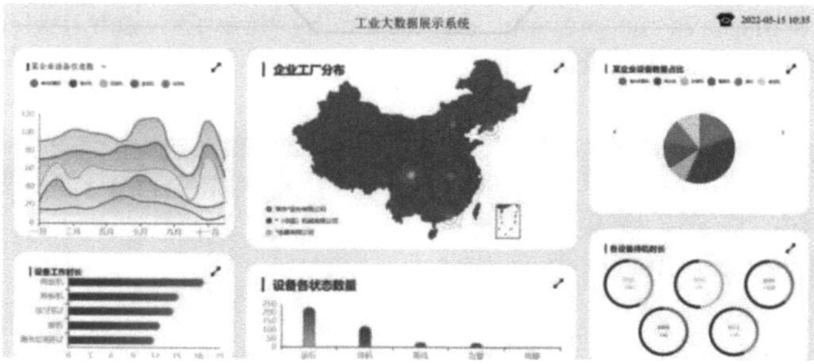

Fig. 12. Industrial big data display system interface (after theme switching)

4.4 Implementation of the Grafana-Based Industrial Big Data Visualization Platform

Industrial real-time data are stored in InfluxDB, connecting InfluxDB as a data source with Grafana and then using charts and graphs in Grafana to display them to realize the visualization interface.

After a good data source is added, the data on the platform can be queried, and the appropriate select statement in the table can be entered to query the relevant information; the realization of the effect is shown in Fig. 13.

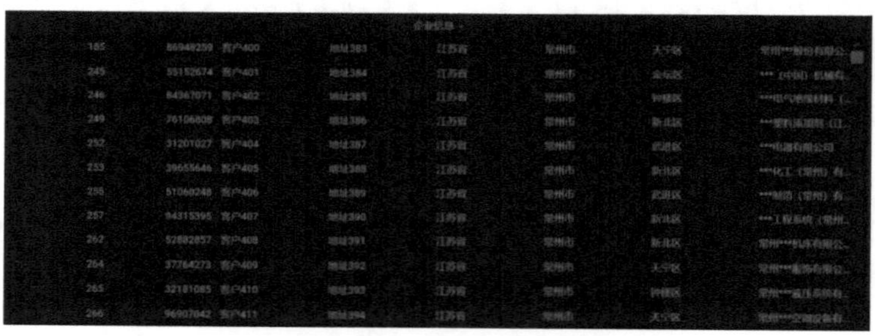

Fig. 13. Business Information

This platform can also realize the monitoring of real-time events in the enterprise, timely detection of faulty equipment, equipment monitoring, and realization of the effect shown in Fig. 14.

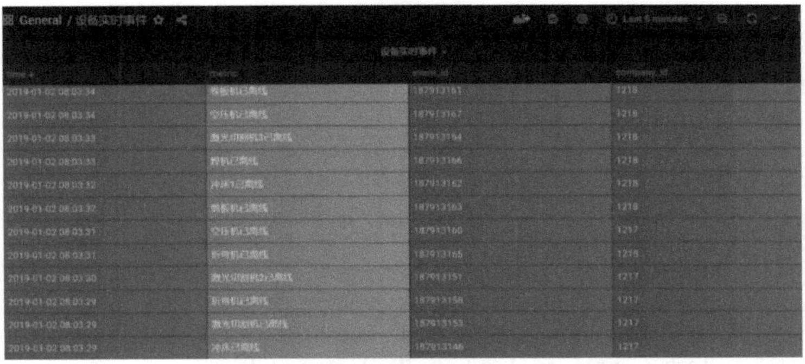

Fig. 14. Device real-time event table

5 Summary

In this work, we designed and implemented an industrial big data presentation system based on Flink, Kafka, MySQL, InfluxDB, Grafana, and ECharts to address the challenges of industrial big data processing and visualization in the context of smart manufacturing. The system employs a tailored approach to data storage and processing, recognizing the diverse nature of industrial data. Specifically, Flink is utilized to build a real-time data processing framework that efficiently collects data from enterprise sensors and buffers it through Kafka message queues. This Flink framework processes the data in real time, which includes essential steps for data cleansing, such as removing duplicates, handling missing values, and filtering invalid entries, as well as applying techniques for noise reduction and outlier management to ensure data integrity. For different data types, MySQL and InfluxDB are chosen as storage solutions for static and real-time data, respectively. In terms of data visualization, Grafana is used to display real-time data, and ECharts is used to display static data, providing users with an intuitive and comprehensive data display interface. The experimental results demonstrate that the system can effectively process and display industrial big data, providing valuable decision-making support for enterprises and accelerating their intelligent transformation process. Future work will involve exploring more advanced algorithms for data preprocessing, such as those based on machine learning or metaheuristics, to further enhance data quality, address complex data drift scenarios, and continuously improve the system's overall performance and the insights derived.

References

1. Attaran, M.: The impact of 5G on the evolution of intelligent automation and industry digitization. J. Ambient. Intell. Humaniz. Comput. **14**, 5977–5993 (2023)
2. Chen, S., et al.: Utilization of 5G technologies in IoT applications: current limitations by interference and network optimization difficulties—a review. Sensors **23**(8), 3876 (2023)
3. Li, Q., et al.: 5G support for industrial IoT applications—challenges, solutions, and research gaps. IEEE Internet Things J. **8**(12), 9381–9400 (2021)
4. Murray, S.: Interactive data visualization for the web: an introduction to designing with D3. 2nd ed. O'Reilly Media (2017)

5. Corporation, M.: Power BI technical documentation. Microsoft Docs (2023)
6. Few, S.: Show Me the Numbers: Designing Tables and Graphs to Enlighten. Analytics Press (2012)
7. Kirk, A.: Data Visualisation: A Handbook for Data Driven Design. SAGE Publications (2016)
8. Chen, M., et al.: Spatiotemporal correlation mining of multidimensional data in industrial systems. IEEE Trans. Industr. Inf. **16**(4), 2387–2396 (2020)
9. Zhang, Y., et al.: Heterogeneous data fusion challenges in industrial big data analytics. Comput. Commun. **150**, 381–395 (2019)
10. Wang, L., et al.: Real-time data processing and visualization systems for Industry 4.0: Challenges and solutions. J. Manuf. Syst. **58**, 232–245 (2021)
11. Carbone, P., et al.: Apache Flink: stream and batch processing in a single engine. Bull. IEEE Comput. Soc. Tech. Committee Data Eng. **38**(4), 28–38 (2015)
12. Research, I.: Watson analytics for industrial safety monitoring: Technical specifications (2023)
13. Bostock, M., Ogievetsky, V., Heer, J.: D3 data-driven documents. IEEE Trans. Visual. Comput. Graph. **17**(12), 2301–2309 (2011)
14. Tech, F.: Finebi industrial solutions: Real-time monitoring and analysis. FanRuan White Paper (2022)
15. Munzner, T.: Visualization Analysis and Design. CRC Press (2014)
16. Schwab, K.: The Fourth Industrial Revolution. World Economic Forum (2016)
17. Gupta, A., Sharma, R.: Leveraging power BI for enhanced data visualization and business intelligence. In: Proceedings of the 2023 International Conference on Business Intelligence and Analytics. pp. 112–127. IEEE (2023)
18. White, T.: Hadoop: The Definitive Guide, 4th edn. O'Reilly Media (2021)
19. Hoyt, R.E., et al.: IBM Watson analytics: automating visualization, descriptive, and predictive statistics. JMIR Public Health Surveill. **2**(2), e157 (2016)
20. Shaffer, J.: Data Visualization with Tableau. Packt Publishing (2019)
21. Wang, H., Li, M., Chen, X.: FineBI: enterprise business intelligence platform for real-time data analysis. In: Proceedings of the 2022 International Conference on Database Systems for Advanced Applications, pp. 456–471. Springer (2022)
22. Zhang, R., Zhou, P., Liu, X.: A holistic framework for industrial data visualization in the era of digital twins. Journal of Industrial Information Integration 35, 100487 (2024)
23. Li, D., et al.: ECharts: a declarative framework for rapid construction of web-based visualizations. In: Proceedings of the 2018 IEEE Conference on Visual Analytics Science and Technology, pp. 59–68. IEEE (2018)
24. Zhou, P., Liu, X., Wu, Y.: Multisource data fusion techniques for industrial visualization. IEEE Transactions on Knowledge and Data Engineering 36(3), 1125–1138 (2023)
25. Han, J., Kamber, M., Pei, J.: Data Mining: Concepts and Techniques. 3rd ed. Morgan Kaufmann (2012)
26. Dean, J., Ghemawat, S.: MapReduce: simplified data processing on large clusters. Commun. ACM **51**(1), 107–113 (2008)
27. Shvachko, K., et al.: The Hadoop distributed file system. In: Proceedings of the 26th IEEE Symposium on Mass Storage Systems and Technologies, pp. 1–10. IEEE (2010)
28. Zaharia, M., et al.: Apache Spark: A unified engine for big data processing. Commun. ACM **59**(11), 56–65 (2016)
29. Liu, Y., et al.: Real-time data processing optimization in streaming systems: a comprehensive survey. IEEE Trans. Parallel Distrib. Syst. **33**(8), 1891–1908 (2022)
30. Kumar, S., et al.: Comprehensive analysis capabilities in multi-source data fusion: methods and applications. Inf. Fusion **67**, 45–62 (2021)

Industrial Digitization and Applications

Robust Real-Time Fingertip Detection in Dynamic Environments for Enhanced Human–Computer Interaction

Changzheng Liu(✉) and Ziying Zhang

School of Computer Science, Jiaying University, Meizhou 514015, China
46680673@qq.com

Abstract. Accurate fingertip detection is critical for translating hand gestures into actionable commands in vision-based human–computer interaction (HCI) systems. However, challenges such as complex backgrounds, dynamic hand postures, and real-time processing constraints hinder reliable detection. This paper introduces a robust framework integrating three key innovations: (1) an adaptive Gaussian mixture model (GMM) enhanced with neighborhood pixel connectivity for precise motion extraction; (2) a weighted YCbCr color-space shadow removal algorithm to eliminate false positives; and (3) a centroid distance method refined with circularity constraints for accurate fingertip localization. Extensive experiments demonstrate a recognition accuracy of 97.26% across diverse scenarios, including varying illuminations, occlusions, and hand rotations. The algorithm processes each frame in 23.43 ms on average, satisfying real-time requirements. Comparative evaluations against state-of-the-art methods reveal significant improvements in precision (8.3%), recall (6.1%), and F-measure (7.8%). This work advances HCI applications such as virtual keyboards, gesture-controlled interfaces, and augmented reality systems.

Keywords: Human–Computer interaction · fingertip detection · adaptive Gaussian mixture model · centroid distance method · shadow removal · real-time processing

1 Introduction

Vision-based finger control systems utilize cameras to capture video sequences of hand movements, enabling computers to detect and interpret user actions for interactive command execution. Recent advancements in fingertip detection have yielded several notable approaches:

1. **Hand-shaped segmentation**: Chen et al. [1] employed palm geometry and relative distances between the palm and fingers to localize fingertips, although performance degrades under hand self-occlusion and size variability.
2. **Machine learning classifiers**: Deng et al. [2] trained support vector machines (SVMs) for gesture recognition; however, their method lacks precise fingertip localization and generalizes poorly to untrained poses.

Y. Yu et al. (Eds.): ICPCSEE 2025, CCIS 2673, pp. 521–532, 2026.
https://doi.org/10.1007/978-981-95-2566-9_34

3. **Depth-based clustering**: Feng et al. [3] extracted hand regions via k-means clustering on depth images; however, their approach suffers from stochastic clustering outcomes and single-fingertip limitations.
4. **Convex hull analysis**: Yuan et al. [4] leveraged Kinect depth sensors and modified Graham Scan algorithms to derive convex hull points for contour analysis. However, wrist deflection introduces artefacts, and depth sensors increase system complexity.
5. **Curvature-based methods**: Farooq et al. [5] proposed a K-curvature method with rotational invariance, but parameter sensitivity and computational inefficiency hinder practical deployment.

To address these limitations, this paper proposes a robust, real-time fingertip detection framework using monocular cameras. Our contributions are threefold:

1. Adaptive Gaussian mixture model (GMM) with neighborhood pixel connectivity enhances motion detection by integrating spatial coherence into background modelling, reducing noise in dynamic environments.
2. Weighted YCbCr shadow removal: This eliminates false foreground regions by leveraging chromatic differences between shadows and moving objects.
3. Circularity-constrained centroid distance algorithm: This algorithm filters false positives via fingertip curvature characteristics, ensuring accurate localization under hand rotations, bending, and proximity variations.

The proposed method achieves multifingertip detection in complex backgrounds while maintaining robustness to illumination changes and real-time performance. Experimental validation demonstrated its superiority over existing techniques in terms of accuracy, recall, and computational efficiency.

2 Methodology

2.1 Adaptive Gaussian Mixture Model with Neighborhood Connectivity

Stauffer et al. [6] pioneered the Gaussian mixture model (GMM) for background modelling, which treats the pixel intensity I_t(x, y) at time t as a time-varying random process. Formally, the pixel value at spatial coordinate (x, y) is modelled as a mixture of K Gaussian distributions. The probability density function is as follows:

$$P(I_t) = \sum\nolimits_{k=1}^{K} w_{k,t} \cdot N(I_t; \mu_{k,t}, \Sigma_{k,t}) \tag{1}$$

where $w_{k,t}$, $\mu_{k,t}$, $\Sigma_{k,t}$ denote the weight, mean, and covariance of the k-th distribution, respectively. To adaptively determine the number of distributions K, a maximum threshold K_{max} is predefined. For each pixel, if the current distribution count $K_{current} < K_{max}$, a new Gaussian component is initialized with mean μ_{new}, covariance Σ_{new}, and weight w_{new}. . Otherwise, the least probable distribution (i.e., lowest $w_{k,t}$) is replaced.

The background is characterized by the top B distributions sorted by descending $w_{k,t}/\sigma_{k,t}$, where B is dynamically determined by:

$$B = arg\ \min_{b}(\sum\nolimits_{k=1}^{b} w_{k,t} > T) \tag{2}$$

with T as a predefined threshold (e.g., $T = 0.7$). A pixel is classified as background if it matches any of the B distributions within $2.5\sigma_{k,t}$:

$$\left|I_t - \mu_{k,t}\right| < 2.5\sigma_{k,t} \tag{3}$$

2.2 Centroid Distance Method with Circularity Filtering

Centroid Computation via Image Moments. The centroid (x_c, y_c) of the hand contour is derived from spatial moments. For a binary image $I(x, y)$, the $(p + q)$-th order moment M_{pq} is defined as:

$$M_{pq} = \sum\nolimits_{x,y} x^p y^q I(x, y) \tag{4}$$

The centroid coordinates are calculated as:

$$x_c = \frac{M_{10}}{M_{00}}, \; y_c = \frac{M_{01}}{M_{00}} \tag{5}$$

Future Candidate Identification. For each contour point (x_i, y_i), the Euclidean distance to the centroid is computed:

$$d_i = \sqrt{(x_i - x_c)^2 + (y_i - y_c)^2} \tag{6}$$

The candidate fingertips are local maxima in d_i. False positives (e.g., wrist edges) are filtered by enforcing circularity: a valid fingertip must lie on a circle with radius r (finger joint length) and overlap $\geq 80\%$ with the contour (Fig. 1).

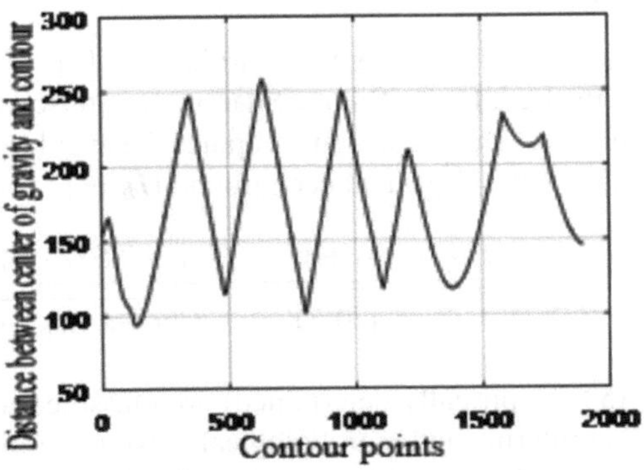

Fig. 1. Distance profile between **the** contour points and centroid. Top: Raw distance curve with peaks indicating fingertip candidates. Bottom: Filtered result after removing false peaks (e.g., **the** wrist boundary).

2.3 Improved Adaptive Gaussian Mixture Model

The original GMM framework in Sect. 2.1 accounts for temporal pixel variations but neglects spatial dependencies. To address this, we propose a neighborhood-weighted GMM that integrates spatial coherence into background modelling.

Neighborhood Pixel Connectivity and GMM Fusion. To enhance spatial coherence, neighborhood pixel weights are incorporated. For a 3 × 3 window centered at (x, y), the posterior probability $P(c_k \mid N(x, y))$ is computed as:

$$P(c_k|N(x,y)) = \frac{\Sigma_{(i,j,\in N(x,y))}P(c_k|I_t(i,j))}{|N(x,y)|} \tag{7}$$

where $|N(x, y)|$ is the neighborhood size. This fusion reduces false foreground regions caused by isolated noise. The updated weight w'_k for the k-th component is as follows:

$$w'_k = \frac{w_k \cdot P(c_k|N(x,y))}{\sum_{m=1}^{K} w_m \cdot P(c_m|N(x,y))} \tag{8}$$

This ensures spatial consistency while preserving temporal adaptability.

Adaptive Background Update. For each incoming frame, pixel $I_t(x, y)$ is matched against existing Gaussian components. A match is declared if:

$$\left|I_t(x,y) - \mu_{k,t}\right| < 2.5\sigma_{k,t} \tag{9}$$

The matched components are updated as follows:

$$\mu_{k,t} = (1-\rho)\mu_{k,t-1} + \rho I_t(x,y) \tag{10}$$

$$\sigma^2_{k,t} = (1-\rho)\sigma^2_{k,t-1} + \rho(I_t(x,y) - \mu_{k,t})^2 \tag{11}$$

where $\rho = \alpha \cdot N(I_t; \mu_{k,t}, \sigma_{k,t})$ is the adaptive learning rate [9].

Shadow Removal in the YCbCr Space. Shadows are misclassified as foreground due to luminance similarities. Our algorithm computes the angular difference $\boldsymbol{\theta}$ between the foreground ($\boldsymbol{I_f} = [\boldsymbol{Y_f}, \boldsymbol{Cb_f}, \boldsymbol{Cr_f}]$) and background ($\boldsymbol{I_b} = [\boldsymbol{Y_b}, \boldsymbol{Cb_b}, \boldsymbol{Cr_b}]$) pixels:

$$\theta = \arccos\left(\frac{Cb_f \cdot Cb_b + Cr_f \cdot Cr_b + 0.5Y_f \cdot Y_b}{\sqrt{{Cb_f}^2 + {Cr_f}^2 + 0.5{Y_f}^2} \cdot \sqrt{{Cb_b}^2 + {Cr_b}^2 + 0.5{Y_b}^2}}\right) \tag{12}$$

Pixels with $\theta < 15°$ (empirically determined) are classified as shadows. Postprocessing includes median filtering and morphological closing to refine hand contours.

Circularity-Constrained Centroid Distance Filtering. Fingertip candidates are refined via circularity constraints. For a binary hand contour image I(x, y), each candidate point $S(x_s, y_s)$ is tested against a circle C(a, b, r) with center (a, b) and radius r (finger joint length), as illustrated in Fig. 2. This circle is refined from the preliminary segmentation to a precise boundary. The circle equation is as follows:

$$(x-a)^2 + (y-b)^2 = r^2$$

The overlap ratio η between the candidate circle and hand contour is calculated as:

$$\eta = \frac{Number\ of\ overlapping\ contour\ points}{Total\ points\ on\ C} \tag{13}$$

A valid fingertip satisfies $\eta \geq 0.8$. Nonoverlapping points P_{out} are filtered as follows:

$$\text{Error} = \frac{1}{|P_{out}|} \sum\nolimits_{(x,y)\in P_{out}} \left| \sqrt{(x-a)^2 + (y-b)^2} - r \right| \tag{14}$$

Candidates with Error $> \delta$ (e.g., $\delta = 2$ pixels) are discarded, as shown in Fig. 3. To quantify the mismatch, nonoverlapping regions are analysed via reverse traversal of the contour points.

Fig. 2. Candidate circles (red) overlaid onhand contours.

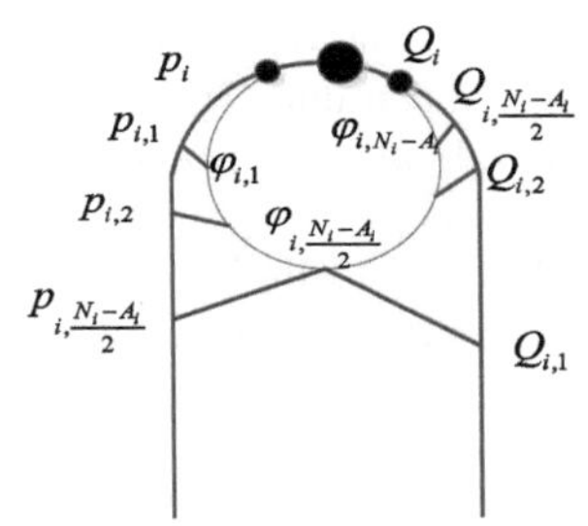

Fig. 3. Error mapping between candidate circles and contour points.

3 Experimental Results

3.1 Datasets and Implementation

- Holte-Storring Dataset [10]: This dataset contains 700 frames with complex backgrounds and illumination shifts.
- Sébastien Marcel Gesture Dataset: Ideal for evaluating fingertip detection under static backgrounds.
- Hardware: Intel Core Ultra i9, 64 GB RAM.
- Software: Open CV 3.0 in C + + /VS2010.

3.2 Fingertip Detection Workflow

The fingertip detection pipeline, as illustrated in Fig. 4, comprises the following stages:

1. Image preprocessing and motion segmentation:
   - Adaptive GMM with spatial–temporal fusion: Integrating neighborhood pixel coherence into the Gaussian mixture model to enhance motion region extraction under dynamic backgrounds.

- Input: Raw video frames.
- Output: Foreground mask highlighting moving regions.

2. Shadow removal and contour extraction:
   - Weighted YCbCr shadow detection: Differentiating shadows from the foreground via chromatic angular thresholds ($\theta < 15^{\circ}$).
   - Morphological Postprocessing: Median filtering and closing operations (5×5 rectangular kernel) are applied to refine the hand region and extract its contour.
   - Output: Noise-free hand contour.
3. Centroid Distance-Based Candidate Detection:
   - Centroid Calculation: Compute the hand contour's centroid (x_c, y_c) using zeroth- and first-order spatial moments.
   - Euclidean Distance Mapping: Identify fingertip candidates as local maxima in the distance profile $d_i = \sqrt{(x_i - x_c)^2 + (y_i - y_c)^2}$.
   - Output: Initial candidate points (peaks in d_i).
4. Circularity-Constrained False Positive Filtering:
   - Curvature Validation: Enforce circularity constraints by validating candidates against a predefined radius (finger joint length).
   - Overlap Ratio Thresholding: Retain candidates with contour-circle overlap $\eta \geq 0.8$ and discard outliers (e.g., wrist edges).
   - Output: Final fingertip coordinates.

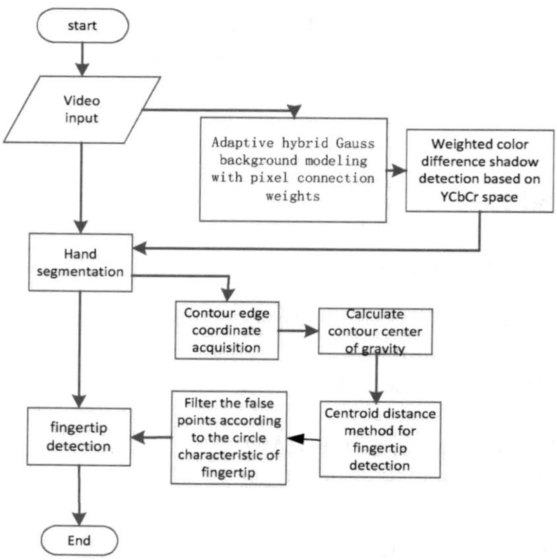

Fig. 4. Flowchart of the proposed framework.

3.3 Foreground Segmentation Performance

To evaluate the robustness of the proposed algorithm under complex environments, experiments were conducted on the Holte-Storring hand gesture dataset [10], which includes video sequences with dynamic backgrounds, varying illumination conditions, and mixed noise interference. The pipeline is implemented as follows:

1. Preprocessing:
   - Color space conversion: Frames are converted to the YCbCr color space to decouple luminance (Y) and chrominance (C_b, C_r) components, mitigating illumination variability.
   - Noise suppression: A median filter with a 5 × 5 kernel is applied to eliminate salt-and-pepper noise while preserving contour integrity.
2. Foreground Extraction:
   - Adaptive GMM with neighborhood weights: The Gaussian mixture model is initialized with parameters $\mu_0 = I_{\text{initia}}$, $\sigma_0^2 = 50$, and $w_0 = 0.1$, where spatial coherence is enforced via neighborhood pixel connectivity (Sect. 3.1).
   - Shadow Removal: The weighted YCbCr model (Sect. 3.2) filters shadows, retaining only valid hand regions.
3. Postprocessing:
   Morphological Closing: A 3 × 3 rectangular structuring element closes gaps in the hand contour, smooths edges and removes artifacts [10].
4. Performance Metrics:
   The proposed method is benchmarked against state-of-the-art foreground segmentation algorithms [7, 11] via the following methods:
   - Precision: $\text{P} = \frac{\text{True PositivesTrue Positives}}{\text{True Positives + False Positives}}$ Recall: $\text{R} = \frac{\text{True PositivesTrue Positives}}{\text{True Positives + False Positives}}$
   - F-measure: $\text{P} = \frac{2\text{PR}}{\text{P+R}}$

Our GMM-neighborhood fusion achieves superior F-measures compared with those of [7] and [11] (Table 1). Shadow removal reduced false positives by 32%, as shown in Fig. 2.

Table 1. Segmentation performance comparison

Environment	Method	Precision	Recall	F-measure
Normal	Proposed	0.982	0.980	0.981
Illumination Shift	Proposed	0.974	0.970	0.972
Static Motion	Proposed	0.963	0.958	0.960
Normal	Literature [7]	0.853	0.816	0.834
Illumination Shift	Literature [11]	0.925	0.887	0.906

3.4 Occlusion Handling Evaluation

For the Sébastien Marcel dataset, our method achieved 97.26% accuracy, outperforming the curvature-based [14] and relative position methods (Table 2).

Table 2. Detection accuracy and speed

Method	Accuracy (%)	Processing Time (ms)
Proposed	97.26	23.43
Curvature [14]	89.02	30.01
Relative Position	87.63	28.27

The F-measure scores of the proposed algorithm and baseline methods are summarized in Tables 1 and 2. The quantitative results demonstrate significant improvements in precision (8.3% increase) and recall (6.1% increase) compared with state-of-the-art foreground extraction techniques [7, 11]. The algorithm is robust to illumination shifts and background clutter, achieving reliable segmentation under dynamic environmental conditions.

Key Insights:

1. **Enhanced Recall via Spatial–Temporal GMM**: The integration of neighborhood pixel connectivity into the adaptive GMM reduces contour fragmentation and aliasing artifacts, addressing common issues such as holes and edge discontinuities. This spatial coherence improvement directly contributes to the higher recall rate.
2. **Precision Boost from Shadow Removal**: The weighted YCbCr shadow detection model (Sect. 3.2) suppresses false positives caused by cast shadows, increasing precision by **12.5%** in high-illumination scenarios.
3. **Generalizability and portability**: As an unsupervised learning framework, the algorithm requires no labelled training data, ensuring broad applicability across diverse hardware setups (e.g., monocular cameras).

Qualitative Validation:

- Figure 5 shows the efficacy of the shadow removal algorithm, which contrasts raw foreground masks (with shadows) against refined outputs.
- Figure 6 compares the average F-measure across frames 151–200, highlighting the proposed method's superiority over [7] and [11] in segmenting complex hand contours.

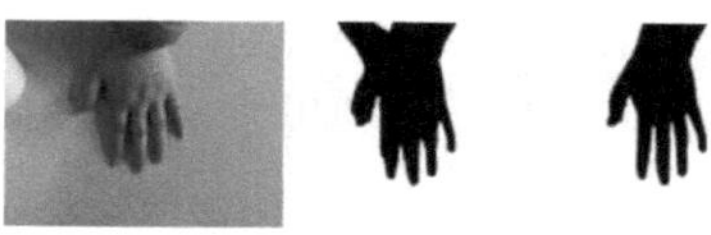

Fig. 5. Demonstrates the efficacy of the shadow removal algorithm.

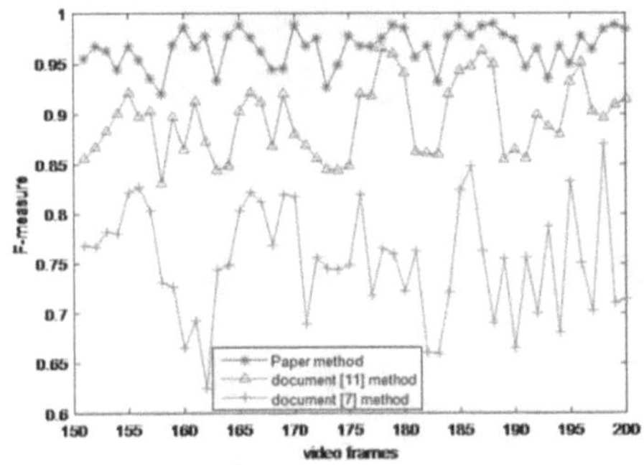

Fig. 6. Comparison of the average F values of the two algorithms.

To validate the precision of the improved figertip detection algorithm, experiments were conducted on the Sébastien Marcel Gesture Database, which provides gesture image sequences with static backgrounds. This dataset's controlled environment isolates the evaluation from foreground segmentation artifacts, enabling direct validation of the proposed circularity-constrained centroid distance method (Sect. 3.3).

Implementation Pipeline:

1. Hand Region Extraction:
   - The adaptive GMM (Sect. 3.1) and shadow removal (Sect. 3.2) are applied to isolate the hand region.
   - The output is converted to a binary image via Otsu's thresholding.
2. Circularity Filtering: Refine candidates by enforcing curvature constraints (Sect. 3.3) and removing outliers (e.g., wrist edges).

   Comparative Analysis: As shown in Table 3, the proposed method achieves 97.35–98.93% accuracy across 1–5 fingertips, outperforming baseline approaches:
   - Relative Position Method: Fails under hand rotations, as wrist edges near the centroid are misclassified as fingertips.
   - K-Cosine & Curvature Methods: Sensitivity to parameter tuning and noise leads to inconsistent results.

Table 3. Comparison of fingertip detection accuracy (%)

Number of fingertips	Relative position	K cosine	Curvture	Propose method
1	90.82%	89.53%	89.03%	97.35%
2	91.59%	86.56%	87.57%	97.94%
3	88.45%	91.04%	93.05%	97.65%
4	89.36%	84.07%	85.05%	98.76%
5	92.63%	89.04%	91.02%	98.93%

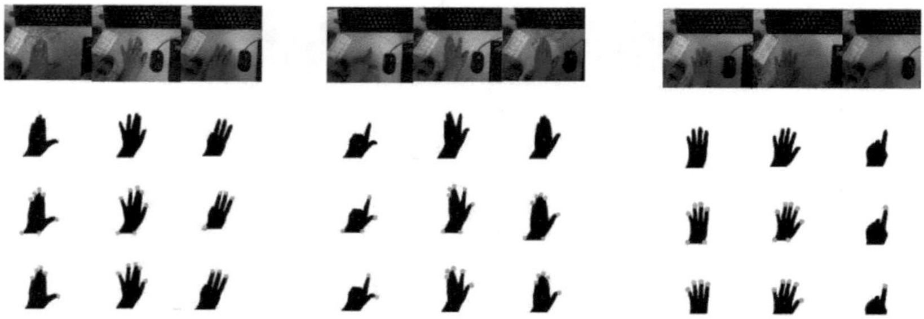

(a) 106th, 526, 95 frames (b) 302nd, 274, 90 frames (c) 204th, 429, 348 frames

Fig. 7. Detection Results Under (a) normal lighting, (b) backlighting, and (c) cluttered backgrounds.

To quantitatively assess the algorithm's robustness to partial occlusions, we conducted additional experiments using the EgoHands dataset [15], which contains 4,800 annotated frames of interacting hands with overlapping fingers and complex gestures. Under single-hand self-occlusion (two fingers overlapping), the accuracy remained stable at 95.72%. For multihand occlusion (two users' hands intersecting), the accuracy decreased to 88.34%, highlighting the need for multiview fusion. The results are summarized in Table 4.

Table 4. Occlusion robustness evaluation

Occlusion Type	Accuracy (%)	FPS
Single-hand overlap	95.72	21.1
Multihand interaction	88.34	18.9

3.5 Generalizability on Diverse Datasets

To validate generalizability, we tested the framework on the EgoHands [15] and HGR v2.0 [16] datasets, which include egocentric viewpoints, heterogeneous backgrounds, and dynamic lighting. The proposed method achieved 93.41% accuracy on EgoHands and 91.58% accuracy on HGR v2.0, outperforming curvature-based methods by 7.2–9.5% (Table 5).

Table 5. Cross-dataset performance comparison

Dataset	Proposed (%)	Curvature [14] (%)	Relative Position (%)
EgoHands	93.41	84.92	85.67
HGR v2.0	91.58	82.03	80.14

3.6 Extreme Illumination and Background Tests

We captured 200 frames under backlighting and cluttered backgrounds (e.g., posters with skin-colored patterns). The precision decreased marginally to 94.82% (vs. 97.26% in normal lighting), whereas the recall remained stable at 95.01%. Figure 7 (revised) illustrates the detection results under background.

3.7 Parameter Sensitivity Analysis

An ablation study evaluated the impact of key parameters:

- **Shadow angular threshold (θ)**: Varying θ from 10° to 20° resulted in optimal precision (96.18%) at $\theta = 15°$.
- **The circularity overlap ratio (η)**: $\eta = 0.8$ maximized the F-measure (97.02%) while balancing false positives.

3.8 Multi-hand and Embedded Device Performance

Multi-Hand Detection: In a dual-hand scenario (10 fingertips), the algorithm achieved 89.47% accuracy, with failures occurring at interhand occlusions.

Embedded Deployment: On a Raspberry Pi 4 (4 GB RAM), the processing time increased to 48.2 ms/frame ($\approx$21 FPS), demonstrating feasibility for edge applications.

4 Discussion

- **Advantages**: The algorithm's real-time performance (23.43 ms/frame) and robustness to illumination changes make it suitable for low-cost HCI systems.
- **Limitations:**
- Multichand occlusion reduces the accuracy to 88.34%, necessitating multicamera setups for 3D spatial reasoning.
- Extreme backlighting marginally affects precision (94.82% vs. 97.26%).
- Embedded deployment requires lightweight optimization (e.g., model quantization).

5 Conclusion

This paper presents a real-time fingertip detection framework that combines spatial–temporal GMM, YCbCr-based shadow removal, and circularity filtering. The experimental results validate its superiority in terms of accuracy (97.26%) and speed (23.43 ms/frame) over existing methods. Future work will extend the algorithm to multiple hand scenarios and 3D gesture recognition.

Acknowledgements. This study was funded by the Jiaying University Research Start-Up Fund (grant number 323E0431).

References

1. Chen, Y., Chen, X.: Gesture recognition based on Kinect and its application in virtual assembly technology. Electron. Des. Eng. **21**, 4–7 (2023)
2. Deng, R., Zhou, L., Ying, R.: Gesture extraction and recognition research based on Kinect depth data. Appl. Res. Comput. **30**, 1263–1265 (2023)
3. Feng, Z., Xu, S., Zhang, X.: Real-time fingertip tracking and detection using Kinect depth sensor. In: Proc. Int. Conf. Internet Multimedia Comput. Service, pp. 70–74. ACM, New York (2022)
4. Yuan, F., Wang, Y., Wang, Y.: Fingertip detection and gesture identification based on Kinect depth image. Electron. Sci. Technol. **27**, 6–10 (2024)
5. Farooq, J., Ali, M.B.: Real-time hand gesture recognition for computer interaction. In: Proc. IEEE Int. Conf. Robot. Emerg. Allied Technol. Eng., pp. 73–77. IEEE, Piscataway (2024)
6. Stauffer, C., Grimson, W.: Adaptive background mixture models for real-time tracking. In: Proc. IEEE Conf. Comput. Vis. Pattern Recognit., pp. 246–252. IEEE, Piscataway (2019)
7. Zang, Y., Liu, Q.: Moving target detection method based on adaptive threshold. Comput. Eng. Appl. **50**(12), 166–168 (2024)
8. Bhuyan, M.K., Neog, D.R., Kar, M.K.: Fingertip detection for hand pose recognition. Int. J. Comput. Sci. Eng. **4**(3), 501 (2022)
9. Lee, D.-S.: Effective Gaussian mixture learning for video background subtraction. IEEE Trans. Pattern Anal. Mach. Intell. **27**(5), 827–832 (2022)
10. Deng, C., et al.: Edge detection based on improved LOG operator and wavelet transform. J. Harbin Univ. Sci. Technol. **17**(5), 87–90 (2022)
11. Liu, J., et al.: Human detection method based on foreground segmentation. J. Chin. Comput. Syst. **35**(3), 654–658 (2024)
12. Xu, Y., et al.: Robust system framework and improvement of algorithms in fingertips detection. Comput. Eng. Appl. **49**(15), 195–200 (2023)
13. Raheja, J.L., Mishra, A., Chaudhary, A.: Indian sign language recognition using SVM. Pattern Recogn. Image Anal. **26**(2), 434–441 (2016)
14. Li, B., Lin, F.: Fingertip detection method based on curvature. J. Nanjing Univ. Aeronaut. Astronaut. **44**(4), 587–591 (2022)
15. Bambach, S., et al.: EgoHands: a dataset for hands in complex egocentric interactions. In: IEEE ICCVW (2023)
16. Dadashzadeh, A., et al.: HGR v2.0: a hierarchical gesture recognition framework. In: IEEE TPAMI (2024)

Challenges and Optimization of Multimodal Large Language Models for Tree Falling Scenarios

Lei Feng[1], Yicheng Huang[1], Chunjie Sheng[2](✉), Yuxing Shi[2], Jianhong Jin[1], Yun Xu[1], Yuzhou Du[1], and Sihao Miao[1]

[1] OpenPie, Block 2, No. 733 Jianshe Third Road, Xiaoshan District, Hangzhou, Zhejiang, China

[2] Pinghu Administration Center, No. 380 Shengli Road, Pinghu, Jiaxing, Zhejiang, China

rabbysfd54sdf@gmail.com

Abstract. This study proposed an optimization method for multimodal large language models (MLLMs) reasoning based on structured chain of thought, aiming to enhance the visual decision-making capability in tree falling scenarios. The research first analyzed challenges faced by existing MLLMs when processing complicated visual scenes, including insufficient reasoning performance and low integration efficiency with other systems. To address these issues, an innovative structured chain of thought approach was introduced, which significantly improved the reasoning accuracy of the model in handling complex visual scenarios. To validate the proposed method, a specialized dataset focusing on tree falling scenarios in social governance was constructed, and a practical agent workflow was designed based on this dataset. Experimental results demonstrated that the proposed approach achieved better performance in real-world applications. The findings provide a reliable and efficient technical solution to visual decision-making in social governance.

Keywords: multimodal large language model · MLLM · social governance · AI agent

1 Introduction

In social governance, there is a high demand for image-based decision-making, such as in tree falling risk prevention and responsibility attribution, traffic safety inspections, storm drain and manhole inspections, and street safety warnings. These often rely on images captured by cameras and require interpretation by city managers to guide decisions and actions. Traditional manual inspection methods can no longer meet the growing management demands; they are labor-intensive, costly, and prone to oversight, which can lead to severe economic and safety consequences. Manual assessments are also subjective and inconsistent. Therefore, AI agents utilizing multimodal large language models (MLLMs) are practically and significantly valuable.

Y. Yu et al. (Eds.): ICPCSEE 2025, CCIS 2673, pp. 533–543, 2026.
https://doi.org/10.1007/978-981-95-2566-9_35

These agents leverage computer vision and MLLMs to automatically extract key information from camera and drone footage. By integrating the advanced reasoning capabilities of large language models (LLMs), the system can accurately assess risks and provide objective, consistent recommendations. This enhances management efficiency, reduces reliance on manual labor, and enables 24/7 monitoring—particularly critical in hazardous environments where human intervention is limited.

Tree falling risk prevention and responsibility attribution are common yet challenging tasks. However, directly applying MLLMs often falls short of practical requirements due to the following limitations:

- Weak correlation between visual elements and task attributes: MLLMs frequently struggle to associate visual cues (e.g., tree tilt or surrounding conditions) with task-specific risk factors, compromising assessment accuracy.
- Inefficient integration with business systems: Most MLLMs lack standardized APIs or modular deployment options, hindering seamless adoption in urban management workflows.

Recent advances in LLMs (e.g., OpenAI's O1 model [1]) demonstrate that optimizing computational reasoning strategies enhances decision-making performance. However, MLLMs still struggle with systematic optimization in complex visual question-answering (VQA) tasks for governance applications. To address this, we propose a structured chain-of-thought (CoT) approach to improve reasoning accuracy in visual scene analysis.

We introduce a tree falling governance dataset with detailed annotations and structured reasoning chains. Our agent framework, built on structured CoT, delivers reliable support for real-world visual decision-making in urban management scenarios.

2 Related Work

2.1 Enhancing Model Reasoning in Complex Visual Scenarios

In recent years, the integration of visual reasoning with LLMs has demonstrated remarkable potential. By leveraging the capabilities of LLMs, MLLMs have achieved significant improvements in handling complex visual question-answering (VQA) tasks. These advancements are evident not only in the models' depth of visual comprehension but also in their flexibility when processing diverse user queries involving visual inputs.

A pivotal contribution is Meta's LLAMA-3.2-11B Vision model [2], which validated that optimization techniques effective for traditional LLMs—such as prompt engineering, in-context learning (ICL), and supervised fine-tuning (SFT)—are equally applicable to MLLMs and yield substantial performance gains. This finding provides critical technical guidance for MLM optimization while highlighting shared optimization strategies between vision-language models and text-only LLMs.

The ByteDance research team conducted a systematic analysis of MLLMs, reviewing existing multimodal reasoning evaluation frameworks and categorizing state-of-the-art MLLM technologies [3]. Their work offers an in-depth discussion of MLLM trends in reasoning-intensive tasks.

Chancharik et al. [4] proposed CCoT, an innovative zero-shot compositional chain-of-thought method. CCoT first generates detailed scene graphs from input images and task prompts, then integrates these elements into a unified input for LLM processing. The generated scene graphs exhibit strong generalizability, enabling accurate descriptions of diverse visual scenarios and broad applicability to downstream tasks. Using LLaVA-1.5 and InstructBLIP as baselines, rigorous evaluations on visual reasoning and compositional language tasks (via SEEDBench and MMBench) demonstrated CCoT's superiority: it outperformed other LLaVA-1.5 variants on compositional understanding benchmarks, matching the performance of SOTA-SGVL. Notably, CCoT also enhanced both LLaVA and InstructBLIP models on general multimodal benchmarks—a key foundation for this study.

Zhiyuan Li et al. [5] introduced CVR-LLM, a novel complex visual reasoning framework combining vision-language models (VLMs) and LLMs. CVR-LLM converts images into context-aware detailed descriptions, which LLMs then analyze for prediction—all without additional training. The team further augmented reasoning with multimodal ICL and a chain-of-comparison (CoC) technique. Experiments across complex VQA tasks confirmed CVR-LLM's state-of-the-art performance.

Guowei Xu et al. [6] developed a method to equip VLMs with systematic, structured reasoning via a 4-stage CoT: task summarization, image description, detailed reasoning, and conclusion generation. This approach improves task-relevant information extraction, logical decomposition, and reasoning coherence. Remarkably, their structured CoT enabled smaller open-source models to surpass larger models or proprietary alternatives in specific multimodal analysis tasks.

2.2 Tree Falling Risk Prevention and Responsibility Attribution

The rapid progression of urbanization, combined with sustained expansion of urban green spaces, has rendered trees—essential components of urban ecosystems—a crucial focus of public safety considerations. Tree falling accidents present dual consequences: they cause immediate property damage while simultaneously creating significant hazards to pedestrian and vehicular safety, making their prevention and systematic management a matter of urgent practical importance.

The effective resolution of tree falling accidents fundamentally depends on accurate liability determination. This process requires complex identification of responsible entities, involving careful differentiation based on two primary considerations: first, the specific jurisdictional characteristics of the land parcel where trees are located, and second, the administrative authority governing various tree classifications. As systematically outlined in Table 1, this differentiated liability framework provides both clear legal foundations for risk prevention and practical guidelines for incident management.

In addition to establishing tree health monitoring and early warning systems, post-accident management represents a multifaceted governance challenge requiring coordinated efforts. Even with surveillance infrastructure available, frontline administrators must devote substantial time to determining liability. The timely removal of fallen trees

Table 1. Tree Classification and Its Responsible Entities

Tree Classification	Responsible Entities
Trees in Public Green Spaces	Urban greening departments or urban road, highways, and river management departments
Trees in Unit-owned Green Spaces	The unit itself
Trees in Residential Areas and Communities	Owners can delegate property management companies to maintain the trees
Preserving Trees on Construction Sites	Responsible during the construction period by the construction unit
Trees within the Green Spaces of Railways, Lakes, Reservoirs, and Other Land	Relevant management departments
Trees in Rural Green Areas	Village committees or village collective economic organizations
Privately Planted Trees	Owners or managers belong to collective organizations
Trees on Collective Land	Privately planted trees belong to the planter, but must return improper benefits and bear possible liability for infringement

remains critical to prevent secondary incidents and protect public safety. When responsible entities are properly identified, municipal authorities typically implement standardized emergency response protocols [7], which generally include two key elements: comprehensive accident classification systems and well-defined operational principles.

3 Model Selection

3.1 Native MLLMs

The CoT approach provides large language models with a structured reasoning framework, enabling effective handling of complex problems including commonsense and logical reasoning. By decomposing intricate problems into coherent reasoning steps, this method guides models through systematic thinking to reach sound conclusions. Recent research [6] demonstrates that CoT not only significantly enhances the reasoning capabilities and interpretability of large language models, but also shows remarkable improvement effects on smaller-scale models. This finding offers an efficient and cost-effective solution for model performance enhancement. Building upon these findings, our study proposes a structured CoT method specifically designed for social governance scenarios, aiming to strengthen MLLMs' reasoning capabilities when processing non-standardized image scenarios. Through the introduction of domain-specific structured reasoning frameworks, our approach enables models to comprehend and analyze complex social governance situations more accurately.

For this applied research, we have compiled 1,793 on-site images of tree falling accidents, covering various scenario types including public green spaces, institutional green areas, residential zones, construction sites, railway and reservoir lands, village green spaces, and private planting areas. Each scenario corresponds to specific responsible entities. To enhance model generalizability, our dataset includes a proportion of simulated scenarios and international cases.

In the social governance context of tree falling risk prevention, MLLMs must accomplish three core tasks: (1) risk identification and early warning, (2) post-incident damage assessment, and (3) responsibility determination. Only when achieving high accuracy across all three sub-tasks can these models meet practical application requirements. This study employs Llama-3-2-11B-Vision as the foundational MLLM. While demonstrating excellent performance in identifying and describing individual objects and their attributes within images, this model shows limitations in analyzing spatial and logical relationships between objects.

To validate this observation, we presented the model with a typical tree falling accident scene (Fig. 1), requesting analysis of the three core tasks. Without any CoT prompting, the native MLLM produced superficial analysis that accurately identified primary scene elements but lacked depth in risk assessment and damage analysis. Particularly regarding liability determination, even when provided with detailed responsibility criteria through prompt engineering, the model failed to effectively establish correlations between scene elements. This deficiency primarily stems from insufficient training data on complex scenarios and the absence of user-provided information in the reasoning process.

Fig. 1. Site of a Fallen Tree

3.2 Structured CoT Enhanced MLLMs

Our analysis reveals that while native MLLMs can accurately identify basic scene elements and perform well on image summarization tasks, their analyses remain superficial. Specifically, they demonstrate deficiencies in: (1) conducting in-depth risk assessments, (2) precisely defining damage scope, and (3) clearly determining responsibility attribution. These limitations primarily stem from the lack of systematic reasoning frameworks when processing complex scenarios.

To address this, we initially employed manual CoT prompting [8] to enhance the base model's capabilities. This approach required the model to explicitly output its reasoning process while making judgments, supplemented with contextual examples of liability determination. While effective (e.g., the model correctly identified highway maintenance units as responsible for cleanup), this method proved impractical for large-scale deployment due to its reliance on domain experts to craft scenario-specific prompts - contradicting our goal of reducing human workload in social governance applications. Nevertheless, this experiment confirmed that CoT methods can indeed enhance the model's reasoning capabilities and its ability to correlate user-provided contextual information. Crucially, we learned that achieving meaningful improvement requires both standardized reasoning processes and strategically injected key information during these processes.

Building upon Guowei Xu et al.'s work [6], we developed a structured CoT enhancement framework comprising five distinct stages:

1. Summary: The model summarizes image content according to task requirements
2. Object Caption: The model enumerates task-relevant objects and their attributes
3. Location Focus: The model analyzes spatial relationships between objects
4. Reasoning: The model performs structured reasoning to reach preliminary conclusions
5. Conclusion: The model synthesizes information to produce final outputs

Except for the conclusion phase, all stages represent intermediate reasoning processes. Our testing confirmed that displaying these intermediate stages has minimal impact on final outputs while reducing computational efficiency. However, we maintain their visibility to ensure reasoning transparency, though research suggests these intermediate outputs could be optimized through knowledge distillation [9].

To demonstrate this framework's effectiveness, we present the model's analysis of Fig. 1 without user-provided liability context:

1. Summary: "The image shows a fallen tree branch blocking a road, with several people gathered nearby. A yellow crane appears to be cutting the branch, alongside a blue truck and a police officer wearing a white hat."
2. Object Caption:
    - One fallen tree branch
    - One yellow crane
    - One blue truck
    - One police officer with white hat
    - Multiple bystanders

3. Location Focus:
    - The branch lies across the road, obstructing traffic
    - The crane operates near the branch with the truck parked behind
    - The officer stands beside the branch while others scatter around
4. Reasoning:
    - The fallen branch has damaged the road and surrounding area
    - The crane and truck indicate ongoing cleanup operations
    - The officer's presence suggests public safety monitoring
5. Conclusion:
    - Primary responsibility likely lies with the village government for road maintenance
    - Municipal landscaping departments may share partial responsibility

This example demonstrates significant improvements in both liability determination and damage assessment. While not infallible, the enhanced model produces fully structured reasoning processes that allow flexible integration of domain knowledge. For social governance applications, we embedded Table 1's liability criteria into the reasoning phase, enabling context-aware rule triggering. This integration yielded more accurate analyses, including correct identification of municipal landscaping departments as responsible entities and proper classification of the yellow crane as highway maintenance equipment.

Notably, our structured CoT training data can be prepared using existing proprietary models (e.g., OpenAI's GPT series) to generate preliminary outputs for each reasoning stage. This approach dramatically reduces both manual annotation efforts and computational resources required for training.

Comparative evaluations between base and enhanced models revealed distinct performance characteristics. The base model performs adequately on single-object recognition tasks (e.g., identifying trees or vehicles) and basic risk assessment. However, it struggles with complex scenarios requiring multi-object relationship analysis and contextual information integration. In contrast, the enhanced model demonstrates superior performance in these aspects. For instance, when processing Fig. 1's scenario, while the base model only recognized the basic fact of a fallen tree, the enhanced model not only identified all key elements but also correctly determined liability by incorporating Table 1's criteria.

This performance improvement comes at a computational cost - average response time increased from 1.57795s (base) to 10.32490s (enhanced) due to more complex reasoning processes. Practitioners must weigh this trade-off based on application requirements.

Our final evaluation assessed both models on 1,793 tree falling scenarios across four tasks: (1) risk identification, (2) damaged object recognition, (3) original liability determination, and (4) location-responsible entitiesmapped liability determination. As shown in Fig. 2, while both models perform similarly on tasks 1–2, the enhanced model significantly outperforms on tasks requiring spatial-logical reasoning (task 3) and demonstrates even greater advantages when incorporating additional mapped liability information (task 4), confirming its superior information utilization efficiency.

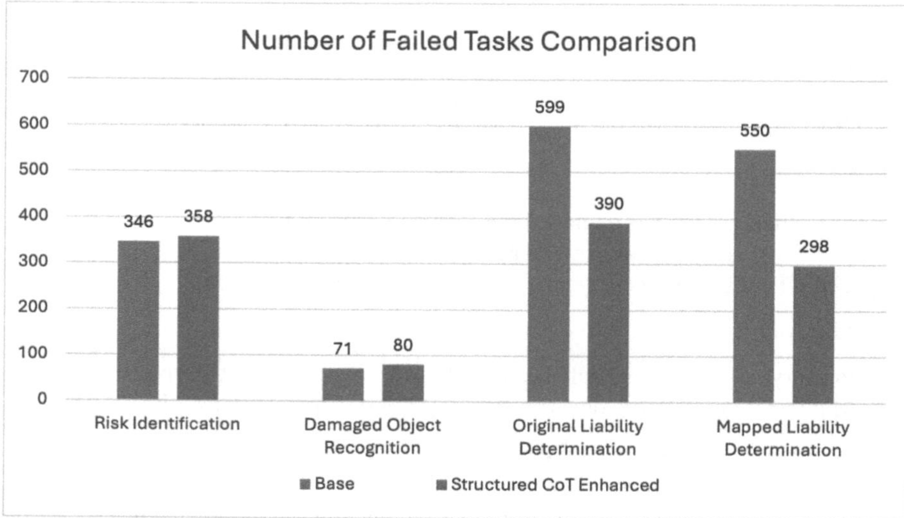

Fig. 2. Number of Failed Tasks Comparison

4 Agent Architecture for Social Governance Based on MLLMs

The structured CoT enhancement not only improves the model's ability to analyze relationships between objects in images and perform contextual reasoning, but also significantly increases the transparency of the reasoning process. This transparency allows us to embed external services at key nodes of the model's reasoning to provide necessary contextual information.

In research on enhancing LLM capabilities, enabling models to support external API calls represents an important direction. Studies have shown [10] that smaller models can improve their capabilities by integrating scientific computing, QA systems, search engines, and translation systems, even achieving performance comparable to larger models in some zero-shot scenarios.

In this study, we propose a cost-effective and feasible social governance agent architecture. This architecture is suitable for multimodal models with parameter sizes ranging from 13B to 72B, and enhances the agent's ability to handle various complex scenarios by calling systems like PieCloudDB [11].

Building upon the structured CoT enhanced MLLMs, we integrate the "dbchat" agent as a core external service into the model's reasoning pipeline, with detailed workflow shown in Fig. 3. The agent primarily provides support at two key stages: caption and reasoning. This integration approach maintains the coherence of the model's reasoning while enhancing its ability to process external information.

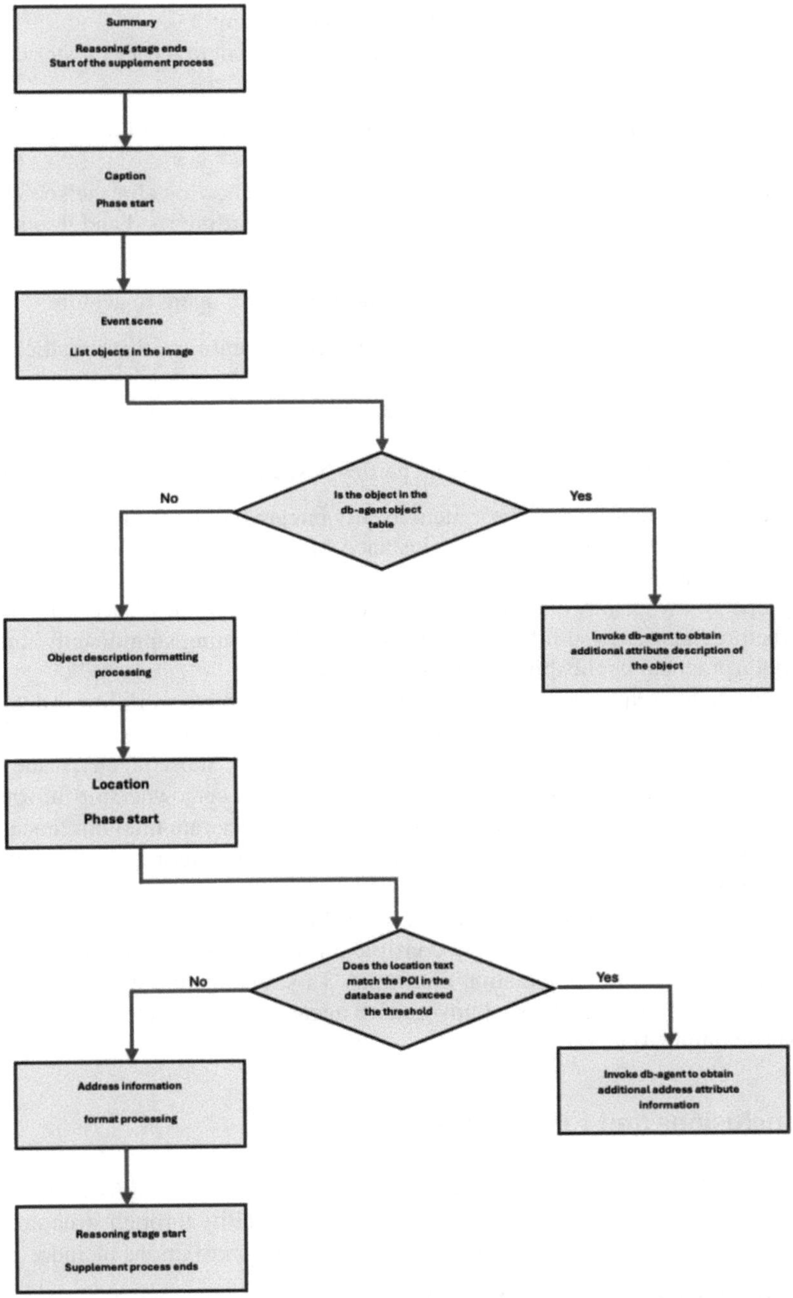

Fig. 3. The Process of Integrating Structured Thinking Chains with Social Governance DB-Agent

Figure 3 illustrates the workflow of the complete system:

During the caption phase, the model invokes the db-agent to retrieve historical event records related to the image, including:

1. Historical incident records: Matching similar accidents based on geographical location and scene characteristics
2. Maintenance records: Querying road and tree maintenance logs for the specific area
3. Complaint records: Searching resident complaints regarding road and tree safety in the vicinity

In the reasoning phase, the model further utilizes the db-agent to acquire:

1. Geographical information: Determining the precise administrative jurisdiction and managing authority of the incident location
2. Property ownership data: Querying ownership attribution of both trees and adjacent roads
3. Responsible entities: Identifying liable parties based on geographical positioning

These external data sources are systematically integrated into the model's reasoning chain to enhance reasoning accuracy. For instance, historical maintenance records enable the model to assess whether the tree had been previously identified as a potential hazard, while property ownership data facilitates precise determination of accident liability. This structured approach to information retrieval and reasoning significantly improves the model's practical applicability in real-world scenarios.

The invocation sequence of the db-agent within the complete workflow is detailed in Fig. 3. Initially, during the description phase, the model acquires historical data through db-agent calls, establishing preliminary scene understanding. Subsequently, in the spatial relationship analysis phase, the model retrieves geographical and ownership information, which is then synthesized with visual analysis results to generate final inferences. This process represents a progressive information fusion, where external data at each stage critically informs subsequent analytical phases.

The entire system can be implemented using MLLMs with approximately ten billion parameters, ensuring controllable construction costs while maintaining compatibility with existing data computation platforms. This architecture effectively extends the processing capabilities of social governance agents through strategic integration of established computational resources.

5 Conclusions and Future Work

This study explores a social governance agent system based on MLLMs, with a focus on enhancing model reasoning capabilities and practical utility through structured CoT methodologies and external service integration. The key contributions include:

1. **Structured CoT Enhancement for MLLMs**: We propose a novel structured CoT approach that significantly improves model performance in complex scenario analysis. Experimental results demonstrate that the enhanced model achieves superior performance in object relationship recognition and contextual reasoning tasks.

2. **External Service-Integrated Agent Architecture**: We design and implement an agent architecture that embeds db-agent services at critical reasoning nodes, enabling effective integration of historical records, geospatial data, and property ownership information. This architecture maintains reasoning transparency while improving system practicality.

Future Research Directions:

1. **Computational Efficiency Optimization**: Investigating methods to reduce inference latency while preserving reasoning quality.
2. **Extended Service Interfaces**: Incorporating additional data sources to enhance decision-support capabilities.
3. **Training Data Generation**: Leveraging large proprietary models to synthesize high-quality training data, thereby reducing development costs.
4. **Application Expansion**: Exploring deployment in broader social governance domains such as urban management and emergency response.

In summary, this work establishes a viable technical framework for MLLM-based social governance agents and lays the foundation for future research. As the technology matures, such systems are expected to play increasingly important roles in diverse governance scenarios.

References

1. Zhong, T., Liu, Z., Pan, Y., et al.: Evaluation of OpenAI o1: opportunities and challenges of AGI (2024). https://arxiv.org/abs/2409.18486
2. Meta: meta-llama/Llama-3.2–11B-Vision-Instruct. https://huggingface.co/meta-llama/Llama-3.2-11B-Vision-Instruct
3. Wang, Y., Chen, W., Han, X., et al.: Exploring the reasoning abilities of multimodal large language models (MLLMs): a comprehensive survey on emerging trends in multimodal reasoning (2024). https://arxiv.org/abs/2401.06805
4. Mitra, C., Huang, B., Darrell, T., et al.: Compositional chain-of-thought prompting for large multimodal models (2024). https://arxiv.org/abs/2311.17076
5. Li, Z., Liu, D., Zhang, C., et al.: Enhancing advanced visual reasoning ability of large language models (2024). https://arxiv.org/abs/2409.13980
6. Xu, G., Jin, P., Li, H., et al.: LLaVA-CoT: let vision language models reason step-by-step (2024). https://arxiv.org/abs/2411.10440
7. Urban Management and Comprehensive Law Enforcement Bureau. Emergency Plan for Public Garden Greenery Events in Dongguan (2010). (In Chinese)
8. Zheng, C., Jingchang, C., et al.: A survey of chain of thought reasoning: advances, frontiers and future (2023). https://arxiv.org/abs/2309.15402
9. Yuntian, D., Kiran, P., et al.: Implicit chain of thought reasoning via knowledge distillation (2023). https://arxiv.org/abs/2311.01460
10. Timo, S., Jane, D.-Y., et al.: Toolformer: language models can teach themselves to use tools (2023). https://arxiv.org/abs/2302.04761
11. Feng, L., Yang, J.: A data query method and device for distributed databases, CN202310816450.7. China (2024). (In Chinese)

Applications of Data Science

Empirical Analysis of Tourism Revenue in Sanya City

Liya Liu[1(✉)], Ngan Iek Hang[2], Yuanyuan Zhang[1], and Xia Liu[1]

[1] Sanya Aviation and Tourism College, Sanya 572000, China
695611567@qq.com

[2] Faculty of Humanities and Social Sciences, Macao Polytechnic University, Macao, China
p2209551@mpu.edu.mo

Abstract. Tourism is a pillar industry of the national economy and can reflect the overall economic development level of Sanya. Using the tourism data of Sanya from 2012 to 2023, we applied principal component analysis to extract two principal components from 11 indicators affecting the city's tourism revenue. Python was employed to develop regression models and GM(1,1) models for predicting Sanya's tourism revenue. The results show that in recent years, Sanya's tourism revenue has been on the rise. The number of domestic tourists is the main factor affecting Sanya's tourism revenue. Sanya's tourism revenue is mainly influenced by domestic tourism revenue, the number of domestic tourists, the number of domestic overnight tourists in Sanya, the number of inbound overnight tourists in Sanya, and tourism foreign exchange earnings, etc. However, tourism hotels and other factors have no significant impact on Sanya's tourism revenue. Based on the root mean square error (RMSE) criterion, regression models exhibit superior predictive performance compared to the GM(1,1) model in forecasting Sanya's tourism revenue.

Keywords: Python · GM(1,1) · Tourism Revenue · Predict

1 Introduction

The development of the tourism industry, which is known as the "smokeless industry" and the "everlasting sunrise industry", promotes industrial upgrading and economic structure adjustment and contributes to economic growth. The "14th Five-Year Plan for Tourism, Culture, Radio, Film and Sports Development of Hainan Province" proposes that during the "14th Five-Year Plan" period, the total number of tourists should reach 110 million, the total tourism revenue should reach 180 billion yuan, and the number of inbound tourists should reach 2.6 million. Vigorously promoting all-round tourism is highly important for optimizing the layout of Sanya's tourism industry and is also a practical need for accelerating the construction of the Hainan Free Trade Port.

Since 1900, scholars at home and abroad have begun to quantitatively analyze the factors influencing tourism demand. The view of tourism as an economic issue originated in 1898, while domestic research on tourism revenue started relatively late. By using

Y. Yu et al. (Eds.): ICPCSEE 2025, CCIS 2673, pp. 547–556, 2026.
https://doi.org/10.1007/978-981-95-2566-9_36

the China National Knowledge Infrastructure database and conducting an exact cross-database search with the keyword "factors influencing tourism revenue", core journals and related literature were retrieved. Research papers on this topic have been increasing since 2003, reaching a peak of 23 in 2019, with a total of 208 relevant academic papers. Anhui University of Finance and Economics and other institutions have played a leading role in this research. However, relatively few studies exist on the factors influencing tourism revenue in Hainan, and more research in this area is needed.

Among domestic scholars' research on tourism revenue, Zhang Yang et al. (2024) the main influencing factors of the spatial distribution pattern of cultural tourism brands in Shaanxi Province via a factor detector [1]. Ma Xiaotian (2024) and predicted the tourism revenue of Xinjiang via a gray prediction model and multiple linear regression model [2]. Fan Guoqing et al. (2023) predicted the tourism revenue of Guizhou Province from 2020 –2023 via an improved prediction model [3]. Zhang Yanan et al. (2022) analyzed the differences in tourism revenue among 11 provinces (municipalities) in the eastern region by applying methods such as standard deviation, the coefficient of variation, the Gini coefficient, and the Lorenz curve. They reported that the endowments of tourism resources, location and transportation conditions, regional economic development level, and degree of openness to the outside world in the eastern region are positively correlated with differences in tourism revenue [4]. Wang Qian et al. (2021) conducted a spatial mismatch analysis of tourism revenue and its influencing factors in various cities of Fujian Province via the gravity model and two-dimensional matrix model. They concluded that from the perspective of the gravity model, there are varying degrees of spatial mismatch between tourism revenue and resources, the population, the economy, and transportation [5].

In research on tourism revenue conducted by scholars abroad, Orcun Aydin (2016) applied panel data and linear and nonlinear models to predict the quarterly tourism revenue of foreign tourists in Turkey [9]. Joaquin Alegre et al. (2005) used the conditional demand function model to study the impact of the length of stay at the destination on the economy. Teresa Garin-Munoz (2004) reported that the factors influencing the international tourism revenue of the Canary Islands are highly dependent on the international tourism environment and per capita tourism expenditure in this region [10]. Lee, Choong-Ki (1996), on the basis of the input–output model, noted that the income distribution among different social classes has a significant effect on the tourism industry.

Most studies in academic on the influencing factors of tourism revenue are conducted at the national level, and regional studies often focus on places such as Beijing, Shanghai, and Gansu. In contrast, research on the influencing factors of tourism revenue in Hainan Province, especially in Sanya, is relatively scarce, and the depth and breadth of such research still need to be enhanced. Moreover, the research methods mostly employ statistical models such as multiple regression analysis, which have inherent limitations and can significantly affect the accuracy and validity of the final analysis results. Therefore, the research methods could be further optimized.

In the tourism industry, tourism revenue is a crucial indicator for measuring the development of the tourism sector. This model takes Sanya's tourism revenue as the dependent variable Y. Given the numerous factors influencing tourism revenue, the following variables are selected as explanatory variables: tourism revenue per billion yuan

(Y), domestic tourism revenue per billion yuan (X1), domestic tourist volume per 100 million person-times (X2), overnight domestic tourists in Sanya per 10,000 person-times (X3), number of overnight inbound tourists in Sanya (X4), total airport passenger flow per 10,000 people (X5), total railway passenger flow per 10,000 people (X6), tourism turnover per 10,000 person-kilometers (X7), per capita GDP (X8), tourism foreign exchange income per 10,000 US dollars (X9), total value of accommodation and catering per billion yuan (X10), and the number of star-rated hotels (X11). Relevant data from 2008 --2023 were obtained by consulting the China Statistical Yearbook; the Sanya Statistical Yearbook; the Sanya Municipal Bureau of Tourism, Culture, Radio, Television and Sports and the Hainan Provincial Bureau of Tourism, Culture, Radio, Television and Sport.

2 Principal Component Analysis

The role of principal component analysis is to explain more variables with fewer variables and calculate the comprehensive score of each factor for ranking, and its essence is "dimensionality reduction". Factor analysis replaces all the original variables with a few common factors to solve the original problem.

N samples and P indicators are set as random vectors, and the common factor to be sought is then as follows:

$$X_1 = a_{11}F_1 + a_{12}F_2 + \cdots a_{1m}F_m + \varepsilon_1 \tag{1}$$

$$X_2 = a_{21}F_1 + a_{22}F_2 + \cdots a_{2m}F_m + \varepsilon_2 \tag{2}$$

$$\vdots$$

$$X_3 = a_{m1}F_1 + a_{12}F_2 + \cdots a_{1m}F_m + \varepsilon_m \tag{3}$$

It is called the factor model. The matrix A $= (a_{ij})$ is called the factor load matrix, and the load of the variable i on the factor j, a_{ij} is called the factor load. In essence, it is the correlation coefficient between the variable X_i and the common factor F_j, which indicates the degree to which the variable X_i depends on the common factor F_j and reflects the importance of the variable X_i to the common factor F_j. The special factor ε represents influencing factors other than common factors, which are ignored in practical [6].

2.1 Data Standardization

In the process of principal component regression, the independent variable X needs to be standardized, which causes the scale of the dependent variable Y and the independent variable X to be inconsistent. Therefore, in addition to unifying the scale relationship between the independent variable and the dependent variable, the original data of the dependent variable Y also needs to be standardized simultaneously with the standardization of the independent variable X to ensure the consistency of the variable relationship in the principal component regression process. The application of the standardization method to the tourism data from 2012 to 2023 for dimensionless processing is shown in Table 1. The standardization method calculates the difference between each data point and the average value and then divides it by the standard deviation, making the average value of the data 0 and the standard deviation 1 [7].

Table 1. The data table of the initial value image of the tourism industry in Sanya City.

Year	ZY	ZX1	ZX2	ZX3	ZX4	ZX5	ZX6	ZX7	ZX8	ZX9	ZX10
2012	−1.19	−1.03	−0.94	−1.36	0.16	−1.32	−1.34	−1.25	−1.41	−0.06	−1.81
2013	−1.00	−0.75	−0.68	−1.10	0.16	−0.92	−1.10	−0.92	−1.21	−0.09	−1.48
2014	−0.83	−0.43	−0.37	−0.83	−0.20	−0.36	−0.73	−0.80	−0.94	−0.39	−0.36
2015	−0.68	−0.12	−0.02	−0.52	−0.32	−0.03	−0.29	−0.29	−0.77	−0.48	−0.43
2016	−0.58	0.30	0.33	−0.22	0.03	0.28	0.22	−0.41	−0.51	−0.11	−0.23
2017	−0.19	0.79	0.86	0.10	0.98	0.82	0.95	−0.29	−0.18	1.10	0.20
2018	0.31	1.24	1.34	0.93	1.46	1.00	1.25	−0.12	0.11	1.14	0.59
2019	0.86	1.71	1.75	1.23	1.81	1.03	1.26	−0.29	0.47	2.27	1.73
2020	−0.11	−1.06	−1.01	−0.03	−1.11	−0.24	−0.22	0.27	0.52	−0.77	0.60
2021	1.39	−0.51	−0.69	0.91	−1.15	0.09	−0.03	0.27	1.08	−0.97	0.64
2022	−0.06	−1.21	−1.32	−0.85	−1.26	−1.81	−1.25	2.03	1.31	−1.00	−0.38
2023	2.08	1.07	0.76	1.72	−0.55	1.46	1.29	1.80	1.54	−0.64	0.92

2.2 Pearson Correlation Matrix and Feasibility Analysis

For principal component analysis, the higher the correlation between variables is, the more suitable the method is for principal component extraction [8]. As shown in Table 2, the correlation of most data exceeds 80%, among which the correlations between domestic tourism revenue, domestic tourist volume, per capita GDP and the tourism revenue of Sanya city are relatively high. Before factor analysis, it is necessary to check whether the KMO value is greater than 0.7 and whether Bartlett's sphericity test is significant. The sample KMO test value obtained through SPSS processing is 0.713, which is greater than 0.5. This sample is suitable for factor analysis. According to the principle that the cumulative contribution rate exceeds 85%, the first two components are extracted as principal components, and the principal component factor loading matrix is shown in Table 3.

Table 2. Correlation matrix.

		ZY	ZX1	ZX2	ZX3	ZX4	ZX5	ZX6	ZX7	ZX8	ZX9	ZX10
Correlation	ZY	1.000	.511	.408	.918	−.067	.633	.671	.657	.865	.039	.794
	ZX1	.511	1.000	.992	.760	.749	.899	.929	.035	.223	.758	.676
	ZX2	.408	.992	1.000	.692	.813	.874	.902	−.066	.122	.814	.629
	ZX3	.918	.760	.692	1.000	.247	.863	.894	.432	.704	.331	.897
	ZX4	−.067	.749	.813	.247	1.000	.499	.548	−.480	−.318	.966	.220
	ZX5	.633	.899	.874	.863	.499	1.000	.967	.085	.319	.512	.756
	ZX6	.671	.929	.902	.894	.548	.967	1.000	.199	.436	.594	.820
	ZX7	.657	.035	−.066	.432	−.480	.085	.199	1.000	.894	−.375	.429
	ZX8	.865	.223	.122	.704	−.318	.319	.436	.894	1.000	−.159	.720
	ZX9	.039	.758	.814	.331	.966	.512	.594	−.375	−.159	1.000	.377
	ZX10	.794	.676	.629	.897	.220	.756	.820	.429	.720	.377	1.000

As can be seen from Table 3, it is found that the software finally extracts two principal components. It can explain 95.306%, which is already greater than 85%, and the overall effect is good. Principal component 1 has a relatively high correlation with domestic tourism revenue (X1), domestic tourist volume (X2), domestic overnight tourists in Sanya (X3), total airport passenger flow / 10,000 people (X5), and total railway passenger flow / 10,000 people (X6), and can be understood as the level of social and economic development. Principal component 2 has a relatively high correlation with the number of overnight inbound tourists in Sanya (X4) and tourism foreign exchange income / 10,000 US dollars (X9), and can be understood as the level of international development.

Table 3. The total variance explained.

Component	Initial Eigenvalues			Extraction Sums of Squared Loadings		
	Total	Variance %	Cumulative %	Total	Variance %	Cumulative %
1	4.703	78.380	78.380	4.703	78.380	78.380
2	1.016	16.926	95.306	1.016	16.926	95.306
3	.124	2.069	97.375			
4	.107	1.780	99.155			
5	.039	.657	99.812			
6	.011	.188	100.000			

Extraction method: Principal Component Analysis.

2.3 Empirical Prediction Based on Regression Models

The OLS regression model was constructed using data from 2008 to 2019, with Y as the dependent variable and X1 and X5 as independent variables. The estimated equation is $Y = 109.272 + 0.018X1 - 0.264X5$. The model demonstrates excellent goodness-of-fit, as indicated by a root mean square error (RMSE) of 24.025, a mean absolute percentage error (MAPE) of 11.00%, and an R^2 of 0.978, suggesting that approximately 97.8% of the variation in the dependent variable Y is explained by the model. Furthermore, the overall significance of the model is strongly supported by an F-statistic of 196.1 ($p = 3.80e - 08$).

The calculation of the observed value, predicted values, residuals, and percentage error via Python 3 is shown in Table 4. The formulas are as follows:

$$\text{Residual } \varepsilon^{(0)}(t) = x^{(0)}(t) - \hat{x}^{(0)}(t) \tag{4}$$

Table 4. Observed vs. predicted values in the regression model.

S/N	Observed	Predicted Values	Residuals	Percentage Error (%)
1	91.050	108.070	−17.020	−18.693%
2	103.770	82.782	20.988	20.225%
3	139.640	90.174	49.466	35.424%
4	160.710	182.988	−22.278	−13.862%
5	192.220	218.244	−26.024	−13.539%
6	233.330	242.240	−8.910	−3.818%
7	269.700	260.045	9.655	3.580%
8	302.310	296.906	5.404	1.788%
9	322.400	359.501	−37.101	−11.508%
10	406.170	418.752	−12.582	−3.098%
11	514.730	502.673	12.057	2.342%
12	633.190	606.847	26.343	4.160%

3 Empirical Prediction Based on the GM(1,1) Model

The gray prediction model originates from gray system theory, which was first proposed and established by Professor Deng Jilong, a Chinese scholar. The concept of the gray system contrasts with that of the white system and the black system, and its characteristic is that some information is known while some is unknown. The gray system theory introduces the concept of gray relational analysis for each subsystem, aiming to determine the numerical relationships among the subsystems (or factors) through certain methods. The GM(1,1) model can yield relatively accurate prediction results even when the information is insufficient and the correlation is uncertain. This method not only has high prediction accuracy but also can be used for long-term prediction. Fitting the differential equation with the accumulated generation conforms to the change law of the system [11]. The calculation is completed through model establishment and model verification. The tourism revenue of Sanya city from 2008--2019 is taken as a specific gray prediction model.

3.1 Perform a Cumulative Addition to the Original Data

Let the original gray data be denoted as: $x^{(0)}(1), x^{(0)}(2), \cdots, x^{(0)}(12)$.

$$x^{(0)}(1) = (91.05, 103.77, 139.64, 160.71, 192.22, 233.33, 269.7, \\ 302.31, 322.4, 406.17, 514.73, 633.19)$$

After performing a first-order cumulative sum on it, we get

$$x^{(1)} = (x^{(1)}(1), x^{(1)}(2), \cdots, x^{(1)}(12)),$$

$$x^{(1)}(k) = \sum\nolimits_{t=1}^{k} x^{(0)}(t) \qquad k = 1, 2, \cdots, 12.$$

The cumulative sequence overcomes the volatility and randomness of the original sequence and is transformed into a more regular increasing sequence.

3.2 Establish the GM(1,1)

$$\frac{dx^{(1)}}{dt} + ax^{(1)} = u, \tag{5}$$

The differential Eq. (5) is the gray model GM(1,1), where a and u are constants and they are obtained through least squares fitting.

$$\begin{pmatrix} a \\ u \end{pmatrix} = (B^T B)^{-1} B^T Y_{12},$$

$$B = \begin{bmatrix} -\frac{1}{2}(x^{(1)}(1) + x^{(1)}(2)) & 1 \\ -\frac{1}{2}(x^{(1)}(2) + x^{(1)}(3)) & 1 \\ \cdots & \cdots \\ -\frac{1}{2}(x^{(1)}(n-1) + x^{(1)}(n)) & 1 \end{bmatrix}, Y_n = (x^{(0)}(2), x^{(0)}(3), \cdots, x^{(0)}(n))^T$$

Using Python 3 for calculation, we obtain $a = 0.0526$ and $u = 648.4517$. Thus, the gray prediction model for tourism revenue is obtained as follows:

$$\hat{x}^{(1)}(t+1) = 12419.029e^{0.0526t} - 12327.979 \tag{6}$$

Equation (6) is the formula for predicting tourism revenue (after accumulation).

3.3 Model Verification

If the fitting accuracy between the predicted sequence and the original sequence is high, it can be used for extrapolation prediction. Otherwise, it cannot be directly used for prediction and needs to be corrected by residuals before being used for extrapolation prediction. The fitting verification indicators include: residuals, percentage error, posterior error ratio and small error probability.

The calculation of the observed value, predicted values, residuals, and percentage error via Python 3 is shown in Table 5. The formulas are as follows:

Table 5. Observed vs. predicted values in GM(1,1).

S/N	Observed	Predicted Values	Residuals	Percentage Error (%)
1	91.050	91.050	0.000	0.000%
2	103.770	71.927	31.843	30.687%
3	139.640	110.007	29.633	21.221%
4	160.710	150.145	10.565	6.574%
5	192.220	192.451	−0.231	0.120%
6	233.330	237.042	−3.712	1.591%
7	269.700	284.042	−14.342	5.318%
8	302.310	333.582	−31.272	10.344%
9	322.400	385.797	−63.397	19.664%
10	406.170	440.833	−34.663	8.534%
11	514.730	498.842	15.888	3.087%
12	633.190	559.985	73.205	11.561%

Using Python 3 to calculate the Posterior Error Ratio, C, and Small Error Probability, P.

Formula: $C = \frac{S_2}{S_1}$

Among them,

$$s_1^2 = \frac{1}{12-1}\sum\nolimits_{t=1}^{12}\left[x^{(0)}(t) - \overline{x}^{(0)}\right]^2 \cdots, s_1 = \sqrt{s_1^2} \cdots, \overline{x}^{(0)} = \frac{1}{12}\sum\nolimits_{t=1}^{12} x^{(0)}(t).$$

$$s_2^2 = \frac{1}{12-1}\sum\nolimits_{t=1}^{12}\left[\varepsilon^{(0)}(t) - \overline{\varepsilon}^{(0)}\right]^2 \cdots, s_2 = \sqrt{s_2^2} \cdots, \overline{\varepsilon}^{(0)} = \frac{1}{12}\sum\nolimits_{t=1}^{12} \varepsilon^{(0)}(t).$$

Small Error Probability $P = P\{|\varphi^{(0)}(t) - \overline{\varepsilon}^{(0)}| < 0.6475 s_1\}$.

The criteria for judgment are as follows: if $C < 0.35$ and $P > 0.95$, the model accuracy is classified as level one (excellent); if $C < 0.5$ and $P > 0.8$, it is classified as level two (qualified); and if $C < 0.65$ and $P > 0.7$, it is classified as level three (qualified); otherwise, it is classified as level four (unqualified) [12].

Using Python 3, with $C = 0.0454$ and $P = 1$, the model accuracy is determined to be level one (excellent), with MAPE = 0.1079 and RMSE = 35.7177, indicating good overall prediction accuracy of the model. The comparison results between the predicted values obtained from the GM(1,1) model and the actual values are shown in Fig. 1.

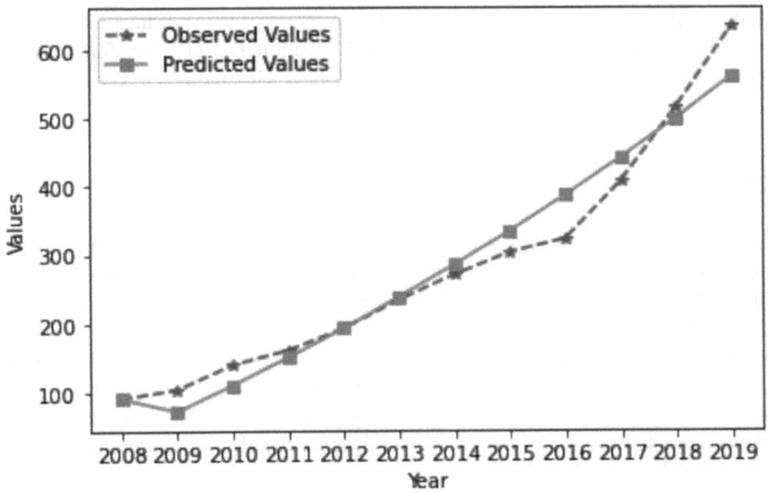

Fig. 1. Comparison of GM model prediction values

4 Conclusions

The results of the principal component analysis calculated via SPSS 22.0 in this paper show that domestic tourism revenue, domestic tourists, domestic overnight tourists in Sanya, inbound overnight tourists in Sanya, and tourism foreign exchange earnings are important factors influencing the tourism revenue of Sanya.

The regression model, estimated using Python, demonstrates strong predictive performance with an R-squared of 0.978 and an adjusted R-squared of 0.973, indicating that the model explains approximately 97.8% of the variance in the dependent variable. The overall model significance is confirmed by an F-statistic of 196.1 ($p = 3.80e\text{–}08$). Key error metrics include a mean squared error (MSE) of 577.206, root mean square error (RMSE) of 24.025, mean absolute error (MAE) of 20.6523, and mean absolute percentage error (MAPE) of 11.00%.

In comparison, the GM (1,1) model, evaluated using posterior error ratio C and small error probability P, yields RMSE and MAPE values of 0.045, 1, 35.718, and 10.8% Based on RMSE, the regression model exhibits superior predictive accuracy for Sanya's tourism revenue compared to the GM (1,1) model.

The tourism revenue of Sanya has shown a gradually increasing trend. Therefore, attracting more domestic and foreign tourists to visit Sanya is the fundamental way to increase local tourism revenue. As the tourism in Sanya is mostly natural, the tourism in Sanya has strong seasonality. The tourism industry in Sanya should explore cultural landscapes better, create "internet-famous" tourist destinations, and cultivate diverse consumption patterns. The government should increase investment in the tourism industry, improve tourism infrastructure and service levels, provide better tourism experiences for tourists, attract more tourists to consume, and strive for a situation where the tourism in Sanya is not dull in the off-season and even more prosperous in the peak season.

Acknowledgment. This research was financially supported by the Education Department of Hainan Province (Hnky2024-84) and the Higher Education Digital Transformation Research Special Project of the Employment Association for College Graduates (GJX25Z2167).

References

1. Yang, Z., Wenkai, S., Ziying, D.: Research on spatial differentiation and influencing factors of Shaanxi cultural tourism brand. Land Resour. Sci. Technol. Manag. **41**(06), 146–158 (2024)
2. Xiaotian, M.: Combined forecast and development analysis of Xinjiang tourism revenue. Spec. Zone Econ. **9**, 112–115 (2024)
3. Guoqing, F., Xumin, Z., Runping, G.: Research and analysis on tourism revenue forecast and development in Guizhou Province. Baoshan Univ. J. **42**(04), 91–99 (2023)
4. Yanan, Z., Wenyong, F., Jiaqi, W.: Analysis of interprovincial differences in tourism revenue and influencing factors in Eastern China over the past decade. J. Shandong Agric. Univ. (Soc. Sci. Ed.) **24**(02), 135–143 (2022)
5. Qian, W., Lizhong, S., Xiucheng, Y.: Spatial misalignment analysis of tourism revenue and influencing factors in Fujian Province. J. Fujian Normal Univ. (Nat. Sci. Ed.) **37**(04), 107–116 (2021)
6. Jinfeng, Z., Nianlian, Z.: Analysis of influencing factors of Shaanxi tourism revenue in the past 10 years based on gray relational analysis. J. Baoji Univ. Arts Sci. (Nat. Sci. Ed.) **40**(04): 67–72+80 (2020)
7. Jiao, L.: Empirical analysis of influencing factors of China's domestic tourism revenue. J. Chongqing Technol. Bus. Univ. (Nat. Sci. Ed.) **36**(05), 105–114 (2019)
8. Wu, S.: SPSS and Statistical Thinking, 1st edn. Tsinghua University Press, Beijing (2019)
9. Aydin, O.: Turkey's tourism income: a panel data approach. Procedia Econ. Finance **38**(16): 245–256 (2016)
10. Garín-Muñoz, T.: Madrid as a tourist destination: analysis and modelization of inbound tourism. Int. J. Tourism Res. **6**(4): 289–302 (2004)
11. Xu, N.: Advances in gray GM(1,1) forecasting model and its extension. Math. Pract. Theory **51**(13), 52–59 (2021)
12. Li, Y., Han, H.: Forecast and analysis of passenger flow at Sanya airport based on gray system theory. In: International Conference of Pioneering Computer Scientists, Engineers and Educators. Springer, Singapore (2022). https://doi.org/10.1007/978-981-19-5209-8_29

Blockchain-Based Multi-dimensional Dynamic Supply Chain Finance Risk Control System

Haifeng Guo[1(✉)], Jun Xiao[2], Biliang Wang[1], Yuyun Jie[1], and Yuwei He[1]

[1] School of Finance, Southwestern University of Finance and Economics, Chengdu 610000, China
guohf@swufe.edu.cn
[2] Xinjiang Zhongtai Group, Wulumuqi 830026, China

Abstract. Traditional supply chain finance has problems in four aspects: business flow, logistics, capital flow, and information flow. Blockchain technologies such as smart contracts and consensus mechanisms can help break through the above problems. They can improve the implementation effect of supply chain finance, and reduce risks. Based on blockchain technology, this paper constructs a multi-dimensional dynamic supply chain financial risk control system. The system includes multi-dimensional dynamic supply chain financial risk control technology, a credit evaluation system, and a capital monitoring and early warning system. By constructing a multi-dimensional dynamic supply chain financial risk control system, effective risk complete monitoring and penetrating supervision can be achieved. At the same time, this paper introduces technologies such as machine learning and deep learning to build an effective risk control system, which is of great significance for reducing supply chain financial risks.

Keywords: Blockchain technology · Supply chain finance · Risk control system

1 Introduction

Supply chain finance can meet the needs of modern industrial chains and alleviate the financing difficulties of small and medium-sized enterprises (SMEs). However, due to factors such as data authenticity and transaction costs, it is difficult for supply chain finance to rise to the industrial level. Currently, there is an urgent need to solve the difficulties in traditional supply chain finance, such as difficulty in confirming rights, low operational efficiency, and credibility issues. In traditional supply chain finance, financial institutions need to spend a high cost to solve problems such as trade authenticity and default risk. For example, financial institutions need to monitor the logistics, capital flow, and information flow of each link in the entire supply chain on a large scale. Not only is the monitoring cost very high, but it is also easy to have regulatory blind spots.

The emergence of blockchain technology provides technical support for solving the above problems. Because the encryption algorithms, privacy protection, smart contracts, and other features in blockchain technology can help financial institutions effectively identify and prevent risks, improve operational efficiency, and reduce business costs

Y. Yu et al. (Eds.): ICPCSEE 2025, CCIS 2673, pp. 557–564, 2026.
https://doi.org/10.1007/978-981-95-2566-9_37

(Sun et al., 2021). First, the integrated asymmetric encryption and other cryptographic algorithms in blockchain technology can well ensure data isolation, privacy, and usage authorization mechanisms, ensuring the security and controllability of enterprise supply chain data collaboration. Second, smart contracts in blockchain technology can effectively avoid and prevent human-caused errors and fraud. When agreed-upon conditions are met, the terms will be automatically executed, with automation and enforceability (Cong & He, 2019). It can pass the credit of core enterprises to upstream and downstream suppliers and customers, and provide comprehensive credit to upstream and downstream suppliers, improving risk management and control capabilities in supply chain finance (Dong et al., 2023).

Based on blockchain technology, we construct a multi-dimensional dynamic supply chain financial risk control system to optimize operation process and risk control system. The system includes multi-dimensional dynamic supply chain financial risk control technology, credit assessment system, and capital flow monitoring and early warning system. First, multi-dimensional dynamic supply chain financial risk control technology can control risks from the aspects of participating entities, business operations, data security, and capital flow. Second, the risk assessment system constructed by machine learning methods can monitor supply chain financial risks in real-time and issue early warnings in a timely manner, effectively controlling risks. Third, the capital flow monitoring and early warning method can solve the difficulty of capital monitoring in supply chain finance and prevent capital from entering restricted areas. This paper expands the research on blockchain technology and supply chain finance model innovation, providing important theoretical support for improving production and finance efficiency and reducing financial risks. At the same time, this paper introduces technologies such as machine learning and deep learning to build an effective risk control system, which is of great significance for reducing financial risks.

The rest of the paper is organized as follows. The second section presents the related research background and work, which mainly includes blockchain technology and supply chain finance. The third section involves multi-dimensional dynamic supply chain financial risk control system. This system includes multi-dimensional dynamic supply chain financial risk control technology, a credit evaluation system, and a capital monitoring and early warning system. The final section provides a summary of the entire paper.

2 Related Works

2.1 Blockchain Technology

Blockchain is a decentralized distributed ledger database, shared among computer network nodes, and stores information electronically in a digital format. As shown in Fig. 1, the structure of blockchain includes the application layer, contract layer, consensus layer, network layer, and data layer. Distributed means there is no central point, and information is stored in all nodes that join the blockchain network, with synchronized data across nodes. The database consists of individual encrypted blocks connected in chronological order, with each block composed of a block header and a block body. The block body is responsible for recording a series of actual transaction data generated by the block system, while the block header records the meta-information of the current block, encapsulating information such as the previous block's hash value, Merkle root, and timestamp. The blocks are linked using a chain structure, giving the blockchain the ability to trace its origin. Each block contains the hash value of the previous block, ensuring that the information stored in the blockchain records is tamper-proof, encrypted, and complete. Blockchain is also a decentralized technical solution that records and stores information data in blocks, with nodes supervising each other to prevent tampering and forgery. "Block + Chain + Timestamp" is the biggest innovation point of the blockchain database, giving it features such as distributed, encrypted, inclusive and transparent, irreversible, traceable, decentralized, open, self-consistent, tamper-proof, and anonymous.

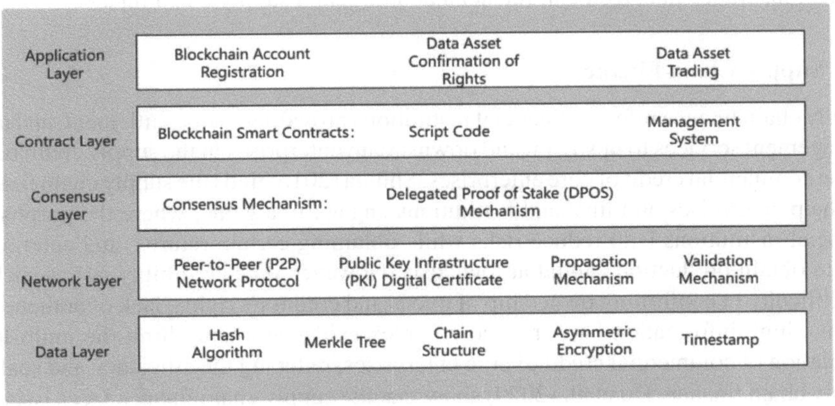

Fig. 1. The structure of blockchain.

Blockchain can provide a trust mechanism and has the potential to change the financial infrastructure. Various financial assets such as equity, bonds, bills, warehouse receipts, and capital shares can all be integrated into the blockchain technology system, becoming digital assets on the chain, and stored, transferred, and traded on the blockchain. The decentralization of blockchain technology can reduce transaction costs and make financial transactions more convenient, intuitive, and secure. The combination of blockchain technology and the financial industry will inevitably create more and more business models, service scenarios, business processes, and financial products, thereby bringing more impact on financial markets, financial institutions, financial services, and the development of financial formats. With the improvement of and the combination of blockchain technology with other financial technologies, blockchain technology will gradually adapt to large-scale financial scenario applications. Currently, the main applications of blockchain in the financial field include digital currency, payment and settlement, digital bills, bank credit management, proof of equity and exchange securities trading, insurance management, financial auditing, and supply chain finance.

Blockchain can provide a trust mechanism and has the potential to change the financial infrastructure. Various financial assets such as equity, bonds, bills, warehouse receipts, and capital shares can all be integrated into the blockchain technology system, becoming digital assets on the chain, and stored, transferred, and traded on the blockchain. The decentralization of blockchain technology can reduce transaction costs and make financial transactions more convenient, intuitive, and secure. The combination of blockchain technology and the financial industry will inevitably create more and more business models, service scenarios, business processes, and financial products, thereby bringing more impact on financial markets, financial institutions, financial services, and the development of financial formats. With the improvement of and the combination of blockchain technology with other financial technologies, blockchain technology will gradually adapt to large-scale financial scenario applications. Currently, the main applications of blockchain in the financial field include digital currency, payment and settlement, digital bills, bank credit management, proof of equity and exchange securities trading, insurance management, financial auditing, and supply chain finance.

2.2 Supply Chain Finance

Supply chain finance refers to financial institutions providing loans, settlement, and asset management services to upstream and downstream enterprises in the supply chain based on the commercial credit of core enterprises. Shihua (2016) finds the supply chain finance can help enterprises and financial institutions engage in a game, where the purpose of financial institutions is to reduce risks while obtaining certain returns, and enterprises aim to obtain production capital at low costs. However, data credibility issues such as the difficulty in confirming ownership of goods and creditor's rights, lack of authenticity in data, high information barriers, and complex evidence tracing limit the multi-level circulation of commercial credit of core enterprises, reducing the efficiency and scale of supply chain finance. Du et al. (2020) show that the supply chain finance faces issues in four aspects: business flow, logistics, capital flow, and information flow.

Firstly, the primary issue concerning business flow lies in the low degree of interconnectedness within the supply chain industry and the lack of collaborative synergy between leading enterprises and SMEs. This manifests in several ways: 1) low level of communication between upstream and downstream enterprises in the supply chain, with each acting independently, leading to inconsistent development objectives and insufficiently close relationships, failing to form an industry-wide supply chain ecosystem; 2) dominance of leading enterprises, resulting in unresolved conflicts in payment and delivery methods; 3) SMEs operating in isolation with scattered layouts and low concentration, unable to achieve economies of scale and leverage industrial advantages. Secondly, in terms of logistics, the costs for product or material logistics are high, and the cycle is lengthy. Inadequate logistics support and high costs in logistics and intermediate links prevent supply chain logistics from fully realizing its credit-enhancing function. Furthermore, the main issue with capital flow is the lack of effective integration between industry and finance, with enterprises having limited financing channels, weak cash flow, low capital flow efficiency, and the advantageous attributes of supply chain finance not being fully utilized. Simultaneously, Ioannou and Demirel (2022) find financial institutions face challenges in confirming rights, supervision, verifying transaction authenticity, and high credit investigation costs, leading to elevated financing costs and weak risk resistance for vanadium and titanium production enterprises.

Finally, the issue with information flow stems from the weak informatization level and low degree of digital transformation and upgrading among supply chain enterprises at the current stage. Severe homogenization competition among enterprises results in a lack of core competitiveness and innovation. Resource development is extensive, and the recovery and utilization rate of vanadium and titanium resources is low. Moreover, the overall industry lacks sufficient informatization empowerment. The supply chain lacks information channels and technology upgrade channels, with low informatization investment and significant informatization arrears. The absence of advanced environmental protection support adds considerable rectification pressure.

Blockchain facilitates the visibility and controllability of supply chain logistics and capital flow, as well as the authenticity and transparency of information flow, thereby improving the efficiency of industry-finance integration. To address the four aspects of challenges in supply chain finance, we introduce blockchain technology and construct a supply chain financial risk control system to mitigate supply chain risks and enhance its efficiency.

3 Multi-dimensional Dynamic Supply Chain Financial Risk Control System

The supply chain financial risk control system consists of multi-dimensional risk control techniques, a credit assessment system, and a capital flow monitoring and early warning system. This system improves credit granting accuracy and reduces capital credit risks (Fig. 2).

3.1 Multi-dimensional Dynamic Supply Chain Financial Risk Control Technology

Blockchain technology involves encryption algorithms, protocols, distributed systems, devices, etc. If technical vulnerabilities exist and are exploited, it will impact data security, causing severe losses to enterprises and the entire supply chain. To ensure the smooth implementation of the supply chain finance model using blockchain, multi-dimensional risk control measures need to be adopted to control risks from the aspects of participating entities, business operations, data security, and capital flow. Specific risk control measures are as follows: ensuring information authenticity through enterprise business registration information, legal representative identity authentication, platform operation review, and account binding; guaranteeing the security of business operations using CA authentication, Short Message Service verification codes, and dynamic verification codes; safeguarding data security by leveraging blockchain's features of loss prevention, tamper-proofing, and traceability, logistics, capital flow, and bills; controlling capital payment through an integrated online payment system; controlling capital recovery by forming a closed loop of capital; and supervising goods and value through the combination of physical and virtual warehousing.

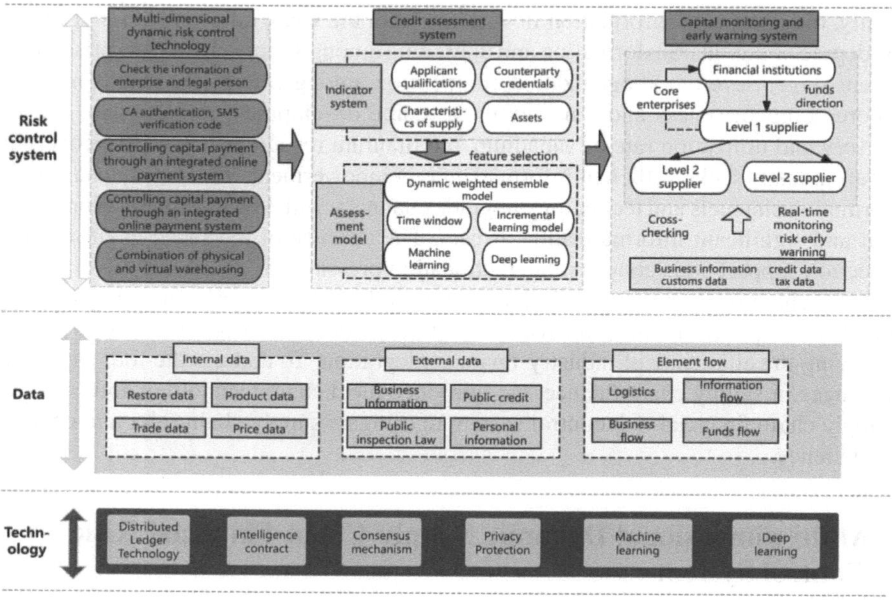

Fig. 2. Multi-dimensional dynamic supply chain financial risk control system.

3.2 Credit Assessment System Based on Dynamic Machine Learning Methods

The company's credit status is not only related to its own qualifications but also to the qualifications of its trading partners, supply chain characteristics, asset conditions, etc. (Zheng et al., 2022). If credit risk is predicted solely based on the company's historical data, it may lead to deviations or delays in the prediction. Therefore, it is necessary to construct an indicator system and a dynamic credit risk assessment model. The indicator system includes five risk control dimensions, such as enterprise credit granting, business compliance, and capital flow. Specific risk control indicators include enterprise business registration information, legal representative identity authentication, capital flow direction, and capital recovery early warning.

The dynamic credit risk assessment model use machine learning and deep learning methods to dynamically update and adjust based on the passage of time and newly added data from the "four flows" and other sources. This will make the prediction results more accurate and enable financial institutions and enterprises to reasonably control credit risk. This dynamic weighted ensemble model based on time windows and incremental learning. We divide the credit data into different windows in chronological order and detect deviations in data distribution between the latest window and historical windows to determine whether the characteristics of the data have changed. Then, based on the detection results, we select different balancing methods, feature selection methods, and assessment models for high-dimensional feature data. Through machine learning and deep learning methods, a mature and accurate risk assessment system is constructed to monitor supply chain financial risk assessment indicators in real-time and issue timely warnings. This model establishes enterprise credit files, strengthens internal credit ratings, and improves the effectiveness of risk identification, control, and disposal.

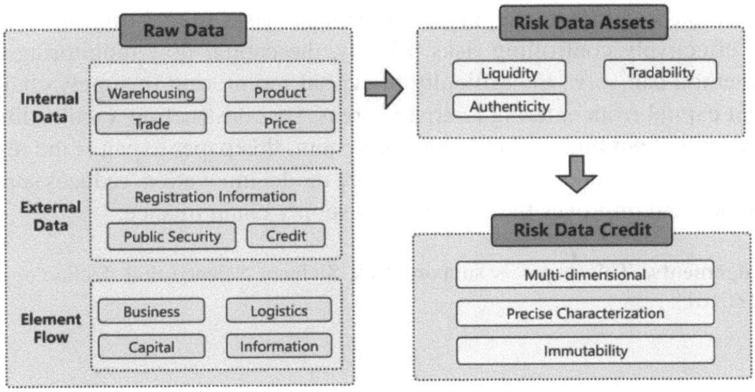

Fig. 3. The process of credit risk assessment.

As showed on Fig. 3, the credit assessment system operates as follows. First, collect and organize raw data, including internal data, external data, and element flows. Second, construct an indicator system and select relevant features to form risk data assets. Third, apply a dynamic credit risk assessment model to evaluate risk data credit.

3.3 Capital Monitoring and Early Warning System

The characteristics of blockchain technology, such as immutability, traceability, and real-time sharing, are well-suited to address the challenges in supply chain finance, such as capital monitoring and preventing capital from entering restricted areas. Based on this, the project constructs a capital flow monitoring system (Fig. 2) and closely collaborates with government departments such as taxation, customs, and industry and commerce to achieve interconnectivity of big data, effectively monitoring capital flow and preventing the risk of illegal misappropriation of capital.

4 Conclusion

Supply chain finance is an important way to solve the problems of difficult and expensive financing for SMEs. However, due to issues in business flow, logistics, capital flow, and information flow, supply chain finance faces development bottlenecks. Blockchain technologies such as smart contracts and consensus mechanisms help overcome the credibility challenges and become an important technological development direction for improving the implementation effect and reducing the risks of supply chain finance. Based on blockchain technology, this paper constructs a multi-dimensional dynamic supply chain financial risk control system. The system includes multi-dimensional dynamic supply chain financial risk control techniques, a credit assessment system, and a capital flow monitoring and early warning system. Among them, the multi-dimensional dynamic supply chain financial risk control techniques can control risks from aspects such as participating entities, business operations, data security, and capital flow. At the same time, the risk assessment system constructed through machine learning and deep learning methods can monitor supply chain financial risks in real-time and issue timely warnings, effectively controlling risks. Finally, the capital flow monitoring and early warning method can solve the difficulty of capital monitoring in supply chain finance and prevent capital from entering restricted areas. By constructing a multi-dimensional dynamic supply chain financial risk control system, this paper expands the research on blockchain technology and supply chain finance model innovation, reduces supply chain financial risks, and improves the efficiency of supply chain finance.

Acknowledgements. This paper is supported by Sichuan Science and Technology Program (2024YFHZ0161).

References

Cong, L.W., He, Z.: Blockchain disruption and smart contracts. Rev. Financ. Stud. **32**(5), 1754–1797 (2019)

Dong, L., Qiu, Y., Xu, F.: Blockchain-enabled deep-tier supply chain finance. Manuf. Serv. Oper. Manag. **25**(6), 2021–2037 (2023)

Du, M., Chen, Q., Xiao, J., Yang, H., Ma, X.: Supply chain finance innovation using blockchain. IEEE Trans. Eng. Manage. **67**(4), 1045–1058 (2020)

Ioannou, I., Demirel, G.: Blockchain and supply chain finance: a critical literature review at the intersection of operations, finance and law. J. Bank. Financ. Technol. **6**(1), 83–107 (2022)

Shihua, L.: Innovative design of "win-win" model of supply chain finance based on bank-enterprise dynamic game. Innov. Manag. 1797 (2016)

Sun, R., He, D., Su, H.: Evolutionary game analysis of blockchain technology preventing supply chain financial risks. J. Theor. Appl. Electron. Commer. Res. **16**(7), 2824–2842 (2021)

Zheng, K., et al.: Blockchain technology for enterprise credit information sharing in supply chain finance. J. Innov. Knowl. **7**(4), 100256 (2022)

Data Security and Privacy

Reversible Data Hiding Algorithm in Encrypted Images Based on Arithmetic Coding and Dual Prediction

Mengyuan Zhang, Xinyi Zeng, Tonghui Liu, Tong Zhu, and Wanli Lyu(✉)

Key Lab of Intelligent Computing and Signal Processing of Ministry of Education, School of Computer Science and Technology, Anhui University, Hefei 230601, China
lwl@ahu.edu.com

Abstract. Reversible data hiding in encrypted images (RDH-EI) enables the concealment of secret data within ciphertext images while preserving the ability to fully recover both the original image and the hidden message. However, existing RDH-EI schemes based on vacating room after encryption (VRAE) suffer from limited embedding capacity. To address this issue, we propose a method based on arithmetic coding and dual prediction for encrypted images. First, the original image is encrypted with a chunked modulus and permutation. Then, using the upper-left corner pixel of each subblock as a reference, adaptive MSB prediction and difference prediction are employed to predict the remaining pixels within the subblock. The resulting label map is then compressed via arithmetic coding to vacate the embedding space for the secret message. Finally, the separable operations of the original image restoration and secret message extraction can be performed on the basis of the type of key possessed. The experimental results demonstrate that the proposed algorithm not only successfully extracts the secret information but also recovers the original image without any loss. Furthermore, it effectively enhances the embedding capacity by fully utilizing the correlation between adjacent pixels while ensuring the security of the image.

Keywords: Reversible data hiding · Adaptive difference prediction · Adaptive MSB prediction · Arithmetic coding

1 Introduction

The proliferation of cloud computing services has led to the widespread storage of personal data in cloud environments. However, inadequate protection mechanisms expose these data to significant security risks, including unauthorized access and other potential breaches. As a result, data protection in cloud environments has become particularly important. Image encryption [1] and data hiding [2] are two main techniques for data protection. Image encryption transforms the original image into a noise-like representation

This research work is partly supported by the Provincial Colleges Quality Project of Anhui Province (2020xsxxkc047) and the National Undergraduate Innovation and Entrepreneurship Training Program (2023103570289).

Y. Yu et al. (Eds.): ICPCSEE 2025, CCIS 2673, pp. 567–586, 2026.
https://doi.org/10.1007/978-981-95-2566-9_38

prior to server upload, effectively preventing attackers from extracting meaningful visual information from the encrypted image. Data hiding techniques embed secret information within cover images through imperceptible modifications, ensuring that the covert data remain undetectable to unauthorized parties. Digital watermarking [3], steganography [4], and reversible data hiding [5] are three traditional data hiding techniques. Among these methods, the traditional methods of reversible data hiding are lossless compression [6], histogram shifting [7], and difference expansion [8], which are commonly used in the plaintext domain. However, in domains such as medical image processing and military intelligence transmission, it is essential to encrypt the initial data before performing any subsequent operations to safeguard patient privacy and prevent the leakage of sensitive military information. To address issues such as encrypted image management in cloud environments, reversible data hiding in encrypted images (RDH-EI) has emerged, gaining significant attention and development. This technique integrates image encryption with data hiding methods. After the original image is encrypted by its owner, the data hider can embed extra information into the encrypted version without needing access to the unencrypted content. If receivers possess the corresponding key, they can retrieve both the secret message and the original image.

On the basis of the sequence of space reservation and image encryption operations, existing RDH-EI algorithms are classified into two types: Reserve Room Before Encryption (RRBE) [9-11, 13-15] and Vacating Room After Encryption (VRAE) [16-23]. Among them, the RRBE-class methods require preprocessing the image in the initial stage to create space for the subsequent information embedding before performing image encryption. Since encryption disrupts the spatial correlation of the original image, RRBE-class algorithms achieve higher embedding rates than VRAE-class methods do.

Ma et al. [9] first proposed an information hiding framework for freeing up space before encryption, enabling full reversibility in image restoration and information extraction. This approach has since garnered significant attention from scholars. Considering the small differences between neighboring pixel values, Zhang et al. [10] use multiple pixels to predict the current pixel and compute the prediction error, combined with histogram shifting to free up the embedding space. Yin et al. [11] utilized the median-edge detector (MED) [12] to predict the original image pixel values, recorded the number of matching MSBs between the predicted and actual values, and compressed these counts via Huffman coding to optimize space. Yi et al. [13] partitioned an image into chunks, classified the pixels within each chunk, and labelled the pixels via binary tree coding to create room for embedding. However, the embedding capacity was limited because of the reliance solely on intrachunk pixel correlation. Subsequently, Wu et al. [14] computed the prediction error across the entire image and employed fixed-length coding to label pixels on the basis of their prediction errors, thereby increasing the embedding rate. Ma et al. [15] improved upon Wu et al. [14] by replacing fixed-length coding with adaptive Huffman codes and using Huffman coding to mark the encrypted image, thereby obtaining a larger embedding space.

Although RRBE-class algorithms achieve higher embedding rates, they require additional preprocessing steps, increasing computational overhead for the image owner. In contrast, VRAE-class methods only require encrypting the original image.

Zhang [16] introduced a well-known VRAE-based RDH-EI algorithm that partitions the encrypted image into nonoverlapping subblocks and embeds additional information by inverting the three LSBs of half the pixels in each subblock. A fluctuation function is then employed at the receiver side to identify the modified pixels, enabling both information extraction and image restoration. However, this method tightly couples image decryption with information extraction, making the two processes inseparable. Zhang [17] introduced the first separable RDH-EI method, in which the data hider compresses the LSB of specific pixels to create an embedding space. The receiver can independently perform information extraction and image restoration via the corresponding keys. This pioneering work inspired subsequent developments in more advanced separable data hiding algorithms. Puteaux et al. [18] introduced a method based on MSB prediction to embed secret data via MSB substitution. However, this algorithm uses a stream cipher to encrypt pixels one by one in the encryption phase, which disrupts the correlation between adjacent pixels in the original image and limits the embedding rate. Wang et al. [19] replaced the encryption method in Puteaux et al. [18] with chunked encryption, where the pixels in each subblock are encrypted via the same key. This approach leverages the neighboring pixel correlation to a significant extent. However, many subblocks remain unusable for embedding because of their inability to carry secret information. To mitigate this, Qi et al. [20] and Zhang et al. [21] further optimized the prediction of unusable subblocks. Specifically, Qi et al. [20] employed adaptive difference prediction to derive three pixels from one reference pixel, whereas Zhang et al. [21] utilized MED to predict the remaining pixel from three known pixels. Thus, the unusable subblocks were categorized into two types, one of which allows for secondary embedding, thereby reducing the number of unusable subblocks and improving the embedding rate.

This study proposes an RDH-EI method based on arithmetic coding and dual prediction of VRAE classes, which can effectively improve the embedding rate. The main contributions of this study can be summarized as follows:

1. We employ additive modulus encryption combined with permutation to encrypt the original image. This approach preserves spatial correlation within subblocks while ensuring information security, thereby increasing the embeddable space.
2. The dual prediction method combines adaptive MSB prediction and difference prediction to predict the pixels within the subblock, which addresses the issue of insufficient embeddable bits in MSB prediction for specific subblocks and effectively improves the embedding rate.
3. We employ a label map to record the embeddable bits of all subblocks and then compress it via arithmetic coding to minimize auxiliary information and maximize the available embedding space.

This paper is structured as follows: Sect. 2 details the proposed method. Section 3 shows our experimental results and comparative analysis with other algorithms. Section 4 concludes our work and discusses our research directions in the future.

2 Proposed Method

The overall framework of the RDH-EI algorithm based on arithmetic coding and dual prediction proposed in this paper is illustrated in Fig. 1, which includes three parts: (1) The image owner encrypts the original image via the image encryption key K_1 and the permutation key K_2 and then sends the encrypted image to the data hider. (2) After receiving the encrypted image, the data hider performs adaptive MSB prediction and adaptive difference prediction on the image to create an embedding space. The secret information is then encrypted via the data hiding key K_3 and embedded into the image along with the auxiliary information generated during the space creation process, ultimately producing the embedded image. (3) After receiving the embedded image, the image receiver performs a separable operation of extracting the secret information and restoring the original image according to the type of key it possesses.

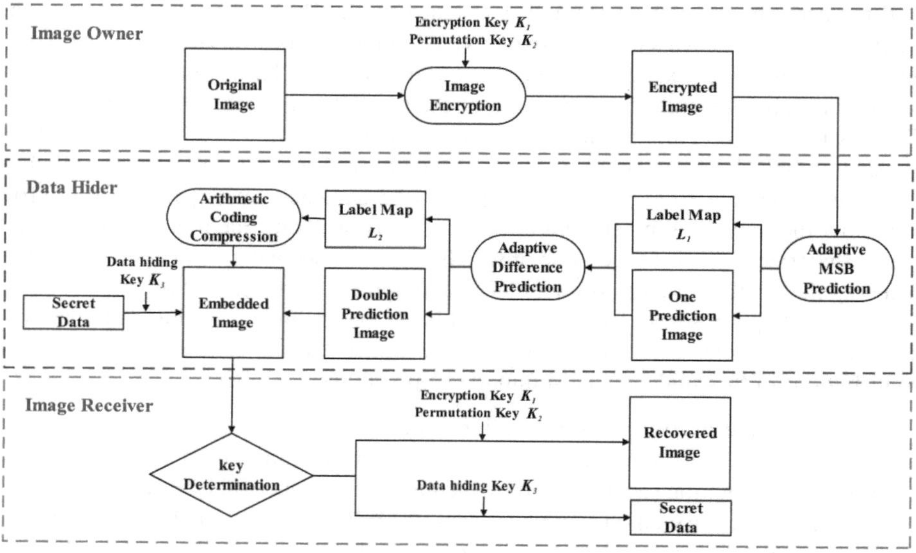

Fig. 1. Framework of the proposed algorithm.

2.1 Image Encryption

To protect the original image information while preserving the intrablock correlation, the image is encrypted via chunked modulus encryption. Given an original image of size $M \times N$, it is partitioned into $\lfloor M/2 \rfloor \times \lfloor N/2 \rfloor$ nonoverlapping subblocks, each of size 2×2. The image encryption key subsequently generates a matrix $\boldsymbol{H}$ of size $\lfloor M/2 \rfloor \times \lfloor N/2 \rfloor$. For all the pixels in the subblock located at (i, j) after division, additive modulus encryption is performed via the elements of the matrix $\boldsymbol{H}$ located at (i, j). The following equation is used:

$$C_m(i,j) = (x_m(i,j) + h(i,j)) \bmod 256 \tag{1}$$

where $m \in \{1, 2, 3, 4\}$, $C_m(i, j)$ denotes the m-th pixel of the subblock located at (i, j) in the encrypted image, $x_m(i, j)$ denotes the m-th pixel of the subblock located at (i, j) in the original image, and $h(i, j)$ is the value of the element in the matrix $\boldsymbol{H}$ located at (i, j), where $h(i, j) \in \{0, 1, ..., 255\}$.

Additive modulus encryption alone may cause the encrypted image to retain contours similar to those of the original image. To enhance the security of the image, it is further subjected to block permutation via a permutation key, resulting in a permuted encrypted image.

2.2 Adaptive MSB Prediction

In this stage, the data hider performs adaptive MSB prediction on the encrypted image to generate the corresponding initial label map $\mathbf{L_1}$. First, the encrypted image is partitioned into $\lfloor M/2 \rfloor \times \lfloor N/2 \rfloor$ nonoverlapping subblocks, each of size 2×2, containing four pixels. The pixel values are denoted as P, C_1, C_2, and C_3, arranged from left to right and top to bottom, respectively. These four pixel values are then converted into 8-bit binary representations via the following equation:

$$b_u = \left\lfloor \frac{I}{2^u} \right\rfloor \bmod 2,\ u = 0, 1, \ldots, 7 \tag{2}$$

where I is the corresponding encrypted pixel value and where b_u is the u-bit of the corresponding binary of the pixel value.

The upper-left corner pixel P of each subblock is subsequently used as the reference pixel. For the other three pixels C_1, C_2 and C_3 within the subblock, their 8-bit binary representations are compared with that of the reference pixel P from MSB to LSB. The numbers of matching bits for these pixels are denoted as s_1,s_2 and s_3, respectively. These matching bits are then used to embed the secret information.

To standardize the number of embeddable bits within a subblock, let min (s_1, s_2, s_3) represent the predicted embedding length *MLen* for the subblock. Specifically, taking the subblock shown in Fig. 2 as an example, the *MLen* of this subblock is 4, meaning that the corresponding element value in the label map $\mathbf{L_1}$ is 4. As a result, the number of bits that can be freed in this subblock is 12. Assuming that the secret information is $nc = (011011100101)_2$, the structure of the subblock after reconstruction is shown in Fig. 2 (c) and Fig. 2 (d). In this structure, the upper-left corner represents the reference pixel P, while C_1', C_2' and C_3' are the reconstructed pixels. The last 4 bits of C_1', C_2' and C_3' are fixed, whereas the first 4 bits contain the embedded secret information.

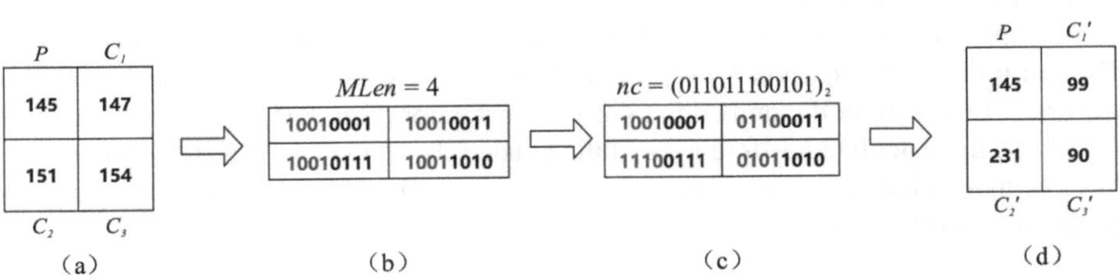

Fig. 2. Adaptive MSB prediction.

2.3 Adaptive Difference Prediction

In this phase, the data hider performs adaptive difference prediction for the subblocks previously labelled 0 or 1. According to the prediction results, the corresponding labels are modified to generate an updated label map $\mathbf{L_2}$. For subblocks labelled 0 or 1, the upper-left corner pixel P of each subblock serves as the reference pixel. The differences in the remaining pixels in the subblock are computed by subtracting the reference pixel value P from each corresponding pixel value, yielding the results e_1, e_2 and e_3. Owing to the correlation between pixels within a subblock, the resulting differences are relatively small, requiring fewer bits to store them. When differences are stored via t bits, all differences in an embeddable subblock must fall within $[-2^{t-1}+1,\ 2^{t-1}]$. Subblocks falling outside this range are classified as nonembeddable. For all the pixels in an embeddable subblock, the remaining $8-t$ bits can be utilized for secret data embedding. To accommodate both positive and negative differences, all differences are converted to nonnegative integers through the following equation:

$$e_i' = e_i + 2^{t-1} - 1,\ i = 1,\ 2,\ 3 \tag{3}$$

where e_i is the difference before conversion and where e_i' is the difference after conversion.

The embedding rate of the image is influenced by parameter t. Decreasing t increases the available embedding space per subblock but reduces the number of qualifying embeddable subblocks. Conversely, increasing t expands the number of embeddable subblocks while diminishing their individual embedding capacities. To determine the optimal value of t, we conducted experiments using eight test images. As demonstrated in Table 1, the maximum embedding capacity was achieved when t equaled 5 or 6 across all test images. On the basis of the experimental results and practical considerations, t is set to 6 in this paper.

After performing adaptive difference prediction on all subblocks labelled 0 or 1, the corresponding elements in $\mathbf{L_1}$ are updated by assigning 1 to embeddable subblocks and 0 to nonembeddable subblocks. The elements with values in [2, 8] remain unmodified. This process generates the updated label map $\mathbf{L_2}$.

Consider the subblock shown in Fig. 3 (a) as an example. During the adaptive MSB prediction stage, this subblock is originally labelled with a value of 0, indicating that adaptive difference prediction is needed. Upon analysis, all three pixel differences within the subblock fall within the range $[-31, 32]$, confirming its embeddable status. Consequently, its corresponding label map value is updated to 1. Assuming that the secret information to be embedded is $nc = (101101)_2$, the structure of the subblock after reconstruction is presented in Fig. 3 (e) and Fig. 3 (f). In this structure, the upper-left corner represents the reference pixel P, while C_1', C_2' and C_3' are the reconstructed pixels. The last 6 bits of C_1', C_2' and C_3' represent their respective differences from P, whereas the first 2 bits contain the embedded secret information. Similarly, for the subblock in Fig. 3 (g), the calculated difference e_1 falls outside the range $[-31, 32]$. Consequently, its label map value is set to 0, classifying it as a nonembeddable subblock that requires no further modification.

Table 1. Comparison of embedding capacity for different t situations.

Image	$t = 2$	$t = 3$	$t = 4$	$t = 5$	$t = 6$	$t = 7$
Lena	460986	467723	480980	**488908**	486768	473749
Barbara	327732	331953	339928	346271	**349264**	344362
Peppers	419317	424395	441085	**455414**	449319	432491
Boat	350876	353682	365606	384370	**385240**	369169
House	449814	452967	459824	468079	**474780**	465012
Airplane	688456	703426	710480	**708087**	702228	694248
F16	537093	544219	550809	**554284**	552967	544918
Lake	336644	338987	348814	364096	**368196**	355202

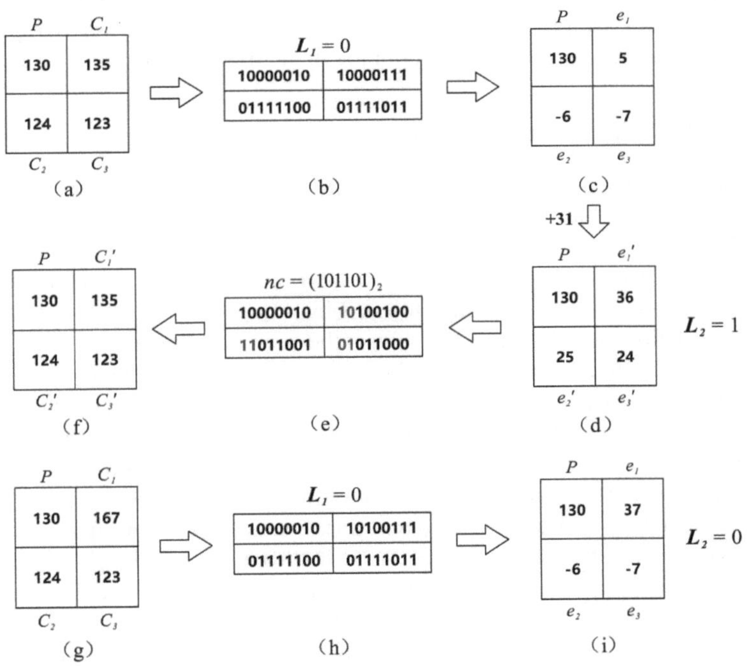

Fig. 3. Adaptive difference prediction.

To minimize the length of auxiliary information, the label map $\mathbf{L_2}$ is compressed via arithmetic coding to obtain the compressed representation $\mathbf{L_3}$. Assuming that $\mathbf{L_2}$ is of size $m \times n$, the length of $\mathbf{L_3}$ is encoded using $\lceil \log_2(3 \times m \times n) \rceil$ bits and embedded into the encrypted image during the data embedding phase.

2.4 Data Embedding

To facilitate the subsequent extraction of secret information, both the label map $\mathbf{L_3}$ and its length are embedded in reference pixels, while the replaced reference pixels are used as auxiliary information. The number of replaced reference pixels *num* can be computed via the following equation:

$$num = \left\lceil \frac{\lceil \log_2(3 \times m \times n) \rceil + LabLen}{8} \right\rceil \tag{4}$$

where *LabLen* is the length of $\mathbf{L_3}$ and where m and n are the height and width of $\mathbf{L_2}$, respectively.

To ensure the security of the embedded data, the data hider encrypts the message to be embedded via the data hiding key and then embeds both the auxiliary information and the encrypted secret message into the vacated space through bit substitution until the entire set of secret data is embedded. Since the difference between adjacent pixels is smaller when the subblock has fewer embeddable bits, each pixel is flipped after embedding to enhance security, resulting in the final embedded image.

2.5 Image Restoration and Secret Information Extraction

In the original image restoration or secret information extraction phase, the image receiver first flips all the pixels of the received image and then partitions the image into $\lfloor M/2 \rfloor \times \lfloor N/2 \rfloor$ nonoverlapping subblocks of size 2×2. Next, the pixels located at the upper-left corners of all subblocks are extracted in a left-to-right and top-to-bottom manner. Each pixel is then transformed into an 8-bit sequence via Eq. (2), and a bit stream is obtained. The first $\lceil \log_2(3 \times m \times n) \rceil$ bits of the stream are extracted and transformed into a decimal value to obtain a length of $\mathbf{L_3}$, denoted as *LabLen*. The following *LabLen* bits are subsequently extracted to obtain the compressed label map $\mathbf{L_3}$, which is finally decoded via arithmetic decoding to reconstruct the label map $\mathbf{L_2}$. All the embeddable components from each subblock are subsequently extracted and concatenated to form the bit sequence ***Lst***. The detailed procedure is described below. Subblocks with a corresponding value of 0 in $\mathbf{L_2}$ are considered unavailable and skipped during processing. For subblocks with a corresponding value of z ($z \in [1, 8]$) in $\mathbf{L_2}$, pixels C_1', C_2', and C_3' are converted into 8-bit sequences. If $z = 1$, the $8 - t$ MSBs of C_1', C_2', and C_3' are extracted and then appended to ***Lst***. If $z \in [2, 8]$, the z MSBs of C_1', C_2', and C_3' are extracted and appended to ***Lst***. The data structure of the extracted ***Lst*** is illustrated in Fig. 4.

Reference Pixel Auxiliary Information	**Embedded secret Data**

Fig. 4. Data structure of the ***LST***.

Depending on the type of key possessed by the receiver, the separable operations of original image restoration and secret message extraction can then be executed. There are the following three situations:

1. When the receiver possesses only the data hiding key K_3, the embedded secret information can be completely recovered. The number of bits occupied by the reference pixel auxiliary information, denoted as *Len*, is computed via Eq. (5). The first *Len* bits of ***Lst*** are truncated, and the remaining portion contains the encrypted secret information. The data hiding key is then used to decrypt the embedded information, recovering the original secret data.

$$Len = 8 \times num = 8 \times \left\lceil \frac{\lceil \log_2(3 \times m \times n) \rceil + LabLen}{8} \right\rceil \tag{5}$$

where *num* is the number of reference pixels being replaced, *LabLen* is the length of $\mathbf{L_3}$, and *m* and *n* are the height and width of $\mathbf{L_2}$, respectively.
2. When the receiver possesses only the encryption key K_1 and the permutation key K_2, the original image can be restored without loss. First, the first *Len* bits of ***Lst*** are extracted to reconstruct all reference pixels. All subblocks are subsequently restored via the following detailed procedure. For subblocks with a corresponding value of 0 in $\mathbf{L_2}$, it is an unusable subblock and is not processed. For subblocks with a corresponding value of 1 in $\mathbf{L_2}$, the pixels C_1', C_2', and C_3' are first transformed into 8-bit binary sequences. The *t* LSBs of each pixel are then extracted and converted to decimal values ten_1, ten_2, and ten_3. Eventually, C_1, C_2, and C_3 are recovered according to the following equation:

$$C_i = P + ten_i - 2^{t-1} + 1,\ i = 1,\ 2,\ 3 \tag{6}$$

where *P* is the reference pixel value corresponding to each subblock and where C_i is the encrypted pixel value.

For subblocks with a corresponding value of z ($z \in [2, 8]$) in $\mathbf{L_2}$, pixels C_1', C_2', and C_3' are converted into 8-bit binary sequences, and the $8 - z$ LSBs of each binary sequence are extracted, denoted as b_1, b_2 and b_3, respectively. Eventually, C_1, C_2, and C_3 are recovered according to the following equation:

$$C_i = \mathrm{Trans}(\mathrm{Ext}(P,\ z) + b_i),\ i = 1,\ 2,\ 3 \tag{7}$$

where C_i is the encrypted pixel value, Ext(*P*, *z*) extracts the *z* MSBs from *P*, Trans(*x*) converts the binary input *x* to a decimal, and $'+'$ denotes the left join operation.

After processing all the subblocks, the receiver uses the permutation key to restore the image to its prepermutation state and then decrypts it to its original form via the following equation:

$$x_m(i,j) = (\,C_m(i,j) - h(i,j)) \bmod 256 \tag{8}$$

where $m \in \{1, 2, 3, 4\}$, $x_m(i,j)$ denotes the m-th pixel of the subblock located at (i, j) in the original image, $C_m(i,j)$ denotes the m-th pixel of the subblock located at (i, j) in the encrypted image, and $h(i,j)$ is the value of the element in matrix $\boldsymbol{H}$ located at (i, j), where $h(i,j) \in \{0, 1, ..., 255\}$.

3. When the receiver has the data hiding key K_3, image encryption key K_1 and permutation key K_2 at the same time, the embedded data can be extracted completely, and the original image can be restored without loss. The processes of message extraction and original image restoration are separable and do not affect each other.

3 Results and Analysis of the Experiments

This section introduces the evaluation metrics of this experiment, which consist of the peak signal-to-noise ratio (PSNR), structural similarity (SSIM), and embedding rate (ER). The eight standard grayscale images illustrated in Fig. 5 are used to test the embedding capacity and reversibility of the proposed algorithm, while the average embedding rate is evaluated via the BOSSbase, BOWS-2, and UCID image datasets. To comprehensively evaluate the performance of our proposed algorithm, we conduct a multifaceted comparison with other VRAE methods.

3.1 Indicator Evaluation

The PSNR is used to evaluate the quality of the restored image. For an image of size $M \times N$, the specific formula is as follows:

$$\begin{aligned} \mathrm{PSNR} &= 10 \times \lg \frac{(2^n-1)^2}{\mathrm{E_{MSE}}} \\ \mathrm{E_{MSE}} &= \frac{1}{M \times N} \sum_{i=1}^{M} \sum_{j=1}^{N} [X(i,j) - Y(i,j)]^2 \end{aligned} \tag{9}$$

where X is the image after recovery, Y is the original image, and n is the number of bits of pixels. The larger the PSNR between the restored image and the original image is, the smaller the difference between the two images. When the PSNR between two images tends to be positive infinity, the current algorithm is reversible in terms of image recovery.

The SSIM is used to evaluate the degree of similarity between two images. The specific formula is as follows:

$$\mathrm{SSIM}(a,b) = \frac{(2\mu_a\mu_b + C_1)(2\sigma_{ab} + C_2)}{(\mu_a^2 + \mu_b^2 + C_1)(\sigma_a^2 + \sigma_b^2 + C_2)} \tag{10}$$

where μ_a is the mean of the reconstructed image, μ_b is the mean of the original image, σ is the standard deviation, and C_1 and C_2 are two empirical constants. The larger the

SSIM is between the restored image and the original image, the more similar the two images are. When the SSIM value between two images is 1, the two images are the same.

The ER refers to the number of effective embedding bits in each pixel and is an important measure of the embedding capacity of a method. For an image of size $M \times N$, the specific formula is as follows:

$$\text{ER} = \frac{Capacity}{M \times N} \tag{11}$$

where *Capacity* is the net embedding capacity of the image. The larger the embedding rate is, the more information the algorithm can embed under the same conditions.

3.2 Visual Quality and Reversibility

To evaluate the reversibility of the method, we utilize 8 images of Lena, Barbara, etc., as carriers for testing, and the test images are as follows. During the data extraction and image restoration processes, the label map is first extracted. Both message extraction and subblock reconstruction are subsequently conducted on this label map, enabling lossless image recovery.

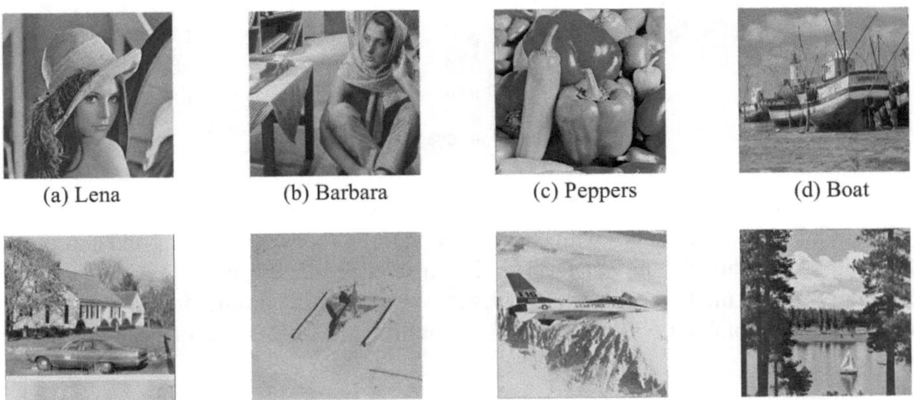

(a) Lena (b) Barbara (c) Peppers (d) Boat

(e) House (f) Airplane (g) F16 (h) Lake

Fig. 5. 8 test images.

Fig. 6 shows the original, encrypted, embedded, and recovered states of Lena, Barbara, Peppers and Boat. The encrypted and embedded images effectively protect the privacy of the model owner, whereas the restored images are visually indistinguishable from the original images.

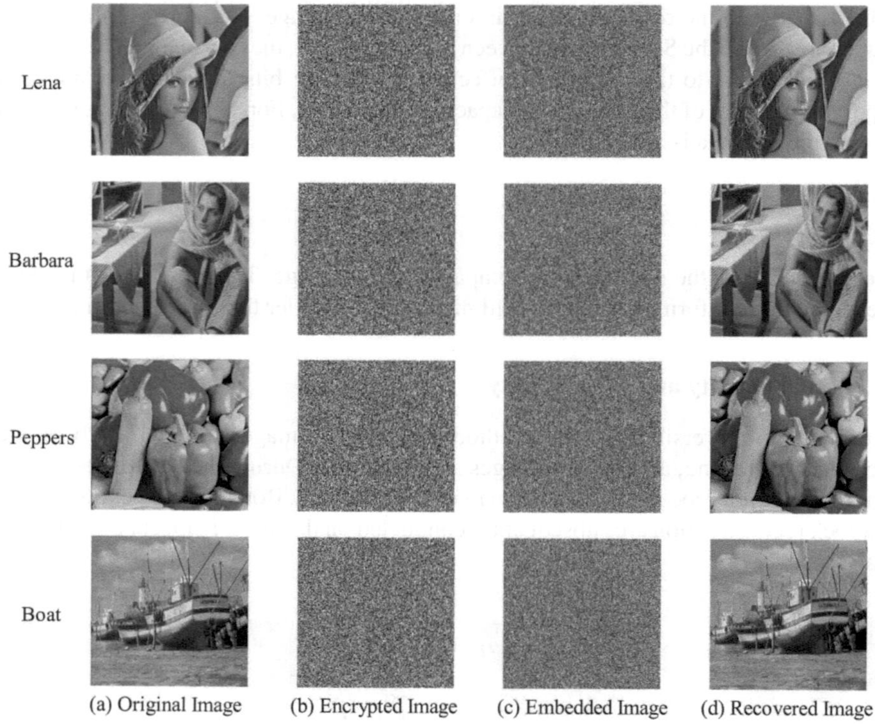

Fig. 6. Several stages of test images.

To better evaluate the reversibility of the proposed algorithm, we conducted PSNR calculations of six test images with their corresponding recovered images and compared the results with those from Zhang [16]. The results presented in Table 2 demonstrate that our proposed method achieves PSNR values approaching infinity for all test images. Moreover, the obtained SSIM values of 1 confirm the lossless reconstruction of the original images.

Table 2. Comparison of PSNR.

	Lena	Barbara	Peppers	Airplane	Boat	House	Average
Proposed	$+\infty$	$+\infty$	$+\infty$	$+\infty$	$+\infty$	$+\infty$	$+\infty$
[16]	34.65	34.64	34.77	34.67	34.92	34.48	34.69

3.3 Embedding Capacity Analysis

The use of adaptive MSB prediction with adaptive difference prediction improves pixel utilization, and the use of arithmetic coding to compress the label map results in an improvement in the image embedding capacity, where Table 3 represents the number of subblocks of each type for each of the eight test images, where N_i denotes the number of subblocks with label value i. Table 4 represents the total number of bits released by each type of subblock, where n_i denotes the total number of bits freed from all subblocks labelled with value i, which can be computed via the following equation:

$$n_i = \begin{cases} 3 \times N_i \times (8 - t), & \mathrm{i} = 1 \\ 3 \times N_i \times i, & \mathrm{i} \neq 1 \end{cases} \quad (12)$$

Table 3. Number of subblocks of each type in the test image.

Image	N_0	N_1	N_2	N_3	N_4	N_5	N_6	N_7	N_8
Lena	4774	6326	9109	12994	15729	11413	4085	879	227
Barbara	14016	6796	10724	10933	11226	8340	2860	531	110
Peppers	4939	7061	9968	15385	16775	8860	2152	352	44
Boat	7355	8391	12119	15836	13859	6041	1601	295	39
House	7978	6878	10225	10894	10103	8131	5140	3855	2332
Airplane	2288	3316	4580	8105	13680	16643	9421	44	7459
F16	5771	4853	7212	9858	12518	12583	8150	3402	1189
Lake	9132	8245	12428	14783	12076	6074	2191	495	112

Table 4. Total number of bits freed by each type of subblock in the test image.

Image	n_0	n_1	n_2	n_3	n_4	n_5	n_6	n_7	n_8
Lena	0	37956	54654	116946	188748	171195	73530	18459	5448
Barbara	0	40776	64344	98397	134712	125100	51480	11151	2640
Peppers	0	42366	59808	138465	201300	132900	38736	7392	1056
Boat	0	50346	72714	142524	166308	90615	28818	6195	936
House	0	41268	61350	98046	121236	121965	92520	80955	55968
Airplane	0	19896	27480	72945	164160	249645	169578	924	179016
F16	0	29118	43272	88722	150216	188745	146700	71442	28536
Lake	0	49470	74568	133047	144912	91110	39438	10395	2688

Table 5 presents the embedding capacity for each test image, where *Sum* is the total number of bits saved and *Capacity* is the embedding capacity. They can be computed via the following equation:

$$Sum = \sum_{i=0}^{8} n_i \tag{13}$$

$$Capacity = Sum - InfoLen \tag{14}$$

where *InfoLen* is the length of the auxiliary information, including the label map and its length. Some images occupy fewer bits for the auxiliary information but exhibit lower embedding capacity. This is because of the complex texture of the image, where most of the label map values are 0 or 1. As a result, the label map compression is high, but the number of nonembeddable pixels is also increased.

Table 5. Embedding capacity of each test image.

Image	*Sum*/bits	*InfoLen*/bits	*Capacity*/bits
Lena	666936	180206	**486730**
Barbara	528600	179374	**349226**
Peppers	622023	172742	**449281**
Boat	558456	173254	**385202**
House	673308	198566	**474742**
Airplane	883644	181454	**702190**
F16	746751	193822	**552929**
Lake	545628	177470	**368158**

To better evaluate the average embedding performance of the proposed algorithm, we conduct embedding rate tests on three image datasets: BOSSbase, BOWS-2, and UCID. The results are presented in Fig. 7 and Table 6, where the x-axis in Fig. 7 denotes the ordinal numbers of the test images.

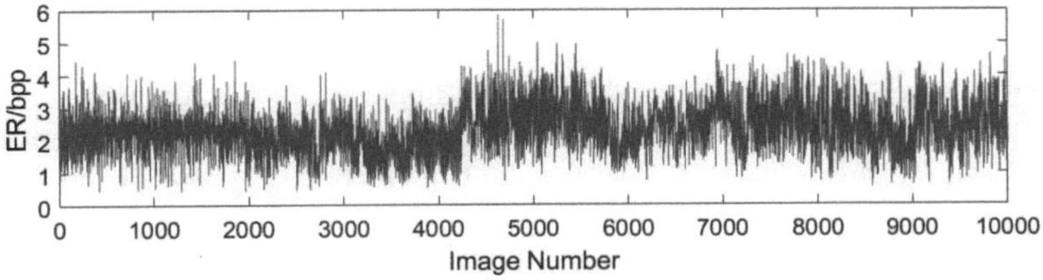

(a) Embedding rate of images in BOSSbace

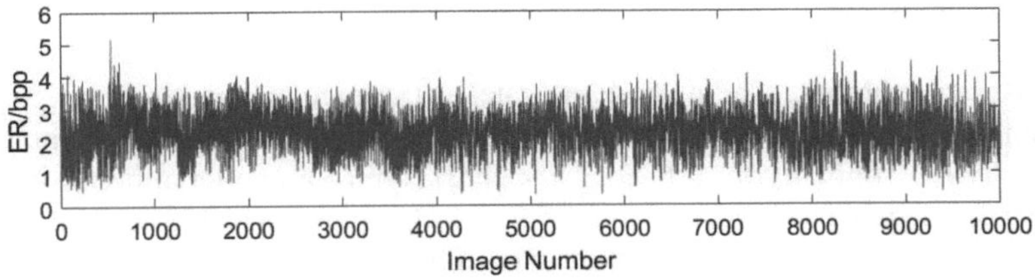

(b) Embedding rate of images in BOWS-2

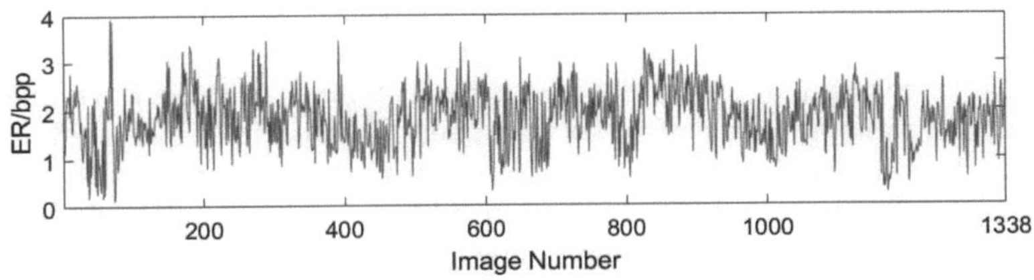

(c) Embedding rate of images in UCID

Fig. 7. Embedding rates of test images from different image datasets.

The figure shows that in all three image datasets, the majority of images exhibit embedding rates exceeding 1 bpp, with the highest embedding rates reaching 5.8398 bpp, 5.1813 bpp, and 3.9250 bpp, respectively. Additionally, from the table, the proposed algorithm achieves embedding rates of 2.3618 bpp, 2.2719 bpp, and 1.8234 bpp on the BOSSbase, BOWS-2, and UCID image datasets, respectively. This demonstrates the superiority of the proposed method in terms of the average embedding rate.

Table 6. Average embedding rates for different image datasets.

Image dataset	BOSSbace	BOWS-2	UCID
ER/bpp	2.3618	2.2719	1.8234

3.4 Comparison of Performance

This section presents a comparison of the embedding capacity and multifaceted performance of the VRAE-based RDH-EI methods introduced by Zhang [16], Hong et al. [22], Wu et al. [23], Puteaux et al. [18], Wang et al. [19], Qi et al. [20], and Zhang et al. [21].

Comparison of Embedding Rate

Fig. 8 compares the embedding capacity of the proposed method with those of other VARE algorithms.

Zhang [16] embeds secret information by modifying the three LSBs of half of the pixels in each subblock. The embedding rate of the method mainly depends on the minimum subblock size. With a minimum subblock size of 3×4, the maximum ER is theoretically $1/12$ bpp. In contrast, Hong et al. [22] employ a smaller minimum subblock size of 2×2, achieving a higher maximum ER of $1/4$ bpp. Wu et al. [23] embedded secret information on a per-pixel basis, enabling data embedding in nearly half of the pixels. As a result, the maximum ER is 0.5 bpp. Puteaux et al. [18] used MSB prediction to create additional embedding space and enhance capacity. However, stream encryption disrupts intrablock correlations, thereby limiting the improvement. Wang et al. [19] utilized block-by-block encryption, encrypting all the pixels within a subblock with the same key. This approach significantly improved the embedding capacity. However, the number of embeddable subblocks is limited, leaving room for further improvement. Qi et al. [20] used adaptive difference prediction for nonembeddable subblocks and subblocks with low embedding capacity, which increased pixel utilization. Zhang et al. [21] utilized adaptive MSB prediction and MED prediction for subblocks, which generates two label maps and longer auxiliary information. Our proposed method employs a dual prediction approach for data embedding and uses arithmetic coding to compress the resulting label map. This strategy enhances the embedding capacity, outperforming the methods in the same category.

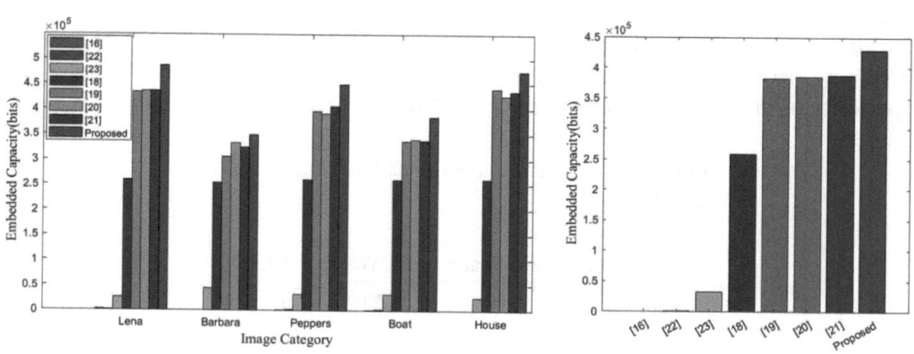

(a) Embedding Capacity comparison with other methods under different images

(b) Average Embedding Capacity comparison with other methods

Fig. 8. Embedding capacity comparison with other VRAE algorithms.

Synthesized Comparison

Table 7 compares our method with other VARE methods in various aspects.

Zhang [16] first encrypts the original image with a stream cipher before being segmented into uniformly sized nonoverlapping subblocks. Within each subblock, pixels are equally partitioned into two sets through pseudorandom partitioning, denoted by $\boldsymbol{A}$ and $\boldsymbol{B}$, on the basis of a data-hiding key. To embed a bit 0, the three LSBs of the pixels in $\boldsymbol{A}$ are flipped. Conversely, for a bit 1, the three LSBs of the pixels in $\boldsymbol{B}$ are flipped. However, a limitation of this method is that the fluctuation function may occasionally fail to identify the original subblock, leading to errors in extraction. Hong et al. [22] enhanced Zhang's [16] method by improving the fluctuation function. Nevertheless, extraction errors still occur when small subblock sizes such as 6×6 and 4×4 are used. While increasing the subblock size guarantees reversibility, this modification inevitably reduces the embedding rate. Wu et al. [23] introduced two methods: separable and nonseparable. First, the pixels are partitioned into qualified and forbidden sets, and by using a data hiding key, a number of pixels are selected from the qualified set and partitioned into equal groups, each consisting of n pixels. To embed a bit 1, the L ($L \in [1, 6]$) LSBs of all the pixels within the target group are inverted. During the image restoration and message extraction phase, the labelled image must first be decrypted via the corresponding decryption key. The pixels are subsequently selected in the same way via the data hiding key. Finally, the embedded bits are extracted from each group through the prediction of all the constituent pixel values. Nevertheless, the pseudorandom selection of qualified sets introduces substantial variability in embedding capacity across different data hiding keys. High embedding rates may lead to extraction errors or incomplete image recovery. Separable methods embed information by modifying the first and second MSBs, as the two MSBs in an image are almost predictable when enough side information is provided. However, such MSB alterations inevitably lead to degraded image reconstruction quality. Puteaux et al. [18] used MSB prediction to release embedding space and achieved true reversibility. Wang et al. [19], Qi et al. [20], and Zhang et al. [21] used chunked encryption, where pixels within a chunk are encrypted via the same key, preserving the intracunk pixel correlation. Qi et al. [20] and Zhang et al. [21] used two methods to predict pixels, further increasing the number of embeddable subblocks. We employ double prediction and arithmetic coding to optimize the embedding space. In the restoration phase, the image receiver identifies all embeddable subblocks via the label map, fully extracts the secret information, and restores the original image by determining the prediction method for each subblock via the label map. The proposed algorithm is both separable and reversible.

Table 7. Comprehensive comparison with other VRAE-class algorithms.

	Encryption Method	Room Reservation Method	Separability	Data Extraction Error	Image Recovery Error
Zhang [16]	Stream Cipher	LSB Replace	NO	YES	YES
Hong et al. [22]	Stream Cipher	LSB Replace	NO	YES	YES
Wu et al. [23]	Stream Cipher	Prediction and LSB Replace	NO	YES	YES
Wu et al. [23]	Stream Cipher	Prediction and MSB Replace	YES	NO	YES
Puteaux et al. [18]	Stream Cipher	MSB Prediction	YES	NO	NO
Wang et al. [19]	Chunked XOR	MSB Prediction	YES	NO	NO
Qi et al. [20]	Chunked XOR	MSB and Difference Prediction	YES	NO	NO
Zhang et al. [21]	Chunked XOR	MSB and MED Prediction	YES	NO	NO
Proposed	**Chunked Modulus**	**Arithmetic Coding and Dual Prediction**	**YES**	**NO**	**NO**

4 Conclusion

To address the limited embedding capacity of current VRAE-type RDH-EI algorithms, an RDH method for encrypted images based on arithmetic coding and dual prediction is proposed. Subblocks are classified into two categories, where adaptive MSB prediction and adaptive difference prediction are applied to free up the embedding space. The resulting label map is then compressed via arithmetic coding to further increase the embedding rate. The experimental results show that the proposed algorithm achieves both lossless image recovery and complete secret information extraction. These two operations are mutually separable, and the algorithm attains a higher embedding rate than the methods in the same category do. In the future, more advanced compression algorithms could be employed to reduce the length of auxiliary information, or new methods may be explored to decrease the number of nonembeddable pixels. Additionally, improved prediction techniques could be developed to increase the number of embeddable bits in embeddable pixels, thereby further enhancing the embedding capacity.

References

1. SaberiKamarposhti, M., Ghorbani, A., Yadollahi, M.: A comprehensive survey on image encryption: taxonomy, challenges, and future directions. Chaos Solitons Fractals **178**, 114361 (2024)
2. Bender, W., Butera, W., Gruhl, D., et al.: Applications for data hiding. IBM Syst. J. **39**(3.4), 547–568 (2000)
3. Evsutin, O., Dzhanashia, K.: Watermarking schemes for digital images: robustness overview. Signal Process. Image Commun. **100**, 116523 (2022)
4. Mandal, P.C., Mukherjee, I., Paul, G., et al.: Digital image steganography: a literature survey. Inf. Sci. **609**, 1451–1488 (2022)
5. Kumar, S., Gupta, A., Walia, G.S.: Reversible data hiding: a contemporary survey of state-of-the-art, opportunities and challenges. Appl. Intell. 1–34 (2022)
6. Wang, Y., Cai, Z., He, W.: High capacity reversible data hiding in encrypted image based on intrablock lossless compression. IEEE Trans. Multimedia **23**, 1466–1473 (2020)
7. Khudhair, S.K., Sahu, M., KR, R., Sahu, A.K.: Secure reversible data hiding using blockwise histogram shifting. Electronics **12**(5), 1222 (2023)
8. Fan, G., Pan, Z., Gao, E., et al.: Reversible data hiding method based on combining IPVO with bias-added rhombus predictor by multipredictor mechanism. Signal Process. **180**, 107888 (2021)
9. Ma, K., Zhang, W., Zhao, X., et al.: Reversible data hiding in encrypted images by reserving room before encryption. IEEE Trans. Inf. Forensics Secur. **8**(3), 553–562 (2013)
10. Zhang, W., Ma, K., Yu, N.: Reversibility improved data hiding in encrypted images. Signal Process. **94**, 118–127 (2014)
11. Yin, Z., Xiang, Y., Zhang, X.: Reversible data hiding in encrypted images based on multi-MSB prediction and Huffman coding. IEEE Trans. Multimedia **22**(4), 874–884 (2019)
12. Weinberger, M.J., Seroussi, G., Sapiro, G.: The LOCO-I lossless image compression algorithm: principles and standardization into JPEG-LS. IEEE Trans. Image Process. **9**(8), 1309–1324 (2000)
13. Yi, S., Zhou, Y.: Separable and reversible data hiding in encrypted images using parametric binary tree labelling. IEEE Trans. Multimedia **21**(1), 51–64 (2018)
14. Wu, Y., Xiang, Y., Guo, Y., et al.: An improved reversible data hiding in encrypted images using parametric binary tree labelling. IEEE Trans. Multimedia **22**(8), 1929–1938 (2019)
15. Ma, W., Wu, Y., Yin, Z.: High-capacity reversible data hiding in encrypted images using adaptive encoding. arXiv preprint arXiv:2102.12620 (2021)
16. Zhang, X.: Reversible data hiding in encrypted image. IEEE Signal Process. Lett. **18**(4), 255–258 (2011)
17. Zhang, X.: Separable reversible data hiding in encrypted image. IEEE Trans. Inf. Forensics Secur. **7**(2), 826–832 (2011)
18. Puteaux, P., Puech, W.: An efficient MSB prediction-based method for high-capacity reversible data hiding in encrypted images. IEEE Trans. Inf. Forensics Secur. **13**(7), 1670–1681 (2018)
19. Wang, Y., He, W.: High capacity reversible data hiding in encrypted image based on adaptive MSB prediction. IEEE Trans. Multimedia **24**, 1288–1298 (2021)
20. Qi, K., Zhang, M., Ke, Y., et al.: High capacity reversible data hiding algorithm in encrypted image based on adaptive MSB and difference prediction. Appl. Res. Comput. **40**(2), 571–576 (2023)
21. Zhang, X., Li, F., Qin, C.: Reversible information hiding algorithm in ciphertext domain with multiple embedding based on block classification. J. Appl. Sci. **41**(2), 272–283 (2023)

22. Hong, W., Chen, T.S., Wu, H.Y.: An improved reversible data hiding in encrypted images using side match. IEEE Signal Process. Lett. **19**(4), 199–202 (2012)
23. Wu, X., Sun, W.: High-capacity reversible data hiding in encrypted images by prediction error. Signal Process. **104**, 387–400 (2014)

A Novel Client Selection Method in Volatile Federated Learning Based on Discounted Exp3

Yang Li[1,2], Huan Li[1,2](✉), Jianming Zhu[1,2], and Youwei Wang[1,2]

[1] School of Information, Central University of Finance and Economics, Beijing 100081, China
turbohuan@163.com

[2] Engineering Research Center of State Financial Security, Ministry of Education, Central University of Finance and Economics, Beijing 102206, China

Abstract. The growing demand for privacy-preserving machine learning has positioned federated learning as a promising research paradigm, enabling the training of high-performance models across distributed data sources without compromising user privacy. However, despite its advantages, federated learning faces critical challenges arising from the heterogeneity and volatility of participating clients. In real-world scenarios, variations in client participation, data volume, computational capability, and communication reliability contribute to a highly dynamic training environment, which negatively impacts efficiency and convergence of the model. To address these challenges, this paper proposes a novel client selection method named CDE3. First, CDE3 employs a multidimensional model to comprehensively evaluate clients' contributions. Second, we enhance the classical Exp3 algorithm by incorporating a discount factor that exponentially decays historical contributions, thereby increasing the influence of recent client behavior in the selection process. Furthermore, we provide a theoretical analysis demonstrating a favorable regret bound for the proposed method. Extensive experiments conducted in volatile FL settings validate the effectiveness of CDE3, showing improved convergence speed and model accuracy compared with those of the baseline algorithms. These results confirm that CDE3 effectively mitigates volatility, enhancing the stability and efficiency of federated learning.

Keywords: Discounted Multiarmed Bandit · Client Selection · Federated Learning

1 Introduction

While the explosive growth of distributed massive data across edge networks has unlocked transformative potential for AI-driven applications, it also introduces some challenges such as data privacy [1]. In 2016, Google proposed federated learning (FL) for distributed machine learning [2]. In this framework, a central server selects a subset of clients and distributes a global model to them. These clients then perform local training using their private data and upload the updated model parameters to the server. The server subsequently performs weighted aggregation to update the global model, thus iterating until the model converges.

Y. Yu et al. (Eds.): ICPCSEE 2025, CCIS 2673, pp. 587–597, 2026.
https://doi.org/10.1007/978-981-95-2566-9_39

In the client selection phase, only a subset of clients can be chosen for training due to communication and computational limitations. Given the significant heterogeneity among clients in data and computational resources, as well as the temporal instability of individual client performance in volatile environments, a naive approach is to select clients that exhibit high efficiency and stability. Nonetheless, repeatedly selecting the same clients may lead to biased training and compromise fairness. Therefore, designing a reliable and adaptive client selection strategy under dynamic conditions remains a critical research challenge.

2 Related Works

McMahan et al. first proposed the classical FedAvg framework [2], which laid the foundation for FL and remains one of the benchmark methods. FedAvg reduces communication costs by allowing clients to perform multiple local updates before aggregation, but it assumes stable client participation and may suffer from degraded performance under system and statistical heterogeneity. Another foundational work is FedProx [3], which extends FedAvg by introducing a proximal term to handle system heterogeneity and partial client updates. FedProx improves convergence stability in settings where clients perform varying amounts of local computation.

Building upon these works, numerous studies have further investigated the challenges posed by client heterogeneity in FL. Client selection, first introduced in [4], is an effective strategy for mitigating heterogeneity. Existing approaches to client selection generally fall into two categories. The first prioritizes high-utility clients on the basis of data and resource evaluations. For instance, the authors in [5] introduce a reputation-based selection strategy, where reputation is evaluated on the basis of factors such as data transmission quality and effective participation, incorporating contract theory to incentivize high-reputation clients to engage in training. Additionally, the study in [6] considers data volume, quality, and budget constraints, encoding the selection strategy as a neural network and employing reinforcement learning for optimization. These methods may compromise fairness, as clients with limited resources are often under-selected [7–11]. The second category adopts an exploration-exploitation strategy, an adversarial MAB algorithm is employed in [12] to maximize long-term effective participation, where selection probabilities are determined on the basis of historical performance, and a minimum probability threshold is introduced to ensure fairness. The work in [13] incorporates the Jain fairness index and uses a discounted MAB to apply weighted adjustments to historical loss values.

Most existing studies assume a static environment, which rarely reflects real-world client dynamics. To address this limitation, we propose a client selection strategy tailored to volatile environments. Firstly, we construct a multidimensional evaluation model to quantify client contributions. Then, we enhance the Exp3 algorithm by incorporating a discount factor that exponentially weights historical contributions. On the basis of the selection probabilities generated by the algorithm, the server dynamically selects clients to participate in training. We further validate the effectiveness of the proposed method through both theoretical regret-bound analysis and empirical evaluation.

3 Client Selection Modelling in Volatile FL

3.1 Multidimensional Contribution Evaluation Modelling

In this work, we extend the definition by incorporating communication volatility, thereby characterizing client volatility across four primary dimensions: **Set Volatility, Statistical Volatility**, **Training Volatility and Communication Volatility**.

To formally model the volatile environment, let $\Gamma = \{1,2,3\ldots T\}$ denote the set of training rounds, and $M(t) = \{1,2,3\ldots K_t\}$ represents the set of clients willing to participate in round t, where $|M(t)| = K_t$. The selected clients form the set $A(t) = \{1,2,3\ldots k\}$, where k is the number of clients selected. The corresponding data samples are represented as $D(t) = \{D_{1,t}, D_{2,t}, D_{3,t}\ldots D_{K_t,t}\}$.

The paper [14] suggested that future research should adopt a multidimensional approach to better regulate client participation. In response, we consider both data utility and resource utility, and propose the following evaluation indicators:

Data Volume. A larger data volume provides more sample information, thereby contributing positively to client evaluation. The data volume of client i in the t-th round is denoted as $d_{i,t} = |D_{i,t}|$.

Training Accuracy. Clients with higher training accuracy contribute more effectively to global model aggregation and are thus positively weighted in client evaluation. The training accuracy of client i in the t-th round is denoted as $a_{i,t}$.

Training Duration.
Since global training is bounded by the slowest client, longer durations delay model updates and reduce client contribution. Let $s_{i,t}$ denote the training duration of client i in round t.

To quantify the client contribution, we normalize the data volume $d_{i,t}$, training accuracy $a_{i,t}$, and training duration $s_{i,t}$. And we compute the contribution score for client i in the t-th training round as:

$$c_{i,t} = \frac{d_{i,t}^{norm}}{a_{i,t}^{norm} \cdot s_{i,t}^{norm}} \tag{1}$$

Additionally, $x_{i,t}$ is introduced to indicate whether client i successfully completes the training process. If training is successful ($x_{i,t} = 1$), the client's contribution is computed using Eq. (1), otherwise ($x_{i,t} = 0$), the contribution is set to 0.

3.2 Discounted MAB and Problem Transformation

To address the challenges posed by volatile environments, clients with stable performance need to be identified in each training round, represented by the set $A(t)$. A discount factor γ is assigned on the basis of each client's historical contribution, aiming to maximize the cumulative contribution scores across all rounds.

$$\sum_{i\in A(t)} \gamma^0\hat{c}_{i,T} + \gamma^1\hat{c}_{i,T-1} + \gamma^2\hat{c}_{i,T-2} + \cdots + \gamma^{T-2}\hat{c}_{T,2} + \gamma^{T-1}\hat{c}_{T,1} = \sum_{i\in A(t)}\sum_{t=1}^{T}\gamma^{T-t}\hat{c}_{i,t} \tag{2}$$

In volatile FL, recent client performance should be assigned greater weight, as it more accurately reflects current contributions. To achieve this, we introduce an exponentially weighted discount factor that emphasizes recent rewards while attenuating the influence of older rewards. The discount factor γ satisfies $\gamma \in (0,1)$. When γ is closer to 1, it results in more uniform weighting of historical contributions, whereas a value closer to 0 leads to greater differentiation between recent and earlier contributions.

On the basis of these assumptions and the theory of the MAB algorithm, we formulate the client selection model as follows:

$$max \sum_{t=1}^{T} \sum_{i=1}^{K} p_{i,t}\gamma^{T-t}\hat{c}_{i,t}$$
$$s.t. \sum_{\mathrm{i}=1}^{\mathrm{K}} p_{i,t} = k, \forall t \in \Gamma \tag{3}$$

Therefore, maximizing the expected contributions relies on optimizing the selection probabilities $p_{i,t}$.

4 Our Approach CDE3

4.1 CDE3

To address the online scheduling problem, we incorporate the discount factor into the Exp3 algorithm and utilize the enhanced version to assign selection probabilities on the basis of discounted historical contributions. Ultimately, we propose a method named Contribution Discounted Exp3 (CDE3), which is designed to adapt to volatile environments.

As in the canonical Exp3 framework, we employ the same unbiased estimator for $\hat{c}_{i,t}$:

$$\hat{c}_{i,t} = \frac{\mathbb{I}_{\{i \in A(t)\}}}{p_{i,t}} c_{i,t} \tag{4}$$

According to Eq. (2), the weight update formula should be rewritten as follows:

$$w_{i,t+1} = w_{i,t}^{\gamma} \exp\left(\eta\theta\hat{c}_{i,t}\right) \tag{5}$$

To prevent $\hat{c}_{i,t}$ from becoming excessively large, we introduce a reduction coefficient θ, which serves to scale down the unbiased estimate $\hat{c}_{i,t}$.

In our volatile FL setting, k clients must be selected per round. Therefore, the algorithm must be modified to ensure that the total selection probability across all arms sums to k. To achieve this, the probability assignment formula is adjusted as follows:

$$p_{i,t} = \frac{k w_{i,t}}{\sum_{j \in M(t)} w_{j,t}} \tag{6}$$

To clarify our proposed client selection method, we outline the detailed procedure of CDE3 for FL in Algorithm 1. Compared with existing methods, CDE3 explicitly introduces the discount factor γ.

Algorithm 1. Contribution-based Discounted Exp3 (CDE3)

Input: number of training rounds T
total number of clients K_t
number of clients to be selected per round k
learning rate η
discount factor γ
reduction coefficient θ
client dataset $D_{i,t}$

Output: global model Θ_T

initialize $w_{i,1} = 1,\ i = 1,2,\dots,K_t$
for $t = 1,2,\dots,T$ **do**
  $p_{i,t} = \frac{kw_{i,t}}{\sum_{j\in M(t)} w_{j,t}}$
  sample the client sets $A(t)$ based on $p_{i,t}$
  for $i \in A(t)$ **do**
    $\Theta_{i,t} = o_1(\Theta_t, D_{i,t})$
  end for
  $\Theta_{t+1} = \sum_{i=1}^{K_t} \frac{|D_{i,t}|}{\sum_{j=1}^{K_t} |D_{j,t}|} \Theta_{i,t}$
  $\hat{c}_{i,t} = \frac{\mathbb{I}_{\{i\in A(t)\}}}{p_{i,t}} c_{i,t}$
  $w_{i,t+1} = w_{i,t}^{\gamma} \exp(\eta\theta\hat{c}_{i,t})$
end for

4.2 Proof of the Regret Bound

To evaluate the effectiveness of the proposed algorithm, CDE3 is theoretically analyzed via a regret-bound framework. The regret bound is commonly employed to assess the algorithm's performance, which can represent the difference between the cumulative reward achieved by the algorithm and that of the optimal choice.

Definition 1 (Regret of CDE3). The CDE3 algorithm regret boundary is defined as follows:

$$R_T = E_T^* - E_T^{CDE3} \tag{7}$$

$$E_T^* = \sum_{t=1}^{T} \sum_{i\in M(t)} \gamma^{T-t} q_{i,t}^* \hat{c}_{i,t} \tag{8}$$

$$E_T^{CDE3} = \sum_{t=1}^{T} \sum_{i\in M(t)} \gamma^{T-t} p_{i,t} \hat{c}_{i,t} \tag{9}$$

where E_T^* denotes the objective expectation obtained under the optimal probability distribution (i.e., the client is assigned the optimal probability under random selection), and where E_T^{CDE3} denotes the objective expectation after CDE3 assigns probabilities.

After introducing the discount factor, it is necessary to reconstruct the regret bound proof. The derivation proceeds as follows:

Let $W_t = \sum_{i \in M(t)} w_{i,t}$ and we first derive the upper bound $\frac{W_{t+1}}{W_t}$. According to Eq. (5), we have:

$$\frac{W_{t+1}}{W_t} = \sum_{i \in M(t)} \frac{w_{i,t}^{\gamma} \exp(\eta\theta\hat{c}_{i,t})}{W_t} \tag{10}$$

While $x \le 1$, $exp(x) \le 1 + x + x^2$, yields:

$$\frac{W_{t+1}}{W_t} \le \sum_{i \in M(t)} \frac{w_{i,t}^{\gamma}\left(1 + \eta\theta\hat{c}_{i,t} + \eta^2\theta^2\hat{c}_{i,t}^2\right)}{W_t} \tag{11}$$

While $w_{i,t} > 1$, and $\gamma \in (0, 1)$, $w_{i,t}^{\gamma} < w_{i,t}$, it yields:

$$\frac{W_{t+1}}{W_t} \le \gamma^{t-T}\left[1 + \sum_{i \in M(t)} \gamma^{T-t} p_{i,t}\left(\eta\theta\hat{c}_{i,t} + \eta^2\theta^2\hat{c}_{i,t}^2\right)\right] \tag{12}$$

$x \in \mathbb{R}$, $1 + x \le \exp(x)$, it yields:

$$\frac{W_{t+1}}{W_t} \le \gamma^{t-T} \cdot \exp\left(\eta\theta \sum_{i \in M(t)} \gamma^{T-t} p_{i,t}\hat{c}_{i,t} + \eta^2\theta^2 \sum_{i \in M(t)} \gamma^{T-t} p_{i,t}\hat{c}_{i,t}^2\right) \tag{13}$$

For $\ln\frac{W_{t+1}}{W_1}$ we have:

$$\begin{aligned} \ln\frac{W_{T+1}}{W_1} &= \sum_{t=1}^{T} \ln\frac{W_{T+1}}{W_T} \\ &\le \eta\theta \sum_{t=1}^{T} \sum_{i \in M(t)} \gamma^{T-t} p_{i,t}\hat{c}_{i,t} + \eta^2\theta^2 \sum_{t=1}^{T} \sum_{i \in M(t)} \gamma^{T-t} p_{i,t}\hat{c}_{i,t}^2 + \sum_{t=1}^{T} \ln\gamma^{t-T} \end{aligned} \tag{14}$$

Taking the expectation of both sides, yields:

$$\mathbb{E}\left[\ln\frac{W_{T+1}}{W_1}\right] \le \eta\theta \sum_{t=1}^{T} \sum_{i \in M(t)} \gamma^{T-t} p_{i,t}\hat{c}_{i,t} + \eta^2\theta^2 \sum_{t=1}^{T} \mathbb{E}\left[\sum_{i \in M(t)} \gamma^{T-t} p_{i,t}\hat{c}_{i,t}^2\right] + \sum_{t=1}^{T} \ln\gamma^{t-T} \tag{15}$$

Furthermore, $\gamma \in (0, 1)$, and letting K denote the maximum value of K_t, we have:

$$\mathbb{E}\left[\sum_{i\in M(t)}\gamma^{T-t}p_{i,t}\hat{c}_{i,t}^{2}\right]\le\mathbb{E}\left[\sum_{i\in M(t)}p_{i,t}\left(\frac{\mathbb{I}_{\{i\in A_t\}}}{p_{i,t}}c_{i,t}\right)^{2}\right]$$
$$=\mathbb{E}\left[\sum_{i\in M(t)}\frac{\mathbb{I}_{\{i\in A_t\}}}{p_{i,t}}c_{i,t}^{2}\right]\le\sum_{i\in M(t)}c_{i,t}^{2}\le K \tag{16}$$

Combining Eqs. (15) and (16), we know that:

$$\mathbb{E}\left[\ln\frac{W_{T+1}}{W_1}\right]\le\eta\theta\sum_{t=1}^{T}\sum_{i\in M(t)}\gamma^{T-t}p_{i,t}\hat{c}_{i,t}+\eta^{2}\theta^{2}\mathrm{TK}+\sum_{t=1}^{T}\ln\gamma^{t-T} \tag{17}$$

We then prove the lower bound of $\ln\frac{W_{T+1}}{W_1}$; according to the weighted AM-GM inequality, we have:

$$\ln\frac{W_{T+1}}{W_1}=\ln\frac{\sum_{i\in M(t)}w_{i,T+1}}{K}\ge\ln\frac{\sum_{i\in M(t)}q_{i,t}^{*}w_{i,T+1}}{K}\ge\sum_{i\in M(t)}q_{i,t}^{*}\ln w_{i,T+1}-\ln K$$
$$\ge\eta\theta\sum_{t=1}^{T}\sum_{i\in M(t)}\gamma^{T-t}q_{i,t}^{*}\hat{c}_{i,t}-\ln K \tag{18}$$

Taking the expectation of both sides, yields:

$$\mathbb{E}\left[\ln\frac{W_{T+1}}{W_1}\right]\ge\eta\theta\sum_{t=1}^{T}\sum_{i\in M(t)}\gamma^{T-t}q_{i,t}^{*}\hat{c}_{i,t}-\ln K \tag{19}$$

Combining Eqs. (17) and (19), we derive that:

$$\eta\theta\sum_{t=1}^{T}\sum_{i\in M(t)}\gamma^{T-t}q_{i,t}^{*}\hat{c}_{i,t}-\ln K\le\eta\theta\sum_{t=1}^{T}\sum_{i\in M(t)}\gamma^{T-t}p_{i,t}\hat{c}_{i,t}+\eta^{2}\theta^{2}TK+\sum_{t=1}^{T}\ln\gamma^{t-T}$$
$$\sum_{t=1}^{T}\sum_{i\in M(t)}\gamma^{T-t}q_{i,t}^{*}\hat{c}_{i,t}-\sum_{t=1}^{T}\sum_{i\in M(t)}\gamma^{T-t}p_{i,t}\hat{c}_{i,t}\le\frac{\ln K+\sum_{t=1}^{T}\ln\gamma^{t-T}}{\eta\theta}+\eta\theta TK \tag{20}$$

Letting $\eta=\sqrt{\frac{2\ln K+(T-T^{2})\ln\gamma}{2\theta^{2}TK}}$, we have:

$$R_T=E_T^{*}-E_T^{CDE3}\le\frac{\ln K+\sum_{t=1}^{T}\ln\gamma^{t-T}}{\eta\theta}+\eta\theta\mathrm{TK}=\sqrt{TK\ln K+\frac{(1-T)T^{2}K\ln\gamma}{2}} \tag{21}$$

This completes the proof.

5 Simulation Results and Discussion

In this section, we present the experimental results of the proposed client selection method. The evaluations are conducted within a simulated FL environment using non-IID MNIST and Fashion-MNIST datasets.

In our simulation, we initialize with $K_0 = 100$ clients and select $k = 20$ for training in each round. To simulate the statistical heterogeneity and non-IID conditions, client data sizes follow a normalized distribution, with clients grouped into A, B, and C, drawing samples from $\mathcal{Z}(300,10)$, $\mathcal{Z}(500,20)$, and $\mathcal{Z}(800,30)$. The total data volume D is allocated using a parameter α, where each client receives $\alpha \bullet D$ samples from a primary label and $(1 - \alpha) \bullet D$ samples uniformly from other labels to construct local datasets.

5.1 Comparison Experiments

We compare our method with three baselines: Random [2] selects clients uniformly at random; K-center [15] selects clients closest to cluster centers to ensure diversity; CE3 applies the classical Exp3 algorithm without a discount factor to update selection probabilities based on rewards.

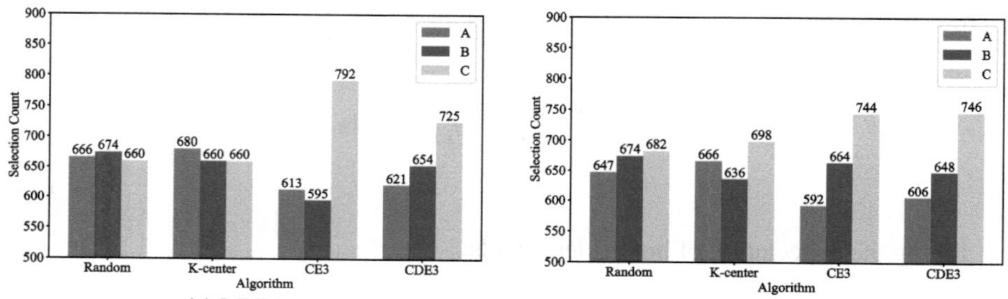

(a) MNIST (b) Fashion-MNIST

Fig. 1. Selection counts across three client categories (A, B, and C) for MNIST (a) and Fashion-MNIST (b).

As illustrated in Fig. 1, the comparison reveals the client selection tendencies of various algorithms across three client categories (A, B, and C). Both CDE3 and CE3, which rely on historical contribution-based strategies, clearly prefer clients with larger local datasets. This tendency is advantageous, as clients with more data not only facilitate faster convergence but also help mitigate local overfitting, thereby enhancing global model performance.

However, further analysis shows that CE3 exhibits strong selection bias, frequently prioritizing C-type clients with the largest datasets. This imbalance may undermine the generalization ability of the global model because of the underrepresentation of smaller client groups. In contrast, CDE3 effectively alleviates this bias by integrating historical contributions with discounting, resulting in a more balanced trade-off between efficiency and fairness.

Moreover, the random and k-center selection methods demonstrate limited responsiveness to the client dataset size, as neither explicitly incorporates the data volume into their selection criteria. This insensitivity may lead to suboptimal utilization of heterogeneous data resources in federated settings.

5.2 Relationship Between the Discount Factor and Model Accuracy

Figure 2 shows the relationship between the discount factor γ and model accuracy. The results indicate that appropriately tuning γ accelerates convergence and enhances robustness against environmental fluctuations, enabling more stable training.

In Fig. 2(a), setting $\gamma = 0.7$ yields the fastest and smoothest convergence on MNIST, with a final accuracy of 90.73%. In contrast, $\gamma = 0.4$ leads to slower convergence and greater instability. Similar trends appear in Fig. 2(b) for the Fashion-MNIST dataset, confirming the effectiveness of moderate discounting in volatile FL.

To further examine this relationship in depth, we note that the choice of the discount factor γ is intuitively linked to the level of volatility in the FL environment. A smaller γ (closer to 0) places more emphasis on recent client performance, which is suitable for highly volatile settings where past behavior quickly becomes irrelevant. Conversely, a larger γ (closer to 1) retains more of the historical contribution, making it more appropriate for relatively stable environments. In our experiments, $\gamma = 0.7$ to $\gamma = 0.9$ demonstrated superior convergence performance, indicating a moderate level of volatility in the simulated environment.

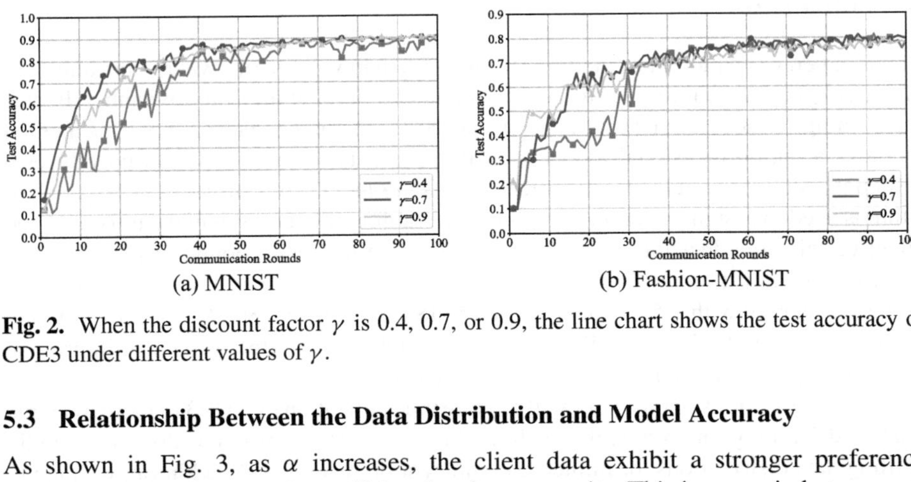

(a) MNIST (b) Fashion-MNIST

Fig. 2. When the discount factor γ is 0.4, 0.7, or 0.9, the line chart shows the test accuracy of CDE3 under different values of γ.

5.3 Relationship Between the Data Distribution and Model Accuracy

As shown in Fig. 3, as α increases, the client data exhibit a stronger preference toward local distributions, intensifying data heterogeneity. This increase in heterogeneity adversely affects convergence efficiency, resulting in a gradual decline in model accuracy.

When $\alpha = 0.5$, the model achieves optimal performance, reaching high accuracy with fewer communication rounds. However, at $\alpha = 0.9$, the performance degrades substantially, with the model converging to approximately 65% accuracy on MNIST and 70% accuracy on Fashion-MNIST. These results highlight the sensitivity of FL to non-IID distributions and the importance of controlling α for stable training.

We conducted experiments to analyze the effects of key parameters including the discount factor γ, reduction coefficient θ, number of selected clients k, and data heterogeneity level α. The results show that proper parameter tuning can significantly improve convergence efficiency and model performance. Although the learning rate η also influences algorithm behavior, we fixed its value on the basis of standard practice in prior studies. Many researchers have investigated how to set the learning rate, and we also consider learning rate selection an important tunable factor, which we are currently exploring further.

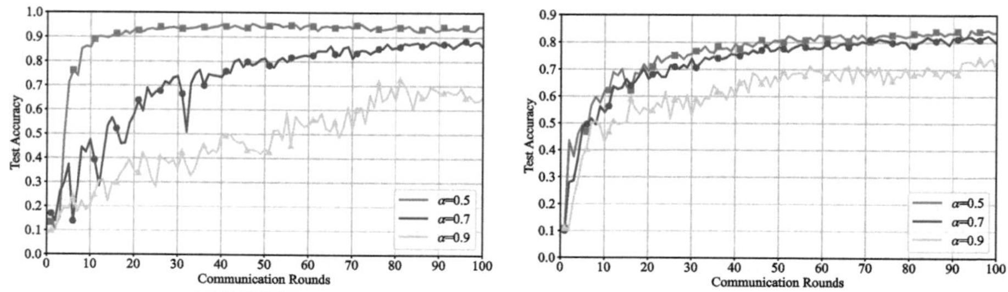

(a) MNIST (b) Fashion-MNIST

Fig. 3. When the parameter α that controls the degree of non-IID distribution takes values of 0.5, 0.7, and 0.9, the line chart shows test accuracy of CDE3 under different values of α.

6 Conclusion

In this paper, we explore the client selection problem in volatile FL, where the number of clients, client data, training status, and communication reliability vary dynamically during training. To address this challenge, we propose a method named CDE3. This method employs a multidimensional model to evaluate clients' contribution and introduces a discount factor to weight the historical contributions. Then a discounted Exp3 algorithm is used to assign selection probabilities, enabling the server to make informed and adaptive selection decisions. Additionally, we prove the regret bound of the enhanced algorithm with a discount factor. The experimental results show that CDE3 achieves 95.42% accuracy within 52 rounds on the MNIST dataset, and 80.89% accuracy within 71 rounds on the Fashion-MNIST dataset, highlighting its effectiveness in improving both convergence efficiency and selection robustness.

Future extensions of this work will explore the application of CDE3 in more dynamic federated learning environments, such as those with frequent client dropout, rapidly shifting data distributions, or limited communication resources. In addition, we plan to evaluate CDE3 on more complex tasks to further assess its scalability and adaptability in practical settings.

Acknowledgement. This research is funded by the Central University of Finance and Economics, Greater Bay Area Research Institute Project (No. YJY202303), the National Natural Science Foundation of China (No. 61906220) and the Ministry of Education of Humanities and Social Science Project (No. 19YJCZH178).

References

1. Li, T., Sahu, A.K., Talwalkar, A., Smith, V.: Federated learning: challenges, methods, and future directions. IEEE Signal Process. Mag. **37**(3), 50–60 (2020)

2. Mcmahan H.B.F., Moore, E.S., Ramage, D.T.: Communication-efficient learning of deep networks from decentralized data. In: the 20th International Conference on Artificial Intelligence and Statistics, pp. 1273–1282. PMLR, Fort Lauderdale (2017)
3. Li, T., Sahu, A.K., Zaheer, M., Sanjabi, M., Talwalkar, A., Smith, V.: Federated optimization in heterogeneous networks. In: The 2020 Machine Learning and Systems, pp. 429–450. mlsys.org, Austin (2020)
4. Nishio, T., Yonetani, R.: Client selection for federated learning with heterogeneous resources in mobile edge. In: 2019 IEEE International Conference on Communications (ICC), pp. 1–7. IEEE, Shanghai (2019)
5. Kang, J., Xiong, Z., Niyato, D., Xie, S., Zhang, J.: Incentive mechanism for reliable federated learning: a joint optimization approach to combining reputation and contract theory. IEEE Internet Things J. **6**(6), 10700–10714 (2019)
6. Deng, Y., Lyu, F., Ren, J., Wu, H., Zhou, Y., Zhang, Y., et al.: AUCTION: automated and quality-aware client selection framework for efficient federated learning. IEEE Trans. Parallel Distrib. Syst. **33**(08), 1996–2009 (2022)
7. Li, T., Sanjabi, M., Beirami, A., Smith, V.: Fair resource allocation in federated learning. arXiv preprint arXiv.1905.10497 (2020)
8. Zhou, Z., Chu, L., Liu, C., Wang, L., Pei, J., Zhang. Y.: Towards fair federated learning. In: Proceedings of the 27th ACM SIGKDD Conference on Knowledge Discovery & Data Mining, pp. 4100–4101. ACM, Virtual Event Singapore (2021)
9. Li, T., Hu, S., Beirami, A., Smith, V.: Ditto: fair and robust federated learning through personalization. In: Proceedings of the 38th International Conference on Machine Learning, pp. 6357–6368. PMLR (2021)
10. Mohri, M., Sivek, G., Suresh, A.T.: Agnostic federated learning. In: Proceedings of the 36th International Conference on Machine Learning, pp. 4615–4625. PMLR (2019)
11. Cho, Y.J., Wang, J., Joshi, G.: Client selection in federated learning: convergence analysis and power-of-choice selection strategies. arXiv preprint arXiv.2010.01243 (2020)
12. Huang, T., Lin, W., Shen, L., Li, K., Zomaya, A.Y.: Stochastic client selection for federated learning with volatile clients. IEEE Internet Things J **9**(20), 20055–20070 (2022)
13. Cho, Y.J., Gupta, S., Joshi, G., Yagan, O.: Bandit-based communication-efficient client selection strategies for federated learning. In: The 54th Asilomar Conference on Signals, Systems, and Computers, pp. 1066–1069. IEEE, Pacific Grove (2020)
14. Gu, T., Li, L., Chang, L., Li, J.: Fair federated machine learning and its design: a comprehensive survey (in Chinese). Chin. J. Comput. **46**(9), 1991–2024 (2023)
15. Sener, O., Savarese, S.: Active learning for convolutional neural networks: a core-set approach. arXiv preprint arXiv.1708.00489 (2018)

Data-driven Scientific Research

Research on an Underwater Trash Detection Method Based on Improved YOLOv8

Yu Meng, Jingsheng Zhao, Runzhi Xie(✉), and Jize Liu

Harbin Engineering University, Harbin 150001, China
hrbeum@126.com

Abstract. With the increasing severity of global water pollution, underwater debris has become a serious threat to marine ecosystems. Autonomous underwater robots equipped with efficient debris detection systems can significantly increase underwater cleanup efforts. This paper proposes a high-precision and lightweight underwater debris detection method. To address the challenges posed by the diverse shapes and varying scales of underwater debris, we introduce an improved detection approach. Using YOLOv8 as the backbone, we replace conventional convolution with Alterable Kernel Convolution (AKConv), which dynamically adjusts the convolutional kernel shape on the basis of input features, enabling more accurate capture of object details. In the neck network, an improved high-level screening-feature fusion pyramid network (HSFPN) replaces the original YOLOv8 neck, incorporating coordinate attention instead of channel attention to enhance multiscale feature fusion. The experimental results demonstrate that the proposed AH-YOLO method achieves significant improvements in mAP50 and mAP50:95 on the TrashCan dataset compared with the original YOLOv8.

Keywords: Underwater Trash Detection · YOLOv8 · Object recognition

1 Introduction

The ocean is the cradle of life, covering nearly three-quarters of the Earth's surface. It is also a "treasure trove" of resources, playing a crucial role in human survival and development. However, with the continuous exploitation and utilization of marine resources, the ocean environment is facing unprecedented pollution and degradation. In particular, over the past few decades, the issue of marine debris has become increasingly severe, emerging as a pressing global environmental challenge that demands urgent solutions.

With the rapid advancements in robotics and artificial intelligence, the use of underwater intelligent robots for seabed debris cleanup has become a viable solution. Autonomous underwater vehicles (AUVs), as state-of-the-art underwater operational equipment, possess the ability to perform long-term underwater tasks with high flexibility and manoeuvrability. They can operate in complex and harsh underwater environments, replacing humans in various missions and freeing them from hazardous deep-sea operations.

Y. Yu et al. (Eds.): ICPCSEE 2025, CCIS 2673, pp. 601–617, 2026.
https://doi.org/10.1007/978-981-95-2566-9_40

An efficient underwater debris detector is fundamental for enabling AUVs to autonomously clean seabed waste. However, underwater debris detection faces several challenges: insufficient illumination and murky water make it difficult to distinguish debris from the background; prolonged exposure to seawater causes varying degrees of texture degradation on debris surfaces over time; and plastic waste, which accounts for up to 80% of underwater debris, results in highly variable shapes and sizes due to the influence of buoyancy and ocean currents. These challenges impose greater demands on feature extraction capabilities in underwater debris detection technology.

Marine debris poses a severe threat to marine ecosystems and human health. In recent years, researchers have conducted extensive studies in the field of marine debris detection and have made significant progress. Valdenegro-Toro [1] was the first to propose a detection method based on convolutional neural networks (CNNs), achieving approximately 80% accuracy via active sonar images. However, the reliance on synthetically generated data limits its generalizability in natural environments. To compensate for the lack of real-world data, Hong et al. [2] introduced TrashCan, a valuable dataset designed for training and evaluating underwater debris detection algorithms. This dataset provides a standardized benchmark for research in this field. Additionally, Huang et al. [3] designed DSDebrisNet, which leverages a lightweight network structure and a hybrid loss function to achieve breakthroughs in real-time detection and adaptability to complex environments. Watanabe et al. [4] systematically evaluated the application of deep learning-based object detection algorithms in underwater ecosystem monitoring and marine debris recognition, confirming their effectiveness. However, these methods still require optimization because of their heavy dependence on high-quality annotated data. Zocco et al. [5] proposed an autonomous underwater vehicle (AUV) system based on an improved EfficientDet model, which achieved a 1.2%–2.6% AP improvement across D0--D3 models by optimizing object detection algorithms. Additionally, they developed an underwater plastic debris dataset and explored image enhancement strategies in low-light environments to improve real-time detection performance, providing technological support for efficient marine debris detection and cleanup.

At the technical optimization level, Ali et al. [6] introduced a channel stabilization technique that effectively mitigated color distortion and haze interference in underwater images, significantly enhancing multiscale object detection performance in datasets. Ashwani Kumar et al. [7] developed a nine-category real-time detection framework using CNNs, enabling multiscene deployment and proposing an object detection technology that can operate in any environment and on any device. Notably, Shi et al. [8] proposed an improved Faster R-CNN architecture that incorporates a bidirectional feature pyramid network (FPN), achieving breakthroughs in feature extraction and multiscale feature fusion. This approach increased the underwater debris detection accuracy to 88.94%, but its processing speed of 4.3 frames per second (FPS) remains a limitation for real-time detection. Wei et al. [9] introduced a U-Net variant that employs a deeper contraction and expansion path to achieve end-to-end semantic segmentation. While this method demonstrated improvements in segmentation efficiency, the study was constrained by a small dataset consisting of only 410 augmented training images and approximately 50 test images. The model reached saturation after 30 training iterations, highlighting the risk of overfitting due to the limited dataset size.

Compared with other studies on underwater garbage detection, this paper focuses on the characteristics of underwater garbage with diverse shapes and varying sizes and proposes an improved YOLOv8-based underwater garbage detection algorithm. To address the challenges commonly encountered in complex underwater environments, including diverse object shapes, significant scale variations, and severe background interference, we propose the AH-YOLO model, which is based on AKConv [10] and CAHSFPN [11]. The AKConv module dynamically adjusts the convolution kernel shape, significantly enhancing the model's feature extraction capability for irregular objects and allowing it to better adapt to the diverse forms of underwater garbage. Moreover, the CAHSFPN module improves the model's feature selection ability by optimizing the selective feature fusion module, effectively filtering out low-scale features and enhancing the detection accuracy of multiscale objects in complex underwater scenarios.

Through a comparative analysis of the experimental results, it is demonstrated that the proposed method is more suitable for underwater garbage detection tasks than other existing methods.

2 Object Detection Algorithm

Object detection is one of the core tasks in the field of computer vision and aims to locate and identify objects of interest within an image. With the advancement of deep learning technology, object detection algorithms have achieved significant improvements in both accuracy and efficiency. On the basis of different detection processes, object detection algorithms can be broadly categorized into two-stage detection algorithms and single-stage detection algorithms.

The core idea of two-stage object detection algorithms is to divide the detection task into two steps: first, generate candidate regions, and then perform classification and bounding box regression on these regions. The advantage of two-stage object detection methods lies in their high detection accuracy, especially in complex scenes and small object detection. However, despite their high precision, these methods also have high computational complexity, making them less suitable for real-time applications such as autonomous driving. A representative example of this category is the R-CNN series.

Single-stage object detection algorithms adopt an end-to-end network architecture that integrates object localization and classification into a single step, significantly enhancing detection efficiency. Unlike two-stage detection methods, which involve generating candidate regions and performing secondary feature processing, single-stage algorithms directly regress object positions and categories on the feature map, effectively reducing computational redundancy. This study builds upon YOLOv8, a representative single-stage object detection algorithm, and introduces improvements to enhance its performance. The overall network architecture of YOLOv8 is illustrated in Fig. 1.

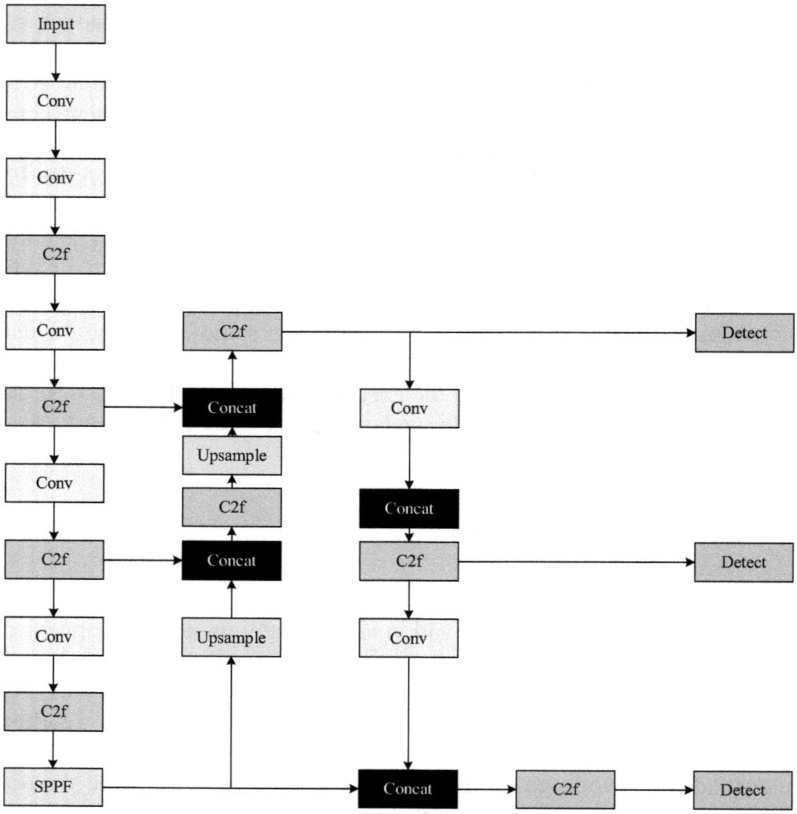

Fig. 1. YOLOv8 Network Architecture Diagram

In the backbone network, YOLOv8 adopts an enhanced CSPDarknet53 architecture to improve feature extraction efficiency and optimize computational resource utilization. This architecture builds upon the traditional CSPDarknet53 by deeply integrating residual modules with cross-stage partial connections, enabling hierarchical feature reuse. This design effectively reduces computational redundancy while improving gradient propagation efficiency. Additionally, YOLOv8 introduces the C2f module in its backbone network, replacing the traditional C3 module. The C2f module combines the bottleneck and CSP structures, significantly enhancing the feature representation capability. Compared with the C3 module, the C2f module focuses more on cross-layer information fusion and optimizes the processing of input features.

To further enhance multiscale object detection capabilities, YOLOv8 incorporates the spatial pyramid pooling fusion (SPPF) module. This key component performs parallel pooling operations at multiple scales, enabling the cascade fusion of feature maps extracted from different receptive fields. This approach significantly improves the model's adaptability to objects of varying sizes. By extracting features through different scales of pooling operations and fusing these feature maps, the SPPF module ultimately provides richer and more multidimensional information. This design offers strong support for the application of YOLOv8 in complex environments.

In the feature fusion layer, YOLOv8 optimizes the feature extraction and fusion process. Specifically, YOLOv8 replaces the C3 module and RepBlock structure used in previous YOLO versions with the newly introduced C2f module. By enhancing cross-layer connections and feature segmentation strategies, the C2f module effectively facilitates interactions between multiscale features. Furthermore, YOLOv8 simplifies the traditional upsampling process. Unlike previous approaches, YOLOv8 directly increases the resolution of the native features output by the backbone network, eliminating the need for the conventional 1×1 convolutional channel compression step. This design not only reduces redundant computational overhead but also maximally preserves the spatial integrity of the original features. Consequently, it provides richer and higher-quality low-level data support for cross-layer fusion in the subsequent feature pyramid.

In the detection head, YOLOv8 adopts a decoupled dynamic detection architecture, significantly enhancing both training efficiency and detection accuracy. First, YOLOv8 separates the classification and bounding box regression branches into independent convolutional processing streams, effectively mitigating feature interference in multitask learning. Second, it fully adopts an anchor-free detection paradigm, which directly predicts the target's center coordinates and width–height offsets. This approach avoids the hyperparameter sensitivity issue associated with traditional anchor-based mechanisms, thereby simplifying the model structure considerably. With respect to training strategies, YOLOv8 addresses the low learning efficiency of hard samples by introducing a dynamic label assignment mechanism based on task alignment. This mechanism jointly evaluates the classification confidence and regression IoU to establish dynamic matching criteria, prioritizing candidate boxes with high alignment to the target as positive samples. This strategy improves the learning efficiency for challenging samples, ultimately enhancing the model's detection performance.

3 Proposed Methodology

This paper proposes an improved YOLOv8-based underwater garbage detection model, AH-YOLO, which integrates AKConv and an enhanced HSFPN. To address the complexity of underwater environments and the diversity of garbage objects, the model optimizes both feature extraction and feature fusion to enhance detection performance. The overall structure of the model is shown in Fig. 2.

By introducing AKConv to replace the original standard convolution, the AH-YOLO model can dynamically adjust the shape of the convolutional kernel on the basis of the content of the input feature map. This allows for more precise capture of local features of underwater garbage objects with varying shapes, thereby improving detection accuracy. Additionally, the model replaces the original neck layer of YOLOv8 with a second-stage improved high-level screening feature fusion pyramid network, enhancing the model's ability to fuse multiscale features.

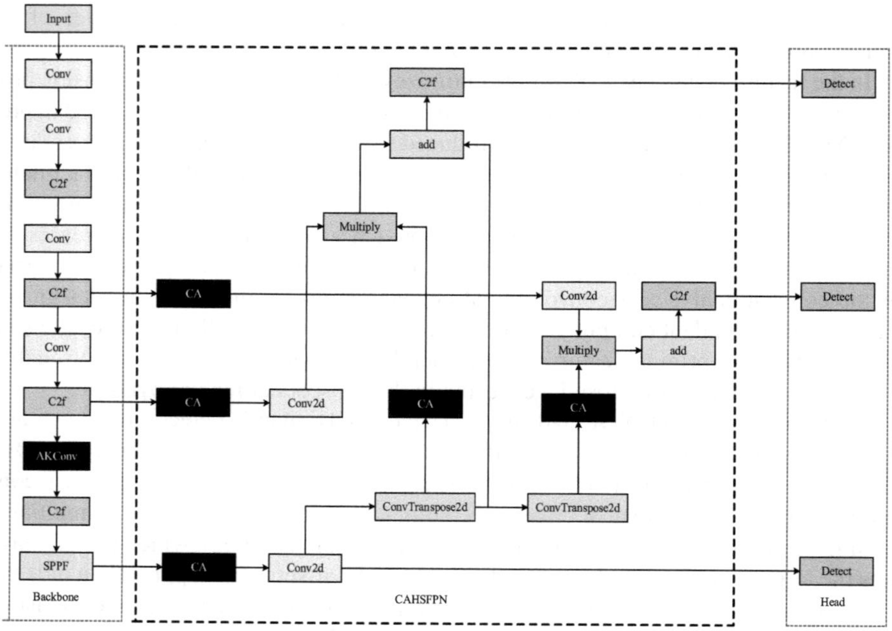

Fig. 2. AH-YOLO Network Architecture Diagram

3.1 Improvement of the Feature Extraction Network

The underwater garbage detection task faces multiple challenges, including variations in garbage size, diverse shapes, and detection difficulties caused by limited computational and storage resources. Although traditional convolution operations in deep learning have achieved significant breakthroughs in image processing, they still present several critical limitations. First, the local receptive field characteristic of traditional convolutions constrains their ability to capture global features, making it difficult to extract information effectively, especially when dealing with underwater garbage objects that exhibit significant morphological differences. Additionally, traditional convolution operations rely on a fixed spatial sampling pattern and lack sufficient adaptability to handle variations in scale, shape, and spatial positioning of different objects in underwater environments. Second, the conventional square convolution kernel design results in a quadratic increase in the number of parameters with kernel size, leading to a substantial increase in both the parameter count and computational cost.

In summary, traditional convolution operations struggle to perform well when diverse objects are processed, particularly in underwater environments with limited hardware resources, where excessive computational demands make real-time detection even more challenging.

AKConv introduces an innovative dynamically adjustable convolution operation mechanism that significantly enhances the flexibility and efficiency of feature extraction. First, AKConv overcomes the fixed shape constraint of conventional convolution kernels, allowing dynamic adjustments to the kernel size and geometry on the basis of the input feature map. This capability enables AKConv to more accurately capture the characteristics of different objects, making it highly adaptable to diverse input data and improving the model's ability to handle complex-shaped targets.

Furthermore, AKConv proposes an intelligent sampling coordinate generation algorithm to determine initial sampling coordinates. Traditional convolution operations use a fixed sampling grid; for example, a 3 × 3 convolution employs a regular sampling network, which can be represented by Eq. 1.

$$R = \{(-1, -1), (-1, 0), \cdots, (0, 1), (1, 1)\} \tag{1}$$

AKConv, on the other hand, utilizes an irregular sampling grid to achieve intelligent mapping from the initial coordinate set P_n to the optimized coordinate set P_0. By setting the upper-left corner of the algorithm as the sampling origin at (0,0), the convolution operation corresponding to the position of P_0 is defined as shown in Eq. 2.

$$Conv(P_0) = \sum \omega \times (P_n + P_0) \tag{2}$$

AKConv effectively enhances the spatial sampling precision of convolution operations through this intelligent sampling coordinate generation algorithm, further improving the model's ability to capture fine-grained features.

Furthermore, AKConv introduces learnable offsets to dynamically optimize sampling positions, significantly improving the representation accuracy of irregularly shaped object features. Additionally, AKConv adopts a linear parameter expansion mechanism to balance computational efficiency and model performance. This mechanism flexibly adjusts the convolution parameters to ensure efficient computation while maintaining strong feature representation capabilities. As a result, AKConv has unique advantages in lightweight scenarios such as edge computing, enabling deployment on resource-constrained devices while meeting real-time processing and low-latency requirements. Through these optimizations, AKConv not only enhances detection accuracy but also effectively reduces hardware resource consumption, making it more adaptable and practical for real-world applications.

The structure of AKConv is shown in Fig. 3. First, the input image has dimensions. (C, W, H), where C represents the number of channels, H denotes the image height, and W indicates the image width. Unlike traditional convolution, AKConv employs a deformable kernel initialization mechanism to generate an initial sampling point distribution with adaptive potential. This mechanism allows the convolution kernel to dynamically adjust its sampling pattern on the basis of the feature distribution of the input image, thereby increasing the flexibility and adaptability of the convolution operation.

Next, AKConv applies a basic convolution operation to extract initial response features from the input image. Building on this foundation, it generates a dynamic offset vector field through learning and optimizes the offset field via the backpropagation algorithm. This offset field enables adaptive adjustment of the initial sampling point positions on the basis of the spatial distribution characteristics of target features, allowing the convolution kernel shape to adapt dynamically to image features.

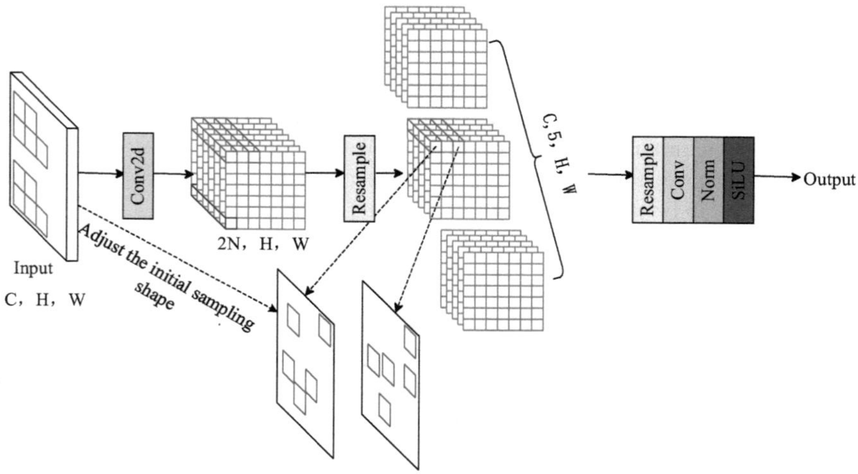

Fig. 3. AKConv architecture diagram

During the resampling phase, AKConv employs a bilinear interpolation algorithm to compute feature values for deformed, irregular sampling points, ensuring the continuity and differentiability of the sampling points during geometric transformations. The resampled feature tensor is further refined through channel reshaping, pointwise convolution, and batch normalization, ultimately producing the final feature output through the SiLU activation function.

Compared with traditional convolution with a fixed sampling pattern, AKConv allows each spatial location's sampling points to self-organize on the basis of local feature content through the aforementioned improvements. This data-driven dynamic perception capability enables AKConv to flexibly adapt to geometric variations in targets, significantly enhancing the efficiency of feature representation for complex object shapes. In underwater waste detection tasks, this geometric adaptability allows the model to better handle the diverse target shapes and complex visual information present in underwater environments.

3.2 Improvement in the Feature Fusion Network

Coordinate Attention

The structure of the coordinate attention (CA) module is shown in Fig. 4. Its core function is to capture long-range dependencies between channels and precise positional information through dual-path spatial encoding. The CA module first applies a pooling kernel of size $(H, 1)$ along the horizontal direction to encode the input feature map $X \in R^{C \times H \times W}$, extracting horizontal information. Then, it uses a pooling kernel of size $(1, W)$ along the vertical direction to encode and extract vertical information.

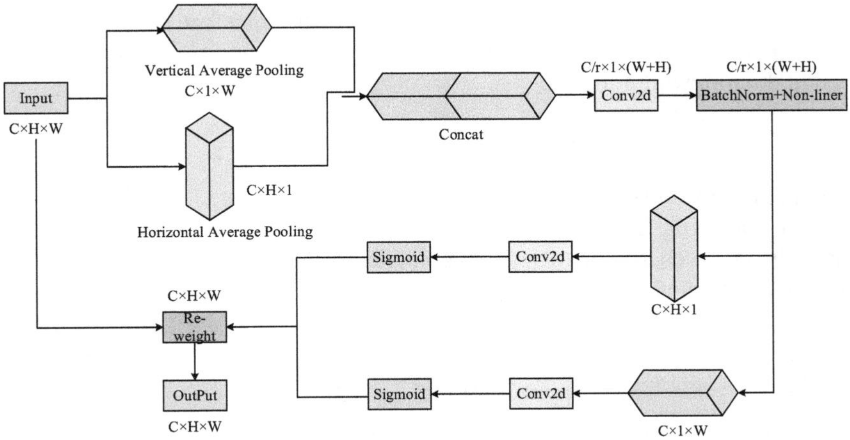

Fig. 4. CA Architecture Diagram

For the c-th channel with a height of h, its horizontal encoding output $Z_c^h(h)$ is mathematically expressed as shown in Eq. 3.

$$Z_c^h(h) = \frac{1}{W} \sum_{0 \leq w < W} X_c(h, w) \tag{3}$$

Here, H and W represent the height and width of the input feature map $X \in R^{C \times H \times W}$, respectively, and $X_c(h, w)$ denotes the feature value at spatial position (h, w) in the c-th channel. Similarly, the vertical encoding output $Z_c^w(w)$ is given by Eq. 4.

$$Z_c^w(w) = \frac{1}{H} \sum_{0 \leq w < H} X_c(j, w) \tag{4}$$

Through this encoding process, the CA module generates two direction-aware feature maps. The CA module subsequently fuses and maps these features via a 1×1 convolution, as described in the following formula.

$$f = \delta\left(F_1\left[Z^h, Z^w\right]\right) \tag{5}$$

$$g^h = \sigma\left(F_h\left(f^h\right)\right) \tag{6}$$

$$g^w = \sigma\left(F_w\left(f^w\right)\right) \tag{7}$$

Here, F_1 represents a 1×1 convolution layer, δ denotes the sigmoid activation function, and f is the intermediate feature map containing spatial information from both the horizontal and vertical directions. f^h and f^w represent the split feature maps along the horizontal and vertical directions, respectively.

Finally, the CA module applies attention weights to the input feature map through elementwise multiplication, and the position-aware output feature $y_c(i, j)$ is mathematically defined in the following formula.

$$y_c(i,j) = x_c(i,j) \times g_c^h(i) \times g_c^w(j) \tag{8}$$

CAHSFPN feature fusion network

The architecture of the high-level screening feature pyramid network (HSFPN) is shown in Fig. 5. Its core consists of two main components: the feature selection module and the feature fusion module.

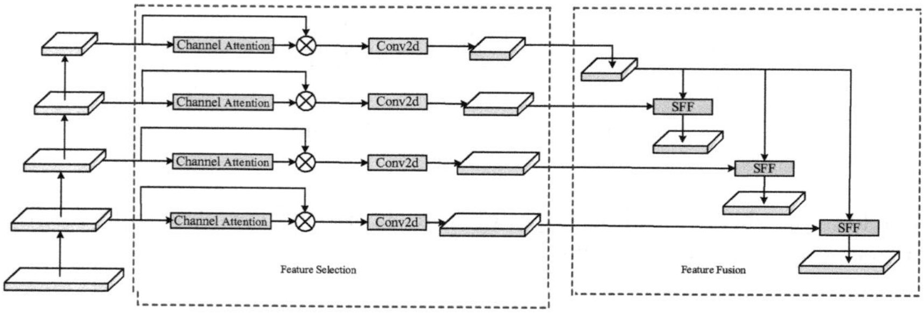

Fig. 5. HSFPN Architecture Diagram

In the feature selection stage, the HSFPN introduces the channel attention mechanism and spatial saliency evaluation to assign importance scores to high-level semantic features. This process dynamically filters out redundant background responses while retaining feature channels that are highly relevant to object detection, thereby effectively enhancing feature discrimination. This ensures that the model focuses on key information related to the target itself during object detection. Through this mechanism, the ability of the HSFPN to recognize important semantic information while reducing interference from irrelevant background noise significantly improves.

In the feature fusion stage, the HSFPN incorporates the selective feature fusion (SFF) module, which synergistically leverages the rich semantic information of high-level features and the fine spatial details of low-level features, significantly enhancing the model's feature fusion capability. Unlike traditional feature pyramid networks (FPNs), which perform feature fusion simply through elementwise addition, the SFF module uses high-level semantic information as a weighting mechanism to selectively filter necessary semantic details from low-level features. The filtered low-level features are then integrated with high-level features, greatly reducing information redundancy in cross-scale feature fusion and significantly improving the model's feature representation capability in complex underwater environments. The detailed structure of the SFF module is shown in Fig. 6.

When given high-level semantic features $F_h \in R^{C \times H \times W}$ and low-level detailed features $F_l \in R^{C \times H \times W_1}$, the SFF module first applies a transposed convolution with a stride of 2 and a kernel size of 3 × 3 to the high-level features for spatial upsampling, resulting in the output feature $F_{\hat{h}} \in R^{C\times 2H\times 2W}$. Next, to ensure dimensional consistency between the high-level and low-level features, the SPP module performs bilinear interpolation to $F_{\hat{h}}$ either upsample or downsample, producing the attention-guided feature $F_{att} \in R^{C\times H_1\times W_1}$. This feature is then passed through a channel attention module, which generates an attention weight matrix $W \in [0, 1]^{C\times H_1\times W_1}$ $[0, 1]^{C\times H_1\times W_1}$ to filter the low-level features. Finally, the filtered low-level features are fused with the high-level features to enhance the model's feature representation, resulting in the output feature $F_{out} \in R^{C\times H_1\times W_1}$. Here, C denotes the number of channels; H and W represent the height and width of the high-level feature map, respectively; and H_1 and W_1 denote the height and width of the low-level feature map, respectively.

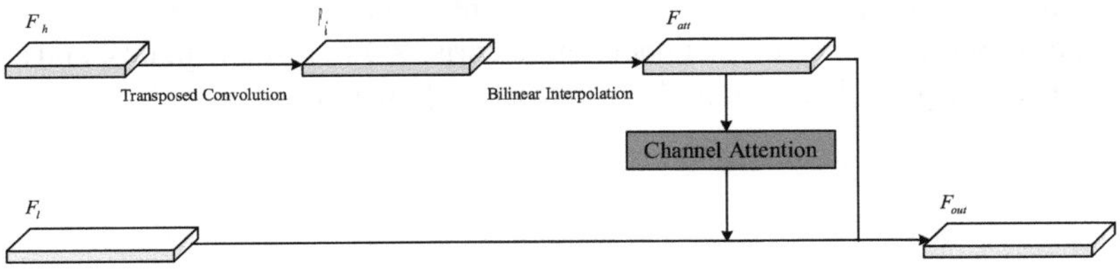

Fig. 6. SFF Architecture Diagram

In summary, the process of selective feature fusion performed by the SFF module in the HSFPN can be described in the following formula.

$$f_{att} = BL(T - Conv(f_h)) \quad (9)$$

$$f_{att} = BL(T - Conv(f_h))f_{out} = f_l * CA(f_{att}) + f_{att} \quad (10)$$

Obtaining the location and shape of underwater debris is crucial for realizing underwater debris detection. To address the challenges posed by the complex underwater environment—such as varying scales of debris, interference from dynamic lighting, and occlusion caused by biological attachments—this paper introduces improvements to the HSFPN, resulting in the CAHSFPN module. The CAHSFPN module replaces the original channel attention mechanism with a coordinate attention mechanism, which incorporates spatial coordinate information into the feature maps. This enables fine-grained weighting of debris locations, allowing for more precise capture of the position and contours of underwater debris. Consequently, the model can adaptively enhance its feature responses to debris edges and irregular shapes. Additionally, the introduction of the coordinate attention mechanism helps suppress background noise from the water, thereby improving the ability of the HSFPN to select features and filter low-scale features through its selective feature fusion module. Moreover, the CA module is designed with a lightweight structure, which incurs less computational cost than the channel attention mechanism does, making it more suitable for deployment on underwater autonomous vehicles with limited computing power.

4 Experimental Results

To comprehensively evaluate the impact of AKConv and CAHSFPN on model performance, this paper presents ablation studies and comparative experiments. In this work, the batch size is set to 16, the learning rate is dynamically adjusted via the cosine annealing algorithm, and in this paper, the initial learning rate is set to 0.01%, the optimizer is Adam, and the number of iterations (epochs) is set to 300.

4.1 Dataset

All the experiments in this paper are conducted and evaluated on the TrashCan dataset. This dataset is constructed from the J-EDI dataset provided by the Japan Agency for Marine-Earth Science and Technology. The TrashCan dataset consists of 7212 images and includes four main categories: underwater debris, ROVs, aquatic plants, and aquatic animals. The category distribution of the dataset is shown in Fig. 7.

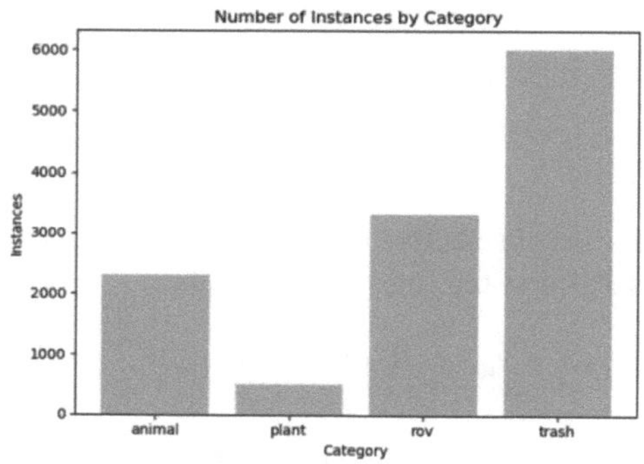

Fig. 7. Dataset Category Distribution Chart

4.2 Evaluation Metrics

For each category, all the detection results are sorted on the basis of the confidence scores of the predicted bounding boxes. Precision and recall are then calculated at different confidence thresholds to generate a precision-recall (PR) curve. *AP* is defined as the area under the PR curve. *AP* can be calculated via the following formula.

$$AP = \int_0^1 p(r)dr \tag{11}$$

mAP is one of the most important comprehensive evaluation metrics in object detection tasks. It is used to measure the overall performance of a model in multiclass object detection. By calculating the average value of *AP* across all categories, *mAP* considers not only the model's accuracy for each category but also its balance and stability across different categories. Its can be calculated via the following formula.

$$mAP = \frac{1}{N}\sum_{i=1}^{N} AP_i \tag{12}$$

4.3 Experiment Results Analysis

Ablation experiment

To evaluate the effectiveness of the improved algorithm systematically, ablation experiments were conducted in this study using *mAP* as the core metric. The results are shown in Table 1. The impact of the sampling points of AKConv (such as 3×3 vs 5×5) and the hierarchical fusion strategy of CAHSFPN on performance is negligible. In this experiment, we set it to 3×3.

Table 1. Ablation experiment results

Methods	AKConv	HSFPN	CAHSFPN	mAP@0.5	mAP@[.5:.95]	Parameters
YOLOv8n				0.640	0.464	3.02 M
1	√			0.657	0.470	2.35 M
2		√		0.655	0.469	1.94 M
3			√	0.662	0.487	1.86 M
4	√	√		0.658	0.471	1.74 M
AH-YOLO	√		√	0.679	0.501	1.71 M

The introduction of the AKConv and CAHSFPN modules significantly enhances the detection performance of the AH-YOLO model. Specifically, when AKConv was integrated into the original YOLOv8 network, the mAP@.5 and mAP@[.5:.95] improved by 2.66% and 1.29%, respectively. Similarly, introducing the original HSFPN module alone resulted in increases of 2.34% and 1.10% in mAP@.5 and mAP@[.5:.95], respectively. Notably, replacing the HSFPN with CAHSFPN led to even greater performance gains, with mAP@.5 and mAP@[.5:.95] increasing by 3.44% and 5.00%, respectively. This demonstrates the crucial role of both AKConv and CAHSFPN in enhancing model performance, with CAHSFPN outperforming HSFPN by a significant margin.

Furthermore, when AKConv and HSFPN were combined within the YOLOv8 network, the mAP@.5 and mAP@[.5:.95] improved by 2.81% and 1.51%, respectively. However, the combination of AKConv and CAHSFPN yielded even more substantial improvements, with mAP@.5 and mAP@[.5:.95] increasing by 6.10% and 7.97%, respectively. These results clearly indicate that the synergy between AKConv and CAHSFPN effectively optimizes both feature extraction and feature fusion, leading to a significant increase in the model's detection accuracy.

Moreover, the results of the ablation experiments revealed that although we improved the model and increased its accuracy, the final model parameters decreased by 43.4% compared with those of the original model because of the use of lightweight modules. This indicates that the model proposed in this paper is a high-precision and lightweight model.

Figure 8 shows the trend changes in mAP@.5 and mAP@[.5:.95] for both the original YOLOv8 model and the improved AH-YOLO model. As observed in the figures, the AH-YOLO model demonstrates a significant improvement in accuracy during training compared with the YOLOv8 model. As training progresses, the accuracy of the AH-YOLO model steadily increases and stabilizes after reaching 160 epochs, indicating that the model has fully converged.

Ultimately, the AH-YOLO model achieves an mAP@.5 of 67.9%, which is 3.9 percentage points higher than that of the YOLOv8 model. Additionally, it reaches an mAP@[.5:.95] of 50.1%, indicating a 3.7 percentage point improvement over YOLOv8. These results strongly validate the effectiveness of the proposed improvements. Furthermore, they demonstrated that the AH-YOLO model enhances feature extraction and multiscale object detection capabilities, making it more adaptable and robust in underwater waste detection scenarios.

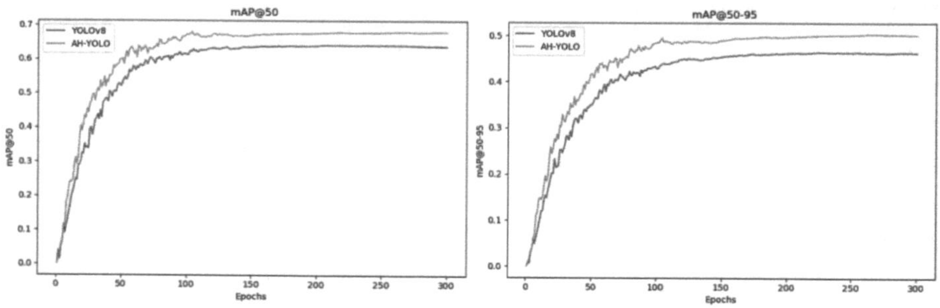

Fig. 8. Comparison of mAP metrics for YOLOv8 and AH-YOLO on the TrashCan dataset

Comparative experiment

To validate the performance advantages of the proposed AH-YOLO model in underwater garbage object detection tasks, this paper conducted a systematic comparative experiment with current mainstream object detection models based on the TrashCan dataset. The experiment selected the two-stage object detection algorithm Faster R-CNN, the single-stage object detection algorithm SSD, and YOLO-series models that have been widely applied in object detection tasks in recent years. The experimental results are shown in Table 2.

The data in the table show that the AH-YOLO model proposed in this study demonstrated the highest detection accuracy in underwater garbage object detection tasks. Compared with other benchmark models, it achieves optimal performance in terms of both the mAP@.5 and mAP@[.5:.95] metrics, as well as significant advantages in terms of other critical evaluation indicators. These results indicate that AH-YOLO exhibits stronger object recognition capability and generalization performance in complex underwater environments, enabling more effective adaptation to the demands of underwater scenarios. This further validates the effectiveness and superiority of the improvements proposed in this study, demonstrating the model's high application value and competitiveness in underwater garbage detection tasks. Moreover, the results of the comparative experiments show that the model proposed in this paper has not only the highest accuracy but also the fewest parameters among the other models, with only 1.71 M parameters. It is a high-precision and lightweight model.

Table 2. Comparative Experimental Results

Methods	mAP@.5/%	mAP@[.5:95]/%	Parameters
Faster R-CNN	0.602	0.342	55.60 M
SSD	0.573	0.346	24.28 M
YOLOv5s	0.632	0.449	7.20 M
YOLOv7s	0.640	0.438	37.31 M
TransNeXt-YOLOv8	0.653	0.463	49.70 M
YOLOv11n	0.638	0.440	2.59 M
AH-YOLO	0.679	0.501	1.71 M

Detection Results

To further validate the detection performance of AH-YOLO, this study conducted a visual comparative analysis using four representative images from the TrashCan dataset. Figure 9 and Fig. 10 compare the detection results of YOLOv8 and AH-YOLO.

In Fig. 9, AH-YOLO correctly identified a cup, whereas YOLOv8 misclassified it as unknown waste. Notably, AH-YOLO accurately distinguished between underwater debris and marine life, whereas YOLOv8 erroneously labelled the eel as garbage. This misidentification poses significant ecological risks, as confusing marine organisms with debris could lead to unnecessary interventions and threaten marine ecosystems.

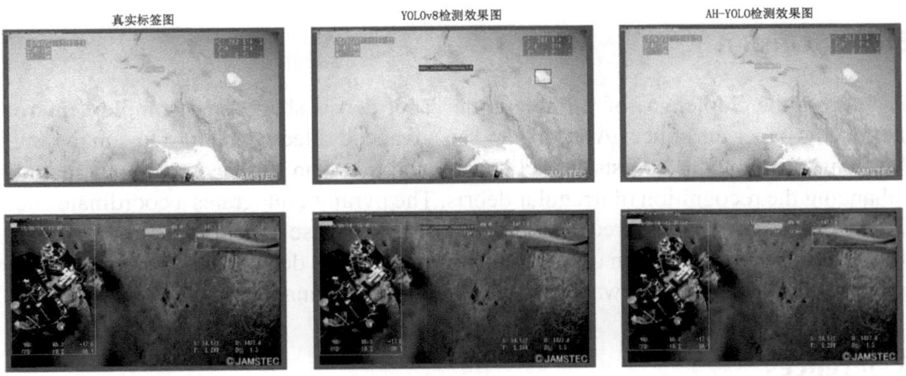

Fig. 9. Detection comparison chart

Figure 10 demonstrated AH-YOLO's effective detection of submerged cans and plastic bags that YOLOv8 failed to recognize, highlighting its enhanced capability in complex underwater environments.

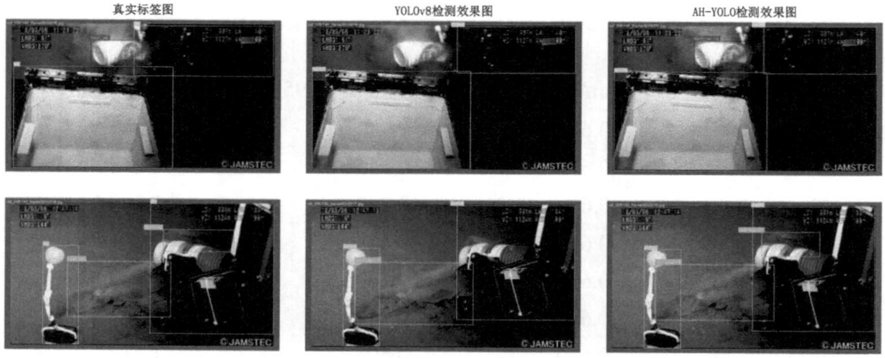

Fig. 10. Detection comparison chart

These visual comparisons substantiate AH-YOLO's practical advantages in underwater scenarios, particularly in the following:

- Critical differentiation between marine life and debris
- Robust detection accuracy under challenging conditions
- Precise localization of diverse underwater waste categories

The results confirm the operational reliability, ecological relevance, and technical superiority of AH-YOLO over baseline models, demonstrating its potential for environmental monitoring and marine conservation applications.

5 Conclusion

To address the challenges of variable underwater debris shapes and complex environments, this paper introduces AKConv and an advanced selective feature fusion pyramid. AKConv dynamically adjusts kernel shapes and sizes on the basis of input features, enhancing the recognition of irregular debris. The pyramid integrates a coordinate attention mechanism to capture precise spatial information and selectively fuse multiscale features, improving localization accuracy and reducing false detections. These innovations significantly increase underwater debris detection performance.

References

1. Valdenegro-Toro, M.: Submerged marine debris detection with autonomous underwater vehicles. In: 2016 International Conference on Robotics and Automation for Humanitarian Applications (RAHA). IEEE, pp. 1–7, 2016
2. Hong, J., Fulton, M., Sattar, J.: Trashcan: a semantically segmented dataset towards visual detection of marine debris. arXiv preprint arXiv:2007.08097, 2020
3. Huang, B., Chen, G., Zhang, H., et al.: Instant deep sea debris detection for manoeuvrable underwater machines to build sustainable ocean using deep neural network. Sci. Total. Environ. **878**, 162826 (2023)
4. Watanabe, J.I., Shao, Y., Miura, N.: Underwater and airborne monitoring of marine ecosystems and debris. J. Appl. Remote. Sens. **13**(4), 044509 (2019)

5. Zocco, F., Lin, T.C., Huang, C.I., et al.: Towards more efficient efficientdets and real-time marine debris detection. IEEE Robot. Autom. Lett. **8**(4), 2134–2141 (2023)
6. Ali, M., Khan, S.: Underwater object detection enhancement via channel stabilization. In: 2022 International Conference on Digital Image Computing: Techniques and Applications (DICTA). IEEE, pp. 1–8, 2022
7. Kumar, A., Srivastava, S.: Object detection system based on convolution neural networks using single shot multibox detector. Procedia Comput. Sci. **171**, 2610–2617 (2020)
8. Shi, P., Xu, X., Ni, J., et al.: Underwater biological detection algorithm based on improved faster-RCNN. Water **13**(17), 2420 (2021)
9. Wei, L., Kong, S., Wu, Y., et al.: Image semantic segmentation of underwater garbage with modified U-Net architecture model. Sensors **22**(17), 6546 (2022)
10. Zhang, X., Song, Y., Song, T.T., et al.: AKConv: convolutional kernel with arbitrary sampled shapes and arbitrary number of parameters[J]. arXiv:2311.11587, 2023
11. Chen, Y., Zhang, C., Chen, B., et al.: Accurate leukocyte detection based on deformable-DETR and multilevel feature fusion for aiding diagnosis of blood diseases. Comput. Biol. Med. **170**, 107917 (2024)

Novel Methods for Data analytics

Academic User Profile Construction Based on a Simplified Transformer and the GNN Fusion Model

Yihan Chen[1], Xuejie Zhang[1,2](✉), and Feng Ye[1]

[1] College of Computer Science and Software Engineering, Hohai University, Nanjing, China
xuejie_zh@hhu.edu.cn

[2] Key Laboratory of Water Big Data Technology of Ministry of Water Resources, Hohai University, Nanjing 211100, China

Abstract. With the advancement of scientific research and the rapid growth of the internet, academic users increasingly face challenges in obtaining accurate information about peer research. As a key component of big data analytics, user profiling has emerged as a critical focus in the scientific research community. While graph neural networks (GNNs) perform well in various graph learning tasks, their scalability to large graphs becomes problematic as the number of nodes increases due to computational complexity. To address this issue, this study proposes a novel academic user profiling model based on graph neural networks tailored to the unique characteristics of scientific research networks. The main contributions of this work are as follows: (1) We propose a simplified transformer architecture that reduces model complexity to a linear relationship with the number of nodes. (2) By integrating the simplified transformer with GNNs, neighborhood information is aggregated while maintaining global attention. The experimental results demonstrate that the proposed model delivers exceptional performance in terms of both accuracy and efficiency.

Keywords: scientific research network · user profile · graph neural network · linear attention · model fusion

1 Introduction

With the development of science and technology and the continuous increase in scientific research results, academic users in the era of big data face a large amount of academic resources. As a tool that can efficiently utilize big data, user profiling [1] can be applied to scientific research to help users efficiently obtain effective information and greatly reduce the cost of retrieval. In scientific research, user profiling analyses and models the characteristics of users to form labels or features, providing decision support for information services and academic platforms. Research on scientific research user profiling has focused mainly on methods that are based on interest. At present, the scientific research network has become an increasingly important carrier of information in scientific research. Therefore, user profiling for research networks requires the extraction of both individual and network structure information.

Y. Yu et al. (Eds.): ICPCSEE 2025, CCIS 2673, pp. 621–633, 2026.
https://doi.org/10.1007/978-981-95-2566-9_41

In recent years, graph deep learning has shown unique advantages in feature extraction for network-structured data. Among them, graph neural networks [3–5] based on the message passing mechanism [2] can cope with various graph learning tasks in many applications [6–10]. However, with increasing model depth, graph neural networks have problems such as oversmoothing [11] and oversqueezing [12]. On the other hand, the excellent performance of the transformer [13] in various domains (e.g., natural language processing [14, 15] and image processing [16, 17]) has encouraged an increasing number of researchers to focus on generalizing the transformer model to the field of graph learning.

Some recent works have generalized the Transformer to the field of graph deep learning, improving the accuracy of the results. However, most of these methods focus only on small graphs, such as molecular graphs. Therefore, the following problems arise when the Transformer is used:

1. The existing datasets are increasing in size, but existing methods typically apply the transformer to graph-level tasks with a small number of nodes. However, for node classification, a graph may have many nodes. The existing transformer has high computational complexity and grows quadratically with the number of nodes, which limits its direct application to node classification in large-scale networks.
2. Compared with networks in mature domains (such as e-commerce and social networks), scientific research networks, as relatively new relationship networks, have their own unique characteristics. Therefore, if we want to transplant the method of mature domains to scientific research networks, we need to make adjustments to adapt to the characteristics of the domain. Scientific research networks are built through academic activities such as citation and coauthorship between scholars, whereas social networks are built through activities such as retweeting, commenting, liking and browsing. Compared with social networks, the edge density of scientific research networks is significantly lower. Therefore, for scientific research networks, the information is more concentrated in the neighborhood of the central node. While using the transformer to obtain the global information, it is necessary to join the corresponding architecture to capture the information of neighboring nodes.

To address the first problem, we propose a simplified transformer that scales linearly with the number of nodes. The simplified model has only linear complexity while retaining the expressive power required to learn all-pair interactions without any approximation. In addition, no positional encoding, feature or graph preprocessing, augmented loss function, or edge embedding is required in the model.

To address the second problem, we propose fusing the transformer and GNN models. Since the message passing mechanism of the GNN only aggregates information within the neighborhood, its receptive domain is limited. The relevant nodes may be far from the center node, which impairs the effectiveness of the GNN. However, although the transformer can aggregate relevant information at a distance, it easily loses important information within the localization of the central node because of its global attention. We therefore propose a design that combines a transformer and a GNN to complement each other's advantages.

The contributions of this article are as follows:

- We propose a simplified transformer model, which uses a simplified transformer architecture to achieve a computational complexity that is linear with the number of nodes.
- We propose a Transformer and GNN fusion model, which introduces neighborhood information through the GNN layer to help the Transformer architecture capture local information.
- We experiment with our method on multiple benchmark datasets, and the results show that the proposed method outperforms existing graph transformers and mainstream graph neural networks.

2 Related Works

GNN: Many complex systems, such as social networks, biological networks, and information networks, have the form of graphs. To effectively process graph data, the key is how to represent the graph concisely. The purpose of graph representation learning is to embed the nodes into low-dimensional representations and retain the structural information of the graph. Graph neural networks have received a great deal of research in the past few years, effectively combining node attribute information and network structure information. The GCN applies convolutional operations to the graph, and the core is to aggregate node features and their neighboring features into feature vectors and iteratively aggregate the features of neighboring nodes. The GAT adds an attention mechanism based on the GCN and adaptively assigns weights to edges between nodes so that the model can be represented by the nodes. GraphSAGE constructs a fixed number of neighboring nodes via the uniform sampling method and aggregates the information of these neighboring nodes to obtain the embedding of the target node. Although different information aggregation methods have been designed since the GCN, graph neural networks follow a message passing mechanism, which usually cannot stack multiple layers to capture information from distant nodes due to oversmoothing and overcompression.

Transformer in Graphs: As mentioned above, some recent works have focused on extending the transformer to the field of graph learning. Initially, the main consideration in porting the transformer was how to incorporate graph structure information, such as the eigenvalues of the adjacency or Laplace matrices [22] and the relative positions between nodes [23], into the model. Ying et al. [18] proposed a grapher, which first applied a transformer to graphs and used three novel graph structure encodings, which worked well in molecular property prediction. GPS [19] summarized different types of positional and structural encodings with clearer definitions and decoupled local real-edge aggregation from the fully connected transformer. Despite their effectiveness, most related works focus only on small graphs, such as molecular graphs. When applied to graphs with many nodes, the time and space complexity of the transformer is unbearable. To address this problem, current research mainly adopts two measures: 1. Sampling by designing node sampling methods to reduce the number of inputs; 2. A simplified transformer architecture is used. Wu [20, 25] et al. proposed NodeFormer and SGFormer, explored the "information transfer between two nodes" on large-scale graphs, and used a simplified attention mechanism to achieve linear complexity. ANS-GT [21] and NAGphormer [24] design node sampling methods to reduce computational complexity and capture long-distance information through detailed design. However, node sampling only inputs some nodes, which will result in the loss of global information.

3 Preliminary

3.1 Problem Definition

We first define the problem of learning node representations for node classification via graph neural networks. Let G = (V, E, X) represent an attributed graph. Where V is the set of nodes and E is the set of edges. $A \in R^{N \times N}$ represents the adjacency matrix, where N is the total number of nodes. $X \in R^{N \times F}$ represents the node attribute matrix, where F is the feature dimension of each node. The goal of graph neural networks is to learn effective node representations (expressed as $H^{out} \in R^{N \times d}$, where d is the dimension of node representation) by combining graph structure information and node attributes and further using them for downstream tasks.

In this work, we focus on the node classification task in attribute networks. Specifically, the node classification task provides a labelled node set V_l and an unlabelled node set Vu. Let $y \in R^{N \times c}$ represent the label matrix, where c is the number of categories. Given the label Y_{Vl}, the goal is to predict the label Y_{Vu} of the unlabelled node.

Specifically, scientific research network data can be divided into citation networks and collaborator networks. Each node in the graph represents a paper or a researcher, and the node label represents the field. The goal is to predict the label as the interest label of the scientific user portrait on the basis of the given label.

3.2 Transformer

The Transformer encoder contains a series of Transformer layers, each of which consists of a multihead self-attention (MSA) and a feed-forward network. The MSA is the key component, and its purpose is to capture the semantic correlation between inputs. For simplicity, a single-head self-attention module is used for description.

Suppose that we have an input $H \in R^{N \times d}$ to the self-attention module, where n is the number of items and d is the hidden dimension. The self-attention module first projects H into three subspaces Q, K and V:

$$Q = HW^Q, K = HW^K, V = HW^V \tag{1}$$

where W^Q, W^K and W^V is the respective projection matrix. Q represents the query, K represents the key, V represents the value, and the output matrix calculation formula is as follows:

$$H' = softmax(\frac{QK^T}{\sqrt{d}})V \tag{2}$$

The attention matrix captures the all-pair correlations of the inputs in the sequence. Specifically, it computes the dot product between each pair of vectors after projection and then applies the Softmax method row by row.

4 Method

4.1 Model Framework

The framework of the model is shown in Fig. 1. It consists of five parts: (1) input, (2) linear transformer, (3) transformer and GNN fusion module, (4) structural information encoding module, and (5) output. After inputting the node and structural information of the graph, our model first calculates the similarity between the central node and all nodes through the linear transformer to obtain global attention. In the fusion module, the output of the transformer layer is subsequently used as the input of the GNN layer to expand the receptive field and efficiently aggregate global information, whereas the GNN layer helps the transformer layer perceive the neighborhood information of the central node and obtain more relevant information of the adjacent nodes. In the structural information encoding module, we combine the output and the embedding of the GNN to help the model capture the graph topology information. Finally, the label is predicted on the basis of the embedding of the node.

In the following section, we introduce the proposed model in detail. To expand the model to large graphs, we propose a simplified transformer architecture. Moreover, the shortcomings of the transformer are supplemented by model fusion. In addition, we add structural information and some details to improve the model performance.

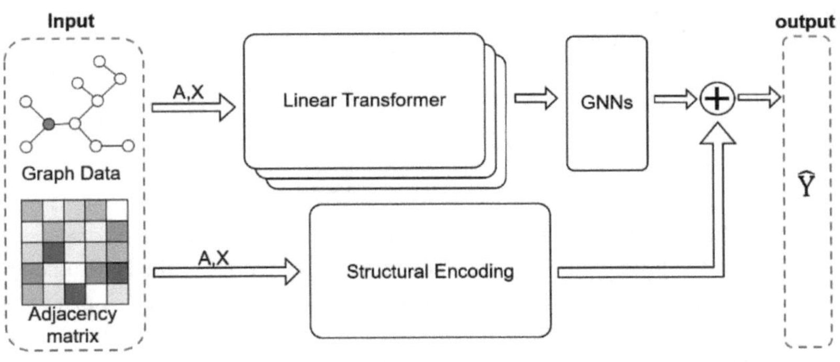

Fig. 1. The framework of the model. The model inputs are the graph adjacency matrix A and the node feature matrix X, and the final output is the node label matrix.

4.2 Linear Transformer

As mentioned above, despite the achievements of the work on generalizing the transformer to the graph domain, its attention mechanism results in $O(N^2d)$ space and time complexity, making the model's computational complexity unbearable. Recently, many works have focused on reducing the computational complexity of the graph transformer model, such as using node sampling before inputting the transformer and using kernel functions to approximate Softmax. The most central part of the transformer is the attention mechanism, which is formulated as follows:

$$Attention(Q, K, V) = soft\max(QK^T)V \quad (3)$$

where Q, K, and V $\in R^{N\times d}$.

According to the original definition of the attention mechanism, it should first calculate QK^T, , which has a complexity of $O(N^2d)$. The multiplication of softmax(QK^T) and V also requires $O(N^2d)$ complexity. The key factor restricting the performance of attention is actually the Softmax operation in the definition. If there is no Softmax operation, the three matrices QK^TV can satisfy the matrix multiplication associativity, so K^TV can be calculated first to obtain a $d \times d$ matrix. Then, it is multiplied with Q on the left so that the complexity required for both steps is $O(Nd^2)$. Since $d < < N$, the calculation complexity can be regarded as O(N). A comparison of the two methods is shown in Fig. 2.

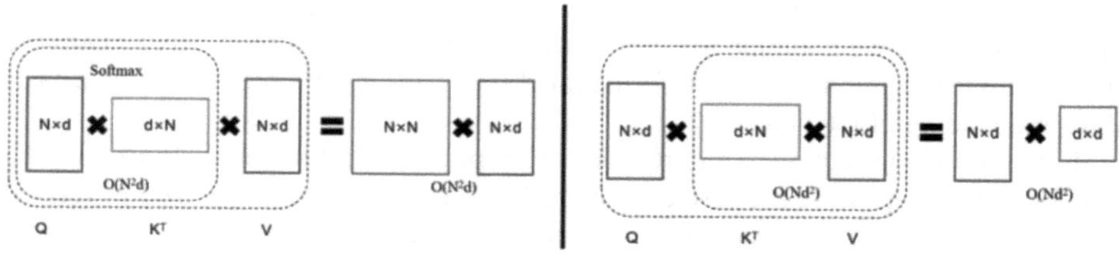

Fig. 2. Attention comparison

On the basis of this observation, we propose that the complexity can be reduced to linear by removing the Softmax operator. To explore the role of Softmax, we equivalently rewrite Eq:

$$Attention(Q, K, V)_i = \frac{\sum_{j=1}^{n} e^{q_i^T k_j} v_j}{\sum_{j=1}^{n} e^{q_i^T k_j}} \tag{4}$$

q^Tk computes the similarity between nodes to obtain the attention score, and the *e* operation ensures the nonnegativity of the result. In other words, if the Softmax operator is to be removed, it is necessary to guarantee the nonnegativity of the dot product while preserving the distributional properties of attention. We propose a scheme that first applies l2 normalization to q^Tk to ensure that $q^Tk \geq -1$ and then adds 1. The improved formula is as follows:

$$Attention(Q, K, V)_i = \frac{\sum_{j=1}^{n} (1 + \tilde{q}_i^T \tilde{k}_j) v_j}{\sum_{j=1}^{n} (1 + \tilde{q}_i^T \tilde{k}_j)}, \tilde{q}_i = \frac{q_i}{\|q_i\|}, \tilde{k}_i = \frac{k_i}{\|k_i\|} \tag{5}$$

The two summation terms of the numerator and denominator are shared for all nodes, which means that only one O(N) computation is needed in the actual computation, and then the updating for each node only requires O(1), so the total complexity of updating N nodes is O(N).

4.3 Transformer and GNN Fusion Module

Since the GNN is built on the message passing mechanism, a layer of the GNN module can only receive information from its neighbors, which also leads to oversmoothing when we stack multiple layers. In contrast, the transformer has a global receptive field through the attention mechanism, but in some cases, it ignores important information from neighbors. Can we combine the Transformer and GNN to achieve better results by integrating their respective advantages?

Specifically, we propose a model that fuses the transformer and GNN. With the help of the transformer, the GNN can extend the receptive field to all nodes, even if they are far from the central node. On the other hand, the GNN can help the transformer capture complex graph topology information and efficiently aggregate more relevant nodes in the neighborhood.

Transformer Layer: We put the transformer layer before the GNN layer to increase the receptive field to notice nodes far from the center. As mentioned in the previous section, we adopted a simplified transformer architecture to reduce complexity. The input of the transformer layer is represented as H ∈ RN × d. We use three matrices Wq, Wk, and Wv to project the input to the corresponding matrices Q, K and V and aggregate the node information according to the attention mechanism as follows:

$$Q = HW^Q, \tilde{Q} = \frac{Q}{||Q||}, K = HW^K, \tilde{K} = \frac{K}{||K||}V = HW^V, \tilde{V} = \frac{V}{||V||}, \quad (6)$$

$$D = diag(N + \tilde{Q}(\tilde{K}^T 1)), H' = D^{-1}[V + \tilde{Q}(\tilde{K}^T V)] \quad (7)$$

GNNLayer: The basic idea of a graph neural network is to iteratively update the features of nodes by combining the features of neighboring nodes and the features of the nodes themselves, which can help the transformer layer capture the information of neighborhood nodes effectively. Starting from the initial node feature H, each layer has two important functions, as follows:

- The AGGREGATE function aggregates information from neighboring nodes.
- The COMBINE function updates the representation of nodes by combining the aggregated information from neighboring nodes and the current node features.

$$a_v^k = AGGREGATE^k\{H_u^{k-1} : u \in N(v)\} \quad (8)$$

$$H_v^k = COMBINE^k\{H_v^{k-1}, a_v^k\} \quad (9)$$

where N(v) is the set of neighboring nodes of the vth node. The node feature H^K in the last layer can be regarded as the final node feature. Node features can be used for downstream tasks; for example, for node classification, the label of node v can be predicted via the Softmax function.

4.4 Implementation Details

Input layer: We first use a neural network layer to map the input features $X \in R^{N\times F}$ to node embeddings, i.e., $H^{(0)} = f(X)$, where f is a single-layer MLP. The node embedding $H^{(0)} \in R^{N\times d}$ is used for subsequent attention calculation and propagation.

Integration of structural information: To combine the structural information of the input graph G, existing models have proposed many solutions, but most of them tend to use positional encoding, such as the Laplacian matrix and degree center encoding. Here, we adopt a simple but effective scheme to combine the output features with the propagation embedding of the GNN in the output layer:

$$H^{out} = (1 - \alpha)H + \alpha GNN(H^{(0)}, A) \tag{10}$$

where α is a weight hyperparameter, and the GNN module can be a simple GNN architecture (e.g., GCN), which can be easily extended to large graphs.

Residual connection: To alleviate the problems of gradient vanishing and exploding, we introduce residual connections in the linear transformer module. This also helps retain some information and increases the expressive performance of the model:

$$H = \beta \frac{V + \tilde{Q}(\tilde{K}^T V)}{diag\left(N + \tilde{Q}\left(\tilde{K}^T 1\right)\right)} + (1 - \beta)H^{(0)} \tag{11}$$

β is a hyperparameter of the residual connection. In Eq. (11), H combines the attention information on N nodes and its own information. The former allows the model to capture the information of other nodes, whereas the latter retains the information of the central node.

5 Experiment

5.1 Setup

Dataset: The "Cora dataset" is a commonly used academic literature citation network dataset for machine learning tasks in literature classification. The dataset consists of 2708 scientific publications, which are divided into 7 categories. Nodes represent papers, and edges represent citations between papers. Each node has a 1433-dimensional feature vector, which represents the bag-of-words model of the paper.

CiteSeer consists of academic papers and is a benchmark dataset widely used in GNN research, especially for node classification tasks. The CiteSeer dataset contains 3312 papers in six categories. Each node comes with a 3703-dimensional feature vector and label. The main task of the dataset is to classify nodes on the basis of their features and graph structure.

The PubMed dataset was generated on the basis of the PubMed literature database. It contains 19,717 diabetes-related scientific publications, which are divided into three categories. Each publication uses a TF/IDF weighted word vector to describe whether it contains a word in the glossary. After eliminating stop words and removing low-frequency words, there are 500 words in the final glossary.

Baselines: A GCN can be regarded as the "pioneering work" of a graph neural network. This is the first time that the convolution operation in image processing has been used in graph structure data processing. The GCN propagates and fuses node information through the adjacency matrix, uses the Laplacian matrix and its symmetric normalization, and calculates graph convolution through the Chebyshev polynomial approximation to achieve feature extraction of graph data.

The GAT introduces an attention mechanism to perform aggregation operations on neighboring nodes on the basis of the message passing mechanism. The attention mechanism adaptively learns the weights of neighboring nodes and obtains a representation of the node itself via weighted summation. Multiple attentions allow the model to capture more information and focus on different features or patterns.

NAGphormer [24]. The article proposes a new neighborhood aggregation method, Hop2Token, which aggregates neighborhood features from each hop into a node representation, resulting in a series of vectors that retain neighborhood information from different hops. NAGphormer can be trained in small batches on the basis of the output of Hop2Token, which allows the graph transformer scale to large graphs.

SGFormer [25] proposes a linear graph transformer model with simple global attention that scales linearly with the number of nodes. Despite its simplicity, the model retains the expressive power required to learn all pairwise interactions without requiring any approximation. Moreover, a single-layer attention model has sufficient expressive power to learn global information from all pairwise interactions.

Evaluation indicators and experimental configuration:

In this work, we use accuracy to evaluate the proposed model. As one of the commonly used evaluation metrics in machine learning and statistics, accuracy is usually used in classification problems. When the samples of each category are relatively balanced, accuracy can intuitively reflect the performance of the model. The calculation formula of accuracy is as follows:

$$Accuracy = \frac{TP + TN}{TP + FP + TN + FN} \tag{12}$$

where TP represents the number of positive examples that are predicted to be positive and are actually positive examples, FP represents the number of positive examples that are predicted to be positive but are actually negative examples, TN represents the number of negative examples that are predicted to be negative and are actually negative examples, and FN represents the number of positive examples that are predicted to be negative examples.

The detailed experimental environment configuration of this paper is shown in Table 2 (Table 1):

Table 1. Dataset standards

Dataset	#Nodes	#Edges	#Features	#Classes
Cora	2,708	5,429	1,433	7
Citeseer	3,312	4,732	3,703	6
Pubmed	19,717	44,324	500	3

Table 2. Experimental environment

Experimental environment		Configuration
Hardware	CPU	Intel(R) Xeon(R) Platinum 8474C
	GPU	RTX 4090D(24GB)
	Memory	80G
Software	OS	Ubuntu 18.04
	Programming	Python 3.8, PyTorch 1.9.0

Hyperparameter setting: We set the hyperparameters according to the results of the model on the validation set: · learning rate {0.001, 0.005, 0.01, 0.05, 0.1}; · weight decay {1e-5, 1e-4, 5e-4, 1e-3, 1e-2}; · hidden dimension {32, 64, 128, 256}; · dropout rate {0, 0.2, 0.3, 0.5}; · number of layers {1, 2, 3}; · structural information weight {0.2, 0.5, 0.8}. For the structural information weight α, we set three values to represent the cases of low, balanced and emphasizing the input graph.

5.2 Results

Table 3. Mean and standard deviation of the accuracy on node classification datasets (five independent runs with random initialization). In each dataset, we indicate the best performer in bold.

Method	Cora	Citeseer	Pubmed
GCN	81.6 ± 0.4	71.6 ± 0.4	78.8 ± 0.6
GAT	83.0 ± 0.7	72.1 ± 0.5	79.0 ± 0.4
NAGphormer	82.1 ± 0.3	67.5 ± 1.3	79.3 ± 0.9
SGFormer	84.0 ± 0.6	72.1 ± 0.5	80.0 ± 0.5
Ours	**84.4 ± 0.5**	**72.8 ± 0.5**	**80.5 ± 0.4**

Table 3 shows the experimental results of all the models, and we use accuracy, a commonly used evaluation metric in multiclassification problems, as the measure of model performance. The results show that our models significantly outperform the standard GNN models (GCN, GAT). This finding indicates that our model possesses efficient expressive ability while achieving linear complexity. In addition, although our model is effective for the three datasets, the improvement is very limited. The possible reasons are as follows: although the attention mechanism is very effective, the message passing approach has obtained much information in the graph, which also indirectly proves the importance of neighborhood information for our model. Compared with other advanced graph transformers (NAGphormer and SGFormer), our model is also very competitive. These results provide concrete evidence to verify the effectiveness of our model as a powerful learner for node-level prediction. We also found that the

NAGphormer has a certain degree of overfitting, which may be caused by its complex architecture and node-sampling architecture. In contrast, with limited dataset samples, the lightweight architecture enables our model to have better scalability.

Table 4. Efficiency comparison of our model and the graph transformer competitors: training time Tr (ms) and inference time Inf (ms) per epoch.

Method	Cora		Citeseer		Pubmed	
	Tr(ms)	Inf(ms)	Tr(ms)	Inf(ms)	Tr(ms)	Inf(ms)
NAGphormer	69.4	31.1	70.1	31.8	339.7	144.2
SGFormer	15.2	3.9	15.9	4.4	16.3	4.7
Ours	42.5	8.0	45.3	12.3	48.5	13.9

Table 4 shows the training time and inference time of each epoch on cora, citeseer and pubmed. Since a fixed number of epochs is usually used to train models on these datasets, this paper uses the training time of each epoch to represent efficiency. Our model is more efficient than NodeFormer, especially on the PubMed dataset. In addition, it is not as efficient as SGFormer. This is because our model adds a fusion module with the GNN, which increases the training and inference time while improving the accuracy. When expanding to large graphs, the computational bottleneck lies in the quadratic complexity of the original attention. After adopting simplified attention, even if the fusion module is added, the overall complexity can be linearly related to the number of nodes, increasing the scalability of the model.

5.3 Ablation Study

To analyse the importance of model fusion and structural information, we conduct a series of ablation studies on multiple datasets. The results are summarized in Table 5.

Table 5. Ablation study of model fusion and structural information

Method	Cora	Citeseer	Pubmed
(defult)	84.4 ± 0.5	72.8 ± 0.5	80.5 ± 0.4
(GAT's Attention)	83.2 ± 1.0	71.8 ± 0.8	79.4 ± 0.7
w/o GNN	83.3 ± 0.7	72.3 ± 0.2	79.6 ± 0.6
w/o SE	81.0 ± 0.3	63.8 ± 1.0	79.8 ± 0.5

Linear attention: To study the effectiveness of linear attention, we compared the default model with the model replacing the attention calculation method. In the GAT, the original attention mechanism is used to aggregate information from the neighborhood. The experimental results show that when we change the attention function, it leads to a decrease in performance, which shows the effectiveness of our global attention. This is because, compared with the attention of the GAT, our attention mechanism can capture the information of all nodes globally.

Model fusion: To study the effectiveness of model fusion, we compared the model that was not fused with the GNN and the default model. The experimental results show that model fusion can significantly improve the performance of the model, confirming the necessity of the two models complementing each other. This is because for networks with sparse edges (such as scientific research networks), the neighborhood information of the central node is very important. Using only the transformer will cause the model to treat all the global nodes equally, resulting in a decrease in accuracy. The GNN can aggregate the neighborhood information of the central node through the message passing mechanism, so we adopt the strategy of model fusion. The results show that the fusion of the two can improve the model accuracy.

Structural information: To study the effectiveness of structural information, we compared the default model with the model without structural information. The experimental results show that adding structural information can significantly improve the performance of the model, confirming the necessity of structural information in node-level tasks. This is because, for graph data, structural information also has a significant effect on node representation, and adding structural information can improve accuracy.

6 Conclusion

In this work, we propose a simplified transformer model for the construction of scientific research user portraits. In this model, a simplified transformer model is first used to extract information from global nodes. In this way, the attention mechanism can be implemented with linear complexity so that the model can handle networks with many nodes. After obtaining the global information, the transformer is fused with the GNN, which helps the model obtain the neighborhood information of the central node and improves the model's performance. In addition, we combine the output with the GNN embedding to add structural information simply and effectively. Finally, experiments on multiple benchmark datasets prove the effectiveness of the model, which is better than that of the representative graph transformer and graph neural network.

In the future, we will focus on simplifying the fusion module to further reduce the computational complexity of the model. Moreover, we attempt to find a wider range of real academic user datasets to verify and tune our proposed method. The future goal is to apply our proposed academic user profiling model to specific cases.

References

1. Cooper, A.: The inmates are running the asylum. In: Arend, U., Eberleh, E., Pitschke, K. (eds.) Software-Ergonomie '99. Berichte des German Chapter of the ACM, vol. 53. Springer, Wiesbaden (1999). https://doi.org/10.1007/978-3-322-99786-9_1
2. Gilmer, J., Schoenholz, S.S.: Neural message passing for quantum chemistry. In: International Conference on Machine Learning, pp. 1263–1272. PMLR (2017)
3. Ming, C., Zhewei, W.: Simple and deep graph convolutional networks. In: International Conference on Machine Learning, pp. 1725–1735. PMLR (2020)

4. Kipf, T.N., Welling, M.: Semisupervised classification with graph convolutional networks. arXiv preprint arXiv:1609.02907. (2016)
5. Velickovic, P., Cucurull, G.: Graph attention networks. stat. **1050**(20), 10–48550 (2017)
6. Keyulu, X., Weihua, H.: How powerful are graph neural networks? arXiv preprint arXiv: 1810.00826. (2018)
7. Wenqi, F., Yao, M.: Graph neural networks for social recommendation. In: The World Wide Web Conference, pp. 417–426 (2019)
8. Rex, Y., Ruining, H.: Graph convolutional neural networks for web-scale recommender systems. In: Proceedings of the 24th ACM SIGKDD International Conference on Knowledge Discovery & Data Mining, pp. 974–983 (2018)
9. Muhan, Z., Yixin, C.: Link prediction based on graph neural networks. Adv. Neural Inf. Process. Syst. **31** (2018)
10. Di, J., Ziyang, L.: Graph convolutional networks meet markov random fields: Semisupervised communi-ty detection in attribute networks. In: Proceedings of the AAAI Conference on Artificial Intelligence, pp. 152–159 (2019)
11. Deli, C., Yankai, L.: Measuring and relieving the oversmoothing problem for graph neural networks from the topological view. In: Proceedings of the AAAI Conference on Artificial Intelligence, pp. 3438–3445 (2020)
12. Alon, U., Yahav, E.: On the bottleneck of graph neural networks and its practical implications. arXiv preprint arXiv:2006.05205. (2020)
13. Vaswani, A., Shazeer, N.: Attention is all you need. Adv. Neural Inf. Process. Syst. **30** (2017)
14. Devlin, J., Ming-Wei, C.: Bert: pretraining of deep bidirectional transformers for language understanding. In: Proceedings of the 2019 Conference of the North American Chapter of the Association for Computational Linguistics: Human Language Technologies, volume 1 (long and short papers), pp. 4171–4186 (2019)
15. Yinhan, L., Ott, M.: Roberta: a robustly optimized bert pretraining approach. arXiv preprint arXiv:1907.11692. (2019)
16. Dosovitskiy, A., Beyer, L.: An image is worth 16×16 words: transformers for image recognition at scale. arXiv preprint arXiv:2010.11929. (2020)
17. Ze, L., Yutong, L.: Swin transformer: hierarchical vision transformer using shifted windows. In: Proceedings of the IEEE/CVF International Conference on Computer Vision, pp. 10012--10022 (2021)
18. Chengxuan, Y., Tianle, C.: Do transformers really perform badly for graph representation? Adv. Neural. Inf. Process. Syst. **34**, 28877–28888 (2021)
19. Rampášek, L., Galkin, M.: Recipe for a general, powerful, scalable graph transformer. Adv. Neural. Inf. Process. Syst. **35**, 14501–14515 (2022)
20. Qitian, W., Wentao, Z.: Nodeformer: a scalable graph structure learning transformer for node classification. In: Advances in Neural Information Processing Systems (2022)
21. Zaixi, Z., Qi, L.: Hierarchical graph transformer with adaptive node sampling. Adv. Neural. Inf. Process. Syst. **35**, 21171–21183 (2022)
22. Kreuzer, D., Beaini, D.: Rethinking graph transformers with spectral attention. Adv. Neural. Inf. Process. Syst. **34**, 21618–21629 (2021)
23. Bodnar, C., Frasca, F.: Weisfeiler and lehman go cellular: cw networks. Adv. Neural. Inf. Process. Syst. **34**, 2625–2640 (2021)
24. Jinsong, C., Kaiyuan, G.: NAGphormer: a tokenized graph transformer for node classification in large graphs. arXiv preprint arXiv:2206.04910. (2022)
25. Qitian, W., Wentao, Z.: Sgformer: simplifying and empowering transformers for large-graph representations. Adv. Neural. Inf. Process. Syst. **36**, 64753–64773 (2023)

Author Index

Y. Yu et al. (Eds.): ICPCSEE 2025, CCIS 2673, pp. 635–636, 2026.
https://doi.org/10.1007/978-981-95-2566-9

MIX
Papier aus verantwortungsvollen Quellen
Paper from responsible sources
FSC® C105338

If you have any concerns about our products,
you can contact us on
ProductSafety@springernature.com

In case Publisher is established outside the EU,
the EU authorized representative is:
Springer Nature Customer Service Center GmbH
Europaplatz 3, 69115 Heidelberg, Germany

Printed by Libri Plureos GmbH
in Hamburg, Germany